9.9.99

TD1060 .N38 1999
Natural attenuation of
fuels and chlorinated
solvents in the subsurface

Natural Attenuation of Fuels and Chlorinated Solvents in the Subsurface

Natural Attenuation of Fuels and Chlorinated Solvents in the Subsurface

Todd H. Wiedemeier
Hanadi S. Rifai
Charles J. Newell
John T. Wilson

JOHN WILEY & SONS, INC.
New York / Chichester / Weinheim / Brisbane / Singapore / Toronto

D. HIDEN RAMSEY LIBRARY
U.N.C. AT ASHEVILLE
ASHEVILLE, NC 28804

This book is printed on acid-free paper. ♾

Copyright © 1999 by John Wiley & Sons, Inc. All rights reserved.

Published simultaneously in Canada.

No part of this publication may be reproduced, stored in a retrieval system or transmitted in any form or by any means, electronic, mechanical, photocopying, recording, scanning or otherwise, except as permitted under Sections 107 or 108 of the 1976 United States Copyright Act, without either the prior written permission of the Publisher, or authorization through payment of the appropriate per-copy fee to the Copyright Clearance Center, 222 Rosewood Drive, Danvers, MA 01923, (978) 750-8400, fax (978) 750-4744. Requests to the Publisher for permission should be addressed to the Permissions Department, John Wiley & Sons, Inc., 605 Third Avenue, New York, NY 10158-0012, (212) 850-6011, fax (212) 850-6008, E-Mail: PERMREQ@WILEY.COM.

AUTHORS' DISCLAIMER: Although one of the authors of this book is an employee of the United States Environmental Protection Agency (U.S. EPA), this book was not produced as an activity of the U.S. EPA. The contents of this book reflect the professional judgment of the authors. This book does not represent a statement of guidance or policy on the part of the U.S. EPA.

This publication is designed to provide accurate and authoritative information in regard to the subject matter covered. It is sold with the understanding that the publisher is not engaged in rendering professional services. If professional advice or other expert assistance is required, the services of a competent professional person should be sought.

Library of Congress Cataloging-in-Publication Data:

Natural attenuation of fuels and chlorinated solvents in the
subsurface / Todd H. Wiedemeier . . . [et al.].
p. cm.
Includes bibliographical references (p.).
ISBN 0-471-19749-1 (cloth : alk. paper)
1. Hazardous wastes—Natural attenuation. 2. Hazardous wastes—
Natural attenuation—Case studies. I. Wiedemier, Todd H.
TD1060.N38 1999 98-45799
628.5′5—dc21

Printed in the United States of America.

10 9 8 7 6 5 4 3 2 1

To Mom and Dad—Thanks for everything!

I also would like to express my sincere appreciation for the mentoring provided by Drs. John T. Wilson, Salvatore J. Mazzullo, and Frank G. Ethridge.

THW

To Phil Bedient
To Reed and Ryan

HSR
CJN

To Barbara H. Wilson

JTW

CONTENTS

PREFACE

Writing a book about the natural attenuation of fuels and solvents was a daunting task. The available data, information, technical understanding and paradigms changed rapidly during the completion of this book. In many ways this book is more of a detailed snapshot of the current understanding of natural attenuation than the description of a well-understood process. The authors realize, and would like to stress to the reader, that some of the information in this book may be superseded as more is learned. This is particularly true for the biological mechanisms of contaminant attenuation, an area of intense research, theoretical development, and practical application.

This book is intended to present a mixture of the theoretical and practical aspects of understanding, evaluating, and quantifying natural attenuation. Our emphasis is on the application of existing, field-proven techniques, with theoretical discussions when necessary to provide insight into the fate of contaminants. We have tried to summarize the relevant scientific literature, and have included information from various databases. Although this book focuses on fuels and solvents, much of the information presented can be applied to understanding and evaluating the behavior of other contaminants.

Numerous groups and individuals have been instrumental in advancing our understanding of the fate and transport of organic contaminants dissolved in groundwater. Much of the authors' experience in evaluating and understanding natural mechanisms of contaminant attenuation was gained as a direct result of work sponsored by the Air Force Center for Environmental Excellence, Technology Transfer Division (AFCEE/ERT). In 1992, Colonel Ross N. Miller and his team at AFCEE/ERT initiated a nationwide program to evaluate the natural attenuation of fuels and solvents. The intent of this program was to evaluate the behavior of organic contaminants under natural conditions and to develop protocols that would allow detailed evaluation of natural attenuation. Locations were selected to cover a broad range of geographic and hydrogeological conditions. Sites studied under this program were located in regions as far removed and varied as Hawaii, Alaska, Florida, upstate New York, California, and many locations in between. Much of the material in this book is culled from the authors' experience at these and other sites.

The authors would like to thank the following people for their contributions to our understanding of natural attenuation, and the completion of this book; without the help of these individuals this book could not have been written.

- At AFCEE/ERT: Colonel Ross N. Miller, Jerry E. Hansen, Marty Faile, Patrick Haas, Jim Gonzales, and Sam Taffinder.
- At Parsons Engineering Science, Inc.: E. Kinzie Gordon, Matthew A. Swanson, David E. Moutoux, Leigh Alvarado Benson, R. Todd Herrington, Douglas C. Downey, Robert E. Hinchee, and Peter R. Guest.
- At the University of Houston: Monica Suarez, Karen Bennett, and the faculty in Civil and Environmental Engineering for putting value in writing books.
- At Groundwater Services, Inc.: Christina Walsh and Don Beasley (graphics), Dr. Carol Aziz (solvent biodegradation chapter), Rob Balcells, (case studies), Dr. Tom McHugh and Ric Bowers (sources chapter). In addition the other partners at GSI, John Connor, Grant Cox, and Bob Lee, for agreeing to devote company resources to this project.
- At the United States Environmental Protection Agency, Robert S. Kerr Environmental Research Laboratory in Ada, Oklahoma: Don Kampbell, Barbara Wilson, Frank Beck, and Mike Cook.
- At the United States Geological Survey: Dr. Francis H. Chapelle.

CHAPTER 1

OVERVIEW OF NATURAL ATTENUATION

Natural attenuation, also referred to as *intrinsic* or *passive remediation*, has emerged as the approach of choice for remediating groundwater at many contaminated sites. Natural attenuation or the reduction of contaminant concentration and contaminant mass by relying on the assimilative capacity of a groundwater system is nonintrusive, cost-effective, and at its most effective does not involve the transfer of contamination from one medium or phase to another. This approach has been used successfully to control water quality in rivers, streams, lakes, and estuaries. Waste loading permits for industrial and municipal wastewater discharges, for example, are generally based on the ability of the water body to assimilate the proposed waste load in such a way that significant depletion of oxygen is prevented. A number of water quality models that simulate hydrodynamic, physical, chemical, and biological processes to predict the assimilative capacity of surface water systems have been developed and are used routinely by the scientific and regulatory communities. A similar approach has been used for air regulatory programs.

Researchers and practitioners have, in recent years, sought an understanding and acceptance of natural attenuation in groundwater similar to that which exists for surface water. This process is complicated by the nature of groundwater systems and the behavior of the contaminants. Contaminants in the subsurface distribute themselves among the various media (soil, water, and soil vapor) and can exist in different phases (e.g., nonaqueous, aqueous, and vapor phases). Additionally, a large number of processes, such as sorption and biodegradation, control their fate and transport. Many of these processes are still poorly understood. Nonetheless, significant progress in quantifying natural attenuation of contaminants in groundwater systems has been made over the past decade. Field protocols and analytical and numerical models have been developed, and numerous databases have emerged from field characterization studies. These efforts have allowed the scientific, technical, and regulatory communities to make informed decisions regarding the use of natural attenuation as a remedial approach for contaminated groundwater.

1.1 DEFINITION OF NATURAL ATTENUATION

Natural attenuation refers to the observed reduction in contaminant concentrations as contaminants migrate from the source in environmental media. This reduction in concentration in groundwater is due primarily to a number of fate and transport processes, including simple dilution, dispersion, sorption, volatilization, and biotic and abiotic transformations. Naturally attenuating contaminant plumes can take a variety of forms; they might be expanding, stable, or shrinking (Figure 1.1), depending on the trends in the spatial variation of contaminant concentrations with time. Additionally, Rice et al. (1995a) refer to an "exhausted" fuel-hydrocarbon plume as having a short length (less than 70 ft) and an average benzene concentration of less than 10 parts per billion (ppb). Common features of all attenuating plumes are a decline in the dissolved contaminant mass as a function of time, and a decline in contaminant concentrations downgradient from the source (Figure 1.2). In many cases, before the contaminant plume reaches potential receptor exposure points, natural

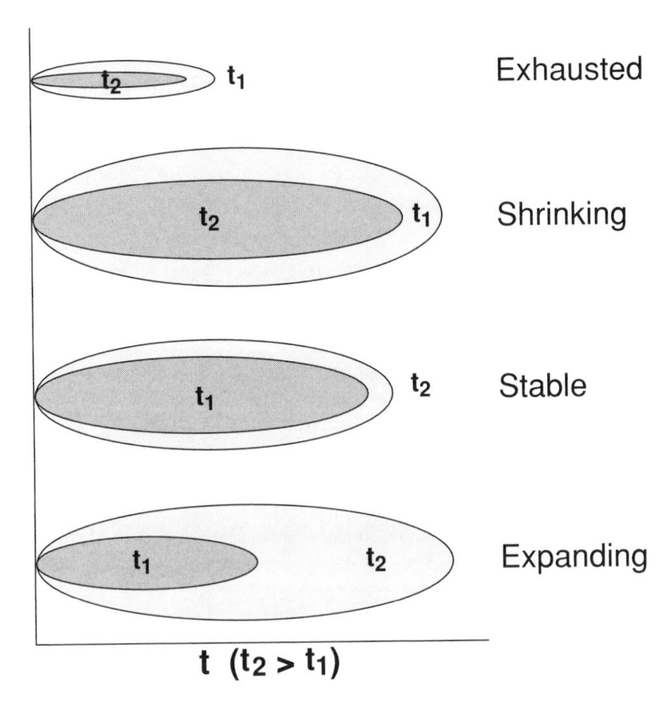

FIGURE 1.1. Types of naturally attenuating plumes.

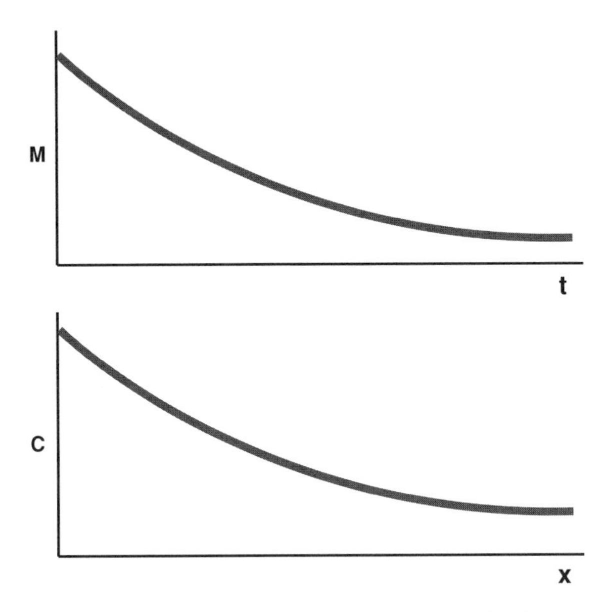

FIGURE 1.2. Decline in plume mass and contaminant concentration in an attenuating plume.

attenuation will reduce dissolved contaminant concentrations below regulatory standards such as maximum contaminant levels (MCLs) or other remediation goals.

In practice, natural attenuation also is referred to as *intrinsic remediation, intrinsic bioremediation, natural restoration, or passive bioremediation.* The use of natural attenuation as a remedial approach, referred to as *monitored natural attenuation* by the U.S. Environmental Protection Agency (U.S. EPA), implies a plume management approach that involves long-term monitoring of the plume. U.S. EPA (1997) defines monitored natural attenuation as follows:

> The term *monitored natural attenuation,* as used in this Directive, refers to the reliance on natural attenuation processes (within the context of a carefully controlled and monitored site cleanup approach) to achieve site-specific remedial objectives within a time frame that is reasonable compared to that offered by other more active methods. The "natural attenuation processes" that are at work in such a remediation approach include a variety of physical, chemical, or biological processes that, under favorable conditions, act without human intervention to reduce the mass, toxicity, mobility, volume, or concentration of contaminants in soil and groundwater. These in-situ processes include biodegradation; dispersion; dilution; sorption; volatilization; and chemical or biological stabilization, transformation, or destruction of contaminants.

In this book the term *natural attenuation* is used to describe all the processes that act to reduce the concentrations of contaminants in groundwater. The term *intrinsic bioremediation* is used to refer to the component of natural attenuation brought about by biological degradation mechanisms.

Natural Attenuation: Advantages and Limitations

In comparison to conventional engineered remediation technologies, natural attenuation offers a number of advantages, especially if intrinsic bioremediation is occurring:

1. During intrinsic bioremediation, contaminants can ultimately be transformed to innocuous by-products (e.g., carbon dioxide, ethene, chloride, and water in the case of chlorinated solvents; and carbon dioxide and water in the case of fuel hydrocarbons), not just transferred to another phase or location within the environment.
2. Natural attenuation is nonintrusive and allows continuing use of infrastructure during remediation.
3. Natural attenuation does not involve generation or transfer of wastes.
4. Natural attenuation is often less costly than other currently available remediation technologies.
5. Natural attenuation can be used in conjunction with, or as a follow-up to, other intrusive remedial measures.
6. Natural attenuation is not subject to limitations imposed by the use of mechanized remediation equipment (e.g., no equipment downtime).

Natural attenuation has the following potential limitations:

1. Time frames for complete remediation may be long.
2. Responsibility must be assumed for long-term monitoring and its associated cost, and the implementation of institutional controls.
3. Natural attenuation is subject to natural and anthropogenic changes in local hydrogeologic conditions, including changes in groundwater flow direction or velocity, electron acceptor and donor concentrations, and potential future releases.
4. The hydrologic and geochemical conditions amenable to natural attenuation are likely to change over time and could result in renewed mobility of previously stabilized contaminants (e.g., manganese and arsenic) and may adversely affect remedial effectiveness.
5. Aquifer heterogeneity may complicate site characterization, as it will with any remedial approach.
6. Intermediate products of biodegradation [e.g., vinyl chloride (VC)] can be more toxic than the original compound [e.g., trichloroethene (TCE)].

1.2 WHY NATURAL ATTENUATION HAS EMERGED AS A FEASIBLE APPROACH TO REMEDIATION

The primary motivation for considering natural attenuation as a potential remediation technology is the lack of efficient, cost-effective remediation technologies that can deal with the large number of contaminated sites across the nation. Pump-and-treat, for example, proved to be an inefficient and costly technology due to "tailing" and "rebound" phenomena (Figure 1.3). In the late 1980s, EPA conducted a study of 19 sites where pump-and-treat technology was being used (U.S. EPA, 1989). The EPA study concluded that contaminant concentrations at pump-and-treat sites usually decrease most rapidly soon after system startup. After the initial reduction, however, the concentrations often tend to level off, thereby significantly slowing the progress toward complete remediation. The main factors impeding the ability to achieve cleanup goals include

- Hydrogeologic factors, such as numerous low-permeability layers or other hydrogeologic complexities
- System design factors, such as poorly designed or improperly located extraction wells
- Contaminant factors, such as continued leaching from source areas and/or the presence of non-aqueous-phase liquids (NAPLs) in the subsurface

In a follow-up study of the same 19 sites, the EPA also concluded that potential NAPL presence was not addressed at the majority of these sites. As a result, a groundwater extraction system may perform as designed but will not achieve the cleanup goals within the predicted time frame (U.S. EPA, 1992a). EPA concluded that NAPLs were "a key factor in the longer-than-anticipated time frames for aquifer restoration." NAPLs were recognized as a continuing source of contaminants for

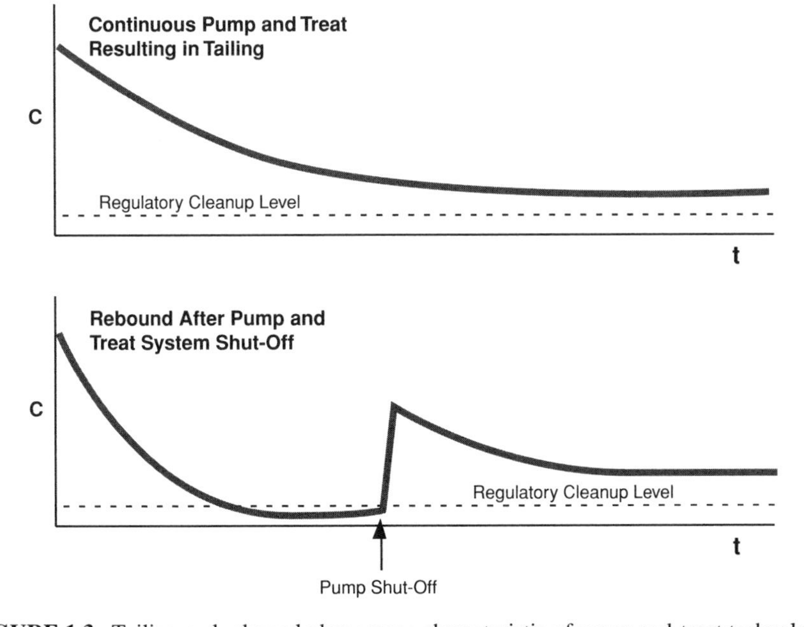

FIGURE 1.3. Tailing and rebound phenomena characteristic of pump-and-treat technology.

the groundwater, sources that typically take "a very long time" to deplete solely by groundwater extraction.

Because of the problems experienced at sites where NAPLs are present, EPA modified the guidelines for remedial actions at Superfund sites affected by such materials in 1992 (U.S. EPA, 1992b). Under a directive, EPA indicated that "the presence of NAPLs will have a significant influence on the time frame required or likelihood of achieving cleanup standards, and should be evaluated when selecting appropriate remedial actions." Furthermore, EPA stated that "actions at each site should be tailored to the specific conditions and applicable requirements at that site." This directive had an important effect on site characterization and remediation strategies, particularly for the Resource Conservation and Recovery Act (RCRA) and Superfund regulatory programs.

A subsequent study conducted by the National Research Council (NRC) Committee on Groundwater Cleanup Alternatives concluded that groundwater pump-and-treat methods could require untenable operating periods (in some cases centuries) to restore groundwater to drinking water standards at sites affected by NAPLs (National Research Council, 1994). To further evaluate the performance of pump-and-treat systems, the NRC Committee established a four-category site rating system characterizing the relative practicability of groundwater cleanup based on site hydrogeologic conditions, contaminant type, and other factors (see Table 1.1). A category 1 site represents ideal conditions for groundwater restoration with a pump-and-treat system (i.e., homogeneous aquifers with a degradable contaminant and no NAPL). A category 4 site, on the other hand, is extremely difficult or impossible to remediate due to the contaminant type, the presence of NAPLs, and complex hydrogeology.

TABLE 1.1. Relative Ease of Cleaning Up Contaminated Aquifers as a Function of Contaminant and Hydrogeologic Factors

	Contaminant Chemistry					
Hydrogeology	Mobile, Dissolved (Degrades/ Volatilizes)	Mobile Dissolved	Strongly Sorbed, Dissolved (Degrades/ Volatilizes)	Strongly Sorbed, Dissolved	Separate-Phase LNAPL	Separate-Phase DNAPL
Homogeneous, single layer	1	1–2	2	2–3	2–3	3
Homogeneous, multiple layers	1	1–2	2	2–3	2–3	3
Heterogeneous, single layer	2	2	3	3	3	4
Heterogeneous, multiple layers	2	2	3	3	3	4
Fractured	3	3	3	4	4	4

Source: Reprinted with permission from *Alternatives for Groundwater Cleanup*. Copyright 1994 by the National Academy of Sciences. Courtesy of the National Academy Press, Washington, DC.

The NRC surveyed 77 sites with operating pump-and-treat systems, classified them in one or two of the four categories, and noted the number of sites where cleanup goals had been achieved. The survey results are summarized below (note that some sites were assigned to two different categories, making the total more than 77):

Category 1 (easiest to remediate)	2 sites	1 had achieved cleanup goals
Category 2	14 sites	4 had achieved cleanup goals
Category 3	29 sites	3 had achieved cleanup goals
Category 4 (hardest to remediate)	42 sites	0 had achieved cleanup goals

According to the NRC, there are several key technical reasons for the difficulty in cleanup:

- Physical heterogeneity, making groundwater migration pathways difficult to predict
- Migration of contaminants to inaccessible regions, such as clays or small pores in aggregates
- Sorption of contaminants to subsurface materials
- Difficulties in characterizing the subsurface, making knowledge of the subsurface incomplete
- Presence of NAPLs, creating long-term continuing sources in the subsurface.

More recently, in a review of the effectiveness of groundwater remediation at 37 sites in the Santa Clara Valley in California, Bartow and Davenport (1995) also concluded that "while pump-and-treat successfully reduced maximum concentrations at most of the sites reviewed, successful attempts to reduce maximum contaminant concentrations to below MCLs are limited."

These studies were instrumental in identifying the limitations of the most frequently applied groundwater remediation alternative, pump-and-treat. As discussed below, other commonly applied technologies have performance limitations as well.

Constraints on Performance of Other Remediation Technologies

Other commonly used remediation technologies, such as enhanced in situ biodegradation and air sparging, also are hampered by technical constraints. For example, enhanced in situ biodegradation employs oxygenated water that is circulated through the affected aquifer zone to enhance localized biodegradation of organic constituents (Lee et al., 1987). Despite extending a pumping system with a biologically destructive process, the mass flux of oxygen that can be delivered is relatively small compared to the potential mass of organics in the source zone. For example, a 20-gallon per minute (gpm; 75-liter per minute) system delivering oxygen at 30 mg/L (pure oxygen feed) would be able to degrade the equivalent of only about 400 kg of biodegradable contaminants per year, making it completely effective only at sites with little or no NAPL. Because enhanced in situ biodegradation requires circulating groundwater through the source zone, better performance is achieved in homogeneous aquifers with relatively high permeabilities. These conditions (no NAPL, favorable hydrogeology) are found at relatively few sites.

In the air sparging process, air is injected directly into the saturated zone, forming small, continuous air channels from the sparge well to the water table (Johnson et al., 1993). Dissolved organics are removed from the groundwater through volatilization, and dissolved oxygen is introduced into the contaminated zone. This technology is generally used to treat fuel-hydrocarbon plumes and is better suited for relatively homogeneous media. The air movement is extremely sensitive to small variations in the permeability of the formation and may cause large portions of the source zone to be bypassed unless very close well spacing is employed (Johnson et al., 1993). For example, if the channels contact only 1% of the NAPL in the source zone, the mass removal of 99% of the contaminants in the source is limited by the transport of oxygen through the water.

Although there are numerous emerging technologies for groundwater remediation, such as oxygen-releasing materials, surfactants, and specialized electron acceptor and electron donor applications, none of these technologies represent a cost breakthrough whereby a large percentage of the sites with contaminated groundwater will be able to be restored to pre-contaminant-release conditions. Therefore, scientific and regulatory attention has been focused on technologies and regulatory approaches that stress managing contaminants in place for some sites. Natural attenuation is one such technology.

Natural Attenuation: How Long and How Far?

Natural attenuation can be thought of as a response to the physical, chemical, hydrogeological, and biological constraints on remediating contaminated aquifers from which we know it is difficult or impossible to remove contaminants but for which it is possible to predict chemical fate and transport in the subsurface. At some sites it may be possible to leave some or all of the contaminants in place and let nature remediate the contaminated groundwater. In such cases, two key questions are raised:

1. How far will the contaminant plume migrate from the source area? Will the contaminants reach potential or actual users of the groundwater, such as human and/or environmental receptors?
2. How long will the contaminant plume persist before the site is restored to pre-contaminant-release conditions? How many years must we wait until cleanup is achieved?

Plume-a-Thon Studies

A new class of studies conducted recently, known as *plume-a-thon studies*, have provided tremendous insights into plume behavior and have supported using natural attenuation as a remediation technology at many hydrocarbon release sites. The California Leaking Underground Fuel Tank (LUFT) Historical Case Analysis (Rice et al., 1995a), for example, reported that plume lengths at 271 fuel hydrocarbon sites in California "change slowly and stabilize at relatively short distances from the release site" (usually less than 250 ft). Of these 271 plumes, 59% were stable, 33% were shrinking, and only 8% were growing. Rice et al. (1995a) also indicated that while active (engineered) remediation may help reduce benzene concentrations, "significant reductions can occur over time, even without remediation."

A Texas Bureau of Economic Geology (BEG) study (Mace et al., 1997), based on 217 sites in Texas, found that most benzene plumes (75%) are less than 250 ft long and have either stabilized or are decreasing in length and concentration. In a manner similar to Rice et al. (1995a), Mace et al. (1997) found no statistical difference between sites where groundwater remediation activities have or have not been implemented. A Florida risk-based corrective action (RBCA) study (Groundwater Services, Inc., 1997) determined a median length of 90 ft for 117 leaking underground storage tank (LUST) sites in Florida (based on a 50-ppb benzene limit).

The Hydrogeologic Database (HGDB) (Newell et al., 1990) was reanalyzed by Newell and Connor (1998), who reported a median BTEX (benzene, toluene, ethylbenzene, and xylene) plume length from 42 service station sites of 213 ft. They also compared the above-mentioned four studies and concluded that most hydrocarbon plumes associated with leaking fuel tanks at service stations are under 200 ft long (Figure 1.4).

Although not as well documented as fuel-hydrocarbon-contaminated sites, similar studies have been completed or are under way for chlorinated solvent sites. The HGDB (Newell et al., 1990), for example, reported a median length of chlorinated ethene [tetrachloroethene (PCE), TCE, dichloroethene (DCE), or VC] plumes of 1000 ft based on 88 sites across the country and a median length of 500 ft for other chlorinated solvent plumes [defined as trichloroethane (TCA) and dichloroethane (DCA) and based on 29 sites].

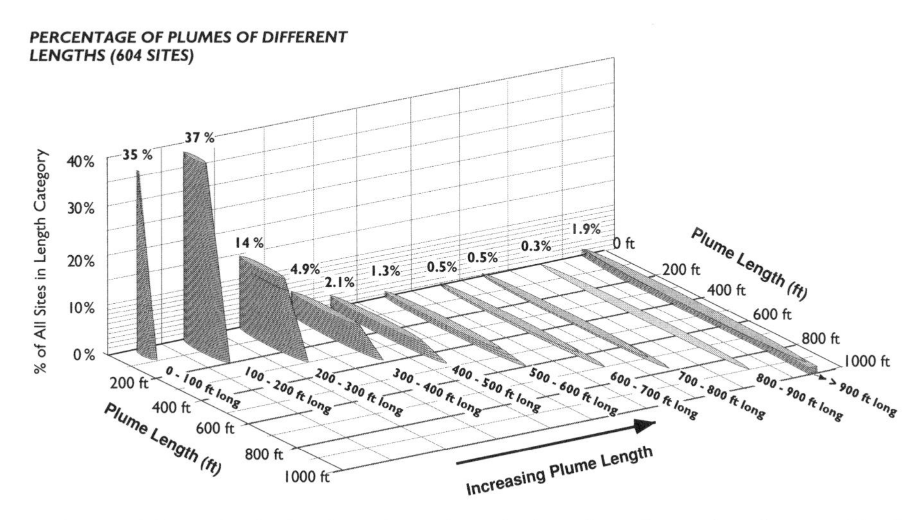

FIGURE 1.4. Limit of migration of petroleum hydrocarbon plumes, based on combined results from four studies: Lawrence Livermore study (Rice et al. 1995a), Texas BEG study (Mace et al., 1997), Florida RBCA study (Groundwater Services, Inc., 1997), and unpublished data from the HGDB database (Newell et al., 1990). (From Newell and Connor, 1998.)

TABLE 1.2. Summary of Important Processes Affecting Solute Fate and Transport

Process	Description	Dependencies	Effect	Quantification
Advection	Movement of solute by bulk groundwater movement	Dependent on aquifer properties, mainly hydraulic conductivity and effective porosity, and hydraulic gradient; independent of contaminant properties	Main mechanism driving contaminant movement in the subsurface	$\upsilon_x = -\frac{K}{n_e}\frac{dH}{dL}$
Dispersion	Fluid mixing due to groundwater movement and aquifer heterogeneities	Dependent on aquifer properties and scale of observation; independent of contaminant properties	Causes longitudinal, transverse, and vertical spreading of the plume; reduces solute concentration	Estimated by: $\alpha_x = 0.83(\log_{10} L_p)^{2.414}$ or $\alpha_x = 0.1L_p$
Diffusion	Spreading and dilution of contaminant due to molecular diffusion	Dependent on contaminant properties and concentration gradients; described by Fick's laws	Diffusion of contaminant from areas of relatively high concentration to areas of relatively low concentration; generally unimportant relative to dispersion at most groundwater flow velocities	$F = -D\frac{dC}{dx}$ $\frac{dC}{dt} = D\frac{d^2C}{dx^2}$
Sorption	Reaction between aquifer matrix and solute whereby relatively hydrophobic organic compounds become sorbed to organic carbon or clay minerals	Dependent on aquifer matrix properties [organic carbon content (f_{oc}) and clay mineral content, bulk density (ρ_b), specific surface area, and total porosity (n) and contaminant properties (solubility, hydrophobicity, octanol–water partitioning coefficient)]	Tends to reduce apparent solute solute transport velocity and remove solutes from the groundwater via sorption to the aquifer matrix	$R = 1 + \frac{\rho_b K_d}{n}$ $K_d = f_{oc}k_{oc}$ $\upsilon_c = \frac{\upsilon_x}{R}$

Recharge (simple dilution)	Movement of water across the water table into the saturated zone	Dependent on aquifer matrix properties, depth to groundwater, surface water interactions, and climate	Causes dilution of the contaminant plume and may replenish electron acceptor concentrations, especially dissolved oxygen	Site specific
Volatilization	Volatilization of contaminants dissolved in groundwater into the vapor phase (soil gas)	Dependent on the chemical's vapor pressure and Henry's law constant	Removes contaminants from groundwater and transfers them to soil gas; typically unimportant for most groundwater systems	$C_a = HC_l$
Biodegradation	Microbially mediated oxidation–reduction reactions that degrade contaminants	Dependent on groundwater geochemistry, microbial population, and contaminant properties; biodegradation can occur under aerobic and/or anaerobic conditions	May ultimately result in complete degradation of contaminants; typically the most important process, acting to truly reduce contaminant mass	$C = C_o e^{-kt}$
Abiotic degradation	Chemical transformations that degrade contaminants without microbial facilitation; only halogenated compounds are subject to these mechanisms in the groundwater environment	Dependent on contaminant properties and groundwater geochemistry	Can result in partial or complete degradation of contaminants; rates typically much slower than for biodegradation	See Chapter 3
Partitioning from NAPL	Partitioning from NAPL into groundwater; NAPL whether mobile or residual, tends to act as a continuing source of groundwater contamination	Dependent on aquifer matrix and contaminant properties, as well as groundwater mass flux through or past NAPL plume	Dissolution of contaminants from NAPL represents the primary source of dissolved contamination in groundwater	See Chapter 2

[a]Notes: υ_x, seepage velocity; K, hydraulic conductivity; n_e, effective porosity; dH/dL, GW gradient; α_x, longitudinal dispervisivity; L_p, plume length; υ_c, contaminant velocity; F, flux; C, contaminant concentration; D, diffusion coefficient; R, retardation factor; ρ_b, bulk density; K_d, distribution coefficient; f_{oc}, fraction organic carbon; k_{oc}, normalized octanol-water partition coefficient; C_a, vapor concentration; C_l, dissolved concentration; H, Henry's law constant; C_o, initial contaminant concentration; k, decay rate.

1.3 SUMMARY OF MAJOR NATURAL ATTENUATION PROCESSES

The environmental fate and transport of an organic compound in the subsurface environment is controlled by the compound's physical and chemical properties and the physical, chemical, and biological nature of the subsurface media through which the compound is migrating. Several processes are known to cause a reduction in the concentration and/or mass of a contaminant dissolved in groundwater (see Table 1.2). Those processes that result only in the reduction of a contaminant's concentration, but not of the total contaminant mass in the system, are termed *nondestructive*. Nondestructive processes include hydrodynamic dispersion (mechanical dispersion and diffusion), sorption, dilution, and volatilization. Those processes that result in the degradation of contaminants are referred to as *destructive*. Destructive processes include biodegradation and abiotic degradation mechanisms. Biodegradation may be the dominant destructive attenuation mechanism acting on a contaminant, depending on the type of contaminant and the availability of electron donors or carbon sources. Abiotic degradation processes (e.g., hydrolysis) are also known to degrade chlorinated solvents; where biodegradation is not occurring, these may be the only destructive processes operating. However, the rates of abiotic processes are generally slow relative to rates of biodegradation for most contaminants.

Natural attenuation results from the integration of all the subsurface attenuation mechanisms (both abiotic and biotic) operating at a given site. Table 1.3 summarizes the biological and abiotic processes that affect fate and transport of chlorinated

TABLE 1.3. Degradation Mechanisms for Selected Organic Compounds

Compound	Degradation Mechanism
PCE	Reductive dechlorination
TCE	Reductive dechlorination, cometabolism
DCE	Reductive dechlorination, direct biological oxidation
Vinyl chloride	Reductive dechlorination, direct biological oxidation
TCA	Reductive dechlorination, hydrolysis, dehydrohalogenation
1,2-DCA	Reductive dechlorination, direct biological oxidation
Chloroethane	Hydrolysis
Carbon tetrachloride	Reductive dechlorination, cometabolism, abiotic
Chloroform	Reductive dechlorination, cometabolism
Methylene chloride	Direct biological oxidation
Chlorobenzenes	Direct biological oxidation, reductive dechlorination, cometabolism
Benzene	Direct biological oxidation
Toluene	Direct biological oxidation
Ethylbenzene	Direct biological oxidation
Xylenes	Direct biological oxidation
1,2-Dibromoethane	Reductive dehalogenation, hydrolysis, direct biological oxidation, and photolysis

solvents and fuel hydrocarbons dissolved in groundwater. Abiotic attenuation mechanisms are discussed in detail in Chapter 3. Biodegradation is discussed in Chapters 4, 5, and 6. It is important to separate abiotic from biotic attenuation mechanisms when evaluating natural attenuation. The methods for correcting apparent attenuation caused by nondestructive attenuation mechanisms are discussed in Chapter 7.

1.4 DEMONSTRATING NATURAL ATTENUATION AT FIELD SITES

Three lines of evidence, in various forms, have been used in recent years to demonstrate intrinsic bioremediation (see, e.g., National Research Council, 1994; Wiedemeier et al., 1995, 1996). These lines of evidence include

1. Historical database showing plume stabilization and/or loss of contaminant mass over time
2. Chemical and geochemical analytical data, including:
   a. Depletion of electron acceptors and donors
   b. Increasing metabolic by-product concentrations
   c. Decreasing parent compound concentrations
   d. Increasing daughter compound concentrations
3. Microbiological data that support the occurrence of biodegradation and give estimates of biodegradation rates.

The first line of evidence involves using historical contaminant data to show that the contaminant plume is shrinking, stable, or expanding at a rate slower than predicted by conservative (slow) groundwater seepage velocity calculations. In some cases a biologically recalcitrant (conservative) tracer present in measurable concentrations within the contaminant plume can be used in conjunction with aquifer hydrogeologic parameters such as seepage velocity and dilution to show that a reduction in contaminant mass is occurring and to estimate biodegradation rate constants. When microorganisms degrade organic contaminants in the subsurface, they cause measurable changes in soil and groundwater chemistry. The second line of evidence involves measurement of this biogeochemical signature.

The first line of evidence is valuable because it shows that the contaminant plume is being attenuated, but it does not necessarily tell the investigator that contaminant mass is being destroyed. The second line of evidence is used to show that contaminant mass is being destroyed, not just being diluted or sorbed to the aquifer matrix. The third line of evidence, microbiological laboratory or field data, can be used to show that indigenous biota are capable of degrading site contaminants at a particular rate. The microcosm study is the most common technique used for this purpose.

Ideally, the first line of evidence should be used to document intrinsic bioremediation (Wiedemeier et al., 1995, 1996). Because of the extensive amount of microbiological research already conducted, microcosm studies typically are not necessary. Several types of soil–sediment and groundwater analytical data are used to obtain the first two lines of evidence, as discussed in Chapter 7. In addition to these lines of evidence, solute transport models can be useful in presenting a weight of evidence supporting natural attenuation.

1.5 NATURAL ATTENUATION STUDIES, PROTOCOLS, AND GUIDANCE

A number of key studies have contributed to our understanding of natural attenuation. Barker and Wilson (1997) and Rifai (1997) described the concept of natural aerobic and anaerobic biological attenuation, respectively, at the Subsurface Remediation Conference held in Dallas in 1992. Wiedemeier et al. (1995, 1996) developed a detailed field protocol for assessing natural attenuation at fuel hydrocarbon and chlorinated solvent sites, respectively. The U.S. EPA (1994a, 1996) conducted two symposia on natural attenuation of fuels and chlorinated solvents, and the U.S. EPA (1994b) Office of Underground Storage Tanks (OUST) has developed an evaluation methodology for natural attenuation. The American Society for Testing and Materials (ASTM) as well as a number of state agencies and private-sector companies have issued guidance on natural attenuation [Florida Department of Environmental Regulation, 1990; Wisconsin Department of Natural Resources, 1993; Amoco (Yang et al., 1995); Chevron (Buschede and O'Reilly, 1995); Minnesota Pollution Control Agency, 1995; Mobil Oil Corporation, 1995; ASTM, 1998].

The number of natural attenuation field assessments also has expanded over the past few years. Of note are the natural attenuation field characterization studies completed by the Air Force Center for Environmental Excellence (AFCEE) for fuel hydrocarbons and chlorinated solvent sites (Wiedemeier et al., 1995, 1996; see Appendix A). These studies provide the most comprehensive database of geochemical characterization collected in support of natural attenuation currently available. Other studies include Barker et al. (1987), Klecka et al. (1990), Caldwell et al. (1992), Borden et al. (1994), Davis et al. (1994), McAllister and Chiang (1994), Rifai and Bedient (1994), Stauffer et al. (1994), Daniel (1995), Wilson et al. (1995), and Wiedemeier and Chapelle (1998).

The numerous studies listed above as well as additional case studies presented at the biennial Battelle Conference Series on In Situ and On Site Bioremediation (Battelle, 1991, 1993, 1995, 1997) and the International Business Communications Annual International Symposia on Intrinsic Bioremediation all demonstrate a number of common themes about natural attenuation:

1. Natural attenuation processes have been observed and documented at numerous sites.
2. Natural attenuation controls the extent of migration of contaminants away from the source and limits the extent of the dissolved plume.
3. Intrinsic bioremediation of fuel hydrocarbons in groundwater occurs universally, while only some sites contaminated with chlorinated solvents exhibit intrinsic bioremediation.

4. Geochemical characterization of the groundwater provides ample data for quantifying the natural attenuation processes that may be important at a given site.
5. Biodegradation is one of the most important attenuation mechanisms that contributes to declines in contaminant concentration and loss of mass in almost all fuel-hydrocarbon plumes and in some chlorinated solvent plumes.
6. Aerobic biodegradation of fuel hydrocarbons is evidenced by depletion of dissolved oxygen inside the plume.
7. Anaerobic biodegradation of fuel hydrocarbons is evidenced by depletion of nitrate and sulfate and production of dissolved iron [Fe(II)] and methane inside the plume.
8. Although slower than aerobic biodegradation, anaerobic biodegradation processes are a significant component of natural attenuation, due to the abundance of anaerobic electron acceptors [NO_3^-, Fe(III), Mn, SO_4^{2-} and CO_2] relative to dissolved oxygen (see data from the Air Force sites in Appendix A).
9. Although some chlorinated solvents (e.g., DCA and VC) will degrade aerobically and most will biodegrade during cometabolism, *halorespiration* is the most important biodegradation reaction involving the common chlorinated solvents. In this reaction, electron donors such as BTEX and methanol are fermented to form the direct electron donor, dissolved hydrogen. Halorespirators then use the hydrogen to dechlorinate the chlorinated solvents.
10. Halorespiration is indicated by (a) dissolved hydrogen concentrations in certain ranges, (b) low reduction–oxidation (redox) potential environments from which electron acceptors such as sulfate have been removed, (c) the presence of methane, which indicates the presence of fermentation reactions and hydrogen, and (d) the presence of biologically produced daughter products that are not typically released at given sites, such as *cis*-1,2-dichloroethene and VC.

1.6 NATURAL ATTENUATION IN THE REGULATORY PROCESS

Starting in 1995, EPA and many states began to develop policies regarding the use of natural attenuation for groundwater remediation. As this book is being written, the process is still ongoing, and the following discussion provides only a brief snapshot of developing regulatory policy.

Federal Policy Regarding Natural Attenuation

Although natural attenuation has been a part of EPA's remediation alternatives since the promulgation of many of the key remediation regulatory programs, a lack of technical understanding and the resulting caution on the part of the regulatory community have greatly limited its application until recent years. Because of increasing interest in its application, EPA issued a memorandum in November 1997 to clarify the use of *monitored natural attenuation* at Superfund, RCRA, underground storage tank (UST), and other contaminated sites. This policy memorandum states that EPA views monitored natural attenuation as only one component of the total remedy at the majority of sites, and it should be "used very cautiously as the sole remedy at contaminated sites." The memorandum stresses that use of monitored natural attenuation did not signify any change in EPA's remediation objectives, including source control and restoration of contaminated groundwater, and that selection of this alternative must be supported by detailed, site-specific technical information that demonstrates the efficacy of this remediation approach. In addition, the progress of natural attenuation must be monitored carefully and compared to expectations.

EPA also stated that while monitored natural attenuation may be used where the circumstances are appropriate, "it should be used with caution commensurate with the uncertainties associated with the particular application." Because of the uncertainties involved, source control and performance monitoring were prescribed as "fundamental components of any monitored natural attenuation remedy." The memorandum provides additional information on several topics associated with the use of monitored natural attenuation (U.S. EPA, 1997), as discussed below.

Role of Monitored Natural Attenuation. The EPA directive states that monitored natural attenuation should not be considered a default or presumptive remedy at any contaminated site. Monitored natural attenuation is an appropriate remediation method only where its use will be protective of human health and the environment and it will be capable of achieving site-specific remediation objectives within a time frame that is reasonable compared to other alternatives.

Demonstrating the Efficacy of Natural Attenuation through Site Characterization. Decisions to employ monitored natural attenuation as a remedy or remedy component should be supported thoroughly and adequately with site-specific characterization data and analysis. Three types of site-specific information or evidence should be used in such an evaluation:

1. Historical groundwater and/or soil chemistry data that demonstrate a clear and meaningful trend of decreasing contaminant mass and/or concentration over time at appropriate monitoring or sampling points. (In the case of a groundwater plume, decreasing concentrations should not be solely the result of plume migration. In the case of inorganic contaminants, the primary attenuating mechanism also should be understood.)
2. Hydrogeologic and geochemical data that can be used to demonstrate indirectly the type(s) of natural attenuation processes active at the site and the rate at which such processes will reduce contaminant concentrations to required levels. For example, characterization data may be used to quantify the rates of contaminant sorption, dilution, or volatilization, or to demonstrate and quantify the rates of biological degradation processes occurring at the site.

3. Data from field or microcosm studies (conducted in or with actual contaminated site media) that directly demonstrate the occurrence of a particular natural attenuation process at the site and its ability to degrade the contaminants of concern (typically used to demonstrate biological degradation processes only).

Unless EPA or the implementing state agency determines that historical data (type 1 above) are of sufficient quality and duration to support a decision to use monitored natural attenuation, EPA expects that data characterizing the nature and rates of natural attenuation processes at the site (type 2 above) should be provided. Where the latter are also inadequate or inconclusive, data from microcosm studies (type 3 above) may also be necessary.

Sites Where Monitored Natural Attenuation May Be Appropriate. EPA expects that monitored natural attenuation will be most appropriate when used in conjunction with active remediation measures (e.g., source control) or as a follow-up to active remediation measures that have already been implemented.

Reasonableness of Remediation Time Frames. Defining a reasonable time frame is a complex and site-specific decision. Factors that should be considered when evaluating the length of time appropriate for remediation include

- Classification of the affected resource (e.g., drinking water source, agricultural water source) and value of the resource
- Relative time frame in which the affected portions of the aquifer might be needed for future water supply (including the availability of alternate supplies)
- Uncertainties regarding the mass of contaminants in the subsurface and predictive analyses (e.g., remediation time frame, timing of future demand, and travel time for contaminants to reach points of exposure appropriate for the site)
- Reliability of monitoring and of institutional controls over long time periods
- Public acceptance of the extended time for remediation
- Provisions by the responsible party for adequate funding of monitoring and performance evaluation over the period required for remediation

Remediation of Contaminated Sources and Highly Contaminated Areas. EPA expects that source control measures will be evaluated for all contaminated sites and that source control measures will be taken at most sites where practicable.

Performance Monitoring. Performance monitoring should continue as long as contamination remains above required cleanup levels.

Contingency Remedies. It is also recommended that one or more criteria ('triggers') be established, as appropriate, in the remedy decision document that will signal unacceptable performance of the selected remedy and indicate when to implement contingency measures. Such criteria might include the following:

- Contaminant concentrations in soil or groundwater at specified locations exhibit an increasing trend.
- Near-source wells exhibit large concentration increases indicative of a new or renewed release.
- Contaminants are identified in sentry or sentinel wells located outside the original plume boundary, indicating renewed contaminant migration.
- Contaminant concentrations are not decreasing at a sufficiently rapid rate to meet the remediation objectives.
- Changes in land and/or groundwater use will adversely affect the protectiveness of the monitored natural attenuation remedy.

State Policy Regarding Natural Attenuation

While the federal government enacted most of the original legislation regarding the remediation of contaminated sites (e.g., Superfund, RCRA, and UST regulations), many states now have been delegated authority by EPA for implementation of their own programs. For example, most UST regulation is performed at the state level, based on general federal policy requirements. State requirements can be more stringent than the federal program but cannot be less stringent than the requirements established by the EPA. Superfund, the Comprehensive Environmental Restoration, Compensation, and Liability Act (CERCLA), is an example of a program that is still implemented largely by the federal government, but many states have their own state Superfund programs that are funded with state monies. The state Superfund programs generally follow the federal Superfund requirements but include provisions unique to the individual states.

In a 1994 publication, the Air Force reviewed state regulations regarding natural attenuation (Air Force Center for Environmental Excellence, 1994) and found eight states with natural attenuation policies. Although these eight states used different terminology for natural attenuation, they all had provisions in their regulations that acknowledged that concentrations of degradable contaminants will naturally decline over time and that public health and the environment will be protected as long as concentrations are reduced to acceptable levels. The states and their key terminology include

Delaware	contamination is controlled under natural conditions
District of Columbia	contamination is controlled under natural conditions
Florida	monitoring only
Iowa	passive cleanup
Michigan	natural attenuation
North Carolina	natural remediation
Ohio	monitoring only
Wisconsin	passive bioremediation with long-term monitoring

The authors concluded that Wisconsin and North Carolina had "progressed further than other states," with Wisconsin having developed natural attenuation guidance and North Carolina providing a detailed description of the state's position.

In 1997, the Partners in RBCA Implementation performed a survey of risk-based corrective action policies being used or considered by state regulators, and included questions regarding the applicability of natural attenuation as a remediation alternative. Of 21 states providing information, 15 states allowed consideration of natural attenuation, 3 allowed consideration of natural attenuation with constraints, and 3 states did not allow natural attenuation (Figure 1.5 and Table 1.4). Summaries of representative state programs are provided below.

Wisconsin. Wisconsin currently requires that at least three cleanup options be evaluated, one of which must be passive bioremediation, and that "if passive bioremediation with long-term monitoring is feasible but not recommended alternative, a clear rationale shall be provided as to why this alternative is not acceptable." Natural attenuation field studies must demonstrate that (1) the plume is stable or receding; (2) natural attenuation processes are reducing the plume mass and concentration; (3) Wisconsin groundwater enforcement standards will be met within a reasonable time period (determined on a site-specific basis); (4) relevant receptors are not being affected; and (5) all other affected media and exposure pathways have been addressed. Groundwater use restrictions are required for the zone containing contaminants in excess of Wisconsin enforcement standards.

Texas. In rules that are still being developed, the proposed Texas Risk Reduction Program will allow natural attenuation remedies at sites in which (1) the plume will not increase in size by more than 500 ft (i.e., 500 ft of additional plume growth is allowed); (2) natural attenuation will result in site cleanup within a 30-year time

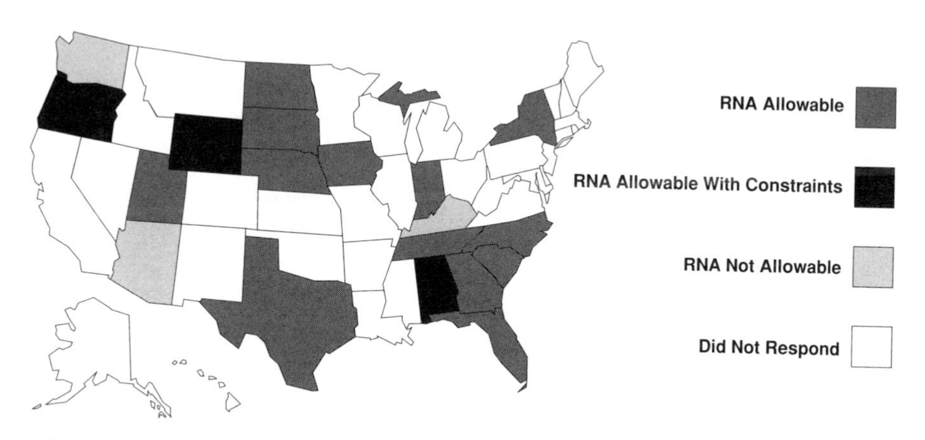

FIGURE 1.5. States where remediation by natural attenuation (RNA) is allowable, allowable with constraints, or not allowable. [From RBCA State Policies Database (www.gsi-net.com/rbcapol).]

TABLE 1.4. Responses of State Regulators on Natural Attenuation Policies as Listed in RBCA State Policies Database (www.gsi-net.com/rbcapol)

State	State Allows Consideration of Natural Attenuation?	Comments
Alabama	Yes	Plume growth limited to 500 ft
Arizona	No	
Delaware	Yes	No unacceptable risks to on-site on off-site persons
Florida	Yes	Default criteria and scientific evaluation process
Georgia	Yes	Not to exceed alternate concentration limits
Indiana	Yes	Subject to postclosure monitoring
Iowa	Yes	Natural attenuation determined by trends in analytical data
Kentucky	No	
Michigan	Yes	No impact to receptor during period of natural attenuation; must provide isolation distance from private and public well water supplies
Nebraska	Yes	Site-by-site basis
New York	Yes	
North Carolina	Yes	
North Dakota	Yes	No restrictions on the use of natural attenuation
Oregon	Yes	Conditional on institutional controls, public hearing, postclosure monitoring
South Carolina	Yes	Prove natural attenuation is occurring (plume stable for 18 months)
South Dakota	Yes	Source removal and groundwater monitoring required
Tennessee	Yes	Not applicable where groundwater is classified as drinking water
Texas	Yes	
Utah	Yes	Case-by-case basis
Washington	No	
Wyoming	Yes	Contamination must be close to remediation standard and the time to complete natural attenuation must approximate time for active remediation without significant spreading

frame; and (3) groundwater use restrictions can be applied to any off-site portion of contaminant plumes. In one of the few instances in which natural attenuation of NAPL layers is allowed, Texas is proposing that removal of product will not be required if (1) the NAPL layer is stable and not moving; (2) the NAPL is not contributing to fire, explosion, or human exposure; and (3) the dissolved contaminant plume is not expanding.

Florida. Florida proposed rules in 1997 in which monitoring of natural attenuation is recognized as a strategy for rehabilitation of hydrocarbon-affected sites. Natural attenuation was determined to be appropriate if (1) there is no free product, (2) no contaminated soils are present that serve as a continuing source of contamination, (3) the plume has not migrated beyond a temporary point of compliance, (4) there is a decrease in the mass of contamination, and (5) Florida groundwater cleanup standards will be met at the temporary point of compliance within five years if natural attenuation is demonstrated to be a protective and cost-effective alternative.

California. California typically is a bellwether state for developing regulations pertaining to pollutants in the environment. The state's position on natural attenuation is no exception. Based on the results of the *California Leaking Underground Fuel Tank* (LUFT) *Historical Case Analyses* (Rice et al., 1995a), researchers at the Lawrence Livermore National Laboratory (LLNL) developed *Recommendations to Improve the Cleanup Process for California's Leaking Underground Storage Tanks* (Rice et al., 1995b). The primary recommendations made by Rice et al. (1995b) include (1) utilizing passive bioremediation as a remedial alternative whenever possible, (2) modifying State Water Resources Control Board (SWRCB) policies to allow risk-based decision making for LUFT cleanups, and (3) requiring risk-based decision making that emphasizes passive bioremediation for LUFT cleanups.

The California SWRCB has since responded to the LLNL recommendations by recommending to "proceed aggressively to close *low risk* soil only cases. For cases affecting *low risk* groundwater (for instance, shallow groundwater with maximum depth to water less than 50 feet and no drinking water wells screened in the shallow groundwater zone within 250 feet of the leak) we recommend that active remediation be replaced with monitoring to determine if the fuel leak plume is stable. Obviously good judgment is required in all of these decisions. However, that judgment should now include knowledge provided by the LLNL report." These recommendations were written as interim guidance pending release of the State Senate Bill 1764 Advisory Committee's Recommendations Report Regarding California's Underground Storage Tank Program and any possible revisions to SWRCB Resolution 92-49 that would result from the advisory committee's findings. Potential problems posed by methyl tertiary butyl ether (MTBE) in groundwater (MTBE was not considered in the LLNL reports) at fuel-hydrocarbon-contaminated sites have delayed changes to SWRCB Resolution 92-49, and at this time the state of California has no specific policy concerning the application of risk-based remediation or natural attenuation at fuel-hydrocarbon-contaminated sites.

Other Approaches for Natural Attenuation. The Air Force review of natural attenuation policies also recognized that two other regulatory approaches could be utilized to implement natural attenuation–based remedies: RCRA *alternative concentration limits* (ACLs) and groundwater mixing zones (Air Force Center for Environmental Excellence, 1994). These provisions cover groundwater contamination in general and are not specific to petroleum products. In both types of provisions, groundwater in a defined area around a site is allowed to be contaminated above water quality standards as long as the water quality at a specified downgradient location conforms to standards. Under RCRA, the regional administrator can "establish an alternative concentration limit for a hazardous constituent," allowing a portion of groundwater to exceed groundwater cleanup standards if it can be demonstrated that human health and the environment are protected. The mixing zone is commonly used in surface water, where small area of degradation is allowed for waste to mix with the receiving water.

Some of the state-specific terminology related to ACLs and mixing zones is as follows:

Alternative Concentration Limits (ACLs)	
Illinois	zone of attenuation, alternate compliance boundary
Kentucky	alternative [waste] boundary
Missouri	alternative criteria
New Hampshire	boundary of the groundwater management zone
North Carolina	compliance boundary
South Dakota	perimeter of operation pollution (POP)
Tennessee	alternative [compliance] boundary
Wisconsin	point of standards application

Mixing Zones	
Florida	zone of discharge
Maryland	mixing zone
South Carolina	mixing zone

Natural Attenuation versus Risk-Based Corrective Action

RBCA and natural attenuation are interrelated management approaches for contaminated groundwater. RBCA is a risk-management approach, where risks from a number of pathways are calculated and the objective is to eliminate excess risk to human receptors and the environment. The RBCA program was devised for state regulatory programs for petroleum release sites (ASTM, 1995) and was extended for chemical release sites (nonpetroleum sites) in 1998. In a general sense, RBCA originated from the regulatory perspective, in which regulators and the regulated community collaborated under the ASTM consensus-building process to write the RBCA standard.

Natural attenuation, on the other hand, has more of its origin in the scientific community, where the goal was to derive a detailed understanding of the fate and transport processes associated with contaminants in groundwater and to transform this understanding into an accepted remediation alternative. AFCEE provided much of the leadership for this program with their initial natural attenuation initiative for petroleum-hydrocarbon sites and subsequent efforts focused on chlorinated solvent plumes.

RBCA and natural attenuation are related programs, as one of the key elements of RBCA is to quantify the risks to receptors located some distance away from the source. To quantify this risk associated with groundwater, usually the critical pathway in a RBCA analysis, the amount of natural attenuation occurring between the source and the receptor exposure point must be quantified. Therefore, many of the state RBCA programs that have been developed to date have some guidelines for quantifying natural attenuation.

1.7 ORGANIZATION OF THIS BOOK

This book contains 11 chapters, including this introductory chapter. In Chapter 2 we consider natural attenuation in the source zone versus attenuation of contaminants in the aqueous phase. The main focus of Chapter 2 is on assessing the impact of NAPLs on site remediation. Hydrodynamic and abiotic processes involved in natural attenuation are discussed in Chapter 3. The processes of advection, dispersion, sorption, dilution, volatilization, hydrolysis, dehydrohalogenation, and reduction reactions are reviewed and evaluated in terms of their relative impact on natural attenuation. Chapters 4, 5, and 6 cover biodegradation. In Chapter 4 we provide an introduction to the biological processes effecting contaminant destruction. In Chapters 5 and 6 we consider the biodegradation of fuel hydrocarbons and chlorinated solvents, respectively.

The assessment and quantification of natural attenuation are discussed in Chapter 7. The simulation of natural attenuation is an important tool for predicting the cleanup time for a contaminant plume. Modeling natural attenuation is considered in Chapter 8. Numerous case studies of sites in which natural attenuation is occurring and in which it is not occurring are considered in Chapters 9 and 10. In Chapter 11 we provide an in-depth discussion covering the design of long-term monitoring programs and some of the key policy decisions related to natural attenuation.

REFERENCES

Air Force Center for Environmental Excellence, 1994, *Review of State Regulations Regarding Natural Attenuation as a Remedial Option*, technical brief, AFCEE, San Antonio, TX, 7 pp.

American Society for Testing and Materials (ASTM), 1995, *Standard Guide for Risk-Based Corrective Action Applied at Petroleum Release Sites*, ASTM E-1739-95, ASTM, Philadelphia.

———, 1998, *ASTM Guide for Remediation by Natural Attenuation at Petroleum Release Sites*, ASTM, Philadelphia.

Barker, J. F., and Wilson, J. T., 1997, Natural biological attenuation of aromatic hydrocarbons under anaerobic conditions, in *Subsurface Restoration*, C. H. Ward, J. A. Cherry, and M. R. Scalf, eds., Ann Arbor Press, Chelsea, MI, pp. 289–300.

Barker, J. F., Patrick, G. C., and Major, D., 1987, Natural attenuation of aromatic hydrocarbons in a shallow sand aquifer, *Ground Water Monit. Rev.*, vol. 7, no. 1, pp. 64–71.

Bartow, G., and Davenport, C., 1995, Pump-and-treat accomplishments: a review of the effectiveness of groundwater remediation in Santa Clara Valley, California, *Ground Water Monit. Rev.*, vol. 15, no. 2, pp. 139–146.

Battelle, 1991, *Proceedings of the First International In Situ and On Site Bioreclamation Symposium*, San Diego, CA, Apr., Battelle Press, Columbus, OH.

———, 1993, *Proceedings of the 2nd International In Situ and On Site Bioreclamation Symposium*, San Diego, CA, Apr., Lewis Publishers, Boca Raton, FL.

———, 1995, *Proceedings of the 3rd International In Situ and On Site Bioreclamation Symposium*, San Diego, CA, Apr., Battelle Press, Columbus, OH.

———, 1997, *Proceedings of the 4th International In Situ and On Site Bioreclamation Symposium*, New Orleans, LA, Apr., Battelle Press, Columbus, OH.

Borden, R. C., Gomez, C. A., and Becker, M. T., 1994, Natural bioremediation of a gasoline spill, in *Hydrocarbon Bioremediation*, R. E. Hinchee, B. C. Alleman, R. E. Hoeppel, and R. N. Miller, eds., Lewis Publishers, Boca Raton, FL, pp. 290–295.

Buschede, T., and O'Reilly, K., 1995, *Protocol for Monitoring Intrinsic Bioremediation in Groundwater*, report by Chevron Research and Technology Company, Chevron, San Francisco, 20 pp.

Caldwell, K. R., Tarbox, D. L., Barr, K. D., Fiorenza, S., Dunlap, L. E., and Thomas, S. B., 1992, Assessment of natural bioremediation as an alternative to traditional active remediation at selected Amoco Oil Company sites, Florida, in *Proceedings of the NGWA/API Conference on Petroleum Hydrocarbons*, pp. 509–525.

Daniel, R. A., 1995, Intrinsic bioremediation of BTEX and MTBE: field, laboratory and computer modeling studies, M.S. thesis, North Carolina State University, Raleigh, NC, 325 pp.

Davis, J. W., Klier, N. J., and Carpenter, C. L., 1994, Natural biological attenuation of benzene in groundwater beneath a manufacturing facility, *Ground Water*, vol. 32, no. 2, pp. 215–226.

Florida Department of Environmental Regulation, 1990, *No Further Action and Monitoring Only Guidelines for Petroleum Contaminated Sites*, FDER, Tallahassee, FL, 6 pp.

Groundwater Services, Inc. (GSI), 1997, *Florida RBCA Planning Study: Impact of RBCA Policy Options on LUST Site Remediation Costs*, a report submitted to the Florida Partners in RBCA Implementation, GSI, Houston, TX, 26 pp.

Johnson, R. L., Johnson, P. C., McWhorter, D. B., Hinchee, R. E., and Goodman, I. 1993, An overview of air sparging, *Ground Water Monit. Remediation*, vol. 13, no. 4, pp. 127–135.

Klecka, G. M., Davis, J. W., Gray, D. R., and Madsen, S. S., 1990, Natural bioremediation of organic contaminants in groundwater: Cliffs–Dow Superfund site, *Ground Water*, vol. 28, no. 4, pp. 534–543.

Lee, M. D., Jamison, V. W., and Raymond, R. L., 1987, Applicability of in-situ remediation as a remediation alternative, *Petroleum Hydrocarbons and Organic Chemicals in Groundwater: A Conference and Exposition*, Houston, TX, Nov. 17–19, National Water Well Association, Westerville, OH, pp. 167–186.

Mace, R. E., Fisher, R. S., Welch, D. M., and Parra, S. P., 1997, *Extent, Mass, and Duration of Hydrocarbon Plumes from Leaking Petroleum Storage Tank Sites in Texas*, Bureau of Economic Geology Geological Circular 97-1, 52 pp.

McAllister, P. M., and Chiang, C. Y., 1994, A practical approach to evaluating natural attenuation of contaminants in groundwater, *Ground Water Monit. Rev.*, vol. 14, no. 2, pp. 161–173.

Mobil Oil Corporation, Environmental Health and Safety Division and Environmental Health Risk Assessment Group, Stonybrook Laboratories, 1995, *A Practical Approach to Evaluating Intrinsic Bioremediation of Petroleum Hydrocarbons in Groundwater*, internal report, Mobil, New York, 29 pp.

Minnesota Pollution Control Agency, 1995, *Assessment of Natural Attenuation at Petroleum Tank Release Sites*, agency guidance, MPCA, St. Paul, MN.

Newell, C. J., and Connor, J. A., 1998, *Characteristics of Dissolved Petroleum Hydrocarbon Plumes: Results from Four Studies*, API technical transfer bulletin, American Petroleum Institute, Washington, DC, 8 pp.

Newell, C. J., Hopkins, L. P., and Bedient, P. B., 1990, A hydrogeologic database for groundwater modeling, *Ground Water*, vol. 28, no. 5, pp. 703–714.

National Research Council, 1994, *Alternatives for Groundwater Cleanup*, National Academy Press, Washington, DC, 315 pp.

Rice, D. W., Dooher, B. P., Cullen, S. J., Everett, L. G., Kastenberg, W. E., Grose, R. D., and Marino, M. A., 1995a, *Recommendations to Improve the Cleanup Process for California's Leaking Underground Fuel Tanks (LUFTs)*, report submitted to the California State Water Resources Control Board and the Senate Bill 1764 Leaking Underground Fuel Tank Advisory Committee, California Environmental Protection Department, Sacramento, CA, 20 pp.

Rice, D. W., Grose, R. D., Michaelsen, J. C., Dooher, B. P., MacQueen, D. H., Cullen, S. J., Kastenberg, W. E., Everett, L. G., and Marino, M. A., 1995b, *California Leaking Underground Fuel Tank (LUFT) Historical Case Analyses*, report submitted to the California State Water Resources Control Board Underground Storage Tank Program and the Senate Bill 1764 Leaking Underground Fuel Tank Advisory Committee, California Environmental Protection Department, Sacramento, CA, 20 pp.

Rifai, H. S., 1997, Natural aerobic biological attenuation, in *Subsurface Restoration*, C. H. Ward, J. A. Cherry, and M. R. Scalf, eds., Ann Arbor Press, Chelsea, MI, pp. 411–427.

Rifai, H. S., and Bedient, P. B., 1994, Field demonstration of natural biological attenuation, in *Hydrocarbon Bioremediation*, R. E. Hinchee, B. C. Alleman, R. E. Hoeppel, and R. N. Miller, eds., Lewis Publishers, Boca Raton, FL, pp. 353–361.

Stauffer, T. B., Antworth, T. B., Boggs, J. M., and MacIntyre, W. G., 1994, A natural gradient tracer experiment in a heterogeneous aquifer with measured in situ biodegradation rates: a case for natural attenuation, in *Proceedings of the Symposium on Intrinsic Bioremediation of Groundwater*, Denver, CO, Aug. 30–Sept. 1, EPA/540/R-94/515, U.S. EPA, Washington, DC, pp. 73–84.

United States Environmental Protection Agency, (U. S. EPA) *Evaluation of Groundwater Extraction Remedies*, Vol. 1, *Summary Report*, EPA/540/2-89/054, U.S. EPA, Washington, DC, 1989.

———, 1992a, *Evaluation of Groundwater Extraction Remedies: Phase II*, Vol. 1, *Summary Report*, Publication 9355.4-05, PB92-963346, 1989, U.S. EPA, Washington, DC, 27 p.

———, 1992b, *Considerations in Ground-Water Remediation at Superfund Sites and RCRA Facilities: Update*, EPA OSWER 9283.1-06, U.S. EPA, Washington, DC, May 27.

———, 1994a, *Proceedings of the Symposium on Intrinsic Bioremediation of Groundwater*, Denver, CO, Aug. 30–Sept. 1, EPA/540/R-94/515, U.S. EPA, Washington, DC.

———, 1994b, *How to Evaluate Alternative Cleanup Technologies for Underground Storage Tank Sites: A Guide for Corrective Action Plan Reviewers*, EPA/510/B-94/003, U.S. EPA, Washington, DC.

———,1996, *Symposium on Natural Attenuation of Chlorinated Organics in Groundwater*, Dallas, TX, Sept. 11–13, EPA/540/R-96/509, U.S. EPA, Washington, DC.

———, 1997, *Use of Monitored Natural Attenuation at Superfund, RCRA Corrective Action, and Underground Storage Tank Sites*, Draft Interim Final, U.S. EPA Office of Solid Waste and Emergency Response Directive 9200.4-17, U.S. EPA, Washington, DC (http://www.epa.gov/oust/directiv/9200417z.htm).

Wiedemeier, T. H., and Chapelle, F. H., 1998, *Technical Guidelines for Evaluating Monitored Natural Attenuation of Petroleum Hydrocarbons and Chlorinated Solvents in Groundwater at Naval and Marine Corps Facilities*, Naval Facilities, Engineering Command, Alternative Restoration Technology Team,

Wiedemeier, T. H., Wilson, J. T., Kampbell, D. H., Miller, R. N., and Hansen, J. E., 1995, *Technical Protocol for Implementing Intrinsic Remediation with Long-Term Monitoring for Natural Attenuation of Fuel Contamination Dissolved in Groundwater*, U.S. Air Force Center for Environmental Excellence, San Antonio, TX.

Wiedemeier, T. H., Swanson, M. A., Moutoux, D. E., Gordon, E. K., Wilson, J. T., Wilson, B. H., Kampbell, D. H., Hansen, J. E., Haas, P., and Chapelle, F. H., 1996, *Technical Protocol for Evaluating Natural Attenuation of Chlorinated Solvents in Groundwater*, Draft, Revision 1, U.S. Air Force Center for Environmental Excellence, San Antonio, TX.

Wilson, J. T., Sewell, G., Caron, D., Doyle, G., and Miller, R. N., 1995, Intrinsic bioremediation of jet fuel contamination at George Air Force Base, in *Intrinsic Bioremediation*, R. E. Hinchee, J. T. Wilson, and D. C. Downey, eds., Battelle Press, Columbus, OH, pp. 91–100.

Wisconsin Department of Natural Resources, 1993, *Natural Biodegradation as a Remedial Action Option: Interim Guidance*, WDNR, Madison, WI, 40 pp.

Yang, X., Jeng, C. Y., Kremesec, V., Fisher, B., and Curran, L., 1995, *Natural Attenuation as a Remedial Alternative,* technical guidance, Amoco Internal Report 29, Amoco, Chicago.

CHAPTER 2

ATTENUATION OF SOURCE ZONES AND FORMATION OF PLUMES

In this chapter we describe the characteristics of the source zones that create dissolved contaminant plumes and discuss the processes that control the partitioning of contaminants into groundwater. Four different source scenarios are presented, with particular emphasis on non-aqueous-phase liquid (NAPL) migration and dissolution. We emphasize the use of a few key relationships and simple models for evaluating source zones. The information presented in the chapter includes a review of key studies performed by U.S. and Canadian researchers and existing databases, as well as new information developed by the authors. One key concept used in this chapter is the definition of a source zone as any area at a site that delivers aqueous-phase contaminants to groundwater. Source areas may include zones with NAPLs, contaminated vadose zone soils, and areas where aqueous-phase releases have been introduced into groundwater.

2.1 DISTRIBUTION OF CONTAMINANTS IN THE SUBSURFACE

Contaminants in the subsurface can be found in up to four different states at sites contaminated by fuel hydrocarbons and chlorinated solvents: dissolved in water, sorbed onto matrix particles, volatilized in soil vapors, or as NAPLs. Soluble fuel compounds and chlorinated solvents in the vadose zone can dissolve in percolating precipitation and be transported in the aqueous phase to the groundwater, usually at concentrations below their free-phase solubility. Almost all dissolved fuel and chlorinated solvents can also sorb onto the aquifer or vadose zone matrix (i.e., to native organic carbon in the matrix; note that the term *sorption* is not used to describe trapping of residual NAPL as described below). Volatile hydrocarbons can partition into the gaseous phase as anthropogenic organic soil vapor in the unsaturated zone. Finally, fuel and chlorinated solvents are present at the majority of contaminated groundwater sites in the form of NAPLs that have migrated into the subsurface. Dissolution of NAPLs (the process by which soluble chemicals partition from the NAPL into the aqueous phase) is probably the most important process that determines how quickly a source will attenuate naturally.

NAPLs are organic, nondissolved liquids in the subsurface. When introduced into the unsaturated zone, both fuel-hydrocarbon NAPLs (e.g., gasoline, diesel fuel, or jet fuels such as JP-4) and chlorinated solvent NAPLs (e.g., TCE) act generally in the same way, and finger down to the saturated zone. At the water table, however, fuel-hydrocarbon NAPLs, which are less dense than water, will float on the water table, forming a thin, pancakelike layer that spreads across the water surface. These products are referred to as light NAPL, or LNAPL. Being denser than water (DNAPL), chlorinated solvents will penetrate the water table and migrate into the saturated zone, sometimes to depths of hundreds of feet. Migration of DNAPLs is strongly influenced by changes in the texture of the aquifer matrix, and DNAPLs will spread laterally where changes in matrix permeability (*capillary barriers*) inhibit vertical NAPL migration.

Although NAPLs can be present as a single chemical (e.g., TCE), they usually are composed of mixtures of chemicals. Almost all fuel hydrocarbons are mixtures; gasoline, for example, is composed of hundreds of different compounds with a wide variety of chemical and physical properties. Many industrial solvents are mixtures of several chlorinated solvents or mixtures of chlorinated solvents and other organics, such as mineral spirits. Therefore, DNAPLs at many sites are composed of mixtures. Sites with pure-phase LNAPLs and DNAPLs have been documented; however, sites with mixed NAPLs are much more common.

When released into the subsurface, NAPL will initially form a large, connected body of continuous NAPL (also called *free-phase NAPL*, *mobile NAPL*, *nonresidual NAPL*, or *free product*) that migrates through the subsurface, invading pore spaces, fractures, and particularly any preferential pathways in the subsurface matrix, and displacing water and air in the unsaturated zone and water alone in the saturated zone. If the source persists over a long period, or if a large volume of NAPL is released, the downward and lateral migration of the NAPL may be extensive, with LNAPL spreading across the water table and DNAPL fingering downward into the saturated zone and spreading laterally across capillary barriers. Once the source loading ceases, however, the mass of continuous NAPL begins to stretch, and processes such as *snap-off* and *bypass* form residual NAPL in the subsurface zones through which the mobile NAPL flowed. Residual NAPL can be thought of as discrete blobs of NAPL in one or a few pores. These individual blobs are not connected to each other and are rendered essentially immobile by capillary forces. Residual NAPL may act as a long-term source of soluble contaminants to groundwater.

2.2 HOW GROUNDWATER PLUMES ARE FORMED

Most dissolved fuel and chlorinated solvent plumes are formed under one of four different source scenarios:

1. *Aqueous-Phase Release to Subsurface.* Under this scenario, water is contaminated at the groundwater surface with organic fuels and/or solvents. This water then migrates to groundwater by means of some transport mechanism, such as recharge from a leaking surface impoundment or discharge from an injection well. Once the delivery mechanism is halted, the contaminant concentrations in the groundwater around the former source decreases rapidly and the dissolved plume appears to separate from the source zone (Figure 2.1*a*). Because only dissolved compounds were introduced to the subsurface, moving groundwater transports them away from the source zone in a distinct, cloudlike plume that disperses slightly as the plume moves away from the source.

2. *Aqueous-Phase and/or NAPL Release to Vadose Zone Only.* Contaminants are released to the subsurface and serve as sources of aqueous-phase contaminants to groundwater. Aqueous-phase contaminants will sorb onto naturally occurring organic carbon in the soil via hydrophobic sorption. Subsequently, clean water (e.g., recharge) passing through the contaminated vadose zone will cause the contaminants to desorb and migrate to the water table as dissolved compounds.

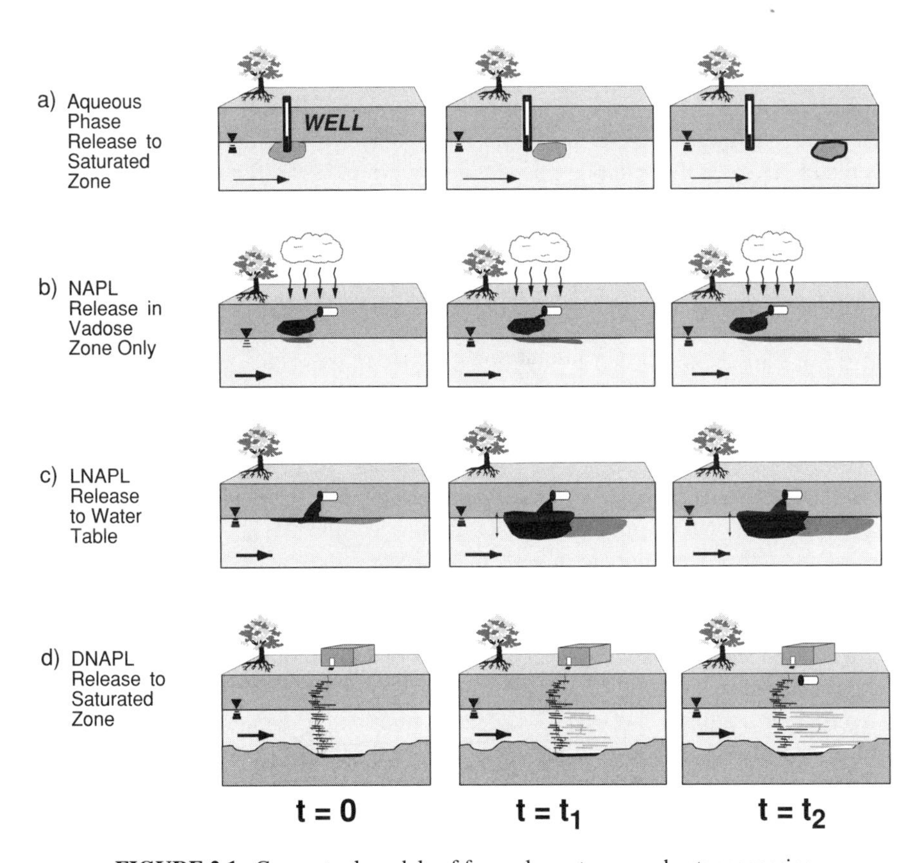

FIGURE 2.1. Conceptual models of four release-to-groundwater scenarios.

More commonly, NAPL is introduced into the vadose zone. Snap-off and bypass processes convert the once-continuous NAPL body into residual NAPL so that all of the NAPL remains in the vadose zone. Rainfall (or human-made recharge from irrigation, drainage, water conveyance, or other systems) infiltrates the subsurface, and the downward-migrating water encounters the residual NAPL. The water flowing past the residual NAPL (and to a lesser extent, sorbed compounds and contaminated soil gas) dissolves the soluble compounds, forming leachate (note that some of the dissolved contaminants in the leachate will sorb to the soil as it migrates toward the water table). The leachate then mixes with the upper few feet of the moving groundwater, forming a thin plume. Because the residual NAPL contains a large mass of dissolvable compounds relative to the rate of dissolution, the NAPL in the vadose zone acts as a long-term, continuous source of dissolved contaminants to groundwater (Figure 2.1*b*). Although this scenario does occur, particularly at locations with very deep groundwater, it is probably more common for NAPLs to penetrate the entire vadose zone to the water table. Under this scenario, discussed below, the NAPL that is located within or immediately above the saturated zone probably will contribute more mass to groundwater contamination than the NAPL in the vadose zone.

3. *LNAPL Release to Water Table.* A NAPL with a density lighter than water, including all common petroleum-based fuels, is released into and penetrates through the vadose zone, delivering LNAPL to the water table (Figure 2.1*c*). The LNAPL spreads and forms a thin layer (or *pool*) across the water table. If the water table fluctuates, the continuous NAPL in the pancakelike layer moves up and down, leaving residual LNAPL in the resulting *smear zone*. As groundwater flows through the zone of residual NAPL, and as it flows underneath any remaining LNAPL, soluble fuel compounds partition from the NAPL and form an aqueous-phase plume. (Note that the word *plume* is used here only for dissolved contaminants, not for NAPL bodies). For gasoline and JP-4 fuels, the soluble components are dominated by the BTEX (benzene, toluene, ethylbenzene, xylene) compounds. Diesel fuels and heating oils will form plumes dominated by dissolved PAHs (polynuclear aromatic hydrocarbons). As described for residual NAPL under the preceding scenario, the continuous LNAPL will also act as a long-term continuous source of dissolved contaminants to groundwater.

4. *DNAPL Release to the Saturated Zone.* Under this scenario, a NAPL containing denser-than-water compounds (e.g., chlorinated solvents) is released to the subsurface in sufficient quantity to migrate through the vadose zone, penetrate the water table, and finger downward into the saturated zone (Figure 2.1*d*). As the DNAPL fingers downward, it can spread laterally on capillary barriers, which may include obvious changes in lithology such as a clay lens in a sand aquifer, or sometimes very subtle changes in structure of a porous medium. In a fractured matrix (either fractured rock or fractured clay), the DNAPL can migrate far in unpredictable distributional patterns. Once the original source is controlled, the continuous mass of DNAPL breaks up and forms residual DNAPL in the vadose and saturated zones, or remains in DNAPL lenses or pools perched on capillary barriers. As groundwater flows past the residual DNAPL in the saturated zone, or flows over the lenses and

pools, soluble chlorinated solvents (e.g., PCE and TCE) will dissolve into the flowing groundwater (and in some cases into the aquifer material itself). The DNAPL will act as a long-term continuous source of dissolved contaminants to groundwater.

Initially, regulators, consultants, and site owners accepted a site conceptual model in which the contaminants were usually released into the vadose zone only, and precipitation-driven infiltration events delivered dissolved contaminants to the saturated zone. Therefore, contaminants in the saturated zone were thought to occur exclusively in the aqueous or sorbed (residual) phases rather than as a NAPL. (This is not surprising, as almost all textbooks written in the 1970s and 1980s had little or no discussion of organic liquids in the subsurface.) Accordingly, groundwater remediation strategies were based on a pump-and-treat philosophy to remove contaminated groundwater. For example, this approach drove almost all Superfund site investigations throughout the 1980s and early 1990s (U.S. EPA, 1993). While practitioners focusing on petroleum-hydrocarbon site remediation encountered and tried to remove LNAPL at many sites, the impact of the residual LNAPL on groundwater remediation was not fully recognized at the time.

In the late 1980s and early 1990s the research community and EPA began to focus on the constraints on aquifer remediation. In 1989, Mackay and Cherry published an overview article on why pump-and-treat systems were not successful at restoring aquifers, and included a detailed discussion of NAPLs as long-term source zones. EPA published a DNAPL Issue Paper (Huling and Weaver, 1991) in 1991, and later that year convened a DNAPL workshop in Dallas, Texas. The workshop proceedings (U.S. EPA, 1992) characterized DNAPLs as difficult-to-detect, difficult-to-remove, long-term sources of groundwater contamination. Additional EPA publications addressed estimating the likelihood of DNAPL presence at sites where DNAPL had not been detected (Newell and Ross, 1992), and the likely presence of undetected DNAPLs at Superfund sites (U.S. EPA, 1993). In 1995, EPA published an LNAPL Issue Paper to address the impact of LNAPL on characterizing and remediating petroleum-hydrocarbon sites (Newell et al., 1995).

Based on this information and a growing body of research, the current conceptual model accepted by the authors is that NAPLs are found at most sites contaminated with petroleum hydrocarbons or chlorinated solvents. Therefore, the last two conceptual plume formation scenarios described above, *LNAPL Release to Water Table and DNAPL Release to the Saturated Zone*, are crucial to understanding sources of dissolved fuel and solvent contamination in groundwater.

2.3 OBSERVATIONS OF SOURCE ZONES

Although considerable effort is devoted to characterizing source zones at contaminated sites, the variability in the distribution of source zone materials (primarily NAPLs) makes detailed delineation difficult. At most sites there will be delineation of (1) contaminated soils in the vadose zone; (2) maps showing where free product (continuous LNAPL or DNAPL) has been observed in monitoring wells, and (3) dissolved contaminant concentration data from the source zone. Less common are measurements of residual NAPL in the saturated zone taken from soil cores or by other methods. In the following sections we summarize information in available plume databases regarding source zone size, presence of NAPL, and source concentrations. These databases represent site information documented during typical site characterization studies and give a general indication of what a typical source zone may be like.

NAPL Chemicals and Abbreviations

As described above, LNAPLs are most often caused by petroleum fuel releases, in which the more soluble, aromatic hydrocarbons, such as benzene, toluene, ethylbenzene, and the xylenes, dissolve from the LNAPL. These four compounds are referred to as the BTEX compounds. The most common chlorinated solvents are the chlorinated ethenes and ethanes and their breakdown products. The chlorinated ethenes are perchloroethene (also called tetrachloroethene), trichloroethene, dichloroethene, and vinyl chloride, and are abbreviated as PCE, TCE, DCE, and VC, respectively. The most common chlorinated ethanes are trichloroethane and dichloroethane, abbreviated as TCA and DCA. Different isomers of the ethenes and ethanes are represented by using the appropriate prefix in conjunction with the abbreviation (e.g., *cis*-1,2-DCE).

Plume Database Studies

Plume database studies are compilations of data from a number of site characterization programs with analyses of key site characteristics. The four studies described below provide summary information about sources, such as the presence of NAPL, estimated amount of free product, extent of soil contamination, and dissolved chemical concentrations.

1. *Texas Bureau of Economic Geology (BEG) Study* (Mace et al., 1997). This database study was based on State of Texas files for 605 LUST (leaking underground storage tank) sites and reviews site, soil, hydrogeologic, and contaminant information. Hydraulic gradients, groundwater flow directions, average plume concentrations, and plume dimensions over time were reported for different climatic and hydrogeologic regions of Texas. Descriptive data were summarized using nonparametric statistics, histograms, and pie charts. Although not reported in this book, a solute-transport relationship was used to estimate plume lengths based on site monitoring data for approximately 220 sites.

2. *Hydrogeologic Database (HGDB)* (American Petroleum Institute, 1989; Newell et al., 1990). The HGDB is a compilation of hydrogeologic and contaminant data for approximately 400 sites around the United States. The data were compiled from responses to a nine-page questionnaire mailed to members of the American Association of Groundwater Scientists and Engineers in 1988. Although the HGDB

was originally intended to be used to determine statistical distributions of hydrogeologic parameters for use in Monte Carlo groundwater modeling by hydrogeologic setting, it also contains information about contaminant plumes and sources such as plume length, source area, presence of NAPL, and maximum concentrations.

3. *Superfund DNAPL Survey* (U.S. EPA, 1993). To evaluate the likelihood of DNAPL presence at Superfund sites, EPA conducted a survey of sites in 1993. Of the 712 Superfund sites surveyed, a subset of 310 provided detailed information about the likelihood of DNAPL presence using indirect measures such as those outlined in an EPA Fact Sheet and other indicators (Newell and Ross, 1992).

4. *Florida Risk-Based Corrective Action (RBCA) Planning Study* (Groundwater Services, Inc., 1997*a*). This study, also based on responses to a questionnaire given to site environmental consultants working on UST (underground storage tank) sites in Florida, reports source, plume, and hydrogeologic information for 117 sites. The data were used to evaluate the impact of various RBCA policy decisions on statewide UST cleanup costs.

Other plume database studies have been performed and provide vital information about dissolved-phase plumes [e.g., the Lawrence Livermore National Laboratory (LLNL) Study (Rice et al., 1995); the San Francisco Bay pump-and-treat study (Bartow and Davenport, 1995)]. These studies provide important information about plumes but do not report details of source characteristics and are not discussed further in this chapter.

Relative Source Area

The plume databases reflect a wide variation in the distribution of NAPLs and characteristics of source zones. For example, the HGDB requested that respondents report the estimated area of the source of contamination causing the dissolved plume. Although this request was very general and open to considerable interpretation on the part of the respondents, the resulting information provides a broad indication of the magnitude of source zones that produce contaminant plumes in groundwater. As shown on Figure 2.2, source areas and the corresponding plumes show considerable variation. Service station fuel-release sites appear to have the smallest source areas and some of the smallest plumes, while industrial chlorinated solvent sites have larger reported source zones and much longer plumes. Non–service station BTEX/fuel-release sites have very large reported source zones with relatively small associated plumes; this is probably due to the respondents reporting the area of large surface impoundments or process units as source zones.

The BEG study presents more focused data on a specific source zone characteristic that is usually measured at UST sites: the extent of contaminated soil in the unsaturated zone. As shown in Table 2.1, the median depth to affected soil at Texas underground storage tank (UST) sites was 4.4 ft BGS (below ground surface), and the median thickness of affected soil was 11 ft, 1 ft less than the reported median depth to water of 12 ft. The reported median for the area of affected soil was 6750 ft^2, the equivalent of a 82- × 82-ft area.

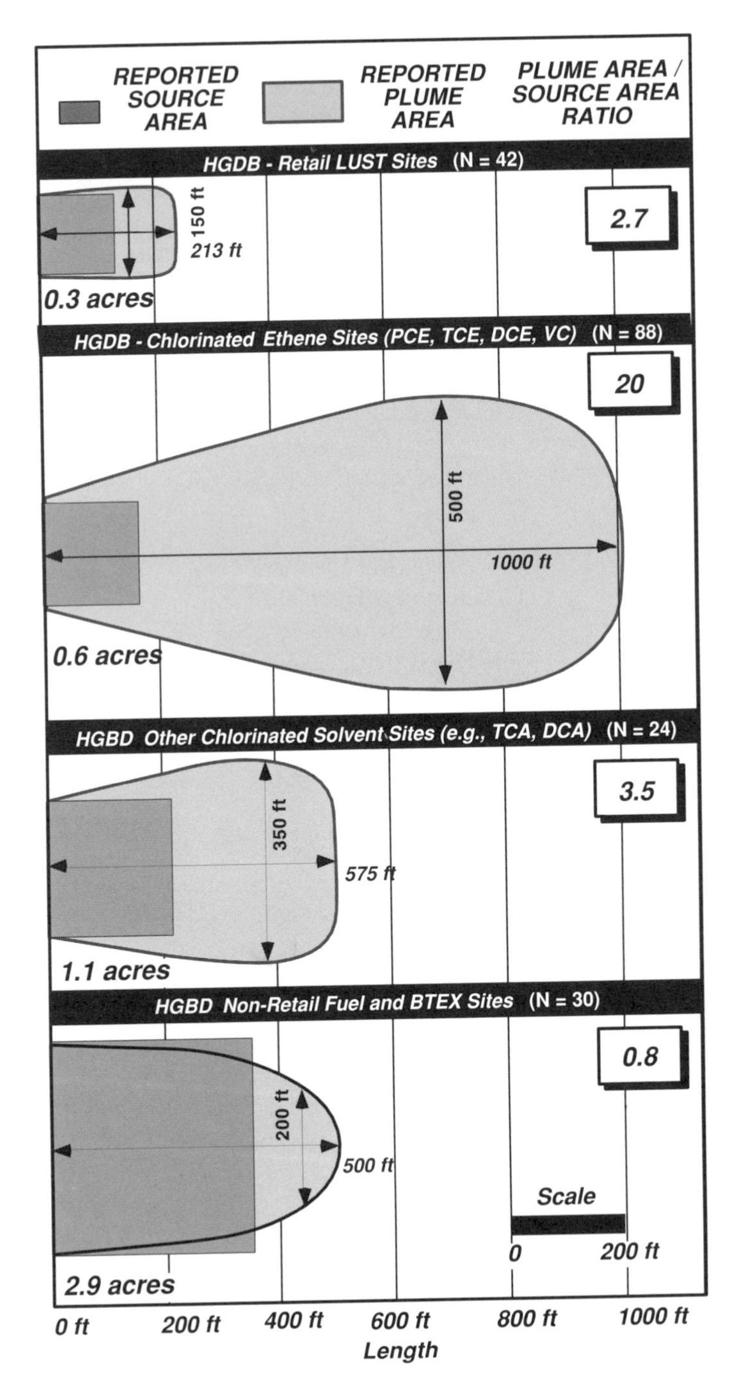

FIGURE 2.2. Relative areas of reported source zones versus plumes from HGDB database for different types of sites. (Data from American Petroleum Institute, 1989.)

TABLE 2.1. Summary of Source Characteristics at Texas LUST Sites

	Depth to Affected Soil (ft)	Thickness of Affected Soil (ft)	Area of Affected Soil (ft^2)	Thickness of Smear Zone[a] (ft)	Maximum Product Thickness (ft)	Thickness Affected Water (ft)
75th percentile	9.0	16	15,000	1.8	4.6	15
Median	**4.4**	**11**	**6,750**	**1.3**	**1.5**	**11**
25th percentile	2.5	7.5	2,546	0.8	0.5	8.0
n	167	161	145	246	102	157

Source: Mace et al. (1997).

[a] Thickness of smear zone estimated from standard deviation of water table fluctuation.

Presence of NAPLs

The presence of NAPLs was reported in the BEG, HGDB, and Superfund DNAPL studies. The HGDB asked respondents if residual NAPLs were present in the vadose zone. As shown on Figure 2.3*a*, a much higher percentage (about 40 to 60%) of petroleum-hydrocarbon sites had residual NAPL delineated in the vadose zone than did chlorinated solvent sites (26%). This may be due to (1) more thorough soil

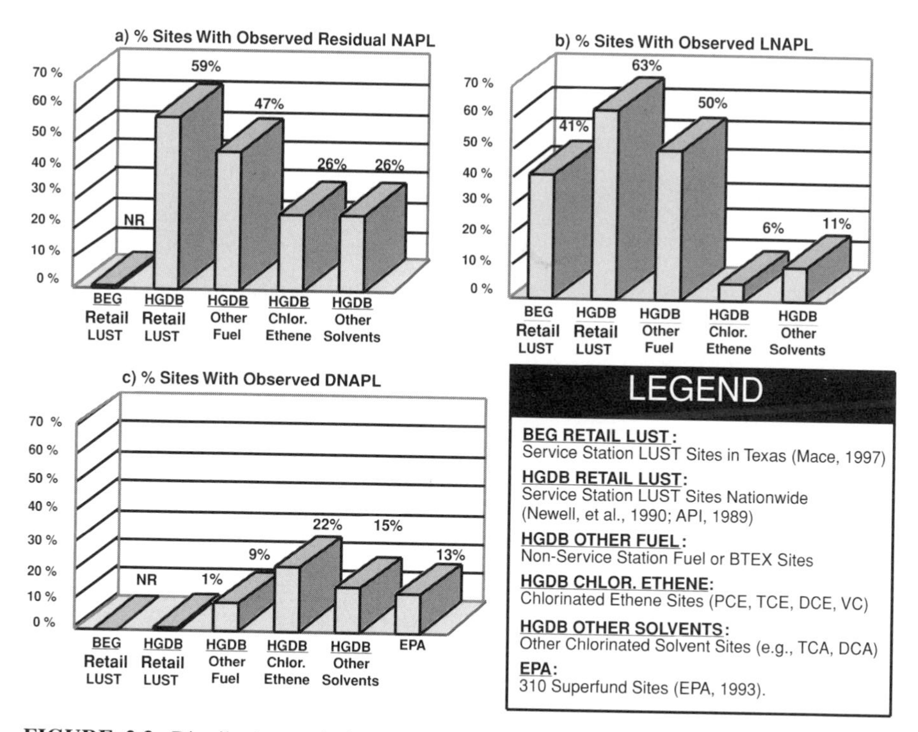

FIGURE 2.3. Distributions of sites with (*a*) observed residual NAPL in vadose zone, (*b*) observed LNAPL, and (*c*) observed DNAPL for different types of sites.

sampling during tank excavations at petroleum sites than is typically conducted at DNAPL sites; (2) smearing of LNAPL near the water table, resulting in a larger residual body of contamination in the vadose zone; and/or 3) a tendency for DNAPL to form fewer fingers and lenses than LNAPL forms in the unsaturated zone.

Continuous LNAPL layers (floating product) were observed at between approximately 40 and 65% of the petroleum-hydrocarbon sites in the BEG and HGDB studies (Figure 2.3*b*). The high percentages of LNAPL sites with floating product demonstrates how the water table intercepts and accumulates LNAPL, making continuous LNAPL easier than DNAPL to detect using a monitoring well network. As would be expected, only 6% of chlorinated ethene–contaminated sites (i.e., PCE, TCE, etc.) and 11% of other chlorinated solvent–contaminated sites had LNAPL layers floating on the water table.

DNAPL was detected much less frequently than LNAPL (Figure 2.3*c*). Only 22% of sites contaminated with chlorinated ethenes (e.g., PCE and TCE) and 15% of sites contaminated with other chlorinated solvents (e.g., TCA and DCA) had observable DNAPL, as reported by HGDB respondents. Low levels of DNAPL detection also were observed in the Superfund DNAPL study (U.S. EPA, 1993), where DNAPL was observed directly at only 40 (13%) of 310 sites.

When different classes of Superfund sites were evaluated by EPA, two patterns were apparent (Figure 2.4): (1) chlorinated solvents were the primary contaminants

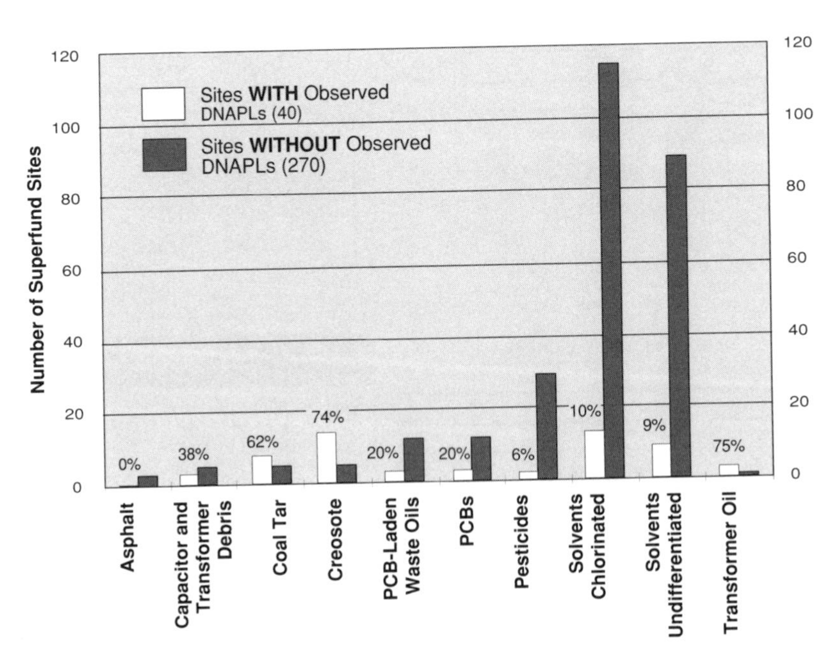

FIGURE 2.4. DNAPL observations at Superfund sites contaminated with different types of hazardous substances. The percentages of sites in each class reporting observed DNAPL are also shown. (From U.S. EPA, 1993.)

at 128 (40%) of 310 Superfund sites studied, and (2) only 10% of the chlorinated solvent sites had observed DNAPL. On the other hand, DNAPL was detected at 74% of the 19 Superfund sites with creosote as the main contaminant. These results are consistent with source loading related to industrial practices and the physical properties of the respective DNAPLs. Creosote sites are characterized by high loading rates for a high-viscosity DNAPL with a specific gravity relatively close to that of water; chlorinated solvent sites, on the other hand, are characterized by lower loading rates for low-viscosity high-density organics that penetrate porous and fractured media more easily.

The Superfund study also indicated that the most common industries associated with DNAPL-contaminated sites are electronics and electrical manufacturing and fabricated metal production. Metal cleaning and degreasing, solvent loading and unloading, storage of drummed solvents, and storage of solvents in USTs were commonly reported industrial practices resulting in releases.

The Superfund DNAPL study focused on a conceptual model that infers that many sites with no observed continuous DNAPL actually have DNAPL present in the subsurface as a immobile (residual) DNAPL, or in lenses or pools that were not indicated by the available monitoring well network. As shown on Figure 2.5, a methodology based on indirect indicators (see discussion below) suggested that 57% of all Superfund sites in the United States have a high or moderate potential for the presence of DNAPL, even though DNAPL was observed at only 5% of all Superfund

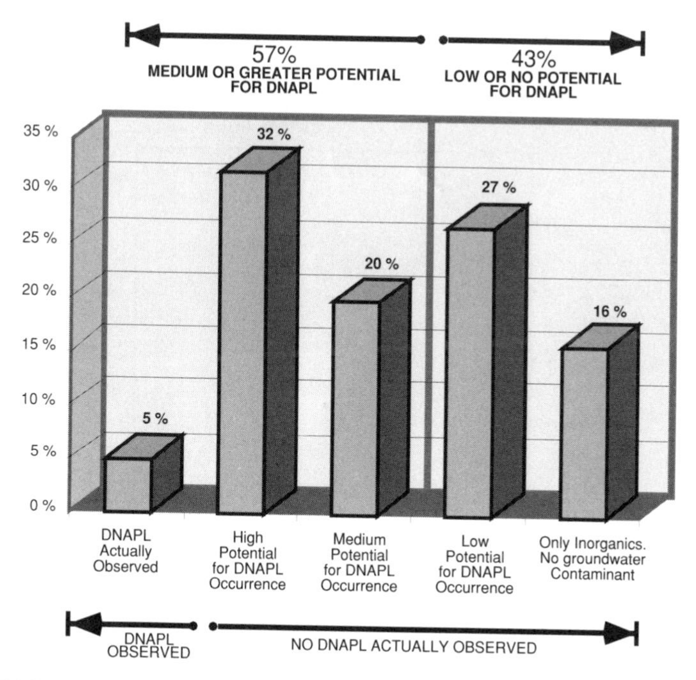

FIGURE 2.5. Potential for occurrence of DNAPL at Superfund sites. (From U.S. EPA, 1993.)

sites in the study. In other words, DNAPL is probably an important source of groundwater contamination at almost all chlorinated solvent–contaminated sites, despite the fact that it is rarely detected in site monitoring wells.

The use of indirect indicators to estimate the potential for the presence of continuous DNAPL at sites where DNAPL has not been observed is now an accepted practice. Site-use indicators are based on the industry type represented, processes used, and types of chemicals that indicate a high probability of DNAPL (e.g., an electronics manufacturing plant with records documenting significant solvent use would have a high potential for DNAPL, based on site historical information). Site-characterization-data indicators are based on generalized criteria, such as soil contaminant concentrations greater than 10,000 mg/kg or dissolved contaminant concentrations exceeding 1% of effective solubility (see the discussion below). Various indirect indicators of DNAPL presence are discussed in more detail by Newell and Ross (1992), Cohen and Mercer (1993), U.S. EPA (1993), and Pankow and Cherry (1996).

Amount of NAPL Present

Although LNAPL is found at almost half of petroleum-hydrocarbon sites, the actual thickness of the LNAPL layer currently found at most LUST sites appears to be relatively small. The Texas BEG study indicated that the median observed maximum LNAPL thickness in monitoring wells at 102 Texas LUST sites with reported LNAPL was 1.5 ft, with 50% of these sites having reported LNAPL thickness between 0.5 and 4.5 ft (Figure 2.6). The relatively small maximum thicknesses at some sites may be related to product-recovery activities; the BEG study noted that

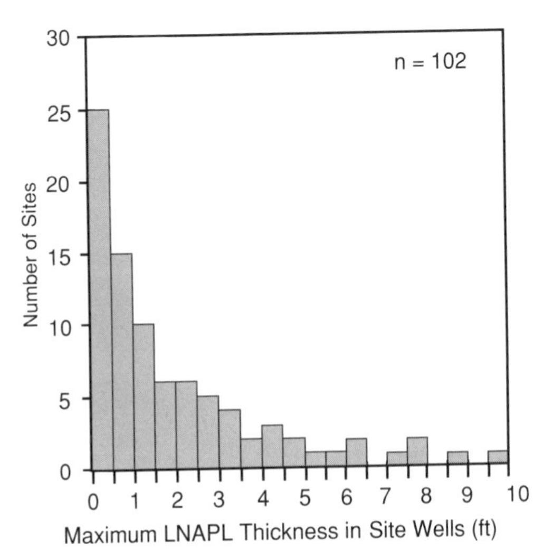

FIGURE 2.6. Maximum LNAPL thicknesses observed at 102 LUST sites in Texas. (From Mace et al., 1997.)

product recovery had been performed at 81 (16%) of 493 sites. The Florida RBCA study indicated a higher percentage of free-phase product recovery projects at 32 (31%) of 102 sites.

Note that the observed LNAPL thickness in a monitoring well is usually 2 to 10 times greater than the actual thickness in the formation, due to capillary effects (Bedient et al., 1994). Therefore, the actual amount of recoverable LNAPL from an aquifer is much lower than the calculated amount would be if the apparent LNAPL layer thickness (i.e., the thickness observed in a monitoring well) is used.

Source Zone Concentrations

As part of the HGDB and Florida RBCA studies, maximum aqueous-phase source concentrations were requested from the respondents. Although these data have considerable scatter, they do illustrate some potential patterns between site type and expected maximum (i.e., source area) concentrations.

For petroleum hydrocarbon sites, both the HGDB and Florida RBCA studies reported maximum concentrations of dissolved benzene found at service station LUST sites. The HGDB survey had higher reported maximum groundwater concentrations of benzene than those in the Florida RBCA study, with median values of 6.5 and 0.36 mg/L, respectively. The most likely reason for the difference is the types of sites reported: The HGDB study, conducted in 1988, probably includes some of the most notable LUST sites across the country at the time, where significant, obvious releases of petroleum hydrocarbons occurred. The Florida RBCA study, on the other hand, represents a cross section of all Florida UST sites being managed in 1996, and sites with relatively small releases are included. Both studies show a large range in observed source concentrations, spanning two orders of magnitude or more.

The Florida RBCA study included data for total volatile organic analytes (TVOAs), which consist primarily of the four BTEX compounds and MTBE. A comparison of these data (Table 2.2) shows several source dissolution trends based on contaminant solubility. As expected, benzene represented only a portion of the

TABLE 2.2. Summary of Maximum Groundwater Concentrations at 117 Florida LUST Sites

	Maximum Source Concentration[a] (mg/L)			
	Benzene	TVOA	PAH	MTBE
90th percentile	5.78	46.04	1.715	6.72
75th percentile	1.300	15.930	0.879	1.220
Median	**0.360**	**1.700**	**0.191**	**0.088**
25th percentile	0.026	0.320	0.034	0.011
n	117	117	50	117

Source: Groundwater Services, Inc. (1997a).

[a] TVOA, total volatile organic analysis; PAH, polynuclear aromatic hydrocarbons; MTBE, methyl *tert*-butyl ether.

TVOA concentration (median values of 0.36 mg/L benzene versus 1.7 mg/L TVOA, due to the contributions of other volatile, soluble organics in gasoline (e.g., the other three BTEX compounds and BTEX breakdown products). Second, the contribution of PAHs to the total dissolved contaminant concentration was relatively small for the median site in Florida (median concentrations of 0.191 mg/L PAHs versus 1.7 mg/L TVOA, or 11%). There appeared to be no obvious concentration trends or changes in the contaminant composition (i.e., benzene versus TVOA versus PAH) over time since release, or when normalized for flushing rate (seepage velocity multiplied by the time since release).

A few sites in Florida exhibited very high concentrations of MTBE, a relatively soluble compound, with most sites having no or very low concentrations. As shown on Figure 2.7, the median MTBE concentration at 117 LUST sites in Florida was only 0.088 mg/L, with the 75th percentile concentration at 1.22 mg/L. Seven sites had maximum MTBE concentrations exceeding 20 mg/L. Benzene concentrations are shown on the figure for comparison.

The HGDB-reported sites contaminated with chlorinated solvents exhibited a wide range of contaminant concentrations. The median TCE concentration at 52 sites reporting TCE as the primary contaminant was 15 mg/L, with the 25th percentile and 75th percentile maximum concentrations being 2.41 and 112.5 mg/L, respectively. One site reported a maximum TCE value of 9000 mg/L, or nine times solubility, indicating either a reporting error or that this groundwater sample contained DNAPL droplets. The 12 sites for which TCA was reported as the primary contaminant had a median maximum TCA concentration of 3.2 mg/L.

Summary

This summary of available source zone data shows the tremendous variability in source conditions between sites and stresses the importance of accounting for the presence of NAPL in site conceptual models. Even well-designed and well-implemented site

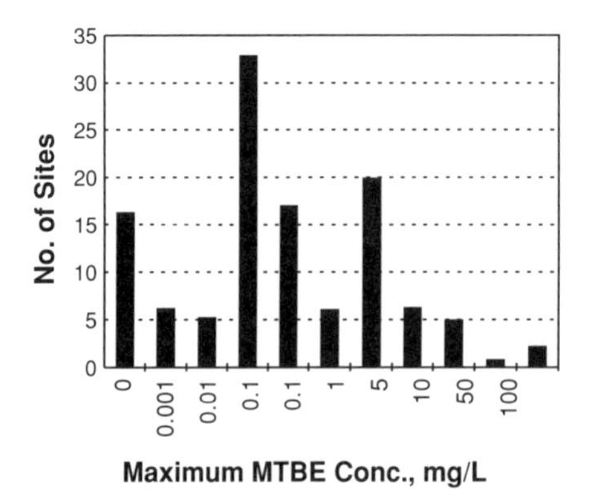

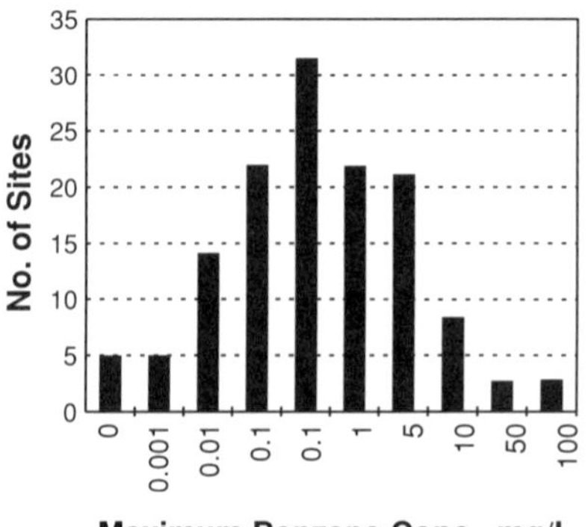

FIGURE 2.7. Maximum MTBE and benzene concentrations reported at Florida retail LUST sites. (From Groundwater Services, Inc., 1997a.)

characterization programs have difficulty in delineating and defining the source of contaminants to groundwater. In general, LNAPL sources are easier to characterize because the LNAPL layer is confined by the water table. Even so, continuous LNAPL is observed in monitoring wells at fewer than half of petroleum-contaminated sites. At DNAPL sites, direct observations of DNAPL are relatively rare (5 to 25% of these sites). The DNAPL that produces the dissolved plume is often never directly observed, but can be inferred through indirect indicators.

2.4 PROCESS DESCRIPTION FOR SOURCE SCENARIO 1: DISSOLVED CONTAMINANT RELEASE TO SUBSURFACE

The goal of this section is to present simple modeling approaches that describe the processes related to scenario 1. In this section, a key parameter, the pore volume, is introduced, and a simple model for representing flushing of dissolved compounds in groundwater is provided. This simple model can be used to estimate how long it will take to remediate an aquifer containing only dissolved compounds via groundwater flushing alone (i.e., in the absence of biodegradation).

As described above, a limited number of plumes are caused by the release of contaminated water containing dissolved organics. Two examples of this source scenario are releases from a process unit where only dissolved organics are documented, or leakage from an industrial wastewater treatment system after oil separation. This type of source (a general term for the area near the release, as the source is actually the original location of the plume) and the resulting plume are relatively easy to remediate, either by active means or through natural attenuation, as there is no long-term continuing source in the subsurface.

When dealing with only dissolved organics, the advection–dispersion equation (see Chapter 6) can be used to describe the change in dissolved concentrations in the source zone at any time after the release. A wide variety of solutions to the advection–dispersion equation, including more complicated solutions that consider the effects of sorption and biodegradation, are discussed in Chapter 6. For source zones containing only dissolved compounds, however, the one-dimensional advection–dispersion equation can be coupled with the concept of the pore volume to provide a useful tool for estimating the volume of water required to flush away dissolved compounds.

Key Parameter: Pore Volume

A *pore volume* is the volume of water required to replace (flush out) water in a unit volume of saturated porous media (see Figure 2.8). For example, for a 1-m^3 (or 1000-L) volume of saturated porous medium with an effective porosity of 0.35, 1 pore volume equals 350 L (note that *effective* porosity is used, as this represents the theoretical amount of water that can be flushed under saturated conditions). In this example, flushing 350 L of water through the control volume represents 1 pore volume. If 3500 L of water were flushed through the control volume, the porous media

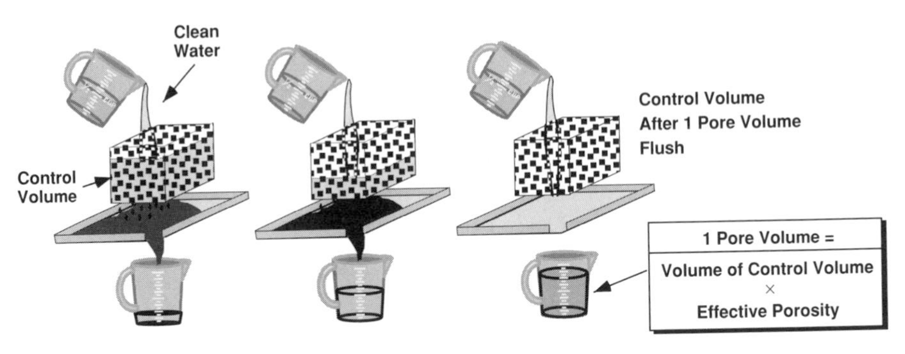

FIGURE 2.8. Schematic of flushing a control volume with 1 pore volume of clean water.

would have been exposed to 10 pore volumes of flushing. If the control volume is 2 m × 1 m × 1 m, 3500 L of flushing would represent only 5 pore volumes. A single pore volume in a control volume (such as a suspected source zone) can be determined using the equation

$$V_{PV} = LWDn_e \tag{2.1}$$

where V_{PV} = volume in control volume (such as a suspected source area) (unitless)
L = length along flow path in control volume (ft)
W = width of control volume (ft)
D = depth of control volume (ft)
n_e = effective porosity (unitless)

The concept of the pore volume can be employed either in an active pumping system (such as a pump-and-treat system) or in a natural attenuation scenario, where natural groundwater flow flushes the source zone. Pore volumes are a common accounting system for both dissolved-contaminant-only systems (such as source scenario 1) and NAPL-dominated systems (source scenarios 2, 3, and 4).

Note that the concept of the pore volume does not mean that all of the water originally in the porous media is removed and replaced by clean water; 1 pore volume is the volume of water that actually flows through the porous media. Porous media (either artificial media such as glass beads, or natural media such as sand grains) contain dead-end pores and restricted flow paths where groundwater flow is greatly reduced. To remove dissolved compounds from these zones, more than 1 pore volume is required. As shown below, achieving a 100-fold reduction in dissolved-contaminant concentration might require 2 to 3 pore volumes of flushing, while a 100,000-fold reduction in concentration would probably require more than 10 pore volumes of flushing. The simple flushing model given below shows the theoretical relationship between pore volumes and contaminant concentration.

Simple Flushing Model

The simple flushing model developed by Newell et al. (1994) is based on an approximation of an analytical solution to the one-dimensional advection–dispersion equation presented by Ogata and Banks (1961):

$$\frac{C_t}{C_0} = \frac{1}{2}\operatorname{erfc}\left(\frac{L - \upsilon_x t}{2\sqrt{\alpha_x \upsilon_x t}}\right) \tag{2.2}$$

where C_t = concentration at any time t at any distance L along control volume (mg/L)
C_0 = original concentration (time = 0) at any point within control volume (mg/L)
L = length along flow path in control volume parallel to groundwater flow (ft)
υ_x = groundwater seepage velocity (ft/yr)
t = time (years)
α_x = longitudinal dispersivity (ft)

Note that the groundwater seepage velocity can be converted to a dissolved compound velocity by dividing by the retardation factor (see Chapter 3). Dispersivity is discussed in Chapter 3, and the complementary error function (erfc) is described in more detail in Chapter 8.

The equation for the number of pore volumes that pass through a control volume during a certain time t is

$$\mathrm{PV} = \frac{\upsilon_x t}{L} \quad \text{or} \quad L = \frac{\upsilon_x t}{\mathrm{PV}} \tag{2.3}$$

where PV = pore volumes (unitless)
υ_x = groundwater seepage velocity (ft/yr)
L = length along flow path in control volume parallel to groundwater flow (ft)
t = time (years)

Using the common assumption that longitudinal dispersivity is equal to 10% of length (see ASTM, 1995; Newell et al., 1996; and Chapter 3), eq. (2.1) becomes

$$\frac{C_t}{C_0} = \frac{1}{2}\operatorname{erfc}\left(\frac{L - \upsilon_x t}{2\sqrt{0.1 L \upsilon_x t}}\right) \tag{2.4}$$

Substituting eq. (2.3) into (2.4) gives

$$\frac{C_t}{C_0} = \frac{1}{2}\operatorname{erfc}\left[\frac{(\upsilon_x t/\mathrm{PV}) - \upsilon_x t}{2\sqrt{0.1(\upsilon_x t/\mathrm{PV})\upsilon_x t}}\right] \tag{2.5}$$

which reduces to

$$\frac{C_t}{C_0} = \frac{1}{2}\operatorname{erfc}\left\{1.58\left[\frac{(1/\mathrm{PV}) - 1}{\sqrt{1/\mathrm{PV}}}\right]\right\} \tag{2.6}$$

Equation (2.6) cannot be solved directly for pore volumes, but can be depicted graphically, as shown in Figure 2.9. The x axis indicates the magnitude of concentration reduction (C_t/C_0) required (e.g., reducing the original contaminant concentration from 5 mg/L to 0.005 mg/L represents a desired unitless cleanup level of 10^{-3}). The worksheet also provides a table and an equation for calculating the retardation factor, which can be multiplied by the pore volume estimate from the chart to provide a total number of pore volumes required to reach a desired cleanup level if the organic compounds sorb to the aquifer matrix (see Chapter 8).

Note that while eq. (2.6) cannot be solved directly for PV, the resulting curve is almost linear throughout the region of interest ($C_t/C_0 < 0.1$), yielding the following approximation:

$$\mathrm{PV} = -0.93 \log_{10}\frac{C_t}{C_0} + 0.75 \qquad \text{for } C_t/C_0 < 0.1 \tag{2.7}$$

This approximation can be generalized further as follows: The number of pore volumes of flushing required to reach a remediation goal (in a no-NAPL zone) is approximately equal to the number of orders of magnitude reduction required to reach the goal plus one (e.g., a 100-fold reduction = 2 + 1 = 3 pore volumes). Although retardation was not included in the derivation, the number of pore volumes can be adjusted to account for retardation by multiplying the total number of pore volumes by the appropriate retardation factor (see Chapter 3 for more information on sorption and retardation).

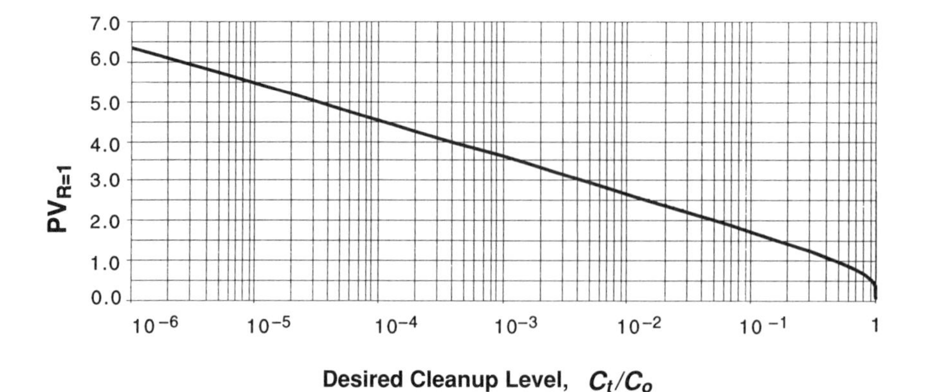

FIGURE 2.9. Pore volumes required to flush dissolved-phase constituents from porous media assuming no retardation ($R = 1$). To account for retardation, multiply pore volumes by the calculated retardation factor. (From Newell et al., 1994.)

When the number of pore volumes is estimated, the time required for natural attenuation to remediate a source zone can be determined by

$$t = \frac{\mathrm{PV} \cdot L}{\upsilon_x} \tag{2.8}$$

where t = time for naturally occurring flow to flush out dissolved compounds and achieve desired cleanup level under natural attenuation (years)
PV = number of pore volumes required to reach desired cleanup level (from Figure 2.9)
L = length of source zone parallel to groundwater flow (ft)
υ_x = groundwater seepage velocity (ft/yr)

This flushing model is a simple planning-level tool that provides estimates of the amount of flushing required to remediate zones containing only dissolved compounds. The key assumptions for applying the flushing model are

- No NAPL is present.
- Flow is uniform throughout the source zone or zone of interest.
- The aquifer is homogeneous.
- The effects of biodegradation are ignored.

The key factors that control the time required for natural attenuation to flush dissolved contaminants from non-NAPL zones are the seepage velocity (greater seepage velocity means more flushing), the length of the source zone parallel to groundwater flow direction (shorter source zones mean that less flushing is required), and the required concentration reduction (less reduction in concentration means less flushing).

One of the most important applications of the simple flushing model is to determine if NAPL is present in the source zone. If aqueous-phase contaminant concentrations in the source zone (the area of the original release) are relatively constant or are not falling as fast as the simple flushing model would indicate, the presence of a continuing source, probably a NAPL, is indicated. The authors' experience is that most sites have source zones that stay contaminated after numerous pore volumes of flushing (either during remediation or from natural groundwater flow), indicating that release events comprised only of aqueous-phase contaminants are relatively rare.

Example 2.1 How long will it take to attain a 5-μg/L cleanup standard in a source zone that contains only dissolved benzene (no NAPL) at a maximum concentration of 5000 μg/L? The source zone is 50 ft long in the direction parallel to groundwater flow and 60 ft long in the direction perpendicular to groundwater flow. The seepage velocity of the aquifer is 100 ft/yr, and the retardation factor is 1.2.

SOLUTION: Calculate C_t/C_0 as 5/5000 or 0.001. Using Figure 2.9, find the number of pore volumes at $R = 1$ as $PV_{R-1} = 3.5$. Adjusting for retardation gives a final pore volume of PV = 4.2. Using eq. (2.8), the time to achieve the cleanup concentration is calculated to be

$$t = \frac{4.2(50)}{100} \quad \text{or} \quad t = 2.1 \text{ years}$$

2.5 PROCESS DESCRIPTION FOR SOURCE SCENARIO 2: CONTAMINANT RELEASE TO VADOSE ZONE ONLY

Under the dissolved-contaminant and/or NAPL-release-to-vadose-zone-only scenario, contaminant in unsaturated soils can leach into infiltrating water, thereby creating a dissolved plume when the percolating water mixes with clean groundwater in the underlying aquifer. Note that this scenario assumes that no continuous NAPL has migrated down to the water table. Although this source scenario was once thought to be the primary mechanism for groundwater contamination (see Section 2.2), an improved understanding of NAPL migration now suggests that vadose zone contamination is not as important as source scenarios 3 and 4 (discussed below). However, there is still considerable regulatory concern about protecting groundwater from soil contamination in the unsaturated zone, and therefore this source scenario must be evaluated and managed at many sites. The following discussion provides simple modeling tools for predicting the resulting concentration of contaminants in an aquifer that underlies contaminated soils in the vadose zone.

Under the scenario in which contaminants are released into the vadose zone only, shallow site geology is idealized as two principal stratigraphic components: (1) a *surface soil column* consisting of unsaturated soils, where contaminant flow is primarily downward, and (2) a saturated, transmissive *water-bearing unit*, where flow is principally horizontal (see Figure 2.1*b*). Contaminant transfer from residual NAPL and/or sorbed contaminants on contaminated soils in the surface soil column to the underlying water-bearing unit produces leachate (contaminated groundwater in the vadose zone) that flows vertically downward. After entering the water-bearing unit, the leachate mixes with horizontally flowing groundwater and spreads horizontally in the downgradient direction of shallow groundwater flow (Figure 2.1*b*).

This scenario can be characterized by two general processes: (1) mass transfer of contaminants from soil and NAPL to the leachate (a physical–chemical process), and (2) infiltration and the subsequent mixing of the resulting leachate with groundwater (a hydrogeologic process). In the following sections we discuss the chemical and hydrogeologic aspects of this source scenario in more detail.

Estimating Leachate Concentrations: Equilibrium Partitioning between Sorbed Contaminants and Water

To estimate the concentration of contaminants in groundwater infiltrating past immobile (sorbed or NAPL) contaminants, the concept of equilibrium partitioning between different phases (e.g., soil to water or NAPL to water) is often employed. This concept assumes that (1) the rate of mass transport within a given phase is slow with respect to the transfer of mass between phases in contact with one another, (2) the equilibrium between any two phases is independent of the presence of additional phases, and (3) physical contact and mixing among the various phases is 100% efficient, neglecting the effects of heterogeneities and preferential pathways. In general, this assumption of instantaneous equilibrium partitioning will tend to overestimate the contaminant mass transferred from the contaminated soil zone to infiltrating water. Nevertheless, equilibrium partitioning is a convenient and common modeling assumption that allows the contaminant concentration in any phase to be expressed as a function of soil concentrations and NAPL and contaminant characteristics.

For example, in the case in which there is no continuous or residual NAPL, equilibrium partitioning assumes that the aqueous-phase concentration (in mg/L) can be related to the soil concentration (in mg/kg) using a mathematical function. The most common mathematical function is the linear, reversible isotherm, in which the dissolved concentration is in direct proportion to the concentration sorbed on the soil (Brusseau et al., 1991). If the soil concentration is 10 mg/kg and the concentration in water in equilibrium with that soil is 5 mg/L, that means that a soil contaminated with 100 mg/kg would produce water with a concentration of 50 mg/L after equilibrium is reached.

Extending the no-NAPL case further, if the soil contaminant described above were present at a concentration of 100 mg/kg but was in contact with infiltrating groundwater with a concentration of 500 mg/L, the equilibrium partitioning approach indicates that there would be mass transfer from the water to the soil as the water–soil system approaches equilibrium. Because the process is assumed to be reversible, when the 100-mg/kg soil is in contact with groundwater contaminated at only 10 mg/L, the mass transfer is from the soil to the water (i.e., leachate is formed). Because of the relatively long time periods associated with water movement through the subsurface, chemical equilibrium is generally assumed to occur when water is in contact with a contaminated media.

The following expression relates the concentration of the soil with no NAPL to the concentration of the leachate (ASTM, 1995; Connor et al., 1995):

$$C_L = \frac{C_T}{(\theta_{ws} / \rho_b) + k_d + H(\theta_{as} / \rho_b)} \tag{2.9}$$

$$k_d = k_{oc} f_{oc} \tag{2.10}$$

where C_L = concentration of contaminant in soil leachate (mg/L)
C_T = bulk contaminant concentration on soil mass for all phases (mg/kg)
θ_{ws} = volumetric water content of surface soils (cm^3 H_2O/cm^3 soil)
ρ_b = soil bulk density (g soil/cm^3 soil)
k_d = soil–water distribution (partition) coefficient (cm^3 H_2O/g soil)
H = Henry's law constant for contaminant (cm^3 H_2O/cm^3 air)
θ_{as} = volumetric air content of vadose zone soils (cm^3 air/cm^3 soil)
k_{oc} = organic carbon partition coefficient for contaminant (cm^3 H_2O/g C)
f_{oc} = fraction of organic carbon (g C/g soil)

For typical soils, conservative default values for equilibrium soil moisture parameters as a function of the Unified Soil Classification System (USCS; Lambe and Whitman, 1969) or U.S. Department of Agriculture textural classification method (Peck et al., 1974) are provided in Table 2.3. To employ these defaults, the user must first determine the appropriate classification for soils within the surface soil column using the predominant soil type observed in site boring logs. The default soil moisture values provided in Table 2.3 represent equilibrium levels associated with an annual rainfall of 30 in. (U.S. median) and are derived using the Brooks and Corey (1964) soil characteristic model and Burdine's (1953) equations for the relative permeability of unsaturated soils. Sensitivity analyses show that use of default soil moisture values based on a median rainfall level for the United States provides model results that are on average within 5% of those obtained using the site-specific, rainfall-dependent soil moisture

TABLE 2.3. Default Moisture Soil Parameters and Saturated Hydraulic Conductivity Values Based on USCS Soil Type

USCS Soil Type[a]		Default Soil Moisture Parameters[a]			Default Hydraulic Conductivity,
Symbol	Description	θ_T	θ_{ws}	θ_{as}	$K_{vs}{}^c$ (cm/s)
SW	Sand, clean, well-graded	0.41	0.08	0.33	10^{-2}
SP	Sand, clean, poorly-graded	0.41	0.08	0.33	10^{-2}
SM	Sand, silty	0.41	0.12	0.29	10^{-3}
SC	Sand, clayey	0.38	0.23	0.15	10^{-5}
ML	Silt, sandy	0.43	0.26	0.17	10^{-5}
ML	Silt	0.46	0.30	0.16	10^{-5}
MH	Silt, clayey	0.36	0.24	0.12	10^{-5}
CL	Clay, sandy, low plasticity	0.38	0.31	0.07	10^{-6}
CL	Clay, silty, low plasticity	0.36	0.34	0.02	10^{-7}
CH	Clay, high-plasticity	0.38	0.38	0	10^{-8}

Source: Connor et al. (1997). Reprinted with permission from NGWA, Westerville, Ohio.

[a] Unified Soil Classification System (USCS) described in Lambe and Whitman (1969) and ASTM (1992).

[b] Default values for volumetric water (θ_{ws}) and volumetric air (θ_{as}) contents are to be matched to predominant soil type in surface soil column.

[c] Typical saturated hydraulic conductivity (K_{vs}) values matched to median values reported by Freeze and Cherry (1979) and Rawls and Brakensiek (1985).

values derived using the Brooks–Corey model. Consequently, use of default soil moisture values represent a reasonable, simplifying measure.

Estimating Leachate Concentrations: Equilibrium Partitioning with NAPL

When infiltrating water comes in contact with residual or continuous NAPL in the soil, a different, approximate relationship based on solubility and the mole fraction of contaminants in the NAPL describes equilibrium partitioning. This approximation, similar to Raoult's law for gases, is known as the *effective solubility relationship* and can be written as

$$C_L \approx XS \tag{2.11}$$

where C_L = concentration of contaminant in soil leachate (mg/L)
X = mole fraction of contaminant in NAPL (unitless)
S = aqueous solubility of contaminant (mg/L)

This equilibrium concentration may also be referred to as the effective solubility of the compound from the mixture. Experimental evidence (Banerjee, 1984; Broholm and Feenstra, 1995) has suggested that eq. (2.10) produces reasonable approximations of effective solubilities for mixtures of structurally similar compounds, and that the relationship works best for binary mixtures of similar compounds. For other mixtures, the error is greater due to the complex solubility relationships created; however, the method is appropriate for many environmental studies for which there are many other uncertainties (Pankow and Cherry, 1996).

For complex mixtures (e.g., multiple identified and unidentified solvents, or mixed fuels and solvents) it is necessary to estimate the weight percent and an average molecular weight of the unidentified fraction of the NAPL before the calculation can be completed. In doing so, it should be remembered that increasing the average molecular weight for the unidentified fraction will produce greater estimated effective solubilities for the identified contaminants. A higher molecular weight for the unidentified fraction will result in a lower mole fraction for those unknown compounds, and therefore higher mole fractions (and solubilities) for the known compounds. Pankow and Cherry (1996) provide an example of these calculations for a mixture of chlorinated and nonchlorinated compounds.

The effective solubility relationship indicates that once leachate is in contact with NAPL, the total concentration of the contaminant dissolved in the leachate remains constant, even if the total concentration of the NAPL in the soil increases. In other words, aqueous-phase concentrations in leachate will increase together with soil concentrations only while the soil contaminants are sorbed (no NAPL). If the soil concentration is high enough to indicate the presence of NAPL, the leachate contaminant reaches a maximum concentration determined by the mole fraction of the contaminant in the NAPL and the aqueous solubility of the contaminant (Figure 2.10). This resulting concentration is known as the *effective solubility* of the contaminant.

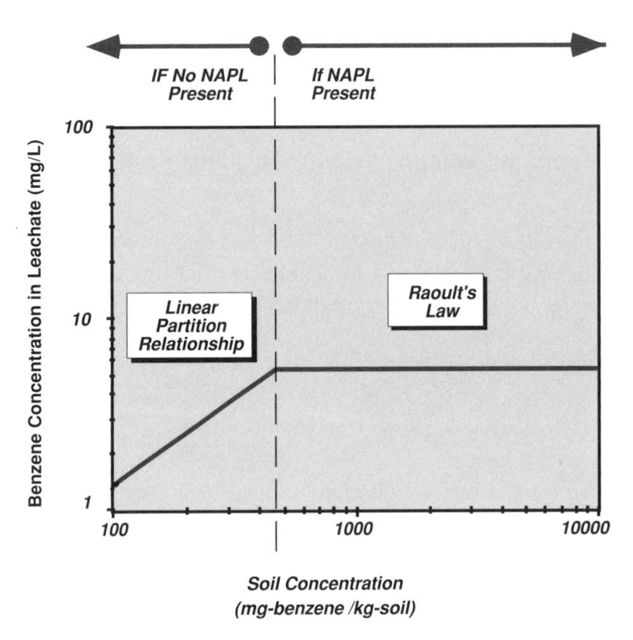

FIGURE 2.10. Typical soil–water partitioning relationships for benzene, showing linear reversible partitioning relationship for soils without NAPL (typically less than a few hundred milligrams of benzene/per kilogram of soil) and Raoult's law for soils containing NAPL.

At field sites, there is considerable heterogeneity in NAPL distributions, and zones with more NAPL often will produce higher concentrations than zones with less NAPL. The effective solubility relationship is theoretical and holds for any flow lines of groundwater (e.g., recharge) that come into contact with NAPL, but many flow lines never come into contact with NAPL. Zones with less NAPL will have more of this dilution, thereby reducing the dissolved contaminant concentrations in the leachate. Because this heterogeneity in NAPL distribution is very difficult or impossible to measure, the theoretical effective solubility relationship is used as a conservative model that will not underpredict leachate concentrations.

In general, NAPLs are almost always present in soils with organic chemical concentrations greater than 10,000 mg/kg, and often in soils with organic concentrations exceeding several hundred mg/kg. Feenstra et al. (1991) provide a method to estimate if NAPLs are present in soils based on equilibrium partitioning relationships (see also Newell and Ross, 1992).

Finally, note that effective solubility is an important process in the saturated zone as well as the unsaturated zone. This is discussed below in more detail.

Hydrogeology of Leaching: Infiltration Estimates and Mixing with Groundwater

Once the concentration of contaminant in the leachate is estimated (using the partitioning relationships presented above), the volume of the leachate that is being generated each year must be calculated to estimate contaminant loadings to an aquifer. In the vadose-to-saturated-zone source scenario, leachate is generated from the net infiltration of precipitation through the surface soil column and then is diluted with fresh groundwater in the water-bearing unit. *Net infiltration* corresponds to total infiltration (precipitation minus runoff) minus the additional loss associated with evapotranspiration. The net infiltration term thereby represents the deep percolation flow through the contaminated vadose zone, which transports contaminants to the groundwater. Another term for net infiltration is *recharge*.

Because net infiltration is very rarely measured, it is a difficult value to estimate accurately. It can be derived from a variety of field and modeling techniques, including soil–water balances, recharge estimated by steady-state yield, streamflow measurements, tracer (e.g., tritium or chloride) studies, water-level fluctuations, soil models (including HELP), Richard's equation, direct measurement with lysimeters, and basin outflow.

One useful study (American Petroleum Institute, 1996) compiled data from more than 100 studies employing various methods to estimate infiltration (Figure 2.11). These data were used by Connor et al. (1997) to develop simple, upper-bound estimates of net infiltration as a function of average rainfall (cm/yr) and the predominant soil type (sand, silt, or clay), assuming a grass ground cover. Connor et al. developed a curve, shown on Figure 2.11, that represents an 80% envelope line for rainfall infiltration data from more than 100 sandy soil sites in 18 geographic regions in the United States as compiled in the American Petroleum Institute (1996) study. This curve provides a conservative (80% upper bound) estimate of net infiltration for sand or gravel soil sites reported in this database. Note that the y axis represents

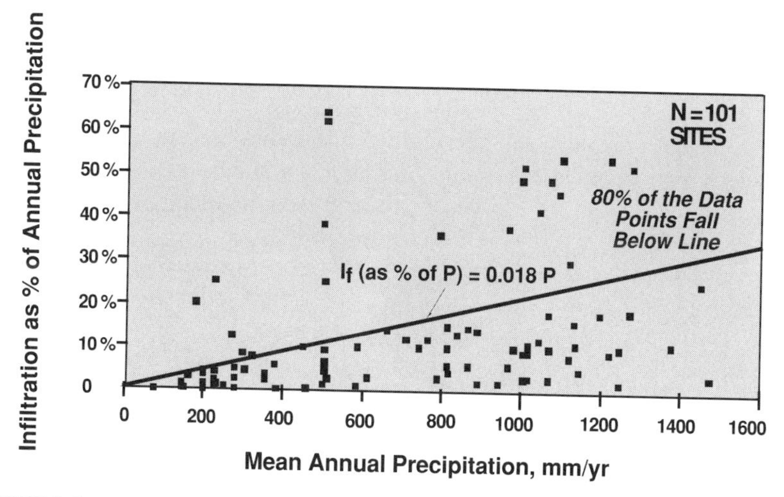

FIGURE 2.11. Annual infiltration as a percentage of annual precipitation as reported from 101 infiltration studies. The line represents an upper-range estimate for infiltration, as 80% of data points fall below the line. (From TNRCC, 1996; original data from American Petroleum Institute, 1996.)

the *percentage* of annual precipitation that becomes net infiltration, so that net infiltration in units of cm/yr becomes a function of annual precipitation squared.

For sandy soils: $$I_f = 0.0018P^2 \tag{2.12}$$

where I_f is the net infiltration per year (cm/yr) and P is the mean annual precipitation (cm/yr).

Curves for silty and clayey soils were then derived from the empirical sandy soil curve based on the relative percent infiltration described by Viessman et al. (1989) for the parameters of the Horton infiltration relationship. This relationship indicates that a silty soil will have 50% of the net infiltration through sandy soil during a theoretical storm event, and clayey soil will have only 10%. The following equations provide upper-bound estimates for net infiltration with K equal to vertical hydraulic conductivity:

For silty soils: $$I_f = 0.0009P^2 \tag{2.13}$$

For clay soils $$I_f = 0.00018P^2 \tag{2.14}$$

Upper-bound net infiltration limit: $$I_{f_{\max}} = K_{\upsilon s} \quad (3.15 \times 10^7 \text{ sec/yr}) \tag{2.15}$$

For more detailed estimates of infiltration, the HELP model, a quasi-two-dimensional deterministic computer-based water budget model can be applied (Schroeder et al., 1988; http://www.ntis.gov/fcpc/cpn7367.htm).

Hydrogeology of Leaching: Mixing with Groundwater

To calculate the final concentration of contaminants in an aquifer underlying unsaturated, contaminated soils, the amount of mixing that occurs between the contaminated leachate and the uncontaminated groundwater must be estimated. The amount of mixing is represented by a *leachate dilution factor* (LDF); in which an LDF of 2, for example, indicates that the leachate will be diluted by a factor of 2, and therefore the leachate concentration will be reduced by a factor of 2.

This dilution factor is based on a simple box model used to estimate mass dilution within a *mixing zone* located in the water-bearing unit directly beneath the contaminated soil mass (Figure 2.12). Dividing the incoming leachate concentration (C_L) by the LDF value yields a steady-state groundwater concentration within the mixing zone. The box model accounts for the volume of leachate that is generated annually by assuming this volume is proportional to (1) the net infiltration rate, times (2) the width in the direction parallel to groundwater flow, times (3) the width perpendicular to groundwater flow. The volume of fresh groundwater mixing with the leachate each year is a product of (1) the groundwater Darcy velocity, times (2) the width perpendicular to groundwater flow, times (3) the mixing zone depth. After canceling out the common term (the width perpendicular to groundwater flow), the

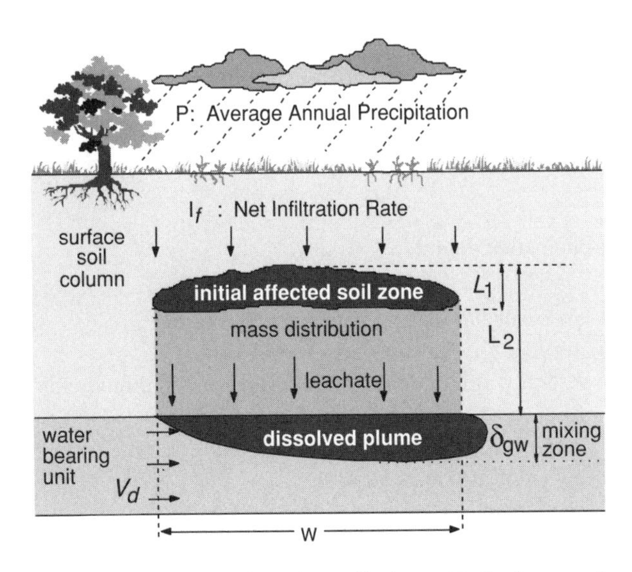

FIGURE 2.12. Conceptual model of leachate dilution with fresh groundwater for SAM model. (From Connor et al., 1997. Reprinted with permission from NGWA, Westerville, Ohio.)

mixing of the leachate and fresh groundwater can be expressed mathematically by (ASTM, 1995)

$$\mathrm{LDF} = 1 + \frac{V_d \delta_{\mathrm{gw}}}{I_f W} \tag{2.16}$$

$$C_{\mathrm{aquifer}} = \frac{C_L}{\mathrm{LDF}} \tag{2.17}$$

where
V_d = groundwater Darcy velocity (cm/yr)
δ_{gw} = groundwater mixing zone thickness (cm)
I_f = net infiltration rate (cm/yr)
W = lateral width of contaminated soil zone parallel to groundwater flow (cm)
LDF = leachate/groundwater dilution factor (unitless)
C_{aquifer} = aquifer concentration (mg/L)
C_L = leachate concentration (mg/L) [typically calculated from eq. (2.9) and/or eq. (2.11)]

The mixing zone depth can be determined from multilevel monitoring well data, estimated [e.g., the ASTM (1995) RBCA standard suggests 2 m as an appropriate mixing zone depth], or modeled using a model such as the one provided in the EPA soil screening guidance (U.S. EPA, 1996):

$$\delta_{gw} = \sqrt{2\alpha_z W} + b\left[1 - \exp\left(\frac{-I_f W}{V_d b}\right)\right] \quad (2.18)$$

$$\alpha_z = 0.0056W \quad (2.19)$$

Upperbound mixing zone depth: $\delta_{gw} \leq b$ (2.20)

where δ_{gw} = groundwater mixing zone thickness (cm)
α_z = vertical groundwater dispersivity (cm)
W = lateral width of contaminated soil zone in direction of groundwater flow (cm)
b = saturated thickness of water-bearing unit (cm)
I_f = net infiltration rate (cm/yr)
V_d = groundwater Darcy velocity (cm/yr)

Volatilization and Biodegradation

Two other processes affect contaminants in soil and leachate: volatilization and biodegradation. Within unsaturated soils, VOCs may partition from the soil, leachate, or NAPL into the soil vapor phase, and subsequently may migrate upward to the ground surface (or into structures) through vapor diffusion. For shallow contaminated soils, such volatilization can significantly reduce the contaminant mass moving downward via leachate migration. The rate of subsurface contaminant volatilization is a function of the chemical characteristics of the contaminants of concern (e.g., the Henry's law constant and gaseous diffusion coefficient) and the volumetric air content of the soil. Due to higher air content, volatilization will be more significant in sandy soils than in clayey soils.

Natural microbial activity within the surface soil column can destroy organic contaminant mass both in the contaminated soil zone and in the migrating leachate. Biodegradation rates are a function of the degradability of the contaminants of concern, as well as the availability of electron acceptors (oxygen, iron, sulfate, etc.), critical parameters for the biological degradation. Biodegradation rate data for the unsaturated zone are extremely limited in the literature. An analysis of rates of aerobic biodegradation of BTEX compounds in the vadose zone is given by DeVaull et al. (1997).

Simple Vadose Zone Leaching Models

Relatively simple manual calculations and/or simple analytical models have been prescribed for many risk assessment programs [such as the ASTM (1995) RBCA process] because these methods have minimal data requirements, are easy to perform and easy to review, and provide reproducible results when different users apply the model to the same site. Two such simple vadose zone leaching models are the ASTM

(1995) soil leaching equation and the Soil Attenuation Model (SAM; Connor et al., 1997). Both models are applicable to analysis of porous media affected by either organic or inorganic compounds, in the absence of mobile NAPLs. Using either site-specific or generic site properties, these models can be used either (1) to predict upper-bound compound concentrations in groundwater, based on an observed soil concentration, or (2) to back-calculate a lower-bound soil cleanup level, based on the applicable groundwater standard at the receptor exposure point.

The ASTM soil leaching model (ASTM, 1995; Connor et al., 1995), developed for use in RBCA studies, is based on the equilibrium partitioning relationship for sorbed contaminants [eq. (2.9)] and the leachate–groundwater mixing algorithm [eq. (2.16)]. The ASTM soil model assumes that all the contaminant mass that is measured in soil samples from a site is present as a sorbed non-NAPL phase and not present as NAPL. This assumption is conservative (i.e., overpredicts actual leachate concentrations), as the linear isotherm relationship [eq. (2.9)] is used so that leachate concentrations increase with increasing soil concentrations for site contaminants. In other words, all the contaminant mass that is actually measured in soil samples from a site (even samples that contain NAPL) is assumed to be in the sorbed phase because eq. (2.9) will typically yield higher (i.e., worst-case) leachate concentrations.

For several parameters, default values are suggested if site-specific parameters are not available (e.g., net infiltration of 30 cm/yr; mixing zone depth of 2 m). Because the model developers wanted the model to remain very simple and conservative (i.e., to have the tendency to overpredict the impacts of contaminated soils), contaminant spreading, volatilization, and biodegradation processes are not incorporated into the model. In practice, however, this model was found to overpredict the impact of contaminated soils in the vadose zone, and consequently, model developers have worked to refine this simple model.

One such refinement is the SAM (Connor et al., 1997), which was directed toward providing a more rigorous characterization of the soil-to-groundwater leachate process and assisting the user in estimating critical input parameters. To accommodate use in risk-based site evaluation efforts, SAM retains the format of a simple, screening-level analytical model requiring limited site-specific data input. SAM includes the following modifications of the ASTM soil leaching model:

1. *Net Infiltration Estimate.* SAM incorporates an empirical relationship for estimation of net rainfall infiltration based on mean annual rainfall and soil type [see eqs. (2.12) through (2.15)].

2. *Sorptive Mass Redistribution.* SAM incorporates depth effects by accounting for the sorptive redistribution of contaminants from the leachate onto soils underlying the contaminated soil zone. This sorptive redistribution reduces contaminant concentrations delivered to the underlying groundwater.

Contaminants dissolved within the advancing leachate front will redistribute onto clean underlying soils through sorption. Such sorptive loss serves to slow the rate of downward contaminant migration and provide additional time for volatilization or biodegradation prior to leachate discharge to underlying groundwater. Sorptive effects typically are characterized using the same equilibrium partitioning relationship as that used for mass transfer from the contaminated soils to infiltrating rainwater, except that for clean underlying soils, the direction of contaminant partitioning is reversed.

SAM corrects the equilibrium soil leachate concentration for the effect of sorptive mass loss as the leachate percolates downward toward the underlying water-bearing unit. This adjustment can prove significant in deep groundwater systems, wherein a significant thickness of uncontaminated soils underlies the contaminated soil zone. For SAM, the contaminated soil zone is characterized as a finite source mass equivalent to the contaminated soil mass times the representative compound concentration. Prior to reaching groundwater, percolating rainwater serves to redistribute this finite source mass among soil, air, and pore fluids throughout the full thickness of the surface soil column.

As a result of this sorptive redistribution of contaminant mass onto intervening soils, the ratio of the initial equilibrium leachate concentration to the leachate concentration that will reach the water-bearing unit is the initial contaminated soil zone thickness (L_1) divided by the total soil column thickness from the top of the contaminated soils to the top of the water-bearing unit (L_2):

$$C_{L2} = C_{L1} \frac{L_1}{L_2} \tag{2.21}$$

where C_{L2} = concentration of contaminant in soil leachate discharged to underlying water-bearing unit (mg/L)
C_{L1} = concentration of contaminant in soil leachate in initial contaminated soil zone (mg/L)
L_1 = thickness of contaminated soil zone (cm)
L_2 = distance from top of contaminated soil zone to top of water-bearing unit (includes both contaminated soil zone and clean zone where contaminants will redistribute) (cm)

3. *Finite Source Mass.* SAM assumes a finite contaminant source mass rather than the infinite source assumed in the ASTM model. Using the relationship (2.22), SAM does not allow the mass flux of contaminants in the leachate to exceed the total mass of contaminants in the vadose zone soils over an exposure duration (typically, a 25- or 30-year period for evaluating chronic effects in a risk assessment). The upper-bound leachate concentration for mass conservation is calculated as

$$C_{L2} \leq \frac{C_T \rho_b L_1}{I_f \text{ED}} \tag{2.22}$$

where C_T = bulk contaminant concentration on soil mass (mg/kg)
ρ_b = soil bulk density (g soil/cm^3 soil)
I_f = net infiltration (cm/yr)
ED = exposure duration (years)

4. *Default Soil Moisture Parameter.* SAM employees default soil moisture parameters (volumetric water and air contents) consistent with the predominant soil type in the surface soil column (see Table 2.3).

5. *Leachate–Groundwater Dilution.* SAM estimates leachate dilution in the groundwater flow system using the same box model incorporated in the ASTM expression. However, a supplementary algorithm, published as part of the EPA soil screening guidance (U.S. EPA, 1996), has been added for estimation of the groundwater mixing zone depth [see eqs. (2.18) through (2.20)].

Example 2.2 What is the resulting dissolved contaminant concentration in an aquifer below contaminated soils in the vadose zone that have benzene concentrations of 200 mg/kg? Assume that no NAPLs are present, 100 cm of annual precipitation (about 39 in/y), and a soil bulk density of 1.7 g/mL. The contaminated soils are clean well-graded sands (see Table 2.3). The top 5 ft of the soil is contaminated and is underlain by 10 ft of clean unsaturated soils before reaching the water table. The measured f_{oc} of the soil is 0.006, and the log K_{oc} value for benzene is 1.58. The Henry's law constant for benzene is 0.22 (cm^3 H_2O/cm^3 air; unitless). The vertical mixing zone is estimated to be 6 ft thick. The source zone is 50 ft long in the direction parallel to groundwater flow and 60 ft long in the direction perpendicular to groundwater flow. The aquifer has a Darcy velocity for groundwater of 70 ft/yr.

SOLUTION: First, calculate the concentration of benzene in the leachate from equilibrium partitioning relationships for soil (no NAPLs) using eq. (2.9).

$$C_L = \frac{200}{(0.08/1.7) + 0.006(10^{1.58}) + 0.22(0.33/1.7)} = 629 \text{ mg/L}$$

Next, redistribute the mass of contamination in the initial contaminated soils over the vertical soil column extending to the water table using eq. (2.21):

$$C_{L2}(\text{mg/L}) = 629\left(\frac{5}{15}\right) = 210 \text{ mg/L}$$

Next, estimate the annual infiltration using eq. (2.12). For sandy soils

$$I_f(\text{cm/yr}) = 0.0018(100)^2 = 18 \text{ cm/yr} \quad \text{or} \quad 0.6 \text{ ft/yr}$$

Next, calculate the leachate dilution factor using eq. (2.16):

$$\text{LDF} = 1 + \frac{70(6)}{0.6(50)} = 15$$

Using eq. (2.17), the concentration of the groundwater contamination in the aquifer will be 15 times lower than the leachate concentration after source mass redistribution, or

$$C_{\text{aquifer}} = \frac{C_{L2}}{15} = \frac{210}{15} = 14 \text{ mg/L}$$

Complex Vadose Zone Models

The ASTM and SAM models are based on relatively simple equations to describe the leaching of contaminants from soils to groundwater, reducing the complexity and the data required to run these models. Other models include additional processes besides equilibrium partitioning and mixing with groundwater. Three more commonly used complex vadose zone models (SESOIL, VADSAT, and Jury's model) are described below (see Table 2.4 for a comparison). These models represent the leaching process in more detail but typically require more site data than that required by the simple models described above.

TABLE 2.4. Comparison of Four Vadose Zone Leaching Models: SESOIL, VADSAT, Jury's Model, and SAM[a]

Model Feature	SESOIL	VADSAT	Jury's	SAM
Net infiltration estimate	×			×
Equilibrium partitioning				
NAPL-to-leachate	×	×		
Soil-to-leachate	×		×	×
Mass transport				
Advection	×	×	×	×
Dispersion		×		
Diffusion	×	×	×	
Sublayer discretization	×			
Sorption	×	×	×	×
Volatilization	×	×	×	
Biodecay	×	×	×	×
Chemical processes: cation exchange, hydrolysis, metal complexation	×			
Leachate–groundwater dilution		×		×
Source depletion	×	×	×	×

Source: Connor et al. (1997). Reprinted with permission from NGWA, Westerville, Ohio.

[a]For model information call:
SESOIL: Scientific Software Group, 703-620-9214
VADSAT: ES&T, 540-552-0685
Jury's: In API DSS, Geraghty and Miller, 410-987-0032
SAM: Groundwater Services, Inc., 713-522-6300

SESOIL. The Seasonal Soil Compartment (SESOIL) model is a numerical code designed to simulate one-dimensional vertical contaminant fate and transport in the unsaturated zone based on soil, chemical, and meteorological input values (General Sciences Corporation, 1996). SESOIL simulates water transport based on the hydrologic cycle and soil properties of the unsaturated soil column to estimate (1) time-varying contaminant concentrations for various depths, and (2) the net loss of contaminant mass from the unsaturated zone in terms of percolation to groundwater, surface runoff, biodegradation, and volatilization.

SESOIL characterizes the site hydrology by accounting for rainfall, surface runoff, infiltration, soil moisture content, and evapotranspiration. The climatic data required by the model consist of monthly climatic statistics for a typical year (e.g., average rainfall, temperature, cloud cover, storm duration and frequency). Contaminant fate and transport are based on chemical mass balances, equilibrium partitioning among four contaminant states, including a residual NAPL and mass transport of dissolved contaminants in groundwater by means of advection and diffusion. SESOIL also accounts for the loss of contaminant mass due to volatilization, biodegradation, and other chemical processes. Chemical data input requirements include solubility, diffusion coefficients, sorption constant, biodegradation rates, and so on. SESOIL does not account for the dilution of leachate in the receiving water-bearing unit.

The unsaturated soil column extends from the ground surface to the groundwater table and is characterized in SESOIL by up to four soil layers, which may be further divided into discrete sublayers. Required soil data for each layer consist of soil parameters averaged over the entire layer, such as dry bulk density, intrinsic permeability, disconnectedness index, and so on. In all, 30 to 40 site- and chemical-specific input values are required for leachate modeling using SESOIL.

VADSAT. VADSAT is a semianalytical code designed to simulate one-dimensional vertical pollutant fate and transport to assess effects of land disposal practices or chemical spills on groundwater quality (American Petroleum Institute, 1995). This model consists of unsaturated and saturated zone components based on analytical solutions of the flow and transport equations, although for leachate modeling, only the unsaturated portion is needed. VADSAT can produce time-varying pollutant concentrations for various depths based on the release and vertical transport of contaminants in the unsaturated zone.

Vertical unsaturated contaminant flow in VADSAT is represented by a steady-state net infiltration rate, input by the user. Soil data requirements include soil moisture parameters, bulk density, dispersivity, and so on. These properties are averaged over the surface soil zone and are assumed to remain constant with depth. Required chemical input parameters include molecular weights, solubility, diffusion coefficients, partitioning parameters, biodegradation rates, and so on, to account for dissolution from NAPLs, advective and dispersive transport of the dissolved contaminant, sorption, volatilization, and biodegradation. In all, about 25 site- and chemical-specific input values are required for leachate modeling using VADSAT.

Jury's Model. Jury's model, implemented in the American Petroleum Institute's (1994) Risk/Exposure Assessment Decision Support System (DSS), is an analytical screening-level model that was developed to evaluate the extent of volatilization to the atmosphere from organic contaminants located below the ground surface. It is not, however, intended for use in simulating volatilization for a specific site, but rather, was designed to assess the volatilization potential of a large number of chemicals under various conditions (Jury et al., 1990). Although primarily a volatilization model, Jury's model does include an expression for estimating the time-varying concentration profile of a soil contaminant in a homogeneous unsaturated zone. Jury's model does not, however, account for leachate dilution in the receiving water-bearing unit.

Vertical unsaturated flow in Jury's model is represented by a steady-state net infiltration rate, input by the user. For the equilibrium partitioning relationship, soil data requirements include soil moisture parameters, bulk density, and so on. These properties are averaged over the surface soil zone and are assumed to be constant with depth. Required chemical input parameters include partitioning parameters, molecular weights, solubility, diffusion coefficients, biodegradation rates, and so on, to account for advective and diffusive transport of the dissolved contaminant, sorption, volatilization, and biodegradation. A total of 19 site- and chemical-specific input values are required in order to predict leachate concentrations using Jury's model.

Model Comparison. A comparison of these three models and the SAM model is provided by Connor et al. (1997) based on hypothetical site data. Predicted soil leachate concentrations and back-calculated soil cleanup values exhibited significant variability among the four models. For a sandy loam soil, initial peak leachate concentrations range over six orders of magnitude, with somewhat less scatter observed in concentrations predicted after 10 years. Soil leachate values and soil cleanup limits determined for a silty clay soil were in somewhat better agreement, ranging over two to three orders of magnitude. These results show the significant uncertainties involved in leachate modeling.

The significant difference between the SAM results and those of the other three models (particularly for the sandy soil case) was due principally to effects of contaminant mass loss via volatilization, which is neglected in SAM. Differences among SESOIL, VADSAT, and Jury's model relate to their differing rates of vertical leachate flow and the consequent variation in mass lost due to volatilization and biodegradation during the leachate travel time through the surface soil column. In all cases, soil cleanup values were less variable than the predicted soil leachate concentrations, with less scatter observed for the clay than for the sandy soil (Connor et al., 1997).

2.6 PROCESS DESCRIPTIONS FOR SOURCE SCENARIOS 3 AND 4: LNAPL AND DNAPL DISSOLUTION IN THE SATURATED ZONE

The most common source of dissolved compounds in groundwater is the presence of NAPLs in the saturated zone. The NAPLs dissolve slowly into groundwater as the groundwater migrates past NAPL in residual blobs or continuous NAPL pools. Dissolution is a very complicated process that can be difficult to quantify and understand. In this section we answer some commonly asked questions regarding the dissolution of NAPLs in the saturated zone by the use of conceptual models (simple diagrams), data from field sites, results from controlled research experiments, information from laboratory studies, and simple dissolution models. Some of the key questions of interest are

- What is the distribution of NAPLs in the subsurface? Do continuous NAPL pools contain 100% NAPL? Does residual NAPL consist of individual blobs in individual pores?
- How does the NAPL distribution change over time?
- If a nondissolved organic compound is present in the subsurface, why aren't dissolved concentrations in source zone monitoring wells at or near the solubility limit for the dissolved compounds?
- Why do we focus primarily on BTEX compounds when characterizing dissolved contaminant plumes resulting from releases of gasoline in the subsurface, when BTEX compounds comprise only a small fraction (about 5 to 15%) of the gasoline itself?
- Is there always recoverable DNAPL in pools at sites with dissolved chlorinated solvent plumes? Why is DNAPL so hard to find compared to LNAPL?
- Does biodegradation affect NAPLs? Does biodegradation affect the dissolution rate?
- How long does dissolution take? How is the lifetime of the source zone predicted?

Although not providing definitive answers to all of these questions, in this chapter, we describe the key conceptual models, calculations, and quantitative models that can be used to describe the dissolution process.

Conceptual Models of NAPL Distribution and Dissolution

The distribution of NAPL in the subsurface has a strong influence on the rate and the nature of the dissolution process in subsurface source zones. At many DNAPL sites and many LNAPL sites, however, the actual extent of NAPL migration is never known precisely. Although evaluation of monitoring well data and soil cores may indicate where continuous NAPL or zones with large amounts of residual NAPL are located, the detection of small or moderate amounts of residual NAPL is problematic with most commonly used site characterization techniques (Cohen and Mercer, 1993; Pankow and Cherry, 1996).

To illustrate how NAPL is distributed in the saturated zone at contaminated sites and how dissolution occurs, a series of conceptual models are presented below. The first group of conceptual models telescope from a large, site-wide scale (typically, hundreds of feet) to the subpore scale (typically, submillimeter). Additional conceptual models illustrate the effects of heterogeneities on NAPL distribution and the migration and dissolution of NAPLs over time.

Site-Wide Model. As shown on Figure 2.13*a*, the site-wide models for LNAPL and DNAPL start similarly, as both LNAPL and DNAPL migrate down through the vadose zone, spread out at changes in soil texture, and then penetrate farther downward in fingers. When the LNAPL hits the water table, however, buoyancy forces cause the LNAPL to spread out in a thin, pancakelike layer, and later to

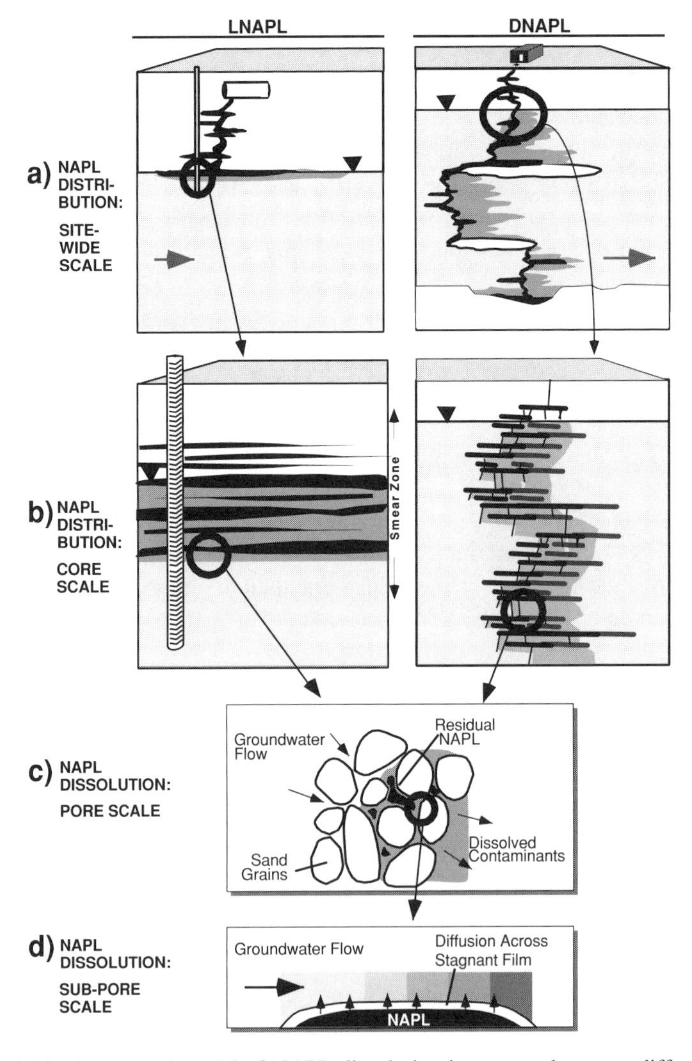

FIGURE 2.13. Conceptual model of NAPL dissolution into groundwater at different scales. (Pore-scale and subpore-scale models adapted with permission from Pankow and Cherry, 1996.)

smear vertically with the changing water table, leaving residual LNAPL throughout the smear zone. An aqueous-phase plume flows from the smear zone, migrating with the moving groundwater close to the water table.

DNAPL penetrates the water table and fingers into the saturated zone, forming lenses at subtle changes in aquifer material and more major, macroscale heterogeneities. In the conceptual model (Figure 2.13*a*), the DNAPL spills down from the lenses and eventually forms a pool on a capillary barrier (in this case an unfractured clay layer) at the bottom of the water-bearing unit. Based on a theoretical treatment of fingering phenomena, Kueper and Frind (1988) suggested that, in general, the fingers formed by chlorinated solvent DNAPLs in saturated sands or finer media should be less than 10 cm wide. Dissolved plumes are created by every finger and pool, with the plumes associated with the discrete fingers being very thin in the *y* direction (the direction coming out of the page), and the plumes created by the pools being very thin in the *z* direction (vertically up and down).

Core-Scale Model. In Figure 2.13*b*, closer, core-scale views on the order of a few feet are shown for the LNAPL and DNAPL releases, corresponding to the size observed when a typical soil core is brought to the surface for evaluation. Note that NAPL can be present, but not observed, in a core, as the amount of NAPL can be small and lack a strong color contrast with the soil. With the LNAPL release, the entire smear zone is visible, and it is apparent that subtle changes in the aquifer material cause a nonuniform distribution of residual LNAPL. In the DNAPL conceptual model, a similar process occurs in which subtle changes in the aquifer materials cause the DNAPL to form an apparently random sequence of fingers leading down to thin horizontal lenses from which more fingers emanate.

The amount of dissolution that occurs at this scale is dependent on (1) the surface area of the NAPL that is exposed to moving groundwater, (2) the size of the NAPL blobs, and (3) the chemical characteristics of the NAPL (e.g., the mole fraction of soluble components and the solubility of NAPL compounds). Note that the presence of residual NAPL in a porous medium will reduce the relative permeability of the residual NAPL zone to some extent. In one laboratory experiment performed by Anderson et al. (1992a), the flow through the residual DNAPL zone was reduced by about 20% by NAPL at a residual saturation of 13%. Kueper and Frind (1992), using a range of residual saturations between 14 and 40%, estimated the corresponding reduction in relative permeability (and therefore groundwater flow) for a PCE–water system to be between 20 and 70%.

Pore-Scale Model. In the pore-scale (millimeters or less) conceptual model, either residual LNAPL or DNAPL blobs are shown (Figure 2.13*c*). Both the LNAPL and DNAPL residual blobs have the same general appearance, and groundwater flow between the sand grains of the porous medium will come into contact with a NAPL blob and dissolve the soluble contaminants. Note that the LNAPL and DNAPL will usually be "nonwetting" in the saturated system. In this case, water (the wetting agent) will be in direct contact with the solid (sand grains in this case) and occupy the smallest pore spaces, whereas the NAPL blobs will occupy the larger pores. In a LNAPL smear zone or a DNAPL pool, the percentage of the pore space occupied by the NAPL (called *saturation*) will be greater than that occupied by water, such that the NAPL may form a continuous mass that could migrate as a body. Under most conditions, however, water will continue to occupy some fraction of the pore space (called the *residual water content*), and it is rare for large volumes of the porous media to be occupied completely by the NAPL at most contaminated sites.

Subpore-Scale Model. At the sub-pore scale (Figure 2.13*d*), the surface of the LNAPL or DNAPL blob is in contact with groundwater, and dissolution is actively taking place. Here a thin film of stagnant water is formed on the blob, and diffusion occurs through the stagnant film to transport soluble contaminants to the film's surface. Moving groundwater mixes with the dissolved compounds leaving the film, forming a dissolved plume. At many parts of each residual NAPL blob, however, there is no moving groundwater near the blob (e.g., the part of the blob adjacent to sand grains) and therefore there is little or no mass transport across the film.

If the NAPL consists of a single compound, the NAPL blob will slowly shrink over time as dissolution continues. If the NAPL consists largely of nonsoluble components with a small fraction of soluble compounds (such as gasoline), dissolution will be affected by effective solubility considerations (see below), and the residual NAPL blob will not shrink appreciably.

Effect of Subsurface Heterogeneities on NAPL Distribution. Large-scale heterogeneities can have different effects on LNAPL and DNAPL migration through the saturated zone. As shown on Figure 2.14, the downward migration of DNAPL is heavily influenced by the presence of microscale heterogeneities in the water-bearing unit and by macroscale heterogeneities such as aquitards and clay lenses. Migration of LNAPLs can be controlled by geologic heterogeneities, for example, where LNAPL is released below the water table (e.g., from a leaking

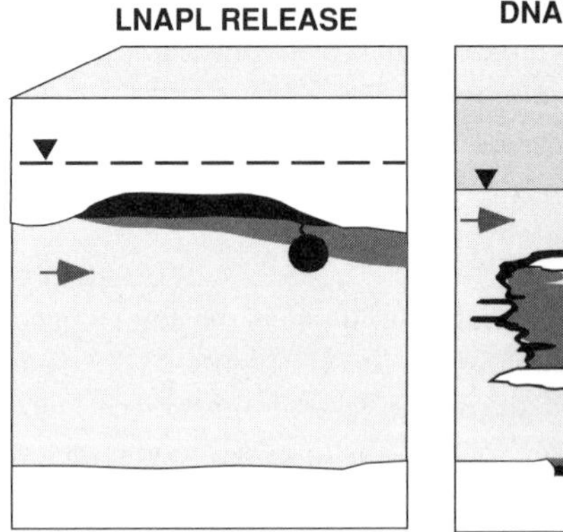

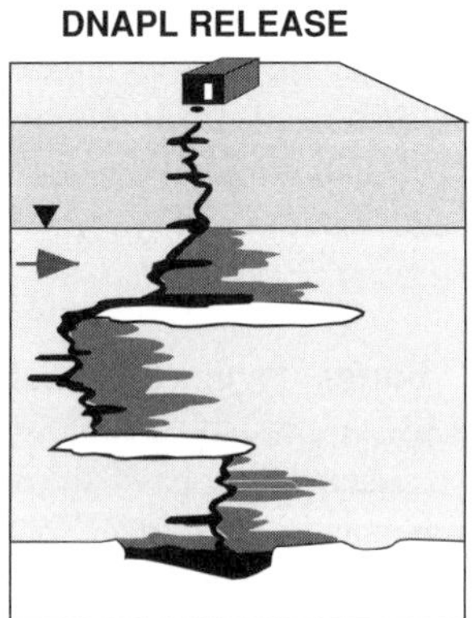

FIGURE 2.14. Potential effects of subsurface heterogeneities on LNAPL and DNAPL releases. (DNAPL figure after Waterloo Centre for Groundwater Research.)

pipeline). In this case, the LNAPL will float upward (leaving a trail of residual LNAPL) and accumulate underneath any stratigraphic traps in the water-bearing unit. This process, familiar to petroleum engineers, can control the migration of an LNAPL layer.

For more detailed presentation of conceptual models of NAPL migration, see the studies by the U.S. EPA (1992), Newell et al. (1995), and Pankow and Cherry (1996). These models describe other hydrogeologic regimes (e.g., fractured rock), source release scenarios, and other processes.

Migration of NAPLs over Time. Simple conceptual models of how NAPL source zones change over time are shown in Figure 2.15*a*, which illustrates how a

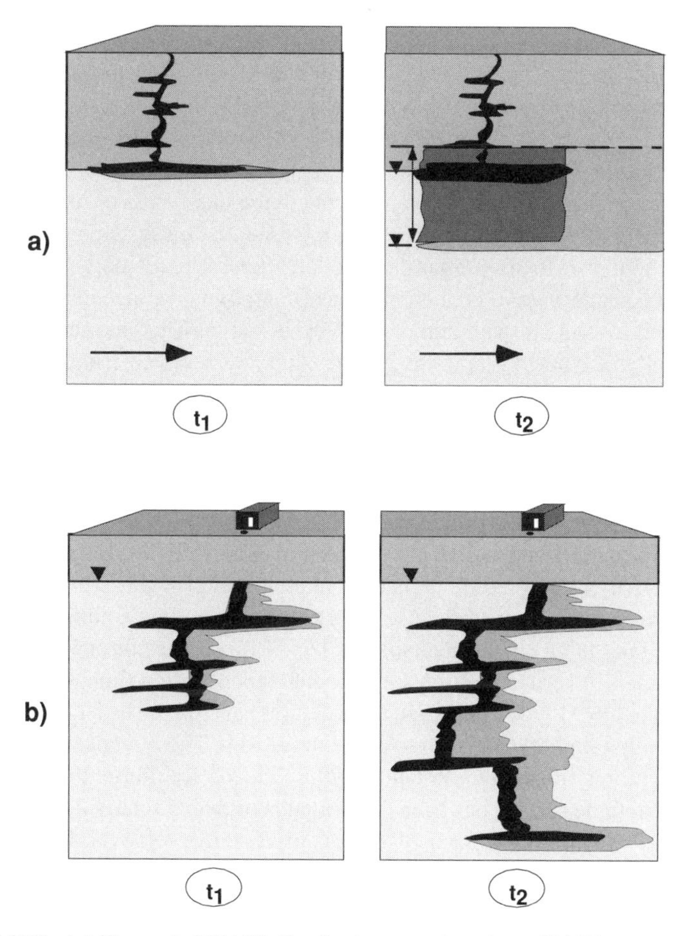

FIGURE 2.15. (*a*) Change in LNAPL distribution over time due to NAPL movement, smearing due primarily to fluctuating water table. Typical time scales: probably weeks or months, with additional smearing during extreme hydrologic events. (*b*) Change DNAPL distribution over time due to downward migration of DNAPL.

thin continuous LNAPL layer produces a thicker residual NAPL smear zone over time as the water table fluctuates up and down. Time t_1 represents conditions just after the release, and time t_2 represents a relatively short time after the release (months or a year). Typical values for the thickness of the smear zone are probably on the order of a couple of feet, as indicated by the Texas BEG and Florida RBCA studies.

In the BEG study (see Section 2.3), the standard deviation of water-level elevations from individual wells at a site was used as one method for estimating smear zone thickness, as it represents the typical range of the water table. Using this method, the median smear zone thickness at Texas UST sites was 1.3 ft (see Table 2.1), with 75% of the sites having smear zone thicknesses below 1.8 ft (Mace et al., 1997). In the Florida RBCA study, site hydrogeologists reported smear zone thicknesses for more than 80 sites, with the median smear zone thickness being 2 ft and with 75% of the sites have a smear zone thickness of 3 ft or less (Groundwater Services, Inc., 1997a).

Water table fluctuation can be seasonal or can extend over a number of wet and dry years. As long as there is a mobile layer of LNAPL at the site, the smearing process will continue. Note that at some sites, the water table will be lowered artificially, due to remediation-related pumping. This can complicate the overall remediation at a site, as groundwater is lowered beneath the main mass of residual LNAPL, greatly reducing the dissolution rate. When the remediation system is turned off, the water table will rise to its prepumping level, thereby resubmerging the residual LNAPL and causing a spike in aqueous-phase contaminant concentrations.

In Figure 2.15*b*, the distribution of DNAPL is shown when the surface supply of DNAPL is active (time t_1) and some time after the removal (time t_2). The time required for the DNAPL distribution to reach steady-state conditions (no further migration) once the surface supply of DNAPL has been removed is important at many sites. If the DNAPL continues to migrate even after the surface source is removed, there is significant benefit from active DNAPL removal to prevent further contamination of the aquifer. On the other hand, if DNAPL migration stops quickly after the primary source is removed, active DNAPL removal may not be as critical.

Although difficult to predict, numerical models can provide estimates of the rate of DNAPL migration. Kueper and McWhorter (1991) applied a numerical model to estimate the rate that a chlorinated solvent DNAPL migrates through fractures in a capillary barrier. In vertical fractures the predicted penetration time was on the order of hours (if the DNAPL can penetrate the fractures) and on the order of 10 days for a simulation with a slightly sloping fracture with a 2° dip. These model results for chlorinated solvent migration suggest that the time to reach equilibrium once the supply of DNAPL from the surface has been controlled is probably relatively short, and that migration probably does not persist for years after source removal. LNAPL models also are available to estimate how long LNAPL layers will continue to spread over time (Charbeneau et al., 1989).

Dissolution of NAPL Zones over Time. At LNAPL sites, the nature of the smearing will control which areas of the smear zone dissolve first (Figure 2.16*a*). Areas of the smear zone with more residual LNAPLs (see residual saturation, defined

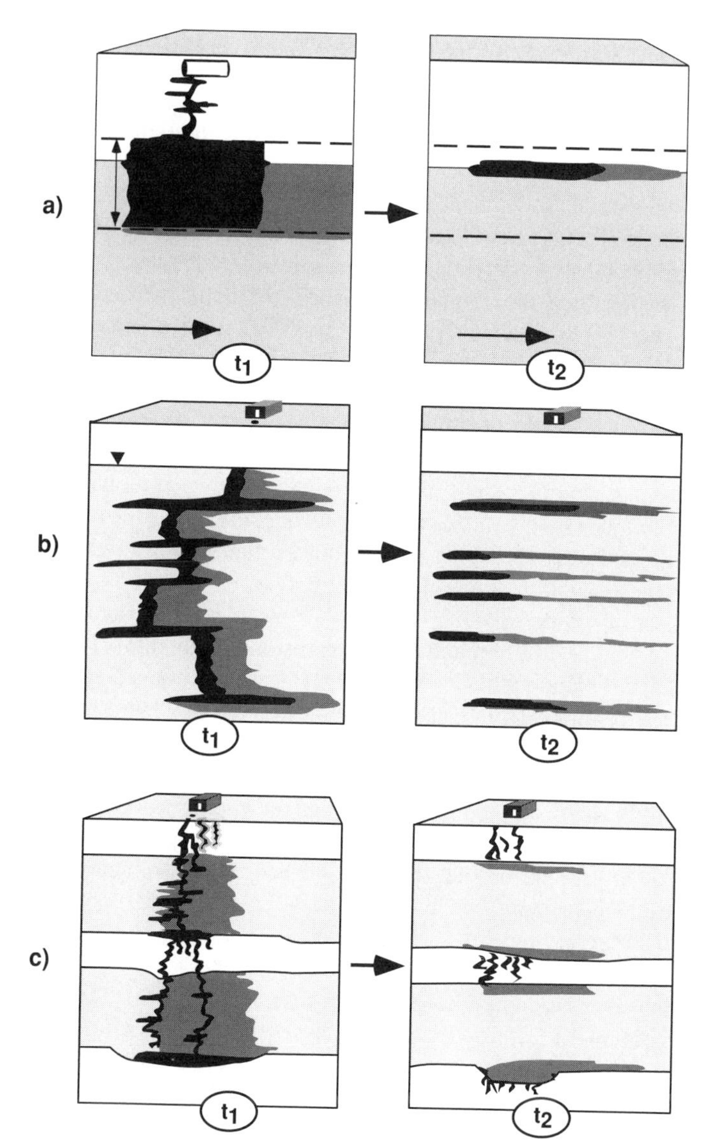

FIGURE 2.16. Conceptual models of NAPL dissolution over time. (*a*) LNAPL is smeared in a zone around the top of the water table, with a zone with higher LNAPL saturation. The zones of lower saturation dissolve first, leaving the zone of higher LNAPL saturation to dissolve last. (*b*) DNAPL forms fingers and pools, and the fingers (being very narrow in the direction of groundwater flow) dissolve first, leaving the pools (being very thick in the direction of groundwater flow). (*c*) DNAPL penetrates a surface clay, a low-permeability unit between two water-bearing units, and finally forms a pool on an unfractured clay. After considerable time, the DNAPL in the water-bearing unit dissolves away, leaving DNAPL in the fractures of the clay and aqueous-phase contaminants that have diffused in the clay that are slowly released to groundwater.

below) will dissolve most slowly. Water quality at sites with more water table fluctuation may be restored more quickly than at sites with little water table fluctuation. This is because a large smear zone means the LNAPL has been spread out, allowing more contact with air in the vadose zone and with groundwater flow in the saturated zone.

In Figure 2.16*b*, a DNAPL zone is shown shortly after a mostly soluble DNAPL has distributed itself into fingers, lenses, and pools (time t_1) and then again after an extensive period of dissolution (time t_2 = years, tens of years, or hundreds of years). In this case, the relatively thin fingers of residual DNAPL have dissolved, leaving the lenses and pools. The reason for the relatively rapid dissolution of the fingers versus the lenses and pools is the flux of clean water flowing through each DNAPL member. Fingers are relatively narrow zones of residual DNAPL that are exposed to a relatively large number of pore volumes over time [see eq. (2.3), where the definition of pore volume includes a length term]. Lenses caused by subtle changes in the texture of the aquifer present only a small cross section to groundwater flow and are exposed to considerably fewer pore volumes than fingers over the same time period. Pools are exposed to groundwater flow only on their top surface, greatly reducing the mass transfer (see the discussion of dissolution from pools, below).

Diffusion of dissolution products into low-permeability zones adjacent to NAPL zones can lengthen the time required for aquifer restoration (Figure 2.16*c*). In this conceptual model, dissolution products from a DNAPL diffuse into a clay near the top of the upper water-bearing unit, into an underlying fractured clay layer, and then into the bottom capillary layer. Once the continuous DNAPL in the higher-permeability zones has dissolved away, the concentration gradient is reversed, and the contaminants in the low-permeability zones start to diffuse out, a very slow process (Mackay and Cherry, 1989; National Research Council, 1994; B. Parker et al., 1994; Pankow and Cherry, 1996). In other words, the clay acts like a giant sponge that will hold NAPL dissolution products (aqueous-phase contaminants), and once the NAPL is gone, will slowly release the dissolution products back into groundwater. This process is particularly important for fractured clay systems (B. Parker et al., 1994).

At the pore scale, dissolution can have two different effects on the residual NAPL, as shown on Figure 2.17. If the NAPL is composed primarily of soluble components, such as pure TCE, pure xylenes, or a mixture of soluble components, the residual NAPL blobs become smaller over time and will eventually disappear (Figure 2.17*a*). On the other hand, if the NAPL is composed of both soluble and effectively insoluble components, dissolution will reduce, but will not eliminate, the NAPL blobs (Figure 2.17*b*). Examples of the latter case are an LNAPL composed of fresh gasoline, in which over 80% of the organic mass has relatively low solubility (mostly straight-chained hydrocarbons such as octane), or a DNAPL composed of a chlorinated solvent mixed with an insoluble mineral oil or some other carrier. In one modeling study performed by Borden and Kao (1992), the volume of residual gasoline in the simulated porous media was reduced by only 2%, even after flushing reduced the aqueous-phase concentrations to very low levels.

In some cases, residual NAPL blobs will be positioned out of the flow path of moving groundwater (e.g., into a dead-end pore) as the blobs shrink, and the rate

a) NAPL With All Soluble Components (e.g., pure TCE)

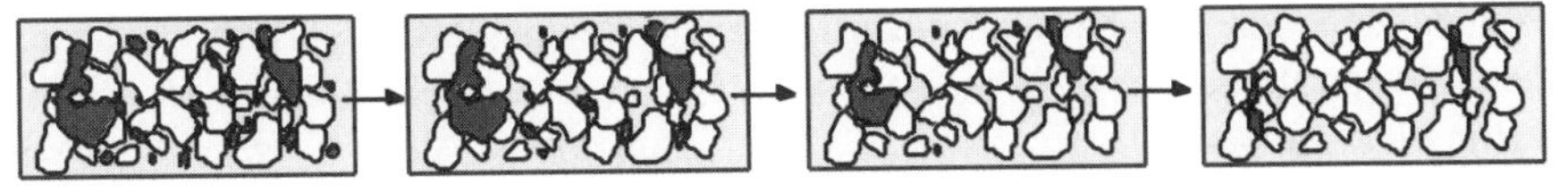

b) NAPL With Mostly Insoluble Components (e.g., gasoline)

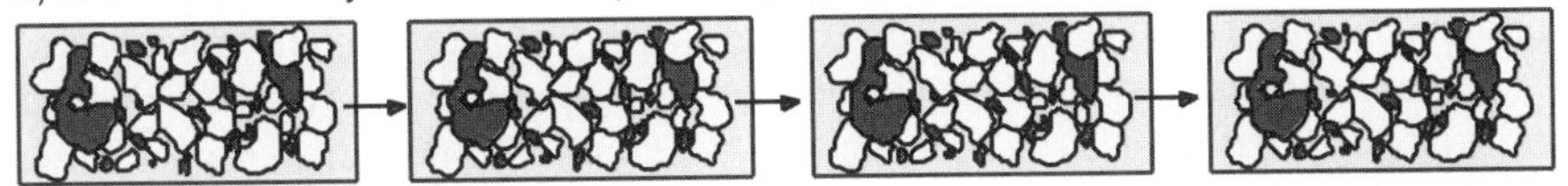

FIGURE 2.17. Effect of dissolution on residual NAPL blob size for two conditions: (*a*) NAPL with all soluble components (e.g., pure TCE) where NAPL blobs get smaller over time; (*b*) NAPL with mostly insoluble components (e.g., gasoline) where NAPL blobs remain close to the same size for a much longer period of time.

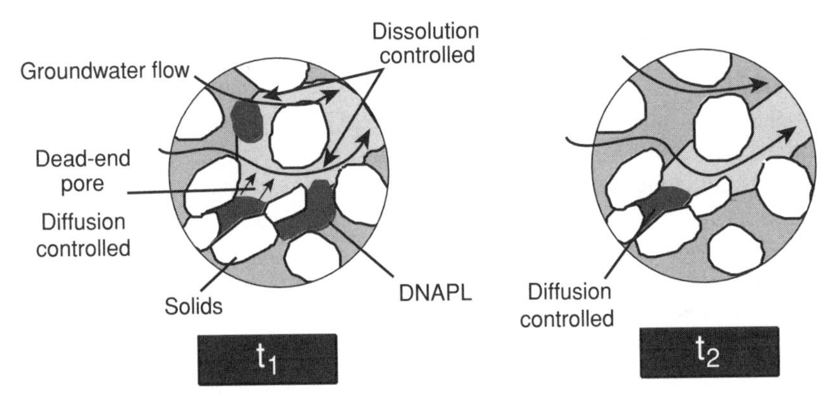

FIGURE 2.18. Effect of dead-end pores on NAPL dissolution. (Reprinted with permission from Pankow and Cherry, 1996.)

of dissolution will be reduced greatly (Figure 2.18). Diffusion is the only active transport mechanism taking contaminants from the residual blob in a dead-end pore to the moving groundwater stream. In this case the NAPL blob will persist for very long time periods, long after residual blobs in the flow stream have been removed. This diffusion-controlled process may be a contributing factor to the observation of dissolution "tails," where additional pore volumes of flushing reduces the dissolved contaminant concentration only very slowly (Borden and Kao, 1992), and the reporting of low-concentration "exhausted" plumes in the BEG and LLNL databases. This concept of tailing is discussed in more detail below from the perspective of laboratory dissolution studies. (Note: an alternative explanation of tailing is from slow-phase desorption processes; see the section on Estimating Leachate Concentration).

The structure of a porous medium may greatly limit the surface area of NAPL blobs that are exposed to moving groundwater. In one study of the unsaturated zone,

Luckner (1991) determined that only 2% of the NAPL blob area was available for oxygen exchange. Etched micromodels and polystyrene cast studies also indicate that some portion of the blob surface area is in contact with immobile water that is wetting the soil particles (Chatzis et al., 1983). Soil column studies support the idea of relatively small contact areas between NAPL blobs and flowing groundwater. Powers et al. (1991), for example, used these studies as the basis for assuming that the effective surface area of residual NAPL blobs was between 1 and 50% of the total surface area. The low surface area of residual NAPL that is exposed to groundwater is another reason why so many remediation efforts have not achieved their cleanup objectives. The subsurface can be thought of as a "reactor" where there is no mixing, and the available surface area for any reaction (e.g, dissolution, biodegradation) is small. With this type of reactor it is very difficult to accelerate the rate at which contaminants are transferred from the NAPL to groundwater. Therefore, natural attenuation, where nature ultimately completes the mass transfer, may be an appropriate strategy at some sites.

Characteristics of Residual NAPL in the Saturated Zone

Most of conceptual models presented above describe NAPL migration at the site level and show how large zones of residual NAPL are formed as LNAPL layers smear on fluctuating water tables and DNAPL fingers its way downward. These residual NAPL zones are important sources for dissolved contaminant plumes, and therefore an appreciation of how the residual NAPL is actually distributed in the subsurface is important.

Wilson and Conrad (1984) presented a prescient overview of residual NAPL as an environmental problem in 1984 when they wrote:

> The migration of hydrocarbon essentially immiscible with water occurs as a continuous multi-phase flow under the influence of capillary, viscous, and gravity forces. Once the source of hydrocarbon is disrupted, and the main body of hydrocarbon displaced, some of it is trapped in the porous media because of capillary forces. Hydrocarbon migration halts as this lower, residual saturation is reached. The trapped hydrocarbon remains as pendular rings and/or isolated, essentially immobile blobs. Residual hydrocarbons act as a continual source of contaminants as, for example, water coming into contact with the trapped immiscible phase leaches soluble hydrocarbon components.

Wilson and Conrad (1984) continued with a discussion of how two trapping mechanisms, *snap-off* and *by-pass*, form residual NAPL. As shown in Figure 2.19, water has advanced into the upper pore, with the arrow signifying a moving interface. A residual NAPL blob snaps off the retreating NAPL and is trapped in the pore by capillary forces. Snap-off is controlled by the body/throat pore size ratio and the wettability of the NAPL. In systems with a low body/throat size ratio (where pore throats are almost as large as the pores), hydrocarbon can be almost completely displaced, and snap-off is relatively uncommon. In most natural porous media, however, a high body/throat size ratio (pore throats much smaller than the pore) is observed, and snap-off will be an important NAPL-trapping mechanism.

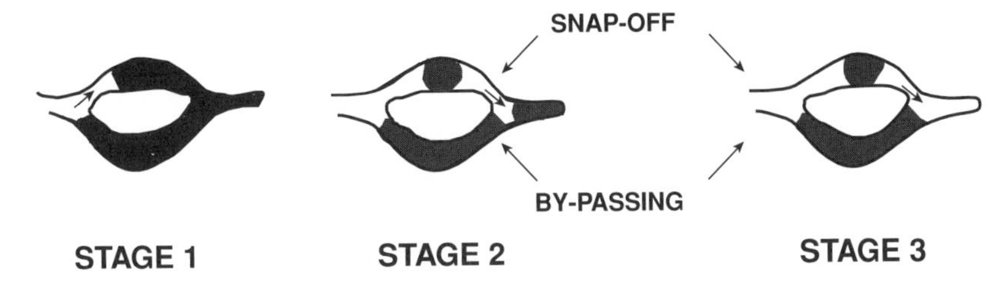

FIGURE 2.19. Mechanisms for formation of residual NAPL. Groundwater advances into the upper pore (moving interfaces are indicated by the small arrows in the pores) and NAPL is trapped by snap-off. The NAPL in the lower pore is trapped by bypassing. (From Wilson and Conrad, 1984. Reprinted with permission from NOWA, Westerville, Ohio, © 1986.)

As shown in Figure 2.19, NAPL is trapped via bypassing when water coming from the upper pore will form an interface between the NAPL being displaced downstream and the NAPL in the lower pore. In this fashion the NAPL in the lower pore is isolated (bypassed) and will remain trapped as residual NAPL in the porous media by capillary forces.

Note that after the continuous NAPL body has been converted to a residual form, the individual NAPL blobs are held very tightly in the porous media by capillary forces. Wilson and Conrad (1984) evaluated the force required to mobilize and completely sweep away residual blobs in porous media in terms of the hydraulic gradient a pumping system would have to generate to either (1) begin blob mobilization, or (2) mobilize all blobs in a porous medium. This relationship, presented as a graph of hydraulic gradient versus required hydraulic conductivity, indicates that mobilization of NAPL blobs by pumping will occur only in very coarse porous media with a very high hydraulic gradient (see Figure 2.20). The rest of the blobs will stay trapped in the porous media, serving as a long-term source of dissolved contaminants.

Although many regulatory programs mandate the complete removal of NAPL, or removal "to the extent practicable," the physical forces that trap residual NAPL mean that only continuous NAPL can be mobilized through the subsurface at most sites, and that the fraction of NAPL represented by the residual saturation (see below) cannot be practicably removed. Therefore, in a practical sense, NAPL removal translates to recovery of a small percentage of NAPL at a site (i.e., whatever continuous NAPL can be collected). Recovery of residual NAPL will only be practicable if it can be excavated, if the porous medium is a coarse gravel (or coarser material), or if an experimental technology such as surfactant–cosolvent addition are applied.

Residual Saturation as Observed in Laboratory Studies. Residual saturation is the fraction of the pore space filled with hydrocarbon that is essentially immobile (i.e., trapped by very strong capillary forces) after snap-off, bypass, and other processes have broken up continuous NAPL. Several sources provide general ranges for residual saturations. Mercer and Cohen (1990) have compiled data from

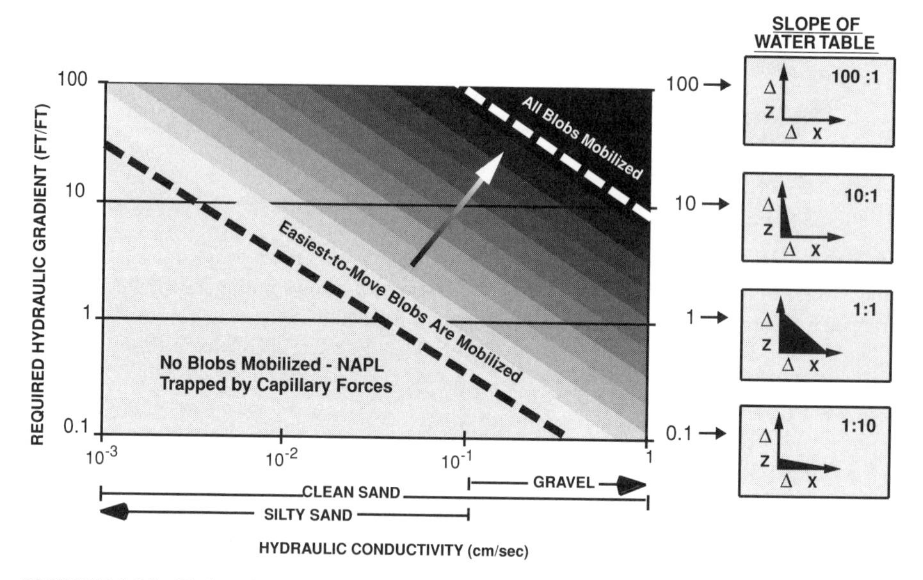

FIGURE 2.20. Hydraulic gradient required to initiate or completely mobilize residual NAPL blobs versus hydraulic conductivity of porous media. Note the median hydraulic gradient from a nationwide survey of 400 aquifers was 0.007 ft/ft, and the median hydraulic conductivity was 0.005 cm/s (Newell et al., 1990). These conditions are well below the values required for blob mobilization. (Adapted with permission from Wilson and Conrad, 1984.)

a number of sources and reported general ranges for residual saturation between 10 and 20% in the vadose zone and between 15 and 50% in the saturated zone. As discussed below, however, the scale of the measurement must be considered before applying concepts regarding residual saturation.

In the laboratory, some additional patterns of residual NAPL behavior can be observed. For example, laboratory studies have shown that residual NAPL saturation will increase with decreasing porosity and increasing body/throat pore size ratios and pore size heterogeneity (Chatzis et al., 1983). Kueper et al. (1993) observed that in a laboratory column, the residual saturation was related to the initial saturation (fraction of pore space occupied by continuous NAPL), as shown in Figure 2.21, although correlations using theoretical oil field models did not fit the data well. The authors explained the observed correlations by observing that a greater percentage of pores are invaded by NAPL, thereby providing a greater number of pores to form residual NAPL. Wilson et al. (1990), however, stressed how soil properties dominate any analysis of residual saturation, and concluded that

1. Prediction of residual saturation in a given soil is very uncertain.
2. Similar soils can exhibit very different residual saturations.

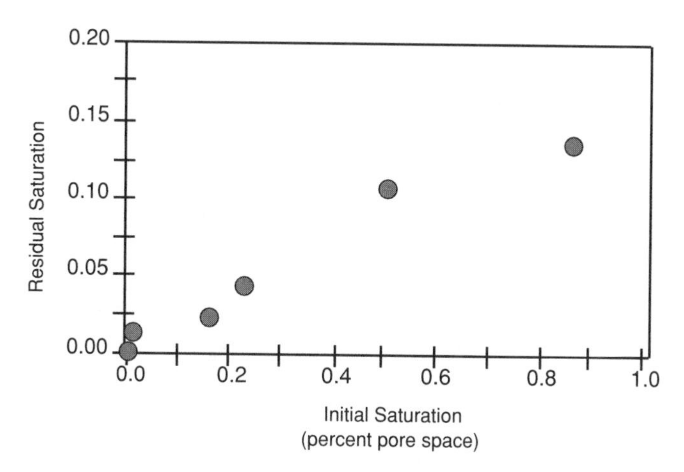

FIGURE 2.21. Relationship between initial saturation and residual saturation from laboratory experiments on Borden sand. (From Kueper et al., 1993. Reprinted with permission from *Journal of Ground Water.*)

3. Even minor amounts of clay or silt in a soil may play a significant role in observed residual saturations.
4. Residual saturations appear to be relatively insensitive to fluid properties and are very sensitive to soil properties (and heterogeneities).

Overall, residual saturations at field sites are difficult to measure and are subject to considerable error (Mercer and Cohen, 1990), and more important, in aquifers with soil heterogeneities (probably almost all sites) they are a function of the scale of measurement (Pankow and Cherry, 1996). For example, a 1-mm-thick finger of DNAPL running vertically downward may have a relatively high residual saturation in the pores where the NAPL invaded and left residual blobs. If a soil sample is taken using a 4-in. core barrel, however, the measured residual saturation will be much lower, due to mixing with adjacent soil material that was never exposed to the NAPL.

The actual distribution of residual NAPL blobs is very complex, as shown in NAPL visualization studies based on pore-and-blob casts of solidifying organic liquid poured into a porous medium and/or etched glass micromodels into which colored organics had been introduced (Chatzis et al., 1983; Wilson et al., 1990; Conrad et al., 1992; Powers et al., 1992). These methods have helped researchers visualize NAPL movement and see the distribution of NAPL in the subsurface after snap-off and bypassing mechanisms have converted it to residual NAPL. For example, Figure 2.22 shows different types of NAPL blob casts, ranging from relatively simple blob shapes to more complex, branching blobs (Conrad et al., 1992).

The numerical distribution and the volume of the residual NAPL in different-sized blobs have been measured in a few laboratory studies. Using uniformly sized glass

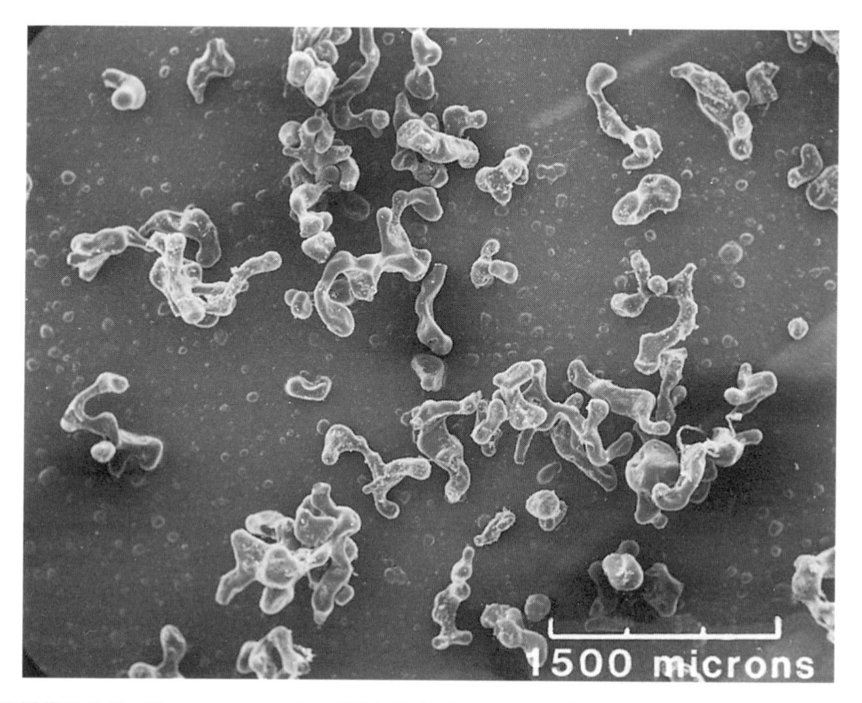

FIGURE 2.22. Photomicrographs of NAPL blob cast from Sevilleta sand. (Reprinted with permission from Conrad et al., 1992.)

beads, Chatzis et al. (1983) found that while most blobs (58%) were trapped as singlets, they held less than 15% of the volume of the residual saturation (Figure 2.23). Blobs occupying five or more pores held at least 50% of the residual, although they represented less than a quarter of the total number of blobs. Although not counting the blobs, Conrad et al. (1992) reported that the most of the residual NAPL blobs in a Sevilleta dune sand (a uniform medium quartz sand) were larger than a doublet or a singlet, and that complex blobs "clearly contain the majority of the residual saturation volume." Powers et al. (1994) reported that a larger percentage of NAPL was entrapped as large blobs in a graded Wagner mix sand than in a uniform Ottawa sand (Figure 2.24).

Measuring NAPL Saturation. Several laboratory methods have been described in the literature for determining NAPL saturation in porous media (American Petroluem Institute, 1960; Amyx et al., 1960; Core Laboratories, 1989; Cohen and Mercer, 1993). Common methods used in the oil industry include a retort fluids distillation method and a solvent extraction method that provide approximations of the relative volumes of NAPL, water, and air in soil samples. Chemical analysis of total petroleum hydrocarbons also may be used to estimate NAPL saturation in a sample if certain fluid and soil parameters are known or may be estimated. Residual saturation values are obtained if the samples at residual saturation (i.e., no continuous-phase NAPL) are analyzed with one of these methods.

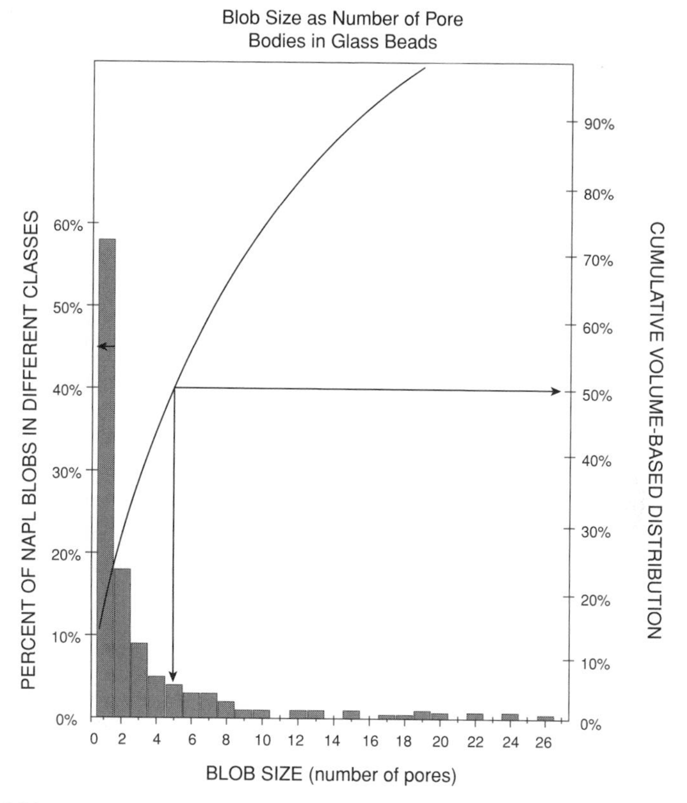

FIGURE 2.23. Distribution of NAPL blob size by number of pore bodies and distribution of NAPL volume by blob size in glass beads. Note that 50% of NAPL volume is in blobs spanning 5 pore volumes or greater, although about 90% of all NAPL blobs span 5 pore volumes or less. (Reprinted with permission from Morrow and Chatzis, 1982.)

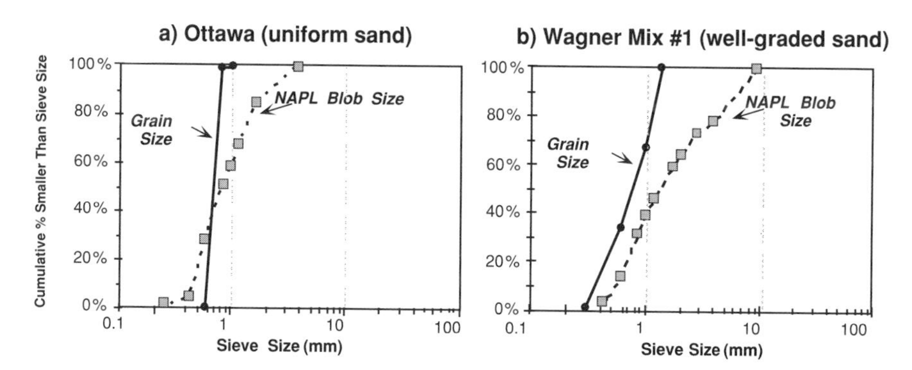

FIGURE 2.24. Distribution of NAPL blob sizes for (a) uniform sized sand and (b) a well-graded sand. Note that larger blob sizes are associated with the well-graded (nonuniform) sand. (Adapted with permission from Powers et al., 1992.)

The retort method, most commonly used for evaluation of sidewall cores from petroleum exploration or production wells, consists of the distillation of NAPL and water from the sample and gravimetric comparison of relative volumes of pore fluids. The water content of a sample is determined by atmospheric distillation concurrently with NAPL content determination. The sample temperature is raised to 100°C until all pore water is removed. At that point, the retort temperature is raised to 650°C to remove the NAPL. The NAPL distilled from the sample is collected in a calibrated receiving tube, in which its volume is measured. Gas content is determined through the injection of mercury into the gas-filled pore space. Measurement of the volume of mercury forced into a sample is a measure of the gas content of the sample. A pressure of 750 psi (50 atm) is normally used for friable or unconsolidated samples. Each of the measured fluid volumes is divided by the total pore volume to obtain the degree of saturation of that fluid or gas (Core Laboratories, 1989). The method is expeditious and inexpensive, with costs under $50 per sample, but lacks precision. It is good for determining whether or not NAPL is present in a sample, and provides a rough estimate of saturation. However, the retort process is hampered by substantial mass losses and cumulative error, resulting in saturations measured only approximately.

The Dean–Stark method, a standardized ASTM procedure, also employs a distillation process, wherein a solvent (usually toluene) in a heated reservoir is vaporized and then passed through a core (or plug) of porous media. As the vapor passes through the core, water and NAPL are evaporated from the core and carried off by the solvent vapor. Water vapor is condensed and drained into a graduated receiving tube. The solvent (including any dissolved NAPL) is refluxed back into the reservoir over the core sample. The core can be completely cleaned of water and NAPL in this apparatus and analyzed further using other porosity and permeability tests. The NAPL saturation, S_{napl}, of the sample is calculated by determining the difference between the water recovered and the total weight loss after extraction and drying.

The Dean–Stark method is more time consuming and somewhat more expensive than the retort process but has the advantage of more accurate results. Errors in saturation determinations are fairly low for soil samples that are obtained from near-surface formations, for which laboratory temperature and pressure conditions are not too dissimilar from the field environment. However, accurate results require some vigilance in the handling and packaging of soil samples. Typically, samples are obtained in brass or aluminum Shelby tubes, each end of which should be sealed carefully to prevent fluid loss or disaggregation of the soil. Sample tubes then should be kept on ice, or preferably frozen, until delivered to the laboratory. Laboratory personnel prepare samples for analysis by drilling horizontal or vertical plugs from the core sample, as specified by the client, and inserting the plugs into the distillation apparatus. Water and NAPL saturations typically are provided along with gas permeability and porosity determinations for approximately $60 per sample, plus a preparation cost of $20 per sample (as of 1997). Most petroleum analytical laboratories are unable to manage toxic or hazardous NAPL-bearing samples, such as samples containing chlorinated solvents or PCBs. A few specialized laboratories are equipped to handle such materials. At the time of this writing, one such laboratory is Core Laboratories' Advanced Technology Center in Dallas, Texas.

A chemical analysis method for NAPL saturation determination that can be inexpensive and simple to obtain for nonchlorinated NAPL depends on analysis of the TPH content of the soil sample and determination or estimation of soil porosity and bulk soil density (Cohen and Mercer, 1993; J. Parker et al., 1994). Determination of soil TPH involves extraction of soil samples and analytical quantification of hydrocarbons in terms of mass of hydrocarbon per mass of dry soil. TPH extractions yield a total hydrocarbon mass that includes NAPL, dissolved components in the aqueous phase, adsorbed components, and components in the gas phase (to the extent that the gas phase is not lost in sample processing). Calculations based on equilibrium partitioning indicate that when a nonaqueous phase exists in a sample, most of the hydrocarbon mass is in the separate phase (i.e., NAPL) unless the natural organic content of the soil is high. Therefore, estimating non-aqueous-phase volume directly from TPH incurs a very small error (J. Parker et al., 1994).

The NAPL saturation, S_{napl}, which is the ratio of NAPL volume to porosity volume, is related to TPH by (J. Parker et al., 1994)

$$S_{\text{napl}} = \frac{\rho_b \cdot \text{TPH}}{\rho_n n (10^6)} \tag{2.23}$$

where S_{napl} = NAPL saturation (unitless)
ρ_b = soil bulk density (g/cm^3)
TPH = total petroleum hydrocarbons, expressed as milligrams of hydrocarbon per kilogram of dry soil
ρ_n = NAPL density (g/cm^3)
n = porosity

Thus NAPL saturation may be determined from TPH analytical results, in conjunction with measurements or estimates of NAPL density, soil bulk density, and porosity. The 1997 cost for spectrophotometric TPH analysis (using EPA Method 418.1) is approximately \$50 per sample. In the case of a chlorinated NAPL, it is necessary to perform gas chromatography/mass spectrometry (GC/MS) hydrocarbon scans that involve summation of GC response over nonoverlapping elution time intervals for volatile and semivolatile hydrocarbon analytical methods such as EPA Methods 8260 and 8270 (Cohen and Mercer, 1993). List costs in 1997 for Methods 8260 and 8270 were typically \$125 and \$300 per sample, respectively.

Example 2.3 What is the approximate NAPL saturation of a soil with a TPH value of 10,000 mg/kg? Assume that the soil bulk density is 1.7 g/cm^3 and the porosity is 0.40. The NAPL, gasoline, has a density of 0.75 g/cm^3. Disregard the small concentrations of hydrocarbon present in the dissolved or sorbed phases in the sample.

SOLUTION: Apply eq. (2.23) to calculate the NAPL saturation of this sample:

$$S_{\text{napl}} = \frac{1.7(10,000)}{0.75(0.40)(10^6)} = 0.06 \text{ (6\% of pore space is filled with NAPL)}$$

Observation of NAPL Zones in the Field. Even when relatively small volumes of soil are used to estimate residual saturation, there is variation in the subsurface over relatively small distances. In a controlled-release experiment at the Borden Landfill at Canadian Forces Base Borden in Ontario, Kueper et al. (1993) released 231 L of PCE (a DNAPL chemical) below the water table in a 3- × 3- × 3.4-m sheet-pile cell. The aquifer material consisted of fine- to medium-grained sands. After 28 days, time-domain reflectometry (TDR) monitoring indicated that DNAPL movement had ceased. The upper 0.9 m of the cell was excavated, and three cores extending to the bottom of the cell were collected. On the basis of visual observations, sampling during the excavation, sampling of the three cores, TDR monitoring, and groundwater sampling using piezometers, the probable distribution of PCE along two perpendicular cell cross sections was developed (Figure 2.25). This distribution shows surprisingly extensive lateral spreading of this DNAPL throughout the cell, even though the aquifer material consisted of sand only (no silt or clay). The DNAPL migrated preferentially through the coarser-grained sands, and the thickness of the laminations and beds containing the DNAPL ranged from a few millimeters to 3 to 5 cm thick. Finally, it was evident that the DNAPL made extensive use of three piezometers as a preferential downward flow path.

Extensive saturation data were collected from the three cores using 2-cm^3 soil samples, and show saturations ranging from 1 to 38% of the pore space, with almost all of the measurements below 20% (Figure 2.26). Laboratory measurements suggested that all saturation values above 15% represented "pooled" PCE, while values

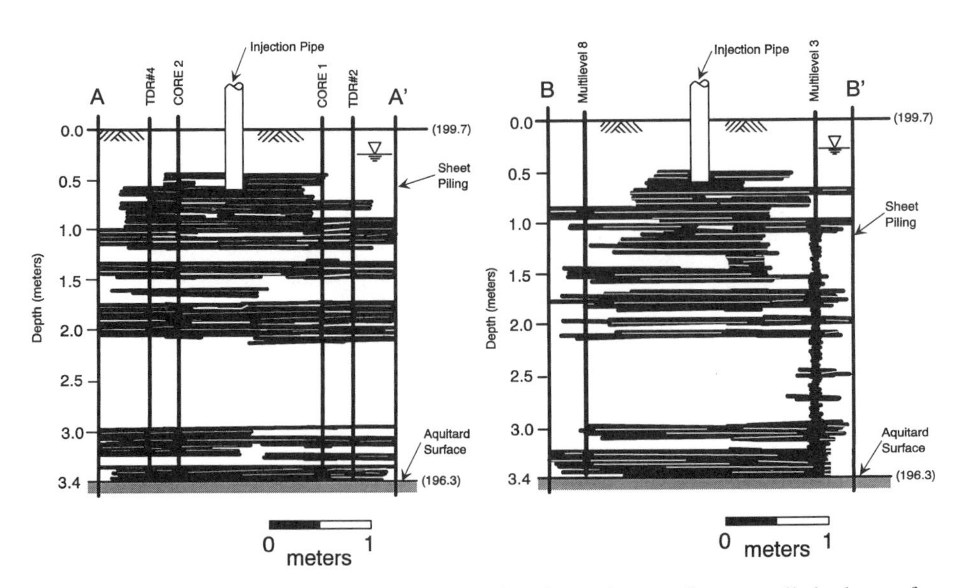

FIGURE 2.25. Probable distribution of PCE migration pathways after controlled release of 230.9 L of PCE into Borden sand test cell. Shown are two perpendicular vertical cross sections (cross sections intersect in middle of each cross section). (From Kueper et al., 1993. Reprinted with permission from *Journal of Ground Water.*)

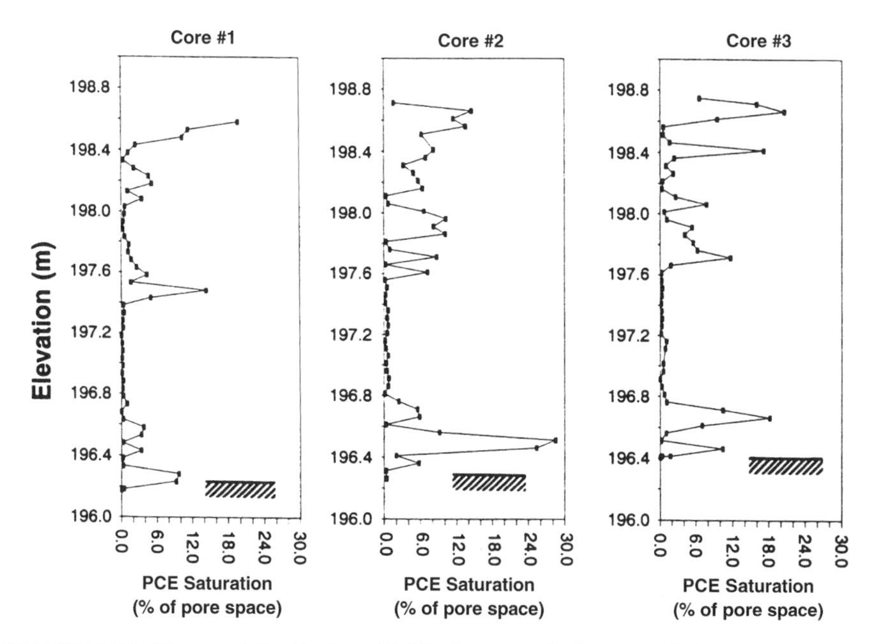

FIGURE 2.26. Measured distribution of PCE after controlled release of 230.9 L of PCE into Borden sand test cell. Shown are three soil cores from surface to bottom of test pit. (From Kueper et al., 1993. Reprinted with permission from *Journal of Ground Water.*)

under 15% represented residual DNAPL (Pankow and Cherry, 1996). The cores also illustrate how subtle changes in aquifer texture on the order of centimeters thick can direct the lateral and downward migration of DNAPL. These subtle changes in texture are almost invisible to a typical hydrogeologic investigation, making following and finding DNAPL in the subsurface extremely difficult.

Huntley et al. (1994) present similar saturation versus depth data for a series of boreholes, with a key difference being that LNAPL from an actual LUST site in San Diego was measured. In Figure 2.27, both LNAPL saturation and grain size are shown, along with the thickness of the observed LNAPL layer in the subsequent well (the difference between the oil–air interface and the oil–water interface). Soil samples were collected every 6 cm using a sample volume of approximately 60 cm^3. The results show typical LNAPL saturations between 5 and 20%, with one core showing a limited zone of up to 50% saturation. In the two boreholes with no LNAPL accumulation in the well (boreholes 9 and 13), saturations of up to 15% were observed below the water table. In boreholes 10 and 11, approximately 100 to 115 cm of LNAPL accumulation was noted, while the maximum saturation below the oil–air interface was less than 25%. In borehole 12, 160 cm of LNAPL was observed in the well, compared to a maximum saturation of 50% slightly above the oil–air interface.

These results from Huntley et al. (1994) indicate that LNAPL and water coexist in the pore spaces between the oil–water interface and the oil–air interface in a

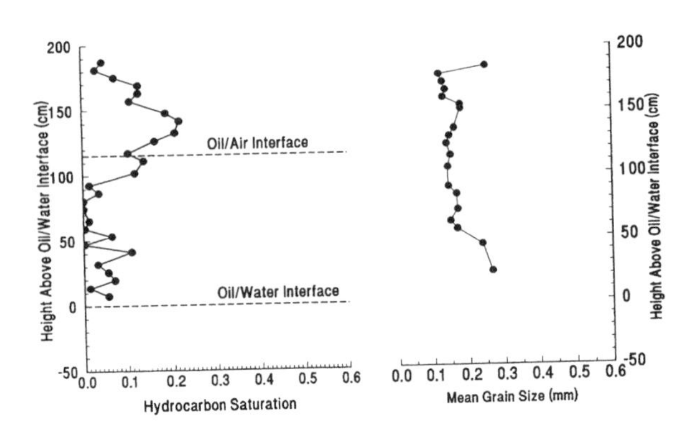

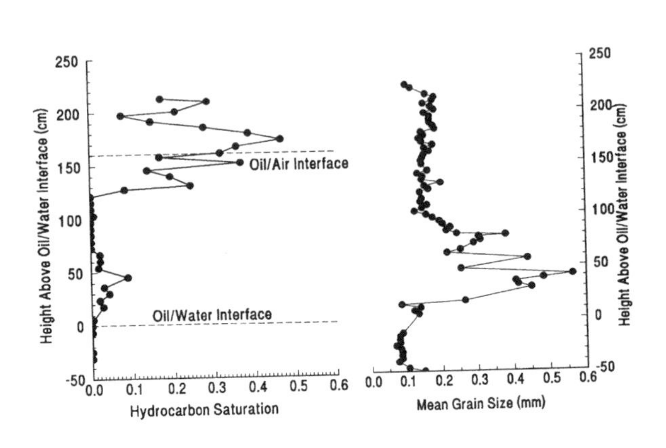

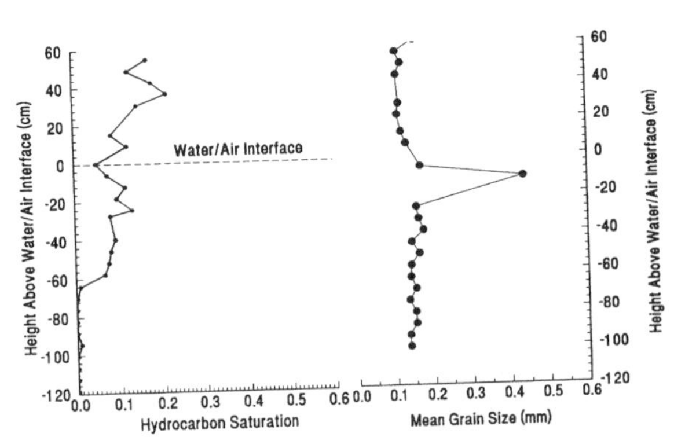

FIGURE 2.27. Grain size and distribution of LNAPL saturation in soil cores from a gasoline LUST site in San Diego, California. (From Huntley et al., 1994. Reprinted with permission from *Journal of Ground Water*, July/August 1994.)

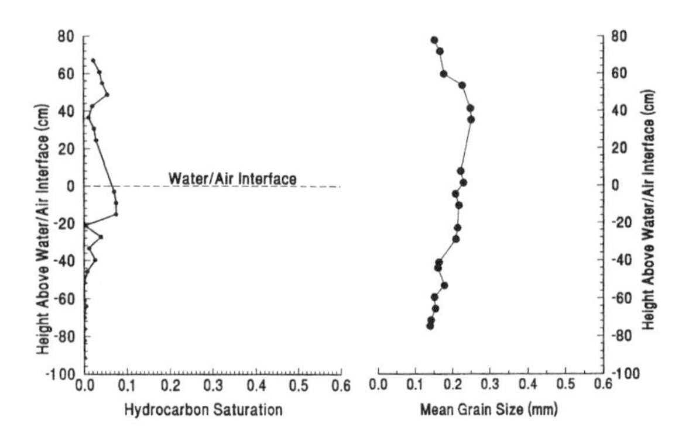

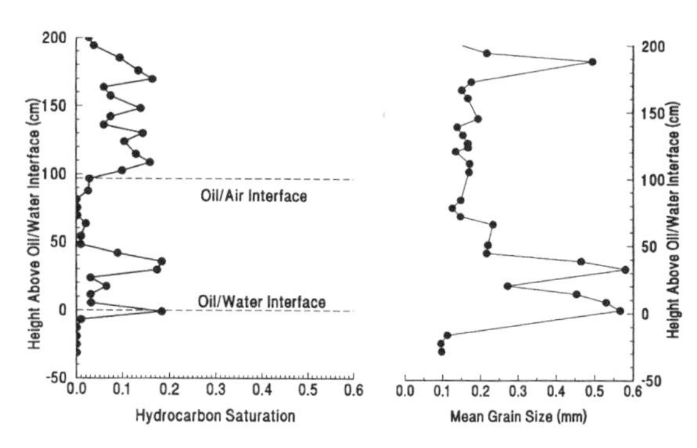

FIGURE 2.27. *(Continued)*

monitoring well, and that there are not measurable zones that are completely saturated with product. Capillary pressure models developed by Farr et al. (1990) and Lenhard and Parker (1990) describe the correlation between saturation and the thickness of LNAPL in a well. Finally, the relatively low LNAPL saturations at this site suggest that the mobility of the LNAPL is limited. If these conditions are similar to those at other LNAPL-contaminated sites, concerns about large expansions of LNAPL source zones over time may not be warranted, even if the amount of continuous LNAPL observed in monitoring wells is relatively large.

The results from the two studies discussed above indicate that the distinction between a residual zone and a continuous NAPL zone is subtle, and that zones where most of the pore space is completely saturated with NAPL may be relatively rare at most sites. In the case of the DNAPL release experiment, saturations below 15% were considered to be residual DNAPL zones, while saturations above 15% indicated pools. At the San Diego LNAPL site, the maximum observed saturation was only 50%, and that was over a very limited range. In other words, many or most LNAPL and DNAPL "pools" are probably characterized as relatively thin layers that contain mostly water in the pore spaces, and therefore have water flowing through them. This distinction becomes more important when choosing between a residual NAPL dissolution model, where the water flows through zones with NAPL blobs, and a NAPL pool dissolution model, where the water flows over the pool (in the case of DNAPL) or under the pool (in the case of an LNAPL) but not through the pool.

Simulating NAPL Dissolution with Process-Based Models

Although there is considerable uncertainty in trying to delineate and understand NAPL zones, there are several quantitative tools that can be used to estimate the rate of dissolution that may occur at a site. Note that most application of these models will involve considerable uncertainty about several key parameters, and therefore the results will be affected by this uncertainty. The following models are provided: (1) a simple pore-scale model, which provides some of the theoretical basis for other models but is typically not applied to solve field-scale problems; (2) simple dissolution models for single-component residual NAPL zones; (3) effective-solubility relationships and simple rules of thumb for multicomponent NAPL dissolution; and (4) models for dissolution from LNAPL and DNAPL pools.

Pore-Scale Dissolution Models. Using concepts from the field of chemical engineering, Feenstra and Guiguer (in Pankow and Cherry, 1996) note that for a single-component NAPL, simple dissolution of the compound may be described by

$$N = K_C(C_W - C_{\text{sat}}) \tag{2.24}$$

where N = flux of species of interest [M/L^{2}T]
K_C = mass transfer coefficient [L/T]
C_W = concentration of compound in bulk aqueous phase [M/L^3]
C_{sat} = concentration of compound at NAPL–water interface (taken as the solubility of the compound) [M/L^3]

The mass transfer coefficient may be calculated various ways, but in all cases the diffusivity of the species of interest is a factor. For example, the stagnant-film model (see Figure 2.28) is based on the relationship where K_c is equal to the diffusion coefficient (units of L^2/T) divided by the thickness of the film δ (units of L). Pankow

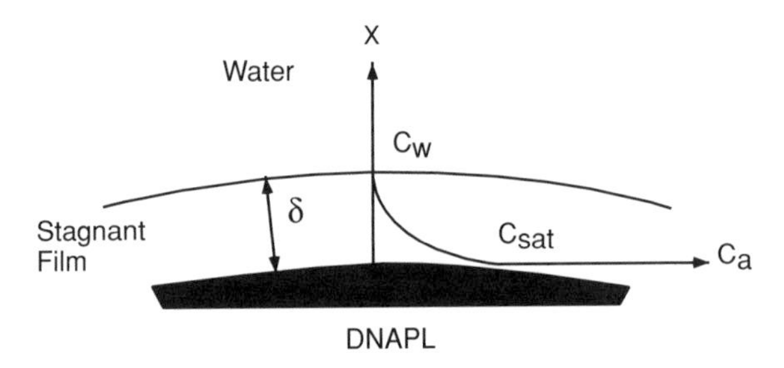

FIGURE 2.28. Schematic of stagnant-film model. (Reprinted with permission from Pankow and Cherry, 1996.)

and Cherry (1996) and Powers et al. (1991) review additional methods for determining mass transfer coefficients.

In a porous medium, the mass transfer rate per volume of porous media is often defined by multiplying the mass flux by the ratio of NAPL surface contact area to the unit volume of porous media, yielding

$$N^* = \lambda(C_W - C_{sat}) \quad (2.25)$$

where N^* = flux of species of interest per unit volume of porous media $[M/L^2T]$
λ = lumped mass transfer coefficient $[L/T]$
C_w = concentration of compound in bulk aqueous phase $[M/L^3]$
C_{sat} = concentration of compound at NAPL–water interface (taken as the solubility of the compound) $[M/L^3]$

The lumped mass transfer coefficient is the product of K_c and the ratio of the NAPL surface contact area and the unit volume of the porous medium:

$$\lambda = K_c \frac{A_s}{V} \quad (2.26)$$

where λ = lumped mass transfer coefficient $[T^{-1}]$
K_c = mass transfer coefficient $[L/T]$
A_s = effective surface contact area between NAPL and groundwater $[L^2]$
V = unit volume of porous medium (unitless) $[L^3]$

It bears repeating that at the field scale, measurement of many of the parameters used for these calculations is not possible, and therefore great uncertainty is introduced. Source terms calculated using these or any other methods should be presented in that light, and if used for solute transport modeling, should be accompanied with a sensitivity analysis.

Dissolution from Single-Component NAPL Blobs. Laboratory studies of dissolution from NAPL blobs in saturated soil columns usually is characterized by three stages (Borden and Kao, 1992): (1) the concentration is stable near the compound's solubility, as dissolution rates are relatively high; (2) the concentration drops exponentially with additional pore volumes; and (3) a long "tail" where the concentration drops very slowly with additional pore volumes. Figure 2.29 shows these three patterns where the TCE concentration is very close to solubility (1000 mg/L) in the stable-concentration stage that extends from 0 to about 100 pore volumes. Next, the concentration declines from 1100 mg/L to about 1 mg/L over the next 400 pore volumes. Finally, there is a long tail, where an additional 1000 pore volumes do not reduce the concentration below 0.200 mg/L. Numerous researchers have used laboratory studies to characterize some or all of the processes behind these three stages (e.g., Hunt et al., 1988; Miller et al., 1990; Mackay et al., 1991; Borden and Kao, 1992; Powers et al., 1992, 1994; Geller and Hunt, 1993; Imhoff et al., 1994).

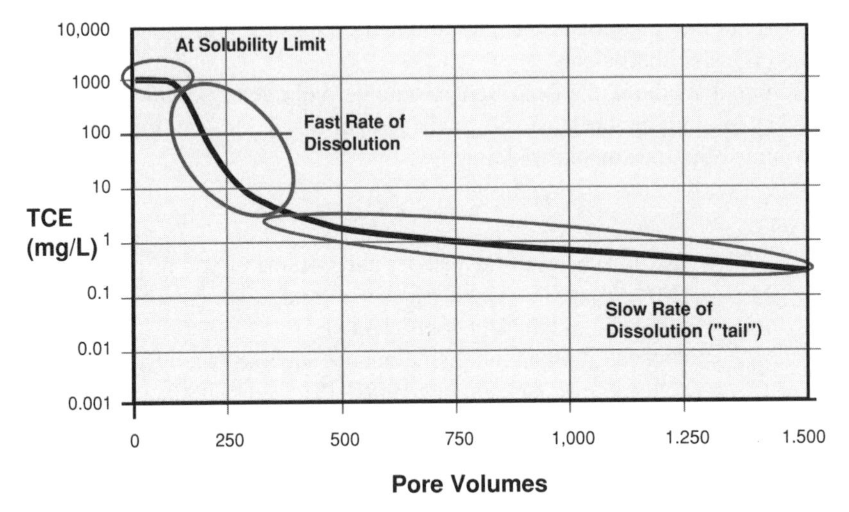

FIGURE 2.29. Dissolution curve for pure TCE in a laboratory column showing different phases of dissolution process. (Data from LaMarche, 1991. Reprinted with permission from Pankow and Cherry, 1996.)

In one of the initial papers addressing dissolution modeling, Hunt et al. (1988) evaluated the dissolution characteristics of NAPL blobs assuming that the blobs were relatively large spheres, and the mass transfer characteristics were obtained from studies of the dissolution of a solid organic (Wilson and Geankopolis, 1966). Using these assumptions, a dissolution model for single-component spherical NAPL blobs was developed. The model is used to illustrate the time required to dissolve pure TCE blobs ("ganglia") of different sizes in Figure 2.30.

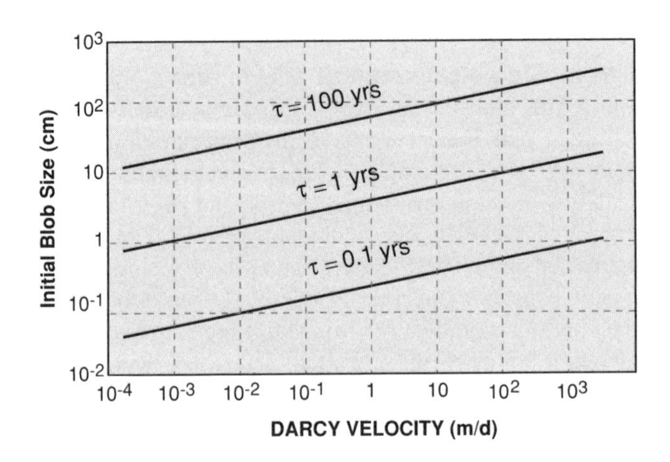

FIGURE 2.30. Time τ required to dissolve TCE blobs of different sizes when dissolved-phase concentrations of groundwater are much lower than the aqueous solubility of TCE. (Reprinted with permission from Hunt et al., 1988.)

Subsequent work identified the structure of the porous media as a critical element in the dissolution of residual NAPL. Powers et al. (1994) focused on how grain size and grain-size distribution affect dissolution concentrations over time. In their column studies, more flushing was required to reduce concentrations from coarser or graded (nonuniform) material than from finer material because of the presence of larger and more amorphous NAPL blobs (Figure 2.31). The data suggested that the concentration

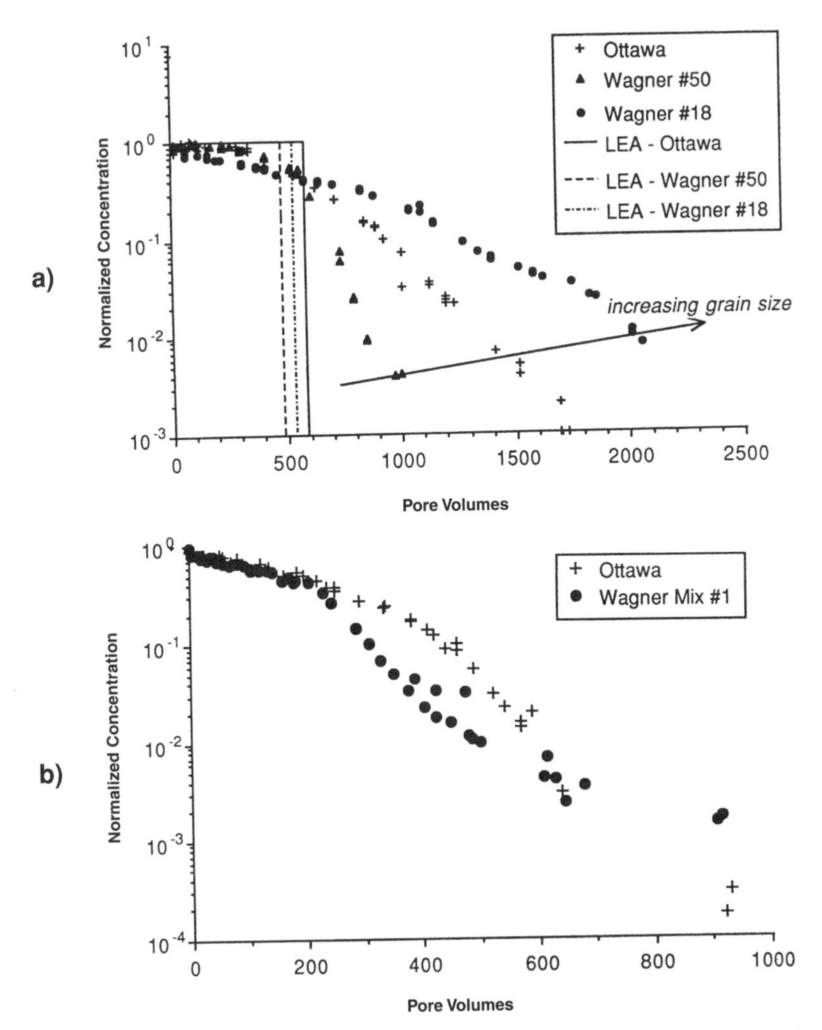

FIGURE 2.31. (*a*) Dissolution curves as a function of grain size (median grain size of Ottawa, Wagner No. 50, and Wagner No. 18: 0.071, 0.045, and 0.120 cm, respectively). Number of pore volumes required for dissolution increases with increasing grain size. LEA indicates predicted dissolution curve if local equilibrium assumption holds for entire soil core. (*b*) Dissolution curves as a function of grain-size distribution [uniformity index (d_{60}/d_{10}) of Ottawa and Wagner No. 1: 1.21 and 2.42, respectively]. Note the more pronounced dissolution "tail" in the less uniform Wagner Mix No. 1 sand. (Reprinted with permission from Powers et al., 1994.)

"tail" was the result of slower dissolution rates from larger, multipore blobs associated with graded sands, and that uniform sands exhibited much less tailing.

Powers et al. (1994) developed a mathematical model of the dissolution from NAPL blobs that incorporated porous-medium properties, the NAPL saturation, and the groundwater flow regime. The model assumed that the NAPL was composed of only one soluble component and therefore did not account for the effects of multiple components or nonsoluble compounds in NAPL. The resulting *theta model* was founded on the assumption that for a given porous medium, changes in the surface area of a NAPL blob as it dissolves can be related to the amount of NAPL trapped in the medium. Two porous-medium properties, median grain size and grain-size distribution, were used as surrogate measures of NAPL blob shape and size distributions. The model matched well the dissolution data from the column studies (Powers et al., 1994); however, the model did not appear to be able to predict the concentration tail for the graded sand.

Although the theta model is not commercially available, Newell et al. (1994) simplified it to produce a simple design tool for predicting the amount of water required to dissolve NAPLs from uniform and graded sands. By ignoring dispersion, the formulation of the theta model was simplified and the resulting dissolution curves were used to develop six curves relating the desired concentration reduction (C_t/C_0) versus α (a measure of how soil type affects dissolution) for three different soil hydraulic conductivities (a function of grain size) and two uniformity indexes (Figure 2.32). The α term is then applied to calculate the required pore volumes using

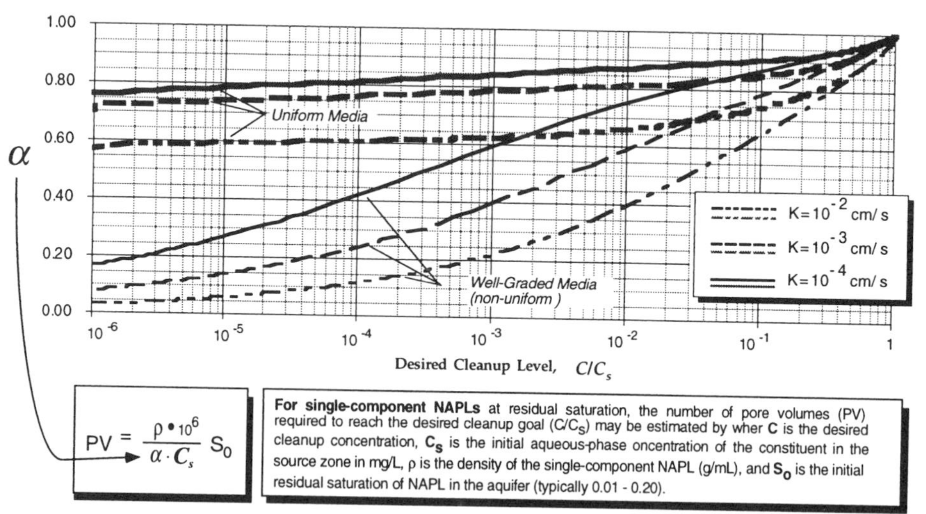

FIGURE 2.32. Pore volumes required to flush aqueous-phase components from single-component residual NAPL in homogeneous soils. α is a dimensionless scaling function for estimating pore volumes based on aquifer hydraulic conductivity and uniform sand (assumed uniformity index of 1.2) versus well-graded sands (assumed uniformity index of 2.4). Curves are based on simplification of the theta model developed by Powers et al. (1994). (From Newell et al., 1994.)

$$PV = \frac{\rho_n S_0 (10^6)}{\alpha C_0} \qquad (2.27)$$

where PV = number of pore volumes required to reach desired cleanup level (from Figure 2.32)
ρ_n = density of NAPL fluid (g/mL)
S_0 = initial NAPL saturation in porous media (fraction of pore space)
α = empirical measure of the effects of soil permeability and uniformity index (see Figure 2.32) (unitless)
C_0 = initial dissolved-phase concentration prior to flushing (mg/L)

Note that this model is very sensitive to the initial aqueous-phase concentration value (C_0). If it is assumed that the source zone behaves like a laboratory column, and C_0 is equal to the solubility of the compound of interest, this simple model will reproduce the behavior of column studies. If C_0 is set equal to an observed concentration from a monitoring well in the field, much higher pore volume estimates will result. The recommended course of action is to make certain that the S_{0r} (residual saturation) value represents the same scale as the C_0 term. In other words, if a field value of C_0 is used, one should ensure that the S_{0r} value represents actual field conditions and not a high-end residual saturation that is used in column studies.

Example 2.4 How many pore volumes will it take to reach a cleanup standard of 5 μg/L in a single-component NAPL zone with an initial concentration of 50,000 μg/L? Assume a NAPL density of 1.5 g/mL, a residual saturation of 10%, and a uniform sand with a hydraulic conductivity of 10^{-3} cm/s.

SOLUTION: First, calculate $C_t/C_0 = 5/50{,}000$ or 10^{-4} and use Figure 2.32 to calculate $\alpha = 0.76$. Using eq. (2.27), calculate the number of pore volumes:

$$PV = \frac{1.5(0.1)(10^6)}{0.76(50)} = 4000 \text{ pore volumes}$$

The number of pore volumes can then be used with eq. (2.2) to estimate how long it will take natural groundwater flow to reduce the source concentration. Important assumptions regarding the theta model are: (1) the NAPL is in the form of blobs rather than pools; (2) the NAPL is composed of a single, soluble component, and therefore the model is not strictly applicable to multicomponent NAPLs or NAPLs like gasoline with insoluble components; (3) there is no biodegradation in the source zone, just flushing with groundwater; and (4) the model does not account for the tailing effect observed in laboratory studies of graded (nonuniform) sands. Important assumptions regarding the simple design tool include all the assumptions listed above, plus the fact that (1) the dispersion is relatively unimportant as water flows through the source zone, and (2) the starting concentration C_0 is usually represented by the pure-phase solubility in laboratory studies but can be replaced with the observed concentration from a monitoring well in a source zone for applications to field sites (see the section on monitoring well concentrations, below).

Dissolution from Multicomponent NAPL Blobs. Dissolution from NAPL blobs with more than one component (such as gasoline) is more complex than dissolution from single-component blobs, as effective solubility effects must be considered [see eq. (2.11)]. Effective solubility is one of the key factors explaining why dissolved concentrations in NAPL zones very rarely approach the solubility of the dissolved compound (see the discussion below) and explains why most of the aqueous-phase mass in gasoline plumes is represented by BTEX compounds/biodegradation products, although they represent only a small fraction of the NAPL mass. In other words, most of the non-BTEX compounds of fuels are relatively insoluble, creating dissolved plumes that are dominated by the soluble BTEX compounds. The results presented below support this conceptual model of a BTEX-dominated plume source in gasoline LNAPLs. For additional supporting data and calculations, see Weidemeier et al. (1995).

Effective Solubility Calculations for Gasoline. Gasoline composition data presented by Johnson et al. (1990a,b) were used to determine the effective solubility of these hydrocarbon mixtures in equilibrium with water [eq. (2.11)]. The total effective solubility of all the compounds was then compared to the effective solubility of the BTEX compounds. Tables 2.5 through 2.7 show these calculations for fresh gasoline and two weathered gasolines. In each of these three fuel samples, BTEX compounds comprise the majority of the dissolved organic mass in equilibrium with water. The non-BTEX components represent a much smaller portion of the dissolved mass. As expected, the theoretical dissolved concentrations from these samples are much higher than what is typically observed in groundwater samples, due to factors such as dilution, the heterogeneous distribution of NAPLs, and the low level of mixing occurring in aquifers [see Bedient et al. (1994) for a more complete discussion].

Note that the total effective solubility of weathered gasoline 1 (126 mg/L) is greater than the total effective solubility of the fresh gasoline (93 mg/L). A comparison of the two samples indicates that the fresh gasoline includes a significant mass of light, volatile compounds that have pure-phase solubilities that are much lower than those of the BTEX compounds (e.g., isopentane with a vapor pressure of 0.78 atm and a

TABLE 2.5. Effective Solubility Data for Fresh Gasoline

Constituent	Mass Fraction	Mole Fraction	Pure-Phase Solubility (mg/L)	Effective Solubility (mg/L)
Benzene	0.0076	0.0093	1780	17
Toluene	0.055	0.0568	515	29
Ethylbenzene	0.0	0.0	152	0
Xylenes	0.0957	0.0858	198	17
Total BTEX	0.16	0.15	152–1780 (range)	63
58 Compounds	0.84	0.85	0.004–1230 (range)	30
	1.00	1.00		93 (68% BTEX)

Source: Data from Johnson et al. (1990a).

TABLE 2.6. Effective Solubility Data for Weathered Gasoline 1

Constituent	Mass Fraction	Mole Fraction	Pure-Phase Solubility (mg/L)	Effective Solubility (mg/L)
Benzene	0.01	0.0137	1780	24
Toluene	0.1048	0.1216	515	63
Ethylbenzene	0.0	0.0	152	0
Xylenes	0.1239	0.1247	198	25
Total BTEX	0.24	0.26	152–1780 (range)	112
58 Compounds	0.76	0.74	0.004–1230 (range)	14
	1.00	1.00		126 (89% BTEX)

Source: Data from Johnson et al. (1990a).

TABLE 2.7. Effective Solubility Data for Weathered Gasoline 2

Constituent	Mass Fraction	Mole Fraction	Pure-Phase Solubility (mg/L)	Effective Solubility (mg/L)
Benzene	0.0021	0.003	1780	5
Toluene	0.0359	0.043	515	22
Ethylbenzene	0.013	0.014	152	2
Xylenes	0.080	0.084	198	15
Total BTEX	0.13	0.14	152–1780 (range)	44
64 Compounds	0.87	0.86	0.004–1230 (range)	21
	1.00	1.00		65 (63% BTEX)

Source: Data from Johnson et al. (1990b).

solubility of 48 mg/L compared to solubilities of 152 to 1780 mg/L for the BTEX compounds). When these light compounds are weathered (probably volatilized), the mole fractions of the BTEX components (the only remaining components with any significant solubility) increase, thereby increasing the total effective solubility of the weathered gasoline. On the other hand, weathered gasoline 2 has a total effective solubility that is lower than fresh gasoline (65 mg/L versus 93 mg/L), suggesting that this gasoline has weathered to such a degree that there has been significant removal of both volatile and soluble components from the gasoline. Newell et al. (1997) extended this calculation for virgin JP-4 jet fuel and found the same pattern: BTEX dominated the aqueous phase, although it represented only a small fraction of the NAPL mass.

Finally, it is important to note that there is considerable variability among different fresh fuels and even more variation among weathered fuels. Therefore, these results should only be used as general indicators that the BTEX compounds and BTEX biodegradation products comprise the majority of the soluble components originating directly from the LAPNL at gasoline release sites. These results should not be used as absolute values for all sites.

Example Multicomponent Dissolution Curves. Borden and Kao (1992) measured the dissolution of BTEX components from a column where 13.5% of the pore space contained residual gasoline. The resulting dissolution curves (Figure 2.33) illustrate another phenomenon related to effective solubility, where the more soluble compounds (benzene at 1750 mg/L and toluene at 530 mg/L) dissolve out of the gasoline blobs faster than the less soluble compounds (xylenes at 170 mg/L and ethylbenzene at 152 mg/L). All of the BTEX compounds exhibited an initial stable high-concentration phase, with benzene having the shortest phase (stable concentration over only the first 30 pore volumes) and ethylbenzene having the longest (stable concentration over the first 300 pore volumes). Concentrations then dropped exponentially, until a tailing effect was observed for most of the compounds. After 880 pore volumes, for example, the benzene concentration was between 1 and 10 parts per billion (ppb); and the toluene and *meta-* + *para*-xylene tail ranged between 10 and 100 ppb.

Borden and Kao (1992) also developed a mathematical model of the process, in which the effects of blob size and the area available for dissolution were simulated assuming that the residual NAPL is present in two forms: (1) a "fast" oil phase that readily exchanges mass with the dissolved phase, consisting of small blobs with a high area/volume ratio; and (2) a "slow" oil phase, for which dissolution is relatively slow, consisting of large blobs with low area/volume ratios. In computer simulations performed in which the model was calibrated to the dissolution data from the column studies, the initial saturation of the fast oil phase was 4.1% of the pore space, while the initial saturation of the slow oil phase was 0.12% of the pore space (about 1⁄35 of the fast oil phase).

The Borden and Kao model is not available commercially, but an analytical solution (described as a rule of thumb) developed by Mackay et al. (1991) can be applied to evaluate dissolution of soluble components from NAPLs comprised of a mixture of components. This model assumes that the NAPL and entire aqueous phase are homogeneous and in chemical equilibrium, with no large NAPL blobs or dead-end pores that cause tailing. The model is a simplification of the dissolution process that occurs in natural porous media but focuses on a key process involving multicomponent NAPLs: effective solubility. The Mackay model simulates how effective solubility affects the amount of flushing that is required to remove various components.

Multicomponent Dissolution Rule of Thumb. Mackay et al. (1991) developed a rule of thumb for predicting the mole fraction of any compound in the NAPL mixture after a given amount of flushing:

$$x_i = x_{i0} \exp(-Q_{\upsilon i} K_{\upsilon i}) \tag{2.28}$$

$$K_{\upsilon i} = \frac{C_{si}}{M_i} \frac{M_n}{\rho_n} \tag{2.29}$$

$$\mathrm{PV} \approx Q_{\upsilon i} S_0 \tag{2.30}$$

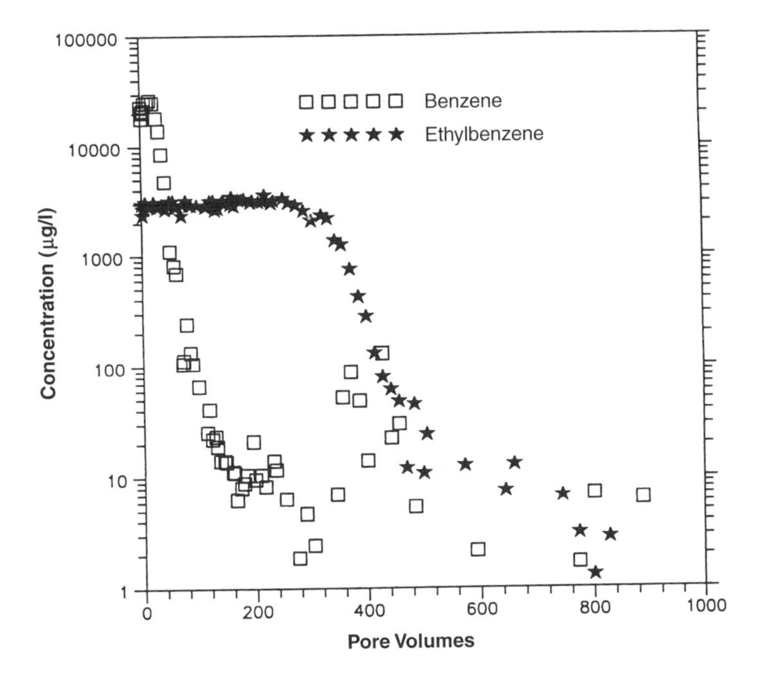

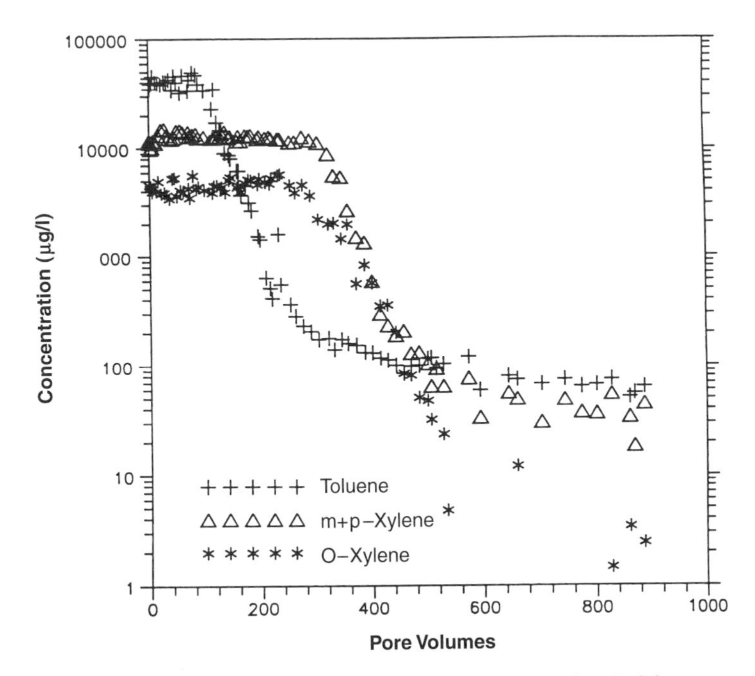

FIGURE 2.33. Dissolution curve for BTEX compounds in gasoline in laboratory column. Notice dissolution tails for all four BTEX compounds. (Reprinted with permission from Borden and Kao, 1992.)

where x_i = mole fraction after flushing with volume $Q_{\upsilon i}$ (unitless)
x_{i0} = initial mole fraction (unitless; see Tables 2.6 to 2.8 for typical values for gasoline)
$Q_{\upsilon i}$ = ratio of clean water volume flushed through NAPL zone to NAPL volume in NAPL zone
$K_{\upsilon i}$ = volumetric partition coefficient (unitless)
C_{si} = aqueous pure-phase solubility of compound i (in units of kg/m^3; e.g., benzene's pure-phase solubility of 1780 mg/L = 1.78 kg/m^3)
M_i = molecular weight of compound i (in units of kg/mol; e.g., benzene's molecular weight of 78 g/mol = 0.078)
M_n = molecular weight of NAPL [in units of kg/mol; Mackay et al. used an example where the molecular weight of a fuel NAPL is 0.1 kg/mol; Pankow and Cherry (1996) used an assumed molecular weight of 0.15 for the unidentified fraction of a DNAPL at a Superfund site]
ρ_n = density of the NAPL (units of kg/m^3; e.g., gasoline might have a typical density of 0.72 to 0.76 g/mL = 720 to 760 kg/m^3)
PV = number of pore volumes of flushing with clean water [unitless; see eq. (2.3)]
S_0 = initial residual saturation of NAPL blobs in porous media

This model was applied to a NAPL comprised of an equimolar mixture of benzene, toluene, and p-xylene, which resulted in a NAPL density of 869 kg/m^3 and a mean NAPL molecular weight of 0.093 kg/mol. The calculated value for benzene's $K_{\upsilon i}$ was 0.00243, giving a value of $Q_{\upsilon i}$ = 948 for a 90% reduction in the NAPL mole fraction (x_i/x_{i0} = 0.1). This means that with an initial residual saturation of 10%, about 95 pore volumes would be required to achieve a 90% reduction in benzene mass in the NAPL. Using this same approach, 390 pore volumes would be required for a 90% reduction of toluene mass in the NAPL, and 1260 pore volumes would be required for a 90% reduction of p-xylene mass.

Note that this expression is an approximation, as the actual mole fractions will be affected by the dissolution of other soluble compounds, whereas the modeled mole fractions are not. In the case where the solute of interest is one of the more soluble components in the NAPL and represents a relatively small amount of the NAPL (e.g., benzene in gasoline), the error will be relatively small. Mackay et al. (1991) provide additional methods for adjusting for errors associated with dissolution of NAPLs comprised mostly of soluble components.

Dissolution from NAPL Pools and Lenses. Several studies have focused on dissolution from NAPL pools, where groundwater flows across a plane where the NAPL saturation is assumed to be 100% of the pore space (i.e., no groundwater flow through the pool). Voudrias and Yeh (1994) reported results from a physical model with a 85-cm-long, 30-cm-high, 20-cm-wide pool of pure toluene. The pool was carefully constructed so that the entire pore volume of the LNAPL pool contained toluene and no water. Steady-state flow conditions at 60.9 cm/day (729 ft/yr) produced a very steep vertical gradient in concentration below the floating LNAPL pool, where concentrations of only 24 mg/L were observed 3 cm below the pool compared to toluene's aqueous solubility of 515 mg/L (the assumed concentration at the pool–water interface). At 8 cm below the pool the concentrations were less than 0.1 mg/L. Experiments performed with a slower groundwater velocity of 28.2 cm/day (338 ft/yr) showed similar results.

Hunt et al. (1988) presented the first dissolution model for pooled NAPLs, in which the advection–dispersion equation was used to estimate the concentration at the *end* of the pool of length L at any distance z above the pool:

$$C(L,z) = C_s \operatorname{erfc}\left[\frac{z}{2(D_\upsilon L/\upsilon_d)^{1/2}}\right] \tag{2.31}$$

$$D_\upsilon = D_e + \upsilon_d \alpha_\upsilon \tag{2.32}$$

where $C(L,z)$ = aqueous-phase concentration at end of NAPL pool length L at distance z above or below the pool (mg/L)
C_s = aqueous-phase solubility of compound (mg/L)
z = vertical distance above pool (LNAPLs) or below pool (DNAPLs)
D_υ = vertical dispersion coefficient (m^2/day)
L = horizontal length of pool in direction of groundwater flow (m)
υ_d = groundwater Darcy velocity (m/day)
D_e = molecular diffusion coefficient (m^2/day; typical range 1 to 5×10^{-10} m^2/s)
α_υ = vertical dispersivity [m; rule of thumb: α_υ = 0.001 times scale, in this case 0.001 times L; reported values for pools: 0.00022 to 0.00029 m from Johnson and Pankow (1992); 0.00013 to 0.00018 m from Voudrias and Yeh (1994)]

Voudrias and Yeh (1994) were able to match their experimental data from the toluene pool with this relationship by calibrating with the vertical dispersivity term. Note that molecular diffusion, typically ignored in most applications of the advection–dispersion equation in aquifers, can be a factor in vertical transport from the pool. Johnson and Pankow (1992) indicate that mechanical dispersion will dominate only in flow regimes greater than 0.1 m/day (120 ft/yr) and that diffusion should be considered for pool dissolution sites with groundwater velocities of less than 120 ft/yr.

Johnson and Pankow (1992) extended Hunt's model to provide a relationship for M_a, the surface-area-averaged mass transfer rate, or the average number of grams of dissolved compounds leaving each square meter of the pool every day (in mg/$m^2 \cdot$ day):

$$M_a = C_s n \sqrt{4 D_\upsilon \upsilon_x / \pi L} \tag{2.33}$$

where M_a = surface-area-averaged mass transfer rate (mg/$m^2 \cdot$ day)
n = porosity (unitless)
υ_x = groundwater seepage velocity (m/day)

The authors noted that no field information is available regarding the geometry of NAPL pools, but cited physical model work performed by Schwille (1988) in which a 24-L release of PCE formed a pool 5 m long with an average thickness of about 0.04 m. If one assumes that the length/thickness ratio of an LNAPL pool is 100 and that NAPL fills all the pore space (saturation = 100%), the mass of NAPL in a pool will be

$$\text{mass} = 0.01L^3 n\rho \tag{2.34}$$

where mass is the mass of NAPL in the pool (kg) and ρ is the density of the NAPL (kg/m^3).

Assuming that the length and width of the pool remain unchanged during dissolution, and that only thinning of the pool occurs, the time to complete dissolution can be provided using

$$\tau_p = \frac{0.01 Ln\rho}{365 M_a} \tag{2.35}$$

or by substituting eq. (2.33),

$$\tau_p = \frac{2.43 \times 10^{-5}\rho}{C_s}\sqrt{\frac{L^3}{D_\upsilon \upsilon_x}} \tag{2.36}$$

where τ_p is the time for complete dissolution of the pool (years). Figure 2.34 shows dissolution time (τ_p) versus groundwater seepage velocity for TCE for four pool lengths ranging from 1 to 10 m (assuming a length/thickness ratio of 100:1; Johnson and Pankow, 1992).

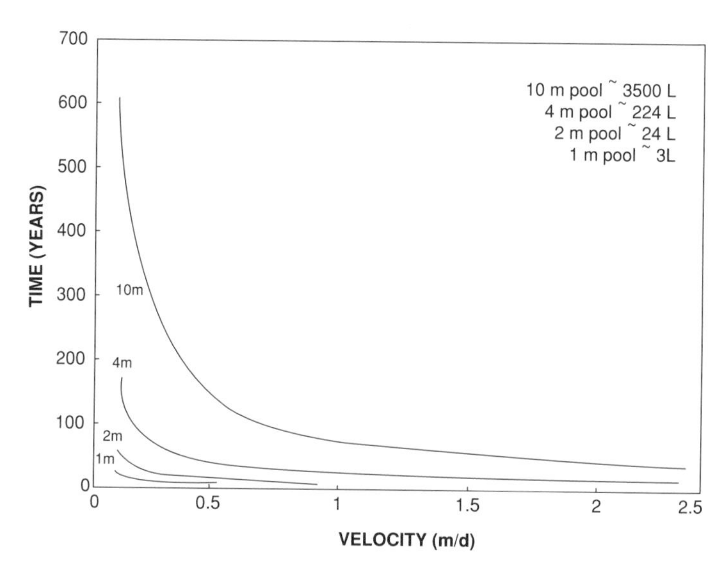

FIGURE 2.34. Dissolution time versus groundwater velocity for TCE pools of four different lengths. (Reprinted with permission from Johnson and Pankow, 1992. Copyright 1992 American Chemical Society.)

Johnson and Pankow concluded that under optimum conditions (homogeneous media, uniform groundwater flow) tens to hundreds of years typically will be required to dissolve pools of chlorinated solvents in the saturated zone. They stress that long-term containment strategies may be the only cost-effective methods to manage DNAPL pools. Note that the pool equations presented above account only for the physical process of dissolution, and that in situ biodegradation, particularly for floating LNAPL pools and potentially for some DNAPL pools, may accelerate the dissolution process by increasing the driving force away from the pool in diffusion-influenced systems.

Using analytical dissolution models applied to a hypothetical 1000-L DNAPL release, Anderson et al. (1992b) evaluated the resulting dissolved concentrations and the time to dissolve the DNAPL. The conceptual release, shown on Figure 2.35, consists of 16 vertical fingers (each have a square cross section ranging between 2 and 20 cm on a side and extending 3 m downward) and 15 horizontal pools (ranging from 2 to 4 m on a side with pool depths in the range 0.1 to 0.3 m). Assuming a porosity of 0.35 and that 15% of the pore space in each finger and pool was occupied by NAPL, only 7.1 L of the 1000-L release was calculated to be present in the fingers, with the vast majority (983 L) residing in the pools.

As shown in Figure 2.35, the estimated source lifetimes for the hypothetical DNAPL release are much longer for the pools than the fingers. The maximum lifetime of a finger is 1.1 years, compared to a maximum lifetime of 273 years for two of the pools. This modeling exercise confirms the conceptual model that groundwater

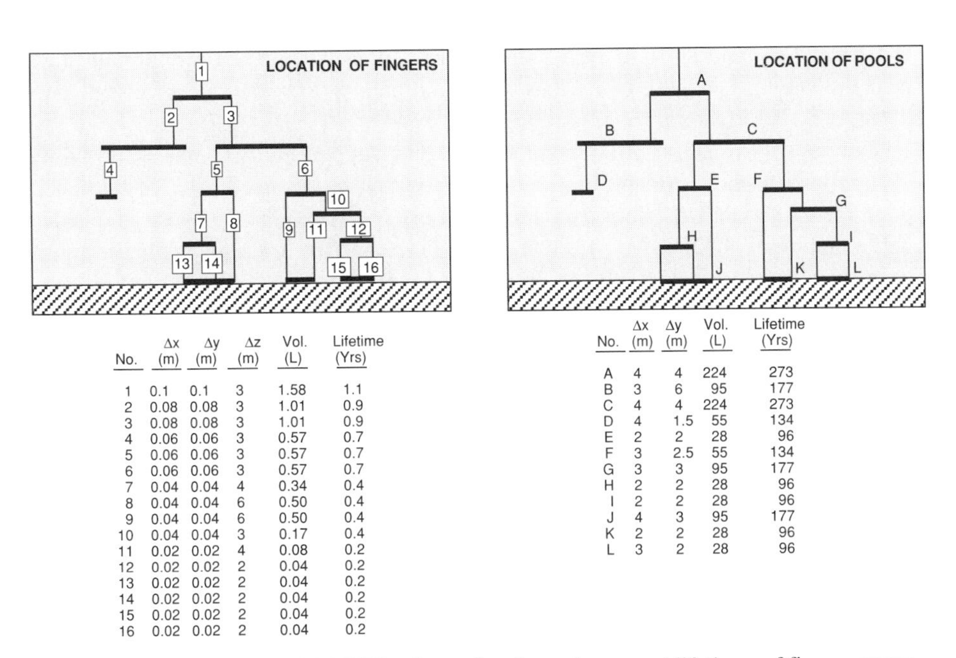

No.	Δx (m)	Δy (m)	Δz (m)	Vol. (L)	Lifetime (Yrs)
1	0.1	0.1	3	1.58	1.1
2	0.08	0.08	3	1.01	0.9
3	0.08	0.08	3	1.01	0.9
4	0.06	0.06	3	0.57	0.7
5	0.06	0.06	3	0.57	0.7
6	0.06	0.06	3	0.57	0.7
7	0.04	0.04	4	0.34	0.4
8	0.04	0.04	6	0.50	0.4
9	0.04	0.04	6	0.50	0.4
10	0.04	0.04	3	0.17	0.4
11	0.02	0.02	4	0.08	0.2
12	0.02	0.02	2	0.04	0.2
13	0.02	0.02	2	0.04	0.2
14	0.02	0.02	2	0.04	0.2
15	0.02	0.02	2	0.04	0.2
16	0.02	0.02	2	0.04	0.2

No.	Δx (m)	Δy (m)	Vol. (L)	Lifetime (Yrs)
A	4	4	224	273
B	3	6	95	177
C	4	4	224	273
D	4	1.5	55	134
E	2	2	28	96
F	3	2.5	55	134
G	3	3	95	177
H	2	2	28	96
I	2	2	28	96
J	4	3	95	177
K	2	2	28	96
L	3	2	28	96

FIGURE 2.35. Conceptual DNAPL release showing volumes and lifetimes of fingers versus pools. (Adapted with permission from Anderson et al., 1992b. Copyright 1992 American Chemical Society.)

flowing through relatively thin fingers will dissolve the fingers relatively quickly, but dissolution from a horizontal DNAPL pool or lens is much slower.

More detailed modeling approaches were provided by Chrysikopoulos (1995), who derived a three-dimensional transient model of dissolution from single-component pools of different shapes, and by Holman and Javandel (1996), who developed a detailed transient model of LNAPL pool dissolution.

Example 2.5 What is the time required to dissolve a pool of TCE 10 m long and 0.1 m thick in an aquifer with a seepage velocity of 1000 ft/yr (0.8 m/day)?

SOLUTION: Using the top curve on Figure 2.34, 0.8 m/day yields 100 years required for dissolution of the pool.

Field-Scale Factors Affecting Dissolution

Although column-scale experiments help explain some of the key processes that control dissolution, several larger-scale factors that are important in the field complicate our ability to quantify the dissolution process. Groundwater sampling resolution and heterogeneities are two such factors.

Sampling Effects on Concentration Measurements. Anderson et al. (1992a) noted that chlorinated solvent concentrations in DNAPL zones were almost always below their expected solubilities, while soil columns containing NAPLs did show concentrations at solubility limits. A 1- × 0.75- × 1-m physical model was constructed to determine if the residual NAPL blobs restricted flow through the NAPL zones in the field, while in soil column experiments the water was forced through the NAPL zone.

The results from the physical model did not support the hypothesis of flow obstructions, as flow through the residual zone was only reduced by ~18% by the presence of the DNAPL blobs located in the center of the physical model at a residual saturation of about 13%. The authors concluded that the low aqueous-phase concentrations of chlorinated solvents in the field were caused by one or more of the following processes: (1) the low cross sections of the DNAPL lenses and pools, where most of the DNAPL resides; (2) dispersion in the zone downgradient from the source; and (3) dilution of thin and/or narrow plumes by clean water in downgradient monitoring wells.

Anderson et al. (1992b) expanded on the last point and modeled the expected dissolved concentrations in simulated monitoring wells with 2-m screens located downgradient from the hypothetical source. The highest model concentration was 12% of solubility of the dissolved solvent in the monitoring wells located 20 m downgradient from the source, 3.8% of solubility at wells located 50 m downgradient from the source, and only 2.8% of solubility in wells located 100 m downgradient from the source. These results clearly show how a conventional monitoring well (with a screen between 5 and 10 ft long) provides a vertical composite sample of groundwater. In addition, DNAPL source zones do not contaminate all of the groundwater flowing through them, and considerable clean water can be mixed with the contaminated water to dilute the resulting groundwater samples. LNAPL source zones are almost certainly affected in this way to some degree, with dilution of contaminated water with fresh water through 5- to 10-ft screen lengths in monitoring wells.

An emplaced-source experiment conducted by the Waterloo Centre for Groundwater Research at the Borden Site illustrates how the use of smaller screening intervals will show that (1) clean flow lines do emerge from DNAPL source zones, and (2) the contaminated flow lines are near solubility. In this experiment, sand from a rectangular block 1.5 m long, 1.0 m high, and 0.5 m wide was excavated, mixed with 18 L of DNAPL, and placed back into the source zone below the water table (Rivett et al., 1994). The DNAPL, consisting of about 10% chloroform (solubility of 8700 mg/L), 42% TCE (reported solubility of 1100 mg/L), and 48% PCE (solubility of 240 mg/L), represented a residual saturation of 6.3% in the source zone (Feenstra, 1997).

A sampling network consisting of more than 2300 multilevel 3.2-mm-diameter sampling tubes was used to measure the dissolution products from the emplaced source, with vertical spacing of 20 to 30 cm and lateral spacing of 0.5 to 4 m. The peak concentrations of TCE (600 to 800 mg/L) observed 1 m downgradient from the source compare well to the calculated effective solubility for TCE (about 0.42 times 1400, or 610, mg/L). Most samples exhibited much lower concentrations, which was attributed to dilution from clean water flowing around the source zone and lower-permeability conditions in the source zone itself due to the presence of trapped air bubbles, DNAPL, and gypsum added to the source as a tracer (Rivett et al., 1994).

Effect of Heterogeneities on Dissolution. Geologic heterogeneities on the micro- and macroscales will significantly affect NAPL migration (see the conceptual models, above) and dissolution. Manivannan et al. (1996) performed studies in columns with NAPL trapped in coarse-grained heterogeneous media and observed an unexpected rise in effluent concentrations after over 300 pore volumes of flushing was performed. Further work with models and two-dimensional flow cells suggested that water was flowing very slowly through this coarse-grained zone at first, but was dissolving NAPL, increasing the relative permeability of the system, and eventually allowing more water with high concentrations of dissolved contaminants to exit the column. The presence of NAPL in the coarse-grained lens increased the total time required to dissolve all of the NAPL compared to that required for a homogeneous packing.

Field-Scale Studies of Dissolution. Dissolution data from actual field sites are relatively rare. In one study, an injection well and a production well were installed 11 ft apart in a silt/silty clay containing a chlorinated solvent DNAPL at a Superfund Site on the Texas Gulf coast (Newell et al., 1991). The test area exhibited the presence of continuous DNAPL, as about 11 ft of DNAPL, primarily bis(2-chloroethyl)ether and 1,2-DCA, was observed in the injection well prior to system startup. The wells were operated at about 0.3 gpm for 30 days, during which five samples were collected from four monitoring points in the test zone. After using a flow model to estimate the number of pore volumes that had passed by each monitoring point, dissolution curves from this field test were constructed (see Figure 2.36). The curves show considerable scatter in the initial concentrations, with

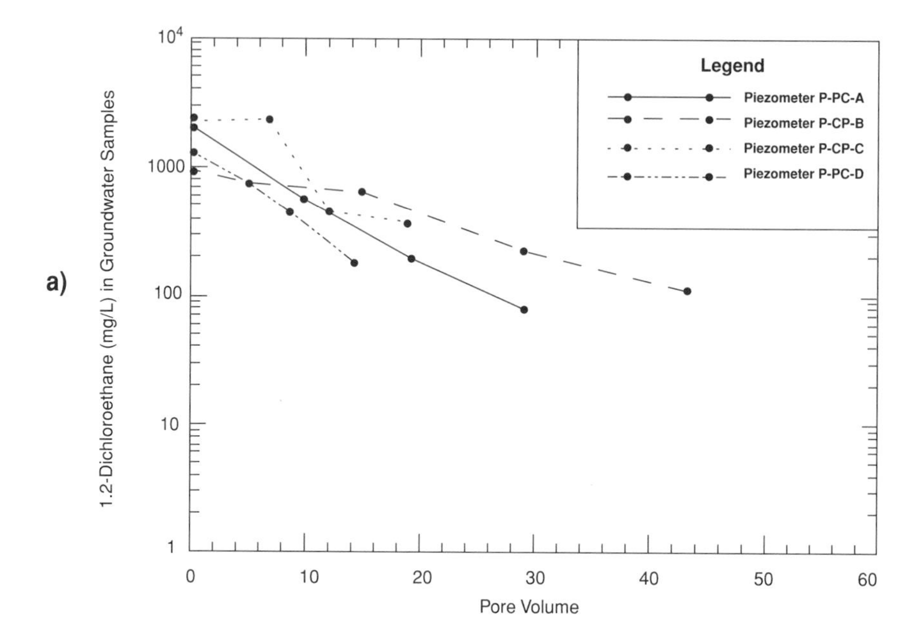

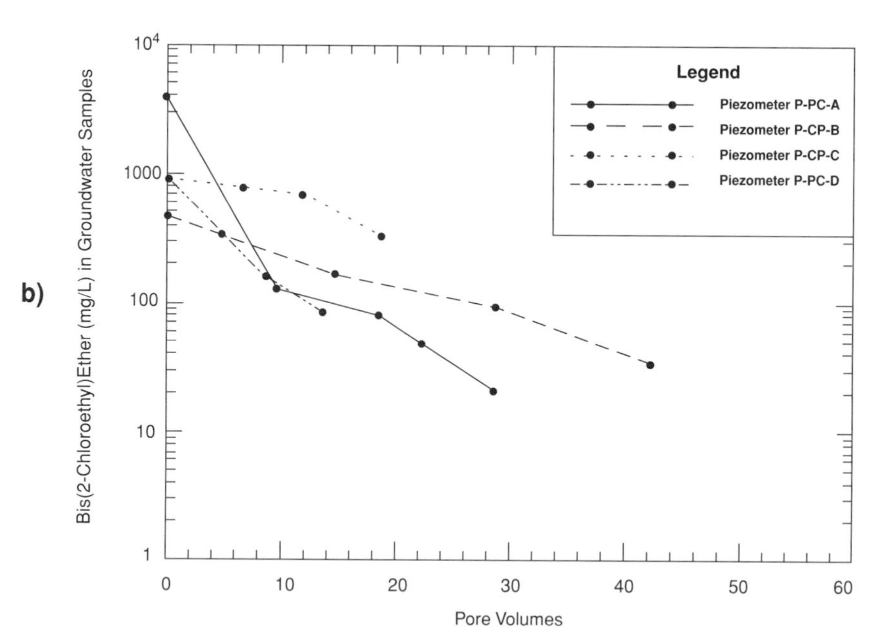

FIGURE 2.36. Dissolution curves from 11-ft test cell, Superfund site on Texas Gulf coast: concentration of (*a*) 1,2-dichloroethane and (*b*) bis(2-chloroethyl ether versus pore volume of injection water. (From Newell et al., 1991. Reprinted with permission from National Ground Water Association.)

the bis(2-chloroethyl)ether concentrations ranging from near 4000 mg/L, compared to a solubility of 10,400 and an estimated effective solubility of 3800 mg/L. The concentrations of both chemicals declined exponentially during the test in most of the monitoring points, with considerable scatter. As expected, no tailing was observed, as the lowest concentrations were still well above 10 mg/L. The conclusion of the test was that no practical pumping system could achieve the target levels for this site.

The field-scale experiment reported by Rivett et al. (1994) provided a vivid example of the effective solubility relationship during the 1128-day experiment, as the most soluble compound (chloroform) almost completely dissolved out of the DNAPL within 600 days. Based on average concentrations from soil cores, only 14% of the chloroform remained in the source after 399 days, compared to 88% of the TCE and about100% of the PCE. At 1128 days, the conclusion of the test, the average concentrations from the soil cores showed no chloroform, 28% of the TCE, and 96% of the PCE remaining in the source zone (Feenstra, 1997).

2.7 SITE-SCALE DISSOLUTION MODEL AND THE EFFECT OF BIODEGRADATION

Although most sites probably have source zones composed of NAPLs in the saturated zone (source scenarios 3 and 4), application of the NAPL dissolution-process models may be too detailed for many sites. Therefore, a more basic approach was adopted in the Air Force's BIOSCREEN Natural Attenuation Model (Newell et al., 1996), where the entire source zone is simulated with a simple box model to provide order-of-magnitude estimates of source lifetimes.

In this simple model, the entire source zone, including continuous NAPLs and dissolved and sorbed contaminants in the saturated and vadose zones, was considered to be located in a box containing a mass of dissolvable contaminants. The rate at which contaminants leave the box is estimated from the rate at which flowing groundwater removes contaminants from the box, plus the amount of saturated-zone biodegradation that occurs in the box. The time required to achieve a cleanup standard can then be estimated by comparing the mass of contaminants in the box versus the time required to remove contaminants from the box.

BIOSCREEN Source Lifetime Model

To use the source lifetime calculation in BIOSCREEN, an estimate for the total mass of dissolvable hydrocarbons in the source zone is developed by evaluating the mass of contaminants in soil samples and/or using estimates of the amount of continuous LNAPL and residual NAPL in the source zone (see Section 2.8). This is usually an order-of-magnitude estimate, yielding an order-of-magnitude estimate for plume lifetime. Next, the mass flux from the source zone can be estimated by multiplying the average source concentration C_s by the groundwater flow rate Q through the source zone (assuming there is no flow or source contribution from the unsaturated zone). If the source concentration is assumed to remain constant during the entire lifetime of the source, the source lifetime (an approximation of the plume lifetime) can be calculated:

$$t_s = \frac{M_0}{QC_s} \tag{2.37}$$

where t_s = source lifetime (years)
M_0 = dissolvable mass in source at time = 0 (mg)
Q = groundwater flow rate through source zone; for an LNAPL site, typically assumed to be (Darcy groundwater velocity) · (width of source zone) · (depth of smear zone) (L/yr)
C_{s0} = observed source concentration at time = 0 (mg/L)

As shown on Figure 2.37, this approach is mathematically simple but is fundamentally inaccurate, as source concentrations will almost certainly decline over time (see Figures 2.29, 2.31, 2.33, and 2.36). If one assumes that the decline will be a first-order process, the source concentration versus time can be simulated with this expression:

$$C_t = C_{s0}e^{-k_s t} \tag{2.38}$$

where C_t = source concentration at time t (mg/L)
C_{s0} = observed source concentration at t = 0 (mg/L)
k_s = source decay coefficient (year^{-1})
t = time (years)

Note that this decay coefficient k_s is not related in any way to first-order decay coefficients for dissolved constituents (also called λ) reported in the literature for

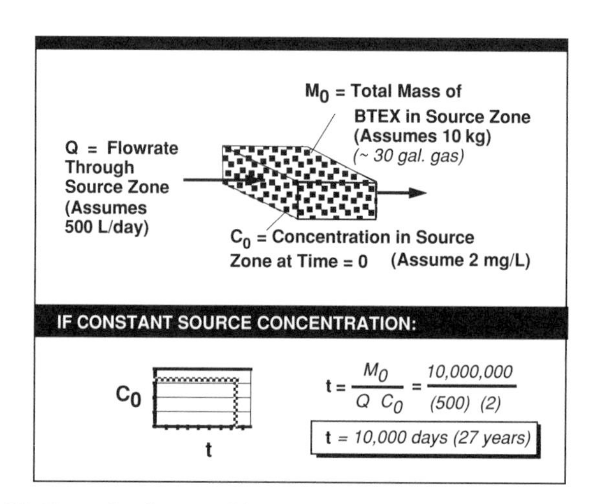

FIGURE 2.37. Example of source lifetime calculation: constant source concentration.

natural attenuation (see Chapters 5 and 6), as the literature values typically represent decay half-lives from 0.1 to 10 years and represent biodegradation of dissolved contaminants in the plume *once they have left the source*. The source decay coefficient values represent how quickly a source zone is being depleted and will usually have much longer half-lives (tens or hundreds of years). It is very important to keep these two decay coefficients, and these concepts, separate.

The source decay coefficient, representing how quickly the source is being depleted, can be derived from site data if one has an estimate of the source mass and flux of contaminants leaving the source (Newell et al., 1996):

$$k_s = \frac{QC_{s0}}{M_0} \tag{2.39}$$

where Q = groundwater flow rate through source zone (L/yr)
C_{s0} = observed source concentration at time = 0 (mg/L)
M_0 = dissolvable mass in source at time = 0 (mg)

As shown in Figure 2.38, this has the effect of greatly lengthening the time required for the source to decay. This model assumes that the only mass leaving the source zone is dissolved in the water flowing through the source zone. Note that Q and C_{s0} are related; the thickness of the source zone should be matched with an appropriate average concentration for that entire depth interval. Because LNAPL source zones are limited in vertical extent and are therefore easier to characterize, the BIOSCREEN method is probably more applicable to LNAPL sites than DNAPL sites.

With a first-order source decay term, the source concentration at any time can be derived [see eq. (2.38)], and the time required to reach any concentration can be calculated using

$$t = -\frac{1}{k_s} \ln \frac{C_t}{C_{s0}} \tag{2.40}$$

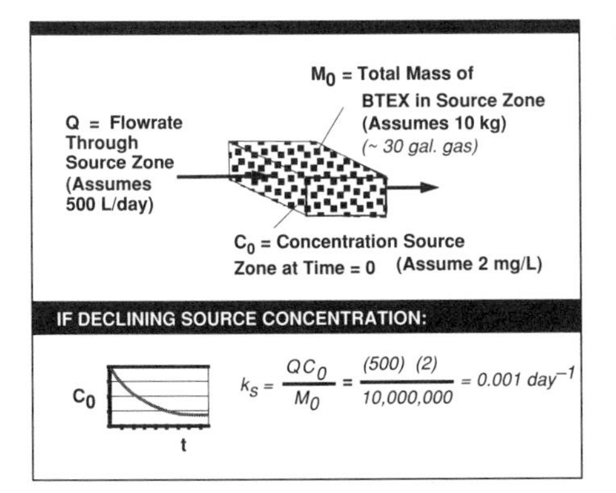

$C_t = C_0\, e^{-0.0001t}$

Time (days)	Time (years)	Source Concentration (mg/L)
0	0	2.0
3,650	10	1.4
7,300	20	0.96
18,250	50	0.32
36,500	100	0.052
54,750	150	0.008
73,000	200	0.001

Rearrange equation to yield time:

$C_t = C_0 e^{-k_s t}$

$t = \frac{\ln(C_t / C_0)}{-k_s}$

$t = \frac{\ln(0.005 / 2)}{-0.0001}$

t = 59,991 days or 164 yrs

FIGURE 2.38. Example of source lifetime calculation: declining source concentration.

where t is the time required to reach concentration C_t (years). Practical application of this method to LNAPL sites, however, provides extremely long estimates for the lifetime of LNAPL source zones (up to hundreds of years). Therefore, the BIOSCREEN model makes an adjustment to account for losses due to biodegradation.

Adjusting for Biodegradation at LNAPL Sites

An evaluation of data from source zone monitoring wells shows that there are other mechanisms whereby contaminants are "leaving the box." At almost all LNAPL sites, groundwater removed from monitoring wells in the source zone will be depleted of electron acceptors (e.g., dissolved oxygen, nitrate, and sulfate) and have elevated concentrations of metabolic by-products [e.g., Fe(II) and methane]. As discussed in Chapter 5, the loss of electron acceptors and the presence of metabolic by-products have been accepted by most groundwater researchers and practitioners to represent the loss of contaminants via biodegradation. Therefore the simple box model for source decay used in BIOSCREEN has an additional experimental term when the loss of electron acceptors and the presence of metabolic by-products are considered.

Equation (2.37) assumes that the biodegradation occurs only *after* the dissolved compounds leave the source zone and therefore the average source concentration observed in the source zone monitoring wells is used for C_{s0}. When electron acceptors and by-products are considered, however, it must be assumed that vigorous aerobic and anaerobic reactions occur directly in the source zone and that monitoring well data from the source zone indicate field concentrations *after* biodegradation has occurred. Therefore, to account for these reactions in the source zone, the C_{s0} term is approximated as the measured source concentration *plus* the amount of biodegradation occurred. This can be accomplished by adding the source concentration and the expressed biodegradation capacity (the amount of organics that can be biodegraded by the electron acceptors in the groundwater flowing through the source zone) to yield an effective source concentration that accounts for biodegradation (see Figure 2.39 for an example):

$$k_s = \frac{Q(C_{s0} + \mathrm{BC})}{M_0} \tag{2.41}$$

where Q = groundwater flow rate through source zone (L/year)
C_{s0} = observed source concentration at time = 0 (mg/L)
BC = expressed biodegradation capacity (see Chapter 5 for information on how to calculate BC) (mg/L)
M_0 = dissolvable mass in source at time = 0 (mg)

A more detailed explanation of expressed biodegradation capacity and how to calculate it is provided in Chapter 5.

The impact of these two different assumptions about the C_s term is significant (see Figure 2.39 for an example). At an actual field site at Hill AFB, for example, the observed source concentration in the field was approximately 15 mg/L total BTEX.

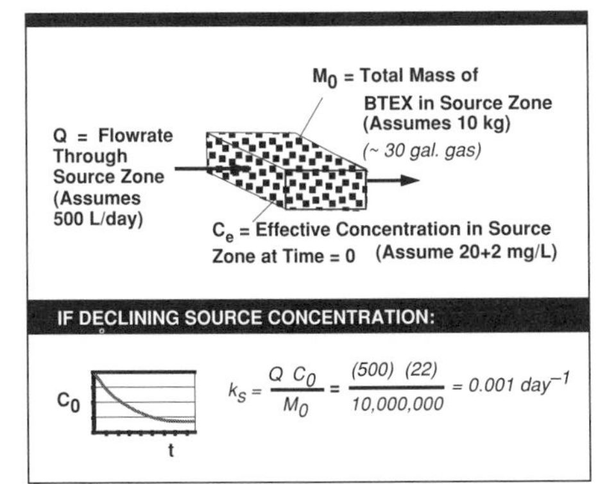

$C_t = C_0 \, e^{-0.001\,t}$

Time (days)	Time (years)	Source Concentration (mg/L)
0	0	2.0
365	1	1.4
700	2	0.96
1825	5	0.32
3650	10	0.052
5475	15	0.008
7300	20	0.001

Rearrange equation to yield time:

$$C_t = C_0 e^{-k_s t}$$

$$t = \frac{\ln(C_t / C_0)}{-k_s}$$

$$t = \frac{\ln(0.005 / 2)}{-0.001}$$

t = 5,991 days or 16.4 yrs

FIGURE 2.39. Example of source lifetime calculation: declining source concentration with biodegradation assuming 20 mg/L biodegradation capacity and 2 mg/L observed concentration.

If one only considers the observed source concentrations, the C_{s0} term used to estimate source lifetime is 15 mg/L. When source zone biodegradation is considered, however, the C_{s0} term is 15 mg/L plus the biodegradation capacity of 31.6 mg/L, or a total of 46.6 mg/L for C_{s0}. Therefore, when biodegradation is considered, the source is actually attenuating three times faster than the dissolution rate predicted by the observed aqueous-phase concentrations alone.

Note that the biodegradation of the LNAPL itself probably is not occurring, but rather, is acting on the dissolved contaminants. Most conceptual models of biodegradation suggest that biological reactions do not occur at any meaningful rate in the NAPL itself. Vigorous aqueous-phase biodegradation will decrease dissolved contaminant concentrations, increasing the driving force of the dissolution reactions, thereby increasing the rate of dissolution [e.g., see eq. (2.22)].

By considering the compounds in source zone groundwater samples, and reflecting on the nature of biological niches, it becomes clear that the biodegradation-before-reaching-the-well scenario is plausible at petroleum release sites. First, almost all the samples collected in fuel-related LNAPL source zones contain BTEX compounds, are depleted in soluble electron acceptors (oxygen, nitrate, and sulfate), and exhibit elevated concentrations of metabolic by-products (iron and even methane, from a relatively slow reaction). Second, the concept of biological niches indicates that BTEX-consuming microorganisms will occupy locations that are (1) near dissolving LNAPL, and (2) near moving groundwater containing electron acceptors. The microorganisms occupy locations immediately adjacent to the LNAPL in the source zone, such as in the pore throats downstream of NAPL blobs, so that the concentrations of BTEX in source zone monitoring wells represents post-biodegradation conditions. An interesting (although not recommended) field test to confirm this hypothesis would be to sterilize the saturated zone of an LNAPL site and observe if the concentrations of BTEX increase significantly.

The same type of adjustment would be required, in theory, at DNAPL sites where significant biodegradation is occurring (see Chapter 6 for a discussion of chlorinated solvent biodegradation). Our inability to estimate the mass of DNAPL at a site makes the source lifetime estimate much less useful at most DNAPL sites compared to LNAPL sites, however.

Example 2.6 How long would it take to reach a 0.001-mg/L cleanup standard for BTEX in a source zone when the site is currently showing 2 mg/L of BTEX in the source zone and 20 mg/L biodegradation capacity and the source has the following characteristics: 10 ft wide, 10 ft deep, and a total of 10 kg of dissolved mass (primarily BTEX) in the source zone (representing about 30 gallons of gasoline)? The groundwater Darcy velocity is 65 ft/yr.

SOLUTION: First calculate Q, the flow rate through the source zone with the following expression:

$$Q = WDV_d = 10(10)(65)(28.3) \approx 184{,}000\ \text{l/yr}\,(500\ \text{L/day})$$

See Figure 2.39 for the rest of the solution using eqs. (2.41) and (2.40).

2.8 ESTIMATING SOURCE MASSES

To estimate the source lifetime for use in soil and groundwater modeling, an estimate of the mass of compounds of concern in the source zone is required. Although five methods are discussed below, note that there is considerable uncertainty with each of these approaches and that estimating source mass will be a probably be an order-of-magnitude endeavor. The first two mass-estimation methods are incorporated into a natural attenuation modeling program for petroleum hydrocarbon sites developed for and with technical guidance from the state of Florida (Groundwater Services, Inc., 1997b) and in natural attenuation software (Groundwater Services, Inc., 1998). The third method, interwell partitioning tracer tests, has been applied at only a handful of field sites, all of which contain DNAPL. The last two methods are based on analyzing the record of aqueous-phase concentration over time to estimate the source mass and have focused on estimating DNAPL mass.

Estimating Source Mass Directly from Sampling Data

The direct method is based on the concept of different compartments (i.e., vadose zone, smear zone, and dissolved in groundwater) at petroleum-hydrocarbon sites as described by Gallagher et al. (1995) and uses actual site contaminant concentration measurements and dimensions to calculate mass. The vadose zone compartment extends vertically from the ground surface down to the current water table elevation and laterally from the zone of highest contamination to where contamination is no longer detectable. The following methodology describes the estimation of the contaminant mass.

1. Draw concentration isocontours for each sample depth, with the outermost boundary representing a 0-concentration or nondetect line.

2. Calculate the average concentration (an area-weighted average is preferable) within each nondetect isocontour for each sampled depth.
3. Multiply (a) the average concentration by (b) the area inside the 0-concentration isocontour for each depth sampled. The resulting units will be concentration-area (e.g., mg/kg · ft^2).
4. Using each value from the preceding step, average two concentration–area results for two different depths and multiply by the thickness. The resulting units will be concentration–volume (e.g., ft^3 · mg/kg). Convert the units to concentration–soil mass (e.g., mg/kg · kg) by multiplying by the soil density. These units simplify further to milligrams.

The second compartment to be considered is the smear zone. This zone is defined vertically to lie between the current water table elevation and the lowest known water table elevation, and horizontally to encompass the area containing or having contained measurable amounts of phase-separated hydrocarbons. This compartment probably contains the bulk of the contaminant mass at LNAPL sites.

1. The horizontal extent of the smear zone is established by one of the following methods: either the area containing groundwater wells having at some time contained measurable amounts of NAPL, or those wells having concentrations of dissolved hydrocarbon compounds greater than some threshold value. For gasoline sites, a value of 3000 ppb total BTEX in groundwater has been proposed by Gallagher et al. (1995) as being representative of the smear zone.
2. Soil samples collected within the smear zone both laterally and vertically are then averaged to determine the average soil concentration in the smear zone.
3. To calculate smear zone mass, multiply (a) the average concentration by (b) the volume inside the assumed smear zone and by (c) the assumed soil density.

The third compartment to be considered is dissolved groundwater contamination located in the source zone. This compartment will typically contain only a small fraction of the mass contained in the other compartments and can frequently be ignored when significant amounts of contaminants are present in other compartments.

1. The source zone for contaminated groundwater can be assumed to be of the same lateral extent as the smear zone if a smear zone exists at the site. If no smear zone is known to exist at the site, the groundwater source zone should be defined as the area inside a contour of known concentration (e.g., 1000 to 3000 ppb total BTEX). If the vertical extent of the dissolved groundwater plume is not known, it may be assumed to extend the full thickness of the aquifer, or 10 ft, whichever is less.
2. Groundwater concentrations located inside the source zone are then averaged both spatially and temporally.

3. To determine the dissolved mass of key compounds, multiply (a) the average concentration, by (b) the area of contaminated groundwater, by (c) the assumed or actual vertical extent, and by (d) the porosity of the aquifer matrix.

Indirect Method to Estimate Smear Zone Mass

If detailed soil contaminant concentration data are not available for the smear zone, where the majority of the source mass is assumed to reside, an indirect method can be applied. Thus an approximation of the contaminant mass in the smear zone is deemed a reasonable value to use for the source mass. The following equation may be used to calculate the source mass:

$$\text{source mass} = \text{SZ} \cdot \text{SA} \cdot nS_0 \cdot \text{FC} \cdot \rho_c C \tag{2.42}$$

where
source mass = mass of dissolvable compound(s) in source zone (kg)
SZ = smear zone thickness (ft)
SA = smear zone area (ft^2)
n = porosity (unitless)
S_0 = residual saturation (unitless)
FC = mole fraction of compound(s) of interest in NAPL (unitless)
ρ_c = pure-phase density of compound of interest (kg/L)
C = conversion = (28.3 L/ft^3)

The indirect method was used to estimate the source mass at 117 sites in Florida as part of a Florida RBCA study (Groundwater Services, Inc., 1997a). Questionnaires were distributed to site consultants, who provided hydrogeological and chemical data about their sites. Because the smear zone area was rarely known, the width of the smear zone was estimated by assuming that it was half the width of the observed dissolved plume at the point of the maximum observed contaminant concentration in groundwater. The length of the smear zone was assumed to be equal to the width, so that the assumed smear zone was square:

$$\text{SA} = (0.5\text{PW})^2 \tag{2.43}$$

where SA is the estimated smear zone area (ft^2) and PW is the plume width at the point with the highest observed dissolved concentration (ft).

Using this method, the RBCA planning study yielded an estimated median benzene mass of 11 kg at Florida UST sites, equivalent to 300 to 500 gallons of gasoline. With source masses calculated in this fashion for each of 117 sites, a median source lifetime for benzene to reach the state's 1-ppb cleanup standard in the source zone was 65 years using the BIOSCREEN method with source biodegradation (note that the time required to reach the 1-ppb cleanup standard at the property boundary was only 34 years). Site-specific values used in this calculation were (1) Darcy velocity of groundwater flow, (2) smear zone depth, (3) plume width [for estimating the area of the smear zone using eq. (2.42)], and (4) the observed maximum source concentration. The 25th percentile values for source mass and source lifetime in the source zone were 3.9 kg and 6 years, respectively. The 90th percentile values for source mass and source lifetime were 160 kg and 703 years, respectively. The reported source lifetimes should be considered as order-of-magnitude estimates, but do match several expectations about source lifetimes at petroleum release sites.

Example 2.7 Determine the source mass of BTEX in a source zone described by the following data: porosity = 0.35 (estimated), residual saturation = 1% (estimated from soil sampling data), fraction BTEX in NAPL = 10% (assumed), density of NAPL = 0.8 g/mL, and smear zone thickness = 5 ft (assumed). There are no detailed soil sampling data from the smear zone, but the width of the dissolved plume in the source zone is 31 ft.

SOLUTION: First estimate the area of the smear zone using eq. (2.43):

$$\text{SA} = [(0.5)(3.1)]^2 = 240\ \text{ft}^2$$

Next use eq. (2.42) to estimate the source mass of BTEX:

$$\text{source mass} = (5)(240)(0.35)(0.01)(0.1)(0.8)(28.3) = 10\ \text{kg}$$

Interwell Partitioning Tracer Tests

Partitioning interwell tracer tests (PITTs) utilize chemical tracers that are pumped through a NAPL zone; tracers that sorb to the NAPL will travel more slowly than tracers that do not sorb to the NAPL (Jin et al., 1995). The resulting data provide (1) a definitive method for detecting DNAPL, (2) a means by which the volume of DNAPL in the subsurface can be determined, and (3) a reliable method for measuring remediation performance (Jin et al., 1995, 1997). Jin et al. (1997) provide a design protocol for performing PITTs and discuss potential interferences with the procedure. Using sensitivity models of a PITT conducted in a hypothetical heterogeneous alluvial aquifer, the authors indicated that PITTs have the ability to measure the volume of residual DNAPL in an aquifer within a few percent. The presence of DNAPL pools increased the error to 30 to 50%, however. The authors concluded that PITTs represent a significant improvement over current DNAPL volume estimation at most sites, and together with core analysis and interpretation of groundwater data, PITTs provide the "only practical approach" for characterizing the potential nature and extent of NAPL contamination.

The tracers are typically organics such as alcohols (Jin, et al., 1997) or gases such as SF6 (Wilson and Mackay, 1995). PITTs have been applied at a few field-scale settings to date. The potential cost of performing these tests, which includes continuous coring, wellfield preparation, pump tests, extensive chemical analysis, computer modeling, and waste disposal of extracted fluids, may limit the widespread application of this approach for many sites, however.

Estimating Source Mass Using Inverse Modeling

Butcher and Gauthier (1994) present a method and equations for estimating DNAPL mass at a site on the measured flux of dissolved contaminants leaving the source zone. They apply a two-step approach, where (1) an inverse modeling solution is developed to estimate flux from the source zone, and (2) the flux is matched to a conceptual model of DNAPL dissolution. Although the authors caution that this method cannot provide useful quantitative estimates without observations over long time periods, it can provide a qualitative indication of the presence of a residual DNAPL source. Application of this method to a industrial site that was in operation from 1956 to 1988 indicated that the volume of residual DNAPL in the source zone was 2144 L, and that 273 years would be required to dissolve the DNAPL completely under current flux rates. Although this estimate had "considerable uncertainty," the accuracy of the approach is expected to improve as more observations are made over time. Therefore, the authors stressed that the method does not provide very useful estimates of remediation early in the life of the site characterization process but should provide a useful framework for evaluating the progress of remediation.

Estimating Source Mass Using Effective Solubility Model

Feenstra (1990) used the preferential dissolution of the more soluble components in a multicomponent NAPL to estimate NAPL mass. With this approach, the change in dissolved component concentration ratios over time from a multicomponent NAPL zone was used to estimate the chemical mass depletion of the NAPL. This approach relies on the effective solubility concept [eq. (2.11)] to evaluate the relative dissolution rate of various NAPL compounds.

The effective solubility model was refined further by Feenstra in 1997 to provide a tool for predicting the change in dissolved concentrations versus NAPL mass without knowledge of the NAPL source zone geometry, groundwater flow rates, or mass transfer coefficients. The model requires groundwater monitoring data collected over a sufficient time period to define changes in aqueous-phase concentration ratios, and information on the cumulative chemical mass dissolved from the source during that time (typically a few years' worth at sites with pump-and-treat systems, perhaps longer for sites undergoing natural attenuation). The groundwater monitoring data must be frequent enough to identify changes in the aqueous-phase concentrations of two or more NAPL components over time without being compromised by variability and scatter in the data. The data must be collected from wells located relatively close to known or suspected NAPL zones, and the NAPL must be a multicomponent NAPL in a residual form (no thick pools of continuous NAPL). Finally, the effects of biodegradation in the source zone are not considered in the current form of the model.

The model is built around a number of *equilibrium cells*, which are used for the following iterative calculation steps made in each cell:

1. Assign a total mass to the NAPL phase.
2. Calculate the mole fractions of components.

3. Calculate the dissolved concentrations based on the effective solubility relationship.
4. Calculate the dissolved-phase mass.
5. Calculate the sorbed mass.
6. Calculate the revised NAPL mass.
7. Calculate the NAPL-filled porosity (saturation).
8. Calculate the water-filled porosity.
9. Repeat calculation steps 2 through 8 until there is little change in dissolved concentrations.

Because water moving through residual NAPL zones will equilibrate rapidly with the NAPL, the composition of the NAPL will change in the downgradient portion of the NAPL. Therefore, multiple equilibrium cells can be used in the model to account for this chromatographic effect. Feenstra (1997) applied the effective solubility model to the emplaced source experiment at the Borden field site described in Section 2.6 (Rivett et al., 1994). The model was able to predict the NAPL mass, consisting of PCE, trichloromethane, and TCE, in the source zone within 20%. When the model was applied to an actual DNAPL site in Dayton, Ohio, the estimated DNAPL mass was 5000 kg.

The effective solubility method has potential for improving source mass estimates. At LNAPL sites, the multicomponent nature of most petroleum-hydrocarbon fuels suggests that this approach could be useful, although the effects of BTEX biodegradation in the source zone (see Section 2.8) need to be addressed before the model can be applied. At many DNAPL sites, biodegradation will be much less vigorous, and the effective solubility model can be applied more directly.

2.9 SOURCE CONTROL AND NATURAL ATTENUATION

Our definition of source control at contaminated sites is the application of an engineered system to either remove source materials (e.g., excavation of contaminated soils, remediation of the vadose zone, NAPL recovery) or prevent the source materials from delivering contaminants to the environment (e.g., installation of a barrier system). The decision to implement source control measures at a site where natural attenuation is being considered can be evaluated from three different perspectives: (1) a regulatory perspective, (2) a technical perspective, and (3) a risk-based perspective.

Source Control from a Regulatory Perspective

In general, the regulatory perspective is that source control measures are necessary at natural attenuation sites. As discussed in Chapter 1, under EPA's regulatory policy, source control is considered a fundamental component of any monitored natural attenuation remedy and it is expected that source control "will be evaluated for all contaminated sites and that source control measures will be taken at most sites."

Many states have similar guidance; for example, rules proposed for Florida state that a natural attenuation site must have no contaminated soil present that serves as a continuing source of contamination. With regard to NAPLs, most states require removal to the extent practicable, often without specific technical guidance on what constitutes the extent practicable. (Currently one state, Texas, is considering allowing NAPLs to remain in place if it can be demonstrated that the NAPLs do not pose a risk or contribute to an expanding groundwater problem.)

Source Control from a Technical Perspective

From a technical perspective, the applicability of source control measures at some natural attenuation sites is less clear. As described in Section 2.2, our conceptual model of contaminated source zones includes the presence of NAPLs in the saturated zone at most sites, even sites where no LNAPL or DNAPL is observed. If continuous (mobile) NAPLs are observed, NAPL recovery will remove some of the source mass. However, the physics of NAPL migration in the subsurface ensures that only a fraction of a continuous (mobile) NAPL can be removed by pumping and that the remainder of the NAPL will remain very tightly trapped in the subsurface (see Section 2.6).

In the vadose zone, excavation can directly remove residual NAPL and other remediation technologies such as soil vapor extraction, and bioventing can remove some of the contaminants of concern. While removing source mass, these technologies, typically do not affect the source material below the water table (e.g., because of technical and cost considerations, excavation typically does not extend below the water table). Consequently, due to source media remaining in place below the water table, efforts to remove contaminants above the water table may not reduce groundwater impacts significantly.

Three studies of UST petroleum release sites support the conclusion that many dissolved-phase plumes at a site are not affected by typical source control measures. First, the California LLNL plume study (Rice et al., 1995) compared plume length versus remediation and concluded: “An analysis of plume length categories shows that none of the remediation treatment variables have a significant impact on the relative frequencies of the different categories” and “We found no difference in plume length between different remediation techniques and sites with no remedial action.” Sites with soil overexcavation (i.e., excavation of contaminated soils and beyond into clean soils) were found to have a slightly higher probability of a declining average plume concentration than sites with no remedial action (52 to 64%). The Texas BEG study (Mace et al., 1997) of UST sites found that “the use of active groundwater remediation has not yet resulted in a lower median plume length at LPST (leaking petroleum storage tank) sites throughout the state where corrective action is under way. This does not mean that remediation does not improve groundwater conditions at individual sites, but that when all LPST sites are reviewed, plume lengths at sites with remediation do not appear different from plume lengths at sites without remediation.” Finally, the Florida RBCA study (Groundwater Services, Inc., 1997a) used an estimation technique based on eqs. (2.42) and (2.43) and concluded that “of the 117 sites included in this study, affected soils have been previously removed at 28 sites.

For these 28 sites, the estimated median groundwater source mass is approximately 34% lower than the median groundwater source mass where overlying soils have not *yet* been removed. These data suggest that, while the soil removal actions have served to reduce groundwater impacts, a significant percentage of the contaminant source (66%) remains in place in the saturated, water-bearing unit." When these source mass estimates were applied to a risk-based cleanup model, the authors found that "soil removal would not significantly affect groundwater remediation requirements."

At sites with DNAPL, the ability to remove source mass is more limited than at sites with LNAPL. First, much more DNAPL mass is expected to be in the saturated zone, as it does not accumulate on the water table. Second, DNAPL is actually observed at only a small fraction of DNAPL-related sites, while LNAPL is observed at 40 to 70% of LNAPL-related sites (see Fig. 2.3), making recovery of continuous DNAPL relatively rare. In addition to difficulties in predicting exactly where DNAPL pools are likely to form, continuous DNAPL recovery can prove extremely difficult, due to technical difficulties in creating sufficient hydraulic gradients (Mercer and Cohen, 1990). Finally, Pankow and Cherry (1996) note that even extensive free-product pumping under the most favorable conditions leaves abundant DNAPL in place to serve as a long-term source of dissolved constituents to groundwater.

In summary, technical constraints will greatly limit the amount of source material that can be removed, and a large volume of source mass will probably be left in the subsurface even after source controls are implemented at many or most sites. With current technology, "removal of NAPL to the extent practicable" does not apply to residual NAPL in the saturated zone. Therefore, at many petroleum-hydrocarbon and chlorinated solvent sites, considerable source material will be left behind even after NAPL has been removed "to the extent practicable."

Source Control from a Risk Perspective

As described in Chapter 1, risk-based corrective action (RBCA) programs for petroleum-hydrocarbon sites have been implemented or considered for implementation by over 40 state LUST remediation programs. As part of the RBCA process, the risk to human health associated with affected surface soils, vadose zone soils, and groundwater is estimated using modeling tools coupled with risk assessment procedures (ASTM, 1995; Connor et al., 1995). At many LUST sites, the risks associated with groundwater contamination may be small and further action may not be required *from a risk perspective* (as opposed to a natural resource protection perspective) when the actual receptors (actual groundwater users or groundwater discharge points) are used in the risk calculation instead of hypothetical receptors (e.g, Groundwater Services, Inc., 1996, 1997a). These results are consistent with the LUST plume studies described in Section 2.3, in which the resulting petroleum-hydrocarbon plumes from these sites are relatively short (typically less than 200 ft) (Rice et al., 1995; Mace et al. 1997), largely due to the effectiveness of aerobic and, particularly, anaerobic biodegradation.

With regard to DNAPL removal, Freeze and McWhorter (1997) developed a conceptual framework to assess risk reduction from DNAPL removal from fractured, low-permeability soils, a relatively common hydrogeologic setting, as chlorinated solvents are often found to penetrate small fractures in clay units at site. They evaluated application of four different "generic" technologies, using assumed mass removal efficiencies for removal of DNAPL, sorbed constituents, and dissolved constituents. The study evaluated contaminant removal from both the porous media itself and the fractures in the media.

After completing this conceptual study, the authors found that "very high mass removal efficiencies are required to achieve significant long-term risk reduction with technology applications of finite duration. Further, it is unlikely that current technologies can achieve such efficiencies in heterogeneous low-permeability soils that exhibit dual porosity properties and preferential pathways." The authors concluded that at the current time, the goal of remediation at these sites should be containment, with the eventual aim of applying emerging technologies with improved mass-removal technologies sometime in the future. Finally, they stated that "given the limited social funds available for environmental protection, it is important to insure that public policies utilize these funds for maximum public benefit, and do not waste them on efforts that are outside the current range of technical feasibility."

In summary, the risk perspective indicates that if LNAPL or DNAPL removal efforts are not likely to achieve a meaningful reduction in plume size or plume lifetime, the value of the removal effort is questionable and resources may better be applied toward other risk control measures.

REFERENCES

American Petroleum Institute, 1960, *Recommended Practice for Core Analysis Procedures*, RP-40, API, Washington, DC.

———, 1989, *Hydrogeologic Data Base for Groundwater Modeling*, API Publication 4476, API, Washington, DC (report and spreadsheet file are available).

———, 1994, *Decision Support System for Exposure and Risk Assessment*, Version 1.0, API, Washington, DC.

———, 1995, *VADSAT: A Vadose and Saturated Zone Transport Model for Assessing the Effects on Groundwater Quality from Subsurface Petroleum Hydrocarbon Releases and Petroleum Production Waste Management Practices*, Version 3.0, API, Washington, DC.

———, 1996, *Estimation of Infiltration and Recharge for Environmental Site Assessment*, Daniel B. Stephens & Associates, Inc., API Publication 4643, API, Washington, DC.

American Society for Testing and Materials (ASTM), 1992, *Standard Classification of Soils for Engineering Purposes (Unified Soil Classification System)*, ASTM D 2487-92, ASTM, Philadelphia.

———, 1995, *Standard Guide for Risk-Based Corrective Action Applied at Petroleum Release Sites*, ASTM E-1739-95, ASTM, Philadelphia.

Amyx, J. W., Bass, D. M., Jr. and Whiting, R. L., 1960, *Petroleum Reservoir Engineering*, McGraw-Hill, New York, 610 pp.

Anderson, M. R., Johnson, R. L., and Pankow, J. F., 1992a, Dissolution of dense chlorinated solvents into ground water: 1. Dissolution from a well-defined residual source, *Ground Water*, vol. 30, no. 2.

Anderson, M. R., Johnson, R. L., and Pankow, J. F., 1992b, Dissolution of dense chlorinated solvents into groundwater: 3. Modeling contaminant plumes from fingers and pools of solvent, *Environ. Sci. Technol.*, vol. 26, pp. 901–908.

Banerjee, S., 1984, Solubility of organic mixtures in water, *Environ. Sci. Technol.*, vol. 18, pp. 587–591.

Bartow, G., and Davenport, C., 1995, Pump-and-treat accomplishments: a review of the effectiveness of ground water remediation in Santa Clara Valley, California, *Ground Water Monit. Rev.*, pp. 139–146.

Bedient, P. B., Rifai, H. S., and Newell, C. J., 1994, *Ground Water Contamination, Transport and Remediation*, Prentice Hall, Upper Saddle River, NJ.

Borden, R. C., and Kao, C., 1992, Evaluation of groundwater extraction for remediation of petroleum contaminated groundwater, *Water Environ. Res.*, vol. 64, no. 1, pp. 28–36.

Broholm, K., and Feenstra, S., 1995, Laboratory measurements of the aqueous solubility of mixtures of chlorinated solvents, *Environ. Toxicol. Chem.*, vol. 14, pp. 9–15.

Brooks, R. H., and Corey, A. T., 1964, *Hydraulic Properties of Porous Media*, Hydrology Paper 3, Colorado State University, Fort Collins, CO.

Brusseau, M. L., Jessup, R. E., and Rao, S. C., 1991. Nonequilibrium sorption of organic chemicals: Elucidation of rate-limiting processes, *Environ. Sci. Technol.*, vol. 25, pp. 134–142.

Burdine, N. T., 1953, Relative permeability calculations from pore-size data, *Trans A.I.M.E.*, vol. 198, pp. 71–77.

Butcher, J. B., and Gauthier, T. D., 1994, Estimation of residual dense NAPL mass by inverse modeling, *Ground Water*, vol. 32, no. 1, pp. 71–78.

Charbeneau, R. J., Wanakule, N., Chiang, C. Y., Nevin, J. P., and Klein, C. L., 1989, A two-layer model to simulate floating free product recovery: formulation and applications, in *National Water Well Association Proceedings of the Petroleum Hydrocarbons and Organic Chemicals in Ground Water Conference*, Houston, TX, Nov., pp. 333–346.

Chatzis, I., Morrow, N. R., and Lim, H. T., 1983, Magnitude and detailed structure of residual oil saturation, *S. Pet. Eng. J.*, vol. 23, no. 2.

Chrysikopoulos, C. V., 1995, Three-dimensional analytical models of contaminant transport from nonaqueous phase liquid pool dissolution in saturated subsurface formations, *Water Resour. Res.*, vol. 31, no. 4, pp.1137–1145.

Cohen, R. M., and Mercer, J. W., 1993. *DNAPL Site Evaluation*, C.K. Smoley, Boca Raton, FL.

Connor, J. A., Nevin, J. P., Fisher, R. T., Bowers, R. L., and Newell, C. J., 1995, *RBCA Spreadsheet System and Modeling Guidelines*, Version 1.0, Groundwater Services, Inc., Houston, TX (www.gsi-net.com).

Connor, J. A., Bowers, R. L., Paquette, S. M., and Newell, C. J., 1997, Soil attenuation model (SAM) for derivation of risk-based soil remediation standards, in *National Ground Well Association Proceedings of the Petroleum Hydrocarbons and Organic Chemicals in Ground Water Conference*, Houston, TX, Nov. (www.gsi-net.com).

Conrad, S. H., Wilson, J. L., Mason, W. R., and Peplinski, W. J., 1992, Visualization of residual organic liquid trapped in aquifers, *Water Resour. Res.*, vol. 28, no. 2, pp. 467–478.

Core Laboratories, 1989, *The Fundamentals of Core Analysis*, Core Laboratories, Houston, TX.

DeVaull, G. E., Ettinger, R. A., Salanitro, J. P., and Gustafson, J. B., 1997, Benzene, toluene, ethylbenzene, and xylenes (BTX) degradation in vadose zone soils during vapor transport:

first-order rate constants, in *Proceedings of the National Groundwater Association Petroleum Hydrocarbons and Organic Chemicals in Ground Water Conference*, Houston, TX, Nov., pp. 365–379.

Farr, A. M., Houghtalen, R. J., and McWhorter, D. B., 1990, Volume estimation of light nonaqueous phase liquids in porous media, *Ground Water*, vol. 28, no. 1, pp. 48–56.

Feenstra, S., 1990, Evaluation of multi-component DNAPL sources by monitoring of dissolved-phase concentrations, in *Proceedings of the International Conference on Subsurface Contamination by Immiscible Fluids*, K. U. Weyer, ed., Calgary, Alberta, A.A. Balkema, Rotterdam, The Netherlands, pp. 65–72.

Feenstra, S., 1997, Aqueous concentration ratios to estimate mass of multi-component NAPL residual in porous media, thesis, University of Waterloo, Waterloo, Ontario.

Feenstra, S., MackKay, D. M., and Cherry, J. A., 1991, A method for assessing residual NAPL based on organic chemical concentrations in soil samples, Ground Water Monit. Rev., vol. 11, no. 2.

Freeze, R. A., and Cherry, J. A., 1979, *Groundwater*, Prentice Hall, Upper Saddle River, NJ.

Freeze, R. A. and McWhorter, D. B., 1997, A framework for assessing risk reduction during dnapl mass removal from low-permeability soils, *Ground Water*, vol. 35, no. 1, pp. 111–123.

Gallagher, M. N., Payne, R. E., and Perez, E. J., 1995, Mass based corrective action, *National Groundwater Association Petroleum Hydrocarbons and Organic Chemicals in Ground Water Conference*, Houston, TX, Nov.

Geller, J. T., and Hunt, J. R., 1993, Mass transfer from nonaqueous phase organic liquids in water-saturated porous media, *Water Resour. Res.*, vol. 29, no. 4, pp. 883–845.

General Sciences Corporation, 1996, *RISKPRO's SESOIL for Windows*, Version 2.6, GSC, Laurel, MD.

Groundwater Services, Inc., 1996, *Implementability of Risked-Based Corrective Action (RBCA) in Wisconsin*, Houston, TX (www.gsi-net.com).

———, 1997a, *Florida RBCA Planning Study*, prepared for Florida Partners in RBCA Implementation, GSI, Houston, TX (www.gsi-net.com).

———, 1997b, *State of Florida Natural Attenuation Decision Support System, Version 0.5 User's Manual*, GSI, Houston, TX (www.gsi-net.com).

———, 1998, *Natural Attenuation Tool Kit*, GSI, Houston, TX (www.gsi-net.com).

Holman, H. Y. N., and Javandel, I., 1996, Evaluation of transient dissolution of slightly water-soluble compounds from a light nonaqueous phase liquid pool, *Water Resour. Res.*, vol. 32, no. 4, pp. 915–923.

Huling, S., and Weaver, J., 1991, *Dense Nonaqueous Phase Liquids*, Ground Water Issue Paper, EPA/540/4-91-002, U.S. EPA, Washington, DC.

Hunt, J. R., Sitar, N., and Udell, K. S., 1988, Nonaqueous phase liquid transport and cleanup: 1. Analysis of mechanisms, *Water Resourc. Res.*, vol. 24, no. 8, pp. 1247–1258.

Huntley, D., Hawk, R. N., and Corley, H. P., 1994, Nonaqueous phase hydrocarbon in a fine-grained sandstone: 1. Comparison between measured and predicted saturations and mobility, *Ground Water*, vol. 32, no. 4, pp. 626–634.

Imhoff, P. T., Jaffé, P. R., and Pinder, G. F., 1994, An experimental study of complete dissolution of a nonaqueous phase liquid in saturated porous media, *Water Resour. Res.*, vol. 30, no. 2, pp. 307–320.

Jin, M., Delshad, M., Dwarakanath, V., McKinney, D. C., Pope, G. A., Sepehrnoori, K., Tilburg, C. E., and Jackson, R. E., 1995, Partitioning tracer test for detection, estimation and remediation performance assessment of subsurface nonaqueous phase liquids, *Water Resour. Res.*, vol. 31, no. 5, pp. 1201–1211.

Jin, M., Butler G. W., Jackson, R. E., Mariner, P. E., Pickens, J. F., Pope, G. A., Brown, C. L., and McKinney, D. C., 1997, Sensitivity models and design protocol for partitioning tracer tests in alluvial aquifers, *Ground Water*, vol. 35, no. 6, pp. 964–972.

Johnson, P. C., Kemblowski, M. W., and Colthart, J. D., 1990a, Quantitative analysis of cleanup of hydrocarbon-contaminated soils by in-situ soil venting, *Ground Water*, vol. 28, no. 3, pp. 413–429.

Johnson, P. C., Stanley, C. C., Kemblowski, M. W., Byers, D. L., and Colthart, J. D., 1990b, A practical approach to the design, operation, and monitoring of in site soil-venting systems, *Ground Water Monit. Remediation*, Spring, pp. 159–178.

Johnson, R. L., and Pankow, J. F., 1992, Dissolution of dense chlorinated solvents in groundwater: 2. Source functions for pools of solvents, *Environ. Sci. Technol.*, vol. 26, no. 5, pp. 896–901.

Jury, W. A., Russo, D., Streile, G., and El Abd, H., 1990, Evaluation of volatilization by organic chemicals residing below the soil surface, *Water Resour. Res.*, vol. 26, no. 1, pp. 13–20.

Kueper, B. H., and Frind, E. O., 1988, An overview of immiscible fingering in porous media, *J. Contam. Hydrol.*, vol. 2.

———, 1992, Two-phase flow in heterogeneous porous media: 1. Model development, *Water Resour. Res.*, vol. 27, pp. 1059–1070.

Kueper, B. H., and McWhorter, D. B., 1991, The behavior of dense, non-aqueous phase liquids in fractured clay and rock, *Ground Water*, vol. 29, pp. 716–728.

Kueper, B. H., Redman, D., Starr, R. C., Reitsma, S., and Mah, M., 1993, A field experiment to study behavior of tetrachloroethylene below the water table: spatial distribution of residual and pooled DNAPL, *Ground Water*, vol. 31, pp. 756–766.

Lamarche, P., 1991, Dissolution of immiscible organics in porous media, Ph.D. thesis, Department of Earth Sciences, University of Waterloo, Waterloo, Ontario, Canada.

Lambe, T. H., and Whitman, R. V., 1969, *Soil Mechanics*, Wiley, New York.

Lenhard, R. J., and Parker, J. C., 1990, Estimation of free hydrocarbon volume from fluid levels in monitoring wells, *Ground Water*, vol. 28, no. 1, pp. 57–67.

Luckner, L., 1991, *Migration Processes in Soil and the Groundwater Zone*, Lewis Publishers, Boca Raton, FL.

Mace, R. E, Fisher, R. S., Welch, D. M., and Parra, S. P., 1997, *Extent, Mass, and Duration of Hydrocarbon Plumes from Leaking Petroleum Storage Tank Sites in Texas*, Geological Circular 97-1, Bureau of Economic Geology, University of Texas at Austin, Austin, TX.

Mackay, D. M., and Cherry, J. A., 1989, Groundwater contamination: pump and treat remediation, *Environ. Sci. Technol.*, vol. 23, no. 6.

Mackay, D. M., Shiu, W. Y., Maijanen, A., and Feenstra, S., 1991, Dissolution of non-aqueous phase liquids in groundwater, *J. Contam. Hydrol.*, vol. 8, pp. 23–42.

Manivannan, I., Powers, S. E., and Curry, G. W., Jr., 1996, Dissolution of NAPLs entrapped in heterogeneous porous media, non-aqueous phase liquids (NAPLs), in *Subsurface Environment: Assessment and Remediation*, L. Reddi, ed., ASCE Specialty Conference, Washington, DC, Nov. 12–14, p. 563.

Mercer, J. W., and Cohen, R. M., 1990, A review of immiscible fluids in the subsurface: properties, models, characterization and remediation, *J. Contam. Hydrol.*, vol. 6.

Miller, C. T., Poirier-McNeill, M. M., and Mayer, A. S., 1990, Dissolution of trapped non-aqueous phase liquids: mass transfer characteristics, *Water Resour. Res.*, vol. 26, no. 11, pp. 2783–2796.

Morrow, N. R. and Chatzis, I., 1982, *Measurement and Correlation of Conditions for Entrapment and Mobilization of Residual Oil*, Report DOE/BC/10310-20, U.S. Department of Energy, Washington, DC.

National Research Council, 1994, *Alternatives for Ground Water Cleanup*, National Academy Press, Washington, DC.

Newell, C. J. and Ross, R. 1992, *Estimating Potential for Occurrence of DNAPL at Superfund Sites*, EPA Quick Reference Fact Sheet, EPA Publication 9355.4-07FS, U.S. EPA, Washington, DC.

Newell, C. J., Hopkins, L. P., and Bedient, P. B., 1990, A hydrogeologic database for groundwater modeling, *Ground Water*, vol. 28, no. 5, pp. 703–714.

Newell, C. J., Connor, J. A., Wilson, D. K., and McHugh, T. E., 1991, Impact of dissolution of dense non-aqueous phase liquids (DNAPLS) on groundwater remediation, in *Proceedings of the National Water Well Association Petroleum Hydrocarbons and Organic Chemicals in Ground Water Conference*, Houston, TX, Nov. (www.gsi-net.com).

Newell, C. J., Bowers, R. L., and Rifai, H. S., 1994, Impact of non-aqueous phase liquids (NAPLs) on groundwater remediation, Summer National AIChE Meeting, Symposium 23, *Multimedia Pollutant Transport Models*, Denver, CO, Aug. 16 (www.gsi-net.com).

Newell, C. J., Acree, S. D., Ross, R. R., and Huling, S. G., 1995, *Light Nonaqueous Phase Liquids*, Ground Water Issue Paper, EPA/540/S-95/500, U.S. EPA, Washington, DC.

Newell, C. J., Gonzales, J., and McLeod, R., 1996, *BIOSCREEN Natural Attenuation Decision Support System*, Version 1.3, EPA/600/R-96/087, U.S. EPA, Washington, DC (www.epa.gov/ada/kerrlab.html).

———, 1997, *BIOSCREEN Natural Attenuation Decision Support System*, Version 1.4 Revisions, U.S. EPA, Washington, DC (www.epa.gov/ada/kerrlab.html).

Ogata, A., and Banks, R. B., 1961, *A Solution of the Differential Equation of Longitudinal Dispersion in Porous Media*, U.S. Geological Survey Professional Paper 411-A, U.S. Government Printing Office, Washington, DC.

Pankow, J. F., and Cherry, J. A., 1996, *Dense Chlorinated Solvents and Other DNAPLs in Groundwater*, Waterloo Press, Portland, OR, 522 pp.

Parker, B. L., Gillham, R.W., and Chemy, J. C., 1996. Diffusive disappearance of immersible-phase organic liquids in fractured geologic media, *Ground Water*, vol. 32, pp. 805–820.

Parker, J. C., Waddill, D. W., and Johnson, J. A., 1994, *UST Corrective Action Technologies: Engineering Design of Free Product Recovery Systems*, prepared for Superfund Technology Demonstration Division, Risk Reduction Engineering Laboratory, Edison, NJ, Environmental Systems & Technologies, Inc., Blacksburg, VA, 77 pp.

Peck, R. B., Hanson, W. H., and Thornburn, T. H., 1974, *Foundation Engineering*, 2nd ed., Wiley, New York.

Powers, S. E., Loureiro, C. O., Abriola, L. M., and Weber, W. J., Jr., 1991, Theoretical study of the significance of nonequilibrium dissolution of nonaqueous phase liquids in subsurface systems, *Water Resour. Res.*, vol. 27, no. 4, pp. 463–477.

Powers, S. E., Abriola, L. M., and Weber, W. J., Jr., 1992, An experimental investigation of nonaqueous phase liquid dissolution in saturated subsurface systems: steady state mass transfer rates, *Water Resour. Res.*, vol. 28, no. 10, pp. 2691–2705.

———, 1994, An experimental investigation of nonaqueous phase liquid dissolution in saturated subsurface systems: transient mass transfer rates, *Water Resour. Res.*, vol. 30, no. 2, pp. 321–332.

Rawls, W. J., and Brakensiek, D. L., 1985, Prediction of soil water properties for hydrologic modeling, in *Proceedings of the Symposium on Watershed Management*, ASCE, New York, pp. 293–299.

Rice, D. W., Grose, R. D., Michaelsen, J. C., Dooher, B. P., MacQueen, D. H., Cullen, S. J., Kastenberg, W. E., Everett, L. G., and Marino, M. A., 1995, *California Leaking Underground Fuel Tank (LUFT) Historical Case Analysis*, California Environmental Protection Department, Sacramento, CA, Nov. 16.

Rivett, M. O., Feenstra, S., and Cherry, J. A., 1994, Transport of a dissolved-phase plume from a residual solvent source in a sand aquifer, *J. Hydrol.*, vol. 159, pp. 27–41.

Schroeder, P. R., Peyton, R. L., McEnroe, B. M., and Sjostrom, J. W., 1988, *The Hydrologic Evaluation of Landfill Performance (HELP) Model*, Vol. 3, *User's Guide for Version 2*, IA DW21931425-01-3, U.S. Army Engineer Waterways Experiment Station, Oct., Vicksburg, MS.

Schwille, F., 1988, *Dense Chlorinated Solvents in Porous and Fractured Media: Model Experiments* (English translation), Lewis Publishers, Boca Raton, FL.

U.S. Environmental Protection Agency (U.S. EPA), 1992, *Dense Nonaqueous Phase Liquids: A Workshop Summary*, EPA/600-R-92/030, U.S. EPA, Washington, DC, Feb.

———, 1993, *Evaluation of the Likelihood of DNAPL Presence at NPL Sites, National Results*, EPA/540-R-93-073, U.S. EPA, Washington, DC.

———, 1996, *Soil Screening Guidance: Technical Background Document*, EPA/504/R-95/128, NTIS No. PP96-963502, U.S. EPA, Washington, DC.

Viessman, W., Lewis, G. L., and Knapp, J. W., 1989, *Introduction to Hydrology*, 3rd ed., HarperCollins, New York.

Voudrias, E. A., and Yeh, M., 1994, Dissolution of a toluene pool under constant and variable hydraulic gradients with implications for aquifer remediation, *Ground Water*, vol. 32, no. 2, pp. 305–311.

Wiedemeier, T. H., Wilson, J. T., Kampbell, D. H., Miller, R. N., and Hansen, J. E., 1995, *Technical Protocol for Implementing Intrinsic Remediation with Long-Term Monitoring for Natural Attenuation of Fuel Contamination Dissolved in Groundwater*, U.S. Air Force Center for Environmental Excellence, San Antonio, TX.

Wilson, E. J., and Geankopolis, C. J., 1966, Liquid mass transfer at very low Reynolds numbers in packed beds, *Ind. Eng. Chem. Fundam.*, vol. 5, no. 1, pp. 9–14.

Wilson, J. L., and Conrad, S. H., 1984, Is physical displacement of residual hydrocarbons a realistic possibility in aquifer restoration? *Proceedings of the National Water Well Association Petroleum Hydrocarbons and Organic Chemicals in Ground Water Conference*, Houston, TX, Nov.

Wilson, J. L., Conrad, S. H., Mason, W. R., Peplinski, W., and Hagan, E., 1990, *Laboratory Investigation of Residual Liquid Organics from Spills, Leaks, and the Disposal of Hazardous Wastes in Groundwater*, EPA/600/6-90/004, U.S. EPA, Washington, DC, Apr.

Wilson, R. D., and Mackay, D. M., 1995, Direct detection of residual nonaqueous phase liquid in the saturated zone using SF6 as a partitioning tracer, *Environ. Sci. Technol.*, vol. 29, no. 5, pp. 1255–1258.

CHAPTER 3

ABIOTIC PROCESSES OF NATURAL ATTENUATION

Many abiotic mechanisms affect the fate and transport of organic compounds dissolved in groundwater. Physical processes include advection and dispersion, while chemical processes include sorption, volatilization, hydrolysis, dehydrohalogenation, hydrogenolysis, and dihaloelimination. Advection transports chemicals along groundwater flow paths and in general does not cause a reduction in contaminant mass or concentration. Dispersion or mixing effects, sorption, and volatilization do not degrade contaminants. Therefore, although these processes will reduce the concentrations of contaminants in groundwater, they will not reduce the total mass of contaminants in the subsurface. On the other hand, hydrolysis and dehydrohalogenation both will result in the destruction of contaminants dissolved in groundwater.

Fuel hydrocarbons and chlorinated solvents are advected, dispersed, and sorbed in groundwater systems. Both classes of compounds will volatilize, although their different components have varying degrees of volatility. Only chlorinated solvents hydrolyze and undergo additional chemical reactions, such as dehydrohalogenation or elimination and oxidation and reduction. As will be seen later in the chapter, these abiotic reactions typically are not complete and often result in the formation of intermediates that may be at least as toxic as the original contaminant.

3.1 ADVECTION

Advective transport refers to the transport of solutes by the bulk movement of groundwater. Advection is the most important process driving the downgradient migration of dissolved contaminants in the saturated subsurface. The seepage velocity of groundwater and dissolved chemicals influenced by advective transport only is given by

$$\upsilon_x = -\frac{K}{n_e}\frac{dH}{dL} \tag{3.1}$$

where
υ_x = seepage velocity [L/T]
K = hydraulic conductivity [L/T]
n_e = effective porosity [L^3/L^3]
dH/dL = hydraulic gradient [L/L]

Newell et al. (1990) developed a statistical estimate of the applicable range of seepage velocities for aquifers in the United States based on data from 400 sites nationwide. Their data (Figure 3.1) indicate that velocities typically range between 10^{-7} and 10^3 ft/day, with an average of 0.24 ft/day. The data shown in Figure 3.1 can be used to estimate the velocity in a given hydrogeologic environment. Seepage velocity is a key parameter in natural attenuation studies because it can be used to estimate the time of travel of a contaminant front:

$$t = \frac{x}{\upsilon_x} \tag{3.2}$$

where
t = time [T]
x = travel distance [L]
υ_x = seepage velocity [L/T]

Solute transport by advection alone yields a sharp solute concentration front. Immediately ahead of the front, the solute concentration is equal to the background concentration (generally not detectable). At and behind the advancing solute front, the concentration is equal to the initial contaminant concentration at the point of release. This is referred to as *plug flow* and is illustrated in Figures 3.2, 3.3, and 3.4. In reality, the advancing front spreads out due to the processes of dispersion and diffusion, and is retarded by sorption, abiotic degradation mechanisms, and biodegradation (see Chapters 4, 5, and 6).

Example 3.1: Estimating Velocity and Time of Travel Given a groundwater aquifer with a hydraulic conductivity of 10^{-3} ft/sec, a gradient of 0.001 ft/ft, and an effective porosity of 0.35, estimate the contaminant seepage velocity and the elapsed time before the contaminant travels 200 ft to a nearby receptor exposure point.

SOLUTION: The seepage velocity can be estimated using eq. (3.1):

$$\upsilon_x = \frac{10^{-3}}{0.35}(0.001) = 2.86(10^{-6}) \text{ ft/sec}$$

The time of travel can be estimated using eq. (3.2):

$$t = \frac{200}{2.86(10^{-6})} = 2.22 \text{ years}$$

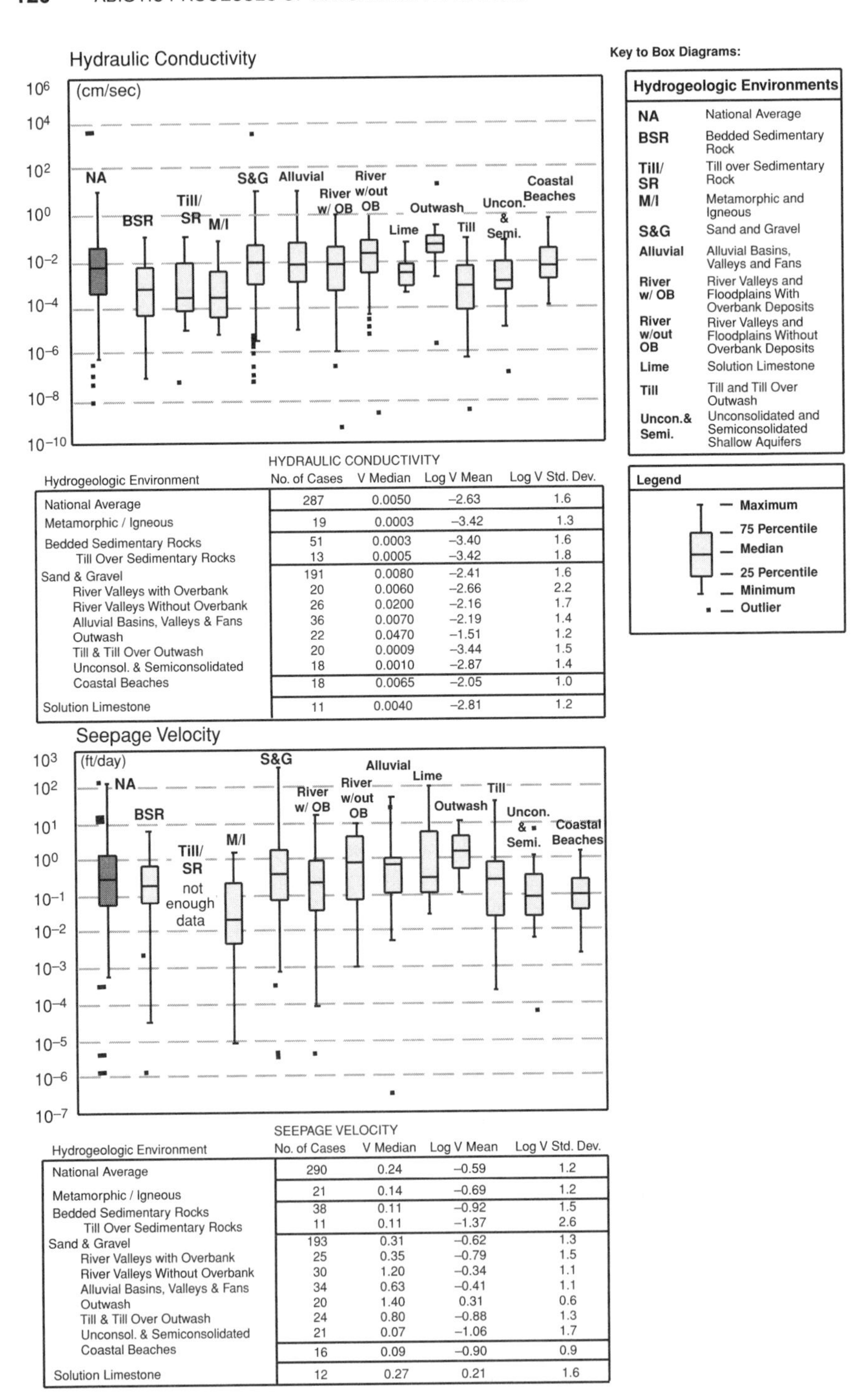

HYDRAULIC CONDUCTIVITY

Hydrogeologic Environment	No. of Cases	V Median	Log V Mean	Log V Std. Dev.
National Average	287	0.0050	−2.63	1.6
Metamorphic / Igneous	19	0.0003	−3.42	1.3
Bedded Sedimentary Rocks	51	0.0003	−3.40	1.6
Till Over Sedimentary Rocks	13	0.0005	−3.42	1.8
Sand & Gravel	191	0.0080	−2.41	1.6
River Valleys with Overbank	20	0.0060	−2.66	2.2
River Valleys Without Overbank	26	0.0200	−2.16	1.7
Alluvial Basins, Valleys & Fans	36	0.0070	−2.19	1.4
Outwash	22	0.0470	−1.51	1.2
Till & Till Over Outwash	20	0.0009	−3.44	1.5
Unconsol. & Semiconsolidated	18	0.0010	−2.87	1.4
Coastal Beaches	18	0.0065	−2.05	1.0
Solution Limestone	11	0.0040	−2.81	1.2

SEEPAGE VELOCITY

Hydrogeologic Environment	No. of Cases	V Median	Log V Mean	Log V Std. Dev.
National Average	290	0.24	−0.59	1.2
Metamorphic / Igneous	21	0.14	−0.69	1.2
Bedded Sedimentary Rocks	38	0.11	−0.92	1.5
Till Over Sedimentary Rocks	11	0.11	−1.37	2.6
Sand & Gravel	193	0.31	−0.62	1.3
River Valleys with Overbank	25	0.35	−0.79	1.5
River Valleys Without Overbank	30	1.20	−0.34	1.1
Alluvial Basins, Valleys & Fans	34	0.63	−0.41	1.1
Outwash	20	1.40	0.31	0.6
Till & Till Over Outwash	24	0.80	−0.88	1.3
Unconsol. & Semiconsolidated	21	0.07	−1.06	1.7
Coastal Beaches	16	0.09	−0.90	0.9
Solution Limestone	12	0.27	0.21	1.6

FIGURE 3.1. Box plots by hydrogeologic environment for hydraulic conductivity (in cm/sec) and seepage velocity (in ft/day). (From Newell et al., 1990.)

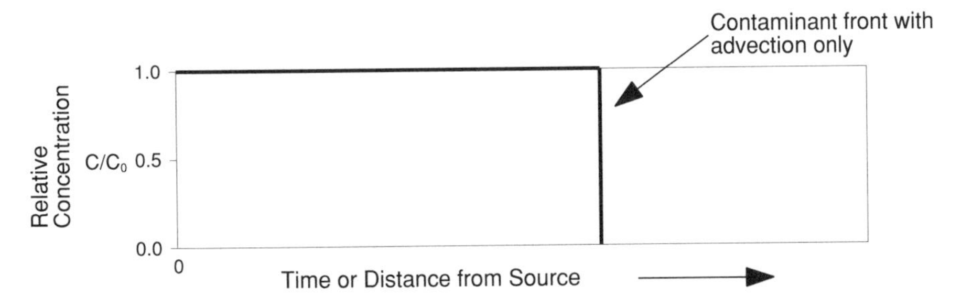

FIGURE 3.2. Breakthrough curve in one dimension showing plug flow with continuous source resulting from advection only.

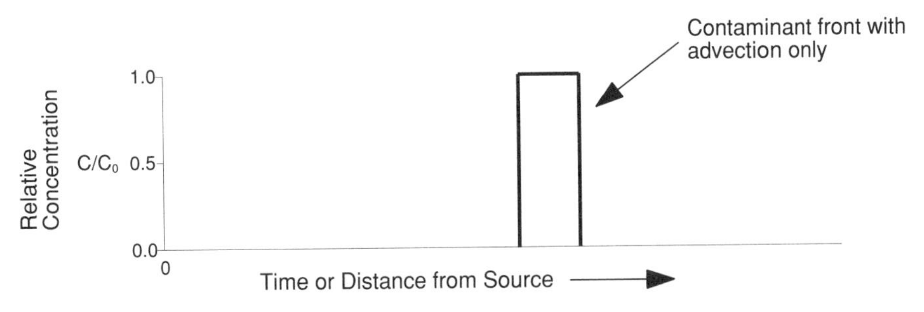

FIGURE 3.3. Breakthrough curve in one dimension showing plug flow with instantaneous source resulting from advection only.

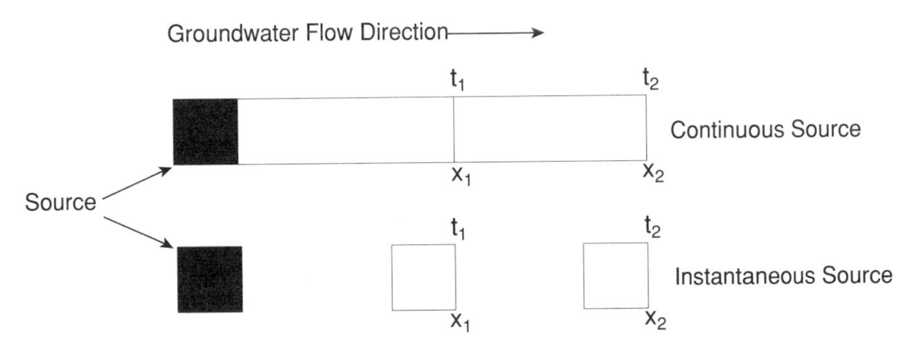

FIGURE 3.4. Plume migration in two dimensions (plan view) showing plume migration resulting from advective flow only with continuous and instantaneous sources.

Potentiometric Surface Maps

A potentiometric surface map is a two-dimensional graphical representation of equipotential lines shown in plan view. Water table elevation maps are potentiometric surface maps drawn for water table (unconfined) aquifers. Potentiometric surface maps for water table aquifers show where planes of equal potential intersect the water table. A potentiometric surface map can be constructed using water-level measurements from piezometers or wells screened in the same relative position within the same hydrogeologic unit. These maps are used to estimate the direction of plume migration and to calculate hydraulic gradients. Seasonal variations in groundwater flow can be assessed by preparing separate potentiometric surface maps using quarterly water level measurements taken over a period of one year or more.

In areas with measurable mobile LNAPL, a correction must be made for the water table deflection caused by the LNAPL. The following relationship, based on Archimedes' principle, provides a correction factor that allows the water table elevation to be adjusted for the effect of floating LNAPL:

$$\text{CDTW} = \text{MDTW} - \frac{\rho_{\text{LNAPL}}}{\rho_w}(\text{PT}) \tag{3.3}$$

where
- CDTW = corrected depth to water [L]
- MDTW = measured depth to water [L]
- ρ_{LNAPL} = density of the LNAPL [M/L^3]
- ρ_w = density of the water, generally 1.0 [M/L^3]
- PT = measured LNAPL thickness [L]

Using the corrected depth to water, the corrected groundwater elevation (CGWE) is given by

$$\text{CGWE} = \text{datum elevation} - \text{CDTW} \tag{3.4}$$

The corrected groundwater elevation value is then used to prepare a potentiometric surface map for the aquifer. Figure 3.5 is an example of a groundwater elevation map for an unconfined aquifer. Water table elevation data used to prepare this map were taken from wells screened across the water table.

Hydraulic Conductivity

Hydraulic conductivity K is a measure of an aquifer's ability to transmit water and is perhaps the most important variable governing fluid flow in the subsurface. Hydraulic conductivity has the units of length over time [L/T]. Observed values of hydraulic conductivity in a variety of aquifer media range over 12 orders of magnitude, from 3×10^{-12} to 3 cm/s (3×10^{-9} to 3×10^3 m/day) (Table 3.1 and Figure 3.6). The data from Newell et al. (1990) indicate a similar range for hydraulic conductivity (see Figure 3.1). In general terms, the hydraulic conductivity for unconsolidated sediments tends to increase with increasing grain size and sorting. The velocity of groundwater and dissolved contaminants is related directly to the

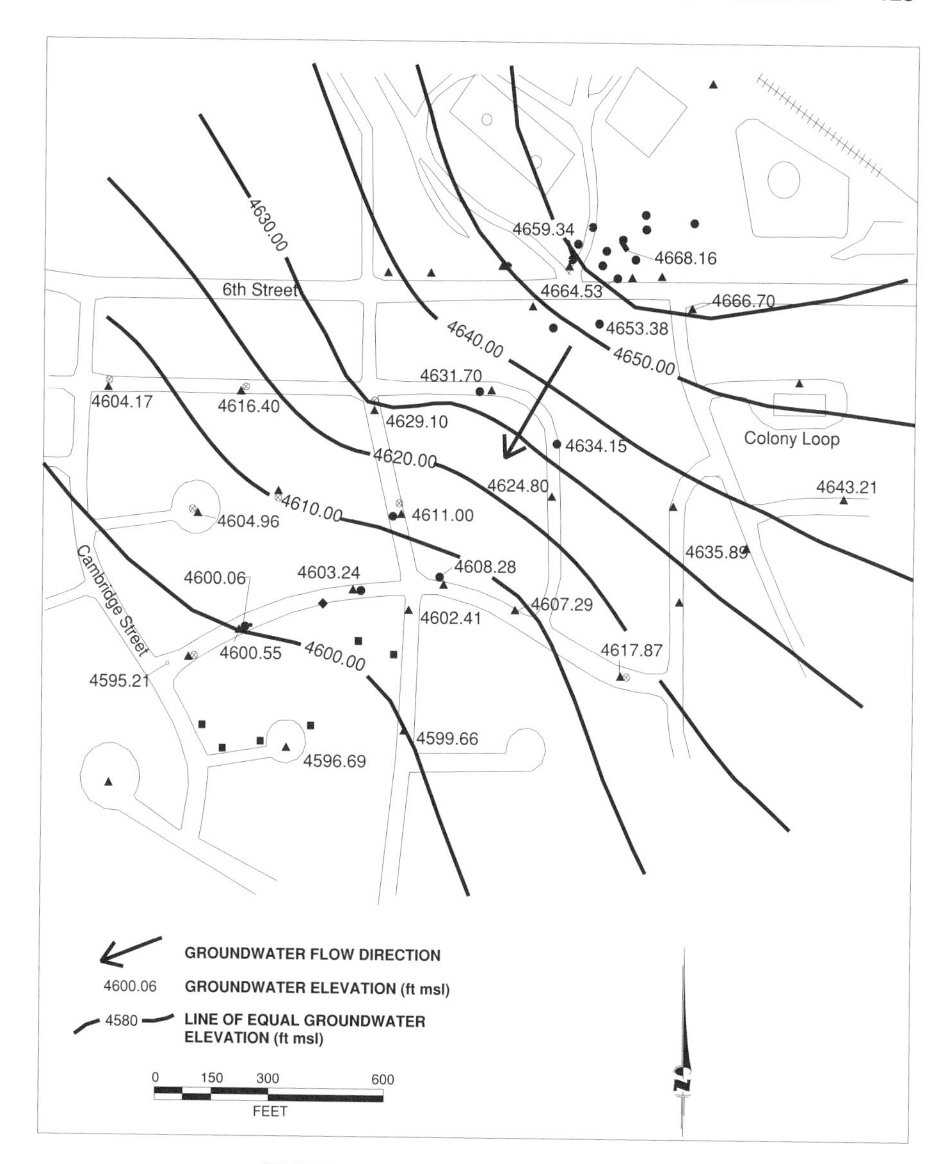

FIGURE 3.5. Groundwater elevation map.

hydraulic conductivity of the saturated zone. Subsurface variations in hydraulic conductivity directly influence contaminant fate and transport by providing preferential pathways for contaminant migration.

The most common methods used to quantify hydraulic conductivity in the subsurface are aquifer pumping tests and slug tests. When analyzed, results from these tests provide an estimate of the transmissivity T of the aquifer such that

$$T = Kb \tag{3.5}$$

TABLE 3.1. Representative Values of Hydraulic Conductivity for Various Sediments and Rocks

Material	Hydraulic Conductivity	
	m/day	cm/s
Unconsolidated sediment		
Glacial till	9×10^{-8}–2×10^{-1}	1×10^{-10}–2×10^{-4}
Clay	9×10^{-7}–4×10^{-4}	1×10^{-9}–5×10^{-7}
Silt	9×10^{-3}–2	1×10^{-7}–2×10^{-3}
Fine sand	2×10^{-2}–2×10^{1}	2×10^{-5}–2×10^{-2}
Medium sand	8×10^{-2}–5×10^{1}	9×10^{-5}–5×10^{-2}
Coarse sand	8×10^{-2}–5×10^{-2}	9×10^{-5}–6×10^{-1}
Gravel	3×10^{1}–3×10^{3}	3×10^{-2}–3
Sedimentary rock		
Karstic limestone	9×10^{-2}–2×10^{3}	1×10^{-4}–2
Limestone and dolomite	9×10^{-5}–5×10^{-1}	1×10^{-7}–6×10^{-4}
Sandstone	3×10^{-5}–5×10^{-1}	3×10^{-8}–6×10^{-4}
Siltstone	9×10^{-7}–1×10^{-3}	1×10^{-9}–1×10^{-6}
Shale	9×10^{-9}–2×10^{-4}	1×10^{-11}–2×10^{-7}
Crystalline rock		
Vesicular basalt	3×10^{-2}–2×10^{3}	4×10^{-5}–2
Basalt	2×10^{-6}–4×10^{-2}	2×10^{-9}–4×10^{-5}
Fractured igneous and metamorphic	7×10^{-4}–3×10^{1}	8×10^{-7}–3×10^{-2}
Unfractured igneous and metamorphic	3×10^{-9}–2×10^{-5}	3×10^{-12}–2×10^{-8}

Source: P. A. Domenico and F. W. Schwartz, *Physical and Chemical Hydrogeology*. Copyright © 1990 John Wiley & Sons, Inc. Reprinted by permission of John Wiley & Sons, Inc.

For a confined aquifer, b is the thickness of the aquifer between confining units; for unconfined aquifers, b is the saturated thickness of the aquifer measured from the water table to the underlying confining layer. Transmissivity has the units of length squared over time $[L^2/T]$.

Hydraulic Conductivity from Pumping Tests. Pumping tests generally provide the most reliable information about aquifer hydraulic conductivity. Pumping tests, in contrast with slug tests, provide a measure of the average hydraulic conductivity in the test area. Pumping test data are most commonly interpreted by graphical techniques. The reader is referred to Walton (1988), Dawson and Istok (1991), and Kruseman and de Ridder (1991) for more information on these techniques.

Hydraulic Conductivity from Slug Tests. Slug tests are relatively easy to conduct and are a commonly used alternative to pumping tests. The biggest advantage of slug tests is that no contaminated water is produced during the test. During

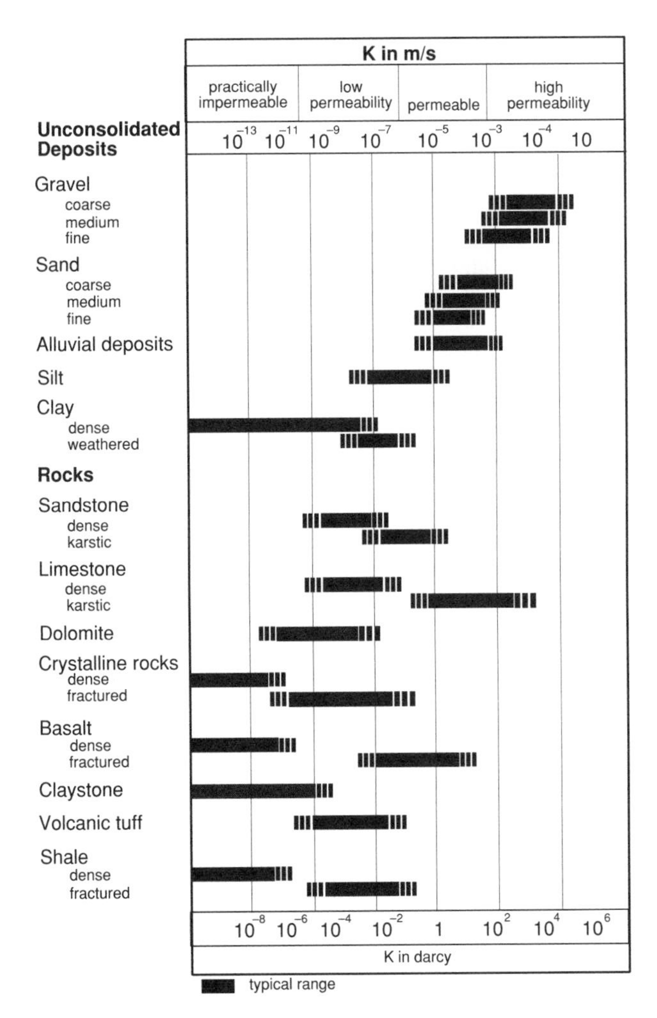

FIGURE 3.6. Range of hydraulic conductivity values. (From K. Spitz and J. Moreno, *A Practical Guide to Groundwater and Solute Transport Modeling*. Copyright © 1996 John Wiley & Sons, Inc. Reprinted by permission of John Wiley & Sons, Inc.)

pumping tests at sites contaminated with organic compounds, large volumes of contaminated water typically are produced. One commonly cited drawback to slug testing is that this method generally gives hydraulic conductivity information only for the area immediately surrounding the monitoring well in which the test was performed. If slug tests are going to be relied on to provide information on the three-dimensional distribution of hydraulic conductivity in an aquifer, numerous slug tests should be performed, both within the same well and in several different monitoring wells at the site. It is not advisable to rely on data from one slug test in a single monitoring well. Data obtained during slug testing generally are analyzed using the

Hvorslev (1951) method for confined aquifers or the Bouwer and Rice (1976) and Bouwer (1989) methods for unconfined conditions.

Hydraulic Head and Gradient

Determining the magnitude of hydraulic gradients is important because gradients influence the direction and rate of contaminant migration. Hydraulic head, specifically variations in hydraulic head within an aquifer, is the driving force behind groundwater movement and solute migration. The hydraulic head H (Figure 3.7) is given by

$$H = z + \frac{P}{\rho g} \tag{3.6}$$

where z = elevation at base of piezometer
P = pressure
ρ = density
g = acceleration due to gravity

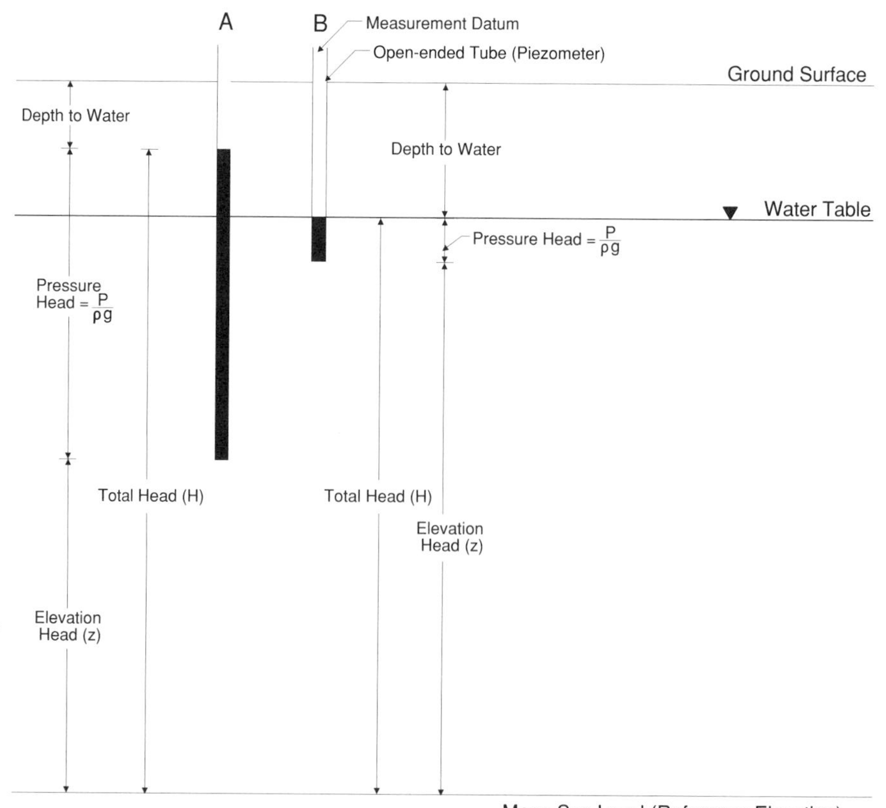

FIGURE 3.7. Hydraulic head.

The hydraulic gradient (dH/dL) is a dimensionless number that represents the change in hydraulic head (dH) between two points divided by the length of groundwater flow between these same two points, parallel to the direction of groundwater flow, and is given by

$$\text{hydraulic gradient} = i = \nabla = \frac{dH}{dL} \tag{3.7}$$

where dH is the change in total hydraulic head between two points [L] and dl is the distance between the two points used for head measurement [L]. Newell et al. (1990) presented a range of values of hydraulic gradient based on their survey of 400 sites in the United States. The data shown in Figure 3.8 indicate that the hydraulic gradients ranged from 10^{-6} to 10^{2} ft/ft with a median national average of 0.006 ft/ft. It should be kept in mind that the majority of sites will experience a seasonal fluctuation in the groundwater table or potentiometric surface. This seasonal fluctuation may cause a variation in the direction of groundwater flow as well as a change in the

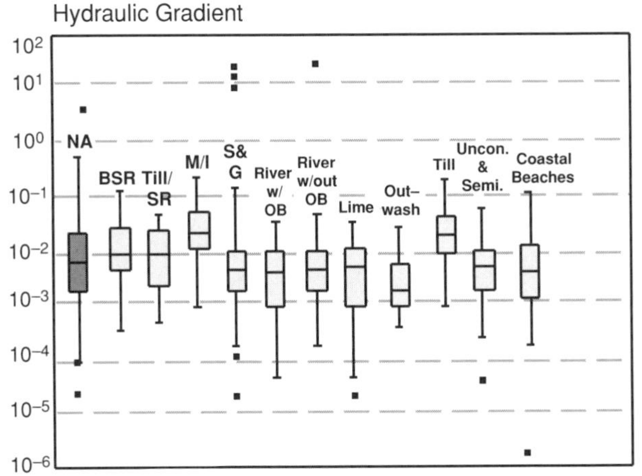

Key to Box Diagrams:

Hydrogeologic Environments	
NA	National Average
BSR	Bedded Sedimentary Rock
Till/ SR	Till over Sedimentary Rock
M/I	Metamorphic and Igneous
S&G	Sand and Gravel
Alluvial	Alluvial Basins, Valleys and Fans
River w/ OB	River Valleys and Floodplains With Overbank Deposits
River w/out OB	River Valleys and Floodplains Without Overbank Deposits
Lime	Solution Limestone
Till	Till and Till Over Outwash
Uncon.& Semi.	Unconsolidated and Semiconsolidated Shallow Aquifers

Hydrogeologic Environment	SEEPAGE VELOCITY No. of Cases	V Median	Log V Mean	Log V Std. Dev.
National Average	346	0.006	−2.22	0.8
Metamorphic / Igneous	23	0.019	−1.77	0.7
Bedded Sedimentary Rocks	52	0.009	−1.96	0.6
Till Over Sedimentary Rocks	17	0.010	−2.11	0.6
Sand & Gravel	223	0.005		
River Valleys with Overbank	25	0.004	−2.52	0.5
River Valleys Without Overbank	30	0.005	−2.25	0.6
Alluvial Basins, Valleys & Fans	38	0.005	−2.29	1.0
Outwash	26	0.002	−2.62	0.5
Till & Till Over Outwash	25	0.010	−1.70	0.7
Unconsol. & Semiconsolidated	25	0.005	−2.42	1.0
Coastal Beaches	25	0.004	−2.37	0.8
Solution Limestone	17	0.006	−2.35	0.8

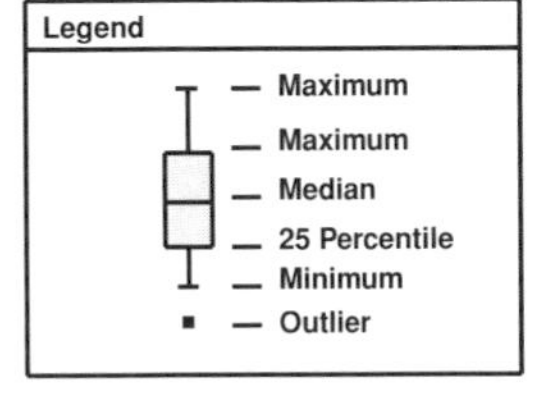

FIGURE 3.8. Box plot by hydrogeologic environment for the hydraulic gradient (in ft/ft). (From Newell et al., 1990.)

hydraulic gradient. These seasonal fluctuations should be quantified, and their impact on contaminant concentrations should be assessed.

Example 3.2: Hydraulic Gradient Calculation Given the water table elevation map shown in Figure 3.9, calculate the hydraulic gradient between points A and B. Assume that all wells are screened across the water table.

SOLUTION: The hydraulic gradient is given by dH/dL. The line connecting points A and B is parallel to the direction of groundwater flow. The water table elevation is

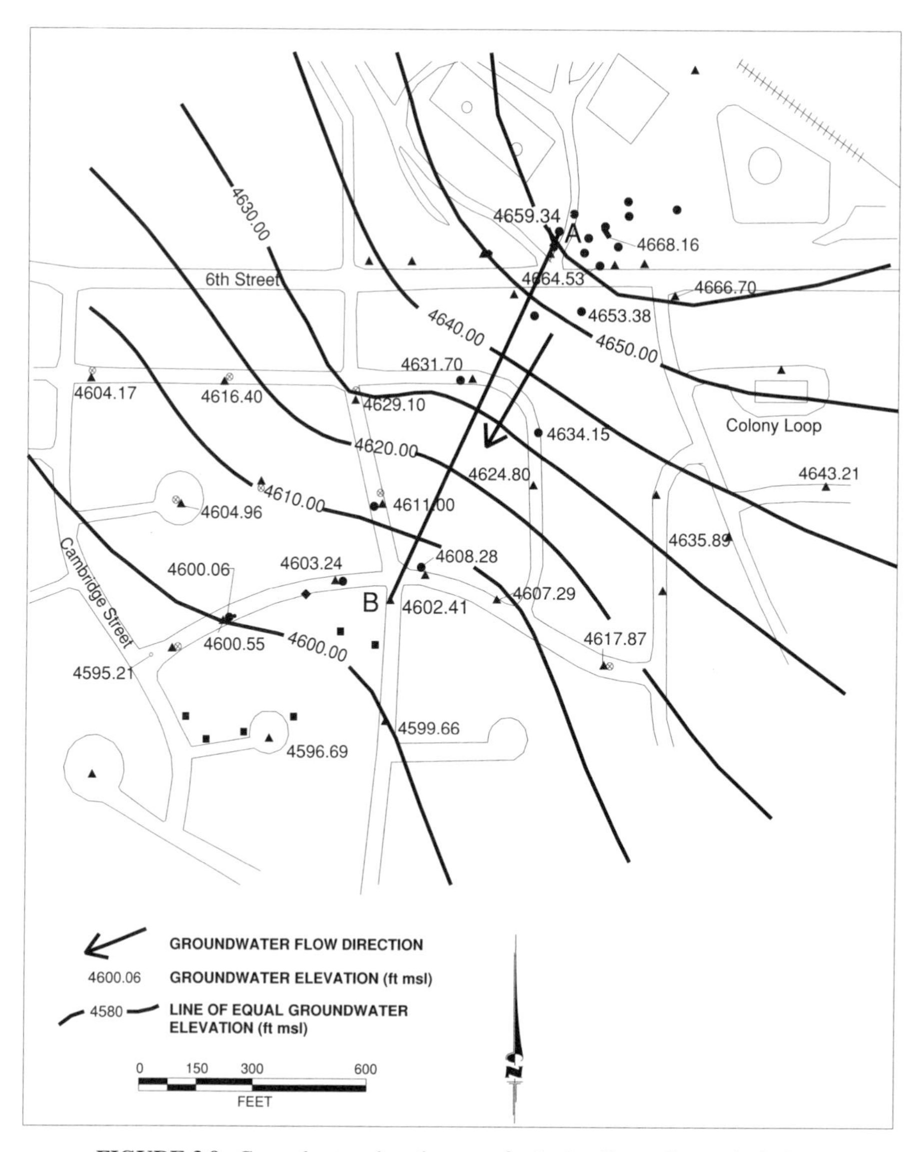

FIGURE 3.9. Groundwater elevation map for hydraulic gradient calculation.

4659.34 ft above mean sea level (MSL) at point A and 4602.41 ft MSL at point B. Therefore, because groundwater flows from areas of high head to areas of lower head,

$$dh = 4602.41 - 4659.34 = -56.93 \text{ ft}$$

The distance between the two points A and B is 936 ft. Therefore;

$$dL = 936 \text{ ft}$$

$$\frac{dH}{dL} = \frac{-56.93 \text{ ft}}{936 \text{ ft}} = -0.06\frac{\text{ft}}{\text{ft}} = -0.06\frac{\text{m}}{\text{m}}$$

Total Porosity and Effective Porosity

Total porosity n is the volume of voids in a unit volume of aquifer. Effective porosity n_e represents the interconnected pore space available for fluid flow. Effective porosity can be estimated using the results of a tracer test. Although this is potentially the most accurate method, time and monetary constraints can be prohibitive. For this reason, the most common technique is to use an accepted literature value for the types of materials making up the aquifer matrix. Table 3.2 and Figure 3.10 present accepted literature values for total porosity and effective porosity

TABLE 3.2. Representative Values of Dry Bulk Density, Total Porosity, and Effective Porosity for Common Aquifer Matrix Materials

Aquifer Matrix	Dry Bulk Density (g/cm^3)	Total Porosity	Effective Porosity
Clay	1.00–2.40	0.34–0.60	0.01–0.2
Peat	—	—	0.3–0.5
Glacial sediments	1.15–2.10	—	0.05–0.2
Sandy clay	—	—	0.03–0.2
Silt	—	0.34–0.61	0.01–0.3
Loess	0.75–1.60	—	0.15–0.35
Fine sand	1.37–1.81	0.26–0.53	0.1–0.3
Medium sand	1.37–1.81	—	0.15–0.3
Coarse sand	1.37–1.81	0.31–0.46	0.2–0.35
Gravely sand	1.37–1.81	—	0.2–0.35
Fine gravel	1.36–2.19	0.25–0.38	0.2–0.35
Medium gravel	1.36–2.19	—	0.15–0.25
Coarse gravel	1.36–2.19	0.24–0.36	0.1–0.25
Sandstone	1.60–2.68	0.05–0.30	0.1–0.4
Siltstone	—	0.21–0.41	0.01–0.35
Shale	1.54–3.17	0.0–0.10	—
Limestone	1.74–2.79	0.0–0.5	0.01–0.24
Granite	2.24–2.46	—	—
Basalt	2.00–2.70	0.03–0.35	—
Volcanic tuff	—	—	0.02–0.35

Source: After Domenico and Schwartz (1990).

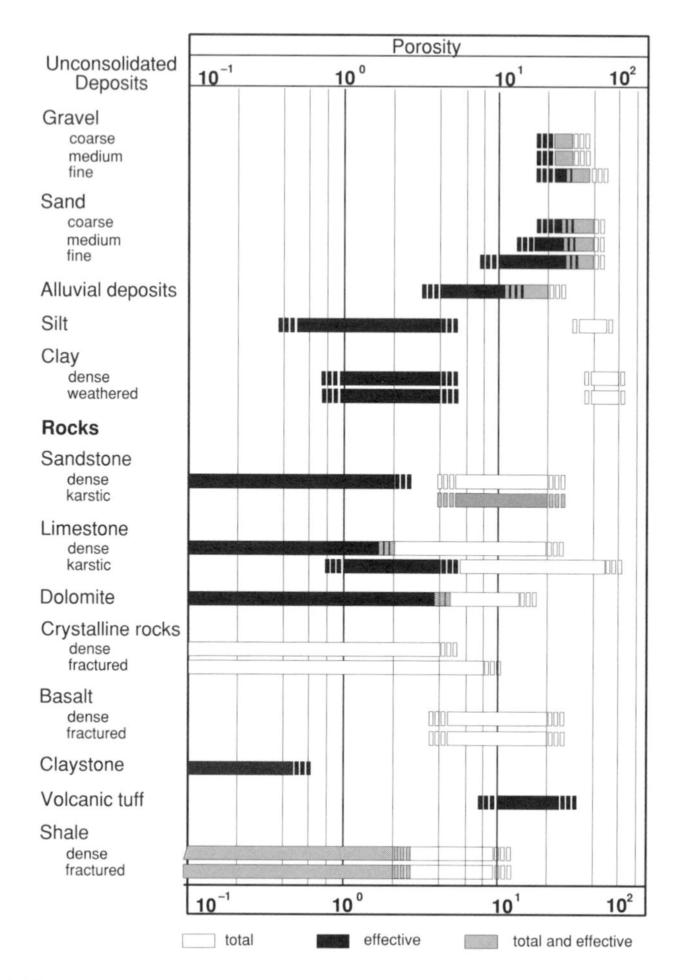

FIGURE 3.10. Range of porosity values for common geologic materials. (From K. Spitz and J. Moreno, *A Practical Guide to Groundwater and Solute Transport Modeling*. Copyright © 1996 John Wiley & Sons, Inc. Reprinted by permission of John Wiley & Sons, Inc.)

3.2 DISPERSION

Hydrodynamic dispersion is the process whereby a contaminant plume spreads out from the main direction of groundwater flow. Dispersion of organic solutes in an aquifer is an important consideration when simulating solute fate and transport. Dispersion results in reduced contaminant concentrations and introduces contaminants into relatively pristine portions of the aquifer, where they may admix with more electron acceptors cross gradient and downgradient from the direction of groundwater flow. Two very different processes cause hydrodynamic dispersion:

mechanical dispersion and molecular diffusion. Mechanical dispersion is the dominant mechanism causing hydrodynamic dispersion at normal groundwater velocities. At extremely low groundwater seepage velocities, molecular diffusion can become the dominant mechanism of hydrodynamic dispersion. Molecular diffusion is generally ignored for most groundwater studies.

Molecular Diffusion

Molecular diffusion occurs when concentration gradients cause solutes to migrate from zones of higher concentration to zones of lower concentration. Molecular diffusion is important only at low groundwater seepage velocities and therefore can be ignored in areas with relatively high groundwater velocities (Davis et al., 1993). The molecular diffusion of a solute in groundwater is described by Fick's laws. Fick's first law applies to the diffusive flux of a dissolved contaminant under steady-state conditions and for the one-dimensional case is given by

$$F = -D\frac{dC}{dx} \tag{3.8}$$

where F = mass flux of solute per unit area of time [M/T]
D = diffusion coefficient [L^2/T]
C = solute concentration [M/L^3]
$\frac{dC}{dx}$ = concentration gradient [M/L^3/L]

For systems where the dissolved contaminant concentrations are changing with time, Fick's second law must be applied. The one-dimensional expression of Fick's second law is

$$\frac{dC}{dt} = D\frac{d^2C}{dx^2} \tag{3.9}$$

where dC/dt is the change in concentration with time [M/T].

The process of diffusion is slower in porous media than in open water because the ions must follow more tortuous flow paths (Fetter, 1988). To account for this, an effective diffusion coefficient D^* is used. Fetter (1988) estimates a range of 1×10^{-9} to 2×10^{-9} m^2/s for D^*. The effective diffusion coefficient is expressed quantitatively as (Fetter, 1988)

$$D^* = wD \tag{3.10}$$

where w is the empirical coefficient determined by laboratory experiments [dimensionless]. The value of w generally ranges from 0.01 to 0.5 (Fetter, 1988).

Mechanical Dispersion

As defined by Domenico and Schwartz (1990), mechanical dispersion is mixing that occurs as a result of local variations in velocity. With time, a given volume of solute gradually will become more dispersed as different portions of the mass are transported

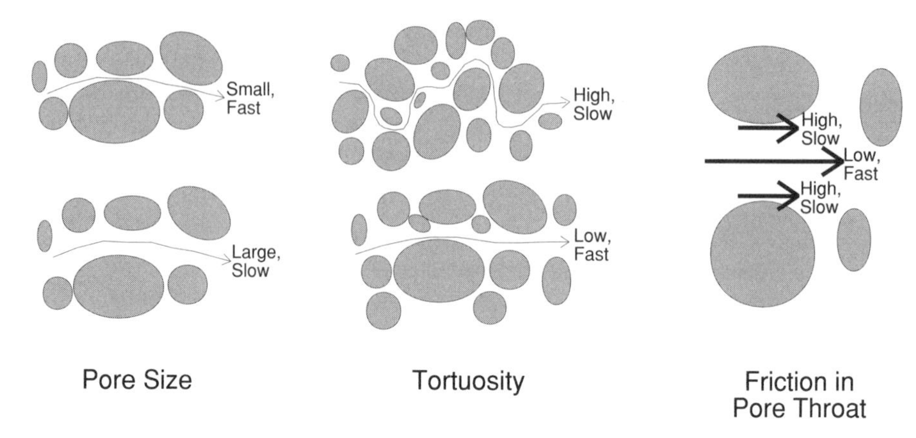

FIGURE 3.11. Physical processes causing mechanical dispersion at the microscopic scale.

at the differing velocities. In general, the main cause of variations of both rate and direction of transport velocities is the heterogeneity of the porous aquifer medium. These heterogeneities are present at scales ranging from microscopic (e.g., pore to pore) to macroscopic (e.g., well to well) to megascopic (e.g., a regional aquifer system).

Three processes are responsible for mechanical dispersion on the microscopic scale (Figure 3.11). The first process is the variation in flow velocity through pores of various sizes. As groundwater flows through a porous medium, it flows more slowly through large pores than through smaller pores. The second cause of mechanical dispersion is *tortuosity*, or *flow path length*. As groundwater flows through a porous medium, some of the groundwater follows less tortuous (shorter) paths, while some of the groundwater takes more tortuous (longer) paths. The longer the flow path, the slower the average seepage velocity of the groundwater and the dissolved contaminant. The final process causing mechanical dispersion is variable friction within an individual pore. Groundwater traveling close to the center of a pore experiences less friction than groundwater traveling next to a mineral grain, and therefore moves faster. These processes cause some of the contaminated groundwater to move faster than the average seepage velocity of the groundwater and some to move slower. This variation in average velocity of the solute causes dispersion of the contaminant.

As a result of dispersion, the solute front travels at a rate that is faster than would be predicted based solely on the average seepage velocity of the groundwater. The overall result of dispersion is spreading and mixing of the contaminant plume with uncontaminated groundwater. Figures 3.12 and 3.13 illustrate the effects of hydrodynamic dispersion on an advancing solute front. The component of hydrodynamic dispersion contributed by mechanical dispersion is given by the relationship:

$$\text{mechanical dispersion} = \alpha_x \upsilon_x \qquad (3.11)$$

where α_x is the dispersivity [L] and υ_x is the average seepage velocity of groundwater [L/T]. *Dispersivity* is a parameter that is characteristic of the porous medium

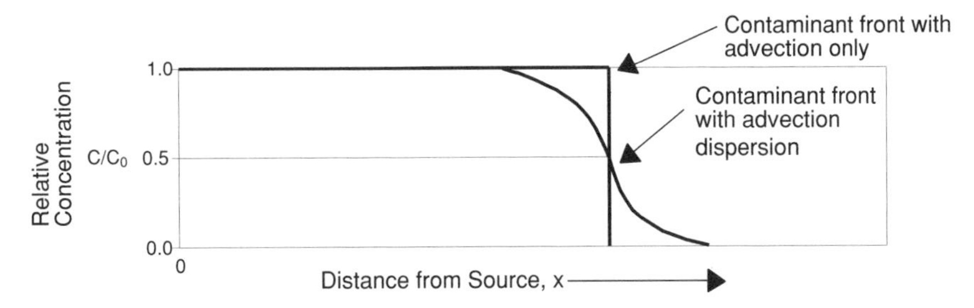

FIGURE 3.12. Breakthrough curve in one dimension showing plug flow with continuous source resulting from advection only and from the combined processes of advection and hydrodynamic dispersion.

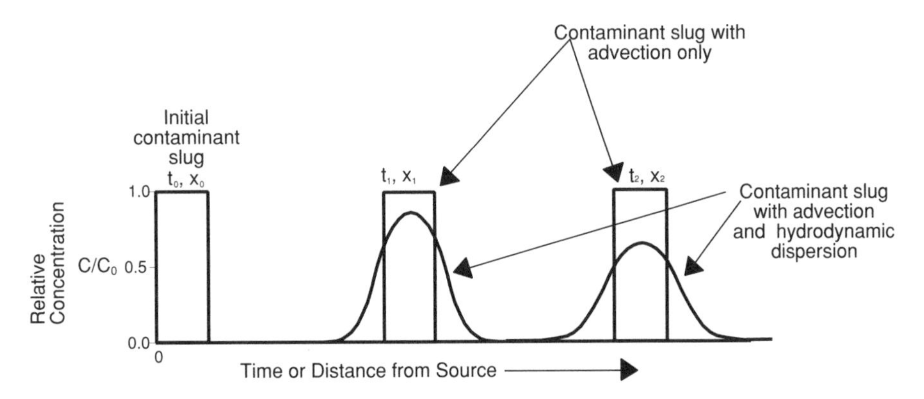

FIGURE 3.13. Breakthrough curve in one dimension showing plug flow with instantaneous source resulting from advection only and from the combined processes of advection and hydrodynamic dispersion.

through which the contaminant migrates. Dispersivity represents the spreading of a contaminant over a given length of flow, and therefore has units of length. It is now commonly accepted (on the basis of empirical evidence) that as the scale of the plume or the system being studied increases, the dispersivity also will increase. Therefore, dispersivity is scale dependent, but at a given scale, data compiled by Gelhar et al. (1985, 1992) show that dispersivity may vary over three orders of magnitude. The data of Gelhar et al. (1992) are presented in Figure 3.14.

Estimating Dispersivity. Several approaches can be used to estimate longitudinal dispersivity α_x at the field scale. One technique involves conducting a tracer test. Although this is potentially the most reliable method, it is time consuming and

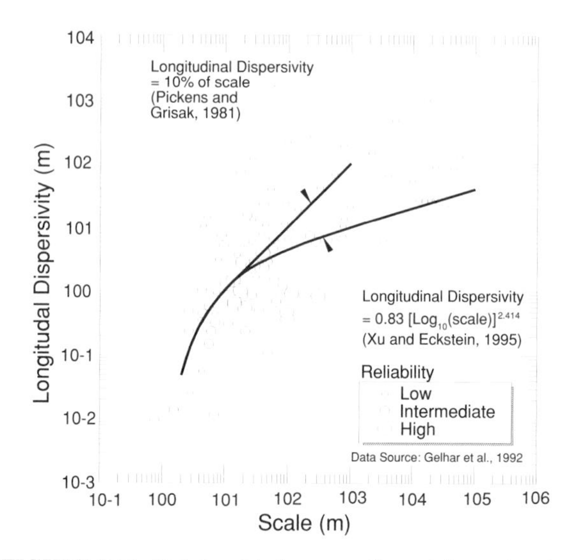

FIGURE 3.14. Relationship between dispersivity and scale.

costly. Another method commonly used to estimate dispersivity when implementing a solute transport model is to start with a longitudinal dispersivity of 0.1 times the plume length (Lallemand-Barres and Peaudecerf, 1978; Pickens and Grisak, 1981; Spitz and Moreno, 1996). This assumes that dispersivity varies linearly with scale. However, Xu and Eckstein (1995) evaluated the same data presented by Gelhar et al. (1992) and by using a weighted least-squares method, developed the following relationship for estimating dispersivity:

$$\alpha_x = 0.83(\log_{10} L_p)^{2.414} \tag{3.12}$$

where α_x is the longitudinal dispersivity [L] and L_p is scale (typically assumed to be plume length [L]). Both relationships are shown on Figure 3.14. In either case, the value derived for dispersivity will be an estimate at best, given the great variability in dispersivity for a given plume length. However, for modeling studies, an initial estimate is needed, and these relationships provide good starting points.

In addition to estimating longitudinal dispersivity, it may be necessary to estimate the transverse and vertical dispersivities (α_T and α_Z, respectively) for a given site. Several empirical relationships between longitudinal dispersivity and transverse and vertical dispersivities have been described. Commonly, α_T is estimated as $0.1\alpha_x$ [based on data from Gelhar et al. (1992)] or as $0.33\alpha_x$ (U.S. EPA, 1986; ASTM, 1995). Vertical dispersivity, α_Z, may be estimated as $0.05\alpha_x$ (ASTM, 1995) or as $0.025\alpha_x$ to $0.1\alpha_x$ (U.S. EPA, 1986) or 0.1 α_T (Newell et al., 1996).

Alternative Approach to Advection–Dispersion Modeling

Molz and Boman (1996) summarize the use of borehole flow meters to characterize the transmissive zones of an aquifer and suggest that the "potential impact of flow meters on site characterization and modeling is dramatic." The borehole flowmeter represents a generalization of a standard, fully penetrating pumping test, in which the standard pump test is extended by measuring the vertical flow distribution within the screened portion of the well. The test is relatively inexpensive, as it can be performed in an hour or two and produces only a few hundred gallons of groundwater. The end result is a detailed measurement of the vertical variation in horizontal hydraulic conductivity. This information can be applied as an alternative approach to conventional solute transport modeling.

In a conventional modeling study, the hydraulic conductivity in a transmissive zone is considered to be uniform, and dispersivity estimates on the order of feet, tens of feet, or more are used to account for the longitudinal spreading of the plume. Typically, the dispersivity estimates are obtained from scale-based rules of thumb or regression of measured dispersivities at sites, as discussed above.

Guven et al. (1992) suggested that plume migration is more accurately characterized by measuring the vertical variation in horizontal hydraulic conductivity, and then modeling the transmissive zone as many separate layers with little or no dispersion but with varying hydraulic conductivities. The authors describe how the separate-layer approach was able to simulate accurately the breakthrough curve generated by a two-well tracer test performed in Alabama, even though no dispersion was incorporated into the model.

With the Molz–Guven approach, knowledge of the vertical variation in horizontal hydraulic conductivity is used to replace dispersivity estimates generated based on plume length or scale, which can introduce inaccuracies during modeling. For example, inspection of the measured dispersivity at field sites versus scale relationship (Figure 3.14) shows greater than an order-of-magnitude scatter in the data and suggests that resolution of the conventional dispersivity modeling approach is limited (unless site-specific calibration is performed). While the Molz–Guven method may increase the accuracy of solute transport modeling, there are no data or studies that indicate the accuracy of the Molz–Guven approach as compared to the conventional advection–dispersion approach.

3.3 ONE-DIMENSIONAL ADVECTION–DISPERSION EQUATION

The one-dimensional advective transport component of the advection–dispersion equation is given by

$$\frac{\partial C}{\partial t} = -\upsilon_x \frac{\partial C}{\partial x} \tag{3.13}$$

where C = contaminant concentration [M/L^3]
t = time [T]
υ_x = average seepage velocity of groundwater [L/T]
x = distance along flow path [L]

Equation (3.13) considers only advective transport of the solute. In some cases this may be a fair approximation for simulating solute migration because advective transport is the main force behind contaminant migration. However, because of dispersion, diffusion, sorption, and biodegradation, this equation generally must be combined with the other components of the modified advection–dispersion equation to obtain an accurate mathematical description of solute transport.

The advection–dispersion equation is obtained by adding hydrodynamic dispersion to eq. (3.13). In one dimension the advection–dispersion equation is given by

$$\frac{\partial C}{\partial t} = D_x \frac{\partial^2 C}{\partial x^2} - \upsilon_x \frac{\partial C}{\partial x} \tag{3.14}$$

where C = contaminant concentration [M/L^3]
t = time [T]
D_x = hydrodynamic dispersion [L^2/T]
x = distance along flow path [L]
υ_x = average seepage velocity of groundwater [L/T]

Example 3.3: Natural Attenuation due to Advection and Dispersion in One Dimension An underground tank leaches an organic compound (benzene) continuously into a one-dimensional aquifer. Estimate the time taken for the benzene concentration to reach 100 mg/L at a distance of 750 ft from the source. Assume a hydraulic conductivity of 10^{-3} ft/sec, an effective porosity of 0.2, and an hydraulic gradient of 0.002 ft/ft. Also assume an initial concentration of 1000 mg/L and a longitudinal dispersivity of 75 ft. Ignore biodegradation.

SOLUTION: Bear (1961) solves the problem of one-dimensional transport assuming an infinite column (see Figure 3.15) with background concentration of zero and an input tracer concentration C_0 at $-\infty \leq x \leq 0$ for $t \geq 0$ at $x = L$ (length of column):

$$C(L,t) = \frac{C_0}{2}\left[\operatorname{erfc}\left(\frac{L - \upsilon_x t}{2\sqrt{D_x t}}\right) + \exp\left(\frac{\upsilon_x L}{D_x}\right)\right]\operatorname{erfc}\left(\frac{L + \upsilon_x t}{2\sqrt{D_x t}}\right)$$

where erfc is the complementary error function shown in Table 3.3.

Note that the second term of this equation can be neglected for most practical problems. The seepage velocity is calculated first using eq. (3.1):

$$\upsilon_x = \frac{10^{-3}(0.002)}{0.2} = 1(10^{-5})\text{ ft/sec} = 0.86\text{ ft/day}$$

The dispersivity is obtained using eq. (3.11):

$$D_x = 75(0.86) = 64.8\text{ ft}^2\text{/day}$$

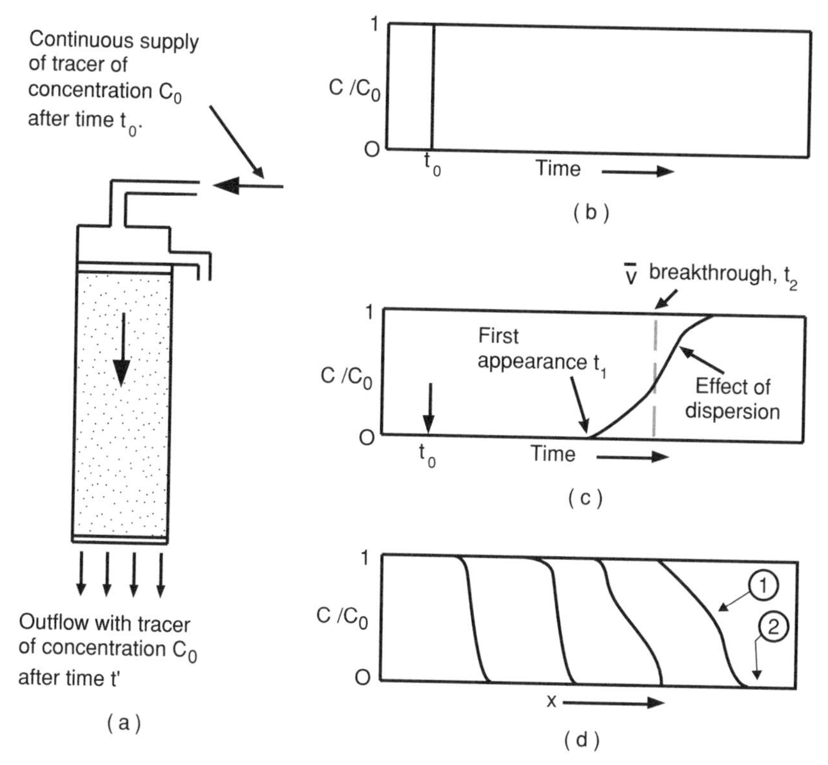

FIGURE 3.15. Longitudinal dispersion of a tracer passing through a column of porous medium: (*a*) column with steady flow and continuous supply of tracer after time t_0; (*b*) step-function tracer input relation; (*c*) relative tracer concentration in outflow from column (dashed line indicates plug flow condition and solid line illustrates affect of mechanical dispersion and molecular diffusion); (*d*) concentration profile in the column at various times. (From R. A. Freeze and J. A. Cherry, *Groundwater*, © 1979. Reprinted by permission of Prentice-Hall, Inc., Upper Saddle River, NJ.)

Using the equation above and ignoring the second term gives

$$C(L,t) = \frac{C_0}{2} \operatorname{erfc}\left(\frac{L - v_x t}{2\sqrt{D_x t}}\right)$$

$$100 = 500 \operatorname{erfc}\left(\frac{750 - 0.86t}{2\sqrt{64.8t}}\right)$$

$$\frac{750 - 0.86t}{2\sqrt{64.8t}} = 0.91$$

Using trial and error, t is approximately 490 days or 1.34 years.

TABLE 3.3. Error Functions[a]

x	erf(x)	erfc(x)	x	erf(x)	erfc(x)
0	0	1	1.1	0.880205	0.119795
0.05	0.056372	0.943628	1.2	0.910314	0.089686
0.1	0.112463	0.887537	1.3	0.934008	0.065992
0.15	0.167996	0.832004	1.4	0.952285	0.047715
0.2	0.222703	0.777297	1.5	0.966105	0.033895
0.25	0.276326	0.723674	1.6	0.976348	0.023652
0.3	0.328627	0.671373	1.7	0.983790	0.016210
0.35	0.379382	0.620618	1.8	0.989091	0.010909
0.4	0.428392	0.571608	1.9	0.992790	0.007210
0.45	0.475482	0.524518	2.0	0.995322	0.004678
0.5	0.520500	0.479500	2.1	0.997021	0.002979
0.55	0.563323	0.436677	2.2	0.998137	0.001863
0.6	0.603856	0.396144	2.3	0.998857	0.001143
0.65	0.642029	0.357971	2.4	0.999311	0.000689
0.7	0.677801	0.322199	2.5	0.999593	0.000407
0.75	0.711156	0.288844	2.6	0.999764	0.000236
0.8	0.742101	0.257899	2.7	0.999866	0.000134
0.85	0.770668	0.229332	2.8	0.999925	0.000075
0.9	0.796908	0.203092	2.9	0.999959	0.000041
0.95	0.820891	0.179109	3.0	0.999978	0.000022
1	0.842701	0.157299			

[a] $\text{erfc}(x) = 1 - \text{erf}(x)$; $\text{erfc}(-x) = 1 + \text{erf}(x)$; $\text{erf}(-x) = \text{erf}(x)$; $\text{erf}(x) = 2/\sqrt{\pi}\int_0^x e^{-u^2}\,du$.

3.4 SORPTION

Many organic contaminants, including chlorinated solvents and BTEX, are removed from solution by sorption onto the aquifer matrix. Sorption is the process in which dissolved contaminants partition from the groundwater and adhere to the particles comprising the aquifer matrix. Sorption of dissolved contaminants onto the aquifer matrix results in slowing (retardation) of the contaminant relative to the average seepage velocity of groundwater and a reduction in dissolved organic concentrations in groundwater. Sorption can also influence the relative importance of volatilization and biodegradation (Lyman et al., 1992). Figures 3.16 and 3.17 illustrate the effects of sorption on an advancing solute front.

Sorption is a reversible reaction; at a given solute concentration, some portion of the contaminant is partitioning out of solution onto the aquifer matrix, and some portion is desorbing and reentering solution. As solute concentrations change, the relative amounts of contaminant that are sorbing and desorbing will change. For example, as solute concentrations decrease (perhaps due to plume migration or solute biodegradation and dilution), the amount of contaminant reentering solution will probably increase. The affinity of a given compound for the aquifer matrix will not

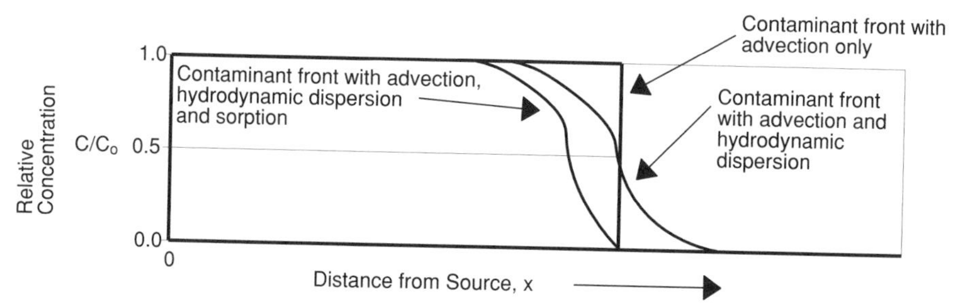

FIGURE 3.16. Breakthrough curve in one dimension showing plug flow with continuous source resulting from advection only; from the combined processes of advection and hydrodynamic dispersion; and from the combined processes of advection, hydrodynamic dispersion, and sorption.

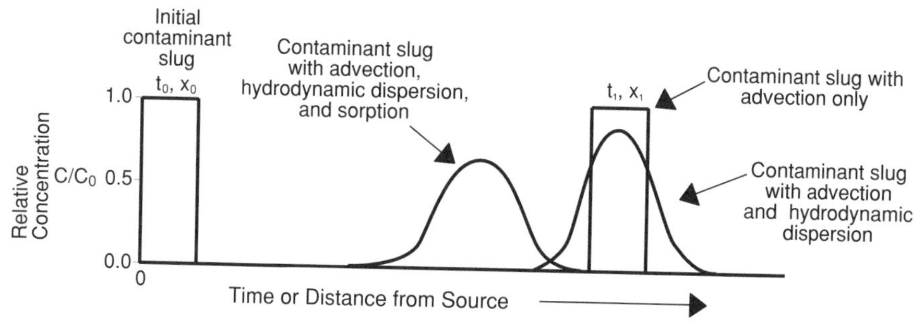

FIGURE 3.17. Breakthrough curve in one dimension showing plug flow with instantaneous source resulting from advection only; from the combined processes of advection and hydrodynamic dispersion; and from the combined processes of advection, hydrodynamic dispersion, and sorption.

be sufficient to isolate it permanently from groundwater, although for some compounds the rates of desorption may be so slow that the loss of mass may be considered permanent for the time scale of interest. Sorption therefore does not permanently remove solute mass from groundwater; it merely retards migration.

Mechanisms of Sorption

Sorption of dissolved contaminants is a complex phenomenon caused by several mechanisms, including London–van der Waals forces, Coulomb forces, hydrogen bonding, ligand exchange, chemisorption (covalent bonding between chemical and aquifer matrix), dipole–dipole forces, dipole-induced dipole forces, and hydrophobic forces. Because of their nonpolar molecular structure, hydrocarbons most commonly exhibit sorption through the process of hydrophobic bonding. When the surfaces comprising the aquifer matrix are less polar than the water molecule, as is generally the case, there is a strong tendency for the nonpolar contaminant molecules to partition from the groundwater and sorb to the aquifer matrix. This phenomenon referred to as *hydrophobic bonding*, is an important factor controlling the fate of many organic pollutants in soils (Devinny et al., 1990). Two components of an aquifer have the greatest effect on sorption: organic matter and clay minerals. In most aquifers, the organic fraction tends to control the sorption of organic contaminants.

Sorption Models and Isotherms

Regardless of the sorption mechanism, it is possible to determine the amount of sorption to be expected when a given dissolved contaminant interacts with the materials comprising the aquifer matrix. Bench-scale experiments are performed by mixing water–contaminant solutions of various concentrations with aquifer materials containing various amounts of organic carbon and clay minerals. The solutions are then sealed with no headspace and left until equilibrium between the various phases is reached. (True equilibrium may require hundreds of hours of incubation, but 80 to 90% of equilibrium may be achieved in one or two days.) The amount of contaminant left in solution is then measured.

The results are commonly expressed as a plot of the concentration of chemical sorbed (μg/g) versus the concentration remaining in solution (μg/L). The relationship between the concentration of chemical sorbed (C_a) and the concentration remaining in solution (C_l) at equilibrium is referred to as the *sorption isotherm* because the experiments are performed at constant temperature. Sorption isotherms generally exhibit one of three characteristic shapes, depending on the sorption mechanism. These isotherms are referred to as the Langmuir isotherm, the Freundlich isotherm, and the linear isotherm (a special case of the Freundlich isotherm).

Langmuir Sorption Model. The Langmuir model describes sorption in solute transport systems in which the sorbed concentration increases linearly with increasing solute concentration at low concentrations and approaches a constant value at high concentrations. The sorbed concentration approaches a constant value because there are a limited number of sites on the aquifer matrix available for contaminant sorption. This relationship is illustrated in Figure 3.18. The Langmuir equation is described mathematically as (Devinney et al., 1990)

$$C_a = \frac{KC_l b}{1 + KC_l} \tag{3.15}$$

where C_a = sorbed contaminant concentration (mass contaminant/mass soil)
K = equilibrium constant for the sorption reaction (μg/g)
C_l = dissolved contaminant concentration (μg/mL)
b = number of sorption sites (maximum amount of sorbed contaminant)

The Langmuir model is appropriate for highly specific sorption mechanisms in which there are a limited number of sorption sites. This model predicts a rapid

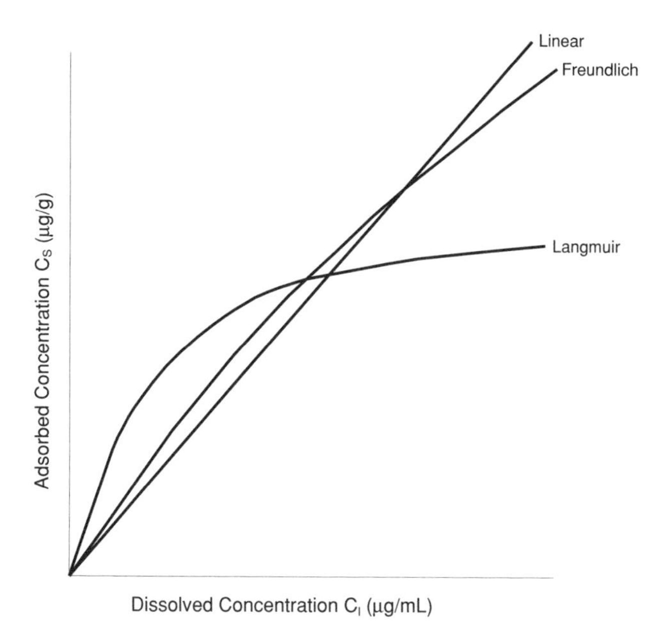

FIGURE 3.18. Characteristic adsorption isotherm shapes.

increase in the amount of sorbed contaminant as contaminant concentrations increase in a previously pristine area. As sorption sites become filled, the amount of sorbed contaminant reaches a maximum level equal to the number of sorption sites, b.

Freundlich Sorption Model. The Langmuir isotherm model can be modified if the number of sorption sites is large (assumed infinite) relative to the number of contaminant molecules. This is generally a valid assumption for dilute solutions (e.g., downgradient from a petroleum-hydrocarbon spill in the dissolved BTEX plume), in which the number of unoccupied sorption sites is large relative to contaminant concentrations. The Freundlich model is expressed mathematically as (Devinny et al., 1990)

$$C_a = K_d C_l^{1/n} \tag{3.16}$$

where C_a = sorbed contaminant concentration (mass contaminant/mass soil, μg/g)
K_d = distribution coefficient
C_l = dissolved concentration (mass contaminant/volume solution, μg/mL)
n = chemical-specific coefficient

The value of n in this equation is a chemical-specific quantity that is determined experimentally. Values of $1/n$ typically range from 0.7 to 1.1 but may be as low as 0.3 and as high as 1.7 (Lyman et al., 1992).

Linear Sorption Isotherm. The simplest expression of equilibrium sorption is the linear sorption isotherm, a special form of the Freundlich isotherm that occurs when the value of n is 1. The linear isotherm is valid for a dissolved species that is present at a concentration less than one-half of its solubility (Lyman et al., 1992). This is a valid assumption for BTEX compounds partitioning from fuel mixtures into groundwater. Dissolved BTEX concentrations resulting from this type of partitioning are significantly less than the pure compound's solubility in pure water. The linear sorption isotherm is expressed as (Jury et al., 1991)

$$C_a = K_d C_l \tag{3.17}$$

where C_a = sorbed contaminant concentration (mass contaminant/mass soil, μg/g)
K_d = distribution coefficient (slope of the isotherm, mL/g)
C_l = dissolved contaminant concentration (mass contaminant/volume solution, μg/mL)

The slope of the linear isotherm is the distribution coefficient, K_d, which is discussed in more detail in subsequent sections. The linear isotherm is the most commonly used model for sorption and natural attenuation studies.

Coefficient of Retardation and Retarded Contaminant Transport Velocity

As mentioned earlier, sorption tends to slow the transport velocity of contaminants dissolved in groundwater. When the average velocity of a dissolved contaminant is less than the average seepage velocity of the groundwater, the contaminant is said to be *retarded.* The coefficient of retardation, R, is used to estimate the retarded contaminant velocity. The difference between the velocity of the groundwater and that of the contaminant is caused by sorption and is described by the coefficient of retardation, R, which is defined as

$$R = \frac{\upsilon_x}{\upsilon_c} \tag{3.18}$$

where R = coefficient of retardation
υ_x = average seepage velocity of groundwater parallel to groundwater flow
υ_c = average velocity of contaminant parallel to groundwater flow

The ratio υ_x/υ_c describes the relative velocity between the groundwater and the dissolved contaminant. When $K_d = 0$ (no sorption), the transport velocities of the groundwater and the solute are equal ($\upsilon_x = \upsilon_c$). If it can be assumed that sorption is described adequately by the distribution coefficient [valid when the fraction of organic carbon (f_{oc}) > 0.001], the coefficient of retardation for a dissolved contaminant (for saturated flow) assuming linear sorption is determined from the distribution coefficient using the relationship

$$R = 1 + \frac{\rho_b K_d}{n} \tag{3.19}$$

where R = coefficient of retardation [dimensionless]
ρ_b = bulk density of aquifer [M/L^3]
K_d = distribution coefficient [L^3/M]
n = porosity [L^3/L^3]

The bulk density, ρ_b, of a soil is the ratio of the soil mass to its field volume. Bulk density is related to particle density by

$$\rho_b = (1 - n)\rho_s \tag{3.20}$$

where n is the total porosity and ρ_s is the density of grains comprising the aquifer. In sandy soils, ρ_b can be as high as 1.81 g/cm^3. In aggregated loams and clayey soils, ρ_b can be as low as 1.1 g/cm^3. Table 3.2 contains representative values of dry bulk density for common sediments and rocks.

The sorption relationship shown in eq. (3.19) expresses the coefficient of retardation in terms of the bulk density and effective porosity of the aquifer matrix and the distribution coefficient for the contaminant. Substitution of this equation into eq. (3.18) gives

$$\frac{\upsilon_x}{\upsilon_c} = 1 + \frac{\rho_b K_d}{n} \tag{3.21}$$

Solving for the contaminant velocity, υ_c, gives

$$\upsilon_c = \frac{\upsilon_x}{1 + \rho_b K_d / n} \tag{3.22}$$

Estimating the Distribution Coefficient

The most commonly used method for expressing the distribution of an organic compound between the aquifer matrix and the aqueous phase is the distribution coefficient, K_d, which is defined as the ratio of the sorbed contaminant concentration to the dissolved contaminant concentration:

$$K_d = \frac{C_a}{C_l} \tag{3.23}$$

where K_d = distribution coefficient (slope of the sorption isotherm, mL/g)
C_a = sorbed concentration (mass contaminant/mass soil or μg/g)
C_l = dissolved concentration (mass contaminant/volume solution or μg/mL)

The transport and partitioning of a contaminant is strongly dependent on the chemical's soil–water distribution coefficient and water solubility. The distribution coefficient is a measure of the sorption/desorption potential and characterizes the tendency of an organic compound to be sorbed to the aquifer matrix. The higher the distribution coefficient, the greater the potential for sorption to the aquifer matrix. The distribution coefficient is the slope of the sorption isotherm at the contaminant concentration of interest. The greater the amount of sorption, the greater the value of K_d. For systems described by a linear isotherm, K_d is a constant. In general terms the distribution coefficient is controlled by the hydrophobicity of the contaminant and the total surface area of the aquifer matrix available for sorption. Thus the distribution coefficient for a single compound will vary with the composition of the aquifer matrix. Because of their extremely high specific surface areas (ratio of surface area to volume), the organic carbon and clay mineral fractions of the aquifer matrix generally present the majority of sorption sites in an aquifer.

Based on the research efforts of Karickhoff et al. (1979), Ciccioli et al. (1980), Rodgers et al. (1980), and Shwarzenbach and Westall (1981), it appears that the primary adsorptive surface for organic chemicals is the organic fraction of the aquifer matrix. However, there is a *critical level of organic matter* below which sorption onto mineral surfaces is the dominant sorption mechanism (McCarty et al., 1981). The critical level of organic matter, below which sorption appears to be dominated by mineral–solute interactions, and above which sorption is dominated by organic carbon–solute interactions, is given by (McCarty et al., 1981)

$$f_{oc_c} = \frac{A_s}{200} \frac{1}{K_{ow}^{0.84}} \tag{3.24}$$

where f_{oc_c} = critical level of organic matter (mass fraction)
A_s = surface area of mineralogical component of aquifer matrix
K_{ow} = octanol–water partitioning coefficient

From this relationship it is apparent that the total organic carbon content of the aquifer matrix is less important for solutes with low octanol–water partitioning coefficients (K_{ow}). Also apparent is the fact that the critical level of organic matter increases as the surface area of the mineralogic fraction of the aquifer matrix increases. The surface area of the mineralogic component of the aquifer matrix is most strongly influenced by the amount of clay. For compounds with low K_{ow} values in materials with a high clay content, sorption to mineral surfaces could be an important factor causing retardation of the chemical.

Several researchers have found that if the distribution coefficient is normalized relative to the aquifer matrix total organic carbon content, much of the variation in observed K_d values between different soils is eliminated (Dragun, 1988). Distribution coefficients normalized to total organic carbon content are expressed as K_{oc}. The following equation gives the expression relating K_d to K_{oc}:

$$K_{oc} = \frac{K_d}{f_{oc}} \tag{3.25}$$

where K_{oc} = soil sorption coefficient normalized for total organic carbon content
K_d = distribution coefficient
f_{oc} = fraction total organic carbon (mg organic carbon/mg soil)

In areas with high clay concentrations and low total organic carbon concentrations, the clay minerals become the dominant sorption sites. Under these conditions, the use of K_{oc} to compute K_d might result in underestimating the importance of sorption in retardation calculations, a source of error that will make retardation calculations based on the total organic carbon content of the aquifer matrix more conservative. In fact, aquifers that have a high enough hydraulic conductivity to spread hydrocarbon contamination generally have a low clay content. In these cases the contribution of sorption to mineral surfaces is generally trivial.

Table 3.4 presents calculated retardation factors for several fuel- and chlorinated solvent-related chemicals as a function of the fraction of organic carbon content of

TABLE 3.4. Calculated Retardation Factors for Several Fuel- and Chlorinated Solvent-Related Chemicals[a]

		Fraction Organic Compound in Aquifer f_{oc}[c]			
Compound[b]	$\log(K_{oc})$	0.0001	0.001	0.01	0.1
Carbon tetrachloride	2.67	1.2	3.3	24.0	231.2
1,1,1-TCA	2.45	1.1	2.4	14.9	139.7
PCE	2.42	1.1	2.3	13.9	130.4
Xylenes (mixed isomers)*	2.38	1.1	2.2	12.8	119.1
Toluene*	2.13	1.1	1.7	7.6	67.4
Ethylbenzene*	1.98	1.0	1.5	5.7	48.0
1,1- or 1,2-DCA	1.76	1.0	1.3	3.8	29.3
Benzene*	1.58	1.0	1.2	2.9	19.7
trans-1,2,-DCE	1.42	1.0	1.1	2.3	13.9
cis-1,2-DCE	1.38	1.0	1.1	2.2	12.8
TCE	1.26	1.0	1.1	1.9	10.0
Chloroethane	1.25	1.0	1.1	1.9	9.8
Methylene chloride	1.23	1.0	1.1	1.8	9.4
MTBE*	1.08	1.0	1.1	1.6	6.9
Vinyl chloride	0.06	1.0	1.0	1.1	1.6

[a] Assumed porosity and bulk density: 0.35 and 1.72, respectively.
[b] Fuel-related chemicals indicated by an asterisk; all others are chlorinated solvents and daughter products.
[c] Units of f_{oc}: grams of naturally occurring organic carbon per gram of dry soil.

the soil. It can be seen from Table 3.4 that R can vary over two orders of magnitude at a site, depending on the chemical in question and the estimated value of porosity and soil bulk density. Earlier investigations reported distribution coefficients normalized to total organic matter content (K_{om}). The relationship between f_{om} and f_{oc} is nearly constant, and assuming that the organic matter contains approximately 58% carbon (Lyman et al., 1992),

$$K_{oc} = 1.724 K_{om} \tag{3.26}$$

Two methods used to estimate the distribution coefficient and amount of sorption (and thus retardation) for a given aquifer–contaminant system are presented below. The first method involves estimating the distribution coefficient by using K_{oc} for the contaminants and the fraction of organic carbon comprising the aquifer matrix. The second method involves conducting batch or column tests to determine the distribution coefficient. Because numerous authors have conducted experiments to determine K_{oc} values for common contaminants, literature values are reliable, and it generally is not necessary to conduct laboratory tests.

Determining the Coefficient of Retardation Using K_{oc}. Batch and column tests have been performed for a wide range of contaminant types and concentrations and aquifer conditions. Numerous studies have been performed using the results of these tests to determine if relationships exist that are capable of predicting the sorption characteristics of a chemical based on easily measured parameters. The results of these studies indicate that the amount of sorption is strongly dependent on the amount of organic carbon present in the aquifer matrix and the degree of hydrophobicity exhibited by the contaminant (Bailey and White, 1970; Karickhoff et al., 1979; Kenaga and Goring, 1980; Brown and Flagg, 1981; Schwarzenbach and Westall, 1981; Chiou et al., 1983; Hassett et al., 1983). These researchers observed that the distribution coefficient, K_d, was proportional to the organic carbon fraction of the aquifer times a proportionality constant. This proportionality constant, K_{oc}, is defined as given by eq. (3.26). In effect, eq. (3.26) normalizes the distribution coefficient to the amount of organic carbon in the aquifer matrix. Because it is normalized to organic carbon, values of K_{oc} are dependent only on the properties of the compound (not on the type of soil). Values of K_{oc} have been determined for a wide range of chemicals. Appendix B lists K_{oc} values for selected chlorinated compounds, BTEX, and TMB.

By knowing the value of K_{oc} for a contaminant and the fraction of organic carbon present in the aquifer, the distribution coefficient can be estimated using the relationship

$$K_d = K_{oc} f_{oc} \tag{3.27}$$

where K_d = distribution coefficient [L^3/M]
K_{oc} = soil adsorption coefficient for soil organic carbon content [L^3/M]
f_{oc} = fraction soil organic carbon (mg organic carbon/mg soil) [M/M]

The fraction of soil organic carbon must be determined from site-specific data. Representative values of the fraction of organic carbon (f_{oc}) in common sediments are given in Table 3.5. When using the method presented in this section to predict sorption of organic compounds, total organic carbon concentrations obtained from the most transmissive aquifer zone unaffected by contamination should be averaged and used for predicting sorption. This is because the majority of dissolved contaminant transport occurs in the most transmissive portions of the aquifer. In addition, because the most transmissive aquifer zones generally have the lowest total organic carbon concentrations, the use of this value will give a conservative prediction of contaminant sorption and retardation.

Determining the Coefficient of Retardation Using Laboratory Tests. The distribution coefficient may be quantified in the laboratory using batch or column tests. Batch tests are easier to perform than column tests. Although more difficult to perform, column tests generally produce a more accurate representation of field conditions than batch tests because continuous flow is involved. Knox et al. (1993) suggest using batch tests as a preliminary screening tool, followed by column studies to confirm the results of batch testing. If conducted properly, batch tests will yield

TABLE 3.5. Representative Values of Fraction of Organic Carbon for Common Sediments

Texture	Depositional Environment	Fraction Organic Carbon	Site Name
Medium sand	Fluvial-deltaic	0.00053–0.0012	Hill AFB, UT
Fine sand		0.0006–0.0015	Bolling AFB, DC
Fine- to medium-grained sand	Back-barrier (marine)	0.0026–0.007	Patrick AFB, FL
Organic silt and peat	Glacial	0.10–0.25	Elmendorf AFB, AK
Silty sand	Glacial	0.0007–0.008	Elmendorf AFB, AK
Silt with sand, gravel, and clay (glacial till)	Glacial	0.0017–0.0019	Elmendorf AFB, AK
Medium sand to gravel	Glacial	0.00125	Elmendorf AFB, AK
Loess (silt)	Eolian	0.00058–0.0016	Offutt AFB, NE
Fine to medium sand	Glaciofluvial or glaciolacustrine	< 0.0006–0.0061	Truax Field, Madison, WI
Fine to medium sand	Glacial	0.00021–0.019	King Salmon AFB, Fire Training Area, AK
Fine to coarse sand	Glacial	0.00029–0.073	Battle Creek ANGB, MI
Sand	Fluvial	0.0057	Oconee River, GA[a]
Coarse silt	Fluvial	0.029	Oconee River, GA[a]
Medium silt	Fluvial	0.020	Oconee River, GA[a]
Fine silt	Fluvial	0.0226	Oconee River, GA[a]
Silt	Lacustrine	0.0011	Wildwood, Ontario[b]
Fine sand	Glaciofluvial	0.00023–0.0012	Various sites in Ontario[b]
Medium sand to gravel	Glaciofluvial	0.00017–0.00065	Various sites in Ontario[b]

[a]From Karickhoff (1981).

[b]From Domenico and Schwartz (1990).

sufficiently accurate results for fate and transport modeling purposes provided that sensitivity analyses for retardation are conducted during the modeling.

Batch testing involves adding uncontaminated aquifer material to a number of vessels, adding solutions prepared using uncontaminated groundwater from the site mixed with various amounts of contaminants to produce varying solute concentrations, sealing the vessel and shaking it until equilibrium is reached, analyzing the solute concentration remaining in solution, and calculating the amount of contaminant sorbed to the aquifer matrix using mass balance calculations. A plot of the concentration of contaminant sorbed versus dissolved equilibrium concentration is then made using the data for each reaction vessel. The slope of the line formed by connecting each data point is the distribution coefficient. The temperature should be held constant during the batch test and should approximate that of the aquifer system through which solute transport is taking place.

Table 3.6 contains data from a hypothetical batch test. These data are plotted (Figure 3.19) to obtain an isotherm unique to the aquifer conditions at the site. A regression analysis can then be performed on these data to determine the distribution coefficient. For linear isotherms, the distribution coefficient is simply the slope of the isotherm. In this example, $K_d = 0.0146$ L/g. Batch-testing procedures are described in detail by Roy et al. (1992).

Column testing involves placing uncontaminated aquifer matrix material in a laboratory column and passing solutions through the column. Solutions are prepared by mixing uncontaminated groundwater from the site with the contaminants of interest and a conservative tracer. Flow rate and time are accounted for, and samples are taken periodically from the effluent end of the column and analyzed to determine contaminant and tracer concentrations. Breakthrough curves are prepared for the contaminants by plotting chemical concentration versus time (or relative concentration versus number of pore volumes). The simplest way to determine the coefficient of retardation (or the distribution coefficient) from the breakthrough curves is to determine the time required for the effluent concentration to

TABLE 3.6. Data from Hypothetical Batch-Test Experiment

Initial Concentration (μg/L)	Equilibrium Concentration (μg/L)	Weight of Solid Matrix (g)	Sorbed Concentration[a] (μg/g)
250	77.3	20.42	1.69
500	150.57	20.42	3.42
1000	297.04	20.42	6.89
1500	510.1	20.42	9.70
2000	603.05	20.42	13.68
3800	1198.7	20.42	25.48
6000	2300.5	20.42	36.23
9000	3560.7	20.42	53.27

[a] Adsorbed concentration = [(initial concentration−equilibrium concentration × volume of solution)]/weight of solid matrix.

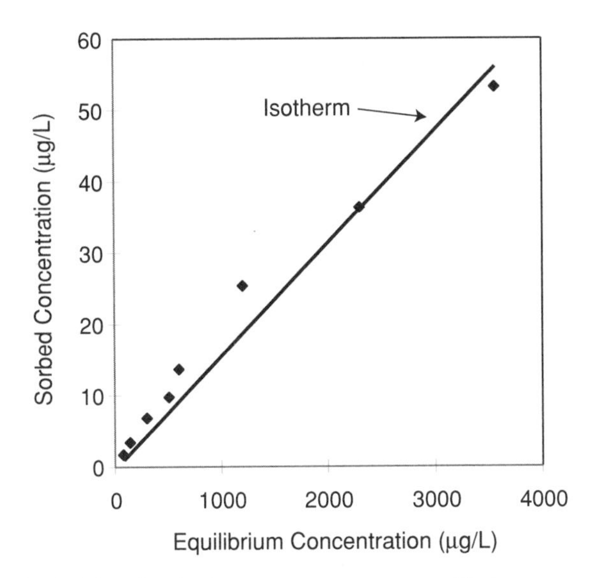

FIGURE 3.19. Plot of sorbed concentration versus equilibrium concentration.

equal 0.5 of the influent concentration. This value can be used to determine average velocity of the center of mass of the contaminant. The retardation factor is determined by dividing the average flow velocity through the column by the velocity of the center of mass of the contaminant. The value thus obtained is the retardation factor. The coefficient of retardation also can be determined by curve fitting using the CXTFIT model of Parker and van Genuchten (1984). Breakthrough curves can also be made for the conservative tracer. These curves can be used to determine the coefficient of dispersion by curve fitting using the model of Parker and van Genuchten (1984).

As described for the K_{oc} method, aquifer samples should be obtained from the most transmissive aquifer zone when using the method presented in this section to predict sorption of the BTEX compounds. This is because the majority of dissolved contaminant transport occurs in the most transmissive portions of the aquifer, and the most transmissive aquifer zones generally have the lowest organic carbon concentrations. Therefore, use of these materials will give a conservative prediction of contaminant sorption and retardation.

Example 3.4: Retarded Solute Transport Velocity Calculation For groundwater flow and solute transport occurring in a shallow, saturated, well-sorted, fine-grained sandy aquifer with an f_{oc} content of 0.7%, a hydraulic gradient of –0.015 m/m, and a hydraulic conductivity of 25 m/day, calculate the retarded contaminant velocity for TCE.

SOLUTION: Because the total porosity, effective porosity, and the bulk density are not given, values for these parameters are obtained from Table 3.2. The median values for total porosity, effective porosity, and bulk density are approximately 0.4, 0.2, and 1.6 kg/L, respectively.

The first step is to calculate the average seepage velocity of groundwater, υ_x:

$$\upsilon_x = -\frac{(25\text{ m/day})(-0.015\text{ m/m})}{0.2} = 1.88\frac{\text{m}}{\text{day}}$$

The next step is to determine the distribution coefficient, K_d. Values of K_{oc} for chlorinated solvents and BTEX are obtained from Appendix B. For TCE, the most conservative value (i.e., that value giving the highest solute velocity) is $K_{oc} = 87$ L/kg, and [using eq. (3.27)]:

$$K_d = \left(87\frac{\text{L}}{\text{kg}}\right)(0.007) = 0.61\frac{\text{L}}{\text{kg}}$$

The retarded contaminant velocity is given by [eq. (3.19)]:

$$\upsilon_c = \frac{1.88\text{ m/day}}{1 + (1.6\text{ kg/L})(0.61\text{ L/kg})/0.4} = 0.55\text{ m/day}$$

This example illustrates that contaminant sorption to TOC can have a profound influence on contaminant transport by slowing the rate of dissolved contaminant migration significantly.

One-Dimensional Advection–Dispersion Equation with Retardation

In one dimension the advection–dispersion equation is given by

$$R\frac{\partial C}{\partial t} = D_x\frac{\partial^2 C}{\partial x^2} - \upsilon_x\frac{\partial C}{\partial x} \tag{3.28}$$

where
- R = coefficient of retardation [dimensionless]
- C = contaminant concentration [M/L^3]
- t = time [T]
- D_x = hydrodynamic dispersion [L^2/T]
- x = distance along flow path [L]
- υ_x = average seepage velocity of groundwater [L/T]

3.5 DILUTION (RECHARGE)

Groundwater recharge can be defined as the entry into the saturated zone of water made available at the water table surface (Freeze and Cherry, 1979). In recharge areas, flow near the water table is generally downward. Recharge defined in this manner may therefore include not only precipitation that infiltrates through the vadose zone, but water entering the groundwater system via discharge from surface water bodies (e.g., streams and lakes). Where a surface water body is in contact with or is part of the groundwater system, the definition of recharge above is stretched slightly. However, such bodies often are referred to as *recharging lakes* or *streams*. Recharge of a water table aquifer has two effects on the natural attenuation of a dissolved contaminant plume. Additional water entering the system due to infiltration of precipitation or from surface water will contribute to dilution of the plume, and the influx of relatively fresh, electron-acceptor-charged water will alter geochemical processes, and in some cases facilitate additional biodegradation.

Recharge from infiltrating precipitation is the result of a complex series of processes in the unsaturated zone. Description of these processes is beyond the scope of this discussion; however, it is worth noting that the infiltration of precipitation through the vadose zone brings the water into contact with the soil and thus may allow dissolution of additional electron acceptors and possibly organic soil matter (a potential source of electron donors). Infiltration therefore provides fluxes of water, inorganic species, and possibly organic species into the groundwater. Recharge from surface water bodies occurs when the hydraulic head of the body is greater than that of the adjacent groundwater. The surface water may be a connected part of the groundwater system, or it may be perched above the water table. In either case, the water entering the groundwater system will not only aid in dilution of a contaminant plume, but it may also add electron acceptors and possibly electron donors to the groundwater.

An influx of electron acceptors will tend to increase the overall electron-accepting capacity within the contaminant plume. In addition to the inorganic electron acceptors that may be dissolved in the recharge (e.g., dissolved oxygen, nitrate, or sulfate), the introduction of water with different geochemical properties may foster geochemical changes in the aquifer. For example, iron(II) will be oxidized back to iron(III). Vroblesky and Chapelle (1994) present data from a site where a major rainfall event introduced sufficient dissolved oxygen into the contaminated zone to cause reprecipitation of iron(III) onto mineral grains. This reprecipitation made iron(III) available for reduction by microorganisms, thus resulting in a shift from methanogenesis back to iron(III) reduction (Vroblesky and Chapelle, 1994). Such a shift may be beneficial for biodegradation of compounds used as electron donors, such as fuel hydrocarbons or vinyl chloride. However, these shifts can also make conditions less favorable for reductive dehalogenation.

Evaluating the effects of recharge can be difficult. The effects of dilution might be estimated if one has a detailed water budget for the system in question. However, if a plume has a significant vertical extent, it cannot be known with any certainty what proportion of the plume mass is being diluted by the recharge. Moreover, because dispersivity, sorption, and biodegradation are often not well quantified, separating out the effects of dilution may be very difficult indeed. Where recharge enters the groundwater from precipitation, the effects of the addition of electron acceptors may be qualitatively apparent due to elevated electron acceptor concentrations or differing patterns in electron acceptor consumption or by-product formation in the area of the recharge. However, the effects of short-term variations in such a system (which are likely due to the intermittent nature of precipitation events in most climates) may not easily be understood. Where recharge is from surface water, the influx of mass and electron acceptors is more steady over time. In this scenario, quantifying the effects of dilution may be less uncertain, and the effects of electron acceptor replenishment may be more easily identified (although not necessarily quantified).

In some cases the effects of recharge-diluting contaminant plumes can be estimated with a simple relationship based on the specific discharge of groundwater passing through the point of interest and the amount of recharge entering the plume area. Note that at most sites, recharge will not actually mix with groundwater in an aquifer but will form a stratified layer on top due to the very low amount of vertical dispersion that is characteristic of aquifer systems. Mixing can be assumed in some cases, such as (1) a very thin, unconfined aquifer (perhaps on the order of a few feet thick); (2) a groundwater well is screened in such a way that its samples are drawn from both the stratified recharge-related aquifer and the deeper zone containing groundwater flowing past a source zone; or (3) the aquifer discharges into a surface water body, and the groundwater associated with the recharge is assumed to be mixed with the original groundwater flowing past a source zone.

The relationship for estimating the amount of dilution caused by recharge is

$$C_L = C_0 \exp\left(-\frac{RWL/V_D}{WT_h V_D}\right) \tag{3.29}$$

Eliminating the width and rearranging gives

$$C_L = C_0 \exp\left(-\frac{RL}{T_h (V_D)^2}\right) \tag{3.30}$$

where C_L = concentration at distance L from origin assuming complete mixing of recharge with groundwater (mg/L)
C_0 = concentration at origin or at distance L = 0 (mg/L)
R = recharge mixing with groundwater (ft/yr)
W = width of area where recharge is mixing with groundwater (ft)
L = length of area where recharge is mixing with groundwater (ft)
V_D = Darcy velocity of groundwater (ft/yr)
T_h = thickness of aquifer where groundwater flow is assumed to mix completely with recharge (ft)

For example, consider a 3-ft-thick aquifer, flowing at 100 ft/yr Darcy velocity, that contains 2 mg/L TCE. If the groundwater in this aquifer is assumed to be mixed with recharge occurring at 0.5 ft/yr, dilution from recharge at L = 1000 ft will result in the following concentration:

$$C_L = 2\exp\left[\frac{-0.5(1000)}{3(100^2)}\right] \quad \text{or} \quad C_L = 1.967 \text{ mg/L}$$

Note that this relationship is an approximation but will give good results if the amount of recharge is less than 50% of original groundwater flow. The key assumption for this relationship is that the recharge water is mixed completely with the groundwater, a process that is limited to a few special cases (see examples above) because dispersion in aquifers is orders of magnitude smaller compared to the dispersion in well-mixed systems (such as rivers or tanks). For a simple relationship to estimate recharge, see Chapter 2.

3.6 VOLATILIZATION

Although not a destructive attenuation mechanism, volatilization does remove contaminants from groundwater. In general, factors affecting the volatilization of contaminants from groundwater into soil gas include the contaminant concentration, the change in contaminant concentration with depth, the Henry's law constant and diffusion coefficient of the compound, mass transport coefficients for the contaminant in both water and soil gas, sorption, and the temperature of the water (Larson and Weber, 1994).

Partitioning of a contaminant between the liquid phase and the gaseous phase is governed by Henry's law. Thus the Henry's law constant of a chemical determines its tendency to volatilize from groundwater into the soil gas. Henry's law states that the concentration of a contaminant in the gaseous phase is directly proportional to the compound's concentration in the liquid phase and is a constant characteristic of the compound. Stated mathematically, Henry's law is given by (Lyman et al., 1992)

$$C_a = HC_l \tag{3.31}$$

where C_a = concentration in air (atm)
H = Henry's law constant (atm · m^3/mol)
C_l = concentration in water (mol/m^3)

Henry's law constants for chlorinated and petroleum hydrocarbons range over several orders of magnitude. For petroleum hydrocarbons at 25°C, Henry's law constants (H values) for the saturated aliphatics range from 1 to 10 atmospheres-cubic meter per mole (atm · m^3/mol); for the unsaturated and cycloaliphatics, H values range from 0.1 to 1 atm · m^3/mol; and for the light aromatics (e.g., BTEX), H's range from 0.007 to 0.02 atm · m^3/mol (Lyman et al., 1992). Values of Henry's law constants for selected chlorinated solvents and the BTEX compounds are given in Appendix B. As indicated in the appendix table, values of H for chlorinated compounds also vary over several orders of magnitude, although most are similar to those for BTEX compounds.

The physiochemical properties of chlorinated solvents, with the exception of vinyl chloride, and the BTEX compounds give them low Henry's law constants. Because of the small surface area of the groundwater flow system exposed to soil gas, volatilization of chlorinated solvents and BTEX compounds from groundwater is a relatively slow process that in the interest of being conservative, generally can be ignored when modeling biodegradation. Chiang et al. (1989) demonstrated that less than 5% of the mass of dissolved BTEX is lost to volatilization in the saturated groundwater environment. Moreover, Rivett (1995) observed that for plumes more than about 1 m below the air–water interface, little, if anything, of solvent concentrations will be detectable in soil gas, due to the downward groundwater gradient in the vicinity of the water table. This suggests that for portions of plumes more than 1 m below the water table, very little, if any, mass will be lost due to volatilization. In addition, vapor transport across the capillary fringe can be very slow (McCarthy and Johnson, 1992), further limiting mass transfer rates. Because of this, the impact of volatilization on dissolved contaminant reduction can generally be assumed to be negligible.

3.7 HYDROLYSIS AND DEHYDROHALOGENATION

Hydrolysis and dehydrohalogenation reactions are the most thoroughly studied abiotic attenuation mechanisms. In general, the rates of these reactions are often quite slow within the range of normal groundwater temperatures, with half-lives of days to centuries (Vogel et al., 1987; Vogel, 1994). Therefore, most information about the rates of these reactions is extrapolated from experiments run at higher temperatures to allow the experiments to be performed within a practical time frame.

Hydrolysis

Hydrolysis is a substitution reaction in which a compound reacts with water and a halogen substituent is replaced with a hydroxyl (OH^-) group. Hydrolysis of organic compounds frequently results in the formation of alcohols and alkenes (Knox et al., 1993 after Johnson et al., 1989):

$$RX + HOH \Rightarrow ROH + HX \tag{3.32}$$

$$H_3C{-}CH_2X \Rightarrow H_2C{=}CH_2 + HX \tag{3.33}$$

Hydrolysis results in reaction products that may be more susceptible to biodegradation, as well as more soluble (Neely, 1985). As such, hydrolysis may be a significant natural attenuation mechanism.

The likelihood that a halogenated solvent will undergo hydrolysis depends in part on the number of halogen substituents. More halogen substituents on a compound will decrease the chance for hydrolysis reactions to occur (Vogel et al., 1987) and will therefore decrease the rate of the reaction. In addition, bromine substituents are more susceptible to hydrolysis than chlorine substituents (Vogel et al., 1987); for example, 1,2-dibromoethane is subject to significant hydrolysis reactions under natural conditions. McCarty (1996) lists 1,1,1-TCA as the only major chlorinated solvent that can be transformed chemically through hydrolysis (as well as elimination), leading to the formation of 1,1-DCE and acetic acid. Based on data from Cline and Delfino (1989) and Haag and Mill (1988), McCarty estimates that 20% of TCA is converted to 1,1-DCE, while 80% is transformed into acetic acid. The 20% converted to 1,1-DCE is significant, however, because 1,1-DCE is considered more toxic than TCA.

Locations of the halogen substituent on the carbon chain also may have some effect on the rate of reaction. The rate also may increase with increasing pH; however, rate dependence upon pH typically is not observed below a pH of 11 (Mabey and Mill, 1978; Vogel and Reinhard, 1986). Rates of hydrolysis may also be increased by the presence of clays, which can act as catalysts (Vogel et al., 1987). Other factors that affect the level of hydrolysis include dissolved organic matter and dissolved metal ions.

Hydrolysis rates can generally be described using first-order kinetics, particularly in solutions in which water is the dominant nucleophile (Vogel et al., 1987). The data shown in Figure 3.20 exhibit a first-order rate of hydrolysis for 1,2,4-trichlorobenzene of 4.3×10^{-3} hr^{-1}. However, this oversimplifies what is typically a much more complicated relationship (Neely, 1985). As noted earlier, reported rates of environmentally significant hydrolysis reactions involving chlorinated solvents are typically the result of extrapolation from experiments performed at higher temperatures (Mabey and Mill, 1978; Vogel, 1994).

Hydrolysis of chlorinated methanes and ethanes has been well demonstrated in the literature. Vogel (1994) reports that monohalogenated alkanes have half-lives on the order of days to months, while polychlorinated methanes and ethanes have half-lives that may range up to thousands of years (e.g., for carbon tetrachloride). As the number of chlorine atoms increases, dehydrohalogenation may become more important (Jeffers et al., 1989). Butler and Barker (1996) note that chlorinated ethenes do not undergo significant hydrolysis reactions (i.e., the rates are slow). Butler and Barker also reported that they were unable to find any studies on hydrolysis of vinyl chloride. A list of half-lives for abiotic hydrolysis and dehydrohalogenation of some

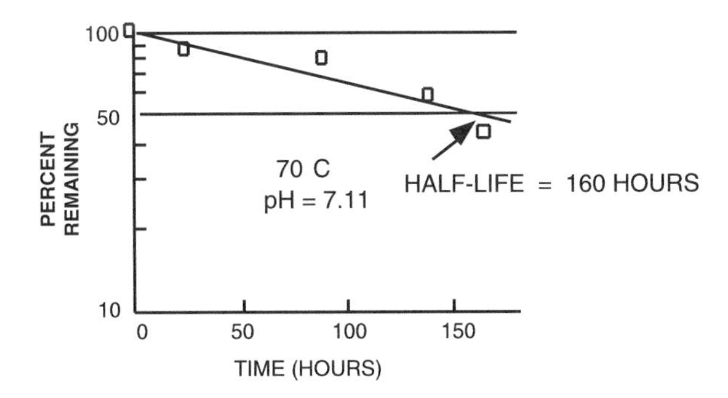

FIGURE 3.20. Hydrolysis data for 1,2,4-trichlorobenzene. (From Johnson et al., 1989.)

chlorinated solvents is presented in Table 3.7. Note that no distinctions are made in the table as to which mechanism is operating; this is consistent with the references from which the table has been derived (Vogel et al., 1987; Butler and Barker, 1996).

One common chlorinated solvent for which abiotic transformations have been well studied is 1,1,1-TCA. 1,1,1-TCA may be transformed to acetic acid abiotically through a series of substitution reactions, including hydrolysis. In addition, 1,1,1-TCA may be reductively dehalogenated to form 1,1-DCA and then chloroethane (CA), which is then hydrolyzed to ethanol (Vogel and McCarty, 1987) or dehydrohalogenated to vinyl chloride (Jeffers et al., 1989). Rates of these reactions have been studied by several parties, as summarized in Table 3.7.

Dehydrohalogenation

Dehydrohalogenation is an elimination reaction involving halogenated alkanes in which a halogen is removed from one carbon atom, followed by subsequent removal of a hydrogen atom from an adjacent carbon atom. In this two-step reaction, an alkene is produced. Although the oxidation state of the compound decreases due to the removal of a halogen, the loss of a hydrogen atom increases it. This results in no

TABLE 3.7. Approximate Half-Lives of Abiotic Hydrolysis and Dehydrohalogenation Reactions Involving Chlorinated Solvents

Compound	Half-Life (yrs)	Product
Chloromethane	No data	
Methylene chloride (dichloromethane)	704[a]	
Trichloromethane (chloroform)	3500[a], 1800[b]	
Carbon tetrachloride	41[b]	
Chloroethane	0.12[c]	Ethanol
1,1-Dichloroethane	61[b]	
1,2-Dichloroethane	72[b]	
1,1,1-Trichloroethane	1.7[a], 1.1[b], 2.5[d]	Acetic acid 1,1-DCE
1,1,2-Trichloroethane	140[b], 170[a]	1,1-DCE
1,1,1,2-Tetrachloroethane	47[b], 380[a]	TCE
1,1,2,2-Tetrachloroethane	0.3[e], 0.4[b], 0.8[a]	1,1,2-TCA TCE
Tetrachloroethene	1.3×10^{6} [b]	
Trichloroethene	1.3×10^{6} [b]	
1,1-Dichloroethene	1.2×10^{8} [b]	
1,2-Dichloroethene	2.1×10^{10} [b]	

[a]From Mabey and Mill (1978).
[b]From Jeffers et al. (1989).
[c]From Vogel et al. (1987).
[d]From Vogel and McCarty (1987).
[e]From Cooper et al. (1987).
[f]From Dilling et al. (1975).

external electron transfer, and there is no net change in the oxidation state of the reacting molecule (Vogel et al., 1987). Contrary to the patterns observed for hydrolysis, the likelihood that dehydrohalogenation will occur increases with the number of halogen substituents. It has been suggested that under normal environmental conditions, monohalogenated aliphatics apparently do not undergo dehydrohalogenation, and these reactions are apparently not likely to occur (March, 1985; Vogel et al., 1987). However, Jeffers et al. (1989) report on the dehydrohalogenation of CA to vinyl chloride. Polychlorinated alkanes have been observed to undergo dehydrohalogenation under normal conditions and extremely basic conditions (Vogel et al., 1987). As with hydrolysis, bromine substituents are more reactive with respect to dehydrohalogenation.

Dehydrohalogenation rates also may be approximated using pseudo-first-order kinetics. Once again, this is not truly a first-order reaction, but such approximations have been used in the literature to quantify the reaction rates. The rates will depend not only on the number and types of halogen substituent, but also on the hydroxide ion concentration. Under normal pH conditions (i.e., near a pH of 7), interaction with water (acting as a weak base) may become more important (Vogel et al., 1987). Transformation rates for dehydrohalogenation reactions are presented in Table 3.7.

The organic compound 1,1,1-TCA is also known to undergo dehydrohalogenation (Vogel and McCarty, 1987). In this case, TCA is transformed to 1,1-DCE, which is then reductively dehalogenated to vinyl chloride. The VC is then either reductively dehalogenated to ethene or consumed as a substrate in an aerobic reaction and converted to carbon dioxide. In a laboratory study, Vogel and McCarty (1987) reported that the abiotic conversion of 1,1,1-TCA to 1,1-DCE has a rate constant of about 0.04 $year^{-1}$. It was noted that this result was longer than indicated in previous studies, but that experimental methods differed. Jeffers et al. (1989) reported on several other dehydrohalogenation reactions; in addition to 1,1,1-TCA and 1,1,2-TCA, both degrading to 1,1-DCE, the tetrachloroethanes and pentachloroethanes degrade to TCE and PCE, respectively. As noted previously, Jeffers et al. (1989) also report that CA may degrade to vinyl chloride, but no information on rates was reported in the general literature.

3.8 REDUCTION REACTIONS

Two abiotic reductive dechlorination reactions that may operate in the subsurface are hydrogenolysis and dihaloelimination. *Hydrogenolysis* is the simple replacement of a chlorine (or another halogen) by a hydrogen, while *dihaloelimination* is the removal of two chlorines (or other halogens) accompanied by the formation of a double carbon–carbon bond. Butler and Barker (1996) review work by Criddle et al. (1986), Jafvert and Wolfe (1987), Acton (1990), and Reinhard et al. (1990). This review suggests that although these reactions are thermodynamically possible under reducing conditions, they often do not take place in the absence of biological activity, even if such activity is only indirectly responsible for the reaction. While not involved in a manner similar to that for cometabolism, microbes may produce reductants that facilitate such reactions in conjunction with minerals in the aquifer matrix, as has been suggested by work utilizing aquifer material from the Borden test site (Reinhard et al., 1990). Moreover, the reducing conditions necessary to produce such reactions most often are created as a result of microbial activity. It is therefore not clear if some of these reactions are truly abiotic, or if because of their reliance on microbial activity to produce reducing conditions or reactants, they should be considered to be a form of cometabolism.

In some cases, truly abiotic reductive dechlorination has been observed; however, the conditions that favor such reactions may not occur naturally. For example, Gillham and O'Hannesin (1994) describe reductive dehalogenation of chlorinated aliphatics using zero-valent iron, in which the iron serves as an electron donor in an electrochemical reaction. However, this is not a natural process. Wang and Tan (1990) reported reduction of TCE to ethene and carbon tetrachloride to methane during a platinum-catalyzed reaction between elemental magnesium and water. Given that the metals involved in these reactions are unlikely to occur naturally in the reduced forms used in the aforementioned work, such processes are not likely to contribute to natural attenuation of chlorinated solvents.

REFERENCES

Acton, D. W., 1990, Enhanced in situ biodegradation of aromatic and chlorinated aliphatic hydrocarbons in anaerobic, leachate-impacted groundwaters, M.Sc. thesis, University of Waterloo, Waterloo, Ontario.

American Society for Testing and Materials (ASTM), 1995, *Emergency Standard Guide for Risk-Based Corrective Action Applied at Petroleum Release Sites*, ASTM E-1739, ASTM, Philadelphia.

Bailey, G. W., and White, J. L., 1970, Factors influencing the adsorption, desorption, and movement of pesticides in soil, in *Residue Reviews*, F. A. Gunther and J. D. Gunther, eds., Springer Verlag, New York, pp. 29–92.

Bear, J., 1961, Some experiments on dispersion, *J. Geophys. Res.*, vol. 66, no. 8, pp. 2455–2467.

Bouwer, H., 1989, The Bouwer and Rice slug test: an update, *Ground Water*, vol. 27, no. 3, pp. 304–309.

Bouwer, H., and Rice, R. C., 1976, A slug test for determining hydraulic conductivity of unconfined aquifers with completely or partially penetrating wells, *Water Resour. Res.*, vol. 12, no. 3, pp. 423–428.

Brown, D. S., and Flagg, E. W., 1981, Empirical prediction of organic pollutant sorption in natural sediments, *J. Environ. Qual.*, vol. 10, no. 3, pp. 382–386.

Butler, B. J., and Barker, J. F., 1996, Chemical and microbiological transformation and degradation of chlorinated solvent compounds, in, *Dense Chlorinated Solvents and Other DNAPLs in Ground Water: History, Behavior, and Remediation*, J. F. Pankow, and J. A. Cherry, eds., Waterloo Press, Waterloo, Ontario, Canada, pp. 267–312.

Chiang, C. Y., Salanitro, J. P., Chai, E. Y., Colthart, J. D., and Klein, C. L., 1989, Aerobic biodegradation of benzene, toluene, and xylene in a sandy aquifer: data analysis and computer modeling, *Ground Water*, vol. 27, no. 6, pp. 823–834.

Chiou, C. T., Porter, P. E., and Schmedding, D. W., 1983, Partition equilibria of nonionic organic compounds between soil organic matter and water, *Environ. Sci Technol.*, vol. 17, no. 4, pp. 227–231.

Ciccioli, P., Cooper, W. T., Hammer, P. M., and Hayes, J. M., 1980, Organic solute–mineral surface interactions: a new method for the determination of groundwater velocities, *Water Resour. Res.*, vol. 16, no. 1, pp. 217–223.

Cline, P. V., and Delfino, J. J., 1989, Transformation kinetics of 1,1,1-trichloroethane to the stable product 1,1-dichloroethene, in *Biohazards of Drinking Water Treatment*, Lewis Publishers, Boca Raton, FL, pp. 47–56.

Cooper, W. J., Mehran, M., Riusech, D. J., and Joens, J. A., 1987, Abiotic transformation of halogenated organics: 1. Elimination reaction of 1,1,2,2-tetrachloroethane and formation of 1,1,2-trichloroethane, *Environ. Sci. Technol.*, vol. 21, pp. 1112–1114.

Criddle, C. S., McCarty, P. L., Elliot, M. C., and Barker, J. F., 1986, Reduction of hexachloroethane to tetrachloroethylene in groundwater, *J. Contam. Hydrol.*, vol. 1, pp. 133–142.

Davis, R. K., Pederson, D. T., Blum, D. A., and Carr, J. D., 1993, Atrazine in a stream-aquifer system: estimation of aquifer properties from atrazine concentration profiles, *Ground Water Monit. Rev.*, Spring, pp. 134–141.

Dawson, K. J., and Istok, J. D., 1991, *Aquifer Testing: Design and Analysis of Pumping and Slug Tests*, Lewis Publishers, Boca Raton, FL, 344 pp.

Devinny, J. S., Everett, L. G., Lu, J. C. S., and Stollar R. L., 1990, *Subsurface Migration of Hazardous Wastes*, Van Nostrand Reinhold, New York, 387 pp.

Dilling, W. L., Tfertiller, N. B., and Kallos, G. J., 1975, Evaporation rates and reactivities of methylene chloride, chloroform, 1,1,1-trichloroethane, trichloroethylene, tetrachloroethylene, and other chlorinated compounds in dilute aqueous solutions, *Environ. Sci. Technol.*, vol. 9, pp. 833–838.

Domenico, P. A., and Schwartz, F. W., 1990, *Physical and Chemical Hydrogeology*, Wiley, New York, 824 pp.

Dragun, J., 1988, *The Soil Chemistry of Hazardous Materials*, Hazardous Materials Control Research Institute, Silver Spring, MD, 458 pp.

Fetter, C. W., 1988, *Applied Hydrogeology*, Merrill, Columbus, OH, 592 pp.

Freeze, R. A., and Cherry, J. A., 1979, *Groundwater*, Prentice Hall, Upper Saddle River, NJ, 604 pp.

Gelhar, L. W., Montoglou, A., Welty, C., and Rehfeldt, K. R., 1985, *A Review of Field Scale Physical Solute Transport Processes in Saturated and Unsaturated Porous Media: Final Project Report*, EPRI EA-4190, Electric Power Research Institute, Palo Alto, CA.

Gelhar, L. W., Welty, L., and Rehfeldt, K. R., 1992, A critical review of data on field-scale dispersion in aquifers, *Water Resour. Res.*, vol. 28, no. 7, pp. 1955–1974.

Gillham, R. W., and O'Hannesin, S. F., 1994, Enhanced degradation of halogenated aliphatics by zero-valent iron, *Ground Water*, vol. 32, no. 6, pp. 958–967.

Guven, O., Molz, F. J., Melville, J. G., El Didy, S., and Boman, G. K., 1992, Three dimensional modeling of a two-well tracer test, *Ground Water*, vol. 30, no. 6, pp. 945–957.

Haag, W. R., and T. Mill, 1988. Transformation kinetics of 1,1,1-trichloroethane to the stable product 1,1-dichloroethene, *Environ. Sci. Tech.*, vol. 22, pp. 658–663.

Hassett, J. J., Banwart, W. L., and Griffin, R. A., 1983, Correlation of compound properties with sorption characteristics of nonpolar compounds by soils and sediments, concepts and limitations, in, *Environment and Solid Wastes*, C. W. Francis and S. I. Auerbach, eds., Butterworth, Boston, pp. 161–178.

Hvorslev, M. J., 1951, *Time Lag and Soil Permeability in Groundwater Observations*, Bulletin 36, U.S. Army Corps of Engineers Waterways Experiment Station Vicksburg, MS, 50 pp.

Jafvert, C. T., and Wolfe, N. L., 1987, Degradation of selected halogenated ethanes in anoxic sediment–water systems, *Environ. Toxicol. Chem.*, vol. 6, pp. 827–837.

Jeffers, P. M., Ward, L. M., Woytowitch, L. M., and Wolfe, N. L., 1989, Homogeneous hydrolysis rate constants for selected chlorinated methanes, ethanes, ethenes, and propanes, *Environ. Sci. Technol.*, vol. 23, pp. 965–969.

Johnson, R. L., Palmer, C. D., and Fish, W., 1989, Subsurface chemical processes, in *Fate and Transport of Contaminants in the Subsurface*, EPA/625/4-89/019, U.S. EPA, Cincinnati, OH and Ada, OK, pp. 41–56.

Jury, W. A., Gardner, W. R., and Gardner, W. H., 1991, *Soil Physics*, Wiley, New York, 328 pp.

Karickhoff, S. W., 1981, Semi-empirical estimation of sorption of hydrophobic pollutants on natural sediments and soils, *Chemosphere*, vol. 10, pp. 833–846.

Karickhoff, S. W., Brown, D. S., and Scott, T. A., 1979, Sorption of hydrophobic pollutants on natural sediments, *Water Resour. Res.*, vol. 13, pp. 241–248.

Kenaga, E. E., and Goring, C. A. I., 1980, ASTM Special Technical Publication 707, ASTM, Philadelphia.

Knox, R. C., Sabatini, D. A., and Canter, L. W., 1993, *Subsurface Transport and Fate Processes*, Lewis Publishers, Boca Raton, FL, 430 pp.

Kruseman, G. P., and de Ridder, N. A., 1991, *Analysis and Evaluation of Pumping Test Data*, International Institute for Land Reclamation and Improvement, The Netherlands, 377 p.

Lallemand-Barres, P., and Peaudecerf, P., 1978, *Recherche des relations entre la valeur de la dispersivite macroscopique d'un milieu aquifère, ses autres caractéristiques et les conditions de mesure*, Etude Bibliographique Bulletin, Bureau de Recherches Géologiques et Minières, Sec. 3/4, pp. 277–287.

Larson, R. A., and Weber, E. J., 1994, *Reaction Mechanisms in Environmental Organic Chemistry*, Lewis Publishers, Boca Raton, FL, 433 pp.

Lyman, W. J., Reidy, P. J., and Levy, B., 1992, *Mobility and Degradation of Organic Contaminants in Subsurface Environments*, C.K. Smoley, Boca Raton, FL, 395 pp.

Mabey, W., and Mill, T., 1978, Critical review of hydrolysis of organic compounds in water under environmental conditions, *J. Phys. Chem. Ref. Data*, vol. 7, pp. 383–415.

March, J., 1985, *Advanced Organic Chemistry*, 3rd ed., Wiley, New York.

McCarthy, K. A., and Johnson, R. L., 1992, Transport of volatile organic compounds across the capillary fringe, *Water Resour. Res.*, vol. 29, pp. 1675–1683.

McCarty, P. L., Reinhard, M., and Rittmann, B. E., 1981, Trace organics in groundwater, *Environ. Sci. Technol.*, vol. 15, no. 1, pp. 40–51.

McCarty, P. L., 1996, Biotic and abiotic transformations of chlorinated solvents in ground water, *Proceedings of the EPA Symposium on Natural Attenuation of Chlorinated Organics in Ground Water*, EPA/540/R-96/509, U.S. EPA, Washington, DC, 169 pp.

Molz, F., and Boman, G., 1996, Site characterization tools: Using a borehole flowmeter to locate and characterize the transmissive zones of an aquifer, *Symposium on Natural Attenuation of Chlorinated Solvents*, U.S. EPA, Washington, DC, EPA/540/R-96/509, pp. 31–34.

Neely, W. B., 1985, Hydrolysis, in *Environmental Exposure from Chemicals*, Vol. 1, W. B. Neely and G. E. Blau, eds., CRC Press, Boca Raton, FL, pp. 157–173.

Newell, C. J., Hopkins, L. P., and Bedient, P. B., 1990, A hydrogeologic database for ground water modeling, *Ground Water*, vol. 28, no. 5, pp. 703–714.

Newell, C. J., McLeod, R. K., and Gonzales, J. R., 1996, *BIOSCREEN Natural Attenuation Decision Support System User's Manual*, Version 1.3, EPA/600/R-96/087, Robert S. Kerr Environmental Research Center, Ada, OK, Aug.

Parker, J. C., and van Genuchten, M. T.,1984, *Determining Transport Parameters from Laboratory and Field Tracer Experiments*, Bulletin 84-3, Virginia Agricultural Experiment Station.

Pickens, J. F., and Grisak, G. E., 1981, Scale-dependent dispersion in a stratified granular aquifer, *Water Resour. Res.*, vol. 17, no. 4, pp. 1191–1211.

Reinhard, M., Curtis, G. P., and Kriegman, M. R., 1990, *Abiotic Reductive Dechlorination of Carbon Tetrachloride and Hexachloroethane by Environmental Reductants: Project Summary*, EPA/600/S2-90/040, U.S. EPA, Washington, DC, Sept..

Rivett, M. O., 1995, Soil-gas signatures from volatile chlorinated solvents: Borden field experiments, *Ground Water*, vol. 33, no. 1, pp. 84–98.

Rogers, R. D., McFarlane, J. C., and Cross, A. J., 1980, Adsorption and desorption of benzene in two soils and montmorillonite clay, *Environ. Sci. Tech.*, vol. 14, no. 4, pp. 457–460.

Roy, W. R., Krapac, I. G., Chou, S. F. J., and Griffin, R. A., 1992, *Batch-Type Procedures for Estimating Soil Adsorption of Chemicals*, Technical Resource Document EPA/530-SW-87-006-F, U.S. EPA, Washington, DC, 100 p.

Shwarzenbach, R. P., and Westall, J., 1981, Transport of nonpolar organic compounds from surface water to groundwater: laboratory sorption studies, *Environ. Sci. Technol.*, vol. G15, pp. 1360–1367.

Spitz, K., and Moreno, J., 1996, *A Practical Guide to Groundwater and Solute Transport Modeling*, Wiley, Inc., New York, 461 pp.

U.S. Environmental Protection Agency (U.S. EPA), 1986, *Background Document for the Ground-Water Screening Procedure to Support 40 CFR Part 269: Land Disposal*, EPA/530-SW-86-047, U.S. EPA, Washington, DC, Jan.

Vogel, T. M., 1994, Natural bioremediation of chlorinated solvents, in *Handbook of Bioremediation*, R. D. Norris et al., eds., Lewis Publishers, Boca Raton, FL.

Vogel, T. M., and McCarty, P. L., 1987, Abiotic and biotic transformations of 1,1,1-trichloroethane under methanogenic conditions, *Environ. Sci. Technol.*, vol. 21, no. 12, pp. 1208–1213.

Vogel, T. M., and Reinhard, M., 1986, Reaction products and rates of disappearance of simple bromoalkanes, 1,2-dibromopropane and 1,2-dibromoethane in water, *Environ. Sci. Technol.*, vol. 20, no. 10, pp. 992–997.

Vogel, T. M., Criddle, C. S., and McCarty, P. L., 1987, Transformations of halogenated aliphatic compounds, *Environ. Sci. Technol.*, vol. 21, no. 8, pp. 722–736.

Vroblesky, D. A., and Chapelle, F. H., 1994, Temporal and spatial changes of terminal electron-accepting processes in a petroleum hydrocarbon-contaminated aquifer and the significance for contaminant biodegradation, *Water Resour. Res.*, vol. 30, no. 5, pp. 1561–1570.

Walton, W. C., 1988. *Ground Water Pumping Tests*, Lewis Publishers, Boca Raton, FL.

Wang, T. C., and Tan, C. K., 1990, Reduction of halogenated hydrocarbons with magnesium hydrolysis process, *Bull. Environ. Contam. Toxicol.*, vol. 45, pp. 149–156.

Xu, M., and Eckstein, Y., 1995, Use of weighted least-squares method in evaluation of the relationship between dispersivity and scale, *Ground Water*, vol. 33, no. 6, pp. 905–908.

CHAPTER 4

OVERVIEW OF INTRINSIC BIOREMEDIATION

In Chapter 3 we presented the most common abiotic mechanisms of natural attenuation and how they work to reduce contaminant concentrations in the subsurface environment. It is important to note that with the exception of degradation mechanisms such as hydrolysis and dehydrohalogenation, these abiotic mechanisms do not remove contaminant mass from the aquifer; they simply transfer organic compounds to another phase or location, or simply cause dilution. In this chapter we provide an introduction to the biological mechanisms of contaminant destruction. Methods for recognizing the occurrence of intrinsic bioremediation in the subsurface are discussed in Chapter 7.

Soils and shallow sediments contain a large variety of microorganisms, ranging from simple prokaryotic bacteria and cyanobacteria to more complex eukaryotic algae, fungi, and protozoa (McNabb and Dunlap, 1975; Ghiorse and Wilson, 1988). Over the past two decades, numerous laboratory and field studies have shown that microorganisms indigenous to the subsurface environment can degrade a variety of organic compounds, including components of gasoline, kerosene, diesel, jet fuel, chlorinated ethenes, chlorinated ethanes, chlorinated methanes, the chlorobenzenes, and many other compounds. Table 4.1 presents a partial list of microorganisms known to degrade organic compounds. Although we now recognize the ubiquitous nature and significance of subsurface microorganisms, study of the microbial ecology and physiology of the subsurface, below the rhizosphere, is still in its infancy. However, great progress has been made in identifying, if not fully understanding, the numerous and diverse types of microbially mediated contaminant transformations that can occur in the subsurface.

To carry out their life functions, microorganisms require a carbon source, electron donors and acceptors, water, and mineral nutrients. Many organic compounds,

TABLE 4.1. Microorganisms Capable of Degrading Organic Compounds

Contaminant	Microorganisms	Biodegradability
Benzene	*Pseudomonas putida, P. rhodochrous, P. aeruginosa, Acinetobacter* sp., *Methylosinus trichosporium* strain OB3b, *Nocardia* sp., methanogens, anaerobes	Moderate to high
Toluene	*Methylosinus trichosporium* strain OB3b, *Bacillus* sp., *Pseudomonas* sp., *P. putida, Cunninghamella elegans, P. aeruginosa, P. mildenberger, P. aeruginosa, Achromobacter* sp., methanogens, anaerobes	High
Ethylbenzene	*Pseudomonas putida*	High
Xylenes	*Pseudomonas putida,* methanogens, anaerobes	High
Jet fuels	*Cladosporium, Hormodendrum*	High
Kerosene	*Torulopsis, Candidatropicalis, Corynebacterium hydrocarboclastus, Candidaparapsilosis, C. guilliermondii, C. lipolytica, Trichosporon* sp., *Rhohosporidium toruloides, Cladosporium resinae*	High
Chlorinated ethenes	*Dehalobacter restrictus, Dehalospirillum multivorans, Enterobacter agglomerans, Dehalococcus entheogenes* strain 195, *Desulfitobacterium* sp. strain PCE1, *Pseudomonas putida* (multiple strains), *P. cepacia* G4, *P. mendocina, Desulfobacterium* sp., *Methanobacterium* sp., *Methanosarcina* sp. strain DCM, *Alcaligenes eutrophus* JMP 134, *Methylosinus trichosporium* strain OB3b, *Escherichia coli, Nitrosomonas europaea, Methylocystis parvus* OBBP, *Mycobacterium* sp., *Rhodococcus erythopolis*	Moderate
Chlorinated ethanes	*Desulfobacterium* sp., *Methanobacterium* sp., *Pseudomonas putida, Clostridium* sp., *Clostridium* sp. strain TCAIIB	Moderate
Chlorinated methanes	*Acetobacterium woodii, Desulfobacterium* sp., *Methanobacterium* sp., *Pseudomonas* sp. strain KC, *Escherichia coli* K-12, *Clostridium* sp., *Methanosarcina* sp., *Hyphomicrobium* sp. strain DM2	Moderate
Chlorobenzenes	*Alcaligenes* sp. (multiple strains), *Pseudomonas* sp. (multiple strains), *P. putida, Staphylococcus epidermis*	Moderate to high

including petroleum hydrocarbons and certain chlorinated solvents, are degraded or transformed by microorganisms through use as a carbon source or as electron acceptors. It appears that those compounds that have been available for microbial metabolism in large quantities and for the longest period of time are the most readily degradable. For example, petroleum hydrocarbons, in the form of crude oil, have resided in the subsurface for hundreds of millions of years. It is these compounds, especially the monoaromatic hydrocarbons, that are the most readily biodegradable. In contrast, consider the chlorinated solvents, which have been produced by human beings for only the last one hundred years or so and were not commonly used until the beginning of World War II. These anthropogenic compounds are much less amenable to biodegradation than are the lighter fractions of petroleum hydrocarbons. Oxygenates such as MTBE have been available for microbial degradation for only about 15 years, and these compounds appear to be the most biologically recalcitrant of common contaminants.

Biological degradation mechanisms can be very important in groundwater systems, depending on the type of contaminant and the prevailing groundwater chemistry. Based on emerging field evidence, it is apparent that intrinsic bioremediation is the dominant process controlling the fate and transport of many organic solutes. For example, Norris (1994) notes that viable petroleum-hydrocarbon-degrading microbial consortia will be present at 99% of sites. During intrinsic bioremediation, organic contaminants are biodegraded by microorganisms indigenous to the subsurface environment. Through a series of oxidation–reduction reactions, dissolved contaminants are ultimately transformed into innocuous by-products such as carbon dioxide, chloride, methane, and water. In some cases, intermediate products of these transformations may be more hazardous than the original compound; however, these intermediates also may be readily degraded. Because indigenous microorganisms are well adapted to the physical and chemical conditions of the subsurface environment in which they reside, they have a distinct advantage over microorganisms introduced artificially into the subsurface to enhance biodegradation. Biodegradation of organic compounds results in a reduction in contaminant concentration (and mass) and a slowing of the contaminant front migration rate relative to the average groundwater seepage velocity. Figures 4.1 and 4.2 illustrate conceptually the effects of biodegradation on advancing solute fronts associated with a continuous source and an instantaneous source, respectively.

To a large degree, the availability of organic compounds for intrinsic bioremediation is dependent on their chemical phase. The most widely accepted conceptual model of biodegradation in groundwater indicates that only dissolved or sorbed organic compounds are available for biodegradation, and will degrade only if they are biodegradable and other compounds (e.g., dissolved oxygen, nitrate, etc.) are available for completing the process. Organic compounds that occur as NAPL (see Chapter 2) are not considered to be available for biodegradation, due to toxic effects and because other necessary reaction compounds (e.g., oxygen, nitrate, etc.) are not present in the NAPL. Mass transfer limitations also will make NAPL biodegradation negligible over the short term. In other words, conditions for biodegradation are so much better for aqueous-phase compounds than for NAPL

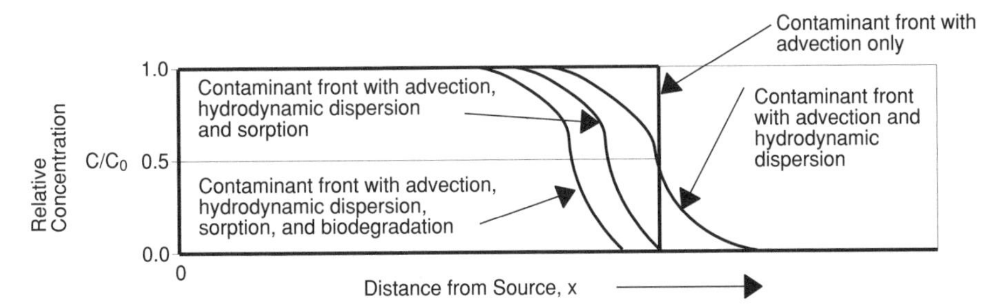

FIGURE 4.1. Breakthrough curve in one dimension showing plug flow with continuous source resulting from advection only, the combined processes of advection and hydrodynamic dispersion, the combined processes of advection, hydrodynamic dispersion, and sorption, and the combined processes of advection, hydrodynamic dispersion, sorption, and biodegradation.

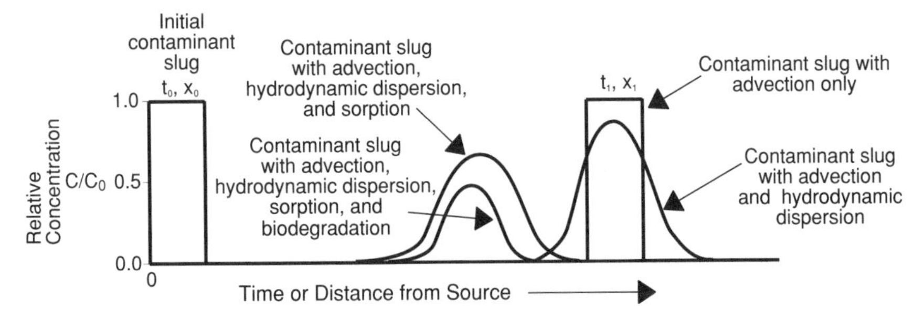

FIGURE 4.2. Breakthrough curve in one dimension showing plug flow with instantaneous source resulting from advection only, the combined processes of advection and hydrodynamic dispersion, the combined processes of advection, hydrodynamic dispersion, and sorption, and the combined processes of advection, hydrodynamic dispersion, sorption, and biodegradation.

that most natural biodegradation of chemicals occurs in the aqueous phase. Over extremely long time periods, such as tens or hundreds of years, NAPL biodegradation may eventually be proven to be important for very low-solubility compounds that do not readily dissolve out of the NAPL. More research is needed to confirm this conjecture, however.

4.1 BIOLOGICAL FATE OF CONTAMINANTS

Field and laboratory research shows that the introduction of soluble organic compounds into groundwater initiates a complex series of responses from subsurface microorganisms that lead to the formation of distinct microbial communities. The dynamics of these communities are driven by the types of compounds harvested by the different communities to obtain energy for growth. Organisms such as humans and most microorganisms obtain energy for growth and activity from physiologically coupling oxidation and reduction reactions. During these growth-promoting reactions, electrons are transferred from one group of compounds called *electron donors* to another group called *electron acceptors*, and energy is released. This process results in the oxidation of the electron donor and reduction of the electron acceptor. Electron donors are compounds that occur in relatively reduced states and include natural organic material, petroleum hydrocarbons (mostly BTEX), and the lightly chlorinated (less oxidized) ethenes, ethanes, methanes, and chlorobenzenes, and dissolved hydrogen. Electron acceptors are elements or compounds that occur in relatively oxidized states and include dissolved oxygen, nitrate, ferric [Fe(III)] iron hydroxide, sulfate, carbon dioxide, and several chlorinated solvents.

Under aerobic (oxic) conditions, many bacteria couple the oxidation of organic compounds (electron donors) to the reduction of oxygen (electron acceptor). All of the primary soluble petroleum hydrocarbon contaminants are biodegraded aerobically, yet only a few of the chlorinated solvents are amenable to aerobic degradation. Under anaerobic conditions, most organic compounds are degraded by groups of interacting microorganisms referred to as a *consortium*. In the consortium, individual types of organisms carry out different specialized reactions which, when combined, can lead to the complete mineralization (conversion to carbon dioxide and water, among other things) of a particular compound. The metabolic interactions between organisms can be complex and may be so tightly linked under a given set of conditions that stable consortia can mistakenly be identified as a single species (Alexander, 1994). Intrinsic bioremediation of dissolved petroleum hydrocarbons is a ubiquitous process and anaerobic consortia are common in these types of dissolved plumes. The intrinsic bioremediation of the common chlorinated solvents is not nearly as pervasive as it is for petroleum hydrocarbons, but an anaerobic consortia must be present for it to occur.

Two other types of reactions are fermentation and cometabolism. During *fermentation*, the compound being biodegraded serves as both the electron donor and the electron acceptor. Fermentation is an important reaction because it supplies other organisms in the consortia with dissolved hydrogen. Dissolved hydrogen is an excellent electron donor that is efficient and that can be used as an electron donor for a surprisingly wide range of reactions, including the in situ biodegradation of chlorinated solvents. During *cometabolism*, an enzyme or cofactor produced by a microbe fortuitously degrades an organic contaminant. During this process the microbe obtains no direct benefit (i.e., does not obtain energy) from destruction of the organic compound. Cometabolic reactions are slow, and they are usually unimportant under natural conditions.

These different reactions can be grouped into the following classification system (after McCarty, 1996):

1. Use of the organic compound as a primary growth substrate
   a. Growth-promoting biological oxidation of the organic compound (e.g., BTEX) through use as an electron donor

   b. Growth-promoting biological reduction of the organic compound (e.g., chlorinated solvents) through use as an electron acceptor (halorespiration)
   c. Fermentation
2. Cometabolism

In the following sections we provide a general description of each reaction. In Chapters 5 and 6 we discuss in depth how each reaction is used to degrade petroleum hydrocarbons and chlorinated solvents, respectively.

Use of an Organic Compound as a Primary Growth Substrate

Growth-Promoting Biological Oxidation of Organic Compounds. The first two biodegradation mechanisms, growth-promoting biological oxidation and reduction, involve the microbial transfer of electrons from electron donors to electron acceptors. During these reactions, the electron donors and electron acceptors are both considered primary growth substrates because they promote microbial growth. Growth-promoting biological oxidation of organic compounds occurs when a microorganism uses the organic compound as an electron donor (primary growth substrate) in a coupled oxidation–reduction reaction.

Biological oxidation can occur under either aerobic or anaerobic conditions. Many organic compounds can be used as electron donors for microbial metabolism, including natural organic carbon, contaminants such as soluble petroleum hydrocarbons (including BTEX), and the less oxidized chlorinated compounds such as DCE, vinyl chloride, DCA, or chlorobenzene. Many of these compounds can be degraded relatively quickly in aerobic environments, as oxygen is an excellent electron acceptor (i.e., provides organisms with large amounts of energy when electrons are transferred to oxygen). Rapid depletion of dissolved oxygen caused by microbial respiration results in the establishment of anaerobic conditions; this is very common in groundwater environments because there is very little mixing in the subsurface compared to most surface environments. When oxygen is consumed in the heart of a contaminant plume, it is replenished only very slowly or not at all because the oxygen is generally consumed at the boundaries until all the contamination has been biodegraded. In fact, once a system becomes anaerobic, it is unlikely that aerobic microorganisms will play a role in the degradation of contaminants in the heart of the plume, as discussed in Chapters 5 and 6. Therefore, anaerobic groundwater conditions are common at contaminated sites, due to the combination of an excess of electron donors (organic compounds such as BTEX) and the insulated, low-mixing environment in the subsurface. When the oxygen is removed by the aerobes, other electron acceptors (sometimes called alternate electron acceptors), such as nitrate, Mn(IV), Fe(III), sulfate, and carbon dioxide, are then used by the anaerobic consortia. The result is the same: Contaminants are oxidized.

In the past few years there has been a growing acceptance that anaerobic reactions at contaminated sites are common, and more important, that these reactions are significant destructive mechanisms in terms of the total mass of contaminants that are biodegraded naturally. Previously, a prevailing rule of thumb had been that the only important oxidizing reaction involved oxygen because the anaerobic reactions were "too slow." This rule has now been supplanted by a new understanding that the amount of mixing of electron donors and electron acceptors is the key factor limiting the amount of biodegradation. In other words, anaerobic reactions are slower than aerobic reactions *when there is an abundant supply of both electron acceptors and electron donors*. In the case of biological oxidation in contaminated aquifers, there is ample time for anaerobic reactions to proceed once the oxygen is gone (the residence time in most plumes is years). Because there is a much larger potential pool of anaerobic electron acceptors [e.g., typically there is more nitrate, Fe(III), sulfate, and carbon dioxide than oxygen] in groundwater systems, much of the contaminant mass removed from these systems is actually accomplished by anaerobes.

Although there is a large body of evidence for direct anaerobic oxidation of soluble petroleum-hydrocarbon contaminants, there is little evidence of such transformations involving chlorinated compounds. In fact, the oxidation of vinyl chloride (Bradley and Chapelle, 1996) is the only known example of oxidation of chlorinated solvents under anaerobic conditions. The biological oxidation of petroleum hydrocarbons under aerobic, denitrifying, Fe(III)-reducing, and sulfate-reducing conditions is discussed further in Chapter 5. The biological oxidation of chlorinated solvents, when it occurs, is discussed in Chapter 6.

Growth-Promoting Biological Reduction of Organic Compounds (Halorespiration). Growth-promoting biological reduction of organic compounds occurs when the compound is used as an electron acceptor (primary growth substrate) during reductive dechlorination. The term *dehalorespiration* was recently coined to describe reductive dechlorination caused by microorganisms that utilize chlorinated compounds as electron acceptors (Hollinger and Schumacher, 1994). In this book the term *halorespiration* is used instead of *dehalorespiration*. Although it appears that reductive dechlorination can occur via cometabolism, this process typically results in the slow and incomplete transformation of chlorinated solvents (Gossett and Zinder, 1996). Halorespiration appears to be the predominate process contributing to the natural biodegradation of the highly chlorinated solvents such as PCE, TCE, TCA, and carbon tetrachloride. This process also can be an important mechanism for the biodegradation of less chlorinated compounds such as DCE and vinyl chloride, depending on environmental factors.

During halorespiration, the chlorinated hydrocarbon is used directly as an electron acceptor and dissolved hydrogen is used directly as an electron donor. Although many laboratory and field studies have reported on the use of a variety of electron donors (e.g., lactate, methanol, toluene) to stimulate halorespiration of chlorinated solvents, it is now recognized that these compounds actually serve as indirect electron donors via fermentation of hydrogen. These compounds do not serve as direct electron donors for halorespiration, but rather, are fermented in situ to produce dissolved hydrogen, which is then consumed by halorespirators. Thus in these reactions, and at sites were halorespiration is occurring naturally, dechlorination is occurring as a two-step process: (1) organic compounds are fermented to produce hydrogen, and (2) the hydrogen is utilized by the dechlorinators for halorespiration. At sites where biological dechlorination is occurring naturally, the fermentation substrates are either biodegradable, nonchlorinated contaminants (e.g., BTEX degradation products, acetone) or naturally occurring, biodegradable organics. Halorespiration affects the widest range of chlorinated aliphatic hydrocarbons under strongly anaerobic conditions. Halorespiration is discussed in more detail in Chapter 6.

Fermentation. Fermentation is a unique reaction in that it requires no external electron acceptors because the organic compound being degraded acts as both an electron donor and an electron acceptor in the first step of the reaction. Fermentation reactions have been used by human beings to make beer and bread since at least 2400 B.C. (Chapelle, 1993). During fermentation, organic compounds are converted to innocuous compounds such as acetate, water, carbon dioxide, and (most important) dissolved hydrogen through a series of internal electron transfers catalyzed by microorganisms.

Dissolved hydrogen is a high-energy electron donor and therefore can be consumed in a wide variety of other reactions. The most obvious fermentation reaction at a contaminated site is methanogenesis, in which organics are consumed and selected fermentation products (including hydrogen) are converted to methane. This reaction is very common, as dissolved methane is observed in groundwater at almost all sites affected by petroleum hydrocarbons and at many sites containing chlorinated solvents. Methanogenesis has proved to be an extremely important mechanism in the complete mineralization of petroleum hydrocarbons and is discussed in Chapter 5.

In addition to methanogenesis and halorespiration, other biological reactions utilize hydrogen. If the required electron acceptors are available, nitrate reducers, iron reducers, and sulfate reducers will preferentially use hydrogen as an electron donor over other electron donors, such as the BTEX compounds. Competition for hydrogen is discussed in Chapter 6.

In summary, fermentation is a very important component of intrinsic bioremediation of organic contaminants because hydrogen is produced by these reactions. This hydrogen is required by other microorganisms, including the methanogens and perhaps more important, by halorespirators, to carry out their life functions.

Cometabolism

Cometabolism is a process in which contaminants (especially chlorinated solvents) or metabolic intermediates are degraded by an enzyme or cofactor that is fortuitously produced by organisms for other purposes. When a compound is biodegraded via cometabolism, it does not serve directly as a primary substrate in a biologically mediated oxidation–reduction reaction. As a result, the organism receives no known benefit (with the possible exception of symbiotic or mutualistic benefits) from the degradation of the compound; in some cases cometabolic degradation may in fact be harmful to the microorganism responsible for the production of the enzyme or cofactor (McCarty and Semprini, 1994). Chlorinated solvents are usually only partially transformed during cometabolic processes, with additional biotic or abiotic degradation generally required to complete the transformation (McCarty and Semprini, 1994).

There are two types of cometabolic reactions observed in the subsurface. Cooxidation occurs when a substrate such as methane is utilized by methanotrophs as an electron donor, and oxygen is used as an electron acceptor. During this reaction an enzyme that degrades some chlorinated solvents (e.g., TCE) is produced. Because oxygen and methane typically occur under vastly different geochemical conditions in the shallow subsurface (i.e., biogenic methane is produced under strongly anaerobic conditions), this type of reaction typically must be engineered. Cooxidation can also affect the biodegradation of intermediates produced during the biodegradation of petroleum hydrocarbons. The second type of cometabolic reaction is a reduction reaction and causes reductive dechlorination of chlorinated solvents. Both of these reactions are very limited under natural conditions, however, and typically are not significant natural attenuation processes.

4.2 THERMODYNAMIC CONSIDERATIONS

Ideally, all biologically mediated reactions produce energy for microbial growth and reproduction. Biologically mediated electron transfer results in oxidation of the electron donor, reduction of the electron acceptor, and the production of usable energy. The energy produced by these reactions can be quantified by the Gibbs free energy of the reaction (ΔG_r), which is given by

$$\Delta G_r^\circ = \sum \Delta G_{f,\,\text{products}}^\circ - \sum \Delta G_{f,\,\text{reactants}}^\circ \tag{4.1}$$

where
ΔG_r° = Gibbs free energy of reaction at standard state
$\Delta G_{f,\,\text{products}}^\circ$ = Gibbs free energy of formation for products at standard state
$\Delta G_{f,\,\text{reactants}}^\circ$ = Gibbs free energy of formation for reactants at standard state

ΔG_r° defines the maximum useful energy change for a chemical reaction at a constant temperature and pressure. The state of an oxidation–reduction reaction relative to equilibrium is defined by the sign of ΔG_r°. Negative values indicate that the reaction is exothermic (energy-producing) and will proceed from left to right (i.e., reactants will be transformed into products and energy will be produced). Positive values indicate that the reaction is endothermic, and in order for the reaction to proceed from left to right, energy must be put into the system. The value of ΔG_r° can be used to estimate how much free energy is consumed or produced during the reaction.

Tables 4.2 and 4.3 present selected electron acceptor and electron donor half-cell reactions and the calculated ΔG_r° values. Table 4.4 gives the Gibbs free energy of formation (ΔG_f°) for species used in these half-cell reactions. Like all living organisms, microorganisms are constrained by the laws of thermodynamics. They can facilitate only those oxidation–reduction reactions that are thermodynamically possible (Chapelle, 1993). That is, microorganisms will facilitate only those oxidation–reduction

TABLE 4.2. Electron Acceptor Half-Cell Reactions

Half-Cell Reaction	ΔG_r° (kcal/mol e$^-$)	E° (V)
$4e^- + 4H^+ + O_2 \Rightarrow 2H_2O$ Aerobic respiration	−18.5	0.80
$5e^- + 6H^+ + NO_3^- \Rightarrow 0.5N_2 + 3H_2O$ Denitrification	−16.9	0.73
$2e^- + 4H^+ + \underline{MnO_2} \Rightarrow Mn^{2+} + 2H_2O$ Pyrolusite dissolution/reduction	−8.6	0.37
$CO_2 + e^- + H^+ + \underline{MnOOH} \Rightarrow MnCO_3 + H_2O$ Manganite carbonation/reduction	−13.3	0.58
$e^- + H^+ + MnO_2 \Rightarrow \underline{MnOOH}$ Pyrolusite hydrolysis/reduction	−12.2	0.53
$e^- + \underline{Fe^{3+}} \Rightarrow Fe^{2+}$ Fe(III) reduction	−17.8	0.77
$8e^- + 9.5H^+ + SO_4^{2-} \Rightarrow 0.5HS^- + 0.5H_2S + 4H_2O$ Sulfate reduction	5.3	−0.23
$8e^- + 8H^+ + CO_{2,g} \Rightarrow CH_{4,g} + 2H_2O$ Methanogenesis	5.9	−0.26
$C_2Cl_{4,g} + H^+ + 2e^- \Rightarrow C_2HCl_3 + Cl^-$ PCE reductive dechlorination	−9.9	0.43
$C_2HCl_3 + H^+ + 2e^- \Rightarrow C_2H_2Cl_2 + Cl^-$ TCE reductive dechlorination	−9.6	0.42
$C_2H_2Cl_2 + H^+ + 2e^- \Rightarrow C_2H_3Cl + Cl^-$ *cis*-DCE reductive dechlorination	−7.2	0.31
$C_2H_3Cl + H^+ + 2e^- \Rightarrow C_2H_4 + Cl^-$ VC reductive dechlorination	−8.8	0.38
$C_2H_2Cl_4 + H^+ + 2e^- \Rightarrow C_2H_3Cl_3 + Cl^-$ PCA reductive dechlorination	−8.7	0.38
$C_2H_3Cl_3 + H^+ + 2e^- \Rightarrow C_2H_4Cl_2 + Cl^-$ TCA reductive dechlorination	−10.3	0.45
$C_2H_4Cl_2 + H^+ + 2e^- \Rightarrow C_2H_5Cl + Cl^-$ DCA reductive dechlorination	−9.0	0.39
$C_2H_5Cl + H^+ + 2e^- \Rightarrow C_2H_6 + Cl^-$ Chlorethane reductive dechlorination	−7.4	0.32
$C_6Cl_6 + H^+ + 2e^- \Rightarrow C_6HCl_5 + Cl^-$ Hexachlorobenzene reductive dechlorination	−9.4	0.41
$C_6HCl_5 + H^+ + 2e^- \Rightarrow C_6H_2Cl_4 + Cl^-$ Pentachlorobenzene reductive dechlorination	−9.7	0.42
$C_6H_2Cl_4 + H^+ + 2e^- \Rightarrow C_6H_3Cl_3 + Cl^-$ Tetrachlorobenzene reductive dechlorination	−8.7	0.38
$C_6H_3Cl_3 + H^+ + 2e^- \Rightarrow C_6H_4Cl_2 + Cl^-$ Trichlorobenzene reductive dechlorination	−8.3	0.36

TABLE 4.3. Electron Donor Half-Cell Reactions

Half-Cell Reaction	ΔG_r° (kcal/mol e⁻)	E° (V)
$1/2 H_2 \Rightarrow H^+ + e^-$ Hydrogen oxidation	−9.9	0.43
$1/4 CH_2O + 1/4 H_2O \Rightarrow 1/4 CO_2 + H^+ + e^-$ Carbohydrate oxidation	−10.0	0.43
$12H_2O + C_6H_6 \Rightarrow 6CO_2 + 30H^+ + 30e^-$ Benzene oxidation	−7.0	0.31
$14H_2O + C_6H_5CH_3 \Rightarrow 7CO_2 + 36H^+ + 36e^-$ Toluene oxidation	−6.9	0.30
$16H_2O + C_6H_5C_2H_5 \Rightarrow 8CO_2 + 42H^+ + 42e^-$ Ethylbenzene oxidation	−6.9	0.30
$16H_2O + C_6H_4(CH_3)_2 \Rightarrow 8CO_2 + 42H^+ + 42e^-$ *m*-Xylene oxidation	−6.8	0.30
$20H_2O + C_{10}H_8 \Rightarrow 10CO_2 + 48H^+ + 48e^-$ Naphthalene oxidation	−6.9	0.30
$18H_2O + C_6H_3(CH_3)_3 \Rightarrow 9CO_2 + 48H^+ + 48e^-$ 1,3,5-Trimethylbenzene oxidation	−6.8	0.30
$18H_2O + C_6H_3(CH_3)_3 \Rightarrow 9CO_2 + 48H^+ + 48e^-$ 1,2,4-Trimethylbenzene oxidation	−6.8	0.29
$4H_2O + C_2H_2Cl_2 \Rightarrow 2CO_2 + 10H^+ + 8e^- + 2Cl^-$ DCE oxidation	−16.1	0.70
$4H_2O + C_2H_3Cl \Rightarrow 2CO_2 + 11H^+ + 10e^- + Cl^-$ Vinyl chloride oxidation	−11.4	0.50
$12H_2O + C_6H_2Cl_4 \Rightarrow 6CO_2 + 26H^+ + 22e^- + 4Cl^-$ Tetrachlorobenzene oxidation	−12.4	0.54
$12H_2O + C_6H_3Cl_3 \Rightarrow 6CO_2 + 27H^+ + 24e^- + 3Cl^-$ Trichlorobenzene oxidation	−10.6	0.46
$12H_2O + C_6H_4Cl_2 \Rightarrow 6CO_2 + 28H^+ + 26e^- + 2Cl^-$ Dichlorobenzene oxidation	−9.2	0.40
$12H_2O + C_6H_5Cl \Rightarrow 6CO_2 + 29H^+ + 28e^- + Cl^-$ Chlorobenzene oxidation	−8.0	0.35

reactions that will yield energy (i.e., $\Delta G_r^\circ < 0$). Microorganisms will not invest more energy into the system than can be released. To derive energy for cell maintenance and production from organic compounds, the microorganisms must couple an endothermic reaction (electron donor oxidation half-cell reaction) with an exothermic reaction (electron acceptor reduction half-cell reaction).

Coupled oxidation–reduction reactions would be expected to occur in order of their thermodynamic energy yield, assuming that there are organisms capable of facilitating each reaction and that there is an adequate supply of organic carbon and electron acceptors (Stumm and Morgan, 1981; Chapelle, 1993). In general, where

TABLE 4.4. Gibbs Free Energy of Formation for Species Used in Half-Cell Reactions and Coupled Oxidation–Reduction Reactions

Species	State[a]	$\Delta G^{\circ}_{f,298.15}$ (kcal/mole)	Source[b]
e^-	i	0	std.
H^+	i	−9.87	std. for 10-7 M
O_2	g	0	std.
H_2O	l	−56.687	Dean (1972)
	Carbon Species		
CO_2	g	−94.26	Dean (1972)
CH_2O, formaldehyde	aq	−31.02	Dean (1972)
C_6H_6, benzene	l	+29.72	Dean (1972)
CH_4, methane	g	−12.15	Dean (1972)
$C_6H_5CH_3$, toluene	l	+27.19	Dean (1972)
$C_6H_5C_2H_5$, ethylbenzene	l	+28.61	Dean (1972)
$C_6H_4(CH_3)_2$, *o*-xylene	l	+26.37	Dean (1972)
$C_6H_4(CH_3)_2$, *m*-xylene	l	+25.73	Dean (1972)
$C_6H_4(CH_3)_2$, *p*-xylene	l	+26.31	Dean (1972)
C_2Cl_4, PCE	l	+1.1	CRC (1990)
C_2HCl_3, TCE	l	+2.9	CRC (1990)
$C_2H_2Cl_2$, 1,1-DCE	l	+5.85	Dean (1972)
$C_2H_2Cl_2$, *cis*-1,2-DCE	l	5.27	CRC (1990)
$C_2H_2Cl_2$, *trans*-1,2-DCE	l	+6.52	CRC (1990)
C_2H_4, ethene	g	+16.28	CRC (1990)
	aq, $m = 1$	+19.43	
C_2H_6, ethane	g	−7.68	CRC (1990)
	aq, $m = 1$	−4.09	
HCl, hydrochloric acid	aq, $m = 1$	−31.372	CRC (1990)
$C_2H_2Cl_4$, 1,1,2,2-PCA	l	−22.73	Dean (1972)
$C_2H_3Cl_3$, 1,1,2-TCA	g	−18.54	Dean (1972)
$C_2H_4Cl_2$, 1,2-DCA	g	−17.68	Dean (1972)
$C_2H_5Cl_1$, Chloroethane	g	−14.47	Dean (1972)
$C_{10}H_8$, naphthalene	l	+48.05	Dean (1972)
$C_6H_3(CH_3)_3$, 1,3,5-TMB	l	+24.83	Dean (1972)
$C_6H_3(CH_3)_3$, 1,2,4-TMB	l	+24.46	Dean (1972)
C_2H_3Cl, vinyl chloride	g	+12.4	Dean (1972)
C_6Cl_6, hexachlorobenzene	l	+0.502	Dolfing and Harrison (1992)
$C_6H_1Cl_5$, pentachlorobenzene	l	+3.16	Dolfing and Harrison (1992)
$C_6H_2Cl_4$, 1,2,4,5-tetrachlorobenzene	l	+5.26	Dolfing and Harrison (1992)
$C_6H_3Cl_3$, 1,2,4-trichlorobenzene	l	+9.31	Dolfing and Harrison (1992)
$C_6H_4Cl_2$, 1,4-dichlorobenzene	l	+14.28	Dolfing and Harrison (1992)
C_6H_5Cl, chlorobenzene	l	+21.32	Dean (1972)
$C_{14}H_{10}$, phenanthrene	l	+64.12	Dean (1972)

continues

TABLE 4.4. *(Continued)*

Species	State[a]	$\Delta G^{\circ}_{f,298.15}$ (kcal/mole)	Source[b]
		Nitrogen Species	
NO_3^-	l	−26.61	Dean (1972)
N_2	g	0	std.
NO_2^-	l	−7.7	Dean (1972)
NH_4^+	aq	−18.97	Dean (1972)
		Sulfur Species	
SO_4^{2-}	i	−177.97	Dean (1972)
H_2S	aq	−6.66	Dean (1972)
H_2S	g	−7.9	Dean (1972)
HS^-	i	+2.88	Dean (1972)
		Iron Species	
Fe^{2+}	i	−18.85	Dean (1972)
Fe^{3+}	i	−1.1	Dean (1972)
Fe_2O_3, hematite	c	−177.4	Dean (1972)
FeOOH, Fe(III) oxyhydroxide	c	−117.2	Naumov et al. (1974)
$Fe(OH)_3$, goethite	a	−167.416	Langmuir and Whittemore (1971)
$Fe(OH)_3$, goethite	c	−177.148	Langmuir and Whittemore (1971)
$FeCO_3$, siderite	c	−159.35	Dean (1972)
		Manganese Species	
Mn^{2+}	i	−54.5	Dean (1972)
MnO_2, pyrolusite	c	−111.18	Stumm and Morgan (1981)
MnOOH, manganite	c	−133.29	Stumm and Morgan (1981)
$MnCO_3$, rhodochrosite	p	−194	Dean (1972)
		Chloride Species	
Cl^-	aq	−31.37	Dean (1972)

[a] c, crystallized solid; l, liquid; g, gaseous; aq, undissociated aqueous species; a, amorphous solid (may be partially crystallized, dependent on methods of preparation); p, freshly precipitated solid; i, dissociated aqueous ionic species (concentration = 1 *m*).

[b] Wherever possible multiple sources were consulted to eliminate the possibility of typographical error. std., accepted by convention.

there are no limitations on electron donors and acceptors, reactions that yield the most energy tend to take precedence over reactions that yield less energy (Stumm and Morgan, 1981). As can be seen in Table 4.2, aerobic respiration, in which microorganisms capable of aerobic respiration (i.e., obligate aerobes or facultative anaerobes) use dissolved oxygen in groundwater as an electron acceptor, is the first reaction to occur. Once the available dissolved oxygen is depleted and anaerobic conditions emerge in portions of the contaminant plume, facultative or obligate anaerobic microorganisms can utilize other electron acceptors, such as nitrate, Fe(III), sulfate, or carbon dioxide.

4.3 OXIDATION–REDUCTION POTENTIAL

The oxidation–reduction potential (ORP) of groundwater is a measure of electron activity and is an indicator of the relative tendency of a solution to accept or transfer electrons. Oxidation–reduction reactions in groundwater containing organic compounds (natural or anthropogenic) are usually biologically mediated, and therefore the ORP of a groundwater system depends on and influences rates of biodegradation. Knowledge of the ORP of groundwater can be used as a qualitative indicator of aerobic versus anaerobic conditions. These measurements, which can be obtained using a portable meter, are also important because some biological processes operate only within a prescribed range of ORP conditions. The ORP of groundwater generally ranges from −400 to 800 mV. It is important to keep in mind that ORP cannot be relied upon to provide concrete evidence for the occurrence of a given terminal electron-accepting process.

As electron acceptors and nutrients are depleted by microbial activity during biodegradation of contaminants, the ORP of groundwater decreases. This results in a succession of bacterial consortia adapted to specific oxidation–reduction regimes and electron acceptors. Metabolic by-products of contaminant biodegradation also exert selective forces, either by presenting different carbon sources or by further modifying the physical and chemical environment of the aquifer. Like organic and inorganic colloids, microorganisms possess complex surface chemistry and can themselves serve as mobile and immobile reactive sites for contaminants. As each geochemical species that can be used to oxidize BTEX is exhausted, the microorganisms are forced to use electron acceptors with a lower oxidizing capacity. When the ORP of groundwater becomes sufficiently low, reactions such as sulfate reduction, methane fermentation, and reductive dechlorination can proceed. Figure 4.3 shows the approximate range of ORPs for selected terminal electron-accepting processes. Figure 4.4 shows the sequence of selected oxidation–reduction reactions based on ORP.

An environment that has a high concentration of electron acceptors has a high electrical potential and is considered *oxidizing*. An environment that has been depleted of electron acceptors has a low electrical potential and is considered *reducing*. In a similar fashion, the reduction potential of individual electron acceptors can be determined based on the amount of energy released when the compound

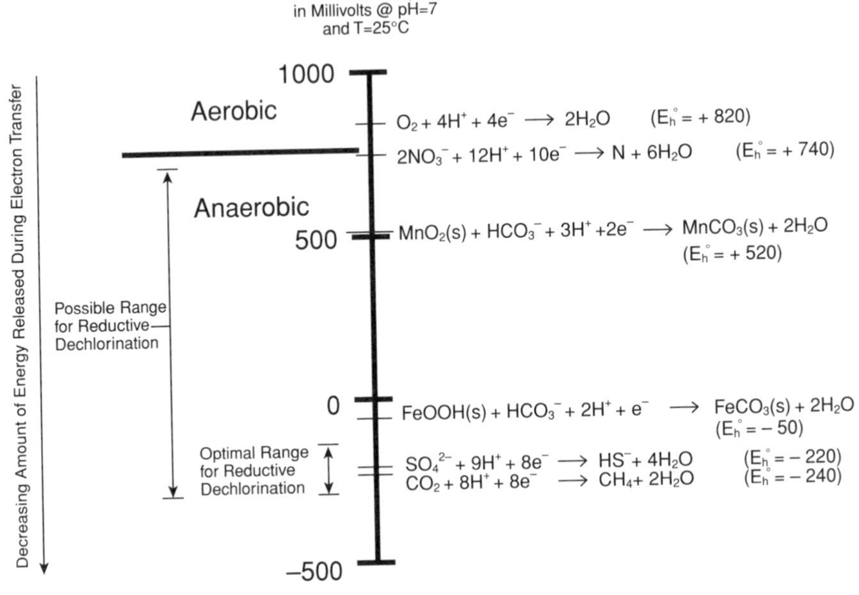

FIGURE 4.3. Oxidation–reduction potentials for various reactions. (Modified from Bouwer, 1994.)

is reduced (i.e., accepts electrons). The reduction potential is related to the energy released by the equation

$$E_H^\circ = \frac{-\Delta G_r^\circ}{n_{\text{electrons}}}(0.0434) \tag{4.2}$$

where E_H° = reduction potential for electron acceptor (V)
ΔG_r° = Gibbs free energy for reduction half-reaction of electron acceptor (kcal/mol)
$n_{\text{electrons}}$ = number of electrons transferred in reaction

and 1 kcal/mol of electrons = 0.0434 V at 25°C.

Thus a negative Gibbs free energy (i.e., energy release) is equivalent to a positive reduction potential. Tables 4.2 and 4.3 show the Gibbs free energy releases for several reduction half-reactions of electron acceptors found in the subsurface. By convention, the energy release is calculated assuming that all reactants and products are present at a concentration of 1 M, except for $[H^+] = 10^{-7}$ (pH 7). These half-reactions illustrate that electrons are combining with the electron acceptors to yield reduced products. However, these are only conceptual reactions because no free electrons exist in solution. To obtain a complete chemical reaction, a reduction half-reaction for an electron acceptor must be combined with an oxidation half-reaction for an electron donor. When two reactions are combined, the Gibbs free

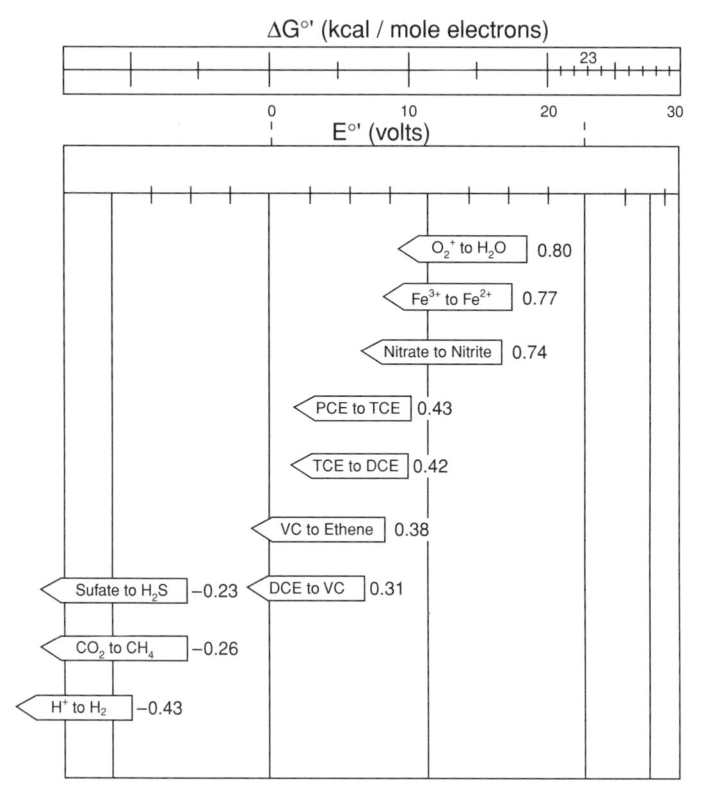

FIGURE 4.4. Reduction potential for various half-cell reactions. (Modified from Stumm and Morgan, 1981.)

energies can be summed. For example, the combination of oxygen reduction with hydrogen oxidation yields

$$\tfrac{1}{2}O_2 + H_2 \Rightarrow H_2O \qquad \Delta G_r^\circ = -57 \text{ kcal/mol } H_2$$

This reaction was obtained by combining two oxygen reduction half-reactions with two hydrogen oxidation half-reactions. From Tables 4.2 and 4.3, $\Delta G_r^\circ = (2 \times -9.9) + (2 \times -18.5)$ kcal/mol. A complete chemical reaction where an electron donor is oxidized and an electron acceptor is reduced is often called an *oxidation–reduction reaction*. An oxidation–reduction reaction typically will occur in nature only if energy is released. Thus the reduction of CO_2 to CH_4 cannot be used to drive the oxidation of Fe(II) to Fe(III) [$\Delta G_r^\circ = +23.7$ kcal/mol Fe(II)]. Note that the Gibbs free energy for an oxidation half-reaction is exactly opposite that of the equivalent reduction half-reaction.

Because the reduction potential of an electron acceptor is proportional to Gibbs free energy associated with the reduction half-reaction, the thermodynamics of a reaction can be determined by comparing the reduction potential of the electron acceptors and donors. The reaction of any electron acceptor with an electron donor that has a lower potential will be energetically favorable, and the larger the difference, the more energy is released. Figure 4.4 shows the reduction potential for a variety of electron acceptors and donors. From this figure it is apparent that the reaction of hydrogen with nitrate (denitrification) is thermodynamically more favorable than with PCE (halorespiration); however, the reaction of hydrogen with PCE (halorespiration) is more favorable than with CO_2 (methanogenesis). Although the thermodynamic favorability of a chemical reaction cannot be linked directly to the rate at which the reaction occurs, Stumm and Morgan (1981) have made the following observations about biologically mediated reactions:

> Although, as stressed, conclusions regarding chemical dynamics may not generally be drawn from thermodynamic considerations . . . it appears that in natural habitats organisms capable of mediating the pertinent oxidation-reduction reactions are nearly always found. . . .
>
> Since the reactions considered are biologically mediated, the chemical reaction sequence is paralleled by an *ecological succession* of microorganism (aerobic heterotrophs, denitrifiers, fermentors, sulfate reducers, and methane bacteria). It is perhaps also of great interest from an evolutionary point of view that there appears to be a tendency for more energy-yielding mediated reactions to take precedence over processes that are less energy-yielding.

In other words, when oxygen is present in the subsurface, reactions utilizing oxygen as the electron acceptor will predominate and aerobic bacteria will dominate. This will be followed by utilization of electron acceptors with decreasing reduction potential [nitrate, Fe(III), sulfate, CO_2] and domination by nitrate reducers, followed by iron reducers, sulfate reducers, and finally, methanogens.

4.4 BIODEGRADATION KINETICS OF DISSOLVED CONTAMINANTS

Biodegradation reactions involving dissolved organic chemicals occur at specific rates. In groundwater these rates are a function of the prevailing environmental conditions, such as temperature and the availability of electron donors and acceptors. Quantifying the biodegradation rate is important for natural attenuation assessments because biodegradation is a key "destructive" mechanism that controls the fate and transport of the organic contaminants. The biodegradation rate can be determined in a variety of ways, including laboratory microcosm studies, literature values, and collection and analysis of field data. In this book we focus on estimating biodegradation rates for contaminants dissolved in groundwater from field data. The reader is referred to the microbiological literature for laboratory estimation procedures of biodegradation rates.

Applicable kinetic models that have been used to represent the biodegradation of organic chemicals in subsurface media are discussed in this section. Various methods for estimating biodegradation rates from field data are discussed in Chapter 7. The models most commonly used to describe the biodegradation of organic compounds dissolved in groundwater include first-order decay, electron-acceptor-limited biodegradation ("instantaneous reaction"), and Monod kinetics. In this section we present a brief overview of these expressions and their impact on the results of natural attenuation modeling. The reader is referred to Bedient et al. (1994) and Rifai and Bedient (1994) for more detailed discussions of this topic.

First-Order Decay Model

One of the most commonly used expressions for representing the biodegradation of an organic compound involves the use of an exponential decay relationship:

$$C = C_0 e^{-kt} \tag{4.3}$$

where C = biodegraded concentration of the chemical
C_0 = starting concentration
k = rate of decrease of the chemical (units of 1/time) $[T^{-1}]$

First-order rate constants are often expressed in terms of a half-life for the chemical:

$$t_{1/2} = \frac{0.693}{k} \tag{4.4}$$

The first-order decay model shown in eq. (4.3) assumes that the solute degradation rate is proportional to the solute concentration. The higher the concentration, the higher the degradation rate. This method is usually used to simulate biodegradation of hydrocarbons dissolved in groundwater. Modelers using the first-order decay model typically use the first-order decay coefficient as a calibration parameter and adjust the decay coefficient until the model results match the field data. With this approach, uncertainties in a number of parameters (e.g., dispersion, sorption, biodegradation) are lumped together in a single calibration parameter.

A simple conceptual model of the first-order decay process in a solute plume is shown in Figure 4.5*a*. In this figure, contaminants are released from a reservoir (representing a NAPL source zone) and dropped on a conveyor belt (representing moving groundwater) that transports the contaminants away from the source. As the contaminants are moving on the conveyor belt, some will react at different times, similar to popcorn popping (indicated by the "zap" labels). The number of reactions that occur on the conveyor depends on (1) the average time it takes for a typical contaminant unit to react (i.e., the half-life of a reaction), and (2) the concentration of contaminant units on the belt. Therefore, more reactions will occur near the reservoir, where there are more contaminants, than farther along on the conveyor, where the concentration of contaminants is lower.

There are three important points illustrated by this simple conceptual model. First, the reaction rate is dependent only on the contaminant itself, more specifically on the

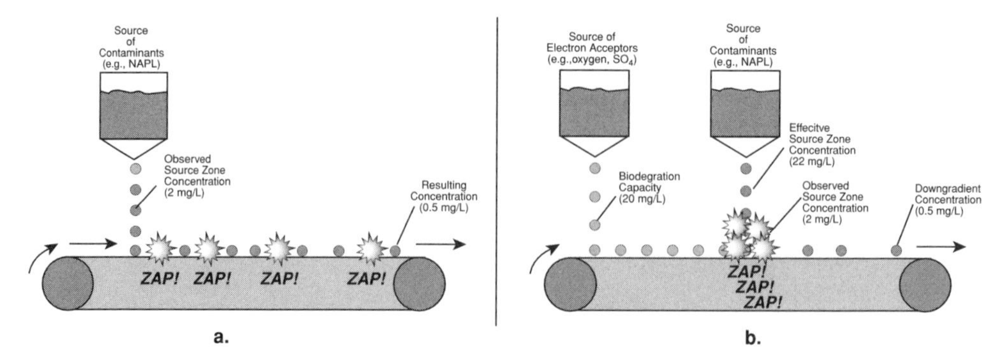

FIGURE 4.5. (*a*) Conceptual model of petroleum-hydrocarbon contaminated site assuming no biodegradation in the source zone and first-order decay of contaminants in plume. Note that this model does not account for observed depletion of almost all electron acceptors and generation of metabolic by-products in source zone. (*b*) Conceptual model of petroleum-hydrocarbon site assuming biodegradation in source zone via electron-acceptor-limited biodegradation (instantaneous reaction). This model does account for observed depletion of almost all electron acceptors and generation of metabolic by-products in source zone. Reaction is relatively "instantaneous" in source zone as the reaction time (after acclimation) is shorter than the residence time in the source area.

average time it takes an individual contaminant to react and the concentration of the contaminant. No other factors, such as the availability of electron acceptors or electron donors, are considered in the first-order decay model. Second, the distance that the contaminant travels down the conveyor belt is controlled by the reaction rate and the speed of the conveyor (i.e., the groundwater seepage velocity). In other words, the first-order decay approach indicates that the faster the seepage velocity, the longer the plume, a conclusion not always true in groundwater systems (see Chapter 8). Third, the rate at which contaminants react on the conveyor (representing contaminants in moving groundwater) is different from the rate at which contaminants leave the reservoir (the source decay rate in the NAPL zone; see Section 2.8).

This conceptual model demonstrates that the first-order decay model is a very simple, easy-to-use approach for quantifying the biodegradation rate of dissolved contaminants as they leave a source zone. Because of its simplicity, it is probably the most widely used method for quantifying biodegradation rates, resulting in an extensive collection of first-order rate constants (typically reported as half-lives) in the literature. For example, literature values for the half-life of dissolved benzene, a readily biodegradable hydrocarbon, range from 10 to 730 days, while the half-life values for TCE, a more biologically recalcitrant constituent, range from 10.7 months to 4.5 years (Howard et al., 1991). Other applications of the first-order decay approach include determining decay rates for radioactive solutes and rates of abiotic hydrolysis of selected organics, such as dissolved chlorinated solvents. One of the best sources for first-order decay coefficients in groundwater systems is the *Handbook of Environmental Degradation Rates* (Howard et al., 1991).

The first-order decay model can easily be integrated into fate and transport analytical and numerical models, and requires the estimation of only a single parameter, the rate constant or decay coefficient. However, this model has the following limitations:

1. The model does not account for site-specific information such as the availability of electron acceptors (for petroleum hydrocarbon contaminants) or electron donors (for chlorinated solvent contaminants). This explains, in part, why the half-lives reported for a given chemical vary over a range of values (see Chapters 5 and 6).
2. Laboratory-determined first-order decay rates are not readily transferable to field situations, due to limitation 1.
3. The model does not assume any biodegradation of dissolved constituents in the source zone. The model assumes that biodegradation starts immediately downgradient from the source.

Example 4.1: Estimating Concentration Reduction with First-Order Model Assuming a benzene concentration in the groundwater of 5 mg/L, estimate the remaining concentrations after 3 years for 1% and 0.5% day^{-1} decay rates.

SOLUTION: We use eq. (4.3), $C = C_0 e^{-kt}$, where $t = 3$ years or 3×365 days/yr $= 1095$ days and $k = 0.01$ day^{-1}, $C = 5e^{-0.01\times1095} = 5(1.76)(10^{-5}) = 8.78(10^{-5})$ mg/L

and for $t = 3$ years or 3×365 days/yr $= 1095$ days and $k = 0.005$ day^{-1},

$$C = 5e^{-kt} = 5(4.19)(10^{-3}) = 0.02 \text{ mg/L}$$

It can be seen from the example calculations that the decay rate has a significant impact on contaminant concentrations and that varying the rate constant will also affect the resulting concentrations. A careful selection of the rate constant is therefore necessary if the first-order decay model is to be used for natural attenuation modeling. In Chapters 5 and 6 we present commonly reported literature values for rate constants for fuels and chlorinated organics, respectively.

Electron-Acceptor-Limited or Instantaneous Reaction Model

The electron-acceptor-limited model (traditionally called the *instantaneous reaction model*) was first proposed by Borden and Bedient (1986) for simulating the aerobic biodegradation of petroleum hydrocarbons. They observed that microbial biodegradation kinetics are fast in comparison with the transport of oxygen and that the growth of microorganisms and utilization of oxygen and organics in the subsurface can be simulated as an electron-acceptor-limited or instantaneous reaction between the organic contaminant and oxygen.

From a practical standpoint, the instantaneous reaction model assumes that the rate of utilization of the contaminant and oxygen by the microorganisms is very high, and that the time required to biodegrade the contaminant is very short, almost instantaneous, relative to the seepage velocity of the groundwater. Using oxygen as an electron acceptor, for example, biodegradation is calculated using the expression

$$\Delta C_R = -\frac{O}{F} \tag{4.5}$$

where ΔC_R = change in contaminant concentration due to biodegradation
O = concentration of oxygen
F = utilization factor, the ratio of oxygen to contaminant consumed

The variable F is obtained from the oxidation–reduction reaction involving the organic and the electron acceptor (see the discussion in Chapter 5).

Based on work conducted by the Air Force Center for Environmental Excellence (Wiedemeier et al., 1995, 1996, 1998; see Appendix A) and summarized by Newell et al. (1995), it appears that the anaerobic biodegradation of petroleum hydrocarbons can be simulated using the assumption of an instantaneous reaction. In addition, Davis et al. (1994) and Kazumi et al. (1997) ran microcosm studies with sulfate reducers and methanogens that confirmed that benzene could be anaerobically degraded in less than 100 days (after acclimation). The instantaneous reaction model has the advantage of not requiring kinetic data, as the microbial kinetics are not what is limiting the reaction. The model, however, is limited to situations in which the microbial biodegradation kinetics are fast relative to the rate of groundwater flow that mixes electron acceptors with dissolved contaminants. The instantaneous reaction for simulating the biodegradation of petroleum hydrocarbons is discussed in more detail in Chapter 5.

A conceptual model for the electron-acceptor-limited/instantaneous reaction model for a petroleum-hydrocarbon site is shown on Figure 4.5*b*. This conceptual model has two reservoirs, with the first reservoir dropping electron acceptors on the conveyor belt (representing naturally occurring electron acceptors in clean, upgradient groundwater), and the second reservoir dropping contaminants on the conveyor belt (representing dissolution of source zone NAPLs). The biological reactions (shown by the "zap" labels) are concentrated in the source zone itself, where the electron acceptors are mixed with the contaminants, because the reactions are fast compared to the residence time in the source zone. Another analogy: dropping burning newspapers into a metal drum, which is then tightly covered. The fire continues until either the electron acceptor (oxygen in the atmosphere) is consumed, leaving only unburned newspaper and no oxygen, or until the newspaper is burned up, leaving only air with a depleted concentration of oxygen. In solute plumes, the NAPL in the source zone is constantly adding new "newspaper" to the fire, therefore consuming all of the "atmosphere" (electron acceptors) and forming a plume. At LNAPL sites the predominant electron acceptors consist of dissolved oxygen, nitrate, Fe(III), sulfate, and carbon dioxide.

There are several key points to be made with this conceptual model. First, the reaction rate depends basically on the amount of electron acceptors that are available. Increasing the amount of contaminants will not increase the reaction rate. Second, increasing the speed of the conveyor belt (i.e., groundwater seepage velocity) will not affect the mass of contaminants leaving the source zone, as electron acceptors and contaminants are mixed together in the same ratio as before the speed-up. Finally, the conveyor-belt approach suggests that reactions are occurring farther along on the conveyor belt. This is caused by the low-level mixing (caused by dispersion) of the contaminant plume with clean groundwater containing electron acceptors. Finally, while the conceptual model represents the reactions of oxygen, nitrate, and sulfate at LNAPL sites relatively well, the reduction of Fe(III) (a solid on the aquifer matrix) and methanogenesis (a reaction based on the fermentation of organics) does not fit this conceptual model well. Reduction of Fe(III) is complicated by the fact that the Fe(III) does not flow with moving groundwater, but is removed from the aquifer matrix, used in the biodegradation reaction, and converted to Fe(II) (which is soluble in groundwater). Methanogenesis is a reaction that depends on fermentation of simple organics (e.g., BTEX breakdown products) to form substrates such as formate or hydrogen. This reaction may be more rate limited than constituent limited.

The electron-acceptor-limited/ instantaneous reaction model has been applied only to petroleum-hydrocarbon sites to date. The kinetics of fermentation and halorespiration, and the observed concentrations of dissolved hydrogen at field sites, suggest that an "instantaneous" reaction kinetics may be appropriate for at least a portion of the biological dechlorination reaction.

Example 4.2: Estimating Concentration Reduction with Instantaneous Reaction Model Assume that BTEX is dissolving into the groundwater in a NAPL zone at a rate that would result in a BTEX concentration of 10 mg/L. Influent groundwater from upgradient of the source is delivering clean groundwater with dissolved oxygen at 6 mg/L. Estimate the resulting concentration of BTEX, assuming a utilization factor of 3 (3 g oxygen consumed for every 1 g of BTEX mineralized) and a relatively fast ("instantaneous") reaction.

SOLUTION: Using the utilization factor $F = 3$, the resulting concentration can be estimated using eq. (4.5) as follows:

$$C = 5 \text{ mg/L} - (6 \text{ mg/L}/3) = 3 \text{ mg/L}$$

Monod Kinetic Model

One of the most common expressions for simulating biodegradation is the hyperbolic saturation function presented by Monod (1942) and referred to as *Monod* or *Michaelis–Menten kinetics*:

$$\mu = \mu_{\max} \frac{C}{K_c + C} \tag{4.6}$$

where μ = growth rate (time^{-1})
$\mu_{\max}$ = maximum growth rate (time^{-1})
C = concentration of growth-limiting substrate (mg/L)
K_c = half-saturation constant

The half-saturation constant, K_c, is also known as the growth-limiting substrate concentration, which allows the microorganism to grow at half the maximum specific growth rate.

The rate equation describing μ as a function of C contains first-order, mixed-order, and zero-order regions. When $C >> K_c$, $K_c + C$ is almost equal to C, and the reaction approaches zero-order with

$$\mu = \mu_{max} \tag{4.7}$$

and μ_{max} becomes the limiting maximum reaction rate. When $C << K_c$, eq. (4.6) reduces to

$$\mu = \frac{\mu_{max}}{K_c} C \tag{4.8}$$

and the reaction approaches first order, with μ_{max}/K_c equal to the first-order rate constant.

In groundwater, the Monod growth function is related to the rate of decrease of an organic compound. This is done by utilizing a yield coefficient, Y, where Y is a measure of the organisms formed per substrate utilized. The change in substrate concentration can then be expressed as follows:

$$\frac{dC}{dt} = \frac{\mu_{max} MC}{Y(K_c + C)} \tag{4.9}$$

where M is the microbial mass in mg/L. Because of the relationship between substrate utilization and the growth of microbial mass, eq. (4.9) is accompanied by an expression of the change in microbial mass as a function of time:

$$\frac{dM}{dt} = \mu_{max} MY \frac{C}{K_c + C} - bM \tag{4.10}$$

where b is a first-order decay coefficient that accounts for cell death.

The advantage of using Monod kinetics is that the constants K_c and μ_{max} uniquely define the rate equation for mineralization of a specific compound. The ratio μ_{max}/K_c also represents the first-order rate constant for degradation when $C << K_c$. This rate constant incorporates both the activity of the degrading population and the substrate dependency of the reaction. It therefore takes into account both population and substrate levels and provides a theoretical basis for extrapolating laboratory rate data to the environment.

The reduction of contaminant concentrations using Monod kinetics can be expressed as

$$\Delta C = M_t \mu_{max} \frac{C}{K_c + C} \Delta t \tag{4.11}$$

where C = contaminant concentration
M_t = total microbial concentration
μ_{max} = maximum contaminant utilization rate per unit mass microorganisms
K_c = contaminant half-saturation constant
Δt = time interval being considered

One of the main difficulties with the Monod kinetic model is estimating the necessary biodegradation parameters for using the model (the maximum growth rate and the half-saturation constants). Monod kinetic parameters have been estimated for only a few compounds, as we see in Chapters 5 and 6.

Example 4.3: Estimating Concentration Reduction with the Monod Kinetic Model Estimate the benzene concentration in groundwater due to biodegradation over a three-year period. Assume a μ_{max} value of 9.3 day^{-1} and a K_c value of 22.16 mg/L. Also assume a benzene source concentration of 5 mg/L and a total microbial concentration of 0.1 mg/L.

SOLUTION: Using eq. (4.11), the change in concentration for benzene over a three-year period can be estimated as follows:

$$\Delta C = M_t \mu_{max} \frac{C}{K_c + C} \Delta t$$

$$= 0.001(9.3)\left(\frac{5}{5 + 22.16}\right)(3)(365)$$

or

$$\Delta C = 1.87 \text{ mg/L}$$

The resulting concentration $C = 5 \text{ mg/L} - 1.87 \text{ mg/L}$, or 3.13 mg/L.

Integrating a Kinetic Biodegradation Model into the Advection–Dispersion Equation

When integrated into the advection/dispersion equation presented in Chapter 3, the mathematical expressions for biodegradation discussed above, allow more accurate simulation of reactive solute transport. For example, substituting first-order decay into eq. (3.14) results in

$$\frac{\partial C}{\partial t} = D_x \frac{\partial^2 C}{\partial x^2} - \upsilon \frac{\partial C}{\partial x} - kC \tag{4.12}$$

where D_x = dispersion coefficient
υ = seepage velocity
k = first-order decay rate

Similarly, incorporating the Monod expression [eq. (4.11)] into the one-dimensional transport equation, results in

$$\frac{\partial C}{\partial t} = D_x \frac{\partial^2 C}{\partial x^2} - \upsilon \frac{\partial C}{\partial x} - M_t \mu_{\max} \frac{C}{K_c + C} \tag{4.13}$$

Integrating the instantaneous reaction into the advection–dispersion equation results in a system of equations, as will be seen in Chapter 8. This is because the instantaneous model will involve a second solute (e.g., oxygen for aerobic biodegradation).

Comparison of First-Order, Instantaneous and Monod Kinetic Models

Although the Monod kinetic model may be the most rigorous of the three models discussed above, the need to determine the Monod kinetic parameters has precluded the use of this model for most field situations. Conceptually, and relative to the fate and transport of a conservative tracer, the instantaneous reaction model represents the maximum concentration reduction possible of a contaminant in a specific groundwater environment. For cases in which concentration reduction never achieves this maximum (i.e., the process is limited by microbial kinetics), the instantaneous reaction model would overestimate the amount of biodegradation. The Monod model or first-order decay model may be more appropriate in such cases.

Rifai and Bedient (1990) compared the instantaneous reaction and the Monod kinetic models. Their results indicate that the instantaneous reaction model is an adequate approximation of Monod kinetics for situations where the rate of the reaction is fast relative to the rate of groundwater transport. Rifai and Bedient (1990) calculated a *Damkohler number*, which is the ratio of the reaction rate to that of the velocity normalized by a characteristic length, and estimate an applicable Damkohler number range for the instantaneous model.

Connor et al. (1994) compare the use of the first-order decay and the instantaneous reaction model and conclude that the first-order model may not be appropriate for simulating aerobic biodegradation of petroleum hydrocarbons because it can overestimate the amount of biodegradation in some cases (i.e., predict more biodegradation than is possible due to electron acceptor limitations).

REFERENCES

Alexander, M., 1994, *Biodegradation and Bioremediation*, Academic Press, San Diego, CA, 302 pp.

Bedient, P. B., Rifai, H. S., and Newell, C. J., 1994, *Groundwater Contamination: Transport and Remediation*, PTR Prentice Hall, Upper Saddle River, NJ, 541 pp.

Borden, R. C., and Bedient, P. B., 1986, Transport of dissolved hydrocarbons influenced by oxygen limited biodegradation: theoretical development, *Water Resour. Res.*, vol. 22, no. 13, pp. 1973–1982.

Bouwer, E. J., 1994, Bioremediation of chlorinated solvents using alternate electron acceptors, in *Handbook of Bioremediation*; R. D. Norris et al., eds., Lewis Publishers, Boca Raton, FL, pp. 149–175.

Bradley, P. M., and Chapelle, F. H., 1996, Anaerobic mineralization of vinyl chloride in Fe(III)-reducing aquifer sediments, *Environ. Sci. Technol.*, vol. 40, pp. 2084–2086.

Chapelle, F. H., 1993, *Groundwater Microbiology and Geochemistry*, Wiley, New York, 424 pp.

Connor, J. A., Newell, C. J., Nevin, J. P., and Rifai, H. S., 1994. Guidelines for use of groundwater spreadsheet models in risk-based corrective action design, *Proceedings of the National Ground Water Association Petroleum Hydrocarbons and Organic Chemicals in Ground Water Conference*, Houston, TX, Nov., pp. 43–55.

CRC, 1990, *CRC Handbook of Chemistry and Physics*, 76th ed., CRC Press, Boca Raton, FL.

Davis, J. W., Klier, N. J., and Carpenter, C. L., 1994, Natural biological attenuation of benzene in groundwater beneath a manufacturing facility, *Ground Water*, vol. 32, no. 2, pp. 215–226.

Dean, J. A., 1972, *Lange's Handbook of Chemistry*, 13th edition, McGraw-Hill, NY.

Dolfing, J., and Harrison, B. K., 1992, The Gibbs free energy of formation of halogenated aromatic compounds and their potential role as electron acceptors in anaerobic environments, *Environ. Sci. Technol.*, vol. 26, pp. 2213–2218.

Ghiorse, W. C., and Wilson, J. T., 1988, Microbial ecology of the terrestrial subsurface, *Adv. Appl. Microbiol.*, vol. 33, pp. 107–172.

Gossett, J. M., and Zinder, S. H., 1996, Microbiological aspects relevant to natural attenuation of chlorinated ethenes, in *Proceedings of the Symposium on Natural Attenuation of Chlorinated Organics in Groundwater*, Dallas, TX, Sept. 11–13, EPA/540/R-96/509, U.S. EPA, Washington, DC.

Holliger, C., and Schumacher, W., 1994, Reductive dehalogenation as a respiratory process, *Antonie Leeuwenhoek*, vol. 66, pp. 239–246.

Howard, P. H., Boethling, R. S., Jarvis, W. F., Meylan, W. M., and Michalenko, E. M., 1991, *Handbook of Environmental Degradation Rates*, Lewis Publishers, Boca Raton, FL.

Kazumi, J., Caldwell, M. E., Suflita, J. M, Lovely, D. R., and Young, L. Y., 1997, Anaerobic biodegradation of benzene in diverse anoxic environments, *Environ. Sci. Technol.*, vol. 31, pp. 813–818.

Langmuir, D., and Whittemore, D. O., 1971, Variations in the stability of precipitated ferric oxyhydroxides, in *Nonequilibrium Systems in Natural Water Chemistry*, J. D. Hem, ed., Advances in Chemistry Series 106, American Chemical Society, Washington, DC.

McCarty, P. L., 1996, Biotic and abiotic transformations of chlorinated solvents in groundwater, in *Proceedings of the Symposium on Natural Attenuation of Chlorinated Organics in Groundwater*, Dallas, TX, Sept. 11–13, EPA/540/R-96/509, pp. 5–9.

McCarty, P. L., and Semprini, L., 1994, Groundwater treatment for chlorinated solvents, in *Handbook of Bioremediation*, R. D. Norris et al., eds., Lewis Publishers, Boca Raton, FL.

McNabb, J. F., and Dunlap, W. J., 1975, Subsurface biological activity in relation to groundwater pollution, *Ground Water*, vol. 13, pp. 33–44

Monod, J., 1942, *Recherches sur la croissance des cultures bactériennes*, Hermann & Cie, Paris.

Naumov, G. B., Ryzhenko, B. N., and Khodakovsky, I. L., 1974, *Handbook of Thermodynamic Data*, (translated from the Russian), USGS-WRD-74-001, U.S. Geological Survey, Washington, DC.

Newell, C. J., Winters, J. A., Rifai, H. S., Miller, R. N., Gonzales, J., and Wiedemeier, T. H., 1995, Modeling of intrinsic remediation with multiple electron acceptors: results from seven sites, in *Proceedings of the Petroleum Hydrocarbons and Organic Chemicals in Groundwater: Prevention, Detection, and Remediation Conference*, Houston, TX, Nov. 29–Dec. 1.

Norris, R. D., 1994, In-situ bioremediation of soils and groundwater contaminated with petroleum hydrocarbons. In Norris, R. D., *Handbook of Bioremediation*, R. D. Norris et al., eds., Lewis Publishers, Boca Raton, FL, pp. 17–37.

Rifai, H. S., and Bedient, P. B., 1990, Comparison of biodegradation kinetics with an instantaneous reaction model, *Water Resour. Res.*, vol. 26, no. 4, pp 637–645.

———, 1994, Field demonstration of natural biological attenuation, in *Hydrocarbon Bioremediation*, R. E. Hinchee, B. C. Alleman, R. E. Hoeppel, and R. N. Miller, eds., Lewis Publishers, Boca Raton, FL, pp. 353–361.

Stumm, W., and Morgan, J. J., 1981, *Aquatic Chemistry*, Wiley, New York.

Wiedemeier, T. H., Wilson, J. T., Kampbell, D. H., Miller, R. N., and Hansen, J. E., 1995, *Technical Protocol for Implementing Intrinsic Remediation with Long-Term Monitoring for Natural Attenuation of Fuel Contamination Dissolved in Groundwater*, U.S. Air Force Center for Environmental Excellence, San Antonio, TX.

Wiedemeier, T. H., Swanson, M. A., Moutoux, D. E., Gordon, E. K., Wilson, J. T., Wilson, B. H., Kampbell, D. H., Hansen, J. E., Haas, P., and Chapelle, F. H., 1996, *Technical Protocol for Evaluating Natural Attenuation of Chlorinated Solvents in Groundwater*, Draft, Revision 1, U.S. Air Force Center for Environmental Excellence, San Antonio, TX.

Wiedemeier, T. H., Swanson, M. A., Moutoux, D. E., Gordon, E. K., Wilson, J. T., Wilson, B. H., Kampbell, D. H., Haas, P. E., Miller, R. N., Hansen, J. E., and Chapelle, F. H., 1998, *Technical Protocol for Evaluating Natural Attenuation of Chlorinated Solvents in Ground Water*, EPA/600/R-98/128, September 1998, U.S. EPA, Washington, DC, ftp://ftp.epa.gov.pub/ada/reports/protocol.pdf.

CHAPTER 5

INTRINSIC BIOREMEDIATION OF PETROLEUM HYDROCARBONS

In Chapter 4 we provided an introduction to the biological mechanisms of contaminant destruction. In this chapter we focus on the biodegradation of the petroleum hydrocarbons with relatively high solubility, mobility, and toxicity that are most commonly found dissolved in groundwater. The most important of these compounds have been the monoaromatic hydrocarbon compounds benzene, toluene, ethylbenzene, and the xylenes (the BTEX compounds), as these monoaromatic hydrocarbon compounds have historically been the regulatory risk drivers at sites contaminated with petroleum hydrocarbons. Because of increased concern regarding methyl *tert*-butyl ether (MTBE), an oxygenate, consideration is also given to the biodegradation of oxygenates such as MTBE. Other compounds, such as polycyclic aromatic hydrocarbons (PAHs), although they are present in fuels and do biodegrade somewhat slowly, are very insoluble and have high sorption coefficients. These properties limit their migration and they typically remain bound to soil–sediment particles and do not partition into groundwater in appreciable quantities. Chlorinated solvent biodegradation is discussed in Chapter 6. In Chapter 7 we present methods for assessment and quantification of intrinsic bioremediation.

Microorganisms perform a unique role in the subsurface in that they are at least partially responsible for producing the crude oil from which benzene, toluene, ethylbenzene, xylene, and most of the other compounds found in petroleum hydrocarbons are derived. As soon as any organism (plant or animal) dies, the remains are attacked by bacteria. Under the right physical, chemical, and biological conditions, this organic matter is transformed but not entirely destroyed and kerogen is formed. Among other things, kerogen contains normal and branched paraffins, naphthenes, and aromatics. Upon continued burial of this material, the temperature in the system increases and crude oil is formed by thermal cracking of kerogen at temperatures between 60 and 150°C (Dickey, 1986). Hydrocarbons used as fuel by humans are produced from the distillation of crude oil. All of the constituents of a refined fuel product (with the exception of oxygenates such as MTBE, which are added to reduce air pollution) are present in the original crude oil; during the oil refining process, the boiling points of the various compounds are used to segregate them into mixtures that are optimal for the use intended. For example, gasoline engines operate most efficiently by burning the lighter hydrocarbons (including BTEX) because they ignite quickly and burn relatively cool. In contrast, jet fuels contain more of the heavier, higher-boiling-point naphthalene compounds; thus, they burn at a higher temperature and produce more energy. Thus bacteria indigenous to the subsurface environment serve a dual role: They are involved in the early stages of oil production, and as will be seen in this chapter, act to degrade the constituents once the oil or refined product is generated, serving to close the loop.

Over the past two decades, numerous laboratory and field studies have shown that hydrocarbon-degrading microorganisms are ubiquitous in the subsurface environment and that these microorganisms can degrade a variety of organic compounds, including components of gasoline, kerosene, diesel, jet fuel, and many other fuel hydrocarbons. Indigenous microorganisms have a distinct advantage over microorganisms injected into the subsurface to enhance biodegradation because they are well adapted to the physical and chemical conditions of the subsurface environment in which they reside. Thus attempts to enhance biodegradation have often met with less-than-anticipated success (Goldstein et al., 1985). Research conducted by the Lawrence Livermore National Laboratories (LLNL; Rice et al., 1995), U.S. Air Force Center for Environmental Excellence (Wiedemeier et al., 1995a), Chevron Research and Technology Company (Buscheck et al., 1996), and the Texas Bureau of Economic Geology (Mace et al., 1997) shows that the majority (greater than about 85 to 90% or more) of the dissolved petroleum-hydrocarbon plumes present in the shallow subsurface of the United States are at steady-state equilibrium, or are receding, probably because of intrinsic bioremediation. The ubiquitous intrinsic biodegradation of petroleum hydrocarbon constituents has caused many states to reevaluate their position concerning the active remediation of these plumes. For example, in some states it is now necessary for the property owner interested in obtaining state funds for active remediation of a fuel spill to show why natural attenuation will not be protective of human health and the environment.

Organic compounds are biodegraded via biological oxidation when electron donors, electron acceptors, and nutrients are combined by microorganisms to produce metabolic by-products and energy for microbial growth. In general, biodegradation proceeds via the following generalized equation:

$$\text{microorganisms} + \text{electron donor} + \text{electron acceptor} + \text{nutrients} \rightarrow \text{metabolic by-products} + \text{energy} + \text{microorganisms}$$

In almost all shallow subsurface environments petroleum hydrocarbons can serve as electron donors in microbial metabolism. When petroleum hydrocarbons are utilized as the primary electron donor for microbial metabolism, they typically are completely degraded or detoxified (Bouwer, 1992). Common electron acceptors include oxygen, nitrate, Fe(III), sulfate, and carbon dioxide. Carbon dioxide, water, nitrogen gas, Fe(II), hydrogen sulfide, and methane are some of the metabolic by-products typically produced from the biodegradation of petroleum hydrocarbons. The biodegradation of petroleum hydrocarbons, especially benzene, toluene, ethylbenzene, and xylenes (BTEX), is limited primarily by electron acceptor availability, and generally will proceed until all of the contaminants that are biochemically accessible to the microbes are oxidized. It is generally observed that an adequate supply of electron acceptors is present in most, if not all, hydrogeologic environments. Table 5.1 summarizes the various mechanisms by which some of the most soluble petroleum hydrocarbons are known to biodegrade.

Subsurface bacteria typically are smaller than many other bacteria, giving rise to a larger surface area/volume ratio. This allows them to utilize nutrients from dilute solutions efficiently (Bouwer, 1992). Bacteria that degrade petroleum hydrocarbons have been known to withstand fluid pressures of hundreds of bars, pH conditions ranging from 1 to 10 standard units, temperatures from 0 to 75°C, and salinities greater than those of normal seawater (Freeze and Cherry, 1979).

Petroleum hydrocarbons biodegrade naturally when an indigenous population of hydrocarbon-degrading microorganisms is present in the aquifer and sufficient concentrations of electron acceptors and nutrients are available to these organisms. In most subsurface environments, both aerobic and anaerobic degradation of petroleum hydrocarbons can occur, often simultaneously in different parts of the plume. For thermodynamic reasons microorganisms preferentially utilize those electron acceptors that provide the greatest amount of free energy during respiration (Bouwer, 1992). The rate of natural biodegradation generally is limited by a lack of electron acceptors rather than by a lack of nutrients such as ammonia, nitrate, or phosphate. Studies at a jet-fuel-contaminated site noted little difference in biodegradation rates in areas with or without nutrient additions (Miller, 1990). These researchers concluded that nitrogen, phosphorus, and other trace nutrients were efficiently recycled by microorganisms at this site.

The driving force for the biodegradation of petroleum hydrocarbons is the transfer of electrons from an electron donor (petroleum hydrocarbon) to an electron acceptor. As discussed in Chapter 4, the energy produced by these reactions is quantified by the Gibbs free energy of the reaction (ΔG_r). Tables 4.2 and 4.3 present half-cell reactions for common electron acceptors and selected petroleum hydrocarbons. These tables also give the calculated ΔG_r values for each of the half-cell reactions. To derive energy for cell maintenance and production from petroleum hydrocarbons, the microorganisms must couple electron donor oxidation with the reduction of an electron acceptor.

Microorganisms are able to utilize electron transport systems and chemiosmosis to combine energetically favorable and unfavorable reactions with the net result of producing energy for life processes. By coupling the oxidation of an electron donor such as BTEX with the reduction of an electron acceptor, the overall reaction becomes energy yielding, as indicated by the negative ΔG_r values. Biodegradation of petroleum hydrocarbons most commonly occurs by means of aerobic, nitrate-reducing, Fe(III)-reducing, sulfate-reducing, and methanogenic respiration. Coupled

TABLE 5.1. Selected Research Supporting the Biodegradation of BTEX

Terminal Electron-Accepting Process	Compound			
	Benzene	Toluene	Ethylbenzene	Xylenes
Aerobic respiration	Marr and Stone (1961), Alvarez and Vogel (1991)	Alvarez and Vogel (1991)	Anecdotal evidence	Alvarez and Vogel (1991)
Denitrification	Major et al. (1988), Kukor and Olsen (1989), Morgan et al. (1993)	Piet and Smeenk (1985), Zeyer et al. (1986), Kuhn et al. (1988), Major et al. (1988), Dolfing et al. (1990), Altenschmidt and Fuchs (1991), Evans et al. (1991a,b, 1992), Schocher et al. (1991), Flyvbjerg et al. (1993), Alvarez and Vogel (1994), Fries et al. (1994), Anders et al. (1995), Chaudhuri and Wiesmann (1995), Colberg and Young (1995), Rabus and Widdel (1995), Zhou et al. (1995), Ball et al. (1996), Ball and Reinhard (1996)	Kuhn et al. (1985), Ball et al. (1991, 1994, 1996), Hutchins (1991a,b), Barbaro et al. (1992), Hutchins (1992), Morgan et al. (1993), Patterson et al. (1993), (1993), Rabus and Widdel (1995), Ball and Reinhard (1996), Ball et al. (1996)	Kuhn et al. (1985, 1988), Major et al. (1988), Evans et al. (1991a), Hutchins (1991a,b, 1992, 1993), Barbaro et al. (1992), Barlaz et al. (1993), Morgan et al. (1993), Alvarez and Vogel (1994), Ball et al. (1994), Rabus and Widdel (1995)

Fe(III) reduction	Cozzarelli et al. (1990), Baedecker et al. (1993), Lovley et al. (1994a, 1996), Kazumi et al. (1997)	Lovley et al. (1989, 1994), Cozzarelli et al. (1990), Lovley and Lonergan (1990), Baedecker et al. (1993), Albrechtsen (1994), Cozzarelli et al. (1994)	Cozzarelli et al. (1990, 1994), Borden et al. (1994)	Borden et al. (1994), Cozzarelli et al. (1994), Barlaz et al. (1995)
Sulfate reduction	Piet and Smeenk (1985), Edwards and Grbić-Galić (1992), Davis et al. (1994), Lovley et al. (1995), Kazumi et al. (1997)	Ward et al. (1980), Piet and Smeenk (1985), Beller et al. (1991, 1992a,b, 1995, 1996), Haag et al. (1991), Edwards et al. (1992), Flyvbjerg et al. (1993), Patterson et al. (1993), Rabus et al. (1993), Thierrin et al. (1993), Ball et al. (1994), Rueter et al. (1994)	Piet and Smeenk (1985), Acton and Barker (1992), Thierrin et al. (1993), Barlaz et al. (1995)	Piet and Smeenk (1985), Haag et al. (1991), Acton and Barker (1992), Edwards et al. (1992), Thierrin et al. (1993), Ball et al. (1994), Rueter et al. (1994), Reinhard et al. (1984), Beller et al. (1995), Ball and Reinhard (1996)
Methanogenesis	Vogel and Grbić-Galić (1986), Wilson et al. (1986, 1990), Grbić-Galić and Vogel (1987), Van Beelen and Van Keulen (1990), Watwood et al. (1991), Ghosh and Sun (1992), Davis et al. (1994), Weiner and Lovley (1998)	Reinhard et al. (1984), Grbić-Galić (1986), Vogel and Grbić-Galić (1986), Wilson et al. (1986, 1990), Beller et al. (1991), Acton and Barker (1992), Liang and Grbić-Galić (1993), Edwards and Grbić-Galić (1994), Grbić-Galić and Vogel (1987)	Wilson et al. (1986), Grbić-Galić (1986), Reinhard et al. (1984), Acton and Barker (1992)	Wilson et al. (1986, 1990), Edwards and Grbić-Galić (1994)

oxidation–reduction reactions would be expected to occur in order of their thermodynamic energy yield assuming that there are organisms capable of facilitating each reaction and that there is an adequate supply of electron donors and acceptors (Stumm and Morgan, 1981; Chapelle, 1993). Aerobic respiration is the first reaction to occur in an aerobic environment that contains microorganisms capable of aerobic respiration (i.e., obligate aerobes or facultative anaerobes) (Bouwer, 1992; Chapelle, 1993). Once the available dissolved oxygen is depleted and anaerobic conditions dominate the interior regions of the dissolved hydrocarbon plume, facultative or obligate anaerobic microorganisms can utilize other electron acceptors in the following order of preference: nitrate, manganese(IV), Fe(III), sulfate, and finally, carbon dioxide. As each electron acceptor being utilized for biodegradation becomes depleted, the next most preferable electron acceptor is utilized. Each successive redox couple provides less energy to the microorganism involved in the reaction.

5.1 BIODEGRADATION PROCESSES FOR PETROLEUM HYDROCARBONS

The degradation of petroleum hydrocarbons and, more specifically, fuel hydrocarbons has been studied and described by numerous researchers for a wide variety of hydrogeologic and physical and chemical conditions. The biodegradation of fuel hydrocarbons is mainly through primary metabolism, in which the hydrocarbon is used as a growth substrate and the degradation yields energy for the organism. Fuel-hydrocarbon biodegradation can be categorized into two different types of biological reactions depending on the presence or absence of dissolved oxygen and the predominant electron acceptors. Aerobic biodegradation refers to the metabolism of petroleum hydrocarbons using oxygen as the terminal electron acceptor, and anaerobic biodegradation encompasses biodegradation reactions that take place in the absence of oxygen.

Aerobic Biodegradation of Petroleum Hydrocarbons

Petroleum hydrocarbons can be biodegraded via the aerobic pathway when indigenous populations of hydrocarbon-degrading microorganisms are supplied with the oxygen and nutrients necessary to utilize petroleum hydrocarbons as an energy source. Much research has been done on the aerobic biodegradation of fuel constituents, and Borden (1994) states that almost all types of petroleum hydrocarbons will biodegrade under aerobic conditions. For example, Jamison et al. (1975) found that a microbial population obtained from a gasoline-contaminated aquifer readily degraded the majority of components found in gasoline. In general, the low- to moderate-molecular-weight hydrocarbons appear to be the most easily degradable, and the resistance to biodegradation increases as the molecular weight of the compound increases. The relative resistance of heavier hydrocarbons to biodegradation is compensated for by the fact that these compounds are more prone to sorb to aquifer solids, so they are not as mobile as the lighter hydrocarbons and thus are not as likely to migrate to potential receptors.

Petroleum-hydrocarbon mixtures with their multitude of potential primary substrates will undergo simultaneous degradation of the aliphatic, aromatic, and alicyclic hydrocarbons. Significantly, low-molecular-weight aromatic hydrocarbons such as BTEX are easily biodegraded in the concentrations found dissolved in groundwater. The mineralization of aromatic hydrocarbons to carbon dioxide and water under aerobic conditions involves the use of oxygen as a cosubstrate during the initial stages of hydrocarbon metabolism, and the later use of oxygen as the terminal electron acceptor for energy production (Higgins and Gilbert, 1978; Gibson and Subramanian, 1984; Young, 1984). The O_2–H_2O oxidation–reduction couple has a high oxidizing potential and when coupled with endothermic reactions involving BTEX can be used by microorganisms to release a large amount of free energy. In fact, reduction of molecular oxygen is one of the most energetically favorable of the oxidation–reduction reactions involved in petroleum-hydrocarbon biodegradation.

Aerobic biodegradation of petroleum hydrocarbons requires the action of oxygenases and therefore the presence of free dissolved oxygen. Subsurface environments quickly become devoid of oxygen if high enough concentrations of organic compounds are present. When this is the case, the rate of aerobic biodegradation will typically be limited by oxygen supply rather than by nutrient concentration (Borden and Bedient, 1986). Although nutrients such as nitrogen and phosphorus are essential for biodegradation of organic contaminants by bacteria, the influence of inorganic and organic nutrients on in situ biodegradation varies substantially, and in some cases, the addition of nutrients into the subsurface environment has been shown to have little or no effect on biodegradation rates of hydrocarbons (Swindoll et al., 1988; Miller, 1990). In any event, biodegradation of petroleum hydrocarbons occurs in most subsurface environments without the addition of supplemental nutrients.

The following equations describe the overall stoichiometry of aromatic hydrocarbon biodegradation. In the absence of microbial cell production, the oxidation (mineralization) of benzene to carbon dioxide and water is given by

$$C_6H_6 + 7.5O_2 \rightarrow 6CO_2 + 3H_2O \tag{5.1}$$

Using this stoichiometry, 7.5 mol of oxygen is required to metabolize 1 mol of benzene. On a mass basis, the ratio of oxygen to benzene is given by:

Molecular weights:	Benzene	$6(12) + 6(1) = 78$ g
	Oxygen	$7.5(32) = 240$ g

$$\text{Mass ratio of oxygen to benzene} = 240:78 = 3.08:1$$

On the basis of this stoichiometry, 3.08 mg of oxygen is required to metabolize 1 mg of benzene completely in the absence of microbial cell production. Similar calculations can be made for toluene, ethylbenzene, and the xylenes. One method that can be used to estimate the amount of oxygen utilized by aerobic bacteria to degrade total BTEX is to average the amount of oxygen consumed during the biodegradation of each compound separately. By doing this, 3.14 mg of oxygen is consumed during the biodegradation of 1 mg of total BTEX (Table 5.2). Because

TABLE 5.2. Mass Ratio of Electron Acceptors Removed or Metabolic By-products Produced to Total BTEX Degraded, BTEX Utilization Factors, and Number of Electrons Transferred for a Given Terminal Electron-Accepting Process[a]

Terminal Electron Accepting Process	Average Mass Ratio of Electron Acceptor to Total BTEX	Average Mass Ratio of Metabolic By-product to Total BTEX	BTEX Utilization Factor, F (mg/mg)
Aerobic respiration	3.14:1	—	3.14
Denitrification	4.9:1	—	4.9
Fe(III) reduction	—	21.8:1	21.8
Sulfate reduction	4.7:1	—	4.7
Methanogenesis	—	0.78:1	0.78

[a] Simple average of all BTEX compounds based on individual compound stoichiometry.

the BTEX compounds are responsible for most of the oxygen consumed in a petroleum-hydrocarbon-contaminated aquifer, it is important to include all these compounds when modeling the environmental fate of petroleum-hydrocarbon plumes under natural conditions.

The stoichiometry presented above does not take into account microbial cell mass production. When cell mass production is accounted for, the mineralization of benzene to carbon dioxide and water is given by

$$C_6H_6 + 2.5O_2 + HCO_3 + NH_4 \rightarrow C_5H_7O_2N + 2CO_2 + 2H_2O \tag{5.2}$$

From this it can be seen that only 2.5 mol of dissolved oxygen is required to mineralize 1 mol of benzene when cell mass production is taken into account. On a mass basis, the ratio of dissolved oxygen to benzene is given by:

Molecular weights:	Benzene	$12(6) + (1)6 = 78$ g/mol
	Oxygen	$2.5(32) = 80$ g/mol

$$\text{Mass ratio of oxygen to benzene} = 80/78 = 1.03{:}1$$

Actual oxygen requirements may vary from those predicted by the stoichiometric relationships presented above because they are dependent on the bacterial yield coefficient (Y_m) that describes the amount of biomass produced per unit mass of substrate biodegraded. Yields of microbial biomass vary depending on the thermodynamics of substrate biodegradation and on the availability of oxygen, nutrients, and substrate concentration (McCarty et al., 1981). Energy for cell maintenance is also needed, and this energy need is not reflected in the stoichiometric relationships presented above. The culmination of these few additional factors suggests that the actual oxygen demand will range from approximately 1 to 3 mg of oxygen per 1 mg of benzene removed through biodegradation. Recent research (Rice et al., 1995; Mace et al.,

1997) shows that the majority of dissolved petroleum-hydrocarbon plumes are at steady-state equilibrium. In fact, most of these plumes are at steady state by the time they are discovered. When this is the case, the microbial death rate is approximately equal to the reproduction rate and the net production of biomass is negligible, so the stoichiometry presented in eq. (5.1) would be appropriate. When dealing with a relatively young plume that has not yet reached steady-state equilibrium, the stoichiometry presented in eq. (5.2) may be more appropriate.

For aerobic respiration to occur, the following conditions must be met: (1) aerobic bacteria must be present, (2) oxygen must be present at concentrations greater than about 0.5 mg/L, (3) biodegradable organic carbon must be present, and (4) oxidizing conditions must prevail. In addition, hydrogen concentrations will be less than about 0.1 M. Most shallow aquifers are characterized by these conditions when petroleum hydrocarbons are introduced into the system. If these conditions are not met, an anaerobic biodegradation pathway probably will result in the biodegradation of petroleum hydrocarbons.

The biodegradation of petroleum hydrocarbons occurs more rapidly under aerobic conditions than during any other terminal electron-accepting process, although denitrification can occur almost as quickly. Because of rapid aerobic biodegradation rates and oxygen's low aqueous solubility, subsurface environments contaminated with petroleum hydrocarbons quickly become devoid of oxygen. A reduction in dissolved oxygen concentrations within an existing plume of dissolved organic carbon is a strong indication that indigenous bacteria have biodegraded this carbon via aerobic respiration. In the experience of the authors, dissolved oxygen is absent in groundwater that contains petroleum hydrocarbons. Although nutrients such as nitrogen and phosphorus are essential for biodegradation of organic contaminants by bacteria, the authors are not aware of a site where nitrogen and phosphorus were limiting factors in petroleum-hydrocarbon biodegradation.

Because of the low aqueous solubility of oxygen, subsurface systems become anaerobic soon after the introduction of petroleum hydrocarbons. Once the dissolved oxygen is depleted, anaerobic microorganisms begin to biodegrade the petroleum hydrocarbons and it is unlikely that the system will become aerobic again until the petroleum-hydrocarbon supply is exhausted.

Anaerobic Biodegradation of Petroleum Hydrocarbons

Although aerobic microbial degradation of aromatic hydrocarbons has been recognized and understood for an extended period, the significance of anaerobic degradation of these compounds was not fully appreciated until recently. Only within the last decade have researchers begun to focus on studying the extent to which anaerobic biodegradation reduces volatile organic compounds such as BTEX and the rate at which such biodegradation occurs. In the last five years there has been a substantial increase in the knowledge of its significance in reducing the mobility and mass of volatile organic compounds dissolved in groundwater systems. In fact, it is now apparent that anaerobic biodegradation is the most significant process working to remove BTEX from groundwater.

One of the first reports of anaerobic benzene oxidation is that by Ward et al. (1980). These researchers reported the formation of $^{14}CO_2$ and $^{14}CH_4$ from radiolabeled [^{14}C]benzene under methanogenic conditions. The materials for the study were obtained from petroleum-contaminated salt marsh and estuarine sediments. Since the work of Ward et al. (1980), numerous researchers have shown that petroleum hydrocarbons can be degraded under anaerobic conditions.

Although we now know that anaerobic microorganisms will degrade organic carbon, including petroleum hydrocarbons, in the absence of molecular oxygen, the biochemical mechanisms underlying the initial enzymatic oxidation reactions are largely unresolved (Ball et al., 1996). Current research, however, is beginning to bring to light the initial reactions in anaerobic biodegradation of petroleum hydrocarbons. For example, Ball et al. (1996) show that the initial reactions in anaerobic ethyl benzene oxidation by denitrifying bacteria involve incorporation of oxygen derived from water into the methylene group, resulting in the formation of 1-phenylethanol. This is in contrast to aerobic biodegradation of aromatic hydrocarbons in which the first oxidation reactions are catalyzed by mono- and dioxygenases (Higgins and Gilbert, 1978; Gibson and Subramanian, 1984; Young, 1984; Smith, 1990).

Soon after petroleum-hydrocarbon contamination enters the groundwater system, rapid depletion of dissolved oxygen caused by increased levels of aerobic microbial respiration results in the establishment of anaerobic conditions within the dissolved contaminant plume. Certain requirements must be met for anaerobic bacteria to degrade petroleum hydrocarbons, including absence of dissolved oxygen; availability of carbon sources such as petroleum hydrocarbons, electron acceptors, and essential nutrients; and proper ranges of pH, temperature, salinity, and redox potential. Depending on the type of electron acceptor present [nitrate, Fe(III), Mn(IV), sulfate, or carbon dioxide], pH conditions, and redox potential, anaerobic biodegradation can occur by denitrification, Mn(IV) reduction, Fe(III) reduction, sulfate reduction, or methanogenesis. Environmental conditions and microbial competition ultimately determine which processes will dominate, but in a typical aquifer denitrification typically occurs first, followed by Mn(IV) reduction, Fe(III) reduction, sulfate reduction, and finally by methanogenesis. Significantly, Vroblesky and Chapelle (1994) show that the dominant terminal electron-accepting process (TEAP) can vary both temporally and spatially in an aquifer with petroleum hydrocarbon contamination. For example, a given area within an aquifer may switch among Fe(III) reduction, sulfate reduction, and methanogenesis, depending on groundwater recharge. With the exception of methanogenesis, all of the TEAPs mentioned above are purely respirative pathways. That is, the microorganisms respire ("breath") nitrate, Mn(IV), Fe(III), and sulfate while oxidizing petroleum hydrocarbons. Methanogenesis is unique in that it combines fermentation with the respiration of carbon dioxide. Each of these TEAPs and their involvement in petroleum-hydrocarbon degradation are discussed below.

An alternative conceptual model has been advanced by Salanitro et al. (1997) in which BTEX degradation in aquifers is controlled by aerobic processes and is limited by the influx of oxygen. They suggest that

> the qualitative and quantitative assessment of field observations regarding the contaminant mass removal . . . is confounded by the continual advective and diffusive influx of O_2 through water table fluctuations, rainfall events, and the vadose zone capillary fringe. At sites in which low DO is measured within the plume soluble hydrocarbons would degrade as rapidly as in ground water with higher DO (e.g., >2 mg/L) if there is a continuous diffusion of O_2. The equilibrium of these low DO plumes would be controlled by abiotic- and microbial-consuming processes. Also, these O_2 diffusion-limited aquifers may not be anaerobic (reducing and at low redox potential) but represent stable, slowly degrading plumes.

However, the results presented by the authors do not necessarily support either the conclusion that degradation of BTEX is dominated by aerobic processes or that most BTEX affected aquifers are not anaerobic. In soil–groundwater microcosms, the authors observed degradation of BTX under both aerobic and a variety of anaerobic conditions. They reported BTX degradation rates of 0.11 to 0.26 day^{-1} in O_2 amended microcosms, 0.06 to 0.22 day^{-1} in NO_3^- amended microcosms, 0.16 to 0.19 day^{-1} in SO_4^- amended microcosms, and 0.04 to 0.20 day^{-1} in anaerobic microcosms without an added electron acceptor. These results are consistent with the observations of numerous other researchers that degradation of BTX compounds can occur under a variety of anaerobic conditions. Salanitro et al. (1997) also reported that CH_4, H_2, and H_2S were produced only in anaerobic microcosms, not in aerobic microcosms or constant low-DO microcosms. Thus the widespread observation of CH_4 production in BTEX-affected aquifers indicates that portions of these aquifers are in fact anaerobic and that degradation of BTEX is occurring through anaerobic processes.

Another conceptual model involves a two-step process, where aerobic bacteria partially degrade the BTEX compounds, and then anaerobic bacteria mineralize the aerobic biodegradation breakdown products. This conceptual model fits the observations of some researchers, who reported no anaerobic biodegradation of benzene at all [see the review in Aronson and Howard (1997)] or only after long lag times (e.g., Kazumi et al., 1997), and could possibly explain the extensive anaerobic reactions occurring in the subsurface where BTEX compounds are present. This conceptual model makes less sense from a thermodynamic point of view, as an aerobe that can break the aromatic ring (which requires energy) would not tend to release the high-energy breakdown products (e.g., pimelate) into the environment for other bacteria to consume. In addition, this mechanism does not match the observations of many other researchers who have reported anaerobic benzene biodegradation (see Table 5.1), or the results of numerous studies where toluene, ethylbenzene, and the xylenes are reported to readily be biodegradable anaerobically (Table 5.1).

Oxidation of Petroleum Hydrocarbons via Denitrification. After almost all dissolved oxygen has been removed from an aquifer via aerobic respiration and anaerobic conditions prevail, nitrate can be used as an electron acceptor by microorganisms that mineralize petroleum hydrocarbons via denitrification. Many species of bacteria are capable of reducing nitrate to produce respirative energy (McCarty, 1972; Riser-Roberts, 1992; Chapelle, 1993) and over the past decade or so, many researchers have shown that benzene, toluene, ethylbenzene, the xylenes, naphthalene, and a variety of other compounds can be degraded when nitrate is used as the terminal electron acceptor (Table 5.1). In fact, many species of bacteria are capable of reducing nitrate to produce respirative energy for organic carbon degradation. Denitrification occurs in the following sequence (Payne, 1981; von Gunten and Zobrist, 1993):

$$NO_3^- \Rightarrow NO_2^- \Rightarrow NO \Rightarrow N_2O \Rightarrow NH_4^+ \Rightarrow N_2$$

Each reaction in this sequence is catalyzed by different microorganisms (Chapelle, 1993).

The following equations describe the overall stoichiometry of benzene oxidation via denitrification. In the absence of microbial cell production, the mineralization of benzene to carbon dioxide and water is given by

$$6NO_3^- + 6H^+ + C_6H_6 \rightarrow 6CO_2 + 6H_2O + 3N_2 \qquad (5.3)$$

Using this stoichiometry, 6 mol of nitrate is required to metabolize 1 mol of benzene. On a mass basis, the ratio of nitrate to benzene is given by:

Molecular weights:	Benzene	$6(12) + 6(1) = 78$ g
	Nitrate	$6(62) = 372$ g

$$\text{Mass ratio of nitrate to benzene} = 372:78 = 4.77:1$$

This shows that 4.77 mg of nitrate is required to metabolize 1 mg of benzene completely. As shown by this stoichiometry, denitrification ultimately results in the formation of carbon dioxide, water, and nitrogen gas (N_2). Similar calculations can be made for toluene, ethylbenzene, and the xylenes. One method that can be used to estimate the amount of nitrate utilized by anaerobic bacteria to degrade total BTEX is to average the amount of nitrate consumed during the biodegradation of each compound separately. Based on these calculations, the average ratio of nitrate consumed per mole of BTEX degraded is 4.9:1 (Table 5.2). Although these calculations do not take into account the production of biomass, it is not likely that this omission will result in significant error (see the discussion under aerobic respiration). Because the BTEX compounds are responsible for much of the nitrate consumed in a petroleum-hydrocarbon-contaminated aquifer, it is important to include all of these compounds when modeling the environmental fate of petroleum-hydrocarbon plumes under natural conditions.

For denitrification to occur, the following conditions must be met (Starr and Gillham, 1993): (1) nitrate-reducing bacteria must be present in the affected aquifer, (2) nitrate must be present, (3) biodegradable organic carbon must be present, and (4) lightly reducing conditions must prevail. Denitrification is favored when all of these conditions are met and hydrogen concentrations are less than about 0.1 nM. If these conditions are not met, another anaerobic biodegradation pathway, such as

Fe(III) reduction, sulfate reduction, or methanogenesis, probably will result in the biodegradation of petroleum hydrocarbons.

Plumes of dissolved petroleum hydrocarbons generally do not contain nitrate, indicating that adaptation of the necessary bacterial population is rapid and that ambient concentrations of nitrate are quickly consumed. This is similar to the relationship observed between dissolved oxygen and BTEX in contaminant plumes, and suggests that nitrate availability, and not microbial processes, may be the limiting factor during denitrification. Thus in areas where denitrification is occurring, there will be a strong correlation between areas with elevated BTEX concentrations and depleted nitrate concentrations relative to measured background concentrations.

Work conducted by several authors suggests that the biodegradation of BTEX under denitrifying conditions occurs in the following order: toluene, *p*-xylene, *m*-xylene, ethylbenzene, and finally *o*-xylene (Norris et al., 1994). The work by Kuhn et al. (1988), Evans et al. (1991a, b), and Hutchins et al. (1991) suggests that benzene is biologically recalcitrant under denitrifying conditions. However, Major et al. (1988) report degradation of benzene under conditions thought to be denitrifying. Kukor and Olsen (1989) and Morgan et al. (1993) also note biodegradation of benzene under denitrifying conditions.

Using sediment collected from the treatment pond of an oil refinery, Ball et al. (1996) showed that strain EB1 was capable of growth using ethylbenzene as the sole electron and carbon source. These researchers also showed that strain EbN1 was also able to use toluene as an electron donor and carbon source under anoxic conditions.

Kuhn et al. (1985) used aquifer material taken from the water–sediment interface of a river in a microcosm study of anaerobic hydrocarbon biodegradation. The indigenous microorganisms were supplied with nitrate. Biodegradation of *p*-xylene and *m*-xylene occurred after three months. Biodegradation of *o*-xylene occurred only after removal of both *p*-xylene and *m*-xylene. The degradation of the xylenes in this experiment was clearly linked to nitrate respiration. In related experiments, *m*-xylene was completely biodegraded under denitrifying conditions in a similar laboratory microcosm, and 80% of ^{14}C-radiolabeled *m*-xylene and 75% of ^{14}C-radiolabeled toluene were mineralized after eight days (Zeyer et al., 1986).

Oxidation of Petroleum Hydrocarbons via Dissimilatory Fe(III) Reduction. After the available dissolved oxygen and nitrate in an aquifer have been depleted, Fe(III) can be used as an electron acceptor. Over the past decade, many researchers have shown that benzene, toluene, ethylbenzene, and the xylenes can be degraded when Fe(III) is used as the terminal electron acceptor (Table 5.1).

The reduction of insoluble iron [Fe^{3+}, ferric iron, Fe(III), or iron(III)] to the soluble form [Fe^{2+}, ferrous iron, Fe(II), or iron(II)] through microbially mediated oxidation of organic matter (including pollutants) in groundwater appears to be a common occurrence. In fact, in typical sedimentary environments, Fe(III) reduction to Fe(II) is primarily the result of the enzymatic activity of specialized Fe(III)-reducing microorganisms (Lovley, 1991). The best forms of Fe(III) for microbiological reduction are poorly crystalline Fe(III) hydroxides, Fe(III) oxyhydroxides, and Fe(III) oxides (Lovley, 1991). Fe(III) in these forms is present in large amounts in some sedimentary systems, thus providing a large reservoir of potential electron acceptors to facilitate biodegradation of petroleum hydrocarbons. Interestingly, studies with iron-reducing isolates show that these microorganisms must be in direct contact with the Fe(III) to facilitate its reduction (Lovley et al., 1991).

The following equations describe the overall stoichiometry of benzene oxidation by iron reduction caused by anaerobic microbial biodegradation. In the absence of microbial cell production, the mineralization of benzene is given by

$$60H^+ + 30Fe(OH)_3 + C_6H_6 \rightarrow 6CO_2 + 30Fe^{2+} + 78H_2O \qquad (5.4)$$

Using this stoichiometry, 30 mol of $Fe(OH)_3$ is required to metabolize 1 mol of benzene. On a mass basis, the ratio of $Fe(OH)_3$ to benzene is given by

Molecular weights:	Benzene	$6(12) + 6(1) = 78$ g
	$Fe(OH)_3$	$30(106.85) = 3205.41$ g

$$\text{Mass ratio of } Fe(OH)_3 \text{ to benzene} = 3205.41{:}78 = 41.1{:}1$$

Thus 41.1 mg of $Fe(OH)_3$ is required to metabolize 1 mg of benzene completely using this stoichiometry. Alternatively, the mass ratio of Fe(II) produced during respiration to benzene degraded can be calculated and is given by

Molecular weights:	Benzene	$6(12) + 6(1) = 78$ g
	Fe^{2+}	$30(55.85) = 1675.5$ g

$$\text{Mass ratio of } Fe^{2+} \text{ to benzene} = 1675.5{:}78 = 21.5{:}1$$

On the basis of this stoichiometry, 21.5 mg of Fe^{2+} is produced during mineralization of 1 mg of benzene. Similar calculations can be made for toluene, ethylbenzene, and xylene. One method that can be used to estimate the amount of Fe(II) produced by anaerobic bacteria to degrade total BTEX is to average the amount of Fe(II) produced during the biodegradation of each compound separately. Based on these calculations, the average ratio of Fe(II) produced per mole of BTEX degraded is 21.8:1 (Table 5.2). Although these calculations do not take into account the production of biomass, it is not likely that this omission will result in significant error (see the discussion under aerobic respiration). These numbers probably represent an extremely conservative estimate of the amount of organic matter assimilated during microbial Fe(III) reduction. The reason for this is that Fe(III) reduction can result in the formation of several Fe(II)-containing minerals such as magnetite (Lovley, 1991), and thus the Fe(II) produced during the reaction will not be measurable in groundwater. In addition, iron can be recycled as the ORP of the groundwater changes.

For Fe(III) reduction to occur, the following conditions must be met: (1) Fe(III)-reducing bacteria must be present in the affected aquifer and in direct contact with the iron, (2) biologically available Fe(III) must be present, (3) biodegradable organic carbon must be present, and (4) reducing conditions must prevail. Fe(III) reduction is favored when all of these conditions are met and hydrogen concentrations range from about 0.2 nM to about 0.8 nM. If these conditions are not met, another anaerobic biodegradation pathway, such as sulfate reduction or methanogenesis, probably will result in the biodegradation of petroleum hydrocarbons. If biologically available Fe(III) is present in the aquifer matrix, the oxidation of BTEX via Fe(III) reduction typically will result in high concentrations of Fe(II) in groundwater within and near the contaminant plume. This being the case, Fe(II) concentrations in groundwater can be measured to determine if Fe(III) is being used as an electron acceptor. If areas with elevated petroleum hydrocarbon concentrations coincide with areas of high Fe(II) relative to measured background concentrations, biodegradation via Fe(III) reduction is probably occurring. Total Fe(III) concentrations as determined through common laboratory techniques do not have much relevance, because without knowing the degree of crystallinity of the iron it is not possible to determine how much of the iron is available to microorganisms.

Although relatively little is known about the anaerobic metabolic pathways involving the reduction of Fe(III), this exothermic process has been shown to be a major metabolic pathway for some microorganisms (Lovley and Phillips, 1987; Chapelle, 1993). In fact, the available evidence suggests that Fe(III) reduction to Fe(II) coupled with the oxidation of petroleum hydrocarbons is an important anaerobic degradation process. High concentrations of dissolved Fe(II) were once attributed to the spontaneous and reversible reduction of Fe(III) oxyhydroxides, which are thermodynamically unstable in the presence of organic compounds under anaerobic conditions (Chapelle, 1993). Yet the work of Lovley et al. (1991) and others suggests that the reduction of Fe(III) cannot occur without microbial mediation. None of the common organic compounds found in low-temperature, neutral, reducing groundwater could reduce Fe(III) oxyhydroxides to Fe(II) under sterile laboratory conditions (Lovley et al., 1991).

Although potentially important in some aquifer systems, not as much is known about the biodegradation of petroleum hydrocarbons under Mn(IV)-reducing conditions. The reader is referred to the excellent summary paper by Lovley (1991) for a review of Fe(III) and Mn(IV) reduction.

Oxidation of Petroleum Hydrocarbons via Sulfate Reduction. Once the available oxygen and nitrate in the groundwater system have been depleted, sulfate-reducing bacteria can begin degrading petroleum hydrocarbons. Decreases in Eh, along with dissolved oxygen and nitrate depletion, will favor sulfate-reducing bacteria. Over the past decade, many researchers have shown that benzene, toluene, ethylbenzene, the xylenes, naphthalene, and a variety of other compounds can be degraded when sulfate is used as the terminal electron acceptor (Table 5.1).

The following equations describe the overall stoichiometry of BTEX oxidation by sulfate reduction caused by anaerobic microbial biodegradation. In the absence of microbial cell production, the mineralization of benzene is given by

$$7.5H^+ + 3.75SO_4^{2-} + C_6H_6 \rightarrow 6CO_2 + 3.75H_2S + 3H_2O \qquad (5.5)$$

Based on this stoichiometry, 3.75 mol of sulfate is required to metabolize 1 mol of benzene. On a mass basis, the ratio of sulfate to benzene is given by

Molecular weights:	Benzene	$6(12) + 6(1) = 78$ g
	Sulfate	$3.75(96) = 360$ g

$$\text{Mass ratio of sulfate to benzene} = 360{:}78 = 4.6{:}1$$

Using this stoichiometry 4.6 mg of sulfate is required to completely metabolize 1 mg of benzene. Similar calculations can be made for toluene, ethylbenzene, and the xylenes. One method that can be used to estimate the amount of sulfate utilized by anaerobic bacteria to degrade total BTEX is to average the amount of sulfate consumed during the biodegradation of each compound separately. Based on these calculations, the average ratio of sulfate consumed per mole of BTEX degraded is 4.7:1 (Table 5.2). Although these calculations do not take into account the production of biomass, it is not likely that this omission will result in significant error (see the discussion under aerobic respiration). Because the BTEX compounds are responsible for most of the sulfate consumed in a petroleum-hydrocarbon-contaminated aquifer, it is important to include all of these compounds when modeling the environmental fate of petroleum-hydrocarbon plumes under natural conditions.

For sulfate reduction to occur, the following conditions must be met: (1) sulfate-reducing bacteria must be present in the affected aquifer, (2) sulfate must be present, (3) biodegradable organic carbon must be present, and (4) reducing conditions must prevail. Sulfate reduction is favored when all of these conditions are met and hydrogen concentrations range from about 1.0 to 4.0 nM. If these conditions are not met, then methanogenesis likely will result in the biodegradation of petroleum hydrocarbons.

Low sulfate concentrations commonly result from the bacterial reduction of sulfate. In waters containing dissolved BTEX, sulfate concentrations typically are depressed, often to the point that they can no longer be measured. Sulfate is reduced to sulfide during the oxidation of BTEX. The presence of decreased concentrations of sulfate and increased concentrations of sulfide relative to background concentrations indicates that sulfate may be utilized in BTEX oxidation reactions at a site. In general, the extent and significance of BTEX biodegradation via sulfate reduction is not well understood (Norris et al., 1994). Although oxidation of benzene by sulfate reduction is thermodynamically favorable, it is not as favorable as aerobic respiration, denitrification, or iron reduction. Additionally, sulfate-reducing microorganisms typically are sensitive to environmental conditions, including temperature, inorganic nutrients, and pH (Zehnder, 1978). An imbalance in suitable environmental conditions could limit the significance of BTEX degradation via sulfate reduction.

According to Hem (1985), low sulfate concentrations commonly result from the bacterial reduction of sulfate. In waters containing dissolved BTEX, sulfate concentrations are further reduced, and these microbes utilize BTEX as the oxidant. Lovley et al. (1995) show that under sulfate-reducing conditions, benzene is mineralized to carbon dioxide and water with no intermediate products such as phenol, benzoate, *p*-hydroxybenzoate, cyclohexane, and acetate. Furthermore, the results of their work show that benzene can be biodegraded in the absence of dissolved oxygen. Thierrin et al. (1992a) described the biodegradation of a BTEX plume from a gasoline spill in Perth, Western Australia, under sulfate-reducing conditions. Edwards et al. (1992) give an example in which toluene and the xylenes are degraded by indigenous microorganisms under sulfate-reducing conditions. Beller et al. (1992a) describe the metabolic by-products of toluene biodegradation under sulfate-reducing conditions.

Oxidation of Petroleum Hydrocarbons via Methanogenesis. Over the past decade, many researchers have shown the benzene, toluene, ethylbenzene, the xylenes, and a variety of other compounds can be degraded when methanogenesis is the terminal electron-accepting process (Table 5.1). Methanogenesis in a contaminated aquifer is a two-step process involving fermentation and respiration. During the first step of methanogenesis, BTEX compounds are fermented by fermentative bacteria to compounds such as acetate and hydrogen. These by-products of fermentation still contain usable energy that is utilized by fermentative and respirative organisms. During methanogenosis, organisms use hydrogen and acetate as metabolic substrates and produce methane, carbon dioxide, and water. According to Chapelle (1993), *methanogenic respiration, in which the carbon present either in* CO_2 *or in acetate serves as the terminal electron acceptor, is one of the most important respirative pathways found in anaerobic subsurface environments.*

The following equations describe the overall stoichiometry of benzene oxidation by methanogenesis. As mentioned above, the first step in methanogenesis involves the fermentation of an organic compound. The fermentation of benzene to acetate and hydrogen is represented stoichiometrically as

Step 1

$$C_6H_6 + 6H_2O \rightarrow 3CH_3COOH + 3H_2 \tag{5.6}$$

The acetate and hydrogen are then utilized by different organisms. In these reactions, carbon dioxide present in the aquifer and the acetate produced in the first reaction are utilized as electron acceptors and the produced hydrogen and acetate are used as electron donors in the following reactions:

Step 2

$$3CH_3COOH \rightarrow 3CH_4 + 3CO_2 \tag{5.7}$$

$$3H_2 + 0.75CO_2 \rightarrow 0.75CH_4 + 1.5H_2O \tag{5.8}$$

Using this stoichiometry, 3.75 mol of methane is produced during the biodegradation of 1 mol of benzene. The mass ratio of methane produced during the oxidation of benzene during methanogenesis can be calculated and is given by

Molecular weights:	Benzene	$6(12) + 6(1) = 78$ g
	CH_4	$3.75(16) = 60$ g

$$\text{Mass ratio of } CH_4 \text{ to benzene} = 60{:}78 = 0.77{:}1$$

Therefore, 0.77 mg of CH_4 is produced during mineralization of 1 mg of benzene. Similar calculations can be made for toluene, ethylbenzene, and xylene. One method that can be used to estimate the amount of methane produced by anaerobic bacteria to degrade total BTEX is to average the amount of methane produced during the biodegradation of each compound separately. Based on these calculations, the average ratio of methane produced per mole of BTEX degraded is 0.78:1 (Table 5.2). Although these calculations do not take into account the production of biomass, it is not likely that this omission will result in significant error (see the discussion under aerobic respiration). Because the BTEX compounds are responsible for much of the methane produced in a petroleum-hydrocarbon-contaminated aquifer, it is important to include all of these compounds when modeling the environmental fate of petroleum-hydrocarbon plumes under natural conditions.

For methanogenesis to occur, the following conditions must be met: (1) methanogenic bacteria must be present in the affected aquifer, (2) carbon dioxide must be present, (3) biodegradable organic carbon must be present, and (4) strongly reducing conditions must prevail. Methanogenesis is favored when hydrogen concentrations are greater than about 5 nM. Because of the relatively small amounts of free energy produced by these reactions, methanogenesis generally is not the thermodynamically favored reaction in the anaerobic environment but will proceed in environments that lack other electron acceptors or after these other electron acceptors (e.g., nitrate, sulfate) have been depleted. Sulfate-reducing bacteria typically outcompete methanogenic bacteria in subsurface environments. Because of this, methanogenesis typically is not observed in groundwater containing high concentrations of sulfate. The presence of methane in groundwater at concentrations elevated relative to background concentrations is a good indicator of methane fermentation because it is the only organic compound in the carbon cycle that is thermodynamically stable. Elevated concentrations of methane correlated with elevated concentrations of total BTEX suggest that methanogenesis is utilized by microbes to biodegrade BTEX.

When the pH of groundwater is buffered by the presence of carbonate in the aquifer, the metabolic activity of the microorganisms becomes the rate-limiting step. The rate of reaction is controlled by the density of the active organisms and by the concentration of metabolizable compounds. Wilson et al. (1986) used a microcosm to show that BTEX was biodegraded under methanogenic conditions. Grbić-Galić and Vogel (1987) show the potential for the complete degradation of benzene and toluene to carbon dioxide and methane by methanogenic bacteria in laboratory studies. Wilson et al. (1987, 1990) described the methanogenic degradation of benzene, toluene, and xylenes in groundwater contaminated by an aviation gasoline spill. Cozzarelli et al. (1990) reported the anaerobic biodegradation of alkylbenzenes in a contaminant plume originating from a spill of crude oil in Minnesota. The groundwater was actively methanogenic and accumulated long-chain volatile fatty acids. Wilson et al. (1993) studied biodegradation of aromatic hydrocarbons in a plume produced from a gasoline spill from a leaking underground storage tank. The water was methanogenic and accumulated acetate. The rates of anaerobic biodegradation in this plume were very similar to the rates in the crude oil spill studied by Cozzarelli et al. (1990) and the aviation gasoline spill studied by Wilson et al. (1990). Using radiolabeled benzene, Weiner and Lovley (1998) show rapid benzene degradation in methanogenic sediments from a petroleum-contaminaed aquifer. The results of this study show conclusively that benzene can be biodegraded under strongly anaerobic conditions.

5.2 PATTERNS OF INTRINSIC BIOREMEDIATION AT FIELD SITES

As mentioned in Chapter 1, several studies published in the past few years (Rice et al., 1995; Mace et al., 1997) show that 85 to 90% (or more) of the petroleum-hydrocarbon plumes present in the United States are at steady-state equilibrium or are receding. There is no doubt at this point that the dominant mechanism limiting the advance of these plumes is biodegradation. The introduction of petroleum hydrocarbons into the subsurface environment causes rapid changes in the prevailing geochemistry of the affected groundwater due to microbially mediated reactions between the contaminants and naturally occurring inorganic electron acceptors. Aerobic respiration, denitrification, Fe(III) reduction, sulfate reduction, and methanogenesis are largely responsible for these changes, and each reaction leaves behind a unique biogeochemical signature. The changes in groundwater geochemistry caused by biodegradation and temporal and spatial variations in dominant terminal electron-accepting processes are considered in this section.

Bulk Changes in Groundwater Affected by Petroleum Hydrocarbons

Figure 5.1 illustrates conceptually the evolution of groundwater geochemistry in an initially pristine aquifer into which petroleum hydrocarbons (or any form of organic carbon that can be oxidized by microorganisms) are introduced. This hypothetical aquifer initially contains dissolved oxygen, nitrate, biologically available Fe(III), sulfate, carbon dioxide, all necessary nutrients, and has a positive oxidation–reduction potential (oxidizing). Shortly after the introduction of organic carbon into an oxygenated aquifer such as this, aerobic bacteria begin to biodegrade soluble organic compounds. Because of the low aqueous solubility of oxygen and the plethora of aerobic bacteria, the system soon becomes anoxic. After dissolved oxygen concentrations in the aquifer fall below about 0.5 mg/L, denitrification will begin if nitrate is present in sufficient concentrations.

During denitrification, nitrate concentrations decrease, typically to the point at which denitrification can no longer be supported. During denitrification, the concentration of hydrogen in groundwater is poised at a concentration of less than about 0.1 nM. This is one factor that results in the competitive exclusion of Fe(III) reducers (and many other anaerobic micoorganisms) during denitrification. After the majority of the available nitrate is consumed, dissolved hydrogen concentrations will begin to increase until the growth of Fe(III) reducers can be supported. Once the dissolved hydrogen concentration reaches about 0.1 nM, Fe(III) reducers will begin utilizing petroleum hydrocarbons if sufficient quantities of biologically available Fe(III) are present. During Fe(III) reduction, biologically available Fe(III) concentrations decrease, typically to the point where Fe(III) reduction can no longer be supported. Also during this period, Fe(II) concentrations in the aquifer increase and the oxidation–reduction potential of the groundwater decreases. During Fe(III) reduction, the concentration of hydrogen in groundwater is poised by the Fe(III)-reducing

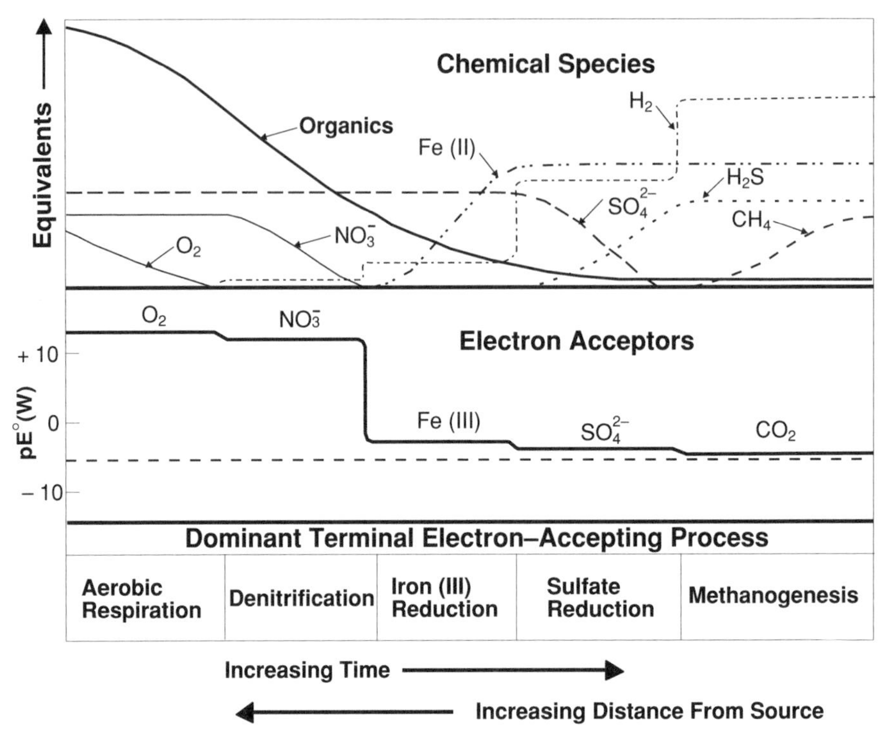

FIGURE 5.1. Conceptual model of the geochemical evolution of groundwater contaminated with petroleum hydrocarbons. (After Bouwer and McCarty, 1984.)

microorganisms at a concentration of between about 0.2 and 0.8 nM. This is one factor that results in the competitive exclusion of sulfate reducers (and many other anaerobic micoorganisms) during Fe(III) reduction. After the majority of the biologically available Fe(III) is consumed, dissolved hydrogen concentrations will begin to increase until the growth of sulfate reducers can be supported. After the concentration of dissolved hydrogen in the aquifer reaches about 1 nM, sulfate reducers will begin utilizing petroleum hydrocarbons if sufficient sulfate is available.

During sulfate reduction, sulfate concentrations decrease, typically to the point at which this terminal electron-accepting process can no longer be supported. Also during this period, H_2S concentrations tend to increase and the oxidation–reduction potential of the groundwater decreases. During sulfate reduction the concentration of hydrogen in groundwater is poised by the sulfate-reducing microorganisms at a concentration of between 1 and 4 nM. This is one factor that results in the competitive exclusion of methanogens. However, after the majority of the available sulfate is consumed, dissolved hydrogen concentrations will begin to increase until the growth of methanogens can be supported. During methanogenesis, methane concentrations increase and the oxidation–reduction potential of the groundwater decreases. In contrast to aerobic respiration, denitrification, Fe(III) reduction, and sulfate reduction, all of which are purely respirative reactions that are limited by the amount of electron acceptors, methanogenesis, in which carbon dioxide is the terminal electron acceptor, is typically a self-perpetuating reaction (as long as sufficient organic carbon is present) because there is an almost inexhaustible supply of carbon dioxide in most aquifer systems and because there is a net production of carbon dioxide during this reaction. During methanogenesis the concentration of hydrogen in groundwater is poised by the methanogens at a concentration of between about 5 and 20 nM.

The sequence starting with aerobic respiration, followed by denitrification, Fe(III) reduction, sulfate reduction, and finally, methanogenesis, represents the ideal case. Many site-specific variables can cause this idealized sequence to be interrupted. For example, if an aquifer has inherently high organic carbon concentrations resulting from depositional processes, all of the dissolved oxygen, nitrate, and biologically available Fe(III) in the system may be depleted and the system is already sulfate-reducing before the introduction of the petroleum hydrocarbons. In such a case, the system will proceed from sulfate reduction to methanogenesis as the biogeochemical system evolves.

In other cases, the system may skip a terminal electron-accepting process entirely. For example, if an aquifer system has inherently low nitrate and biologically available Fe(III) concentrations, the system may go from aerobic respiration directly to sulfate reduction and then methanogenesis. Another possibility that has been observed by the authors on several occasions is that the system will seem to bypass sulfate reduction and proceed directly to methanogenesis. This is evidenced by several sites where sulfate is present, but there are no discernible trends in sulfate concentrations; in fact, in some cases, sulfate concentrations appear to increase in areas with dissolved contamination. These same sites exhibit depleted dissolved oxygen and nitrate concentrations and elevated Fe(II) and methane concentrations. It is not clear if sulfate reduction has been bypassed at these sites or if some unknown geochemical reaction (or perhaps sampling technique or analytical procedure) is masking the trend in sulfate concentrations. The possible scenarios involving the geochemical evolution of groundwater contaminated with petroleum hydrocarbons are as varied as the complexities inherent in the subsurface.

Appendix A summarizes geochemical data obtained from 38 sites across the United States contaminated with petroleum hydrocarbons. Many of these sites are military installations, and these data were collected as part of an evaluation of natural attenuation sponsored by the U.S. Air Force Center for Environmental Excellence (U.S. AFCEE). It was the intent of this evaluation to select sites in a variety of geographic regions so that the effect of variables such as hydrogeologic setting, groundwater temperature, alkalinity, pH, and contaminant concentration could be evaluated. In addition, plumes of many different sizes were evaluated under this program, ranging from large spills at petroleum, oil, and lubricant facilities to small spills at gasoline stations.

The data presented in Appendix A were collected prior to active groundwater remediation, so they represent trends caused mainly by indigenous bacteria. These data show that the groundwater in petroleum-hydrocarbon-affected areas at almost every site had depleted dissolved oxygen, nitrate, and sulfate concentrations, elevated

Fe(II) and methane concentrations, and strongly reducing conditions, as indicated by the highly negative oxidation–reduction potential. The data presented in this table suggest that the groundwater at each of these sites had the necessary organic carbon, electron acceptors, inorganic nutrients, and microbial consortia to facilitate the biodegradation of petroleum hydrocarbons via one or more of the following processes; aerobic respiration, denitrification, Fe(III) reduction, sulfate reduction, and methanogenesis. The data also suggest that the intrinsic bioremediation of petroleum hydrocarbons introduced into groundwater is a universal process. This is in stark contrast to the intrinsic bioremediation of chlorinated solvents introduced into the shallow subsurface. In many cases, chlorinated solvents simply will not biodegrade in the natural environment and may persist for many years (see Chapter 6).

Temporal and Spatial Distribution of Terminal Electron-Accepting Processes in Petroleum-Hydrocarbon-Contaminated Aquifers

We have seen that the intrinsic bioremediation of petroleum hydrocarbons changes the bulk geochemistry of groundwater. Figure 5.2 is an idealized conceptualization of the distribution of TEAPs in space in an aquifer contaminated with petroleum hydrocarbons. As shown by this figure, there is an ideal zonation of TEAPs outward from a source of groundwater contamination. Data collected from field sites do not necessarily conform to this model, as many of the biodegradation processes are occurring concurrently throughout the contamination plume.

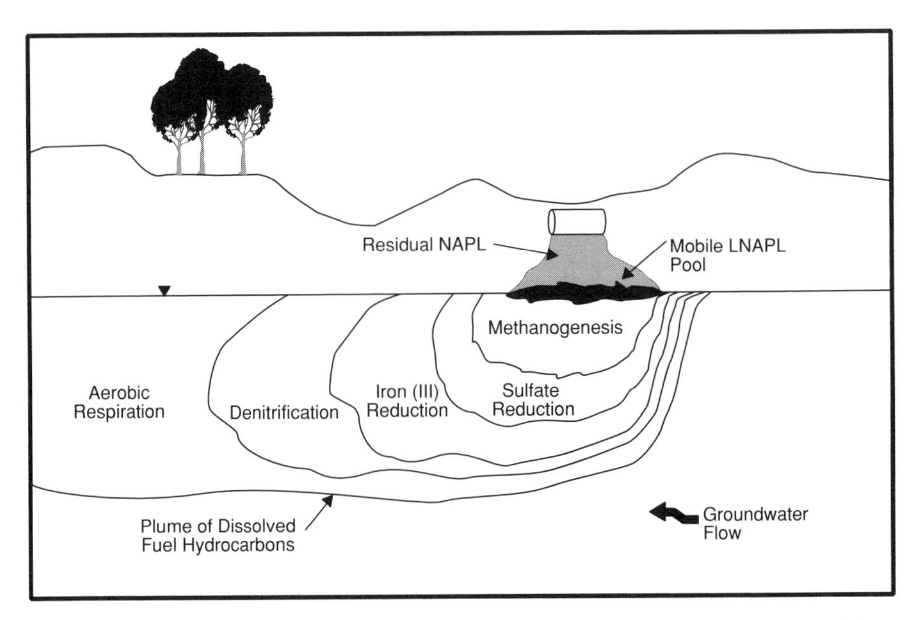

FIGURE 5.2. Conceptualization of electron acceptor zones in the subsurface. (Adapted from Lovley et al., 1994b.)

In addition to spatial changes in the prevailing chemistry of the groundwater in space, field evidence suggests similar changes over time. Vroblesky and Chapelle (1994) used dissolved hydrogen concentrations and other biologically active solutes to study the temporal and spatial changes of terminal electron-accepting processes in a dissolved petroleum-hydrocarbon plume emanating from a petroleum tank farm in Hanahan, South Carolina. These authors found that the dominant terminal electron-accepting process in petroleum-hydrocarbon-contaminated groundwater can change both temporally and spatially in an aquifer depending on recharge (i.e., reaeration and changing water levels). For example, groundwater obtained from a monitoring well at the site in June 1991 was actively methanogenic. In January 1992, the same well was resampled, and the dominant TEAP was found to be Fe(III) reduction.

The authors conclude that a recharge event had introduced sufficient oxygen to change the redox potential of the groundwater such that Fe(III) was precipitated from oxidation of Fe(II) that had initially been reduced by anaerobic processes. Reprecipitation of this iron resulted in a renewed source of Fe(III) for the indigenous Fe(III)-reducing microorganisms. Because of this reprecipitation of biologically available Fe(III), the dominant TEAP shifted from methanogenesis to Fe(III) reduction. In this case, sufficient oxygen had not been introduced to shift the redox potential enough to allow aerobic respiration, even though the groundwater is only about 6 ft below ground surface (BGS), and the shallow subsurface consists of sand with clay lenses (generally unconfined conditions). Vroblesky and Chapelle (1994) conclude that

> during times when little or no sulfate or oxygen is introduced to the groundwater by recharge from rainfall or alternate sources, the available oxygen, Fe(III), and sulfate can be depleted by respiring bacteria, leaving methanogenesis as the dominant terminal electron-accepting process. Introduction of oxygen from rainfall infiltration into methanogenic parts of the aquifer can cause precipitation of Fe(III) as grain coatings. Once the oxygen is depleted, Fe(III)-reducing bacteria can take advantage of the precipitated Fe(III) to sequester most of the electron flow from degradation of organic compounds. If the water table is lowered below the zone of Fe(III) availability, electron flow can be sequestered by less efficient terminal electron-accepting processes, such as sulfate reduction.

This example illustrates that the interaction between petroleum hydrocarbons and the various TEAPs is not stagnant and changes both temporally and spatially. Although the dominant terminal electron-accepting process can change in both time and space, once a system contaminated with petroleum hydrocarbons becomes anoxic, it is unlikely that the oxidation–reduction potential will shift sufficiently to allow aerobic respiration to dominate until the petroleum hydrocarbons are depleted. This is largely due to geochemical scavenging of the dissolved oxygen by compounds such as Fe(II) and Mn(II). The continuing anaerobic conditions prevalent in a petroleum-hydrocarbon-contaminated aquifer, or in an aquifer with naturally high organic carbon concentrations, is important for the intrinsic bioremediation of chlorinated solvents such as perchloroethane, trichloroethene, 1,1,1-trichloroethane, and carbon tetrachloride, as discussed in Chapter 7.

Relative Importance of Biodegradation Mechanisms: A Changing Paradigm

Until about 20 years ago, anaerobic biodegradation of petroleum hydrocarbons, and especially BTEX, was virtually unknown; and up until about five years ago, anaerobic biodegradation mechanisms were not considered important for the biodegradation of organic compounds dissolved in groundwater. This perspective has since changed with the evolution of numerous field studies and laboratory anaerobic biodegradation studies described in Section 5.2. We now know that significant amounts of hydrocarbons dissolved in groundwater can be degraded by anaerobic mechanisms such as denitrification, Fe(III) reduction, sulfate reduction, and methanogenesis (Wiedemeier et al., 1995b).

One method for estimating the relative importance of degradation mechanisms is presented below. This method relies on measured changes in groundwater chemistry to estimate the dominant TEAPs operating at a site. Although there are several limitations to this approach, the importance of anaerobic processes for the biodegradation of petroleum hydrocarbons must not be overlooked.

Changes in groundwater geochemistry can be used in conjunction with the stoichiometric relationships for the various TEAPs to estimate the relative importance of biodegradation mechanisms operating at a site. Biodegradation capacity is the amount of contamination that a given TEAP can "assimilate" or degrade based on the electron-accepting capacity of the groundwater. Expressed biodegradation capacity then is the biodegradation capacity "expressed" by changes in groundwater chemistry caused by a given terminal electron-accepting reaction. A closed system containing 2 L of water can be used to help visualize the physical meaning of biodegradation capacity. Assume that the first liter contains no petroleum hydrocarbons but contains petroleum-degrading microorganisms and electron acceptors and has a biodegradation capacity of exactly x milligrams of petroleum hydrocarbons. The second liter has no biodegradation capacity; however, it contains petroleum hydrocarbons. As long as these 2 L of water are kept separate, the biodegradation of petroleum hydrocarbons will not occur. If these 2 L are combined in a closed system, biodegradation will begin and continue until the petroleum hydrocarbons are depleted, the electron acceptors are depleted, or the environment becomes acutely toxic to the petroleum-degrading microorganisms.

The groundwater beneath a site is an open system, which continually receives additional electron acceptors from the flow of groundwater in the aquifer. This means that the biodegradation capacity is not fixed as it would be in a closed system, and therefore cannot be compared directly to contaminant concentrations in the groundwater. Rather, the expressed biodegradation capacity of groundwater is intended to serve as a qualitative tool. The fate of BTEX in groundwater and the potential impact on receptors is dependent on the complex interaction between the contaminant and solute fate and transport mechanisms such as advection, dispersion, sorption, volatilization, and biodegradation.

Based on the stoichiometry of BTEX biodegradation presented previously, the amount of BTEX biodegraded by a given terminal electron-accepting process can be estimated by

$$EBC_x = \frac{|C_B - C_P|}{F} \tag{5.9}$$

where EBC_x = expressed biodegradation capacity for given terminal electron-accepting process (mg/L)
C_B = average background (upgradient) electron acceptor or metabolic by-product concentration (mg/L)
C_P = lowest measured (generally in NAPL source area) electron acceptor or metabolic by-product concentration (mg/L)
F = BTEX utilization factor (mg/mg, Table 5.2)

Table 5.2 summarizes BTEX utilization factors for aerobic respiration, denitrification, Fe(III) reduction, sulfate reduction, and methanogenesis.

As an example, consider aerobic respiration. The expressed biodegradation capacity of BTEX degraded by respiration of dissolved oxygen is given by

$$EBC_{\text{oxygen}} = \frac{|O_B - O_M|}{3.14}$$

where EBC_{oxygen} = expressed biodegradation capacity for dissolved oxygen
O_B = average background (upgradient) dissolved oxygen concentration (mg/L)
O_M = measured dissolved oxygen concentration in NAPL source area (mg/L)
3.14 = BTEX utilization factor using dissolved oxygen (Table 5.2)

Thus, if groundwater in the contaminant plume is devoid of oxygen and the background dissolved oxygen concentration is 6.0 mg/L, this system has expressed the capacity to assimilate 1.9 mg/L (1900 μg/L) of total BTEX. Similar EAC calculations can be completed for denitrification, manganese(IV) reduction, Fe(III) reduction, sulfate reduction, and methanogenesis.

Table 5.3 presents expressed biodegradation capacity calculations for 38 sites contaminated with petroleum hydrocarbons. Figure 5.3 shows the estimated relative importance of BTEX biodegradation mechanisms based on these calculations. Although this figure represents only an estimate of the relative importance of each mechanism, it is clear that anaerobic biodegradation mechanisms, and specifically sulfate reduction and methanogenesis, are the most important biodegradation mechanisms at many petroleum-hydrocarbon-contaminated sites. This is not surprising, considering the relative abundance of available electron acceptors as shown by Table 5.3. The fact that sulfate reduction and methanogenesis are such important biodegradation mechanisms at sites contaminated with petroleum hydrocarbons is an important observation when one considers the biodegradation of chlorinated solvents mixed with petroleum hydrocarbons. This topic is discussed in Chapter 6.

TABLE 5.3. Expressed Biodegradation Capacity[a]

Site	Aerobic Respiration[b]		Denitrification[c]	
	Dissolved Oxygen Removed (mg/L)	Expressed Biodegradation Capacity (mg/L)	Nitrate Removed (mg/L)	Expressed Biodegradation Capacity (mg/L)
Hill AFB, UT, site 870	5.40	1.72	74.92	15.29
Battle Creek ANGB, MI, site 3	6.81	2.17	14.90	3.04
Madison ANGB, WI	6.30	2.01	43.98	8.98
Elmendorf AFB, AK, hangar 10	0.70	0.22	64.09	13.08
Elmendorf AB, AK, ST-41	12.50	3.98	134.81	27.51
King Salmon AFB, AK, SS-12	10.80	3.44	1.11	0.23
Eglin AFB, FI, police facility	3.70	1.18	0.66	0.14
Patrick AFB, FL, gas station	3.60	1.15	1.06	0.22
MacDill AFB, FL, site 56	2.38	0.76	5.61	1.15
MacDill AFB, FL, pumphouse 75	2.02	0.64	4.15	0.85
Myrtle Beach, SC, police facility	1.86	0.59	5.79	1.18
Langley AFB, VA, site SS04	4.91	1.56	5.97	1.22
Langley AFB, VA, site SS16	6.28	2.00	22.72	4.64
Griffis AFB, NY, pumphouse 5	6.40	2.04	11.67	2.38
Pope AFB, NC, FPTA No. 4	8.60	2.74	8.40	1.71
Seymour Johnson AFB, NC, bldg. 470	8.82	2.81	7.25	1.48
Fairchild AFB, WA, bldg. 1212	8.60	2.74	36.86	7.52
Eaker AFB, AR, gas station	7.80	2.48	2.02	0.41
Dover AFB, DE, site SS27/XYZ	8.20	2.61	81.11	16.55
Bolling AFB, DC, car care center	7.45	2.37	75.76	15.46
Offutt AFB, NE, tank 349	6.00	1.91	32.93	6.72
Westover AFB, MA, Christmas tree fire training area	11.02	3.51	32.09	6.55
Columbus AFB, MS, ST-24	8.33	2.65	10.61	2.16
Shaw AFB, SC, bldg. 1613	7.80	2.48	15.03	3.07
Travis AFB, CA, gas station	3.60	1.15	41.72	8.52
Beale AFB, CA, UST site	8.25	2.63	18.64	3.80
Chanute AFB, IL, 952 site	3.30	1.05	2.34	0.48
Grissom AFB, IN, bldg. 735	8.20	2.61	65.42	13.35
Keesler AFB, MI, SWMU 66 site	1.70	0.54	0.88	0.18
Tyndall AFB, FL, police B site	1.88	0.60	2.65	0.54
Gulf Coast site A	1.60	0.51	—	—
Gulf Coast site B	4.10	1.31	—	—
Gulf Coast site C	4.20	1.34	—	—
Gulf Coast site D	—	—	12.82	2.62
Gulf Coast site E	1.60	0.51	53.92	11.00
Gulf Coast site F	4.80	1.53	—	—
Gulf Coast site I	6.30	2.01	1.46	0.30
Houston site A	1.40	0.45	23.87	4.87
Total expressed biodegradation capacity		65.99		187.19
Percent of total expressed biodegradation capacity		3.00		9.00

[a] Dominant terminal electron-accepting process based on expressed biodegradation capacity.

[b] Electrons transferred during aerobic respiration = $(\Delta C_{\text{oxygen}}/3.14)$.

[c] Electrons transferred during denitrification = $(\Delta C_{\text{nitrate}}/4.9)$.

Fe(III) Reduction[d]		Sulfate Reduction[e]		Methanogenesis[f]	
Fe(II) Produced (mg/L)	Expressed Biodegradation Capacity (mg/L)	Sulfate Removed (mg/L)	Expressed Biodegradation Capacity (mg/L)	Methane Produced (mg/L)	Expressed Biodegradation Capacity (mg/L)
50.95	2.34	98.50	20.10	2.10	2.69
11.95	0.55	24.60	5.02	8.40	10.77
15.59	0.72	88.69	18.10	9.90	12.69
8.95	0.41	24.58	5.02	8.50	10.90
40.45	1.86	59.25	12.09	1.48	1.90
43.95	2.02	8.04	1.64	5.61	7.19
10.45	0.48	30.95	6.32	16.22	20.79
1.85	0.08	999.48	203.98	15.50	19.87
5.06	0.23	109.60	22.37	13.54	17.36
20.88	0.96	64.58	13.18	14.44	18.51
37.45	1.72	83.60	17.06	17.13	21.96
44.89	2.06	864.75	176.48	7.80	10.00
9.29	0.43	236.75	48.32	8.00	10.25
24.40	1.12	19.63	4.01	6.98	8.95
56.25	2.58	12.80	2.61	48.40	62.05
31.59	1.45	46.30	9.45	2.70	3.46
3.10	0.14	59.63	12.17	1.87	2.40
33.79	1.55	88.68	18.10	3.80	4.87
1.96	0.09	60.10	12.27	—	—
20.20	0.98	33.00	6.73	9.10	11.66
0.00	0.00	71.37	14.57	0.01	0.01
23.45	1.08	431.85	88.13	0.01	0.01
63.74	2.92	14.50	2.96	2.05	2.63
44.99	2.06	51.50	10.51	0.99	1.27
14.84	0.68	1995.80	407.31	5.42	6.95
6.83	0.31	25.90	5.29	3.50	4.48
9.17	0.42	145.50	29.69	6.60	8.46
2.77	0.13	108.00	22.04	1.68	2.15
36.14	1.66	—	—	7.28	9.33
1.41	0.06	5.90	1.20	4.48	5.74
4.07	0.19	145.00	29.59	7.00	8.97
—	—	93.00	18.98	0.09	0.12
0.83	0.04	—	—	9.30	11.92
2.88	0.13	—	—	9.40	12.05
56.17	2.58	167.35	34.15	4.92	6.31
—	—	376.00	76.73	1.45	1.86
298.97	13.71	732.00	149.39	0.10	0.12
0.28	0.01	60.80	12.41	1.30	1.67
	47.69		1517.96		342.29
	2.00		70.00		16.00

[d]Electrons transferred during Fe(III) reduction = $(\Delta C_{Fe(II)}/21.8)$.

[e]Electrons transferred during sulfate reduction = $(\Delta C_{sulfate}/4.7)$.

[f]Electrons transferred during methanogenesis = $(\Delta C_{methane}/0.78)$.

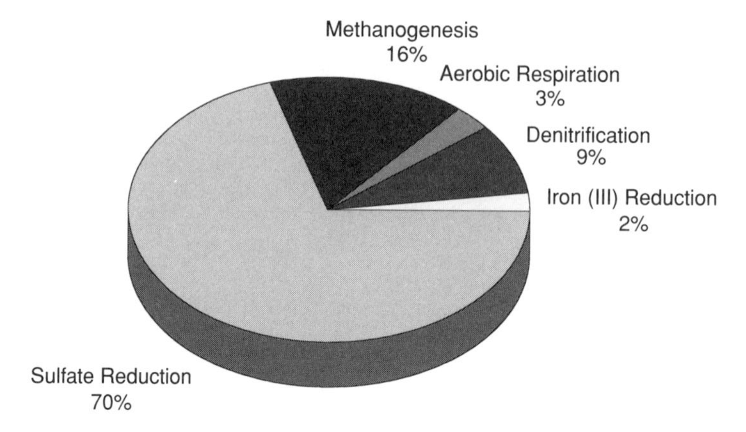

FIGURE 5.3. Relative importance of BTEX biodegradation mechanisms as determined from expressed biodegradation capacity.

The same type of calculations comparing the relative importance of various electron acceptors can be performed on a mass basis. Using the BIOSCREEN model (Chapter 8), the actual mass of electron acceptors consumed and by-products produced was estimated for a gasoline service station site at Keesler AFB (Newell et al., 1996). Note that the following values represent "missing" oxygen, nitrate, and sulfate and "excess" Fe(II) and methane in the plume area from one sampling event:

Dissolved oxygen	13.2 kg
Nitrate	5.7 kg
Fe(II)	134.3 kg
Sulfate	181.2 kg
Methane	53.4 kg

When the utilization factors are included, an equivalent amount of BTEX compounds degraded can be calculated for each reaction:

Dissolved oxygen	4.3 kg (4%)
Nitrate	1.2 kg (1%)
Fe(II)	6.3 kg (5%)
Sulfate	39.4 kg (33%)
Methane	68.5 kg (57%)

The mass of sulfate (181 kg) consumed and methane (53 kg) produced significantly exceed the consumption of dissolved oxygen (13 kg). The mass of hydrocarbon required to reduce the sulfate and produce the methane is substantial (e.g., the

TABLE 5.4. Mean and Recommended First-Order Rate Coefficients for Selected Petroleum Hydrocarbons Presented by Aronson and Howard (1997)

	Mean of Field/In Situ Studies			Recommended First-Order Rate Constants			
				Low End		High End	
Compound	First-Order Rate Constant (day^{-1})	Half-Life (day^{-1})	Number of Studies Used for Mean	First-Order Rate Constant (day^{-1})	Half-Life (day^{-1})	First-Order Rate Constant (day^{-1})	Half-Life (day^{-1})
Benzene	0.0036	193	41	0	No degradation	0.0036	193
Toluene	0.059	12	46	0.00099	700	0.059	12
Ethylbenzene	0.015	46	37	0.0006	1155	0.015	46
m-Xylene	0.025	28	33	0.0012	578	0.016	43
o-Xylene	0.039	18	34	0.00082	845	0.021	33
p-Xylene	0.014	49.5	26	0.00085	815	0.015	46

equivalent of over 100 kg of BTEX). Others have suggested that the consumption of sulfate and the production of methane could be accounted for through consumption of nonhydrocarbon electron donors such as organic carbon, nonhydrocarbon substrates, or "stored" organic polymer nutrients (Salanitro et al., 1997). However, it seems unlikely that these sources could provide the large mass of carbon needed, and crediting nonhydrocarbon sources for the consumption of sulfate and production of methane raises the question of why these concentration changes are observed only in the dissolved BTEX plume, not in background areas.

5.3 PETROLEUM HYDROCARBON RATE DATA FROM THE LITERATURE

A large number of laboratory and field studies in the general literature have reported biodegradation rates for aerobic and anaerobic biodegradation of BTEX. Rifai and Bedient (1994), Rifai et al. (1995), and Aronson and Howard (1997) provide excellent summaries of the majority of these studies. Aronson and Howard (1997), for example, recommended first-order anaerobic decay rates ranging between 0 and 0.0036, 0.00099 and 0.059, 0.0006 and 0.015, and 0.00082 and 0.021 day^{-1} for benzene, toluene, ethylbenzene, and xylene, respectively (Table 5.4). Aronson and Howard additionally reported anaerobic decay rates for other hydrocarbons. Rifai et al. (1995)

TABLE 5.5. Selected Field Biodegradation Rates Presented by Rifai et al. (1995)[a]

	First-Order Decay Rate Constant (day^{-1})				
Site	Benzene	Toluene	Ethylbenzene	Xylene	Reference
Rocky Point, NC	0.0002	0.0021	0.0015	0.0013, 0.0021	Borden et al. (1995)
Kalkaska, MI	0.0095	NR	NR	NR	Chiang et al. (1989)
Columbus, MS	0.007	NR	NR	0.0107	MacIntyre et al. (1993)
Sleeping Bear, MI	NS	0.0045	0.007	0.005–0.009	Wilson et al. (1994), Schafer (1994)
Indian River, FL	0.0085	NR	NR	NR	Kemblowski et al. (1987)
Morgan Hill, CA	0.0035	NR	NR	NR	Kemblowski et al. (1987)
Eglin AFB, FL	BD	0.05–0.013	0.03–0.05	0.02–0.21	Wilson et al. (1994)
San Francisco, CA	0.0028	0.0022	0.0033	0.0023	Buscheck et al. (1993)
Alameda County, CA	0.002	0.0017	0.002	0.0017	Buscheck et al. (1993)
Elko County, NV	0.001	NR	NR	NR	Buscheck et al. (1993)
Sampson County, NC	0.0006	0.0	0.0023	0.0009, 0.0016	Borden et al. (1995)
Traverse City, MI	0.007	0.2	NR	0.004	Wilson et al. (1990)
Perth, Australia	NS	0.006	0.003	0.004, 0.006	Thierrin et al. (1993)

[a]BD, below detection ; NR, not recorded; NS, not significant.

indicate overall (aerobic and anaerobic) rate data from field and microcosm studies ranging between 0.0002 and 0.21 day^{-1} (Table 5.5). Table 5.6 presents first-order rate data collected from several of the 38 Air Force sites mentioned earlier in the chapter. As can be seen from Table 5.6 as well as the summary data discussed above, the rate constants for BTEX can vary over three orders of magnitude, depending on the condition under which the rates were estimated.

Field-reported rate constants generally represent an overall estimate for aerobic and anaerobic reactions and will generally incorporate the specific nature of the site environmental conditions. Laboratory data, on the other hand, are generally derived under controlled environmental and electron acceptor conditions. This implies that laboratory data may not be directly transferable to the field. Furthermore, and

TABLE 5.6. Biodegradation Rate Constants for Selected Air Force Sites

	First-Order Rate Constant (day^{-1})		
Site	Conservative Tracer [a]	Steady-State Plume [b]	Other
Hill AFB, UT, site 870	—	—	0.003
Battle Creek ANGB, MI, site 3	—	—	0.003
Madison ANGB, WI	0.004	—	—
Elmendorf AFB, AK, hangar 10	0.01	—	—
Elmendorf AFB, AK, ST-41	0.005	—	—
King Salmon AFB, AK, SS-12	—	—	0.01
Eglin AFB, FL, police facility	—	—	0.009
Patrick AFB, FL, gas station	—	—	0.0009
MacDill AFB, FL, site 56	—	—	0.003
MacDill AFB, FL, pumphouse 75	0.0005	—	—
Myrtle Beach, SC, police facility	—	0.0021	—
Langley AFB, VA, site SS16	—	—	0.008
Griffis AFB, NY, pumphouse 5	—	—	0.0002
Pope AFB, NC, FPTA No. 4	—	0.0062	—
Seymour Johnson AFB, NC, bldg. 470	—	—	0.003
Fairchild AFB, WA, bldg. 1212	—	0.007	—
Eaker AFB, AR, gas station	—	0.0062	—
Dover AFB, DE, site SS27/XYZ	0.003	—	—
Bolling AFB, DC, car care center	0.003	—	—
Westover AFB, MA, Christmas tree fire training area	—	0.0034	—
Columbus AFB, MS, ST-24	—	0.08	—
Travis AFB, CA, gas station	—	—	0.00049
Beale AFB, CA, UST site	—	—	0.00022

[a] Biodegradation rate constant was estimated using a conservative tracer and the method presented by Wiedemeier et al. (1996).

[b] Biodegradation rate constant was estimated by assuming that the plume is at steady-state equilibrium [see, e.g., Buscheck and Alcantar (1995) and Chapelle et al. (1996a)].

because of the variable geochemistry of groundwater aquifers, field-observed rates are expected to vary across the site and between different sites. Finally, while the first-order decay model is the most commonly used model to describe biodegradation, it has specific limitations as discussed in Chapter 4.

The instantaneous reaction model presented in Chapter 4 offers a more realistic representation of electron acceptor conditions in the field but may not be applicable to all field sites, as discussed in Chapter 4. Recent results from the Air Force Center for Environmental Excellence (AFCEE) Natural Attenuation Initiative (Wiedemeier et al., 1995a,b; see Appendix A) indicate that the anaerobic reactions, which were originally thought to be too slow to be of significance in groundwater, can also be simulated as instantaneous reactions (Newell et al., 1995). For example, Davis et al. (1994) and Kazumi et al. (1997) ran microcosm studies with sulfate reducers and methanogens that confirmed that benzene could be degraded anaerobically in a period of less than 100 days (after acclimation). Considering the time required to replenish electron acceptors in a plume (typically years), it appears appropriate to simulate both aerobic and anaerobic biodegradation of dissolved hydrocarbons as an instantaneous reaction. This conclusion is supported by observing the pattern of anaerobic electron acceptors and metabolic by-products along the plume at natural attenuation research sites (Figures 5.4 and 5.5). From a theoretical basis, the only

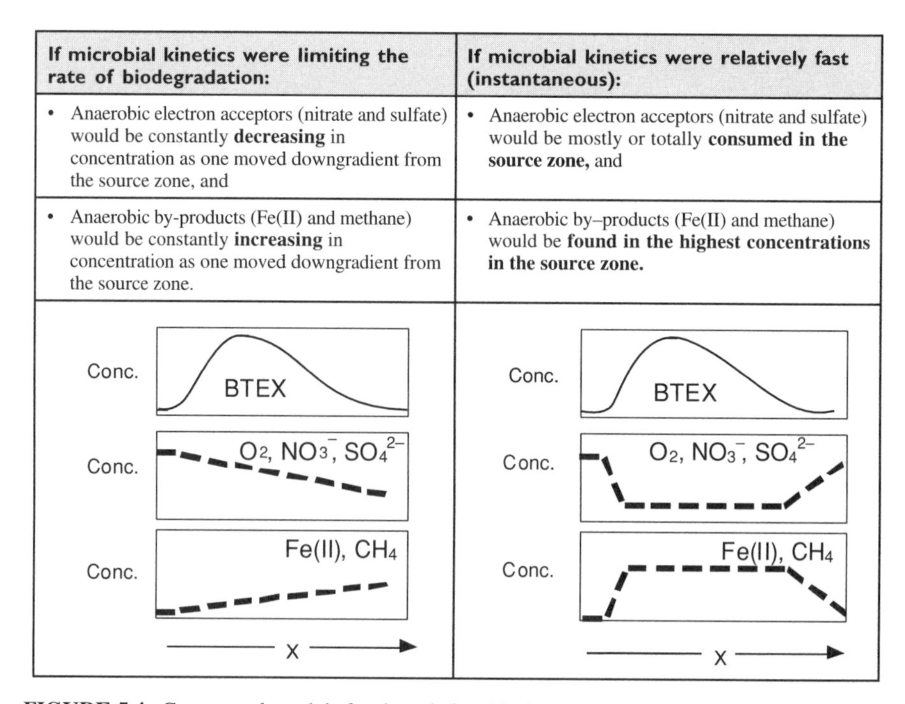

FIGURE 5.4. Conceptual models for the relationship between BTEX, electron acceptors, and metabolic by-products versus distance along centerline of plume. (From Newell et al., 1996.)

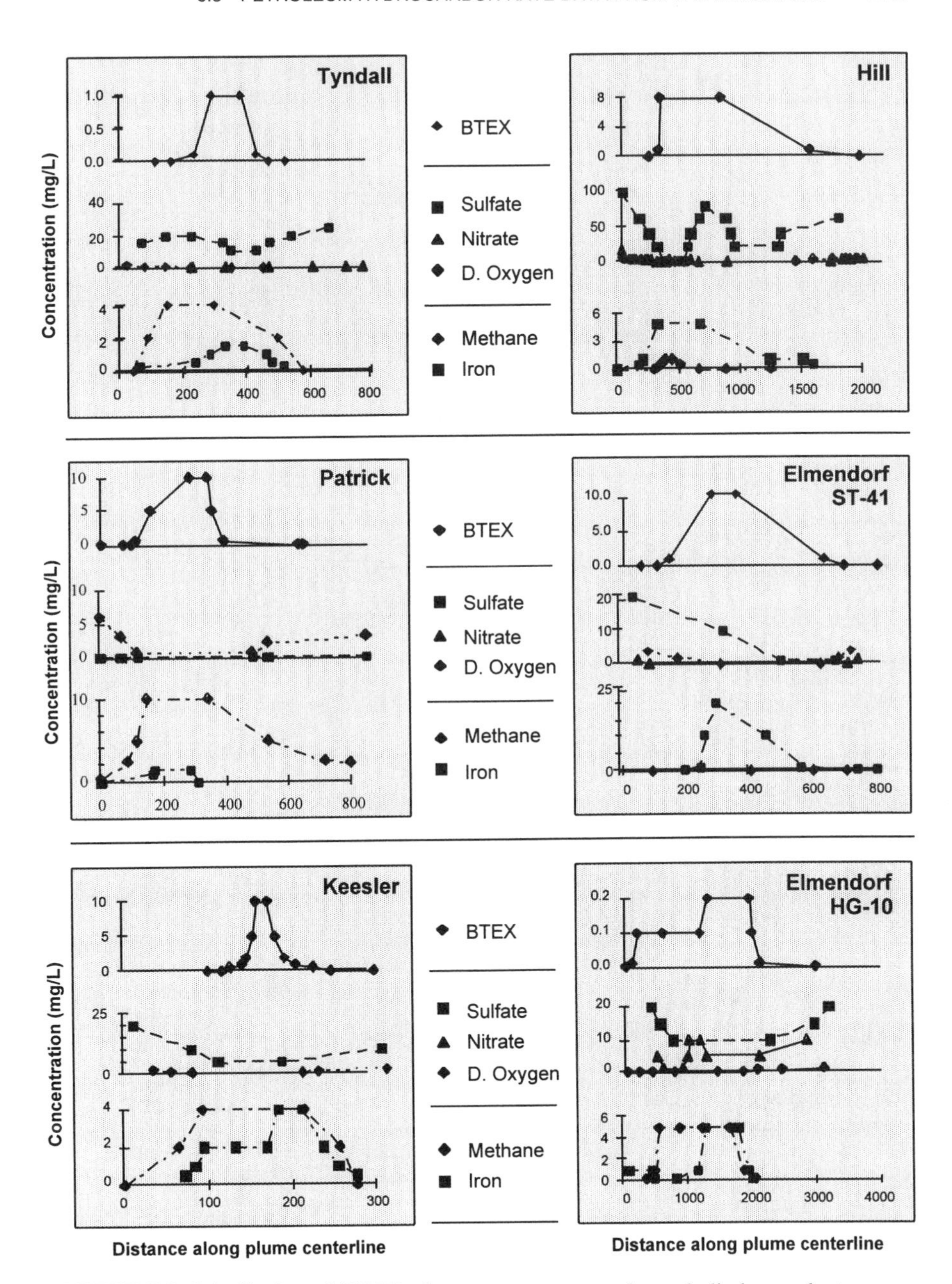

FIGURE 5.5. Distribution of BTEX, electron acceptors, and metabolic by-products versus distance along centerline of plume. (From Newell et al., 1996.) [*Sampling date and source of data:* Tyndall 3/95, Keesler 4/95 (Groundwater Services, Inc.), Patrick 3/94 (*note:* one NO_3 outlier removed, sulfate not plotted), Hill 7/93, Elmendorf site ST41 6/94, Elmendorf site HG 10 6/94 (Parsons Engineering Science).]

TABLE 5.7. Monod Kinetic Parameters for Benzene, Toluene, Ethylbenzene, and the Xylenes

	Benzene		Toluene		Ethylbenzene		Xylenes	
Reference	μmax (mg/mg • day)	K_s (mg/L)	μmax (mg/mg • day)	K_s (mg/L)	μmax (mg/mg • day)	K_s (mg/L)	μmax (mg/mg • day)	K_s (mg/L)
Alvarez et al. (1991)	8.30	12.20	—	—	—	—	—	—
Chang et al. (1993)	7.74	3.17	—	—	—	—	—	—
Kelly et al. (1996)	0.784	0.31	—	—	—	—	—	—
Goldsmith and Balderson (1988)	3.84	20	—	—	—	—	—	—
Alvarez et al. (1991)	—	—	9.9	17.4	—	—	—	—
Allen-King et al. (1994)	—	—	1.3–4.0	< 0.1	—	—	—	—
Chang et al. (1993)	—	—	10.68	1.96	—	—	—	—
Kelly et al. (1996)	—	—	0.649	0.276	—	—	—	—
MacQuarrie et al. (1990)	—	—	0.493	0.655	—	—	—	—
Robertson and Button (1987)	—	—	0.34	0.044	—	—	—	—
Corseuil and Weber (1994)	—	—	—	—	—	—	3.03	0.0007
Kelly et al. (1996)	—	—	—	—	—	—	6.1	8.4
Tabak et al. (1990)	9.3	22.16	9.09	56.74	9	11.81	8.33, 11.50	67, 35.61

Source: Adapted from Bekins et al. (1998).

sites where the instantaneous reaction assumption may not apply are sites with very low hydraulic residence times (very high groundwater seepage velocities and short source zone lengths).

Kinetic-limited sites, however, appear to be relatively rare, as the instantaneous reaction pattern is observed even at sites such as site 870 at Hill AFB, with residence times of a month or less. As shown in Chapter 9, this site has an active sulfate-reducing and methane production zone within 100 ft of the upgradient edge of the plume. When a 1600-ft/yr seepage velocity is considered, this highly anaerobic zone has an effective residence time of 23 days. Despite this very short residence time, significant sulfate depletion and methane production were observed in this zone. If the anaerobic reactions were constrained significantly by microbial kinetics, the amount of sulfate depletion and methane production would be much less pronounced. Therefore, this site supports the conclusion that the instantaneous reaction assumption is applicable to many petroleum release sites.

To apply the instantaneous reaction model, the amount of biodegradation that can be supported by the groundwater that moves through the source zone must be calculated. Using the stoichiometry presented earlier in this chapter, a utilization factor is calculated showing the ratio of the oxygen, nitrate, and so on, consumed to the mass of dissolved hydrocarbon degraded in the biodegradation reactions. Table 5.2 presents utilization factors based on the biodegradation of combined BTEX constituents. For a given background concentration of an individual electron acceptor, the potential contaminant mass removal or *biodegradation capacity* depends on the *utilization factor* for that electron acceptor. Dividing the background concentration of an electron acceptor by its utilization factor provides an estimate (in BTEX concentration units) of the biodegradation capacity (or *assimilative capacity*) of the aquifer by that mode of biodegradation. The total biodegradation capacity is the sum of the biodegradation capacities for each of the individual reactions. The BIOSCREEN and BIOPLUME III models rely on the concept of biodegradation capacity to simulate the fate and transport of dissolved petroleum hydrocarbons.

Finally, although the Monod kinetic model is applicable to most field conditions and accounts for kinetic limitations, the model suffers from a lack of kinetic data. Table 5.7 presents the aerobic Monod rate data for the common dissolved hydrocarbons in groundwater. Anaerobic Monod kinetic data for the majority of petroleum hydrocarbons have not been estimated.

5.4 PETROLEUM-HYDROCARBON PLUME DATABASES

A summary of plume data from several petroleum hydrocarbon databases is presented below. The HGDB (Newell et al., 1990) was reanalyzed by Newell and Connor (1998) to provide estimates of length, width, thickness, and maximum concentration observed at 82 petroleum-hydrocarbon sites. Three of the databases, the LLNL database (Rice et al., 1995), the BEG database (Mace et al., 1997), and the Florida database (Groundwater Services, Inc., 1997), were introduced in Chapter 1 while discussing plume-a-thon studies and discussed in Chapter 2 with regard to contaminant source data. The last database contains extensive natural attenuation data from 38 Air force petroleum hydrocarbon sites (see Appendix A) and was described earlier in this chapter. These five databases illustrate some of the main concepts regarding the natural attenuation of petroleum-hydrocarbon plumes.

Petroleum-Hydrocarbon Sites in the HGDB, LLNL Database, BEG Database, and the Florida Database

The HGDB database (Newell et al., 1990) contains hydrogeologic and chemical information from 400 site investigations across the United States. The data were obtained in a national survey of National Ground Water Association members. This 400-site database includes groundwater plume dimensions for a broad range of groundwater contaminants, including 42 service station BTEX sites, 30 non–service station BTEX sites, 78 chlorinated ethene sites, 25 non–ethene solvent sites, and 21 inorganic sites. As can be seen in Table 5.8, the 42 service station sites show a median benzene/BTEX plume length of 213 ft, while the non–service station sites have a median plume length of 500 ft. This database includes a higher percentage of longer plumes than the LLNL and BEG databases, with six BTEX plume lengths greater than 900 ft.

Data collected for approximately 1200 service station sites in California by LLNL (Rice et al., 1995) indicate that averaged site plume lengths rarely exceed about 250 ft. Spatial and temporal analysis of 271 benzene plumes indicated that benzene plume length tends to change slowly with time, while average concentrations decrease more rapidly than plume lengths. Rice et al. (1995) concluded from their study that significant reduction in benzene concentrations can occur with time at these sites even without active remediation. Based on their findings, Rice et al. (1995) categorized

TABLE 5.8. Forty-two Service Station Sites from HGDB

	Vertical Plume Penetration (ft)	Plume Length (ft)	Plume Width (ft)	Highest Concentration (mg/L)
Service Stations				
90th percentile	37	945	840	112
75th percentile	25	400	250	47.3
Median	**12**	**213**	**150**	**15.0**
25th percentile	5	85	50	2.4
n	37	42	45	43
Other Fuel Release Sites				
90th percentile	62	3000	2160	226,000
75th percentile	32.5	1000	510	885
Median	**17**	**500**	**200**	**55.0**
25th percentile	6	125	100	15.0
n	27	30	29	27

TABLE 5.9. Frequency of LUFT Study Sites That Are Expanding, Shrinking, Stable, or Exhausted

	Plume Length Trend			
Plume Concentration Trend	Positive	None	Negative	Total
Significant positive plume concentration trend	6	13	2	21
Insignificant trend, average plume concentration > 10 ppb	4	34	4	42
Insignificant trend, low average plume concentration < 10 ppb	3	29	15	47
Significant negative plume concentration trend	9	84	68	161
	22	160	89	271
Percent of total	8	59	33	100

Source: Data from Rice et al. 1995.

plumes as *shrinking*, *expanding*, *stable*, and *exhausted*. As shown in Table 5.9, 59% of the sites exhibited a shrinking benzene plume while 33% of the sites were stable or exhausted. Only about 8% of the sites exhibited expanding plumes.

One of the surprising findings reported by Rice et al. (1995) is the lack of correlation between plume length at service stations sites with other measurable hydrogeologic variables, such as groundwater depth, range, and seepage velocity (Figure 5.6). Rice et al. (1995) conclude that plume length may be correlated with unmeasured parameters such as source mass and passive bioremediation rate. Their conclusions are supported by similar research conducted by Newell and Connor (1998) who demonstrated that plume length is affected significantly by source width, depth, concentration, and the biodegradation capacity of the aquifer system.

Rice et al. (1995) also compared concentrations of the benzene, toluene, ethylbenzene, and the xylenes in a large number of monitoring wells at UST sites in California. Table 5.10 shows temporally and spatially averaged BTEX concentrations for all of the cases they studied. Figure 5.7 shows scatter plots of benzene concentration

TABLE 5.10. Summary of Site-, Time-, and Spatially Averaged Groundwater Concentration Data[a]

	Number of Sites	50% Quantile	90% Quantile	99% Quantile
Benzene (ppb)	1,094	22.7	362	3,000
Toluene (ppb)	275	10.4	175	1,213
Ethylbenzene (ppb)	275	17.7	227	1,214
Total xylenes (ppb)	276	24	311	2,088
Total petroleum hydrocarbons (gasoline, ppb)	1,042	871	7,343	36,000
Free product thickness (ft)	450	0.099	0.76	2.18

Source: Rice et al. (1995).

[a]Water chemistry data were temporally averaged over each site's history.

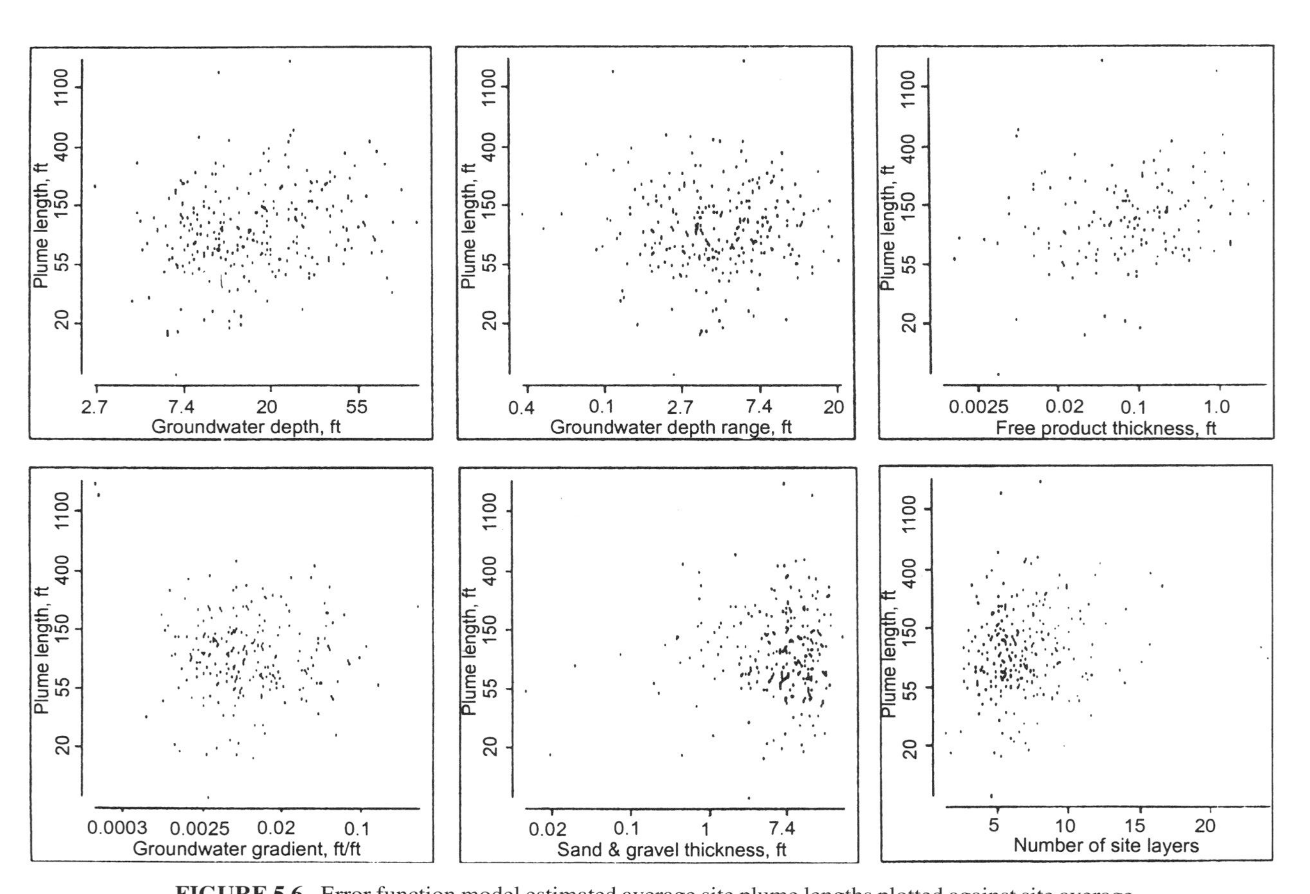

FIGURE 5.6. Error function model estimated average site plume lengths plotted against site average hydrogeologic characteristics. (From Rice et al., 1995.)

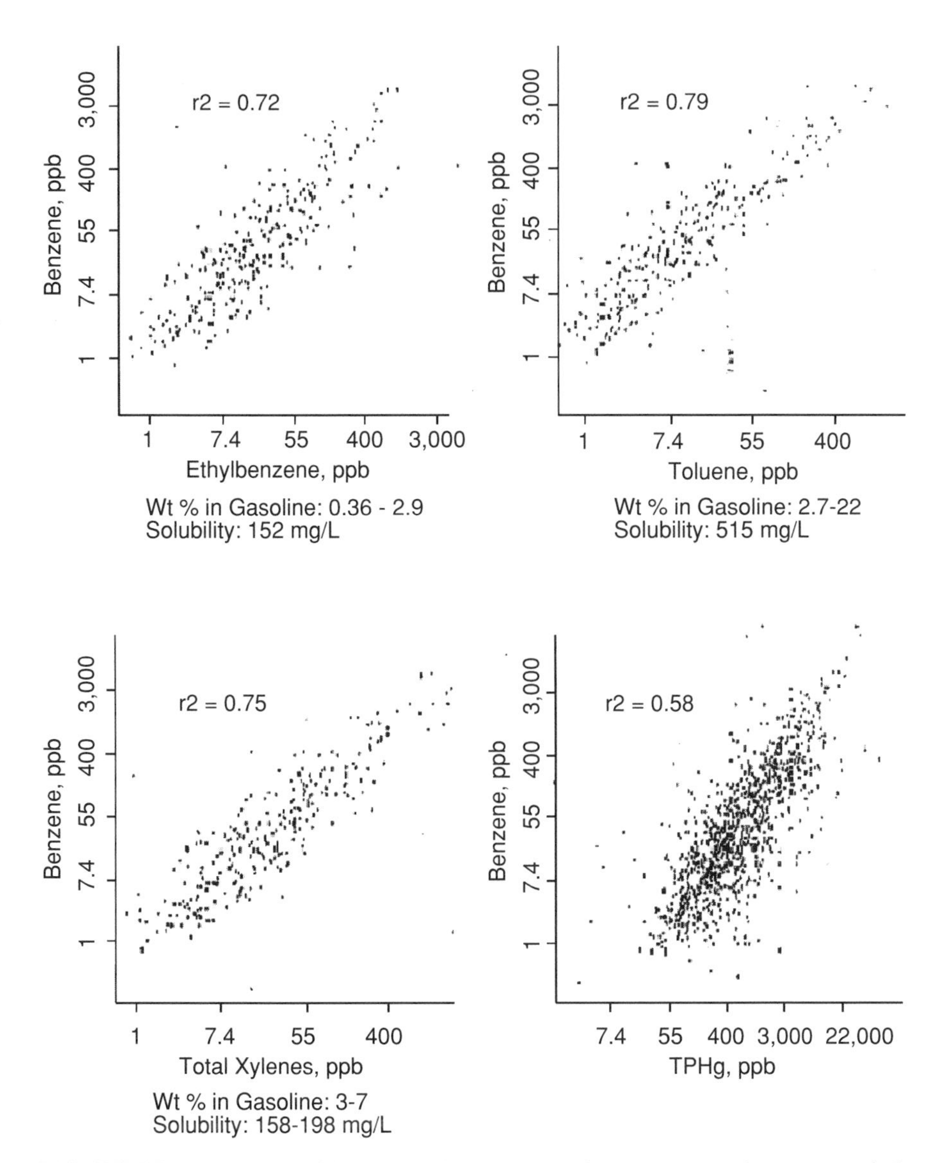

FIGURE 5.7. Scatter plots of site average benzene groundwater concentrations versus ethylbenzene, toluene, total xylenes, and total petroleum-hydrocarbons (gasoline) (TPHg) concentrations. Weight percent benzene in gasoline: 0.12–3.5, solubility of benzene: 1750 mg/L. (From Rice et al., 1995. Solubility data from Appendix B. Weight percent data from California Leaking Underground Fuel Tank Force, 1988.)

versus the concentration of toluene, ethylbenzene, total xylenes, and total petroleum hydrocarbons as gasoline (TPHg). The different data points in Figure 5.7 represent a wide range of hydrogeologic settings, source conditions, and location within the contaminant plume (e.g., source versus fringe). Each individual data point, however, represents the same environmental conditions for each of the four BTEX compounds as they migrated from a source zone to a monitoring well. These data suggest that toluene and ethylbenzene are found at somewhat lower concentrations than benzene and the xylenes, which are comparable in all but the highest concentration wells. This is true even though pure-phase solubility of benzene is 1780 mg/L compared to 500 mg/L for toluene, 150 mg/L for ethylbenzene, and less than 200 mg/L for the xylenes. If samples with lower concentrations are assumed to come from either older plumes or locations farther from the source, the data can be used to look for a *chromatographic effect*, the separation of BTEX constituents in groundwater due to differential rates of degradation. Overall, the data support the observation of many practitioners: There is no obvious chromatographic effect where one constituent (such as benzene) forms a much longer plume than the other constituents. The databases prepared by Mace et al. (1997) and Groundwater Services, Inc. (1997) reported similar findings from their analysis of petroleum sites in Texas and Florida as those reported by Rice et al. (1995) and Newell and Connor (1998).

Petroleum-Hydrocarbon Sites in the Air Force Database

Data compiled from 38 Air Force sites where the AFCEE natural attenuation protocol (Wiedemeier et al., 1998) was applied provide detailed information about plume characteristics, electron acceptors, and general site conditions present at petroleum-hydrocarbon sites (Appendix A). Several overall observations can be made with these data. For example, the Air Force sites were typically better monitored than the service stations in the other plume-a-thon studies. The median number of wells in the Air Force database is 25 wells, in comparison to five or six wells in Rice et al. (1995) and Mace et al. (1997). Detailed geochemical data were collected at the sites such that observations regarding biodegradation capacity can be made. Finally, while no data were collected on source dimensions in the saturated zone, detailed information on LNAPL and soil contamination were collected.

Twenty-four of the sites had reported BTEX plume lengths ranging from 140 to 3000 ft with a median of 555 ft (Appendix A). The median plume width and depth were 310 ft and 17.5 ft, respectively. The median maximum observed total BTEX was approximately 13 mg/L, with the values ranging from <1 to over 405 mg/L (based on 38 sites). Measured soil fraction of organic carbon ranged from 0.00036 to 0.0143, with a median of 0.00145 or 0.15%. Only nine of the sites had detectable LNAPL (liquid non-aqueous-phase liquid), with reported thicknesses ranging from 0.4 to 7 ft, lengths ranging from 80 to 1300 ft, and widths ranging from 30 to 500 ft.

Site-specific biodegradation rate information was developed using the methods discussed by Buscheck and Alcantar (1995), the use of a conservative tracer (the trimethylbenzenes), and based on microcosm experiments. Rates varied from 0.0002 to 0.01 day^{-1}, and the median half-life for 26 of the sites was 0.6 year (Table 5.6).

Electron Acceptor and By-product Data. Electron acceptor data collected from the 38 sites (Table 5.3 and Appendix A) indicated background concentrations of oxygen ranging from 0.8 to 12.6 mg/L with a median of 6.5 mg/L. Nitrate concentrations ranged from 0.19 to 148 mg/L (median value of 3.21 mg/L), and sulfate concentrations ranged from 5.62 to 6020 mg/L (median value of 83.55 mg/L). By-products of the biodegradation reactions included Fe(II), which ranged from 1.9 to 63.75 mg/L (median value of 15.6 mg/L), and methane, which ranged from 0.006 to 48.4 mg/L (median value of 5.61 mg/L).

Linear Regressions of Petroleum-Hydrocarbon Site Data. Correlations between the various measured parameters in the Air Force 38 site database are presented in Figures 5.8 and 5.9. While some of the correlation data confirm results

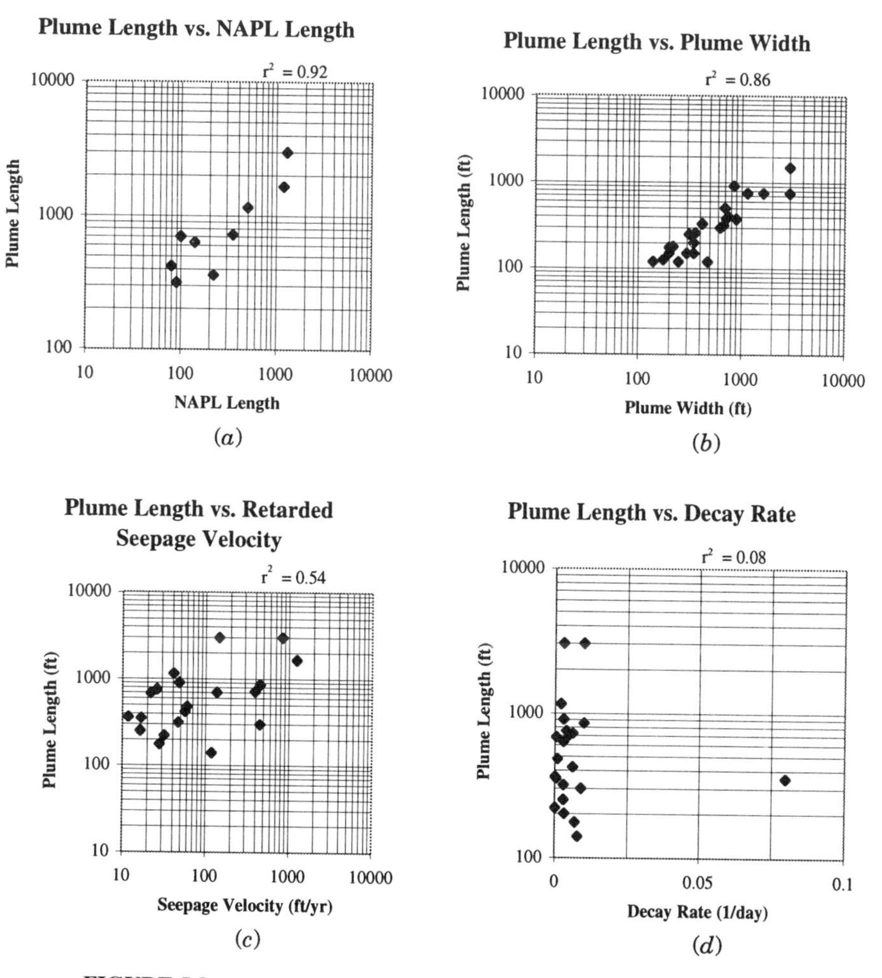

FIGURE 5.8. Correlation between plume length and aquifer parameters.

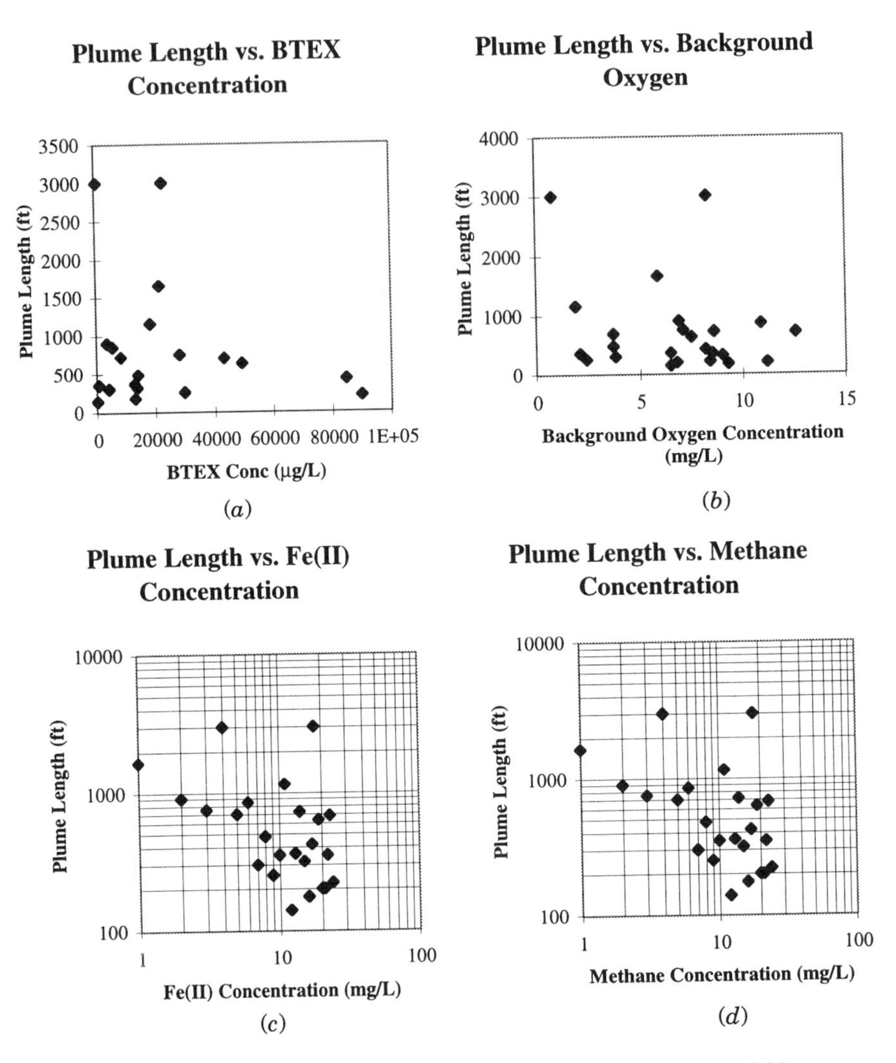

FIGURE 5.9. Correlation between plume length and geochemical variables.

reported by Rice et al. (1995) and Mace et al. (1997), the Air Force data provide a number of different conclusions. For example, a better correlation was found between plume length and the retarded seepage velocity (Figure 5.8*C*, $r^2 = 0.54$) in the aquifer than was observed in the other two studies. Plume length was additionally correlated to mobile NAPL length and width for the sites in which mobile NAPL was observed. Figure 5.8*A* illustrates this relationship, but it should be noted that only nine sites in the database had mobile NAPL; the remainder of the plumes are being sourced by residual NAPL. The Air Force also indicate a correlation between plume length and plume width (Figure 5.8*B*, $r^2 = 0.86$).

Surprisingly, plume length was not correlated to the maximum BTEX concentration measured in the aquifer or to the relative importance of the various terminal electron-accepting processes (Figure 5.9). Additionally, plume length was not correlated to the first-order decay rates reported in the database (Figure 5.8*D*).

5.5 FUTURE CHALLENGES

Although we now know that indigenous bacteria will biodegrade petroleum hydrocarbons without human intervention, one compound that has been added to gasoline since 1979 and appears to be fairly recalcitrant to intrinsic bioremediation is MTBE. This compound is an ether that is made by combining methanol and isobutylene. MTBE was originally added to gasoline mixtures as a replacement for lead to boost the octane rating of the fuel. Since about 1988, MTBE has been used in much higher concentrations (about 15%) to raise the oxygen content of gasoline to reduce emissions. In 1995 about 3.5 billion gallons of MTBE were blended in gasoline in the United States (American Petroleum Institute, 1997). Research showing the biodegradation of MTBE under natural conditions is scarce. Because of its extremely high solubility (miscible with water), it typically is found in fairly high concentrations near facilities with known leaks in which MTBE was used as a fuel additive. In addition, MTBE tends to migrate well beyond the leading edge of the remainder of the fuel hydrocarbon plume. The impacts of MTBE on human health and the environment are unclear at this time. Although not a known carcinogen, the smell and taste of MTBE in groundwater used for drinking water make it undesirable. The presence of MTBE in groundwater is raising new questions concerning the efficacy of natural attenuation for waters in which it is present. Additional research is necessary to determine the toxicity of MTBE and magnitude of the MTBE problem.

In a recent regional study of groundwater in the Denver, Colorado metropolitan area, MTBE was the organic contaminant most often detected (McMahon et al., (1995). More than 80% of groundwater collected from 30 randomly distributed wells in Denver had measurable concentrations of MTBE (McMahon et al., 1994). Perhaps more important than its persistence is the effect that MTBE has on the degradation of other gasoline components in the subsurface. For example, McMahon et al. (1995) added MTBE in concentrations of 0, 200, 2000, and 20,000 ppb to sediment incubations containing 200 ppb of either toluene or benzene. These researchers found that the highest MTBE concentration decreased toluene degradation rates 10% and benzene degradation rates by as much as five times compared to incubations having no MTBE.

REFERENCES

Acton, D. W., and Barker, J. F., 1992, In situ biodegradation potential of aromatic hydrocarbons in anaerobic groundwaters, J. Contam. Hydrol., vol. 9, pp. 325–352.

Albrechtsen, H. J., 1994, Bacterial degradation under iron-reducing conditions, in *Hydrocarbon Bioremediation*, R. E. Hinchee, B. C. Alleman, R. E. Hoeppel, and R. N. Miller, eds., Lewis Publishers, Boca Raton, FL, pp. 418–423.

Allen-King, R. M., Barker, J. F., Gillham, R. W., and Jensen, B. K., 1994, Substrate and nutrient limited toluene biotransformation in sandy soil, *Environ. Toxicol. Chem.*, vol. 13, no. 5, pp. 693–705.

Altenschmidt, U., and Fuchs, G., 1991, Anaerobic degradation of toluene in denitrifying *Pseudomonas* sp.: indication for toluene methylhydroxylation, *Arch. Microbiol.*, vol. 156; pp. 152–158.

Alvarez, P. J., Anid, P. J., and Vogel, T. M., 1991, Kinetics of aerobic benzene and toluene in sandy aquifer material, *Biodegradation*, vol. 2, no. 1, pp. 43–51.

Alvarez, P. J. J., and Vogel, T. M., 1991, Substrate interactions of benzene, toluene, and *para*-xylene during microbial degradation by pure cultures and mixed culture aquifer slurries, *Appl. Environ. Microbiol.*, vol. 57, pp. 2981–2985.

———, 1994, Degradation of BTEX and their aerobic metabolites by indigenous microorganisms under nitrate reducing conditions, *Abstracts of the 94th Annual Meeting*, Q–339, American Society for Microbiology, Washington, DC.

American Petroleum Institute, 1997, *An MTBE Primer*, API, Washington, DC.

Anders, H. J., Kaetzke, A., Kampfer, P., Ludwig, W., and Fuchs, G., 1995, Taxonomic position of aromatic-degrading denitrifying pseudomonad strains K 172 and KB 740 and their description as new members of the genera *Thauera*, as *Thaurea aromatica* sp. nov., and *Azoarcus evansit* sp. nov., respectively, members of the beta subclass of the Proteobacteria, *Int. J. Syst. Bacteriol.*, vol. 45, pp. 327–333.

Aronson, D., and Howard, P. H., 1997, *Anaerobic Biodegradation of Organic Chemicals in Groundwater: A Summary of Field and Laboratory Studies*, Draft Final Report, American Petroleum Institute, Washington, DC.

Baedecker, M. J., Cozzarelli, I. M., Eganhouse, R. P., Siegel, D. I., and Bennett, P. C., 1993, Crude oil in a shallow sand and gravel aquifer: III. Biochemical reactions and mass balance modeling in anoxic groundwater, Appl. Geochem., vol. 8, pp. 569–586.

Ball, H. A., and Reinhard, M., 1996, Monoaromatic hydrocarbon degradation under anaerobic conditions at Seal Beach, California: laboratory studies, *Environ. Toxicol. Chem.*, vol. 15, no. 2, pp. 114–122.

Ball, H. A., Reinhard, M., and McCarty, P. L., 1991, Biotransformation of monoaromatic hydrocarbons under anoxic conditions, in *In Situ Bioreclamation: Applications and Investigations for Hydrocarbon and Contaminated Site Remediation*, R. E. Hinchee and R. F. Olfenbuttel, eds., Butterworth-Heinemann, Boston, pp. 458–463.

Ball, H. A., Hopkins, G. D., Orwin, E., and Reinhard, M., 1994, Anaerobic bioremediation of aromatic hydrocarbons at Seal Beach, CA: laboratory and field investigations, *Proceedings of the Symposium on Intrinsic Bioremediation of Groundwater*, Denver, CO, Aug. 30–Sept. 1, EPA/540/R-94/515, U.S. EPA, Washington, DC.

Ball, H. A., Johnson, H. A., Reinhard, M., and Spormann, A., 1996, Initial reactions in anaerobic ethylbenzene oxidation by a denitrifying bacterium, strain EB1, *J. Bacteriol.*, vol. 178, no. 19, pp. 5755–5761.

Barbaro, J. R., Barker, J. F., Lemon, L. A., and Mayfield, C. I., 1992, Biotransformation of BTEX under anaerobic, denitrifying conditions: field and laboratory observations, J. Contam. Hydrol., vol. 11, pp. 245–272.

Barlaz, M. A., Shafer, M. B., Borden, R. C., and Wilson, J. T., 1993, Rate and extent of natural anaerobic bioremediation of BTEX compounds in groundwater plumes, *Proceedings of the Symposium on Bioremediation of Hazardous Wastes: Research, Development, and Field Evaluations*, Dallas, TX, May 4–6.

Barlaz, M. A., et al., 1995, Intrinsic bioremediation of a gasoline plume: comparisons of field and laboratory results, in *Bioremediation of Hazardous Wastes: Research, Development, and Field Evaluations*, EPA/540/R-95-532, U.S. EPA, Washington, DC.

Bekins, B. A., Warren, E., and Godsy, E. M., 1998, A comparison of zero-order, first-order, and monod biotransformation models, *Ground Water*, vol. 36, no. 2, pp. 261–268.

Beller, H. R., Edwards, E. A., Grbić-Galić, D., Hutchins, S. R., and Reinhard, M., 1991, *Microbial Degradation of Alkylbenzenes Under Sulfate-Reducing and Methanogenic Conditions*, Project Summary, EPA/600/S2-91/027, U.S. EPA, Washington, DC.

Beller, H. R., Reinhard, M., and Grbić-Galić, D., 1992a, Metabolic byproducts of anaerobic toluene degradation by sulfate-reducing enrichment cultures, *Appl. Environ. Microbiol.*, vol. 58, pp. 3192–3195.

Beller, H. R., Grbić-Galić, D., and Reinhard, M., 1992b. Microbial degradation of toluene under sulfate-reducing conditions and the influence of iron on the process, *Appl. Environ. Microbiol.*, vol. 58, pp. 786–793.

Beller, H. R., Ding, W. R., and Reinhard, M., 1995, Byproduct of anaerobic alkylbenzene metabolism useful as indicators of in situ bioremediation, *Environ. Sci. Technol.*, vol. 29, no. 11, pp. 2864–2870.

Beller, H. R., Spormann, A. M., Sharma, P. K., Cole, J. R., and Reinhard, M., 1996, Isolation and characterization of a novel toluene-degrading, sulfate-reducing bacterium, *Appl. Environ. Microbiol.*, vol. 62, pp. 1188–1196.

Borden, R. C., 1994, Natural bioremediation of hydrocarbon-contaminated ground water, in *Handbook of Bioremediation*, R. D. Norris et al., eds., Lewis Publishers, Boca Raton, FL, pp. 177–199.

Borden, R. C., and Bedient, P. B., 1986, Transport of dissolved hydrocarbons influenced by oxygen limited biodegradation: theoretical development, *Water Resour. Res.*, vol. 22, no. 13, pp. 1973–1982.

Borden, R. C., Gomez, C. A., and Becker, M. T., 1994, Natural bioremediation of a gasoline spill, in *Hydrocarbon Bioremediation*, R. E. Hinchee, B. C. Alleman, R. E. Hoeppel, and R. N. Miller, eds., Lewis Publishers, Boca Raton, FL, pp. 290–295.

Borden, R. C., Gomez, C. A., and Becker, M. T., 1995, Geochemical indicators of intrinsic bioremediation, *Ground Water*, vol. 33, no. 2, pp. 180–189.

Bouwer, E. J., 1992, Bioremediation of subsurface contaminants, in *Environmental Microbiology*, R. Mitchell, ed., Wiley-Liss, New York, pp. 287–318.

Bouwer, E. J., and McCarty, P. L., 1984, Modeling of trace organics biotransformation in the subsurface, *Ground Water*, vol. 22, no. 4, pp. 433–440.

Buscheck, T. E., O'Reilly, K. T. O., and Nelson, S. N., 1993, Evaluation of intrinsic bioremediation at field sites, in *Proceedings of the 1993 Petroleum Hydrocarbons and Organic Chemicals in Ground Water: Prevention, Detection, and Restoration*, pp. 367–371, Water Well Journal Publishing Co., Dublin, OH.

Buscheck, T. E. and Alcantar, C. M., 1995, Regression techniques and analytical solutions to demonstrate intrinsic bioremediation, in *Proceedings of the 1995 Battelle International Conference on In-Situ and On Site Bioreclamation*, Apr., Battelle Press, Columbus, OH.

Buscheck, T. E., Wickland, D. C., and Kuehne, D. L., 1996, Multiple lines of evidence to demonstrate natural attenuation of petroleum hydrocarbons, in *Proceeedings of the 1996 Petroleum Hydrocarbons and Organic Chemicals in Ground Water Conference*, Houston, TX, API/NGWA, pp. 445–460.

Chang, M. K., Voice, T. C., and Criddle, C. S., 1993, Kinetics of competitive inhibition and cometabolism in the biodegradation of benzene, toluene, and p-xylene by two Pseudomonas isolates, *Biotechnology and Bioengineering*, vol. 41, no. 11, pp. 1057–1065.

Chaudhuri, B. K. and Wiesmann, U., 1995, Enhanced anaerobic degradation of benzene by enrichment of mixed microbial culture and optimization of the culture medium, *Appl. Microbiol. Biotechnol.*, vol. 43, pp. 178–187.

Chapelle, F. H., 1993, *Groundwater Microbiology and Geochemistry*, Wiley, New York, 424 pp.

Chapelle, F. H., Bradley, P. M., Lovley, D. R., and Vroblesky, D. A., 1996a, Measuring rates of biodegradation in a contaminated aquifer using field and laboratory methods, *Ground Water*, vol. 34, no. 4, pp. 691–698.

Chiang, C. Y., Salanitro, J. P., Chai, E. Y., Colthart, J. D., and Klein, C. L., 1989, Aerobic biodegradation of benzene, toluene and xylene in a sandy aquifer: data analysis and computer modeling, *Ground Water*, vol. 27, no. 6, pp. 823–834.

Colberg, P. J. S., and Young, L. Y., 1995, Anaerobic degradation of nonhalogenated homocyclic aromatic aromatic compounds coupled with nitrate, iron, or sulfate reduction, in *Microbial Transformations and Degradation of Toxic Organic Chemicals*, L. Y. Young and C. E. Cerniglia, eds., Wiley-Liss, New York, pp. 307–330.

Corseuil, H. X., and Weber, W. J., 1994, Potential biomass limitations on rates of degradation of monoaromatic hydrocarbons by indigenous microbes in surface soils, *Water Resources*, vol. 28, no. 6, pp. 1415–1423.

Cozzarelli, I. M., Eganhouse, R. P., and Baedecker, M. J., 1990, Transformation of monoaromatic hydrocarbons to organic acids in anoxic groundwater environment, *Environ. Geol. Water Sci.*, vol. 16, no. 2, pp. 135–141.

Cozzarelli, I. M., Baedecker, M. J., Eganhouse, R. P., and Goerlitz, D. F., 1994, The geochemical evolution of low-molecular-weight organic acids derived from the degradation of petroleum contaminants in groundwater, *Geochim. Cosmochim. Acta*, vol. 58, no. 2, pp. 863–877.

Davis, J. W., Klier, N. J., and Carpenter, 1994, Natural biological attenuation of benzene in groundwater beneath a manufacturing facility, *Ground Water*, vol. 32, no. 2, pp. 215–226.

Dickey, P. A., 1986, *Petroleum Development Geology*, PennWell Publishing Company, Tulsa, OK, 530 pp.

Dolfing, J., Zeyer, J., Binder-Eicher, P., and Schwarzenbach, R. P., 1990, Isolation and characterization of a bacterium that mineralizes toluene in the absence of molecular oxygen, *Arch. Microbiol.*, vol. 154, pp. 336–341.

Edwards, E. A., and Grbić-Galić, D., 1992, Complete mineralization of benzene by aquifer microorganisms under strictly anaerobic conditions, *Appl. Environ. Microbiol.*, vol. 58, pp. 2663–2666.

———, 1994. Anaerobic degradation of toluene and *o*-xylene by a methanogenic consortium, *Appl. Environ. Microbiol.*, vol. 60, pp. 313–322.

Edwards, E. A., Wills, L. E., Reinhard, M., and Grbić-Galić, D., 1992, Anaerobic degradation of toluene and xylene by aquifer microorganisms under sulfate-reducing conditions, *Appl. Environ. Microbiol.*, vol. 58, pp. 794–800.

Evans, P. J., Mang, D. T., and Young, L. Y., 1991a, Degradation of toluene and *m*-xylene and transformation of *o*-xylene by denitrifying enrichment cultures, *Appl. Environ. Microbiol.*, vol. 57, pp. 450–454.

Evans, P. J., Mang, D. T., Kim, K. S., and Young, L. Y., 1991b, Anaerobic degradation of toluene by a denitrifying bacterium, *Appl. Environ. Microbiol.*, vol. 57, pp. 1139–1145.

Evans, P. J., Ling, W., Goldschmidt, B., Ritter, E. R., and Young, L. Y., 1992, Metabolites formed during anaerobic transformation of toluene and *o*-xylene and their relationship to the initial steps of toluene mineralization, *Appl. Environ. Microbiol.*, vol. 58, no. 2, pp. 496–501.

Flyvbjerg, J., Arvin, E., Jensen, B. K., and Olsen, S. K., 1993, Microbial degradation of phenols and aromatic hydrocarbons in creosote-contaminated groundwater under nitrate-reducing conditions, *J. Contam. Hydrol.*, vol. 12, pp. 133–150.

Freeze, R. A., and Cherry, J. A., 1979, *Groundwater*, Prentice Hall, Upper Saddle River, NJ, 604 pp.

Fries, M. R., Zhou, J., Chee-Sanford, J., and Tiedje, J. M., 1994, Isolation, characterization, and distribution of denitrifying toluene degraders from a variety of habitats, *Appl. Environ. Microbiol.*, vol. 60, pp. 2802–2810.

Ghosh, S., and Sun, M. L., 1992, Anaerobic biodegradation of benzene under acidogenic fermentation condition, in *Proceedings of the Conference on Hazardous Waste Research*, L. E. Erickson, S. C. Grant, and J. P. McDonald, eds., Engineering Extension, Manhattan, KS.

Gibson, D. T., and Subramanian, V., 1984, Microbial degradation of aromatic hydrocarbons, in *Microbial Degradation of Organic Compounds*, D. T. Gibson, ed., Marcel Dekker, New York, pp. 181–252.

Goldsmith, C. D., and Balderson, R. K., 1988, Biodegradation and growth kinetics of enrichment isolates on benzene, toluene, and xylene, *Water Sci. Tech.*, vol. 20, no. 11/12, pp. 505–507.

Goldstein, R. M., Mallory, L. M., and Alexander, M., 1985, Reasons for possible failure of inoculation to enhance biodegradation, *Appl. Environ. Microbiol.*, vol. 50, no. 4, pp. 977–983.

Grbić-Galić, D., 1986, Anaerobic production and transformation of aromatic hydrocarbons and substituted phenols by ferulic acid-degrading BESA-inhibited methanogenic consortia, *FEMS Microbiol. Ecol.*, vol. 38, pp. 161–169.

Grbić-Galić, D., and Vogel, T. M., 1987, Transformation of toluene and benzene by mixed methanogenic cultures, *Appl. Environ. Microbiol.*, vol. 53, pp. 254–260.

Groundwater Services, Inc., 1997, Florida RBCA Planning Study, Impact of RBCA Policy Options on LUST Site Remediation Costs, report prepared for Florida Partners in RBCA (PIRI), 24 pp.

Haag, F. M., Reinhard, M., and McCarty, P. L., 1991, Degradation of toluene and *p*-xylene in anaerobic microcosms: evidence for sulfate as a terminal electron acceptor, *Environ. Toxicol. Chem.*, vol. 10, pp. 1379–1389.

Hem, J. D., 1985, *Study and Interpretation of the Chemical Characteristics of Natural Water*, Water Supply Paper 2254, U.S. Geological Survey, Washington, DC, 264 pp.

Higgins, I. J., and Gilbert, P. D., 1978, The biodegradation of hydrocarbons, in *The Oil Industry and Microbial Ecosystems*, K. W. A. Chator and H. J. Somerville, eds., Heyden and Son, London, pp. 80–114.

Hutchins, S. R., 1991a, Biodegradation of monoaromatic hydrocarbons by aquifer microorganisms using oxygen, nitrate, or nitrous oxide as the terminal electron acceptor, *Appl. Environ. Microbiol.*, vol. 57, pp. 2403–2407.

———, 1991b, Optimizing BTEX biodegradation under denitrifying conditions, *Environ. Toxicol. Chem.*, vol. 10, pp. 1437–1448.

———, 1992, Inhibition of alkylbenzene biodegradation under denitrifying conditions by using the acetylene block technique, *Appl. Environ. Microbiol.*, vol. 58, no. 10, pp. 3395–3398.

———, 1993, Biotransformation and mineralization of alkylbenzenes under denitrifying conditions, *Environ. Toxicol. Chem.*, vol. 12, pp. 1413–1423.

Hutchins, S. R., Sewell, G. W., Sewell, D. A., Kovacs, D. A., and Smith, G. A., 1991, Biodegradation of aromatic hydrocarbons by aquifer microorganisms under denitrifying conditions, *Environ. Sci. Technol.*, vol. 25, pp. 68–76.

Jamison, V. W., Raymond, R. L., and Hudson, J. O. Jr., 1975, Biodegradation of high-octane gasoline in groundwater, *Dev. Ind. Microbiol.*, vol. 16.

Kazumi, J., Caldwell, M. E., Suflita, J. M., Lovley, D. R., and Young, L. Y., 1997, Anaerobic degradation of benzene in diverse anoxic environments, *Environ. Sci. Technol.*, vol. 31, no. 3, pp. 813–818.

Kelly, W. R., Hornberger, G. M., Herman, J. S., and Mills, A. L., 1996, Kinetics of BTX biodegradation and mineralization in batch and column studies, *J. Contam. Hydrol.*, vol. 23, no. 1/2, pp. 113–132.

Kemblowski, M. W., Salanitro, J. P., Deeley, G. M., and Stanley, C. C., 1987, Fate and transport of residual hydrocarbon in groundwater: a case study, in *Proceedings of the Petroleum Hydrocarbons and Organic Chemicals in Ground Water: Prevention, Detection and Restoration*, pp. 207–231, National Water Well Association, Dublin, OH.

Kuhn, E. P., Colberg, P. J., Schnoor, J. L., Wanner, O., Zehnder, A. J. B., and Schwarzenbach, R. P., 1985, Microbial transformations of substituted benzenes during infiltration of river water to groundwater: laboratory column studies, *Environ. Sci. Technol.*, vol. 19, pp. 961–968.

Kuhn, E. P., Zeyer, J., Eicher, P., and Schwarzenbach, R. P., 1988, Anaerobic degradation of alkylated benzenes in denitrifying laboratory aquifer columns, *Appl. Environ. Microbiol.*, vol. 54, pp. 490–496.

Kukor, J. J., and Olsen, R. H., 1989, Diversity of toluene degradation following long-term exposure to BTEX in situ, in *Biotechnology and Biodegradation*, Portfolio Publishing, The Woodlands, TS, pp. 405–421.

Liang, L. N., and Grbić-Galić, D., 1993, Biotransformation of chlorinated aliphatic solvents in the presence of aromatic compounds under methanogenic conditions, *Environ. Toxicol. Chem.*, vol. 12, pp. 1377–1393.

Lovley, D. R., 1991, Dissimilatory Fe(III) and Mn(IV) reduction, Microbiol. Rev., June, pp. 259–287.

Lovley, D. R., and Lonergan, D. J., 1990, Anaerobic oxidation of toluene, phenol, and *p*-creosol by the dissimilatory iron-reducing organisms, GS-15, *Appl. Environ. Microbiol.*, vol. 56, no. 6, pp. 1858–1864.

Lovley, D. R., and Phillips, E. J. P., 1987. Competitive mechanisms for inhibition of sulfate reduction and methane production in the zone of ferric iron reduction in sediments, *Appl. Environ. Microbiol.*, vol. 53, pp. 2636–2641.

Lovley, D. R., Baedecker, M. J., Lonergan, D. J., Cozzarelli, I. M., Phillips, E. J. P., and Siegel, D. I., 1989, Oxidation of aromatic contaminants coupled to microbial iron reduction, *Nature*, vol. 339, pp. 297–299.

Lovley, D. R., Phillips, E. J. P., and Lonergan, D. J., 1991, Enzymatic versus nonenzymatic mechanisms for Fe(III) reduction in aquatic sediments, *Environ. Sci. Technol.*, vol. 26, no. 6, pp. 1062–1067.

Lovley, D. R., Woodward, J. C., and Chapelle, F. H., 1994a, Stimulation of anoxic aromatic hydrocarbon degradation with Fe(III) ligands, *Nature (London)*, vol. 370, July 14, pp. 128–131.

Lovley, D. R., Chapelle, F. H., and Woodward, J. C., 1994b, Use of dissolved H_2 concentrations to determine distribution of microbially catalyzed redox reactions in anoxic groundwater, *Environ. Sci. Tech.*, vol. 28, no. 7, pp. 1205–1210.

Lovley, D. R., Coates, J. D., Woodward, J. C., and Phillips, E. J. P., 1995, Benzene oxidation coupled to sulfate reduction, *Appl. Environ. Microbiol.*, vol. 61, no. 3., pp. 953–958.

Lovley, D. R., Woodward, J. C., and Chapelle, F. H., 1996, Rapid anaerobic benzene oxidation with a variety of chelated Fe(III) forms, *Appl. Environ. Microbiol.*, vol. 62, no. 1, pp. 288–291.

Mace, R. E., Fisher, R. S., Welch, D. M., and Parra, S. P., 1997, *Extent, Mass, and Duration of Hydrocarbon Plumes from Leaking Petroleum Storage Tank Sites in Texas*, Geological Circular 97-1, Bureau of Economic Geology, University of Texas at Austin, Austin, TX.

MacIntyre, W. G., Boggs, M., Antworth, C. P., and Stauffer, T. B., 1993, Degradation kinetics of aromatic organic solutes introduced into a heterogeneous aquifer, *Water Resour. Res.*, vol. 29, no. 12, pp. 4045–4051.

MacQuarrie, K. T. B., Sudicky, E. A., and Frind, E. O., 1990, Simulation of biodegradable organic contaminants in ground water: 1. Numerical formulation in principal directions, *Water Resources Res.*, vol. 26, no. 2, pp. 207–222.

Major, D. W., Mayfield, C. I., and Barker, J. F., 1988, Biotransformation of benzene by denitrification in aquifer sand, *Ground Water*, vol. 26, pp. 8–14.

Marr, E. K., and Stone, R. W., 1961, Bacterial oxidation of benzene, *J. Bacteriol.*, vol. 81, pp. 4245–430.

McCarty, P. L., 1972, Energetics of organic matter degradation, in *Water Pollution Microbiology*, R. Mitchell, ed., Wiley-Interscience, New York, pp. 91–118.

McCarty, P. L., Reinhard, M., and Rittmann, B. E., 1981, Trace organics in groundwater, *Environ. Sci. Technol.*, vol. 15, no. 1, pp. 40–51.

McMahon, P. B, Crowfoot, R., and Wydoski, D., 1994, Effect of fuel oxidants on the degradation of gasoline components in sediments of the South Platte alluvial aquifer, *Proceedings of the 5th Annual South Platte Forum: Integrated Watershed Management in the South Platte Basin: Status and Practical Implementation*, Greeley, CO, Oct. 26–27, pp. 25.

———, 1995, Effect of fuel oxidants on the degradation of gasoline components in aquifer sediments, in *Proceedings of the 3rd International Symposium on In Situ and On-Site Bioreclamation*, San Diego, CA, Apr. 24–27.

Miller, R. N. 1990, A field-scale investigation of enhanced petroleum hydrocarbon biodegradation in the vadose zone at Tyndall Air Force Base, Florida, in *Proceedings of the Petroleum Hydrocarbons and Organic Chemicals in Groundwater: Prevention, Detection, and Restoration Conference*, NWWA/API, pp. 339–351.

Morgan, P., Lewis, S. T., and Watkinson, R. J., 1993, Biodegradation of benzene, toluene, ethylbenzene, and xylenes in gas-condensate-contaminated groundwater, *Environ. Pollut.*, vol. 82, no. 2, pp. 181–190.

Newell, C. J., Hopkins, L. P., and Bedient, P. B., 1990, A hydrogeologic database for ground water modeling, *Ground Water*, vol. 28, no. 5, pp. 703–714.

Newell, C. J., Winters, J. W., Rifai, H. S., Miller, R. N., Gonzales, J., and Wiedemeier, T. H., 1995, Modeling intrinsic remediation with multiple electron acceptors: results from seven sites, *Proceedings of the Petroleum Hydrocarbons and Organic Chemicals in Ground Water Conference*, Houston, TX, November 1995, National Ground Water Association, pp. 33–48.

Newell, C. J., McLeod, R. K., and Gonzales, J. R., 1996, *BIOSCREEN Natural Attenuation Decision Support System User's Manual*, Version 1.3, EPA/600/R-96/087, August, Robert S. Kerr Environmental Research Center, Ada, OK.

Newell, C. J., and Connor, J. A., 1998, *Characteristics of Dissolved Petroleum Hydrocarbon Plumes: Results from Four Studies*, API Tech Transfer Bulletin, 8 pp.

Norris, R. D., Hinchee, R. E., Brown, R., McCarty, P. L., Semprini, L., Wilson, J. T., Kampbell, D. H., Reinhard, M., Bouwer, E. J., Borden, R. C., Vogel, T. M., Thomas, J. M., and Ward, C. H., eds., 1994, *Handbook of Bioremediation*, Lewis Publishers, Boca Raton, FL, pp. 177–199.

Patterson, B. M., Pribac, F., Barber, C., Davis, G. B., and Gibbs, R., 1993, Biodegradation and retardation of PCE and BTEX compounds in aquifer material from Western Australia using large-scale columns, *J. Contam. Hydrol.*, vol. 14, pp. 261–278.

Payne, W. J., 1981, The status of nitric oxide and nitrous oxide as intermediates in denitrification, in *Denitrification, Nitrification, and Atmospheric Nitrous Oxide*, C. C. Delwiche, ed., Wiley-Interscience, New York, pp. 85–103.

Piet, G. J., and Smeenk, J. G. M. M., 1985, Behavior of organic pollutants in pretreated Rhine water during dune infiltration, in *Groundwater Quality*, C. H. Ward, W. Giger, and P. L. McCarty, eds., Wiley, New York, pp. 122–144.

Rabus, R., and Widdel, F., 1995, Anaerobic degradation of ethylbenzene and aromatic hydrocarbons by a new denitrifying bacteria, *Arch. Microbiol.*, vol. 163, pp. 96–103.

Rabus, R., Nordhaus, R., Ludwig, W., and Widdel, F., 1993, Complete oxidation of toluene under strictly anoxic conditions by a new sulfate-reducing bacterium, *Appl. Environ. Microbiol.*, vol. 59, pp. 1444–1451.

Reinhard, M., Goodman, N. L., and Barker, J. F., 1984, Occurrence and distribution of organic chemicals in two landfill leachate plumes, *Environ. Sci. Technol.*, vol. 18, pp. 953–961.

Rice, D. W., Grose, R. D., Michaelsen, J. C., Dooher, B. P., MacQueen, D. H., Cullen, S. J., Kastenberg, W. E., Everett, L. G., and Marino, M. A., 1995, *California Leaking Underground Fuel Tank (LUFT) Historical Case Analyses*, California State Water Resources Control Board, Sacramento, CA.

Rifai, H. S., and Bedient, P. B., 1994, Modeling contaminant transport and biodegradation in ground water, in *Advances in Environmental Science Groundwater Contamination*, D. C. Adriano, A. K. Iskandar, and I. P. Murarka, eds., Science Reviews, Northwood, OH.

Rifai, H. S., Borden, R. C., Wilson, J. T., and Ward, C. H., 1995, Intrinsic bioattenuation for subsurface restoration, in *In Situ and On-Site Bioreclamation, The 3rd International Symposium*, R. E. Hinchee and R. F. Olfenbuttel, eds., Battelle Memorial Institute, Butterworth-Heinemann, Boston.

Riser-Roberts, E., 1992, *Bioremediation of Petroleum Contaminated Sites*, C. K. Smoley, Boca Raton, FL, 461 pp.

Robertson, B. R., and Button, D. K., 1987, Toluene induction and uptake kinetics and their inclusion in the specific affinity relationship for describing rates of hydrocarbon metabolism, *Appl. Environ. Microbiol.*, vol. 53, pp. 2193–2205.

Rueter, P., Rabus, R., Wilkes, H., Aeckersberg, F., Rainey, F. A., Jannasch, H. W., and Widdel, F., 1994, Anaerobic oxidation of hydrocarbons in crude oil by new types of sulphate-reducing bacteria, *Nature*, vol. 372, pp. 455–458.

Salanitro, J. P., Wisniewski, H. L., Byers, D. L., Neaville, C. C., 1997, Use of aerobic and anaerobic microcosms to assess BTEX biodegradation in aquifers: Groundwater Monitoring and Remediation, vol. 17, no. 3, pp. 210–221.

Schaefer, M. B., 1994, *Methanogenic Biodegradation of Alkylbenzenes in Groundwater from Sleeping Bear Dunes National Lakeshore*, M.S. Thesis, North Carolina State University, Raleigh, NC.

Schocher, R. J., Seyfried, B., Vazquez, F., and Zeyer, J., 1991, Anaerobic degradation of toluene by pure cultures of denitrifying bacteria, *Arch. Microbiol.*, vol. 157, pp. 7–12.

Smith, M. R., 1990, The biodegradation of aromatic hydrocarbons by bacteria, *Biodegradation*, vol. 1, pp. 191–206.

Starr, R. C., and Gillham, R. W., 1993, Denitrification and organic carbon availability in two aquifers, *Ground Water*, vol. 31, no. 6, pp. 934–947.

Stumm, W., and Morgan, J. J., 1981, *Aquatic Chemistry*, Wiley, New York.

Swindoll, M. C., Aelion, C. M., and Pfaender, F. K., 1988, Influence of inorganic and organic nutrients on aerobic biodegradation and on the adaptation response of subsurface microbial communities, *Appl. Environ. Microbiol.*, vol. 54, no. 1, pp. 221–217.

Tabak, H. H., Govind, R., and Desai, S., 1990, *Determination of Biodegradability Kinetics of RCRA Compounds Using Respirometry for Structure Activity Relationships*, EPA/600/D-90/136, 18.

Thierrin, J., Davis, G. B., Barber, C., Patterson, B. M., Pribac, F., Power, T. R., and Lambert, M., 1992a, Natural degradation rates of BTEX compounds and naphthalene in a sulfate reducing groundwater environment, in *In-Situ Bioremediation Symposium '92*, Niagara-on-the-Lake, Ontario, Canada, Sept. 20–24.

Thierrin, J., Davis, G. B., Barber, C., and Power, T. R., 1992b, Use of deuterated organic compounds as groundwater tracers for determination of natural degradation rates within a contaminated zone, *Proceedings of the 6th International Symposium on Water Tracing*, Karlsruhe, Germany.

Thierrin, J., Davis, G. B., Barber, C., Patterson, B. M., Pribac, F., Power, T. R., and Lambert, M., 1993, Natural degradation rates of BTEX compounds and napthalene in a sulphate reducing groundwater environment, *Hydrol. Sci. J.*, vol. 38, no. 4, pp. 309–322.

Van Beelen, P., and Van Keulen, F., 1990, The kinetics of the degradation of chloroform and benzene in anaerobic sediment from the River Rhine, *Hydrobiol. Bull.*, vol. 24, no. 1, pp. 13–20.

Vogel, T. M., and Grbić-Galić, D., 1986, Incorporation of oxygen from water into toluene and benzene during anaerobic fermentative transformation, *Appl. Environ. Microbiol.*, vol. 52, pp. 200–202.

von Gunten, U., and Zobrist, J., 1993, Biogeochemical changes in groundwater-infiltration systems: column studies, *Geochim. Cosmochim. Acta*, vol. 57, pp. 3895–3906.

Vroblesky, D. A., and Chapelle, F. H., 1994, Temporal and spatial changes of terminal electron-accepting processes in a petroleum hydrocarbon-contaminated aquifer and the significance for contaminant biodegradation, *Water Resour. Res.*, vol. 30, no. 5, pp. 1561–1570.

Ward, D. M., Atlas, R. M., Boehm, P. D., and Calder, J. A., 1980, Microbial biodegradation and chemical evolution of oil from the Amoco spill, *Ambio*, vol. 9, pp. 277–283.

Watwood, M. E., White, C. S., and Dahm, C. N., 1991, Methodological modifications for for accurate and efficient determination of contaminant biodegradation in unsaturated calcareous soils, *Appl. Environ. Microbiol.*, vol. 57, no. 3, pp. 717–720.

Weiner, J. M., and Lovley, D. R., 1998, Rapid benzene degradation in methanogenic sediments from a petroleum-contaminated aquifer, *Appl. Environ. Microbiol.*, vol. 64, no. 5, pp. 1937–1939.

Wiedemeier, T. H., Wilson, J. T., Kampbell, D. H., Miller, R. N., and Hansen, J. E., 1995a, *Technical Protocol for Implementing Intrinsic Remediation with Long-Term Monitoring for Natural Attenuation of Fuel Contamination Dissolved in Groundwater*, U.S. Air Force Center for Environmental Excellence, San Antonio, TX.

Wiedemeier, T. H., Wilson, J. T., and Miller, R. N., 1995b, Significance of Anaerobic Processes for the Intrinsic Bioremediation of Fuel Hydrocarbons, in *Proceedings of the Petroleum Hydrocarbons and Organic Chemicals in Groundwater: Prevention, Detection, and Restoration Conference*, Houston, TX, Nov. 29–Dec. 1, NWWA/API.

Wiedemeier, T. H., Swanson, M. A., Wilson, J. T., Kampbell, D. H., Miller, R. N., and Hansen, J. E., 1996, Approximation of biodegradation rate constants for monoaromatic hydrocarbons (BTEX) in ground water, *Ground Water Monitor. Remed.*, vol. 16, no. 3, pp. 186–195.

Wiedemeier, T. H., Hansen, J. E., and Haas, P., 1998, *Cost and Performance Summaries for Natural Attenuation of Fuel Hydrocarbons*, U.S. Air Force Center for Environmental Excellence, San Antonio, TX.

Wilson, B. H., Smith, G. B., and Rees, J. F., 1986, Biotransformations of selected alkylbenzenes and halogenated aliphatic hydrocarbons in methanogenic aquifer material: a microcosm study, *Environ. Sci. Technol.*, vol. 20, pp. 997–1002.

Wilson, B. H., Bledsoe, B., and Kampbell, D., 1987, Biological processes occurring at an aviation gasoline spill site, in *Chemical Quality of Water and the Hydrologic Cycle*, R. C. Averett and D. M. McKnight eds., Lewis Publishers, Boca Raton, FL, pp. 125–137.

Wilson, B. H., Wilson, J. T., Kampbell, D. H., Bledsoe, B. E., and Armstrong, J. M., 1990, Biotransformation of monoaromatic and chlorinated hydrocarbons at an aviation gasoline spill site, *Geomicrobiol. J.*, vol. 8, pp. 225–240.

Wilson, J. T., Kampbell, D. H., and Armstrong, J., 1993, Natural bioreclamation of alkylbenzenes (BTEX) from a gasoline spill in methanogenic groundwater, in *Proceedings of the Environmental Restoration Technology Transfer Symposium*, San Antonio, TX.

Wilson, J. T., Pfeffer, F. M., Weaver, J. W., Kampbell, D. H., Wiedemeier, T. H., Hansen, J. E., and Miller, R. N., 1994, Intrinsic remediation of JP-4 jet fuel, in *Proceedings of the EPA Symposium on Intrinsic Remediation in Ground Water*, EPA/540/R-94/515, 189.

Young, L. Y., 1984, Anaerobic degradation of aromatic compounds, in *Microbial Degradation of Aromatic Compounds*, D. R. Gibson, ed., Marcel Dekker, New York, pp. 487–523.

Zehnder, A. J. B., 1978, Ecology of methane formation, in *Water Pollution Microbiology*, R. Mitchell ed., Wiley, New York, pp. 349–376.

Zeyer, J., Kuhn, E. P., and Schwarzenbach, R. P., 1986, Rapid microbial mineralization of toluene and 1,3-dimethylbenzene in the absence of molecular oxygen, *Appl. Environ. Microbiol.*, vol. 52, no. 4, pp. 944–947.

Zhou, J., Fries, M. R., Chee-Sanford, J. C., and Tiedje, J. M., 1995, Phylogenetic analyses of a new group of denitrifiers capable of anaerobic growth on toluene and description of *Azoarcus tolulyticus* sp. nov., *Int. J. Syst. Bacteriol.*, vol. 45, pp. 500–506.

CHAPTER 6

INTRINSIC BIOREMEDIATION OF CHLORINATED SOLVENTS

Since they were first produced some 100 years ago and came into common use in the 1940s, chlorinated solvents and their natural degradation (daughter) products have become some of the most prevalent organic contaminants found in the shallow groundwater of the United States. The solvent properties of these compounds have resulted in many industrial uses. In addition to being excellent degreasing agents, chlorinated solvents are nearly nonflammable and noncorrosive. They are used to clean and degrease everything from rockets to electronics to clothing. The use of these compounds as dry-cleaning agents, in particular, has resulted in their application at a large number of locations. The most commonly used chlorinated solvents are perchloroethene (PCE), trichloroethene (TCE), 1,1,1-trichloroethane (TCA), and carbon tetrachloride (CT).

Compared to the biodegradation of petroleum hydrocarbons, which is better understood, researchers are just beginning to understand the microbial degradation of chlorinated solvents. The specific details of many degradation pathways remain to be discovered. Unlike petroleum hydrocarbons, which can be oxidized by microorganisms under either aerobic or anaerobic conditions, many chlorinated solvents are degraded only under specific ranges of oxidation–reduction potential. For example, as far as the scientific community knows, PCE is biologically degraded through use as a primary growth substrate only under strongly reducing anaerobic conditions. In this chapter we review the biological processes that bring about the degradation of the most common chlorinated solvents, present conceptual models of chlorinated solvent plumes, and summarize data from field studies with chlorinated solvent plumes.

6.1 OVERVIEW OF CHLORINATED SOLVENT BIODEGRADATION PROCESSES

The degradation of chlorinated solvents has been described under a wide variety of physiochemical conditions. As discussed in Chapter 4, the biodegradation of organic chemicals can be grouped into two broad categories: (1) use of the organic compound as a primary growth substrate, and (2) cometabolism. As discussed below, the use of the organic compound as a primary growth substrate is probably the most important biological mechanism working to degrade the common chlorinated solvents and many of their daughter products in the subsurface. When organic compounds, including chlorinated solvents, are used as primary growth substrates, the mediating organism obtains energy for growth. Some chlorinated solvents are used as electron donors and some are used as electron acceptors when serving as primary growth substrates. When used as an electron donor, the contaminant is oxidized. Conversely, when used as an electron acceptor, the chlorinated solvent is reduced via a reductive dechlorination process called halorespiration (see below). To summarize, chlorinated solvents can be biodegraded through use as a primary growth substrate in the following reactions:

- Halorespiration (a form of reductive dechlorination)
- Oxidation reactions
  - —Aerobic oxidation
  - —Anaerobic oxidation

It is important to note that not all chlorinated solvents can be degraded via all of these reactions. In fact, vinyl chloride is the only chlorinated solvent known to degrade via all of these pathways.

In addition to their use as a primary growth substrate, chlorinated solvents can also be degraded via cometabolic pathways. During cometabolism, microorganisms gain carbon and energy for growth from metabolism of a primary substrate, and chlorinated solvents are degraded fortuitously by enzymes present in the metabolic pathways. The organism obtains no known benefit from the biodegradation of the chlorinated solvent, which in some cases often yields products that are toxic to the microorganism. Cometabolism reactions can be either oxidation or reduction reactions (under aerobic or anaerobic conditions). For example J. T. Wilson and B. H. Wilson (1985) showed that TCE could be cometabolically oxidized under aerobic conditions in soils that were fed methane. During this reaction methanotrophs utilize oxygen as the electron acceptor and methane as the electron donor. TCE is degraded using the same enzyme required for methane oxidation. Although this type of oxidation reaction is possible under natural conditions, methane and oxygen typically are not found together in large quantities in the shallow subsurface, and cometabolic oxidation is not a significant process in many chlorinated solvent plumes.

In addition to oxidative cometabolism, anaerobic reductive dechlorination can also occur via cometabolism. Because the organisms that cause these reactions are

TABLE 6.1. Biological Degradation Processes for Selected Chlorinated Solvents

Compound	Halo-respiration	Direct Aerobic Oxidation	Direct Anaerobic Oxidation	Aerobic Cometabolism	Anaerobic Cometabolism
PCE	×				×
TCE	×			×	×
DCE	×	×	×	×	×
Vinyl chloride	×	×	×	×	×
1,1,1-TCA	×			×	×
1,2-DCA	×	×		×	×
Chloroethane		×		×	
Carbon tetrachloride	×				×
Chloroform	×			×	×
Methylene chloride		×	×	×	

ubiquitous at field sites, anaerobic cometabolic dechlorination is responsible for some of the chloroethene biodegradation observed (Gossett and Zinder, 1996). However, the process of cometabolic reductive dechlorination is "sufficiently slow and incomplete that a successful natural attenuation strategy typically cannot completely rely upon it" (Gossett and Zinder, 1996).

The types of biodegradation reactions that have been observed for different chlorinated solvents are presented in Table 6.1. Because of the slow and incomplete nature of cometabolic biodegradation pathways, biodegradation with chlorinated solvents as a primary growth substrate typically contributes more to natural attenuation processes than cometabolic processes. The role of hydrogen as an electron donor under halorespiration is now widely recognized as the key factor governing the dechlorination of chlorinated compounds (Gossett and Zinder, 1996; Hughes et al., 1997; Newell et al., 1997), and for natural attenuation of ethenes, the "success or failure of natural attenuation can be linked to the specific type of dechlorinator present (i.e., cometabolic or direct), as well as to the relative supply of H_2 precursors compared with the supply of chlorinated ethene that must be reduced" (Gossett and Zinder, 1996). The various mechanisms for biodegradation of chlorinated solvents are discussed below.

Halorespiration of Chlorinated Solvents (Reductive Dechlorination Driven by Hydrogen)

The chemical term *reduction* was originally derived from the chemistry of smelting metal ores. Ores are composed of metal atoms coupled with other elements. As the ores are smelted to the pure metal, the weight is reduced through the removal of the other elements. Chemically, the positively charged metal ions receive electrons to become the electrically neutral pure metal. Chemists adopted the term *reduction* for any chemical reaction that added electrons to an element. In a similar manner, the chemical reaction of pure metals with oxygen results in the removal of electrons from the neutral metal to produce positively charged metal ions. Chemists adopted the term *oxidation* to refer to any chemical reaction that removes electrons from a material. For one compound to be reduced, another compound must be equally oxidized. Such a balanced reaction is called an oxidation–reduction reaction. In aerobic metabolism, organic substrates such as sugars and fatty acids are oxidized and oxygen is reduced yielding energy. For example, in the following reaction, glucose is the electron donor and oxygen is the electron acceptor:

$$C_6H_{12}O_6\,(\text{glucose}) + 6O_2 \Rightarrow 6CO_2 + 6H_2O + \text{energy}$$

In contrast, reductive dechlorination is a reaction in which a chlorinated solvent acts as an electron acceptor and a chlorine atom on the molecule is replaced with a hydrogen atom. This results in the reduction of the chlorinated solvent. When this reaction is biological, and the organism is utilizing the substrate for energy and growth, the reaction is termed *halorespiration*. Only recently have researchers demonstrated the existence of halorespiration (Holliger and Schumacher, 1994). Prior to this research, reductive dechlorination was thought to be strictly a cometabolic process. During halorespiration, hydrogen is used directly as an electron donor. The hydrogen is produced in the terrestrial subsurface by the fermentation of a wide variety of organic compounds, including petroleum hydrocarbons and natural organic carbon. Because of its importance in the microbial metabolism of the halorespirators, the relative supply of hydrogen precursors compared to the amount of chlorinated solvent that must be degraded is an important consideration when evaluating natural attenuation.

Figure 6.1 illustrates the reductive dechlorination of PCE to TCE and the production of a hydrogen ion and a chloride ion. Similar reactions have been shown to reduce DCE to VC and VC to ethene. Figure 6.2 shows the pathway for complete reductive dechlorination of chlorinated ethenes. In general, reductive dechlorination of the ethenes occurs by sequential dechlorination from PCE to TCE to DCE to VC and finally to ethene. Freedman and Gossett (1989) were first to show the complete

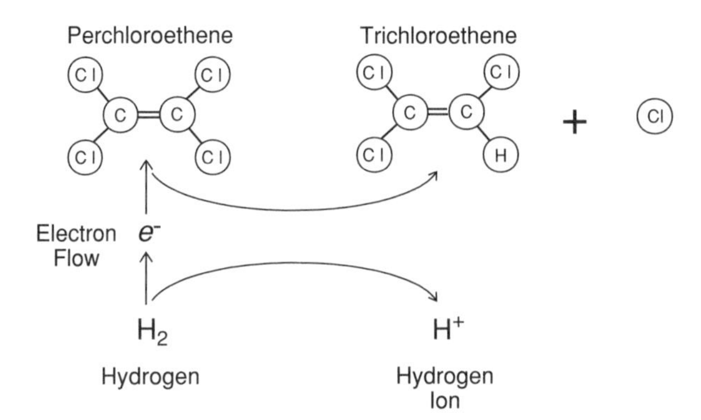

FIGURE 6.1. Reductive dechlorination of perchloroethene to trichloroethene.

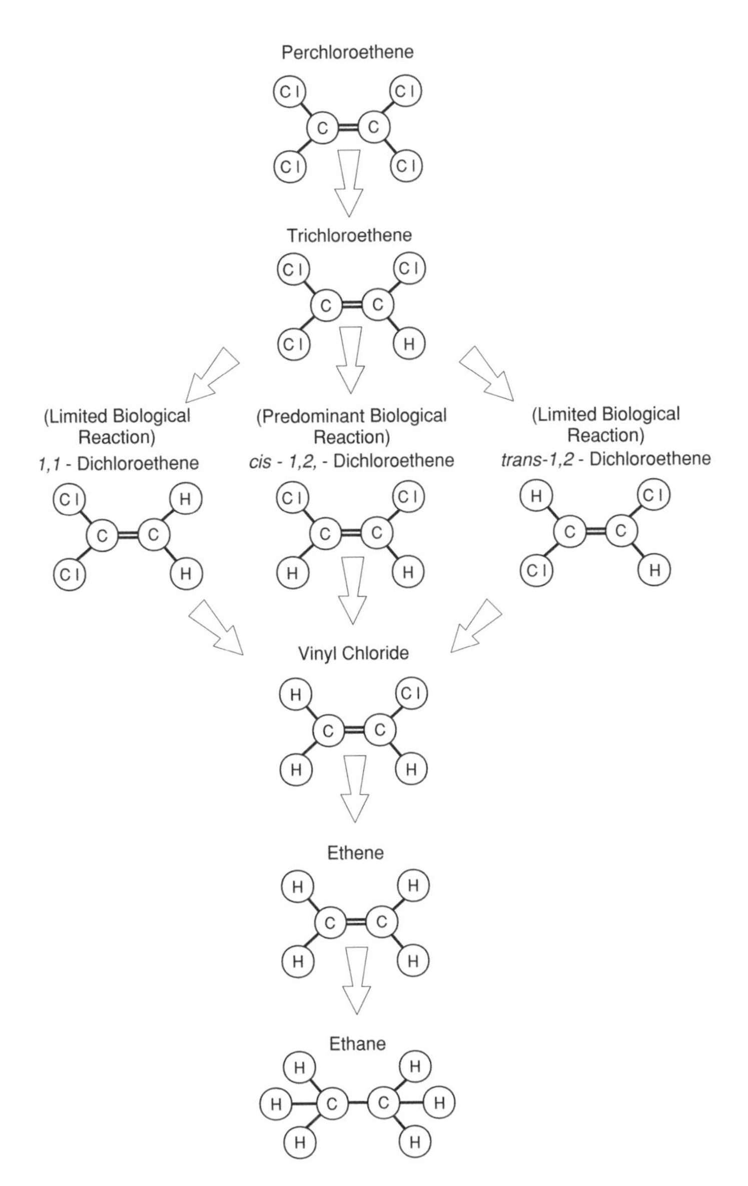

FIGURE 6.2. Reductive dechlorination of chlorinated ethenes. (After Vogel et al., 1987.)

reductive dechlorination of PCE to ethene under methanogenic conditions. Reductive dechlorination of the chlorinated ethanes is shown in Figure 6.3. A summary of key biotic and abiotic reactions for the chlorinated ethenes, ethanes, and methanes first developed by Vogel et al. (1987) is shown in Figure 6.4.

Based on the reaction kinetics and field data analysis, halorespiration probably accounts for the majority of chlorinated solvent biodegradation at many of the sites

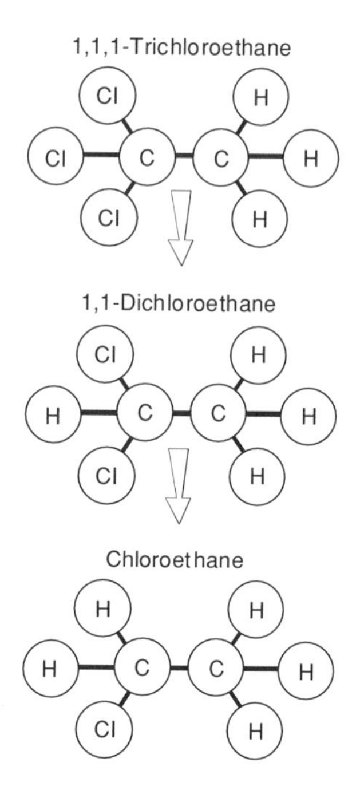

FIGURE 6.3. Reductive dechlorination of chlorinated ethanes.

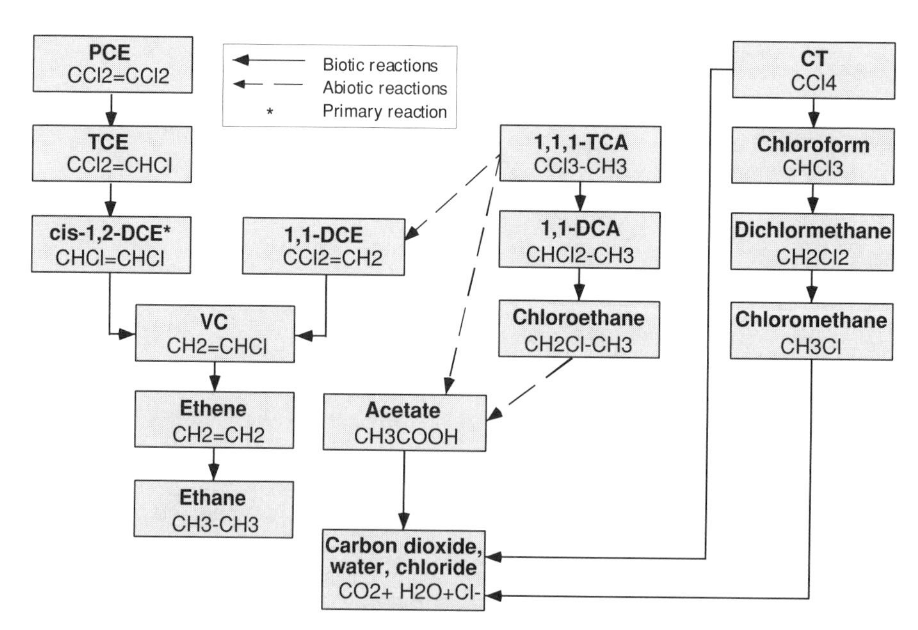

FIGURE 6.4. Abiotic and biological transformation pathways for selected chlorinated solvents. (Adapted from McCarty and Semprini, 1994; after Vogel et al., 1987.)

where biodegradation is significantly attenuating the plume. For halorespiration to occur, the following conditions must exist: (1) the subsurface environmental must be anaerobic and have a low oxidation–reduction potential, (2) chlorinated solvents amenable to halorespiration must be present, and (3) there must be an adequate supply of fermentation substrates for production of dissolved hydrogen.

Oxidation–Reduction Environment Required for Halorespiration. The environmental chemistry and oxidation–reduction potential of a site play an important role in determining if halorespiration will occur. Based on thermodynamic considerations, reductive dechlorination reactions will occur only after both oxygen and nitrate have been depleted from the aquifer. In groundwater, the electrons required for microbial reduction of chlorinated solvents are extracted from dissolved hydrogen. This hydrogen is fermented from native organic matter, contaminants released from fuel spills, such as benzene, toluene, ethylbenzene, and xylene compounds; volatile fatty acids in wastes, such as landfill leachate; and other fermentable organics. Thus dechlorination facilitated by indigenous microorganisms will not occur until the environment becomes sufficiently reduced to support fermentation. As shown by Figure 4.4, reductive dechlorination will not occur in either aerobic environments or conditions where nitrate reduction is occurring. This is because oxygen and nitrate are higher-energy electron acceptors than the chlorinated solvents and the halorespirators cannot compete. However, if sulfate is being consumed at a particular location in a contaminated aquifer, or if methane is being produced, the environment is suitable for reductive dechlorination and the halorespirators can obtain enough energy to compete with the sulfate reducers and the methanogens.

Importance of Dissolved Hydrogen in Halorespiration. Significant progress has been made in recent years in understanding the biochemistry of halorespiration. It is now understood that halorespiration is typically based on the following generalized oxidation–reduction reaction:

$$H_2 + C{-}Cl \Rightarrow C{-}H + H^+ + Cl^-$$

where C−Cl represents a carbon–chloride bond in a chlorinated solvent. In this reaction H_2 is the electron donor, which is oxidized, and the chlorinated solvent is the electron acceptor, which is reduced. Although a few other electron donors besides hydrogen have been identified that can drive halorespiration, these compounds are also fermentation products. The only pure culture isolated to date that can dechlorinate PCE to ethene completely requires hydrogen as the electron donor (Maymo-Gatell et al., 1997). Therefore, hydrogen appears to be the most important electron donor for halorespiration.

Until recently, most laboratory research concerning the anaerobic degradation of chlorinated compounds focused on methanogenic systems. Such systems typically involve the introduction of an organic compound, such as acetate, lactate, methanol, ethanol, or even a co-contaminant such as toluene, which is fermented to produce hydrogen among other things. Although chlorinated compounds have been observed to be degraded in a variety of laboratory systems (Bouwer and McCarty, 1983; Vogel and McCarty, 1985; Bouwer and Wright, 1988; Freedman and Gossett, 1989; Sewell and Gibson, 1991), it is now clear that these systems probably contained at least two distinct strains of bacteria. One strain fermented the organic carbon to produce hydrogen, and another strain utilized the hydrogen as an electron donor for halorespiration. Only in the last four years have researchers begun to fully recognize the role of hydrogen as the electron donor in the reductive dechlorination of PCE and TCE (Holliger et al., 1993; Gossett and Zinder, 1996; Smatlak et al., 1996; Ballapragada et al., 1997).

Prior to understanding the role of hydrogen as the electron donor, there were numerous studies in which the effectiveness of indirect electron donors, such as glucose or toluene, for stimulating chlorinated solvent degradation were investigated. These studies typically did not measure the steady-state flux of hydrogen. For example, Table 6.2 shows how PCE and TCE have been shown to undergo reductive

TABLE 6.2. Sources, Donors, Acceptors, and Products of Reductive Dechlorinating Laboratory Systems

Reference	Source	Donor	Acceptor–Product
Bouwer and McCarty (1983)	Digester	Organic material	PCE-TCE
Vogel and McCarty (1985)	Bioreactor	Acetate	PCE-VC, CO_2
Kleopfer et al. (1985)	Soil	Soybean meal	TCE-DCE
Barrio-Lage et al. (1987)	Swamp muck	Organic material	PCE-VC
	Soil	Methanol (?)	PCE-VC
Fathepure et al. (1987)	Methanosarcina	Methanol	PCE-TCE
	DCB-1	3CB,[a] pyruvate, RF[b]	PCE-TCE
Baek and Jaffe (1989)	Digester	Formate	TCE-VC,CA[c]
		Methanol	TCE-VC,CA
Freedman and Gossett (1989)	Digester	Methanol	PCE-VC, ethene
		Glucose	PCE-VC, ethene
		H_2	PCE-VC, ethene
		Formate	PCE-VC, ethene
		Acetate	PCE-VC, ethene
Scholz-Muramatsu et al. (1990)	Bioreactor	Benzoate	PCE-DCE
Gibson and Sewell (1990)	Aquifer	VFA[d]	PCE-DCE
Sewell and Gibson (1991)	Aquifer	Toluene	PCE-DCE
Sewell et al. (1991)	Aquifer	VFA	PCE-DCE
	Landfill	VFA	PCE-VC

[a] 3-Chlorobenzoate.

[b] Rumen fluid.

[c] Chloroethane.

[d] Volatile fatty acid.

dechlorination in a variety of anaerobic systems with various indirect electron donors. Understanding the results of these studies requires an understanding of the microbial processes that result in the production and utilization of hydrogen. In the following sections we discuss the biological processes that result in the production of hydrogen and the competition among bacteria for the utilization of this hydrogen.

Generation of Dissolved Hydrogen by Fermentation. Under natural conditions, fermentation is the process that generates the hydrogen used in reductive dechlorination. Fermentation is a balanced oxidation–reduction reaction in which different portions of a single substrate are oxidized and reduced, yielding energy. In contrast to the oxidation reactions discussed previously, fermentation does not require an external electron acceptor, such as oxygen, nitrate, or a chlorinated solvent. Fermentation yields substantially less energy per unit of substrate compared to oxidation reactions, which utilize an external electron acceptor; thus fermentation generally occurs when these external electron acceptors are not available. Familiar fermentation reactions include the conversion of glucose to lactic acid in human muscle during periods of intense exercise when oxygen is insufficient to meet energy requirements and the fermentation of glucose to ethanol by yeast, the process that produces the alcohol in beer. However, fermentation by bacteria can also be important in anaerobic aquifers.

Bacterial fermentation can be divided into two categories:

1. *Primary fermentation:* the fermentation of primary substrates such as sugars, amino acids, and lipids to yield acetate, formate, CO_2, and H_2, but also yields ethanol, lactate, succinate, propionate, and butyrate. While primary fermentation often yields H_2, production of H_2 is not required for these reactions to occur.
2. *Secondary fermentation:* the fermentation of primary fermentation products such as ethanol, lactate, succinate, propionate, and butyrate, yielding acetate, formate, H_2, and CO_2. Bacteria that carry out these reactions are called *obligate proton reducers* because the reactions must produce hydrogen in order to balance the oxidation of the carbon substrates. These secondary fermentation reactions are energetically favorable only if hydrogen concentrations are very low (10^{-2} to 10^{-4} atm or 8000 to 80 nM dissolved hydrogen, depending on the fermentation substrate). Thus these fermentation reactions occur only when the produced hydrogen is utilized by other bacteria, such as methanogens which convert H_2 and CO_2 into CH_4 and H_2O. The process by which hydrogen is produced by one strain of bacteria and utilized by another is called *interspecies hydrogen transfer*.

In the absence of external electron acceptors, the hydrogen produced by fermentation will be utilized by methanogens (methane-producing bacteria). In this case, the ultimate end products of anaerobic metabolism of carbon substrates will be CH_4 (the most reduced form of carbon) and CO_2 (the most oxidized form of carbon). Methanogens will carry out the last step in this metabolism, the conversion of H_2 and CO_2 into

CH_4. However, in the presence of external electron acceptors (halogenated organics, nitrate, sulfate, etc.), other products will be formed.

Fermentation of BTEX and Other Substrates. There are a number of compounds besides the ones listed above that can be fermented to produce hydrogen. Sewell and Gibson (1991) noted that petroleum hydrocarbons support reductive dechlorination. In this case, the reductive dechlorination is driven by the fermentation of biodegradable compounds such as the BTEX compounds in fuels.

While anaerobic degradation of benzene has been confirmed (e.g., Kazumi et al., 1997; Weiner and Lovley, 1998), there is still significant controversy as to whether aromatic compounds, such as the BTEX compounds, can be completely mineralized to CO_2 without alternate electron acceptors (i.e., solely by fermentation coupled with a hydrogen-utilizing process such as methanogenesis). Although numerous pure cultures have been isolated that can completely mineralize benzoate or toluene in the absence of oxygen, all of these pure cultures require an external electron acceptor such as nitrate, sulfate, or iron (Harwood and Gibson, 1997). Pure cultures that can ferment aromatics to CO_2 and CH_4 are not expected because the production of CH_4 from complex organics requires at least two strains of bacteria, fermenters and methanogens. As discussed above, the fermentation reactions are not thermodynamically favorable without other bacteria that utilize produced hydrogen and reduce the hydrogen concentrations to a thermodynamically favorable range [i.e., below 1.6×10^{-2} mg/L (8000 nM) for good fermentation substrates or below to 1.6×10^{-4} mg/L (80 nM) for poor fermentation substrates] (Fennell et al., 1997). It is estimated that at least three major groups of bacteria are required for complete mineralization of aromatics by fermentation alone: primary fermenters, obligate proton reducers, and methanogens (Berry et al., 1987). Therefore, experimental determination of aromatic compound fermentation is difficult compared to demonstration of aerobic metabolism or metabolism by nitrate, sulfate, or Fe(III) reducers.

Based on field observations, however, it is known that fermentation occurs at almost all sites where BTEX compounds are found in groundwater. For example, the Air Force Natural Attenuation Initiative (see Chapter 5) resulted in the collection of data from 38 BTEX field sites and 17 chlorinated solvent sites, and all but one site showed measurable concentrations of methane produced. Since methane production requires fermentation products as methanogenic substrates, the presence of methane is clear evidence that fermentation is occurring.

Metabolism of BTEX compounds to produce hydrogen probably requires the involvement of several strains of bacteria. One possible mechanism is a series of reactions, in which other electron acceptors, such as oxygen, nitrate, Fe(III), and sulfate, are used by nonfermenters to break down the aromatics to simpler compounds that can then be used by the fermenters. The following is a possible four-step process:

1. Oxygen is used to oxidize the aromatic ring, which makes it easier for nonfermenting anaerobic bacteria to utilize the aromatic compounds.
2. The nonfermenters [nitrate, Fe(III), and sulfate reducers] break the ring into the carboxylic acid *pimelate* or convert the BTEX compounds directly into these by-products.

3. The pimelate is then used by primary fermenters to produce either hydrogen or secondary fermentation substrates, such as ethanol, lactate, succinate, propionate, and butyrate.
4. The secondary fermentation substrates are used by the obligate proton reducers (secondary fermenters) to form hydrogen.

Note that the four-step reaction above is only one possible reaction sequence. Many laboratory studies have demonstrated the complete mineralization of BTEX compounds under anaerobic conditions (Harwood and Gibson, 1997; Kazumi et al., 1997; Weiner and Lovley, 1998), confirming that step 1 is not necessary for BTEX degradation at field sites. In theory, hydrogen could be produced from primary fermentation alone, but in all likelihood the secondary fermentation reactions are also an important source of dissolved hydrogen.

Although the BTEX compounds are common fermentation substrates at chlorinated solvent sites, there are many other hydrocarbon substrates that are naturally fermented at sites and result in the generation of hydrogen. Examples of easily fermentable organics include acetone, sugars, and fatty acids from landfill leachate. In addition, some groundwaters naturally contain high levels of organic compounds. Fermentation reactions appear to support hydrogen-driven dechlorination at some of these sites. Lyman et al. (1990) provides estimation methods to determine general biodegradability of organic compounds and includes the following rules of thumb:

- Highly branched compounds are more resistant to biodegradation.
- Short-chain compounds are not as quickly degraded as long chains.
- Unsaturated aliphatics (e.g., ethene) are more readily attacked than saturated aliphatics (e.g., ethane).
- Alcohols (e.g., methanol, ethanol), aldehydes, acids, amides, and amino acids are more susceptible to biodegradation than the corresponding alkanes, olefins, ketones, dicarboxylic acids, nitriles, amines, and chloroalkanes.

In summary, hydrogen is generated by fermentation of nonchlorinated organic substrates, including BTEX, acetone, naturally occurring organic carbon, and a variety of other compounds. Methanogens require fermentation products as substrates; therefore, methane production is clear evidence of fermentation in situ, and this fermentation will produce hydrogen. The Air Force Natural Attenuation Initiative found elevated concentrations of methane at 37 of the 38 BTEX field sites, demonstrating that fermentation occurs at almost all field sites with suitable fermentation substrates (e.g., BTEX, acetone, etc.).

Biological Competition for Hydrogen. As hydrogen is produced by fermentative organisms, it is rapidly consumed by other bacteria. The utilization of H_2 by nonfermenters is known as interspecies hydrogen transfer and is required for fermentation reactions to proceed. Although H_2 is a waste product of fermentation, it is a highly reduced molecule, which makes it an excellent, high-energy electron donor (see Figure 4.4). Both organisms involved in interspecies hydrogen transfer benefit from the process. The hydrogen-utilizing bacteria gain a high-energy electron donor, and for the fermenters, the removal of hydrogen allows additional fermentation to remain energetically favorable. This mutually beneficial relationship is known as *syntrophy*. In this section we discuss the competition for hydrogen among hydrogen-utilizing bacteria.

A wide variety of bacteria can utilize hydrogen as an electron donor: denitrifiers, iron reducers, sulfate reducers, methanogens, and halorespirators. Thus the production of hydrogen through fermentation does not, by itself, guarantee that hydrogen will be available for halorespiration. For dechlorination to occur, halorespirators must compete successfully against the other hydrogen utilizers for the available hydrogen. Smatlak et al. (1996) suggest that the competition for hydrogen is controlled primarily by the Monod half-saturation constant $K_s(H_2)$, the concentration at which a specific strain of bacteria can utilize hydrogen at half the maximum utilization rate. They measured a $K_s(H_2)$ value for halorespirators of 100 nM and a $K_s(H_2)$ for methanogens of 1000 nM. Based on this result, they suggested that halorespirators would compete successfully for hydrogen only at very low hydrogen concentrations.

Ballapragada et al. (1997) provided a more detailed discussion of halorespiration kinetics and point out that competition for hydrogen also depends on additional factors, including the bacterial growth rate and maximum hydrogen utilization rate. Thus it is useful to explore a theoretical model of this competition which predicts the ability of halorespirators to compete for hydrogen.

The Monod kinetic model is a relatively simple tool for describing bacterial growth under substrate (hydrogen)-limiting conditions (see Chapter 4). The basic Monod expressions is as follows:

$$\mu = \frac{YkS}{K_s + S} \tag{6.1}$$

where
- μ = growth rate (hr^{-1})
- Y = growth yield (mg biomass/mg substrate)
- k = maximum specific substrate utilization rate (mg substrate/mg biomass/hr)
- S = substrate concentration (mg/L)
- K_s = half-saturation constant (mg/L)

This equation is based on the simplifying assumption that other factors, such as electron acceptors and other nutrients, are not limiting growth. This assumption is often accurate in environments where halorespirators are competing with other organisms for hydrogen. The parameters, growth yield (Y), maximum substrate utilization rate (k), and half-saturation constant (K_s), are characteristic of a given strain of bacteria growing on a given substrate. Values for these parameters can be measured in the laboratory or estimated based on thermodynamic considerations as described by McCarty (1969). The yield and maximum substrate utilization rate are often combined to form the maximum specific growth rate, μ_{max}, with units of time^{-1} (e.g., day^{-1}):

$$\mu_{max} = Yk \tag{6.2}$$

When substituted back into eq. (6.1), the Monod expression becomes

$$\mu = \frac{\mu_{max} S}{K_s + S} \tag{6.3}$$

In this model, bacterial growth rate (μ) is used as a predictor of an organism's ability to compete for the limiting substrate. A strain with a higher growth rate at a given substrate concentration will outcompete a strain of bacteria with a lower growth rate. At high substrate concentrations, the ability to compete is controlled by μ_{max}:

$$\text{If } S >> K_s, \quad \mu \approx \mu_{max} \tag{6.4}$$

In contrast, at low substrate concentrations the ability to compete is controlled by μ_{max}/K_s:

$$\text{If } S << K_s, \quad \mu \approx \frac{\mu_{max} S}{K_s} \tag{6.5}$$

Using this model, the ability of hydrogen-utilizing bacteria to compete for hydrogen can easily be predicted from substrate concentration and two properties of the bacteria, μ_{max} and K_s.

Table 6.3 lists key Monod kinetic parameters for the various hydrogen-utilizing bacteria that have been measured in the laboratory or estimated based on theoretical

TABLE 6.3. Monod Parameters (Calculated and Experimental Values) for Halorespirators versus Denitrifiers, Sulfate Reducers, and Methanogens

Microorganism	Yield, Y (mg VSS/mg H_2)	Maximum Substrate Utilization Rate, k[a] (mg H_2/mg VSS/hr)	Maximum Specific Growth Rate, μ_{max} (hr^{-1})	Half Saturation Constant, K_s (mg/L)
Halorespirator	1.75[b]	0.0114[b] (T = 20°C)	0.01995	0.0002[c]
Denitrifier	1.32[d]	0.044[d] (T = 20°C)	0.05808	—
Sulfate reducer	0.328[e]	0.012[c] (T = 20°C)	0.003936	—
Methanogen	0.237[d]	0.016[f] (T = 20°C)	0.003792	0.0019[c]

[a] Where reported maximum substrate utilization rates, k, were determined experimentally at optimal growth conditions (e.g., 35°C), an Arrehenius-type relationship was used to adjust the values to 20°C to be more representative of conditions in actual subsurface environments. The temperature-correcting expression is $k_2 = k_1\theta^{T_2-T_1}$; T_2 is 20°C and k_2 is the adjusted rate constant at that temperature. k_1 and T_1 are the rate constant and its measurement temperature respectively, as recorded during experimentally controlled optimal growth conditions. θ is an approximation for the temperature correction on the activation energy. Its value ranges, but for these corrections θ was set to 1.042.

[b] From Holliger et al. (1993).

[c] From Smatlak et al. (1960).

[d] Theoretical value calculated using techniques by McCarty (1969, 1972).

[e] From Parkin et al., as referenced by Speece (1996).

[f] From Ballapragada et al. (1997).

considerations. As can be seen, the Monod kinetic model and the μ_{max} term suggest that halorespirators will outcompete methanogens and sulfate reducers at any hydrogen concentration. However, denitrifiers will probably outcompete halorespirators under most conditions as their maximum specific growth rate is approximately three times higher than the halorespirators. Thus this model predicts that high nitrate concentrations in the groundwater will probably be unfavorable for halorespiration, as any hydrogen that is produced by fermentation (which may be minimal if nitrate is still present) would be taken by the nitrate reducers. It should be noted that the kinetic parameters shown in Table 6.3 were taken from a number of different sources; so a direct comparison may yield misleading results. However, the predicted competition for hydrogen is consistent with the simpler model discussed earlier based only on thermodynamic considerations of the various oxidation–reduction reactions mediated by the hydrogen utilizers. The predicted competition is also consistent with field observations, where the relative ability to compete for hydrogen was nitrate reducers > iron reducers > sulfate reducers > methanogens (Chapelle et al., 1995). For autotrophs (including all bacteria utilizing hydrogen as an electron donor), μ_{max} is largely a function of the energy released by metabolism of the substrates (McCarty, 1971). Thus μ_{max} should not vary greatly between different strains of halorespirators.

To summarize the results of the Monod kinetic analyses, halorespirators can probably outcompete methanogens and sulfate reducers for available hydrogen; however, denitrifiers will probably outcompete halorespirators. In other words, the following chain of events probably occurs at most sites undergoing halorespiration:

1. Aerobic bacteria consume nonchlorinated organic substrates until the oxygen is depleted.
2. Nitrate-reducing bacteria consume nonchlorinated organic substrates until the nitrate is exhausted.
3. Iron-reducing bacteria consume nonchlorinated organic substrates until the Fe(III) is expended.
4. Fermentation processes consume nonchlorinated organic substrates and generate hydrogen; dechlorinating bacteria consume hydrogen and dechlorinate the solvents; sulfate-reducing bacteria consume nonchlorinated organic substrates until the sulfate is depleted; and methanogens consume the hydrogen to generate methane.

Smatlak et al. (1996) have suggested that the steady-state concentration of hydrogen will be controlled by the rate of hydrogen production from fermentation. Based on this assumption, they recommend the addition of poor fermentation substrates, such as propionate or butyrate, to an aquifer to stimulate dechlorination. These fermentation substrates will yield low rates of hydrogen production, resulting in low rates of hydrogen utilization by dechlorinators. Both laboratory results and field observations have suggested, however, that the steady-state concentration of hydrogen is controlled by the type of bacteria utilizing the hydrogen and is almost completely independent of the rate of hydrogen production (Chapelle et al., 1995; Ballapragada et al., 1997; Fennell et al., 1997; Carr and Hughes, 1998). For example,

Ballapragada et al., (1997) varied the rate of hydrogen production in mixed methanogenic–halorespirating cultures and saw only a transient increase in hydrogen concentrations. After only 3 to 6 hours, H_2 partial pressures returned to the steady-state concentration observed prior to the increase in hydrogen production. In this culture, the steady-state hydrogen concentration was clearly controlled by the ability of the hydrogen utilizers to use hydrogen at low concentrations and was independent of the rate of hydrogen production. Similarly, researchers have found that steady-state hydrogen concentrations in the field are controlled by the type of bacteria utilizing the hydrogen. Under nitrate reducing conditions, steady-state H_2 concentrations were <0.05 nM; under iron-reducing conditions, they were 0.2 to 0.8 nM; under sulfate-reducing conditions, they were 1 to 4 nM; and under methanogenic conditions, they were 5 to 14 nM (Lovley and Goodwin, 1988; Lovley et al., 1994; Chapelle et al., 1995). Finally, Carr and Hughes (1998) show that dechlorination in a laboratory column is not affected by competition for electron donors at high hydrogen concentrations. Thus it is clear that an increased rate of hydrogen production will result in increased halorespiration without affecting the competition between bacteria for the available hydrogen. Attempting to stimulate halorespiration with poor fermentation substrates may unnecessarily limit the amount of dechlorination and direct hydrogen addition (Newell et al., 1997) may prove to be a viable remedial alternative.

Electron Acceptor/Electron Donor Model for Characterizing Oxidation–Reduction Potential at Chlorinated Solvent Sites. Lovley et al. (1994) and Chapelle et al. (1995) developed a method to characterize the oxidation–reduction conditions in groundwater systems using dissolved hydrogen analyses. This method combines analysis of oxidized and reduced electron acceptor concentrations with analysis of hydrogen concentrations. The relative hydrogen-utilizing efficiency of a microorganism is a very significant factor in microbial ecology because of the competition for limited hydrogen in anaerobic environments. Table 6.4 lists the Gibbs free energy released for various reactions utilizing hydrogen as the electron donor. By convention, standard reaction free energies (G_r°) are calculated assuming that the concentration of all substrates and products is 1 M (except that $[H^+] = 10^{-7}$ M). Although this calculation does not provide the actual energy released from reactions under physiological substrate conditions, it does provide a useful way to compare the relative amount of energy released from different reactions. If the concentration of a substrate (hydrogen) is less than 1 M, energy released from the reaction will decrease. At some low hydrogen concentration, the reaction will become energetically unfavorable. In general, the more energy released by a reaction under standard conditions, the lower the hydrogen concentration at which the reaction will become thermodynamically unfavorable.

From Table 6.4, the organisms that utilize the more oxidized electron acceptors are able to extract more energy per mole of hydrogen consumed. Thus those organisms that gain the most energy from oxidation of hydrogen are able to utilize hydrogen at the lowest concentrations. For example, nitrate reducers are able to extract a relatively large amount of energy from hydrogen and can maintain steady-state hydrogen concentrations below 0.1 nM ($1.5\ 10^{-7}$ atm or 0.0002 μg/L). Figure 6.5

TABLE 6.4. Common Reactions Using Hydrogen as the Electron Donor

Electron Acceptor	Reaction	ΔG_r° (kcal/mol e⁻)
Oxygen	$1/2O_2 + H_2 \Rightarrow H_2O$	–28.5
Nitrate	$2/5NO_3^- + 2/5H^+ + H_2 \Rightarrow 1/5N_2 + 6/5H_2O$	–27.0
Sulfate	$1/4SO_4^{2-} + 3/8H^+ + H_2 \Rightarrow 1/8H_2S + 1/8HS^- + H_2O$	–4.8
CO_2	$1/4CO_2 + H_2 \Rightarrow 1/4CH_4 + 1/2H_2O$	–4.1
Iron(III)	$2Fe^{3+} + H_2^- \Rightarrow 2Fe^{2+} + 2H^+$	–19.6
PCE	$C_2Cl_4 + H_2 \Rightarrow C_2HCl_3 + H^+ + Cl^-$	–19.7
TCE	$C_2HCl_3 + H_2 \Rightarrow C_2H_2Cl_2 + H^+ + Cl^-$	–19.4
DCE	$C_2H_2Cl_2 + H_2 \Rightarrow C_2H_3Cl + H^+ + Cl^-$	–17.1
VC	$C_2H_3Cl + H_2 \Rightarrow C_2H_4 + H^+ + Cl^-$	–18.7

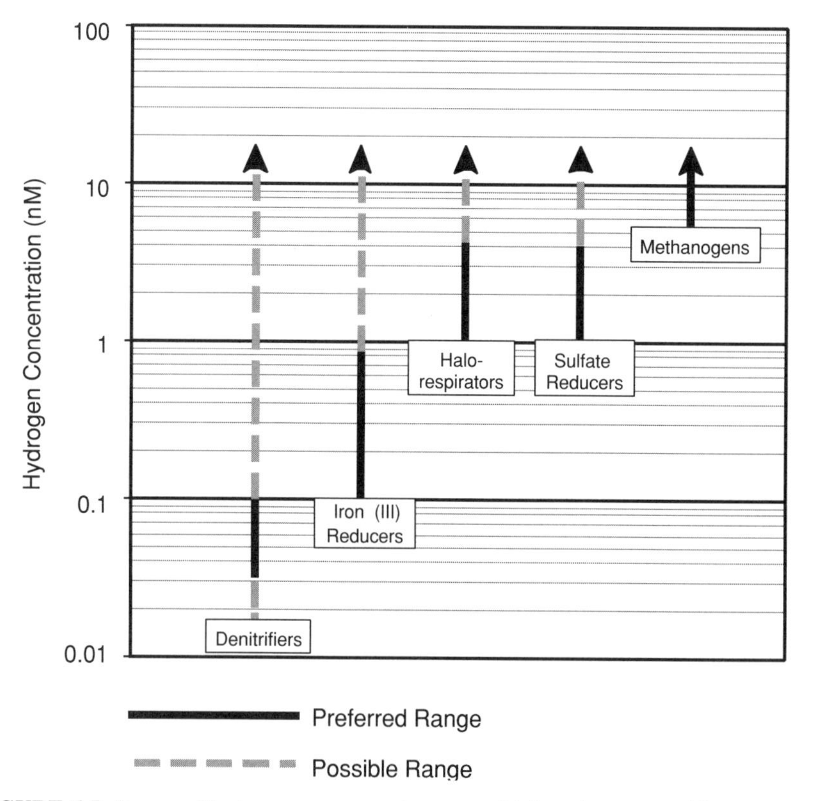

FIGURE 6.5. Range of hydrogen concentration over which various anaerobic reactions are observed. (Data from Lovley and Goodwin, 1988; Chapelle and Lovley, 1992; Vroblesky and Chapelle, 1994.)

graphically illustrates the ability of various microorganisms to utilize hydrogen at low dissolved hydrogen concentrations. Table 6.5 lists the dissolved hydrogen concentrations associated with different oxidation–reduction conditions at field sites.

The concentration ranges shown in Table 6.5 are controlled by the different microbial populations that "poise" (control) the dissolved hydrogen concentration in the aquifer at specific concentrations based on their relative efficiency at utilizing hydrogen. Figure 6.6 shows a conceptualization of the relationship between dissolved hydrogen concentrations and electron acceptors over time at a site, starting when the fermentation substrate is released (fuel gauge full) and ending when the fermentation substrate is depleted and the supply of hydrogen is cut off. The Fe(III) reducers are relatively efficient and therefore require low hydrogen concentrations. As long as Fe(III) is present in the groundwater, the iron reducers rapidly utilize the hydrogen produced, causing dissolved hydrogen concentrations to be maintained at 0.2 to 0.8 nM regardless of the rate of hydrogen production. When the Fe(III) is depleted, dissolved hydrogen concentrations increase to sufficient concentrations so that utilization by sulfate reducers is energetically favorable. When the sulfate is depleted, hydrogen concentrations increase again, so that methanogenic bacteria (which use the relatively poor electron acceptor, CO_2) become competitive. Halorespirators also appear to require hydrogen concentrations on the order of 1 nM for growth to occur (Smatlak et al., 1996). Thus, dechlorination is expected in either sulfate-reducing or methanogenic aquifers.

This conceptual model can be extended to show the flow of electrons from donors to acceptors for competing anaerobic reactions (Figure 6.7). Either organic substrates or hydrogen may act as the primary electron donor, depending on the environmental conditions, but the utilization of hydrogen is more energetically favorable. Solid lines show reactions occurring with hydrogen as the electron donor, while dashed lines indicate reactions using an organic substrate as the electron donor. After the concentration of dissolved oxygen and nitrate in the system falls below some critical concentration (about 0.5 mg/L) because of organic substrate oxidation, fermentation of an organic substrate begins to occur and hydrogen starts to accumulate in the system. Hydrogen accumulation is illustrated conceptually by the level in the hydrogen reservoir.

TABLE 6.5. Range of Hydrogen Concentrations for a Given Terminal Electron-Accepting Process

Terminal Electron-Accepting Process	Dissolved Hydrogen Concentration		
	n*M*	atm[a]	μg/L
Denitrification	< 0.1	$< 1.3 \times 10^{-7}$	$< 0.2 \times 10^{-3}$
Iron(III) reduction	0.2–0.8	$0.26–1.0 \times 10^{-6}$	$0.4–1.6 \times 10^{-3}$
Sulfate reduction	1–4	$1.3–5.0 \times 10^{-6}$	$2.0–8.0 \times 10^{-3}$
Methanogenesis	5–20	$63–250 \times 10^{-6}$	$10–40 \times 10^{-3}$

Source: Adapted from Lovley et al. (1994) and Chapelle et al. (1995).

[a] In gas phase in equilibrium with water containing dissolved hydrogen.

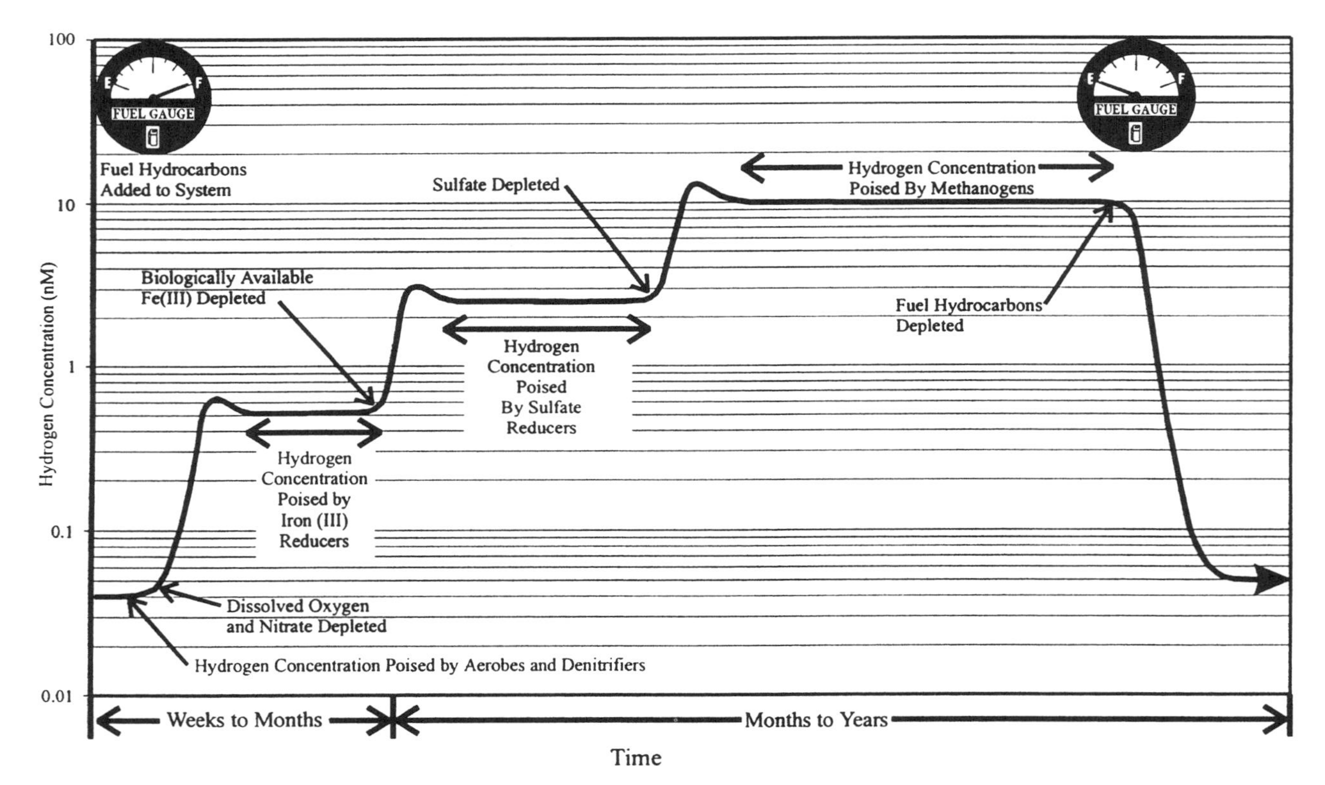

FIGURE 6.6. Conceptualization of the relationship between electron donors, electron acceptors, and dissolved hydrogen concentrations over time. Fuel gauge represents availability of organic substrates for fermentation reactions.

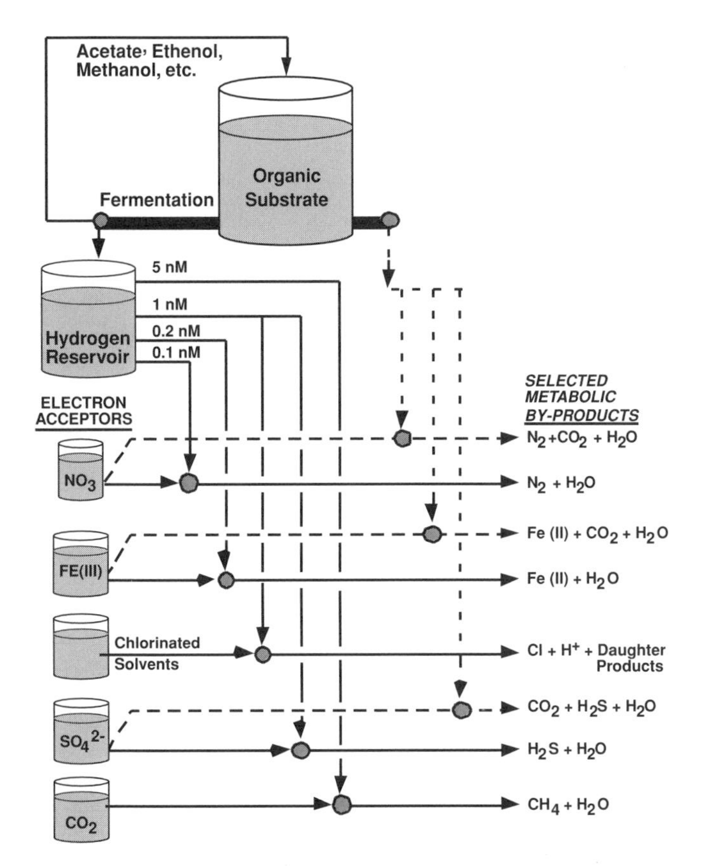

FIGURE 6.7. Hydrogen, organic substrate, chlorinated solvent, electron acceptor, and metabolic by-product diagram. Dashed lines represent reactions involving the organic substrate as the direct electron donor; solid lines represent reactions involving hydrogen as the direct electron donor.

Fe(III)-, chlorinated solvent-, sulfate-, and CO_2-reducing (methanogenic) microorganisms exhibit different efficiencies in utilizing the hydrogen that is continually being produced by fermentation. Hydrogen utilization efficiencies are illustrated conceptually in Figure 6.7 by the pipes coming out of the hydrogen reservoir. Denitrifying bacteria keep the hydrogen concentrations in the aquifer below 0.1 nM. As nitrate becomes depleted, the denitrifying bacterium can no longer function and the hydrogen concentration in the aquifer gradually increases (hydrogen reservoir fills up in our conceptual model). Hydrogen concentrations increase because fermentation continues but there is not enough demand in the system to keep the hydrogen concentration in check. Hydrogen concentrations increase until there is sufficient hydrogen to supply the energy for growth of Fe(III)-reducing bacteria, as shown by the pipe coming out of the hydrogen reservoir that leads to the Fe(III)-reducing reaction. As suggested by this figure, growth-promoting reduction of Fe(III) cannot occur until enough hydrogen accumulates in the reservoir to reach and fill the pipe leading to the Fe(III) reaction. This process occurs at hydrogen concentrations of about 0.2 nM. After the onset of growth-promoting Fe(III) reduction, the hydrogen concentration in the system is poised between 0.2 and 0.8 nM until the supply of biologically available Fe(III) in the aquifer is exhausted. Because fermentation is an ongoing process, hydrogen again begins to accumulate in the system until enough is available to support the growth of sulfate-reducing bacteria. The minimum hydrogen concentration required to support the growth of sulfate reducers is about 1 nM.

Several of the concepts discussed above are combined in the conceptual model shown in Figure 6.8. The figure is arranged thermodynamically, with reactions releasing larger amounts of energy (i.e., aerobic biodegradation and nitrate reduction) shown on the left side of the figure and reactions releasing smaller amounts of energy (such as methanogenesis) shown on the right side of the figure. The figure focuses on the thermodynamic flow of electron acceptors and electron donors in a moderately to highly reduced anaerobic system after oxygen and nitrate are consumed.

A general indication of the *potential* mass flux is shown by the thickness of the pipes connecting each vessel. For each reaction, the flux is inferred based on typical concentrations observed under field conditions. Typical concentrations of the various constituents are shown as the level in the various reservoirs in the process diagram. For example, fermentation produces dissolved hydrogen, which flows through a small pipe into each of the dissolved hydrogen reservoirs. Because the reactions that use hydrogen are much faster than the reactions that generate hydrogen, the pipes leading out from each hydrogen reservoir are much larger than the pipes leading in. Since the inlet pipe is smaller (low capacity) than the outlet pipe (high capacity) at each hydrogen reservoir, the amount in the reservoir (representing dissolved hydrogen concentrations in groundwater) is typically very low (< 0.001 mg/L) at all times.

Fermentation is shown to occur in all three oxidation–reduction environments. Note that the figure suggests that most of the dissolved hydrogen is supplied by fermentation occurring during the methanogenic conditions after sulfate is depleted (Figure 6.8). Although this is conjecture, the ubiquitous nature of methane production at chlorinated solvent sites undergoing halorespiration suggests that these systems do become deeply anaerobic relatively easily and that fermentation of large amounts of dissolved hydrogen occurs after sulfate depletion occurs. While the capacity for iron reduction and sulfate reduction is limited by the amount of Fe(III) and sulfate present, the capacity for fermentation is limited only by the amount of fermentable organic carbon. Thus, at sites with high levels of BTEX or other organics, the majority of fermentation will occur after Fe(III) and sulfate are depleted.

As developed by Lovley and Chapelle, analysis of both electron acceptor levels and hydrogen levels can allow oxidation–reduction potential determination of an aquifer with high confidence (Lovley et al., 1994; Chapelle et al., 1995). The association between hydrogen concentration and oxidation–reduction potential was detailed above. Depletion of Fe(III) and production of Fe(II) indicates iron reduction. Depletion of sulfate and production of sulfide indicates sulfate reduction, and production of methane indicates methanogenesis. If analysis of both electron acceptors and

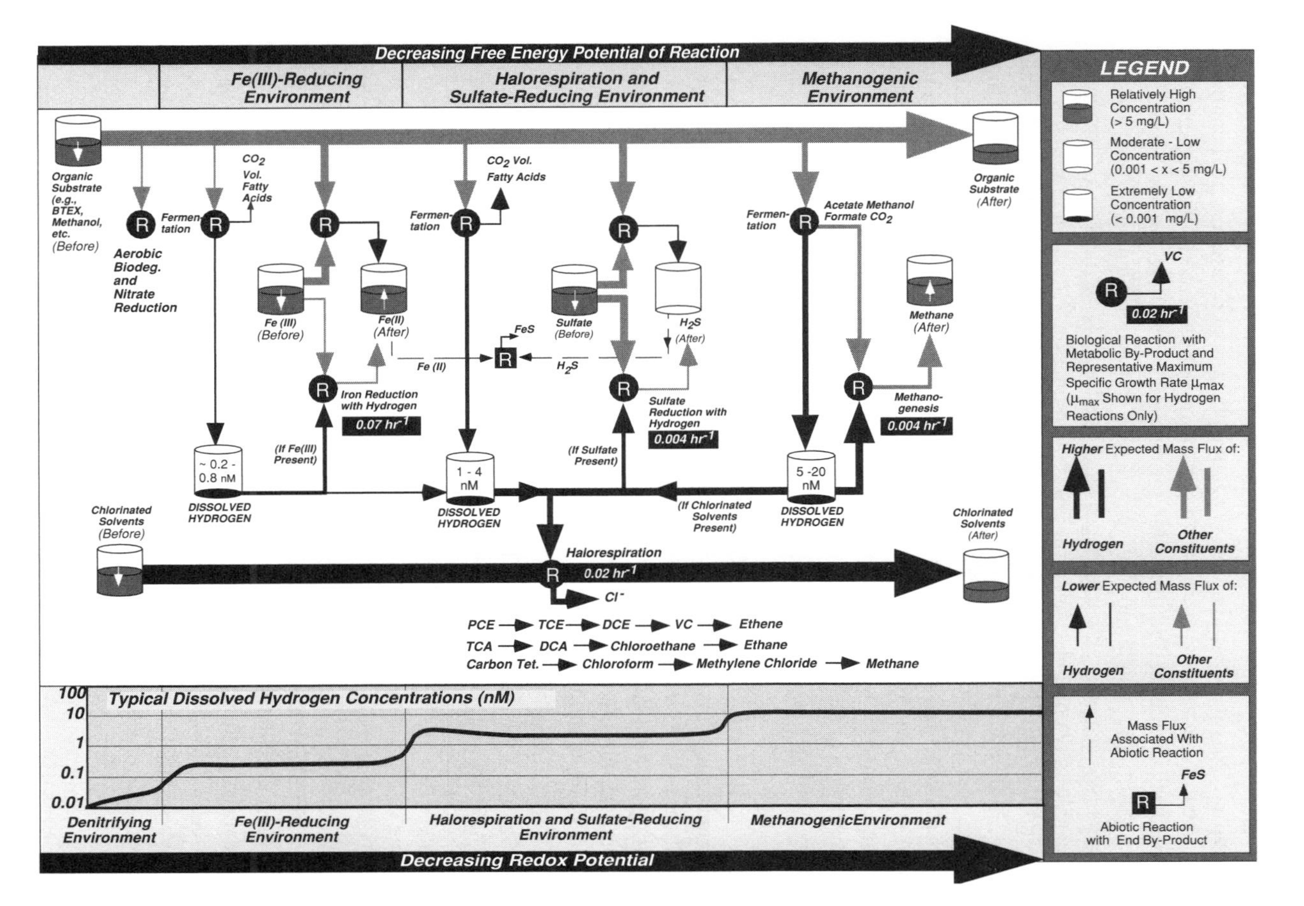

FIGURE 6.8. Thermodynamic flow of electron donors and electron acceptors pathways at chlorinated solvent sites undergoing halorespiration. (From Newell, 1998.)

hydrogen concentration indicate the same oxidation–reduction potential, high confidence can be placed in the assessment.

Although conceptual models of field sites often portray well-defined zones containing different oxidation–reduction environments, in reality aquifers are heterogeneous and poorly mixed; therefore, the oxidation–reduction potential can vary greatly over a small scale within an aquifer. Monitoring wells, in contrast, are low-resolution sampling ports. The typical monitoring well is 2 to 4 in. in diameter, often with a larger borehole diameter, and screened over at least 5 to 10 vertical feet. This composite sample may contain water from multiple oxidation–reduction environments. Therefore, it is possible for a single water sample to contain both oxygen and methane. Although such composite samples can make it more difficult to define boundaries precisely between different oxidation–reduction environments at a field site, the samples still provide a general picture of the range of oxidation–reduction potential present. More precise monitoring of local environments is possible by the use of large numbers of nested monitoring wells that employ small-diameter sample ports screened over narrow intervals. However, the cost of this intensive monitoring is usually justified only at research sites.

Stoichiometry of Reductive Dechlorination. Under anaerobic conditions, Gossett and Zinder (1996) showed that the reductive dehalogenation of the chlorinated ethenes occurs as a series of consecutive irreversible reactions mediated by the addition of 1 mole of hydrogen gas for every mole of chloride removed. Thus the theoretical minimum hydrogen requirement for dechlorination can be calculated on a mass basis as shown below:

- 1 mg of H_2 will dechlorinate 21 mg of PCE to ethene.
- 1 mg of H_2 will dechlorinate 22 mg of TCE to ethene.
- 1 mg of H_2 will dechlorinate 24 mg of DCE to ethene.
- 1 mg of H_2 will dechlorinate 31 mg of VC to ethene.

Complete fermentation of BTEX compounds is expected to yield 0.25 to 0.4 mg H_2 per mg of BTEX. Therefore, for each mg of BTEX consumed, 4.5 to 7 mg of chloride could be released during reductive dechlorination.

However, the utilization of hydrogen for dechlorination will never be completely efficient because of the competition for hydrogen in the subsurface discussed previously. Sewell and Gibson (1991) studied the degradation of PCE in a mixed-culture microcosm fed toluene as a fermentation substrate. They found that less than 10% of the reductive potential produced by this fermentation was utilized for dechlorination. Similarly, Ballapragada et al. (1997) used a mixed-culture fluidized-bed reactor to study degradation of PCE. They fed the reactor with a variety of indirect fermentation-based electron donors and found that 95% of the hydrogen produced was utilized for methane production while only 5% was used for dechlorination.

One rule of thumb that has been proposed is the following: For reductive dechlorination to degrade a plume of dissolved chlorinated solvents completely, organic substrate concentrations greater than 25 to 100+ times that of the chlorinated solvent are required (McCarty, 1997). The reason for this is twofold: (1) the oxidation–reduction potential of the groundwater must be in the sulfate-reducing to methanogenic range, and organic carbon is required to drive the system anaerobic and lower the oxidation–reduction potential; and (2) because of the competition between various microbes for hydrogen, especially between the nitrate reducers and dechlorinators, it is helpful to exert sufficient electron acceptor demand to deplete the system of other electron acceptors. Although the reasoning behind this simple rule of thumb is sound, its veracity has not been demonstrated quantitatively in the field.

Microbiology and Biochemistry of Halorespirators. Since 1993, several pure cultures have been isolated that utilize chlorinated ethenes for halorespiration (see Gossett and Zinder, 1996, for a review). Despite the observation in mixed cultures and at field sites of complete transformation from PCE to ethene, to date only one pure culture has been isolated that is capable of this complete transformation (Maymo-Gatell et al., 1997). The other isolated pure cultures are only capable of transforming PCE to 1,2-DCE.

The bacteria capable of degrading PCE to 1,2-DCE appear to exist in a variety of phylogenetic groups, many of which are genetically related to gram-positive and gram-negative sulfate-reducing bacteria. These partial degraders most commonly utilize hydrogen as the electron donor, but some isolates are able to use other fermentation products, such as pyruvate and acetate. Some of these isolates are also able to utilize electron acceptors other than chlorinated solvents, such as sulfite, thiosulfite, and fumarate. One pure culture capable of halorespirating PCE to 1,2-DCE is a facultative aerobe that can utilize a variety of electron acceptors, including oxygen, and can even ferment a variety of organics (Sharma and McCarty, 1996).

In contrast, the only pure culture capable of complete degradation of PCE to ethene, *Dehalococcus ethenogenes* strain 195, is limited to hydrogen as the electron donor and chlorinated solvents as the electron acceptor (Maymo-Gatell et al., 1997). Chlorinated solvents have been in the environment for only about 50 years; therefore, this bacterium could not have evolved novel enzymes specifically to utilize these compounds as electron acceptors. The "natural" electron acceptor is not known, and either this electron acceptor is uncommon or the strain has lost the ability to utilize it. Electron acceptors that did not support growth include sulfate, sulfite, thiosulfite, nitrate, nitrite, fumarate, and oxygen. However, complex nutritional requirements of this strain make its study in pure culture difficult. Based on analysis of 16S ribosomal DNA, *Dehalococcus ethenogenes* strain 195 should be grouped with eubacteria. However, it does not belong in any known phylogenetic group.

Carr and Hughes (1998) report that PCE dechlorination was initiated in a mixed-culture column experiment after only 24 hours of PCE addition, despite the inoculum having no history of chlorinated ethene exposure. The inoculum was an anaerobic granular sludge obtained from a polyester wastewater containing ethylene glycol. A second mixed culture, derived from solids taken from an aquifer contaminated with chlorinated solvents, was able to begin PCE dechlorination almost immediately after PCE was introduced to the column. The authors reported that PCE and daughter product dechlorination efficiencies improved significantly over the year-long experiment, with final PCE half-lives of 2 to 13 hours observed in the columns.

The biochemistry of halorespiration is also in its infancy. To date, only one enzyme has been purified that is capable of mediating the dehalogenation of a chlorinated compound (Neumann et al., 1996). This enzyme was isolated from *Dehalospirillum mulivorans*, and the reaction mediated by this enzyme involves oxidation and reduction of cobalt in vitamin B_{12}. If the investigation of halorespirators follows the same path as the study of methanogens, a number of novel enzymes and pathways will be revealed.

Chlorinated Solvents That Are Amenable to Halorespiration. Although halorespiration is the most common biodegradation reaction for chlorinated solvents, not all chlorinated solvents can be utilized by bacteria for this purpose. As shown in Table 6.1, all of the chlorinated ethanes (PCE, TCE, DCE, VC) and some of the chlorinated ethanes (TCA, 1,2-DCA) can be degraded via halorespiration; however, methylene chloride has not yet been shown to be degraded by this process. The oxidation state of a chlorinated solvent affects both the energy released by halorespiration and the rate at which the reaction occurs. In general, the more oxidized a compound is (more chlorine atoms on the organic molecule) the more amenable it is for reduction by halorespiration.

As with the ethenes, chlorinated ethanes will also undergo halorespiration. Dechlorination of 1,1,1-TCA has been described by Vogel and McCarty (1987) and Cox et al. (1995), but understanding this pathway is complicated by the rapid hydrolysis (e.g., half-life is 0.5 to 2.5 years) of TCA (Vogel, 1994; see Chapter 3). The speed of these abiotic reactions can make it difficult to discern if TCA breakdown is biological or abiotic in origin.

Halorespiration has been observed with highly chlorinated benzenes such as hexachlorobenzene, pentachlorobenzene, tetrachlorobenzene, and trichlorobenzene (Holliger et al., 1992; Ramanand et al., 1993; Suflita and Townsend, 1995). As discussed by Suflita and Townsend (1995), halorespiration of aromatic compounds has been observed in a variety of anaerobic habitats, including aquifer materials, marine and freshwater sediments, sewage sludges, and soil samples. However, isolation of specific microbes capable of these reactions has been difficult.

Relative Rate of Degradation. Dechlorination is more rapid for highly chlorinated compounds than for compounds that are less chlorinated (Vogel and McCarty, 1985, 1987; Bouwer, 1994). Figure 6.9 qualitatively shows the reaction rate and conditions required for halorespiration of PCE to ethene. PCE (four chlorines) degrades the fastest under all anaerobic environments, while VC (a single chlorine) will degrade only under sulfate-reducing and methanogenic conditions, with a relatively slow reaction rate. Similar reaction summaries are shown for the TCA-to-ethane and the CT-to-methane breakdown sequence in Figures 6.10 and 6.11.

At many chlorinated ethene sites, concentrations of *cis*-1,2-DCE are often higher than any of the parent chlorinated ethene compounds (see the discussion on field

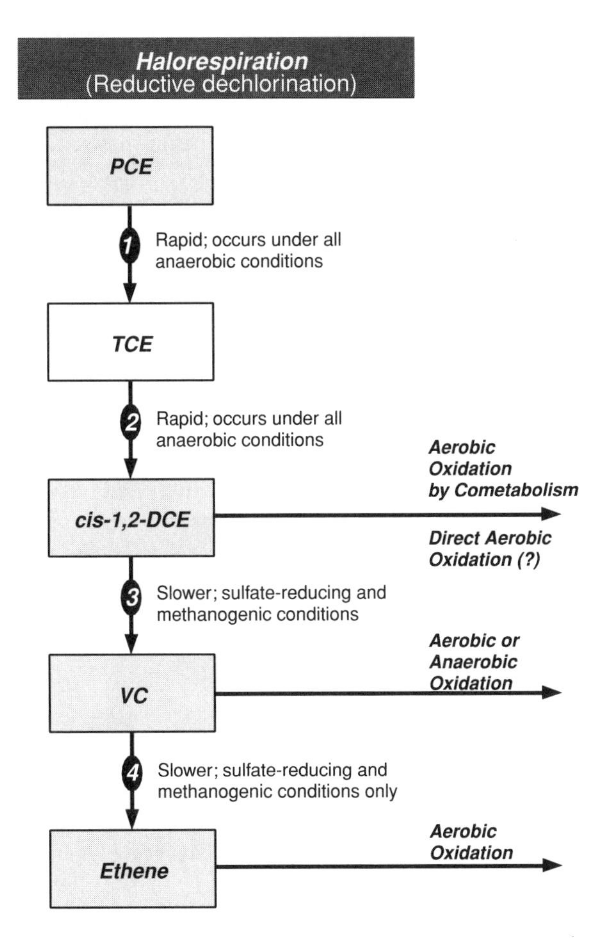

FIGURE 6.9. Reaction sequence and relative rates for halorespiration of chlorinated ethenes, with other reactions shown. (Reaction rate description for halorespiration from Remediation Technologies Development Forum, 1997.)

studies, below, and Table 6.10). The reason for the accumulation of 1,2-DCE may be due to either slower rates of DCE halorespiration or the prevalence of organisms that reduce PCE as far as *cis*-1,2-DCE over organisms that can reduce PCE all the way to ethene (Gossett and Zinder, 1996). Although many researchers have commented that reductive dechlorination will result in the accumulation of VC (see, e.g., Weaver et al., 1995; Bradley and Chapelle, 1996), at many field sites VC accumulation is much lower than *cis*-1,2-DCE (see Table 6.10). This may occur because the vinyl chloride in many chlorinated solvent plumes can migrate to zones that can support direct oxidation of VC, either aerobically and/or anaerobically.

Assessing Reductive Dechlorination at Field Sites. Assessing biological activity at a field site based on monitoring data can be difficult. However, there

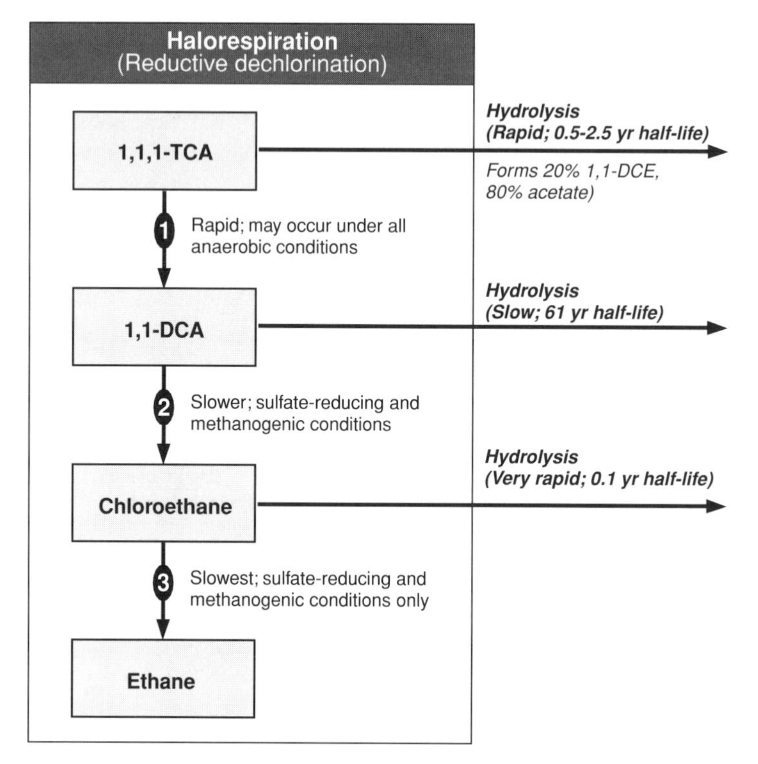

FIGURE 6.10. Reaction sequence and relative rates for halorespiration of chlorinated ethanes associated with 1,1,1-TCA degradation, with other reactions shown. (Reaction rate description for halorespiration from Remediation Technologies Development Forum, 1997.)

are a number of monitoring parameters that can be indicative of halorespiration. First, the presence of methane in the groundwater indicates that fermentation is occurring and that the potential for halorespiration exists. Second, the transformation of PCE and TCE has been studied intensely and many researchers report that of the three possible DCE isomers, 1,1-DCE is the least significant intermediate and that *cis*-1,2-DCE predominates over *trans*-1,2-DCE (Parsons et al., 1984; Parsons and Lage, 1985; Barrio-Lage et al., 1987). This is in line with statements made by the RTDF research consortium (RTDF, 1997), which reported that TCE biodegradation yields almost 100% of the *cis*-1,2-DCE isomer, while manufactured DCE is typically only 10 to 20% *cis*-1,2-DCE. Thus, when groundwater is comprised of more than 80% of the *cis* isomer, the DCE is a likely daughter product of TCE degradation (Wiedemeier et al., 1996a). Third, because chlorinated ethenes are 55 to 85% chlorine by mass, the degradation of these compounds releases a large mass of chloride. Therefore, elevated chloride concentrations also indicate reductive dechlorination.

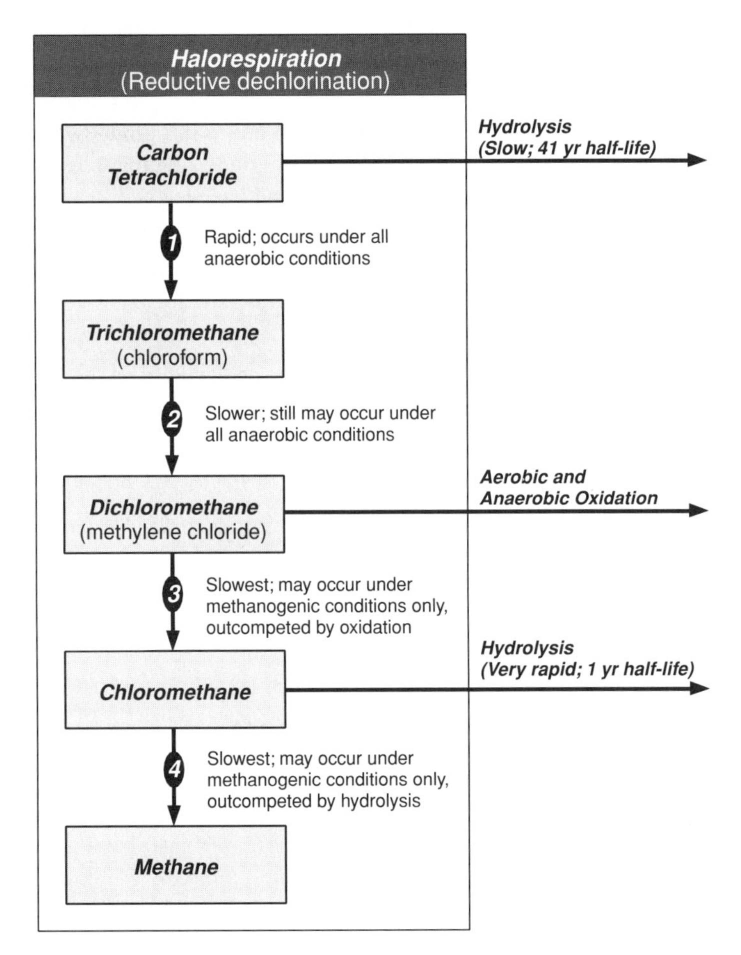

FIGURE 6.11. Reaction sequence and relative rates for halorespiration of carbon tetrachloride and daughter products, with other reactions shown. (Reaction rate description for halorespiration from Remediation Technologies Development Forum, 1997.)

To summarize, the following rules of thumb indicate that reductive dechlorination is occurring:

1. Ethene is being produced (even low concentrations are indicative of biodegradation).
2. Daughter products are being produced (such as *cis*-1,2-DCE or vinyl chloride).
3. Chloride concentrations are elevated.
4. Methane is being produced.
5. Fe(II) is being produced.

6. Hydrogen concentrations are greater than 1 nM.
7. Dissolved oxygen concentrations are low.

As discussed above, the variability associated with collecting groundwater samples makes precise definitions of reactions or zones of differing oxidation–reduction potentially difficult, and the various types of evidence should be weighed together to determine if reductive dechlorination is occurring at a particular site. Wiedemeier et al. (1996a, 1998) present a scoring system that allows this to be done (see Chapter 7).

Oxidation of Chlorinated Solvents

In contrast to halorespiration, direct oxidation of some chlorinated solvents can occur biologically in groundwater systems. In this case, the chlorinated compound serves as the electron donor, and oxygen, sulfate, ferric iron, or other compounds serve as the electron acceptor.

Direct Aerobic Oxidation of Chlorinated Compounds. Under direct aerobic oxidation conditions, the facilitating microorganism uses oxygen as an electron acceptor and obtains energy and organic carbon from degradation of the chlorinated solvent. In general, the more-chlorinated aliphatic chlorinated solvents (e.g., PCE, TCE, and TCA) have not been shown to be susceptible to aerobic oxidation, while many of the daughter products (e.g., vinyl chloride, 1,2-DCA, and perhaps the isomers of DCE) are degraded via direct aerobic oxidation.

Hartmans et al. (1985) and Hartmans and de Bont (1992) show that vinyl chloride can be used as a primary substrate under aerobic conditions, with vinyl chloride being mineralized directly to carbon dioxide and water. Direct vinyl chloride oxidation has also been reported by Davis and Carpenter (1990), McCarty and Semprini (1994), and Bradley and Chapelle (1998). Aerobic oxidation is rapid relative to reductive dechlorination of dichloroethene and vinyl chloride.

Although direct DCE oxidation has not been verified (Bradley and Chapelle, 1998), a recent study has suggested that DCE isomers may be used as primary substrates. Bradley and Chapelle (1998) report that both VC and DCE degraded rapidly in an aerobic streambed environment without accumulation of ethane or ethene, indicating oxidation to CO_2. The authors indicated that the mechanism of oxidation could have been direct or cometabolic oxidation.

Of the chlorinated ethanes, only 1,2-dichloroethane has been shown to be aerobically oxidized. Stucki et al. (1983) and Janssen et al. (1985) show that 1,2-DCA can be used as a primary substrate under aerobic conditions. In this case, the bacteria transform 1,2-DCA to chloroethanol, which is then mineralized to carbon dioxide. Evidence of chloroethane oxidation is scant; however, it appears to degrade rapidly via abiotic mechanisms (hydrolysis) and is thus less likely to undergo extensive, long-term biodegradation. McCarty and Semprini (1994) describe investigations in which 1,2-dichloroethane (DCA) was shown to serve as a primary substrate under aerobic conditions. These authors also document that dichloromethane has the potential to function as a primary substrate in either aerobic or anaerobic environments.

In general, the highly chlorinated ethenes (e.g., PCE, TCE, and TCA) are not likely to serve as electron donors for aerobic microbial degradation reactions. To date, no microorganisms have been identified that can use PCE, TCE, or TCA as primary substrates under aerobic conditions. This is because the highly chlorinated compounds tend to be much more oxidized than most organic carbon present in a groundwater system.

Chlorobenzene and polychlorinated benzenes (up to and including tetrachlorobenzene) have been shown to biodegrade under aerobic conditions. Several studies have shown that bacteria are able to utilize chlorobenzene (Reineke and Knackmuss, 1984), 1,4-DCB (Reineke and Knackmuss, 1984; Schraa et al., 1986; Spain and Nishino, 1987), 1,3-DCB (de Bont et al., 1986), 1,2-DCB (Haigler et al., 1988), 1,2,4-TCB (van der Meer et al., 1987; Sander et al., 1991), and 1,2,4,5-TeCB (Sander et al., 1991) as primary growth substrates in aerobic systems. Nishino et al. (1994) note that aerobic bacteria able to grow on chlorobenzene have been detected at a variety of chlorobenzene-contaminated sites but not at uncontaminated sites. Spain (1996) suggests that this provides strong evidence that the bacteria are selected for their ability to derive carbon and energy from chlorobenzene degradation in situ. The pathways for all of these reactions are similar, bearing a resemblance to benzene degradation pathways (Chapelle, 1993; Spain, 1996).

Aerobic Cometabolism of Chlorinated Compounds. It has been reported that under aerobic conditions, chlorinated ethenes, with the exception of PCE, are susceptible to cometabolic oxidation (Murray and Richardson, 1993; McCarty and Semprini, 1994; Vogel, 1994; Adriaens and Vogel, 1995). Vogel (1994) further elaborates that the oxidation rate increases as the degree of chlorination decreases. Aerobic cometabolism of ethenes may be characterized by a loss of contaminant mass, the presence of intermediate degradation products (e.g., chlorinated oxides, aldehydes, ethanols, and epoxides), and the presence of other products such as chloride, carbon dioxide, carbon monoxide, and a variety of organic acids (Miller and Guengerich, 1982; McCarty and Semprini, 1994). Cometabolism requires the presence of a suitable primary substrate such as toluene, phenol, or methane. For cometabolism to be effective, the primary substrate must be present at higher concentrations than the chlorinated compound, and the system must be aerobic. Because the introduction of high concentrations of oxidizable organic matter into an aquifer quickly drives the groundwater anaerobic, aerobic cometabolism typically must be engineered.

Anaerobic Oxidation of Chlorinated Compounds. Anaerobic oxidation occurs when anaerobic bacteria use the chlorinated solvent as an electron donor by utilizing an available electron acceptor such as ferric iron [Fe(III)]. Bradley and Chapelle (1996) show that vinyl chloride can be oxidized to carbon dioxide and water via Fe(III) reduction. In microcosms amended with Fe(III)-EDTA, reduction of vinyl chloride concentrations closely matched the production of carbon dioxide.

Slight mineralization was also noted in unamended microcosms. The rate of this reaction apparently depend on the bioavailability of Fe(III). In a subsequent paper, Bradley and Chapelle (1997) reported "significant" anaerobic mineralization of both DCE and VC in microcosms containing creek bed sediments. The sediments were taken from a stream where groundwater containing chlorinated ethenes discharges continually. Anaerobic mineralization was observed in both methanogenic and Fe(III)-reducing conditions.

6.2 CLASSIFICATION SYSTEM FOR CHLORINATED SOLVENT PLUMES

Wiedemeier et al. (1996a, 1998) proposed a classification system for chlorinated solvent plumes based on the amount and origin of fermentation substrates that produce the hydrogen that drives halorespiration. Three types of groundwater environments and associated plume behavior—Types 1, 2, and 3—are described below. While the classification system can be used to represent entire plumes, it can also be used to define different zones within a chlorinated solvent plume.

Type 1 Environment: Systems That Are Anaerobic due to Anthropogenic Carbon

For highly chlorinated solvents to biodegrade, anaerobic conditions must prevail within the contaminant plume. Anaerobic conditions are typical at sites contaminated with fuel hydrocarbons, landfill leachate, or other anthropogenic carbon because these organics exert a tremendous electron-acceptor demand on the system. This condition is referred to as a *Type 1 environment*. In a Type 1 environment, anthropogenic carbon is fermented to produce hydrogen, which drives halorespiration.

The geochemistry of groundwater in a Type 1 environment is typified by strongly reducing conditions. This environment is characterized by very low concentrations of dissolved oxygen, nitrate, and sulfate and elevated concentrations of Fe(II) and methane in the source zone. The presence of methane is almost always observed and confirms that fermentation has been occurring at the site, generating hydrogen. If measured, hydrogen concentrations are typically greater than 1 nM. Importantly, a Type 1 environment results in the rapid and extensive degradation of the more highly chlorinated solvents such as PCE, TCE, and DCE.

A conceptual model of the Type 1 environment for a chlorinated ethene plume is shown in Figure 6.12, in which map views and centerline concentration profiles of PCE, TCE, *cis*-1,2-DCE, VC, inorganic electron acceptors (dissolved oxygen, nitrate, sulfate, and carbon dioxide), metabolic by-products (methane and dissolved iron), fermentation substrates (biochemical oxygen demand or BOD), and fermentation products (acetate) are shown. The following sequence of reactions occurs in this type of plume:

$$\text{PCE} \rightarrow \text{TCE} \rightarrow \text{DCE} \rightarrow \text{VC} \rightarrow \text{ethene} \rightarrow \text{ethane}$$

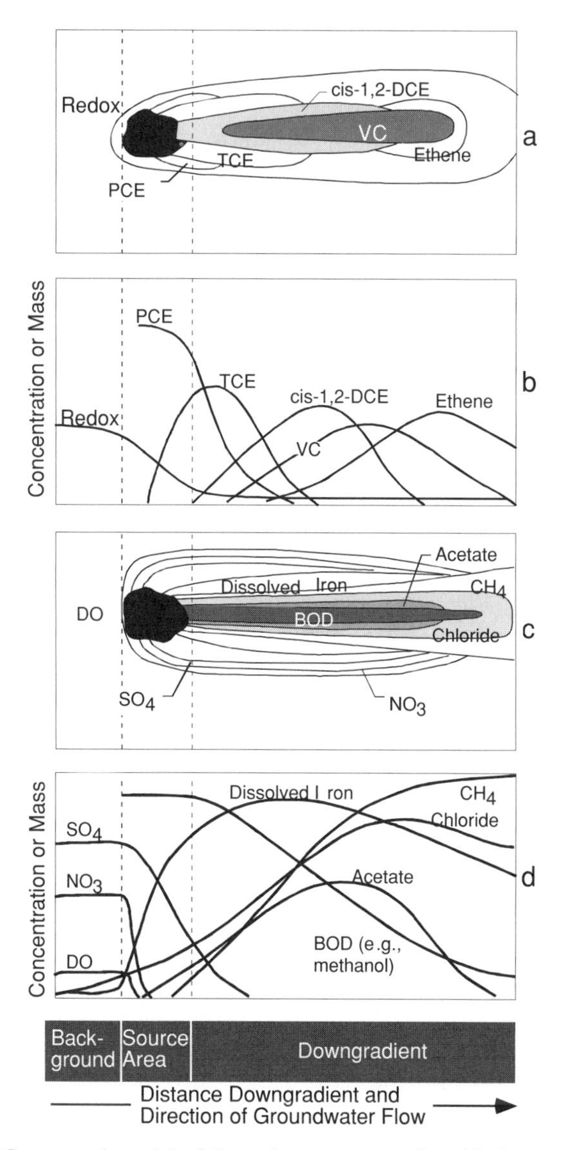

FIGURE 6.12. Conceptual model of Type 1 environment for chlorinated solvent plumes. (From Remediation Technologies Development Forum, 1997.)

In this type of plume, *cis*-1,2-DCE and VC degrade more slowly than TCE; thus they tend to accumulate and form longer plumes (Figure 6.12*a*). In Figure 6.12*b*, the PCE declines to zero and is replaced, in sequence, by a peak in TCE concentrations, followed by a peak in *cis*-1,2-DCE, VC, and ethene. The oxidation–reduction potential is depressed in the source zone by the anthropogenic carbon source and stays depressed throughout the entire plume. Fermentation constituents (BOD and acetate) and inorganics are shown in Figure 6.12*c* and *d*. Figure 6.12*d* illustrates how the fermentation substrate (represented by BOD) extends beyond the source before being consumed. Both panels show long chloride and methane plumes extending far downgradient from the plume area, because chloride is conservative and methane cannot be biodegraded in an anaerobic environment. The acetate curve indicates where active primary fermentation is occurring; declining acetate concentrations are due to consumption by methanogens in the plume area.

Some important questions regarding the long-term behavior of plumes in a Type 1 environment are

1. Is the supply of the fermentation substrate adequate to allow complete microbial reduction of the chlorinated organic compounds, and will the production of hydrogen from fermentation be sufficient to drive dechlorination to completion?
2. What is the role of alternate electron acceptors [e.g., dissolved oxygen, nitrate, Fe(III) and sulfate] in the competition for the hydrogen? In a Type I environment there is usually sufficient anthropogenic carbon delivered to the source zone (typically from dissolution of residual NAPLs or from landfill leachate) to remove all the competing electron acceptors so that fermentation will be a dominant process (as shown by methane production). If the supply of anthropogenic carbon declines (such as through the dissolution of NAPL, removal of a floating product layer, or removal of the landfill leachate), the competing electron acceptors may no longer be depleted.
3. Is vinyl chloride attenuated at rates sufficient to be protective of human health and the environment? Vinyl chloride biodegrades the slowest of the chlorinated ethene compounds in strongly anaerobic environments and may accumulate, forming a longer vinyl chloride plume relative to other chlorinated ethenes (PCE, TCE, or DCE). Based on carcinogen slope factors, vinyl chloride poses a higher potential health risk than the other chlorinated ethenes.

Type 2 Environment: Systems That Are Anaerobic due to Naturally Occurring Carbon

The classification system of Wiedemeier et al. (1996a, 1998) recognized that anaerobic conditions may also result from the fermentation of naturally occurring organic material in the groundwater that flows through chlorinated solvent source zones. This *Type 2 environment* occurs in hydrogeologic settings that have inherently high organic carbon concentrations, such as coastal or stream and river deposits with high concentrations of organics, shallow aquifers with recharge zones in organic-rich environments (such as swamps), or zones affected by natural oil seeps. A Type 2 environment generally results in slower biodegradation of the highly chlorinated solvents compared to a Type 1 environment. However, given sufficient organic loading, this environment can also result in rapid degradation of these compounds.

When evaluating natural attenuation of a chlorinated solvent plume in a Type 2 environment, the same questions apply as for a Type 1 environment. In addition, the same general conceptual model applies (see Figure 6.12). A Type 2 environment typically will not occur in crystalline igneous and metamorphic rock (see the discussion of probable hydrogeologic settings for Type 3 environments, below).

Type 3 Environment: Aerobic Systems due to No Fermentation Substrates

A Type 3 environment is characterized by a well-oxygenated groundwater system with little or no organic matter. Concentrations of dissolved oxygen typically are much greater than 1.0 mg/L. In such an environment, halorespiration will not occur and chlorinated solvents such as PCE, TCE, TCA, and CT will not biodegrade. In this environment, very long dissolved-phase plumes are likely to form. The most significant natural attenuation mechanisms for PCE and TCE will be advection, dispersion, and sorption. However, VC (and possibly DCE) can be rapidly oxidized under these conditions. A Type 3 environment is often found in crystalline igneous and metamorphic rock (fractured or unfractured) such as basalt; granite; schist; phyllite; glacial outwash deposits; eolian deposits; thick deposits of well-sorted clean beach sand with no associated peat or other organic carbon deposits; or any other type of deposit with inherently low organic carbon content if no anthropogenic carbon has been released.

Two conceptual models are provided for environments in which Type 3 behavior occurs. For sources with PCE and TCE, the major natural attenuation processes are dilution and dispersion alone (no biodegradation). As shown in Figure 6.13, the PCE and TCE plumes extend from the source zone and concentrations are slowly reduced by abiotic processes. Chloride concentrations and oxidation–reduction potential will not change as groundwater passes through the source zone and forms the chlorinated ethene plume. If TCA is the solvent of interest, significant abiotic hydrolysis may occur, resulting in a more rapid decrease in TCA concentrations and an increase in chloride concentrations.

In Figure 6.14, a source releases VC and 1,2-DCA into the groundwater at a Type 3 site (an unlikely occurrence, as more highly chlorinated solvents are typically released at sites). Because the VC and 1,2-DCA can be degraded aerobically, these constituents decline in concentration at a significant rate. Chloride is produced, and a depression in dissolved oxygen concentration similar to that occurring at fuel sites, is observed.

Mixed Environments

As mentioned above, a single chlorinated solvent plume can exhibit different types of behavior in different portions of the plume. This can be beneficial for natural biodegradation of chlorinated solvent plumes. For example, Wiedemeier et al. (1996b) describe a plume at Plattsburgh AFB, New York that exhibits Type 1 behavior in the

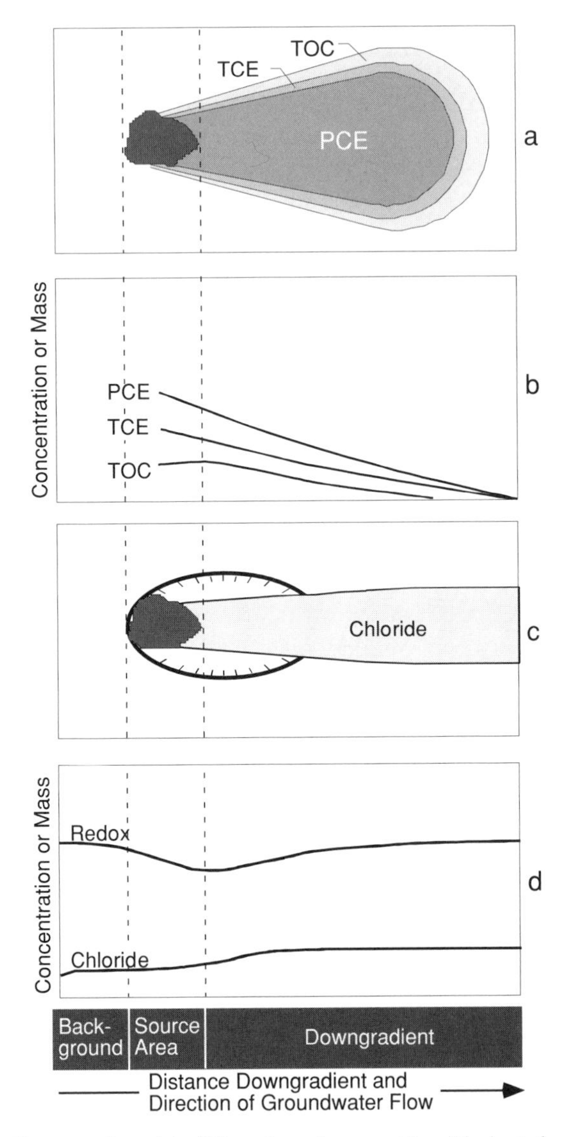

FIGURE 6.13. Conceptual model of Type 3 environment for chlorinated solvent plume due to PCE and TCE release. (From Remediation Technologies Development Forum, 1997.)

source area and Type 3 behavior downgradient from the source (see the case studies in Chapter 10). For natural attenuation, this may be the best scenario. PCE, TCE, and DCE are reductively dechlorinated with accumulation of VC near the source area (Type 1); then VC is oxidized (Type 3) to carbon dioxide, either aerobically or via Fe(III) reduction further downgradient, and does not accumulate. Vinyl chloride is

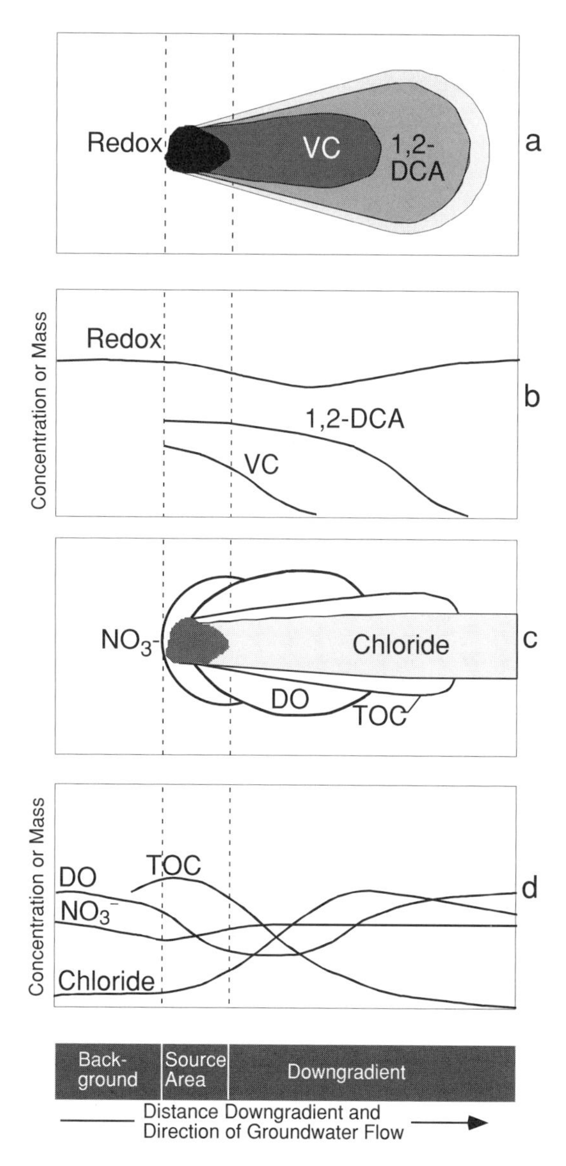

FIGURE 6.14. Conceptual model of Type 3 environment for chlorinated solvent plume with VC and 1,2-DCA release. (From Remediation Technologies Development Forum, 1997.)

removed from the system much faster under these conditions than under strongly reducing conditions.

A less ideal variation of the mixed Types 1 and 3 environments is shown in the conceptual model in Figure 6.15. An extended TCE and 1,2-DCE plume results because insufficient fermentable carbon results in an anaerobic zone that is too short

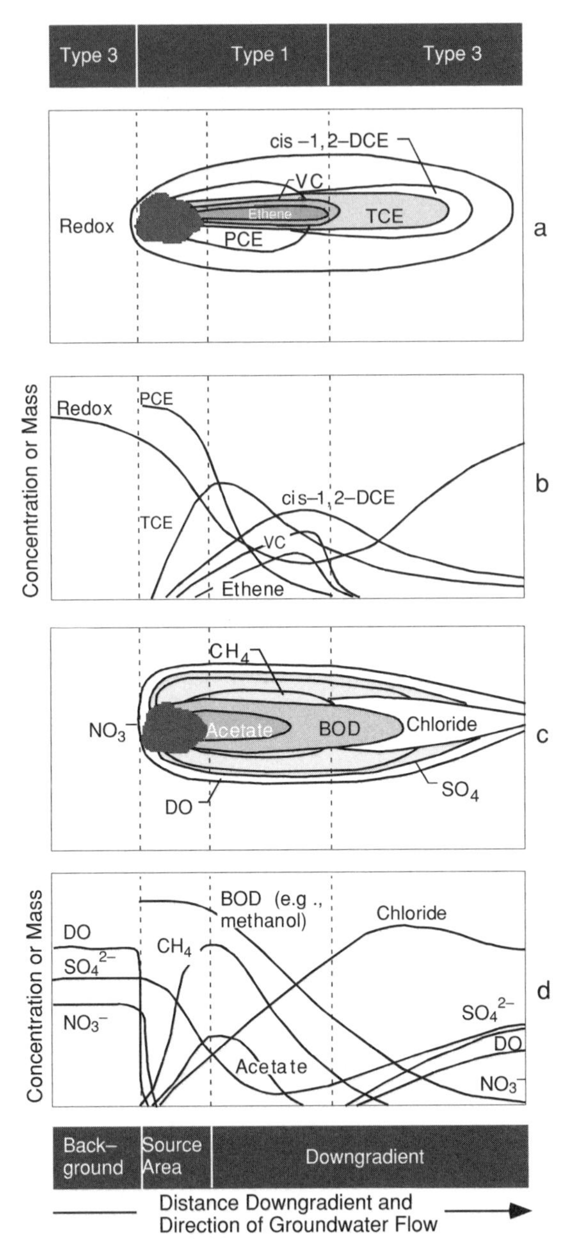

FIGURE 6.15. Conceptual model of mixed environments with Type 1 environment (available fermentation substrates) in source zone and Type 3 environment (no fermentation) in downgradient portion of plume. (From Remediation Technologies Development Forum, 1997.)

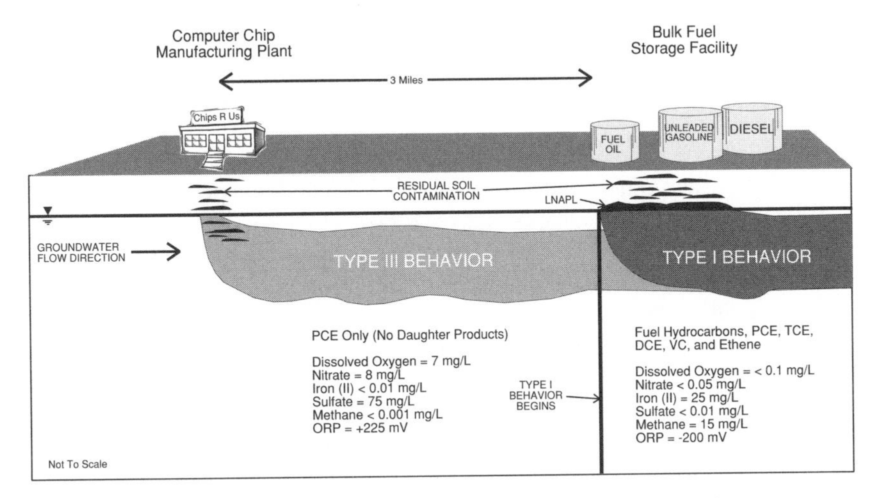

FIGURE 6.16. Conceptual site model of mixed Type 3/Type 1 environments.

for complete biodegradation. Therefore, TCE extends well into the aerobic zone, in which no biodegradation occurs. A long DCE plume also extends into the aerobic zone, indicating insignificant direct aerobic biodegradation was assumed. While a long chloride plume will be observed, the short anaerobic zone means that much less methane is produced, allowing dilution or dispersion to limit the extent of the methane plume.

A less common type of mixed environment occurs in which a chlorinated solvent source zone in a Type 3 environment produces a plume that extends downgradient to a zone in the aquifer where fermentation substrates are supplied (Figure 6.16). For example, a downgradient leaking underground storage tank could introduce BTEX compounds into the chlorinated solvent plume, changing the environment to Type 1. A Type 3 environment would be converted to a Type 2 environment if a downgradient recharge zone introduced naturally occurring fermentation substrates to the subsurface (Figure 6.17).

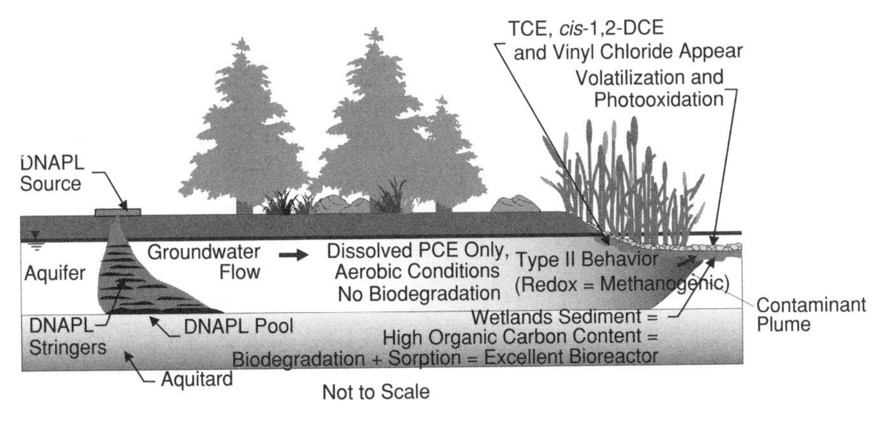

FIGURE 6.17. Conceptual model of a plume discharging to surface water.

6.3 CHLORINATED SOLVENT RATE DATA FROM THE LITERATURE

Biodegradation rate data for various chlorinated solvents are presented in Table 6.6. Note that data are sorted by data source (microcosm, field, or not reported) and by the value of the half-life (where a range was reported, the low end value was used for sorting). As can be seen from these data, there is a considerable amount of variation in the first-order half-lives. Biodegradation rate coefficients for the chlorinated ethenes were generally 0.0001 to 0.01 per day (half-lives of about 20 years to 0.2 year), without any discernible trend in the biodegradation rate coefficients among the different chlorinated compounds.

A groundwater anaerobic biodegradation literature review was performed for several chemicals found in groundwater, including a number of chlorinated solvents (Aronson and Howard, 1997). After reviewing numerous literature references (e.g., data from 16 different field studies/in situ microcosm studies were evaluated for TCE), a range of "recommended values" for the anaerobic biodegradation first-order decay rate coefficients were reported. For many of the chlorinated solvents, the authors defined the low-end rate coefficient based on the lowest measured field value and defined the high-end value as the mean rate coefficient for all the field/in situ

TABLE 6.6. Reported First-Order Biodegradation Reaction Half-Lives, Sorted by Microcosm and Field-Scale Studies and by Lowest Reported Half-Life[a]

Compound	First-Order Biodegradation Half-Life (days)	Microcosm	Field Scale	Reference
PCE	13			Bouwer and McCarty (1983)
	35	×		Parsons et al. (1984)
	877–1019		×	Ellis et al. (1996)
TCE	2	×		Fogel et al. (1986)
	69	×		B. H. Wilson et al. (1996)
	89	×		B. H. Wilson et al. (1986)
	33–90	×		Barrio-Lage et al. (1987)
	141	×		B. H. Wilson et al. (1990)
	141–210	×		Haston et al. (1994)
	231–6930	×		B. H. Wilson et al. (1991)
	116–347		×	Gorder et al. (1996)
	147–845		×	Weaver et al. (1996)
	182–210		×	Ehike et al. (1994)
	198–1100		×	Wiedemeier et al. (1996b)
	231		×	Cox et al. (1995)
	365		×	Lee et al. (1995)
	433–630		×	J. T. Wilson et al. (1996)
	788		×	Benker et al. (1994)
	877–1540		×	Ellis et al. (1996)
	990–1155		×	Dupont et al. (1996)
	1386		×	Swanson et al. (1996)

TABLE 6.6. *(Continued)*

Compound	First-Order Biodegradation Half-Life (days)	Microcosm	Field Scale	Reference
DCE	116	×		Bradley and Chapelle (1997)
	277–4331		×	Wiedemeier et al. (1996b)
	495–1019		×	Ellis et al. (1996)
1,2-DCE	2	×		Fogel et al. (1986)
	77	×		J. T. Wilson et al. (1986)
Chlorobenzene	117			J. T. Wilson et al. (1983)
cis-1,2-DCE	27–82	×		B. H. Wilson et al. (1991)
	77–976		×	Swanson et al. (1996)
	158–495		×	Ehlke et al. (1994)
	289		×	Weaver et al. (1996)
	385–578		×	B. H. Wilson et al. (1996)
	1386		×	Cox et al. (1995)
	89–347			Barbee (1994)
trans-1,2DCE	139	×		J. T. Wilson et al. (1982)
	77			Tabak et al. (1981)
VC	58 (aerobic)	×		Davis and Carpenter (1990)
	23,100	×		Barrio-Lage et al. (1990)
	82		×	Cox et al. (1995)
	98–1690		×	Weaver et al. (1996)
	533–1824		×	Wiedemeier et al. (1996b)
	693–806		×	Ellis et al. (1996)
TCA	70	×		Parsons et al. (1985)
1,1,1-TCA	0.4	×		Bouwer and McCarty (1983)
	< 1	×		Vogel and McCarty (1987)
	198–210	×		Klecka et al. (1990)
1,1-DCA	<< 6 (to CA)	×		Vogel and McCarty (1987)
	63–144	×		Henson et al. (1989)
	158	×		J. T. Wilson et al. (1982)
1,2-DCA	100–365	×		Henson et al. (1989)

[a] Rate constant = 0.693 ÷ half-life.

microcosm studies. Table 6.7 shows the resulting recommended ranges for first-order anaerobic biodegradation rate coefficients for several chlorinated solvents along with the mean value of the field/in situ microcosm studies (note that some minor discrepancies exist between the reported high-end rates and the mean value for the field/in situ microcosm studies).

Data from one site at Plattsburgh AFB were analyzed by Wiedemeier et al. (1996b) to estimate the maximum biodegradation rate coefficients for TCE, TCE daughter products, and BTEX compounds at several zones within the plume. Two methods for

TABLE 6.7. Mean and Recommended First-Order Rate Coefficients for Selected Chlorinated Solvents Presented by Aronson and Howard (1997)

	Mean of Field/In Situ Studies			Recommended First-Order Rate Coefficients				
	First-Order Rate			Low End		High End		
Compound	Coefficient (day^{-1})	Half-Life (day^{-1})	Number of Studies Used for Mean	Coefficient (day^{-1})	Half-Life (day^{-1})	Coefficient (day^{-1})	Half-Life (day^{-1})	Comments
Tetrachloroethylene (PCE)	0.0029	239	16	0.00019	3647	0.0033	210	Lower limit was reported for a field study under nitrate-reducing conditions.
Trichloroethylene (TCE)	0.0025	277	47	0.00014	4950	0.0025	277	Lower limit was reported for a field study under unknown redox conditions.
Vinyl chloride (VC)	0.0079	88	19	0.00033	2100	0.0072	96	Lower limit was reported for a field study under methanogenic/sulfate-reducing conditions.
1,1,1-Trichloroethane (TCA)	0.016	43	15	0.0013	533	0.01	69	Range not appropriate for nitrate-reducing conditions. Expect lower limit to be much less.
1,2-Dichloroethane (DCA)	0.0076	91	2	0.0042	165	0.011	63	Range reported from a single field study under methanogenic conditions.
Carbon tetrachloride (CT)	0.37	1.9	9	0.0037	187	0.13	5	Range not appropriate for nitrate-reducing conditions. Expect lower limits to bed much less.
Chloroform	0.030	23	1	0.0004	1733	0.03	23	Only one field study available. Biodegradation under nitrate-reducing conditions expected.
Dichloromethane	0.0064	108	1	0.0064	108	—	—	Rate constant reported from a single field study under methanogenic conditions.
Trichlorofluoromethane	—	—	—	0.00016	4331	0.0016	433	All studies with very low concentrations of this compound.
2,4-Dichlorophenol	0.014	50	2	0.00055	1260	0.027	26	Range may not be appropriate for nitrate-reducing conditions.

estimating rate coefficients were used: the correction method using chloride and the correction method using trimethylbenzenes, as described in Chapter 7. Both methods gave approximately the same results, allowing the results from the two estimation methods to be averaged. The maximum rate of biodegradation (indicated by half-lives in years) and the location where the maximum degradation occurred are reported below.

TCE half-life	0.6 year in Type 1 zone, 0 to 970 ft on plume centerline
DCE half-life	0.9 year in Type 1 zone, 970 to 1240 ft on plume centerline
VC half-life	1.5 years in Type 3 zone, 1240 to 2560 ft on plume centerline
BTEX half-life	1.5 years in Type 1 zone, 970 to 1240 ft on plume centerline

Note that the 0 point on the plume centerline is the actual source, a former fire training pit. The slowest rate of biodegradation (indicated by half-lives in years) and the location in which the slowest degradation occurred are reported below.

TCE half-life	>100 years in Type 3 zone, 1240 to 2560 ft on centerline
DCE half-life	23 years in Type 1 zone, 0 to 970 ft on centerline
VC half-life	>100 years in Type 1 zone, 0 to 970 ft on centerline
BTEX half-life	7 years in Type 1 zone, 0 to 970 ft on centerline

These data indicate that there are distinct biodegradation zones at chlorinated solvent sites. TCE is rapidly degraded in deeply anaerobic zones near the source (in which the fermentation substrates are located), while DCE and VC are biodegraded via halorespiration more slowly in these areas. In the Type 3 zone located far downgradient of the source, the vinyl chloride is more rapidly biodegraded via anaerobic oxidation and aerobic oxidation, and TCE degradation becomes much slower, as little fermentation is occurring.

6.4 CHLORINATED SOLVENT PLUME DATABASES

A summary of plume data from two different databases containing chlorinated solvent site data is provided below. The first database, the Hydrogeologic Database (HGDB) (Newell et al., 1990), was reanalyzed in 1997 to summarize plume length, plume width, plume thickness, and highest solvent concentration observed at 109 chlorinated solvent sites. The second database condenses extensive site characterization data from 17 Air Force chlorinated solvent sites (see Appendix A), with information on parent compounds versus daughter products concentrations, competing electron acceptors, hydrogen, and metabolic by-products. Together, these databases illustrate many of the processes discussed in this chapter regarding chlorinated plume biodegradation mechanisms, conceptual models, and other processes.

Chlorinated Solvent Sites in the HGDB

The HGDB database, developed from 400 questionnaires completed by National Groundwater Association members, included data for 109 chlorinated solvent sites. The data were broken into two groups: the chlorinated ethenes, in which one or more of the chlorinated ethenes (PCE, TCE, DCE, or VC) was reported to be the major contaminant, and other chlorinated solvent sites, in which all other chlorinated solvents besides the ethenes (e.g., TCA, DCA, chlorobenzene) were lumped together.

As shown in Table 6.8, the median length of the 75 chlorinated ethene plumes was 1000 ft, with one site reporting a plume length of 13,200 ft. These plume lengths are significantly longer than the lengths reported for hydrocarbon plumes (typically less than 250 ft; see Chapter 5). The longer plume lengths for the solvent sites are probably related to some degree to different biodegradation patterns occurring at chlorinated solvent sites than at fuel sites and perhaps due to longer source zones. At fuel hydrocarbon sites, there is a much larger pool of compounds for electron transfer in biodegradation reactions, with BTEX compounds and partially degraded constituents serving as electron donors, and oxygen, nitrate, Fe(III), sulfate, and carbon dioxide serving as electron acceptors. BTEX breakdown products are also fermented to form hydrogen and other constituents that are used by methanogens to produce methane.

TABLE 6.8. Characteristics of Chlorinated Solvent Plumes from HGDB Database[a]

	Plume Length (ft)	Plume Width (ft)	Vertical Penetration (ft)	Highest Concentration (mg/L)
	Chlorinated Ethenes (e.g., PCE, TCE)			
Maximum	13,200	4,950	500	28,000
75th percentile	2,500	1,000	100	72
Median	**1,000**	**500**	**40**	**8.467**
25th percentile	600	200	25	0.897
Minimum	50	15	5	0.001
n	75	75	78	81
	Other Chlorinated Solvents (e.g., TCA, DCA)			
Maximum	18,000	7,500	150	2,500
75th percentile	2,725	1,000	51	13.250
Median	**575**	**350**	**35**	**3.100**
25th percentile	290	188	24	0.449
Minimum	100	100	8	0.016
n	24	24	24	28

Source: Newell et al. (1990); data analyzed in 1997.

[a] Highest concentrations for chlorinated ethenes (28,000 mg/L) was for TCE, which is above the solubility limit. The highest concentration for "other chlorinated solvents" (2500 mg/L) was for chloromethane and toluene.

A key factor limiting BTEX plume length is that BTEX can be degraded by a variety of potential electron acceptors in the plume as it leaves the source zone and disperses into the clean portions of the aquifer.

At chlorinated solvent sites, however, halorespiration is controlled by the amount of fermentation that occurs. Fermentation alone produces the hydrogen that is the required electron donor for the halorespirators. This may be a slower and more spatially limited process than BTEX biodegradation, as there is competition for hydrogen by other bacteria (i.e., sulfate reducers, iron reducers, and methanogens) and conditions favorable for fermentation may be limited to the length of the carbon substrate plume.

In other words, the presence of sulfate, iron, and methane at fuel hydrocarbon sites represent more electron acceptors for BTEX biodegradation and therefore more biodegradation. At chlorinated solvent sites, the presence of sulfate, iron, and (indirectly) methane represents competition for hydrogen, an almost universal electron donor. In source areas, the other electron acceptors are depleted, and hydrogen goes to the dechlorinators. Farther downgradient in the plume, however, electron acceptors that disperse into the plume act to shut down chlorinated solvent biodegradation. This means less biodegradation for the chlorinated solvent parent compounds as they are only degraded by halorespiration.

The other category, other chlorinated solvent sites, had shorter plumes, with a median plume length of 575 ft compared to 1000 ft for the chlorinated ethene sites. Twelve of the 24 plumes were comprised of TCA, which is degraded biologically via halorespiration and other mechanisms and abiotically by hydrolysis (half-life of 0.5 to 2.5 years). Despite the degradability of TCA, the TCA plumes had a median length of 925 ft. The shorter plumes in this database of 24 sites were reported to be comprised of either a general indicator, such as total organic halogens, or individual compounds such as 1,1-dichloroethane, methylene chloride, or chlorobenzene. The median highest concentration at these other chlorinated solvent sites was 3.1 mg/L (see Table 6.8).

Chlorinated Solvent Sites in the Air Force Database

Data compiled from 17 Air Force sites using the AFCEE (Air Force Center for Environmental Excellence) natural attenuation protocol (Wiedemeier et al. 1996a, 1998) provides an in-depth look at natural attenuation processes at chlorinated solvent sites. Note that these data may not be representative of chlorinated plumes in general, as these sites were studied because biodegradation was thought to be occurring.

Plume lengths were reported at 14 sites (Table 6.9), and these data showed a median length of 750 ft. There were significant differences in plume length for different plume classes, with Type 1 plumes (sites with available human-made fermentation substrates such as BTEX) being shorter than Type 3 plumes (sites without available fermentation substrates). Twelve of the sites exhibiting Type 1 plumes had a median plume length of 625 ft, while the two sites with Type 3 plumes had lengths of 1100 and 5000 ft. Four mixed plumes (Type 1 in source zone, Type 3 in the downgradient part of the plume) had a median length of 2538 ft.

TABLE 6.9. Chlorinated Solvent Plume Characteristics

No.	State	Type[a]	Plume Length (ft)	Plume Width (ft)	Plume Thickness (ft)	Total Chlorinated Solvents (mg/L)	Seepage Velocity (ft/yr)	First-Order Biodegradation Rate Coefficient for Solvents (day^{-1})	Half-Life (years)	Method for Calculating Rate Coefficient
13	UT	3	5000	1400	40	4.953	60	0.000006	316.4	Other
9	NY	Mixed	4200[b]	2050	60	774.721	139	0.001[b]	1.9[b]	Other
10	NE	Mixed	3500	1400	50	164.010	152	0.000001	1899	Other
8	MA	1	1800	1200	50	4.340	106	0.0005	3.8	Other
11	FL	Mixed	1575[b]	400	15	1258.842	113	0.0009[b]	2.1[b]	Other
14	AK	3	1100	250	25	4.899	260	0.0065	0.3	Other
1	SC	1	750	550	5	328.208	1600	—	—	—
7	MA	1	750	250	50	50.566	20.8	0.0095	0.2	Conserv. tracer
12	MS	Mixed	750	550	5	0.472	1500	0.01	0.2	Buscheck
4	NE	1	650	450	30	47.909	6.7	0.0006	3.2	Buscheck
3	FL	1	600	350	20	0.429	36	0.0007	2.7	Buscheck
5	WA	1	550	300	10	3.006	32.9	0.001	1.9	Buscheck
2	MI	1	375	100	10	0.397	292	—	—	—
6	OH	1	100	60	10	15.736	25	—	—	—
		Maximum	5000	2050	60	1259	1600	0.01	1899	
		75th percentile	1744	1038	48	136	233	0.00375	3.5	
		Median	**750.0**	**425.0**	**22.5**	**10.3**	**109.5**	**0.0009**	**2.1**	
		25th percentile	613	263	10	3.3	34	0.00055	1.1	
		Minimum	100	60	5	0.397	7	0.000001	0.2	
		n	14	14	14	14	14	11	11	

Source: Data from 14 U.S. Air Force sites compiled by Air Force Center for Environmental Excellence (see Tables 6.10 and 6.11 and Appendix A for additional data).

[a] Plume discharges into stream; may not represent maximum potential plume length.

[b] Mixed refers to Type 1 conditions in source zone, Type 3 conditions in downgradient part of plume. Median length Type 1 sites, 625 ft; mixed sites, 2538 ft; Type 3, 3050 ft (two sites).

Site-specific biodegradation rate information was developed using several methods, including one developed by Buscheck and Alcantar (1995), one based on the use of conservative tracers (the trimethylbenzenes), and other methods, such as model calibration (see Chapter 7 for a more detailed discussion of these methods). Rates varied significantly, with half-lives ranging from over 300 years for a Type 3 site located in Utah to 0.2 year for a Type 1 site (see Table 6.9). The median first-order half-life for the 14 chlorinated plumes was 2.1 years.

Electron Acceptor, Electron Donor, and By-product Data. Extensive chlorinated solvent, electron acceptor, and electron donor data were available for the complete 17-site AFCEE database (see Tables 6.10 and 6.11). The complete data set, shown in Appendix A, was reduced to key indicator data shown in Table 6.10 in units of μg/L and Table 6.11 in units of nmol/L. In these tables, the maximum site concentrations of the chlorinated solvents, ethene and ethane, hydrogen, methane, BTEX, and the change in chloride concentration from background to the interior of the plume are presented. Summary statistics, including the median value for the 17 sites, are also shown.

The data illustrate that halorespiration is or was occurring at these sites on a large scale. For example, the median concentration of the TCE daughter product, *cis*-1,2-DCE, was almost equal to the median concentration of TCE (434 μg/L versus 446 μg/L, respectively) (Table 6.10). There was very little PCE observed, with only eight sites reporting PCE at all. The *cis*-1,2-DCE isomer comprised almost all of the DCE observed at these sites (median concentration of 424 μg/L compared to 2 μg/L for 1,1-DCE and 2 μg/L for *trans*-1,2-DCE). Because *cis*-1,2-DCE is the predominant fraction of biologically produced DCE and represents only a small fraction (about 20%) of manufactured DCE, the data indicate that there were very high concentrations of daughter products at most of the sites. A simple comparison of daughter products (assumed to be all the DCE compounds plus vinyl chloride) to parent compounds (assumed to be PCE + TCE) showed that the median site had twice the amount of daughter products compared to parent compounds (Table 6.10). Finally, the median site had BTEX at twice the concentration as the total chlorinated solvent concentration, indicating that many of these sites had an ample pool of substrates for fermentation of dissolved hydrogen.

When the data were analyzed on a molar basis (units of nanomoles per liter, Table 6.11), the median molar concentrations of parents, daughter compounds, chloride, and methane for the 17 sites were:

Parent compounds (PCE + TCE)	3392	nmol/L (nM)
Daughter compounds (all DCE compounds + VC)	4856	nM
Ethene + ethane	286	nM
Methane	434,500	nM
Excess chloride	4,317,797	nM
Hydrogen	5	nM

TABLE 6.10. Chlorinated Solvent Plume Data[a]

No.	State	Plume Type	PCE (μg/L)	TCE (μg/L)	1,1-DCE (μg/L)	*trans*-1,2-DCE (μg/L)	*cis*-1,2-DCE (μg/L)	VC (μg/L)	Ethene + Ethane (μg/L)	TCA (μg/L)	1,1-DCA (μg/L)	Total Chlorinated Solvents (μg/L)	Hydrogen (nmol/L)	Methane (μg/L)	BTEX (μg/L)	Delta Chloride (μg/L)	*cis*-DCE / all DCE (%)	(DCE + VC) / (PCE + TCE)	Total Solvent / BTEX
1	SC	1	91	718	748	2	4,590	416	15	23,000	8,700	38,265	19.02	6,952	3,295	174,700	86	7.1	11.61
2	MI	1	6	8	0	0	29	0	0	—	—	43	0	3,550	2,421	24,250	100	2.0	0.02
3	FL	1	1	3	0	0	7	21	—	—	—	31	—	9,890	2,840	108,490	100	7.5	0.01
4	NE	1	0	9	0	2	273	817	895	3	—	1,104	—	22,450	3,230	209,400	99	127.0	0.34
5	WA	1	0	63	0	3	124	38	17	—	—	229	—	19,060	5,221	92,600	97	2.6	0.04
6	OH	1	0	9	0	0	1,500	0	6	—	—	1,509	0.648	19,200	440	51,800	100	166.7	3.43
7	MA	1	0	1,600	4	3	434	2,098	8	—	—	4,139	—	14,630	26,125	131,000	98	1.6	0.16
8	MA	1	13	28	1	0	390	1	—	0	0	433	—	2,200	15,601	56,700	100	9.6	0.03
9	NY	Mixed	0	25,280	0	233	51,400	2,080	474	0	0	78,993	11	1,500	16,790	221,000	100	2.1	4.70
10	NE	Mixed	2	17,500	29	9	1,230	1	—	0	1	18,772	9.55	12,290	1	74,520	97	0.1	18,772
11	AS	Mixed	56	15,800	200	389	98,500	6,520	55,225	0	443	121,908	10	6,970	331	324,000	99	6.7	368
12	AS	Mixed	3	39,400	238	145	4,120	1,350	114	130	26	45,413	6	4,000	6	229,000	91	0.1	7,569
13	OK	Mixed	0	12,700	4	33	1,340	0	0	0	0	14,078	2.24	80	43	1,116,000	97	0.1	327
14	MS	Mixed	0	7	3	0	15	10	—	4	1	40	0	6,500	48	180,000	85	3.7	0.8
15	UT	Mixed	0	446	13	8	7,083	469	274	257	187	8,463	1.4	10,400	1,557	—	100	17.0	5.4
16	UT	3	253	355	2	1	16	0	0	64	6	696	5	428	0	179,900	85	0.0	No BTEX
17	AK	3	0	636	0	0	0	0	—	8	0	644	—	390	8,620	3,950	85	0.0	0.07
		Maximum	253	39,400	748	389	98,500	6,520	55,225	23,000	8,700	121,908	19	22,450	26,125	1,116,000	100	167	18,772
		75th percentile	6	12,700	13	9	4,120	817	324	81	107	18,772	10	12,290	5,221	212,300	100	7	90.6
		Median	**0**	**446**	**2**	**2**	**434**	**21**	**16**	**4**	**1**	**1,509**	**5**	**6,952**	**2,421**	**152,850**	**98**	**2.6**	2.1
		25th percentile	0	9	0	0	29	0	4	0	0	433	1	2,200	48	70,065	91	0	0.067
		Minimum	0	3	0	0	0	0	0	0	0	31	0	80	0	3,950	85	0	0.011
		n	17	17	17	17	17	17	12	12	11	17	11	17	17	16	17	17	16

Source: Data from 17 U.S. Air Force sites compiled by Air Force Center for Environmental Excellence (see Tables 6.9 and 6.11 and Appendix A for additional data).

[a]0 indicates nondetectible, value assumed to be 0; — indicates not analyzed.

TABLE 6.11. Chlorinated Solvent Plume Data[a]

No.	State	Type	PCE (nmol/L)	TCE (nmol/L)	1,1-DCE (nmol/L)	*trans*-1,2-DCE (nmol/L)	*cis*-1,2-DCE (nmol/L)	VC (nmol/L)	Ethene + Ethane (nmol/L)	TCA (nmol/L)	1,1-DCA (nmol/L)	Total Chlorinated Solvents (nmol/L)	Hydrogen (nmol/L)	Methane (nmol/L)	BTEX (nmol/L)	Delta Chloride (nmol/L)	Methane / Solvents	Delta Chloride / Solvents	Delta Chloride / Methane
		Mol. wt.:	165.8	131.5	97	97	97	62.5	28	133.4	99	—	2	16	96 (avg)	35.4			
1	SC	1	547	5,460	7,711	22	47,320	6,656	536	172,414	87,879	328,008	19.0	434,500	34,323	4,935,028	1.3	15.0	11
2	MI	1	37	62	0	0	299	0	0	—	—	397	0.0	221,875	25,219	685,028	558.4	1,724.0	3
3	FL	1	7	20	0	0	70	333	0	—	—	429	—	618,125	29,583	3,064,689	1,439.8	7,138.6	5
4	NE	1	0	65	0	25	2,814	13,072	31,964	25	—	16,001	—	1,403,125	33,646	5,915,254	87.7	369.7	4
5	WA	1	0	482	0	33	1,278	606	607	—	—	2,400	—	1,191,250	54,385	2,615,819	496.4	1,090.0	2
6	OH	1	0	68	0	0	15,464	0	204	—	—	15,532	0.6	1,200,000	4,583	1,463,277	77.3	94.2	1
7	MA	1	0	12,167	38	33	4,474	33,568	286	—	—	50,281	—	914,375	272,135	3,700,565	18.2	73.6	4
8	MA	1	78	213	5	0	4,021	22	0	0	0	4,340	—	137,500	162,510	1,601,695	31.7	369.1	12
9	NY	Mixed	0	192,243	0	2,402	529,897	33,280	16,929	0	0	757,822	11.0	93,750	174,896	6,242,938	0.1	8.2	69
10	NE	Mixed	11	133,080	295	97	12,680	21	—	0	10	146,194	9.6	768,125	10	2,105,085	5.3	14.4	3
11	AS	Mixed	338	120,152	2,062	4,010	1,015,464	104,320	1,972,321	0	4,475	1,250,821	10.0	435,625	3,448	9,152,542	0.3	7.3	21
12	AS	Mixed	20	299,620	2,454	1,495	42,474	21,600	4,071	975	267	368,904	6.0	250,000	63	6,468,927	0.7	17.5	26
13	OK	Mixed	0	96,578	44	343	13,814	0	0	0	0	110,780	2.2	5,000	448	31,525,424	0.0	284.6	6,305
14	MS	Mixed	0	56	28	0	155	160	—	28	12	439	0.0	406,250	500	5,084,746	924.7	11,573.2	13
15	UT	Mixed	0	3,392	131	86	73,021	7,504	9,786	1,927	1,889	87,948	1.4	650,000	16,219	—	7.4	—	—
16	UT	3	1,526	2,700	24	5	161	0	0	481	57	4,953	5.0	26,750	0	5,081,921	5.4	1,026.0	190
17	AK	3	0	4,837	0	0	0	0	0	62	0	4,899	—	24,375	89,792	111,582	5.0	22.8	5
	Maximum		1,526	299,620	7,711	4,010	1,015,464	104,320	1,972,321	172,414	87,879	1,250,821	19.02	1,403,125	272,135	31,525,424	1,439.8	11,573	6,305
	75th percentile		37	96,578	131	97	42,474	13,072	6,929	605	1,078	146,194	9.775	768,125	54,385	5,997,175	87.7	1,042	22
	Median		**0**	**3,392**	**24**	**25**	**4,474**	**333**	**286**	**27**	**12**	**16,001**	**5.0**	**434,500**	**25,219**	**4,317,797**	**7.4**	**189.4**	**3.2**
	25th percentile		0	68	0	0	299	0	0	0	0	4,340	1.024	137,500	500	1,979,237	1.3	17	4
	Minimum		0	20	0	0	0	0	0	0	0	397	0.0	5,000	0	111,582	0.0	7	1
	n		17	17	17	17	17	17	15	12	11	17	11	17	17	16	17	16	16

Source: Data from 17 U.S. Air Force sites compiled by Air Force Center for Environmental Excellence (see Tables 6.9 and 6.11 and Appendix A for additional data).

[a] 0 indicates nondetectible, value assumed to be 0; — indicates not analyzed.

When these data were normalized to the moles of parent compounds, the following ratios were calculated:

(PCE + TCE)/(PCE + TCE)	1
(DCE + VC)/(PCE + TCE)	1.43
(Ethene + ethane)/(PCE + TCE)	0.08
Methane/(PCE + TCE)	128
(Excess chloride)/(PCE + TCE)	1273
Hydrogen/(PCE + TCE)	0.0015

The molar data show that the number of moles of daughter products was greater than the moles of parent compounds, and the number of moles of excess chloride (source minus background concentrations) was much greater than the moles of total chlorinated solvents. The extremely high chloride concentration indicates that there was much more chlorinated solvent biodegradation at these sites than was reflected in either the parent compound concentrations or the daughter compound concentrations. Based on the median data for the 17 sites and assuming that the most common parent compound was TCE (three chlorines), the 4.3 million nanomoles per liter (4.3 mM) of chloride represents approximately 187 mg/L of TCE that had been degraded completely. This is much higher than the concentrations of chlorinated solvents observed at any of these sites.

The large amount of apparent biodegradation is not completely unexpected. Parent compounds represent reactions that have not yet occurred, the daughter products represent partially completed reactions, and chloride represents completed reactions. As with fuel sites, considerable biodegradation can occur in source zones at chlorinated solvent sites, and groundwater samples extracted from source zones represents groundwater concentrations *after* significant biodegradation has occurred (see Chapter 8). At these 17 chlorinated solvent sites, the median source zone sample had much more excess chloride than chlorinated solvents, indicating significant biodegradation. An alternative explanation, in which chloride contamination occurred along with the chlorinated solvent and BTEX and entered the aquifer through the source zone, is not supported by any evidence but cannot be disproved with the existing site data.

At first glance, the low concentrations of ethene + ethane appear not to support that significant biodegradation occurred in the source zones. However, DCE and vinyl chloride can be oxidized, yielding no ethene or ethane. Alternatively, the ethene could be degraded anaerobically or can degas, preventing its accumulation. These results indicate that ethane and ethene are not conservative indicators of chlorinated solvent biodegradation.

When the relative moles of methane and chloride from Table 6.11 are evaluated to reflect potential number of hydrogen moles consumed, an estimate of relative hydrogen consumption by halorespiration and methanogenesis can be made. If all the methane is produced from hydrogen (a conservative assumption, as some methanogenesis will occur with other fermentation substrates besides hydrogen), 4 mol of hydrogen will be consumed for every mole of methane produced (4:1). For every mole of chloride produced from halorespiration, 1 mol of hydrogen is consumed (1:1), while the median ratio of chloride moles to methane moles was 8:1 (Table 6.11). These ratios indicate that the relative amount of hydrogen consumed by halorespiration/methanogenesis was 8:4 or 2:1.

The 2:1 ratio from this order-of-magnitude accounting system suggests that hydrogen from fermentation is consumed by both dechlorinators and methanogens and that the dechlorinators may be competing effectively for hydrogen against the methanogens at these sites. (Note that if some of the chloride was generated via DCE or VC oxidation, the accounting system would indicate a more equal distribution of hydrogen between the halorespirators and the methanogens.)

Linear Regressions of Chlorinated Solvent Site Data. The ability to predict certain behaviors, such as the amount of biodegradation and the length of the chlorinated solvent plumes, was evaluated using the linear regressions in Figures 6.18 and 6.19. In Figure 6.18, the BTEX concentration was plotted against the *daughter ratio*, where the daughter ratio was calculated as (total DCE + VC)/ (PCE + TCE). In theory, there should be a relationship between the available fermentation substrate and the amount of dechlorination occurring at a site. Although the slope of the regression was positive (more BTEX means higher fraction of daughter products), the coefficient for determination (r^2) was only 0.15, indicating that other factors affect halorespiration. For example, the presence of unmeasured fermentation substrates, such as acetone, or the existence of TCE as a daughter product of PCE would affect the correlation.

Methane should also serve as an indicator of halorespiration, as the presence of methane indicates active fermentation processes occurring at the site and, therefore, hydrogen production. The r^2 value for the BTEX versus daughter ratio (see Figure 6.18) was only 0.37, however, indicating that data scatter and/or other factors were interfering with this relationship. One possible confounding factor is the potential off-gassing of methane. The solubility of methane is approximately 24 mg/L, and some sites approached the solubility limit for methane (Table 6.10). When the presence of other dissolved gases, such as nitrogen, is considered, the bubbling of produced

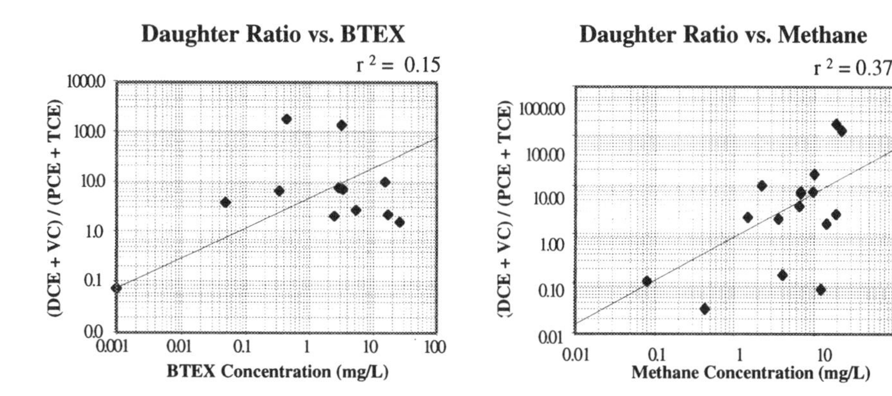

FIGURE 6.18. Comparison of daughter ratio (ratio daughter products DCE and VC to parent compounds PCE and TCE) versus BTEX and methane concentrations. Note that the correlation coefficient (r^2) was performed on the logarithms of the two variables for each graph. [Data from 17 U.S. Air Force sites analyzed as part of the Air Force Center for Environmental Excellence Natural Attenuation Initiative (Wiedemeier et al., 1996a).]

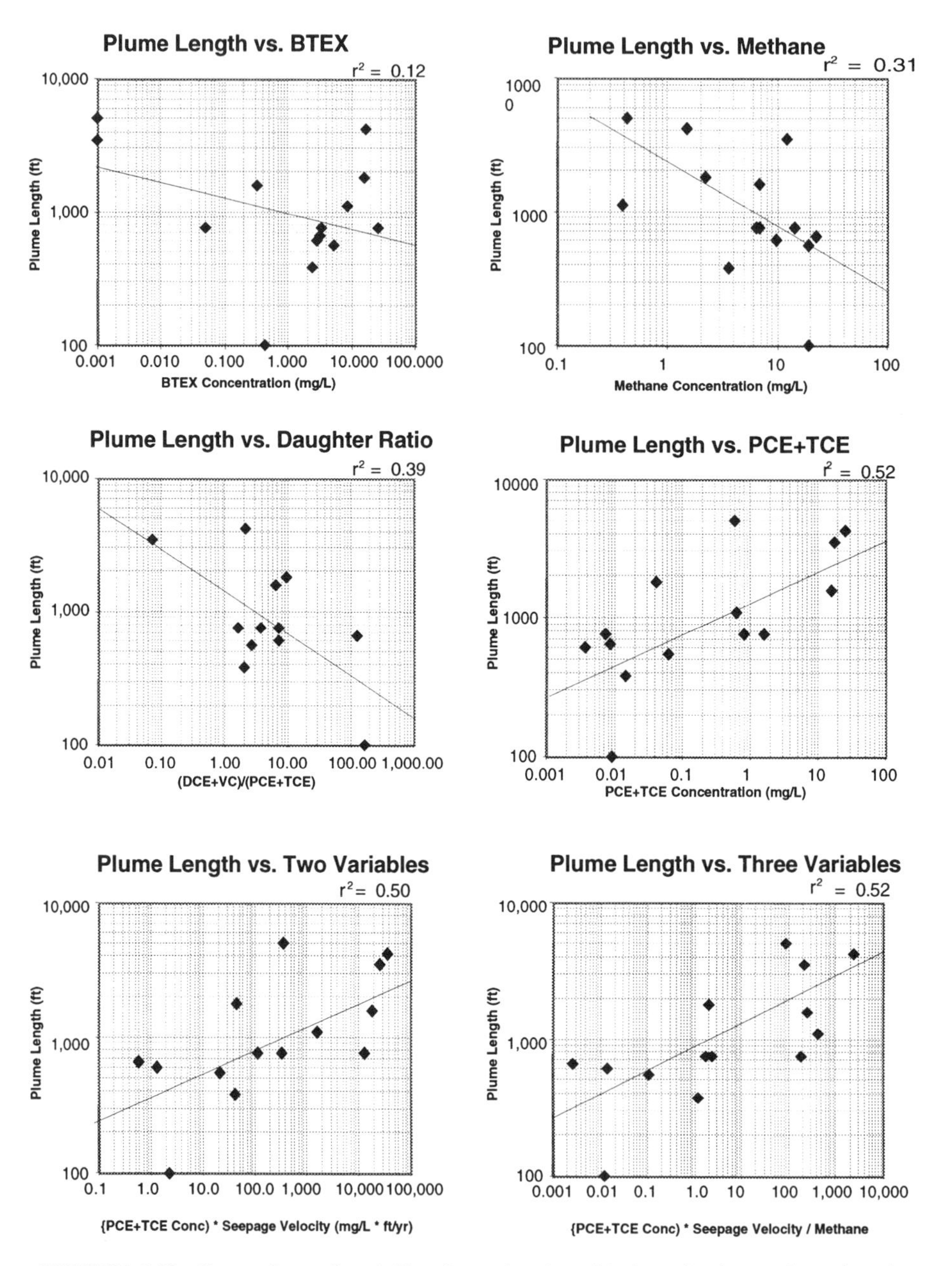

FIGURE 6.19. Comparison of variables for estimating chlorinated solvent plume length. Note that the correlation coefficient (r^2) was performed on the logarithms of the two variables for each graph. [Data from 17 U.S. Air Force sites analyzed as part of the Air Force Center for Environmental Excellence Natural Attentuation Initiative (Wiedemeier et al., 1996a).]

methane may be possible at or near these levels. (*Note:* If methane were bubbled, the hydrogen/methane accounting discussed above would need to be adjusted.)

Six simple models were used to try to predict chlorinated solvent plume length (Figure 6.19). Again, the BTEX concentration proved to be a poor indicator of plume length, as the regression yielded a r^2 of 0.12. Methane concentration and the daughter ratio proved to be better predictors, exhibiting r^2 values of 0.42 and 0.39, respectively. The best correlations were found with parent compound concentrations (PCE + TCE) (r^2 of 0.52) and a two- and three-variable model. The two-variable model incorporated both a source characteristic (PCE + TCE) and a hydrogeologic variable (average seepage velocity) and yielded a r^2 value of 0.50 (note that the resulting product does not represent anything physically, but the two variables reflect both source and hydrogeologic characteristics). A three-variable model accounted for source and hydrogeologic characteristics together with a biodegradation indicator, methane. The first two variables were multiplied together, as they are associated with longer plumes, and divided by the methane concentration, in which higher methane concentrations indicate more biodegradation and shorter plumes. The three-variable model yielded a r^2 value of only 0.54, a slight improvement over the one- and two-variable models (Figure 6.19).

Due to the uncertainties in the data, such as whether the plume has reached its ultimate length and scatter caused by the relatively poor resolution of groundwater monitoring systems, relatively low r^2 values are expected. Nevertheless, the regressions may provide some insight regarding general plume behavior.

REFERENCES

Adriaens, P., and Vogel, T. M., 1995, Biological treatment of chlorinated organics, in *Microbial Transformation and Degradation of Toxic Organic Chemicals*, L. Y. Young and C. E. Cerniglia, eds., Wiley-Liss, New York, 654 pp.

Aronson, D., and Howard, P. H., 1997, *Anaerobic Biodegradation of Organic Chemicals in Groundwater: A Summary of Field and Laboratory Studies*, American Petroleum Institute, Washington DC.

Baek, N. H., and Jaffe, P. R., 1989, The degradation of trichloroethylene in mixed methanogenic cultures, *J. Environ. Qual.*, vol. 18, pp. 515–518.

Ballapragada, B. S., Stensel, H. D., Puhakka, J. A., and Ferguson, J. F., 1997, Effect of hydrogen on reductive dechlorination of chlorinated ethenes, *Environ. Sci. Technol.*, vol. 31, no. 6, pp. 1728–1734.

Barbee, G. C., 1994, Fate of chlorinated aliphatic hydrocarbons in the vadose zone and ground water, *Ground Water Monitor. Remed.*, vol. 14, no. 1, pp. 129–140.

Barrio-Lage, G., Parsons, F. Z., Nassar, R. S., and Lorenzo, P. A., 1987, Biotransformation of trichloroethene in a variety of subsurface materials, *Environ. Toxicol. Chem.*, vol. 6, pp. 571–578.

Beak International, 1992, personal communication with L. Lehmicke, Kirkland, WA.

Benker, E., Davis, G. B., Appleyard, S., Berry, D. A., and Power, T. R., 1994, Groundwater contamination by trichloroethene (TCE) in a residential area of Perth: distribution, mobility, and implications for management, in *Proceedings of Water Down Under '94, 25th Congress of IAH*, Adelaide, South Australia, Australia, Nov.

Berry, D. F., Francis, A. J., and Bollag, J. M., 1987, Microbial metabolism of homocyclic and heterocyclic aromatic compounds under anaerobic conditions, *Microbiol. Rev.*, vol. 51, no. 1, pp. 43–59.

Bouwer, E. J., 1994, Bioremediation of chlorinated solvents using alternate electron acceptors, in *Handbook of Bioremediation*, R. D. Norris et al., eds., Lewis Publishers, Boca Raton, FL, pp. 149–175.

Bouwer, E. J., and McCarty, P. L., 1983, Transformations of 1- and 2-carbon halogenated aliphatic organic compounds under methanogenic conditions, *Appl. Environ. Microbiol.*, vol. 45, no. 4, pp. 1286–1294.

Bouwer, E. J., and Wright, J. P., 1988, Transformations of trace aliphatics in anoxic biofilm columns, *J. Contam. Hydrol.*, vol. 2, pp. 155–169.

Bouwer, E. J., Rittman, B. E., and McCarty, P. L., 1981, Anaerobic degradation of halogenated 1- and 2-carbon organic compounds, *Environ. Sci. Technol.*, vol. 15, no. 5, pp. 596–599.

Bradley, P. M., and Chapelle, F. H., 1996, Anaerobic mineralization of vinyl chloride in Fe(III)-reducing aquifer sediments, *Environ. Sci. Technol.*, vol. 30, pp. 2084–2086.

———, 1997, Kinetics of DCE and VC mineralization under methanogenic and Fe(III)-reducing conditions, *Environ. Sci. Technol.*, vol. 31, pp. 2692–2696.

Bradley, P. M., and Chapelle, F. H., 1998, Effect of contaminant concentration on aerobic microbial mineralization of DCE and VC in stream-bed sediments, *Environ. Sci. Technol.*, vol. 30, no. 5, pp. 553–557.

Buscheck, T. E., and Alcantar, C. M., 1995, Regression techniques and analytical solutions to demonstrate intrinsic bioremediation, in *Proceedings of the 1995 Battelle International Conference on In Situ and On Site Bioreclamation*, San Diego, CA, Apr., Battelle Press, Columbus, OH.

Carr, C., and Hughes, J. B., 1998, High-rate dechlorination of PCE: comparison of lactate, methanol and hydrogen as electron donors, *Environ. Sci. Technol.*, vol. 30, no. 12, pp. 1817–1824.

Chapelle, F. H., 1993, *Groundwater Microbiology and Geochemistry*, Wiley, New York, 424 pp.

Chapelle, F. H., and Lovley, D. R., 1992, Competitive exclusion of sulfate-reduction by Fe(III)-reducing bacteria: a mechanism for producing discrete zones of high-iron ground water, *Ground Water*, vol. 30, no. 1, pp. 29–36.

Chapelle, F. H., McMahon, P. B., Dubrovsky, N. M., Fujii, R. F., Oaksford, E. T., and Vroblesky, D. A., 1995, Deducing the distribution of terminal electron-accepting processes in hydrologically diverse groundwater systems, *Water Resour. Res.*, vol. 31, no. 2, pp. 359–371.

Cox, E., Edwards, E., Lehmicke, L., and Major, D., 1995, Intrinsic biodegradation of trichloroethylene and trichloroethane in a sequential anaerobic–aerobic aquifer, in *Intrinsic Bioremediation*, R. E. Hinchee, J. T. Wilson, and D. C. Downey, eds., Battelle Press, Columbus, OH, pp. 223–231.

Davis, J. W., and Carpenter, C. L., 1990, Aerobic biodegradation of vinyl chloride in groundwater samples, *Appl. Environ. Microbiol.*, vol. 56, pp. 3878.

de Bont, J. A. M., Vorage, M. J. W., Hartmans, S., and van den Tweel, W. J. J., 1986, Microbial degradation of 1,3-dichlorobenzene, *Appl. Environ. Microbiol.*, vol. 52, pp. 677–680.

Dupont, R. R., Gorder, K., Sorenson, D. L., Kemblowski, M. W., and Haas, P., 1996, Case Study: Eielson Air Force Base, Alaska, in *Proceedings of the Symposium on Natural Attenuation of Chlorinated Organics in Ground Water*, Dallas, TX: EPA/540/R-96/509, September 1996.

Ehlke, T. A., Wilson, B. H., Wilson, J. T., and Imbrigiotta, T. E., 1994, In-situ biotransformation of trichloroethylene and *cis*-1,2-dichloroethylene at Picatinny Arsenal, New Jersey, in

Proceedings of the U.S. Geological Survey Toxic Substances Program, Colorado Springs, CO, D. W. Morganwalp, and D. A., Aranson, eds., Water Resources Investigation Report 94–4014, U.S. Geological Survey, Washington, DC.

Ellis, D. E., Lutz, E. J., Klecka, G. M., Pardieck, D. L., Salvo, J. J., Heitkamp, M. A., Gannon, D. J., Mikula, C. C., Vogel, C. M., Sayles, G. D., Kampbell, D. H., Wilson, J. T., and Maiers, D. T., 1996, Remediation technology development forum intrinsic remediation project at Dover Air Force Base, Delaware, in *Proceedings of the Symposium on Natural Attenuation of Chlorinated Organics in Ground Water*, Dallas, TX, Sept. 11–13, EPA/540/R-96/509, U.S. EPA, Washington, DC.

Fathepure, B. Z., and Vogel, T. M., 1991, Complete biodegradation of polychlorinated hydrocarbons by a two-stage biofilm reactor, *Appl. Environ. Microbiol.*, vol. 57, pp. 3418–3422.

Fennell, D. E., Gossett, J. M., and Zindler, S. H., 1997, Comparison of butyric acid, ethanol, lactic acid, and propionic acid as hydrogen donors for the reductive dechlorination of tetrachloroethene, *Environ. Sci. Technol.*, vol. 31, no. 3, pp. 918–926.

Fogel, M. M., Taddeo, A. R., and Fogel, S., 1986, Biodegradation of chlorinated ethenes by a methane-utilizing mixed culture, *Appl. Environ. Microbiol.*, vol. 51, no. 4, pp. 720–724.

Freedman, D. L., and Gossett, J. M., 1989, Biological reductive dehalogenation of tetrachloroethylene and trichloroethylene to ethylene under methanogenic conditions, *Appl. Environ. Microbiol.*, vol. 55, no. 4, pp. 1009–1014.

Gibson, S. A., and Sewell, G. W., 1990, Stimulation of the reductive dechlorination of tetrachloroethene in aquifer slurries by addition of short-chain fatty acids, in *Abstracts of the Annual Meeting of the American Society for Microbiology*, Anaheim, CA, May 14–18.

Gorder, K. A., Dupont, R. R., Sorenson, D. L., Kemblowski, M. W., and McLean, J. E., 1996, Analysis of intrinsic bioremediation of trichloroethene-contaminated ground water at Eielson Air Force Base, Alaska, in *Proceedings of the Symposium on Natural Attenuation of Chlorinated Organics in Ground Water*, Dallas, TX, Sept. 11–13, EPA/540/R-96/509, U.S. EPA, Washington, DC.

Gossett, J. M., and Zinder, S. H., 1996, Microbiological aspects relevant to natural attenuation of chlorinated ethenes, in *Proceedings of the Symposium on Natural Attenuation of Chlorinated Organics in Groundwater*, Dallas, TX, Sept. 11–13, EPA/540/R-96/509, U.S. EPA, Washington, DC.

Haigler, B. E., Nishino, S. F., and Spain, J. C., 1988, Degradation of 1,2-dichlorobenzene by a *Pseudomonas sp.*, *Appl. Environ. Microbiol.*, vol. 54, pp. 294–301.

Hartmans, S., and de Bont, J. A. M., 1992, Aerobic vinyl chloride metabolism in *Mycobacterium aurum li*, *Appl. Environ. Microbiol.*, vol. 58, no. 4, pp. 1220–1226.

Hartmans, S., de Bont, J. A. M, Tamper, J. and Luyben, K. Ch. A. M., 1985, Bacterial degradation of vinyl chloride, *Biotechnol. Lett.*, vol. 7, no. 6, pp. 383–388.

Harwood, C. S., and Gibson, J, 1997, Shedding light on anaerobic benzene ring degradation: a process unique to prokaryotes? *J. Bacteriol.*, vol. 179, no. 2, pp. 301–309.

Haston, Z. C., Sharma, P. K., Black, J. N. P., and McCarty, P. L., 1994, Enhanced reductive dechlorination of chlorinated ethenes, in *Proceedings of the Symposium on Bioremediation of Hazardous Wastes: Research, Development, and Field Evaluations*, Dallas, TX, May 4–6, EPA/600/R-94/075, U.S. EPA, Washington, DC.

Henson, J. M., Yates, M. V., and Cochran, J. W., 1989, Metabolism of chlorinated methanes, ethanes, and ethylenes by a mixed bacterial culture growing off methane, *J. Ind. Microbiol.*, vol. 4., pp. 29–35.

Holliger, C., and Schumacher, W., 1994, Reductive dehalogenation as a respiratory process, *Antonie Leeuwenhoek*, vol. 66, pp. 239–246.

Holliger, C., Schraa, G., Stams, A. J. M., and Zehnder, A. J. B., 1992, Enrichment and properties of an anaerobic mixed culture reductively dechlorinating 1,2,3-trichlorobenzene to 1,3-dichlorobenzene, *Appl. Environ. Microbiol.*, vol. 58, pp. 1636–1644.

———, 1993, A highly purified enrichment culture couples the reductive dechlorination of tetrachloroethene to growth, *Appl. Environ. Microbiol.*, vol. 59, pp. 2991–2997.

Hughes, J. B., Newell, C. J., and Fisher, R. T., 1997, Process for in-situ biodegradation of chlorinated aliphatic hydrocarbons by subsurface hydrogen injection, U.S. patent 5,602,296, Feb. 11.

Janssen, D. B., Scheper, A., Dijkhuizen, L., and Witholt, B., 1985, Degradation of halogenated aliphatic compounds by *Xanthobacter autotrophicus* GJ10, *Appl. Environ. Microbiol.*, vol. 49, no. 3, pp. 673–677.

Kazumi, J., Caldwell, M. E., Suflita, J. M., Lovley, D. R., and Young, L. Y., 1997, Anaerobic degradation of benzene in diverse anoxic environments, *Environ. Sci. Technol.*, vol. 31, no. 3, pp. 813–818.

Klecka, G. M., Gonsior, S. J., and Markham, D. A., 1990, Biological transformations of 1,1,1-trichloroethane in subsurface soils and ground water, *Environ. Toxicol. Chem.*, vol. 9, pp. 1437–1451.

Kleopfer, R. D., Easley, D. M., Hass, B. B., Jr., and Deihl, T. G., 1985, Anaerobic degradation of trichloroethylene in soil, *Environ. Sci. Technol.*, vol. 19, pp. 277–280.

Krumholz, L. R., 1995, A new anaerobe that grows with tetrachloroethylene as an electron acceptor, in *Abstract's of the 95th General Meeting of*, American Society for Microbiology, Washington, DC.

Lee, M. D., Mazierski, P. F., Buchanan, R. J., Jr., Ellis, D. E., and Sehayek, L. S., 1995, Intrinsic and in situ anaerobic biodegradation of chlorinated solvents at an industrial landfill, in *Intrinsic Bioremediation*, R. E. Hinchee, J. T. Wilson, and D. C. Downey, eds., Battelle Press, Columbus, OH, pp. 205–222.

Lovley, D. R., and Goodwin, S., 1988, Hydrogen concentrations as an indicator of the predominant terminal electron-accepting reactions in aquatic sediments, *Geochim. Cosmochim. Acta*, vol. 52, pp. 2993–3003.

Lovley, D. R., Chapelle, F. H., and Woodward, J. C., 1994, Use of dissolved H_2 concentrations to determine distribution of microbially catalyzed redox reactions in anoxic groundwater, *Environ. Sci. Technol.*, vol. 28, no. 7, pp. 1205–1210.

Lyman, W. J., Reehl, W. F., and Rosenblatt, R. H., 1990, *Handbook of Chemical Property Estimation Methods*, American Chemical Society, Washington, DC.

Maymo-Gatell, X., Chien, Y., Gossett, J. M., and Zinder, S. H., 1997, Isolation of a bacterium that reductively dechlorinates tetrachloroethene to ethene, *Science*, vol. 276, pp. 1568–1571.

McCarty, P. L., 1969, Energentics and bacterial growth, Fifth Annual Rudolph Research Conference, Rutgers The State University, New Brunswick, NJ, July 2, 1969.

———, 1971, Energetics and bacterial growth, in *Organic Compounds in Aquatic Environmnets*, S. D. Faust and J. V. Hunter, eds., Marcel Dekker, New York, pp. 495–531.

———, 1972, Energetics of organic matter degradation, in *Water Pollution Microbiology*, R. Mitchell, ed., Wiley-Interscience, New York, pp. 91–118.

———, 1994, An overview of anaerobic transformation of chlorinated solvents, in *Symposium on Intrinsic Bioremediation in Groundwater*, Denver, CO, Aug. 30–Sept. 1, pp. 135–142.

———, 1997, personal communication.

McCarty, P. L., and Semprini, L., 1994, Groundwater treatment for chlorinated solvents, in *Handbook of Bioremediation*, R. D. Norris et al., eds., Lewis Publishers, Boca Raton, FL.

Miller, R. E., and Guengerich, F. P., 1982, Oxidation of trichloroethylene by liver microsomal cytochrome P-450: evidence for chlorine migration in a transition state not involving trichloroethylene oxide, *Biochemistry*, vol. 21, pp. 1090–1097.

Murray, W. D., and Richardson, M., 1993, Progress toward the biological treatment of C_1 and C_2 halogenated hydrocarbons, *Crit. Rev. Environ. Sci. Technol.*, vol. 23, no. 3, pp. 195–217.

Nelson, M. J. K., Montgomery, S. O., O'Neill, E. J., and Pritchard, P. H., 1986, Aerobic metabolism of trichloroethylene by a bacterial isolate, *Appl. Environ. Microbiol.*, vol. 52, no. 2, pp. 383–384.

Neumann, A., Scholz-Muramatsu, H., and Diekert, G., 1994, Tetrachloroethene metabolism of *Dehalorespirillum multivorans*, *Arch. Microbiol.*, vol. 162, pp. 295–301.

Neumann, A., Wohlfarth, G., and Diekert, G., 1996, Purification and characterization of tetrachloroethene reductive dehalogenase from *Dehalorespirillum multivorans*, *J. Biol. Chem.*, vol. 271, no. 28, pp. 16515–16519.

Newell, C. J., 1998. *Direct Hydrogen Addition for the In-Situ Biodegradation of Chlorinated Solvents*, Groundwater Services, Inc., Houston, TX, www.gsi-net.com.

Newell, C. J., Hopkins, L. P., and Bedient, P. B., 1990, A hydrogeologic database for groundwater modeling, *Ground Water*, vol. 28, no. 5, pp. 703–714.

Newell, C. J., Fisher, R. T., and Hughes, J. B., 1997, Direct hydrogen addition for the in-situ biodegradation of chlorinated solvents, in *Proceedings of the Petroleum Hydrocarbons and Organic Chemicals in Ground Water Conference*, NGWA, Houston, TX, Nov.

Nishino, S. F., Spain, J. C., and Pettigrew, C. A., 1994, Biodegradation of chlorobenzene by indigeneous bacteria, *Environ. Toxicol. Chem.*, vol. 13, pp. 871–877.

Parsons, F., and Lage, G. B., 1985, Chlorinated organics in simulated groundwater environments, *J. Am. Water Works Assoc.*, vol. 77, no. 5, pp. 52–59.

Parsons, F., Wood, P. R., and DeMarco. J., 1984, Transformations of tetrachloroethene and trichloroethene in microcosms and groundwater, *J. Am. Water Works Assoc.*, vol. 76, pp. 56–59.

Parsons, F., Barrio-Lage, G., and Rice, R., 1985, Biotransformation of chlorinated organic solvents in static microcosms, *Environ. Toxicol. Chem.*, vol. 4, pp. 739–742.

Ramanand, K., Balba, M. T., and Duffy, J., 1993, Reductive dehalogenation of chlorinated benzenes and toluenes under methanogenic conditions, *Appl. Environ. Microbiol.*, vol. 59, pp. 3266–3272.

Reineke, W., and Knackmuss, H.-J., 1984, Microbial metabolism of haloaromatics: isolation and properties of a chlorobenzene-degrading bacterium, *Eur. J. Appl. Micriobiol. Biotechnol.*, vol. 47, pp. 395–402.

Remediation Technologies Development Forum, 1997, Natural attenuaton of chlorinated solvents in groundwater seminar, Class Notes, RTDF (contacts: Dr. Leo Lehmicke, Beak Consultants, Seattle, WA; Dr. Ron Buchanan, DuPont, Wilmington, DE).

Roberts, P. V., Schreiner, J., and Hopkins, G. D., 1982, Field study of organic water quality changes during groundwater recharge in the Palo Alto baylands, *Water Res.*, vol. 16, pp. 1025–1035.

Sander, P., Wittaich, R.-M., Fortnagel, P., Wilkes, H., and Francke, W., 1991, Degradation of 1,2,4-trichloro- and 1,2,4,5-tetrachlorobenzene by *Pseudomonas* strains, *Appl. Environ. Microbiol.*, vol. 57, pp. 1430–1440.

Scholz-Muramatsu, H., Szewzyk, R., Szewzyk, U., and Gaiser, S., 1990, Tetrachloroethylene as electron acceptor for the anaerobic degradation of benzoate, *FEMS Microbiol. Lett.*, vol. 66, pp. 81–86.

Schraa, G., Boone, M. L., Jetten, M. S. M., van Neerven, A. R. W., Colberg, P. J., and Zehnder, A. J. B., 1986, Degradation of 1,2-dichlorobenzene by *Alcaligenes* sp. strain A175, *Appl. Environ. Microbiol.*, vol. 52, pp. 1374–1381.

Sewell, G. W., and Gibson, S. A., 1991, Stimulation of the reductive dechlorination of tetrachloroethene in anaerobic aquifer microcosms by the addition of toluene, *Environ. Sci. Technol.*, vol. 25, no. 5, pp. 982–984.

Sewell, G. W., Wilson, B. H., Wilson, J. T., Kampbell, D. H., and Gibson, S. A., 1991, Reductive dechlorination of tetrachloroethene and trichloroethene in fuel spill plumes, in *Chemical and Biochemical Detoxification of Hazardous Waste*, Vol. II, J. A. Glaser, ed., Lewis Publishers, Boca Raton, FL.

Sharma, P. K., and McCarty, P. L., 1996, Isolation and characterization of a facultatively aerobic bacterium that reductively dehalogenates tetrachloroethene to *cis*-1,2-dichloroethene, *Appl. Environ. Microbiol.*, vol. 62, pp. 761–765.

Smatlak, C. R., Gossett, J. M., and Zinder, S. H., 1996, Comparative kinetics of hydrogen utilization for reductive dechlorination of tetrachloroethene and methanogenesis in an anaerobic enrichment culture, *Environ. Sci. Technol.*, vol. 30, pp. 2850–2858.

Spain, J. C., 1996, Future vision: compounds with potential for natural attenuation, in *Proceedings of the Symposium on Natural Attenuation of Chlorinated Organics in Groundwater*, Dallas, TX, Sept. 11–13, EPA/540/R-96/509, U.S. EPA, Washington, DC.

Spain, J. C., and Nishino, S. F., 1987, Degradation of 1,4-dichlorobenzene by a *Pseudomonas* sp., *Appl. Environ. Microbiol.*, vol. 53, pp. 1010–1019.

Speece, R. E., 1996, *Anaerobic Biotechnology*, Archae Press, Nashville, TN.

Stucki, G., Krebser, U., and Leisinger, T., 1983, Bacterial growth on 1,2-dichloroethane, *Experientia*, vol. 39, pp. 1271–1273.

Suflita, J. M., and Townsend, G. T., 1995, The microbial ecology and physiology of aryl dehalogenation reactions and implications for bioremediation, in *Microbial Transformation and Degradation of Toxic Organic Chemicals*, L. Y. Young, and C. E. Cerniglia, eds., Wiley-Liss, New York, 654 pp.

Swanson, M., Wiedemeier, T. H., Moutoux, D. E., Kampbell, D. H., and Hansen, J. E., 1996, Patterns of natural attenuation of chlorinated aliphatic hydrocarbons at Cape Canaveral Air Station, Florida, in *Proceedings of the Symposium on Natural Attenuation of Chlorinated Organics in Ground Water*, Dallas, TX, Sept. 11–13, EPA/540/R-96/509, U.S. EPA, Washington, DC.

Tabak, H. H., Quave, S. A., Mashni, C. I., and Barth, E. F., 1981, Biodegradability studies with organic priority pollutant compounds, *J. Water Pollution Cont. Fed.*, vol. 53, pp. 1503–1518.

van der Meer, J. R., Roelofsen, W., Schraa, G., and Zehnder, A. J. B., 1987, Degradation of low concentrations of dichlorobenzenes and 1,2,4-trichlorobenzene by *Pseudomonas* sp. strain P51 in nonsterile soil columns, *FEMS Microbiol. Lett.*, vol. 45, pp. 333–341.

Vogel, T. M., 1994, Natural bioremediation of chlorinated solvents, in *Handbook of Bioremediation*, R. D. Norris et al., eds., Lewis Publishers, Boca Raton, FL.

Vogel, T. M., and McCarty, P. L., 1985, Biotransformation of tetrachloroethylene to trichloroethylene, dichloroethylene, vinyl chloride, and carbon dioxide under methanogenic conditions, *Appl. Environ. Microbiol.*, vol. 49, no. 5, pp. 1080–1083.

———, 1987, Abiotic and biotic transformations of 1,1,1-trichloroethane under methanogenic conditions, *Environ. Sci. Technol.*, vol. 21, no. 12, pp. 1208–1213.

Vogel, T. M., Criddle, C. S., and McCarty, P. L., 1987, Transformations of halogenated aliphatic compounds, *Environ. Sci. Technol.*, vol. 21, no. 8, pp. 722–736.

Vroblesky, D. A., and Chapelle, F. H., 1994, Temporal and spatial changes of terminal electron-accepting processes in a petroleum hydrocarbon-contaminated aquifer and the significance for contaminant biodegradation, *Water Resour. Res.*, vol. 30, no. 5, pp. 1561–1570.

Weaver, J. W., Wilson, J. T., and Kampbell, D. H., 1995, *EPA Project Summary*, EPA/600/SV-95/001, U.S. EPA, Washington, DC.

———, 1996, Extraction of degradation rate constants from the St. Joseph, Michigan, trichloroethene site, in *Proceedings of the Symposium on Natural Attenuation of Chlorinated Organics in Ground Water*, Dallas, TX, Sept. 11–13, EPA/540/R-96/509, U.S. EPA, Washington, DC.

Weiner, J. M., and Lovley, D. R., 1998, Rapid benzene degradation in methanogenic sediments from a petroleum-contaminated aquifer, *Appl. Environ. Microbiol.*, vol. 64, no. 5, pp. 1937–1939.

Wiedemeier, T. H., Swanson M. A., Moutoux D. E., Wilson, J. T., Kambell, D. H., Hansen, J. E., and Haas, P., 1996a, Overview of the technical protocol for natural attenuation of chlorinaed aliphatic hydrocarbons in groundwater under development for the U.S. Air Force Center for Environmental Excellence, *Proceedings of the Symposium on Natural Attenuation of Chlorinated Solvents*, EPA/540/R-96/509, U.S. EPA, Washington, DC, pp. 35–59.

Wiedemeier, T. H., Wilson, J. T., Kambell, D. H., 1996b, Natural attenuation of chlorinated aliphatic hydrocarbons at Plattsburgh Air Force Base, New York, in *Proceedings of the Symposium on Natural Attenuation of Chlorinated Organics in Groundwater*, Dallas, TX, Sept. 11–13, EPA /540/R-96/509, U.S. EPA, Washington, DC.

Wiedemeier, T. H., Swanson, M. A., Moutoux, D. E., Gordon, E. K., Wilson, J. T., Wilson, B. H., Kampbell, D. H., Haas, P. E., Miller, R. N., Hansen, J. E., and Chapelle, F. H., 1998, *Technical Protocol for Evaluating Natural Attenuation of Chlorinated Solvents in Ground Water:* EPA/600/R-98/128, September 1998, U.S. EPA, Washington, DC, ftp://ftp.epa.gov .pub/ada/reports/protocol.pdf.

Wilson, B. H., Smith, G. B., and Rees, J. F., 1986, Biotransformations of selected alkylbenzenes and halogenated aliphatic hydrocarbons in methanogenic aquifer material: A microcosm study, *Environ. Sci. Technol.*, vol. 20, pp. 997–1002.

Wilson, B. H., Wilson, J. T., Kampbell, D. H., Bledsoe, B. E., and Armstrong, J. M., 1990, Biotransformation of monoaromatic and chlorinated hydrocarbons at an aviation gasoline spill site, *Geomicrobiol. J.*, vol. 8, pp. 225–240.

Wilson, B. H., Ehlke, T. A., Imbrigiotta, T. E., and Wilson, J. T., 1991, Reductive dechlorination of trichloroethylene in anoxic aquifer material from Picatinny Arsenal, New Jersey, in *Proceedings of the U.S. Geological Survey Toxic Substances Hydrology Program*, G. E. Mallard, and D. A. Aronson, eds., Monterey, CA, Water Resources Investigation Report 91-4034, U.S. Geological Survey, Washington, DC, pp. 704–707.

Wilson, J. T., and Wilson, B. H., 1985, Biotransformation of trichloroethylene in soil, *Appl. Environ. Microbiol.*, vol. 49, no. 1, pp. 242–243.

Wilson, J. T., McNabb, J. F., Wilson, B. H., and Noonan, M. J., 1982, Biotransformation of selected organic pollutants in groundwater, *Dev. Ind. Microbiol.*, vol. 24, pp. 225–233.

Wilson, J. T., McNabb, J. F., Balkwill, D. L., and Ghiorse, W. C., 1983, Enumeration and characteristics of bacteria indigenous to a shallow water-table aquifer, *Ground Water*, vol. 21, pp. 134–142.

Wilson, J. T., Leach, L. E., Henson, M., and Jones, J. N., 1986, In situ biorestoration as a groundwater remediation technique, *Ground Water Monitor. Rev.*, Fall 1986, pp. 56–64.

CHAPTER 7

EVALUATING NATURAL ATTENUATION

In Chapters 3, 4, 5, and 6 we discussed the most important mechanisms of natural attenuation for petroleum hydrocarbons and chlorinated solvents in groundwater. In this chapter we present various lines of evidence that can be used to determine if natural attenuation is occurring. Because the dominant mechanism working to destroy dissolved organic compounds in the majority of groundwater systems is biodegradation, many of the techniques presented in this chapter are geared toward documenting the occurrence of intrinsic bioremediation. First we review the analytical data that are required to assess natural attenuation. We then discuss the lines of evidence used to demonstrate natural attenuation.

7.1 CHEMICAL AND GEOCHEMICAL ANALYTICAL DATA USEFUL FOR EVALUATING NATURAL ATTENUATION

Several types of chemical and geochemical data are useful for evaluating and quantifying natural attenuation. Tables 7.1 through 7.4 list those chemical and geochemical data and the data quality objectives that are useful for evaluating the natural attenuation of organic contaminants in the subsurface. It must be kept in mind that the data quality objectives listed in Tables 7.2 and 7.4 are only general guidelines that the authors have found useful and may be superseded by local regulations. These tables include the parameters necessary to delineate dissolved contamination and to document natural attenuation, including intrinsic bioremediation.

The analytes listed in these tables and their uses fall under several broad categories, including source term and sorption parameters, contaminants and daughter products, electron acceptors, metabolic by-products, and general water quality parameters. The analytes listed in these tables are useful for (1) estimating the composition and strength

TABLE 7.1. Soil–Sediment Analytical Parameters

Analysis	Data Use	Field or Fixed-Base Laboratory
Aromatic and chlorinated hydrocarbons [benzene, toluene, ethylbenzene, and xylene (BTEX); chlorinated compounds]	Data are used to determine the extent of soil contamination, the contaminant mass present, and the potential need for source removal.	Fixed-base
Total organic carbon (TOC)	The rate of migration of petroleum contaminants in groundwater is dependent on the amount of TOC in the aquifer matrix. High concentrations of organic carbon in the aquifer can support reductive dechlorination.	Fixed-base
Biologically available iron (III)	Optional method that should be used when petroleum hydrocarbons or vinyl chloride are present in the groundwater to predict the possible extent of removal of petroleum hydrocarbons and vinyl chloride via iron reduction. HCl extraction followed by quantification of released iron (III). Under development by the U.S. EPA, Robert S. Kerr Environmental Research Laboratory, Ada, Oklahoma.	Fixed-base
Aromatic and chlorinated hydrocarbons in NAPL	Useful for determining the composition and strength of the contaminant source.	Fixed-base

TABLE 7.2. Soil–Sediment Analytical Data Quality Objectives

Analysis	Minimum Limit of Quantification	Precision	Availability	Potential Data Quality Problems
Aromatic and chlorinated hydrocarbons [benzene, toluene, ethylbenzene, and xylene (BTEX); chlorinated compounds]	1 mg/kg	Coefficient of variation of 20%	Common laboratory analysis	Volatiles lost during shipment to laboratory; prefer extraction in the field
Total organic carbon (TOC)	0.1%	Coefficient of variation of 20%	Common laboratory analysis	Samples must be collected from contaminant-transporting (i.e., transmissive) intervals
Biologically available iron (III)	50 mg/kg	Coefficient of variation of 40%	Specialized laboratory analysis	Sample must not be allowed to oxidize
Aromatic and chlorinated hydrocarbons in NAPL	—	Coefficient of variation of 20%	Common laboratory analysis	Detection limit may be too high

TABLE 7.3. Groundwater Analytical Parameters Useful for Evaluating Natural Attenuation

Analysis	Data Use	Field or Fixed-Base Laboratory
Aromatic and chlorinated hydrocarbons (BTEX, trimethylbenzene isomers, chlorinated compounds)	Method of analysis for BTEX and chlorinated solvents/daughter products, which are the primary target analytes for monitoring natural attenuation; method can be extended to higher-molecular-weight alkylbenzenes; trimethylbenzenes are used to monitor plume dilution if degradation is primarily anaerobic.	Fixed-base
Dissolved oxygen	Concentration below about 0.5 mg/L generally indicate an anaerobic pathway. Measurements made with electrodes; results are displayed on a meter; protect samples from exposure to oxygen during sampling and analysis.	Field
Nitrate	Substrate for microbial respiration in the absence of oxygen.	Fixed-base
Mn(II)	Indication of Mn(IV) reduction during microbial degradation of organic compounds in the absence of dissolved oxygen and nitrate.	Field
Fe(II)	Indication of Fe(III) reduction during microbial degradation of organic compounds in the absence of dissolved oxygen, nitrate, and Mn(IV).	Field
Sulfate (SO_4^{2-})	Substrate for anaerobic microbial respiration.	Field
Hydrogen sulfide	Metabolic by-product of sulfate reduction. The presence of H_2S suggests organic carbon oxidation via sulfate reduction.	Field
Methane, ethane, and ethene	The presence of methane suggests organic carbon degradation via methanogenesis. Ethane and ethene data are used where chlorinated solvents are suspected of undergoing biological transformation. Kampbell et al. (1989) or SW3810 Modified.	Fixed-base
Carbon dioxide	Carbon dioxide is produced during the biodegradation of many types of organic carbon.	Field
Alkalinity	General water quality parameter used (1) to measure the buffering capacity of groundwater, and (2) as a marker to verify that all site samples are obtained from the same groundwater system.	Field
Oxidation–reduction potential (ORP)	The ORP of groundwater reflects the relative oxidizing or reducing nature of the groundwater system. ORP is influenced by the nature of the biologically mediated degradation of organic carbon; the ORP of groundwater may range from more than 800 mV to less than −400 mV. Measurements made with electrodes; results are displayed on a meter; protect samples from exposure to oxygen. Report results against the hydrogen electrode (Eh) by adding a correction factor specific to the electrode used.	Field
pH	Aerobic and anaerobic processes are pH sensitive.	Field
Temperature	Well development.	Field
Conductivity	General water quality parameter used as a marker to verify that site samples are obtained from the same groundwater system.	Field
Major cations	Can be used to evaluate other remedial actions.	Field

TABLE 7.3. *(Continued)*

Analysis	Data Use	Field or Fixed-Base Laboratory
Chloride	General water quality parameter used as a marker to verify that site samples are obtained from the same groundwater system. Final product of chlorinated solvent reduction.	Fixed-base
Total organic carbon	Used to classify plume and to determine if cometabolism is possible in the absence of anthropogenic carbon.	Fixed-base
Hydrogen (H_2)	Sampled at well head requires the production of 100 mL/min of water for 30 min. Equilibration with gas in the field. Determined with a reducing gas detector.	Field

TABLE 7.4. Groundwater Analytical Data Quality Objectives

Analysis	Minimum Limit of Quantification	Precision	Availability	Potential Data Quality Problems
Aromatic and chlorinated hydrocarbons (BTEX, trimethylbenzene isomers, chlorinated compounds)	MCLs	Coefficient of varation of 10%	Common laboratory analysis	Volatilization during shipment and biodegradation due to improper preservation.
Oxygen	0.2 mg/L	Standard deviation of 0.2 mg/L	Common field instrument	Improperly calibrated electrodes or bubbles behind the membrane or a fouled membrane or introduction of atmospheric oxygen during sampling.
Nitrate	0.1 mg/L	Standard deviation of 0.1 mg/L	Common laboratory analysis	Must be preserved.
Mn(II)	0.5 mg/L	Coefficient of variation of 20%	Common field analysis	Possible interference from turbidity (must filter if turbid). Keep out of sunlight and analyze within minutes of collection.
Fe(II)	0.5 mg/L	Coefficient of variation of 20%	Common field analysis	Possible interference from turbidity (must filter if turbid). Keep out of sunlight and analyze within minutes of collection.

continues

TABLE 7.4. *(Continued)*

Analysis	Minimum Limit of Quantification	Precision	Availability	Potential Data Quality Problems
Sulfate (SO_4^{-2})	5 mg/L	Coefficient of variation of 20%	Common laboratory or field analysis	Fixed-base.
Hydrogen sulfide (H_2S)	5 mg/L	Coefficient of variation of 20%	Common field analysis	Possible interference from turbidity (must filter if turbid). Keep sample cool.
Methane, ethane, and ethene	1 μg/L	Coefficient of variation of 20%	Specialized laboratory analysis	SOP must be thoroughly understood by the laboratory before sample submittal. Sample must be preserved against biodegradation and collected without headspace (to minimize volatilization).
Carbon dioxide	5 mg/L	Coefficient of variation of 20%	Common field analysis	—
Alkalinity	50 mg/L	Standard deviation of 20 mg/L	Common field analysis	Analyze sample within 1 hour of collection.
Oxidation-reduction potential (ORP)	± 300 mV	± 50 mV	Common field probe	Improperly calibrated electrodes or introduction of atmospheric oxygen during sampling.
pH	0.1 standard unit	0.1 standard unit	Common field meter	Improperly calibrated instrument; time sensitive.
Temperature	0°C	Standard deviation of 1°C	Common field probe	Improperly calibrated instrument; time sensitive.
Conductivity	50 μS/cm^2	Standard deviation of 50 μS/cm^2	Common field probe	Improperly calibrated instrument.
Major cations	1 mg/L	Coefficient of variation of 20%	Common laboratory analysis	Possible colloidal interferences.
Chloride	1 mg/L	Coefficient of variation of 20%	Common laboratory analysis	—
Chloride (optional; (see data use)	1 mg/L	Coefficient of variation of 20%	Common field analysis	Possible interference from turbidity.
Total organic carbon	0.1 mg/L	Coefficient of variation of 20%	Common laboratory analysis	—
Hydrogen (H_2)	0.1 nM	Standard deviation of 0.1 nM	Specialized field analysis	Numerous; see Appendix A.

of a NAPL source, (2) showing that natural attenuation is occurring, and (3) evaluating the relative importance of the various natural attenuation mechanisms. It should be kept in mind that it may not be necessary to collect all analytes for all sites. The final selection of analytes should be determined on a site-specific basis with the appropriate regulators and will depend on the availability of historical contaminant data, the complexity of the site, and the types of contaminants present. The use of individual analytes is discussed throughout the remainder of this chapter.

Source Term and Sorption Parameters

Mobile and residual NAPL or contaminants sorbed to the aquifer matrix can act as a continuing source of groundwater contamination. To calculate contaminant partitioning into groundwater it is necessary to estimate the location, distribution, concentration, and total mass of contaminants sorbed to soils or present as mobile or immobile NAPL. Obtaining and analyzing samples of mobile NAPL, if present, is an important component of site characterization. Non-aqueous-phase liquid in the subsurface, whether present at residual saturation or in quantities sufficient to cause formation of a mobile NAPL pool, acts as a continuing source of groundwater contamination; as long as NAPL remains in the subsurface at concentrations sufficient to affect groundwater, aqueous-phase contamination will persist. This has several implications for natural attenuation and is an important consideration when developing a model to simulate solute fate and transport. The degree of weathering of the NAPL, and hence its composition and strength, dictates the amount of aqueous-phase contamination at a site. Collection and analysis of NAPL allows better simulation of the influence of the source term in a solute transport model. In some cases, it may be possible to complete equilibrium partitioning calculations to show that the effective solubility of a compound is no longer high enough to affect groundwater at concentrations above regulatory guidelines. Another use of NAPL analyses is to determine if chlorinated compounds present in groundwater came from the NAPL or are daughter products. For example, at Plattsburgh Air Force Base, only petroleum hydrocarbons and trichloroethene (TCE) were detected in the pool of mobile LNAPL. Thus the dichloroethene (DCE) and vinyl chloride (VC) found in groundwater are probably daughter products produced by the reductive dechlorination of TCE. This information can be used to help document the occurrence of reductive dechlorination. The presence of ethene further supports the occurrence of reductive dechlorination.

Knowledge of the total organic carbon (TOC) content of the aquifer matrix is important for sorption and solute-retardation calculations. TOC samples should be collected from a background location in the stratigraphic horizon(s) where most contaminant transport is expected to occur. This will usually be the contaminated portion of the aquifer with the highest hydraulic conductivity.

Contaminants and Daughter Products

The concentrations of contaminants and associated daughter products (which may also be considered contaminants) considered in this book typically are quantified using volatile and semivolatile laboratory analyses. These analytes are used to determine the type, concentration, and distribution of organic contaminants in the aquifer. Method 8020, modified to include the trimethylbenzene isomers, is one method that can be used if site contamination is limited to petroleum hydrocarbons. If chlorinated solvents or mixed petroleum hydrocarbons and solvents are found at the site, the volatile organic compound (VOC) analysis (Method SW8260a) can be useful for evaluating contaminants and daughter products. Addition analyses or different analytical methods may be required, depending on applicable regulations.

The combined dissolved concentrations of BTEX and trimethylbenzenes should not be greater than about 30 mg/L for a JP-4 spill (Smith et al., 1981) or about 135 mg/L for a gasoline spill (American Petroleum Institute, 1985; Cline et al., 1991). If these compounds are found in higher concentrations, sampling errors such as emulsification of NAPL in the groundwater sample probably have occurred and should be investigated. Maximum concentrations of chlorinated solvents dissolved in groundwater from neat solvents should not exceed their solubilities in water. If chlorinated solvents are found in concentrations greater than their solubilities, sampling errors such as emulsification of NAPL in the groundwater sample should be investigated.

Electron Acceptors

Naturally occurring electron acceptors commonly used in microbial metabolism of organic contaminants include dissolved oxygen, nitrate, Mn(IV), Fe(III), sulfate, and in some cases (i.e., during methanogenesis), carbon dioxide. Measurement of these parameters is useful for evaluating the occurrence of intrinsic bioremediation and the relative importance of the various terminal electron-accepting processes. Although microbes utilize Fe(III) as the terminal electron acceptor during Fe(III) reduction, one of the metabolic by-products of this reaction, Fe(II) is measured to confirm the occurrence of Fe(III) reduction. This is because of the difficulty involved in measuring the concentration of biologically available Fe(III) in an aquifer system. A similar relationship exists with manganese. Mn(IV) is used as the electron acceptor and is reduced to Mn(II). Fe(II) and Mn(II) are discussed below.

Metabolic By-products

Readily measurable by-products of microbial metabolism in areas contaminated with organic compounds include Fe(II), carbon dioxide, hydrogen sulfide, methane, ethane, ethene, alkalinity, lowered oxidation–reduction potential, chloride, and hydrogen. There are additional metabolic by-products for which one can analyze but measurement of the parameters listed above has proved useful for evaluating the occurrence of intrinsic bioremediation and the relative importance of the various terminal electron-accepting processes.

General Water-Quality Parameters

Bacteria generally prefer environments with a neutral or slightly alkaline pH. The optimal pH range for most microorganisms is between 6 and 8 standard units; however, many microorganisms can tolerate a pH well outside this range. For example, pH values may be as low as 4.0 or 5.0 in aquifers where sulfides are being oxidized, and pH values as high as 9.0 may be found in carbonate-buffered systems (Chapelle, 1993). In addition, pH values as low as 3.0 have been measured for groundwater contaminated with municipal waste leachates which often contain elevated concentrations of organic acids (Baedecker and Back, 1979). In groundwater contaminated with sludges from cement manufacturing, pH values as high as 11.0 have been measured (Chapelle, 1993).

Groundwater temperature directly affects the solubility of oxygen and other geochemical species. For example, dissolved oxygen is more soluble in cold water than in warm water. Groundwater temperature also affects the metabolic activity of bacteria. Biological activity roughly doubles for every 10°C increase in temperature (Q_{10} rule) over the temperature range between 5 and 25°C.

Conductivity is a measure of the ability of a solution to conduct electricity. The conductivity of groundwater is directly related to the concentration of ions in solution; conductivity increases as ion concentration increases. Because the pH, temperature, and conductivity of a groundwater sample can change significantly within a short time following sample acquisition, these parameters must be measured in the field in unfiltered, unpreserved, "fresh" water collected by the same technique as the samples taken for dissolved oxygen and oxidation–reduction potential analyses. The measurements should be made in a clean glass container separate from those intended for laboratory analysis, and the measured values should be recorded in the groundwater sampling record.

General Groundwater Sampling Considerations

The groundwater sampling procedures presented in this section are important because the quality of several of the biogeochemical indicators used to evaluate intrinsic bioremediation can be affected significantly by poor sampling technique. Poor data quality can result in erroneous conclusions regarding the efficacy, or even the occurrence, of intrinsic bioremediation, so care must be taken during groundwater sample collection. Samples of groundwater can be collected from monitoring wells, monitoring points, or grab sampling locations. Monitoring wells are the most common groundwater sample collection points. Monitoring wells are extremely versatile and can be used for groundwater sampling, aquifer testing, product recovery systems, and long-term monitoring, among other things. Although versatile, monitoring wells are relatively expensive to install and create relatively large quantities of waste during installation, development, and sampling. Perhaps the biggest drawback to conventional monitoring wells is the relatively long screened intervals that typically are used. This tends to cause averaging of groundwater geochemistry over the screened interval. If contamination is limited to a thin portion of the screened interval, measured contaminant concentrations may be biased and groundwater geochemistry may not be representative of the contaminated interval.

Where site conditions and the regulatory environment permit, monitoring points are ideal tools for obtaining groundwater chemical data rapidly and cost-effectively.

As used herein, a monitoring point is a semipermanent small-diameter sampling device that is installed using a Geoprobe sampling device, a cone penetrometer, or using a hand-driven point. Such points can consist of a small section of stainless steel mesh screen or PVC. Monitoring points can be installed and sampled rapidly while generating a minimal volume of waste. In addition, they tend to have relatively short screened intervals that will minimize averaging of groundwater geochemistry over large intervals. Typically, monitoring points are installed in small-diameter boreholes using cone penetrometer testing (CPT), hydraulic percussion, or manually powered equipment.

Grab samples are temporally and spatially discrete samples. Several of the more common instruments used to collect groundwater grab samples include the HydroPunch, Geoprobe, cone penetrometer, or hand-driven points. The collection of grab samples can provide a rapid and cost-effective alternative to the use of monitoring points. Like monitoring points, grab samples generate minimal waste. Because the locations are abandoned upon completion of sampling, analytical results cannot be confirmed and grab sampling probably will not be a cost-effective way to conduct long-term monitoring. Another limitation of grab sampling is that groundwater levels at all sampling locations typically cannot be collected over the space of a few hours. This may make the development of accurate potentiometric surface maps difficult. In addition, if the aquifer is not particularly transmissive, sample collection can require hours, resulting in inefficient equipment utilization.

Because of the accuracy required for many of the analytical procedures listed in Table 7.3, care must be exercised when extracting groundwater from the sampling device. Varied equipment and methods are available for the extraction of groundwater. The approach used should be determined on the basis of application (development, purging, or sampling), hydrogeologic conditions, monitoring location dimensions, and regulatory requirements. Many of the analyses described herein require the use of a flow-through cell.

Portable groundwater extraction devices from three generic classifications are commonly used to collect groundwater samples: grab, suction lift, and positive displacement. The selection of the type of device(s) for the investigation is based on type of activity, well-point dimensions, hydrogeologic conditions, and regulatory considerations. A bailer is perhaps the most common sampling device. Bailers can be used at any depth in wells with a minimum diameter of about 0.5 in.; however, groundwater sample extraction becomes less efficient as the well diameter (and hence the bailer diameter) decreases. Disposable bailers can be used to avoid decontamination expenses and potential cross-contamination problems. Drawbacks of bailers include agitation and aeration of the groundwater and the inability to maintain the steady, nonturbulent flow required to establish a true flow-through cell. Agitation and aeration can be minimized, but not eliminated, through careful immersion into and extraction from the standing column of water in the well and point. Aeration can also be an issue during transfer of the sample from the bailer to the sample container. Once again, this aeration can be minimized but not eliminated. Because of aeration, accurate dissolved oxygen and oxidation–reduction potential measurements can be difficult to obtain.

The suction lift technology is best represented in environmental investigations by the peristaltic pump. A peristaltic pump extracts water using a vacuum created by cyclically advancing a sealed compression along flexible tubing. This pumping technique means that extracted water contacts nothing other than tubing, which can easily be replaced between sampling locations. This reduces the possibility of cross contamination. Furthermore, peristaltic pumps can be used to extract minimally disturbed groundwater from a monitoring location of any size at variable low-flow rates. Because of these features, representative samples are simple to collect, and reliable flow-through cells are simple to establish. The biggest drawback of sampling with a peristaltic pump is the maximum achievable pumping depth, which is equivalent to the height of water column that can be supported by an imperfect vacuum. This effectively limits the use of a peristaltic pump to monitoring locations with groundwater depths of less than approximately 25 ft, depending on the altitude of the site. Also, offgassing can occur in the tubing as a result of the reduced pressures and high rate of cyclical loading. If bubbles are observed in the tubing during purging or sampling, the flow rate of the peristaltic pump must be slowed. If bubbles are still apparent, the tubing should be checked for holes and replaced. The final potential disadvantage with a peristaltic pump is the low flow rate. Although advantageous for sampling, this can be inappropriate during purging or development at locations requiring large extraction volumes.

Positive-displacement pumps, also called submersible pumps, include for example, bladder, Keck, Grundfos Redi-Flo II, Bennett, and Enviro-Tech Purger ES pumps. Each of these pumps operates downhole at depths of up to a few hundred feet and rates of up to several gallons per minute. Therefore, submersible pumps are particularly useful for applications requiring the extraction of large volumes of water or for the extraction of groundwater from depths in excess of 25 ft. Because the pumps operate downhole, they require appropriately sized wells. At a minimum, an inside well diameter of at least 1.5 in. typically is required; however, much larger well diameters can be required, depending on the pump type, extraction depth, and extraction rate selected. Because typical submersible pump design results in contact between the groundwater and internal as well as external surfaces of the pump, rigorous decontamination and quality assurance procedures must be implemented to avoid cross contamination if a pump that is not dedicated to the well is used for sampling.

Purging consists of the evacuation of water from the monitoring location prior to sampling, so that "fresh" formation water will enter the monitoring location and be available for sampling. Because sampling can occur immediately upon completion of purging, it is best to limit groundwater agitation, and consequently, aeration of the groundwater and volatilization of contaminants. Two sources for agitation include the purging device and the cascading of water down the screen as the water level in the well drops. To avoid agitation, a low-disturbance device such as a peristaltic pump or bladder pump is recommended for purging, while equipment such as bailers should be avoided. To avoid aeration, wells and points screened below the water table should be pumped at a rate that prevents lowering of the water table to below the top of the screen, and if practical, wells and points screened across the water table should be pumped at a rate that lowers the total height of the water column no more than about 5%.

A flow-through cell, such as the simple one pictured in Figure 7.1, should be used for measurement of well-head parameters such as pH, temperature, specific conductance, dissolved oxygen, and oxidation–reduction potential. Dissolved oxygen measurements should be taken during well purging and immediately before and after sample acquisition using a direct-reading meter. Because most well purging techniques can allow aeration of collected groundwater samples, it is important to minimize potential aeration by taking the following precautions:

1. Use a peristaltic pump to purge the well when possible (depth to groundwater less than approximately 25 ft). To prevent downhole aeration of the sample in wells screened across the water table, well drawdown should not exceed about 5% of the height of the standing column of water in the well. The pump tubing should be immersed alongside the dissolved oxygen probe beneath the water level in the sampling container (Figure 7.1). This will minimize aeration and keep water flowing past the dissolved oxygen probe's sampling membrane. If bubbles are observed in the tubing during purging, the flow rate of the peristaltic pump must be slowed. If bubbles are still apparent, the tubing should be checked for holes and replaced.
2. When using a bailer, the bailer should be slowly immersed in the standing column of water in the well to minimize aeration. After sample collection, the water should be drained from the bottom of the bailer through tubing into

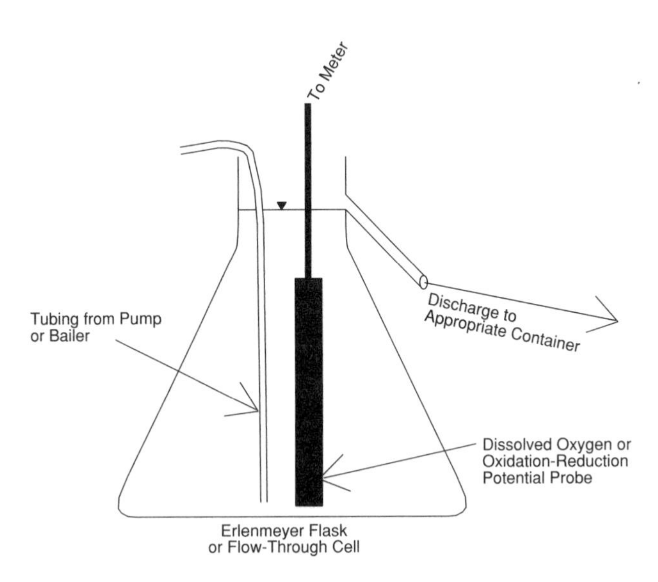

FIGURE 7.1. Suggested procedure for dissolved oxygen and oxidation–reduction potential sampling.

the sampling container. The tubing used for this operation should be immersed alongside the dissolved oxygen probe beneath the water level in the sampling container (Figure 7.1). This will minimize aeration and keep water flowing past the dissolved oxygen probe's sampling membrane.

3. Downhole dissolved oxygen probes can be used for dissolved oxygen analyses, but such probes must be decontaminated thoroughly between wells. In some cases decontamination procedures can be harmful to the dissolved oxygen probe.

Samples should be collected directly from the pump discharge tube or bailer into a sample container of appropriate size, style, and preservation for the desired analysis. Water should be directed down the inner walls of the sample bottle to minimize aeration of the sample. All samples to be analyzed for volatile constituents (e.g., SW8010, SW8020, SW8240, SW8260, and TPH-g) or dissolved gases (e.g., methane, ethane, and ethene) must be filled and sealed so that no air space remains in the container.

7.2 LINES OF EVIDENCE USED TO EVALUATE NATURAL ATTENUATION

Multiple distinct but converging lines of evidence in various forms have been used in recent years to evaluate natural attenuation (see, e.g., National Research Council, 1993; Wiedemeier et al., 1995, 1996a, 1998; U.S. EPA, 1997; ASTM, 1998; Wiedemeier and Chapelle, 1998). The most common lines of evidence used to evaluate natural attenuation of organic compounds dissolved in groundwater include:

1. Historical trends in contaminant data showing plume stabilization and/or loss of contaminant mass over time.
2. Analytical data showing that geochemical conditions are suitable for biodegradation and that active biodegradation has occurred as indicated by the consumption of electron acceptors and/or the production of metabolic by-products. This chemical and analytical data can include evidence of:
   a. Depletion of electron acceptors and donors
   b. Increasing metabolic by-product concentrations
   c. Decreasing parent compound concentrations
   d. Increasing daughter compound concentrations
3. Microbiological data that support the occurrence of biodegradation.

These lines of evidence are summarized in Table 7.5. In addition to these lines of evidence, analytical or numerical solute transport models can be used to examine the processes influencing the transport of organic contaminants in groundwater and to

TABLE 7.5. Summary of the Lines of Evidence Used to Evaluate Natural Attenuation

Line of Evidence	Data Requirements	Comments
Plume stabilization and/or loss of contaminant mass	Historical contaminant data	Can be evaluated using visual techniques or statistical techniques such as those presented by the EPA or the Mann–Whitney *U* test or the Mann–Kendall test. Historical database must be "statistically significant."
Analytical data confirming intrinsic bio-remediation	Data presented in Tables 7.1 and 7.3	One-, two-, and three-dimensional plots plots of contaminants, electron acceptors, and metabolic by-products in time and space. Things to look for include depletion of electron acceptors and donors, increasing metabolic by-product concentrations, decreasing parent compound concentrations, and increasing daughter compound concentrations.
Microbiological laboratory data	Microcosm studies, microbial cell counts, etc.	Should be used only to evaluate the conditions under which biodegradation processes occur or do not occur or when biodegradation rates constants cannot be obtained using other lines of evidence.
Modeling[a]	Model-specific	Useful for examining the relative importance of the processes influencing the transport of organic compounds.

[a]Not a line of evidence but helpful in evaluating natural attenuation.

estimate the influence of natural attenuation mechanisms on solute fate and transport. Solute transport models can also be valuable presentation tools.

The first line of evidence involves using historical contaminant data to show that the contaminant plume is shrinking, stable, or growing at a rate slower than predicted by conservative (slow) groundwater seepage velocity calculations. In some cases a biologically recalcitrant (conservative) tracer present in measurable concentrations within the contaminant plume can be used in conjunction with aquifer hydrogeologic parameters such as seepage velocity and dilution to show that a reduction in contaminant mass is occurring and to estimate biodegradation rate constants. Although the first line of evidence can be used to show that a contaminant plume is being attenuated, it does not necessarily show that contaminant mass is being destroyed.

When microorganisms degrade organic contaminants in the subsurface, they cause measurable changes in soil and groundwater chemistry. The second line of evidence involves documenting these changes and can be used to show that contaminant mass is being destroyed, not just being diluted or sorbed to the aquifer matrix. Chemical changes in groundwater can also be used to infer which contaminant degradation mechanisms are occurring at a site.

The third line of evidence, microbiological laboratory or field data, can be used to show that indigenous biota are capable of degrading site contaminants. The microcosm study is the most common technique used for this purpose. Microcosm studies should only be undertaken when they are absolutely necessary to obtain biodegradation rate estimates that could not be obtained using the other lines of evidence or when the specific mechanism of degradation is not known.

Although not a line of evidence, analytical or numerical solute fate and transport models can prove valuable when evaluating natural attenuation. For example, models can be used to evaluate the relative importance of natural attenuation mechanisms if sufficient historical site data are available. The dominant transport mechanisms affecting the transport of dissolved organic contaminants in many subsurface systems are advection, dispersion, sorption, and degradation (biotic or abiotic). Thus, observed changes in contaminant concentrations downgradient from the NAPL source area will represent the sum of the influences exerted by these processes. Solute transport models that incorporate these mechanisms can be valuable tools for estimating the relative importance of the various attenuation mechanisms. In addition, they can be used to show graphically that without the influence of degradation and mass removal, many contaminants would migrate some distance further than observed. For example, Figure 7.2 shows the observed extent of contamination at a site over the period from August 1993 through September 1995. Also shown on the figure are the results of an analytical model that incorporated advection, dispersion, and sorption only; biodegradation was assumed not to occur. From this figure and what we know about the behavior of petroleum hydrocarbons dissolved in groundwater, it can be inferred that the difference between the observed extent of contamination and the predicted extent of contamination is due largely to biodegradation. To help illustrate this concept further, consider Figure 7.3, which shows model predictions made using measured values of seepage velocity, dispersion, sorption, and biodegradation. Also shown on this figure are measured contaminant concentrations. Because model predictions that incorporate advection, dispersion, sorption, and biodegradation accurately simulate measured contaminant concentrations, it can be concluded that mechanisms of contaminant attenuation at this site are fairly well understood. In this case, the model confirms what is observed at the site and illustrates the relative importance of degradation mechanisms. Models can also be used to evaluate the accuracy of model input parameters. For example, if the sum of advection, dispersion, sorption, and degradation overestimate the loss of contaminant mass downgradient as compared to observed contaminant concentrations, it is likely that one or more model input parameters are incorrect.

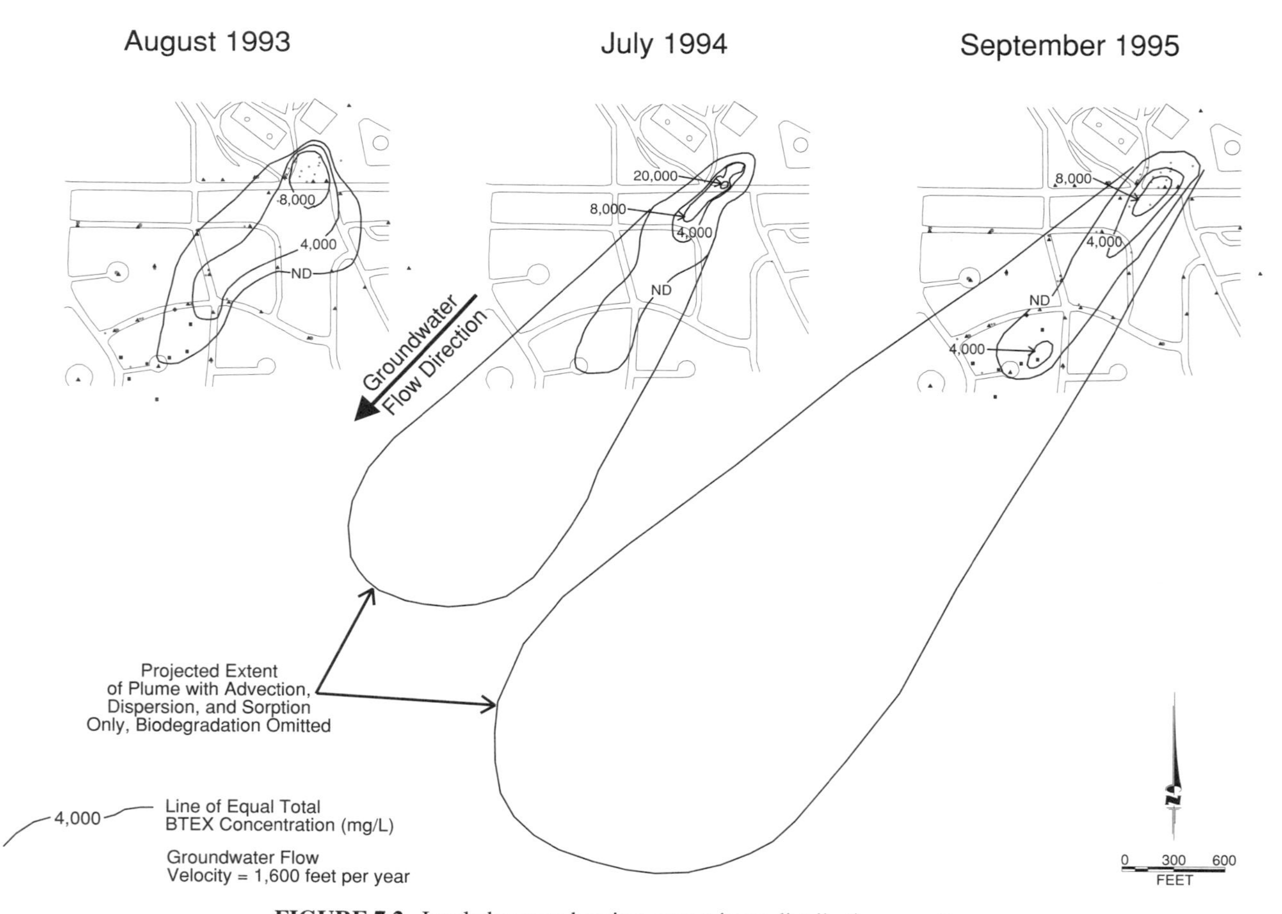

FIGURE 7.2. Isopleth maps showing contaminant distribution over time.

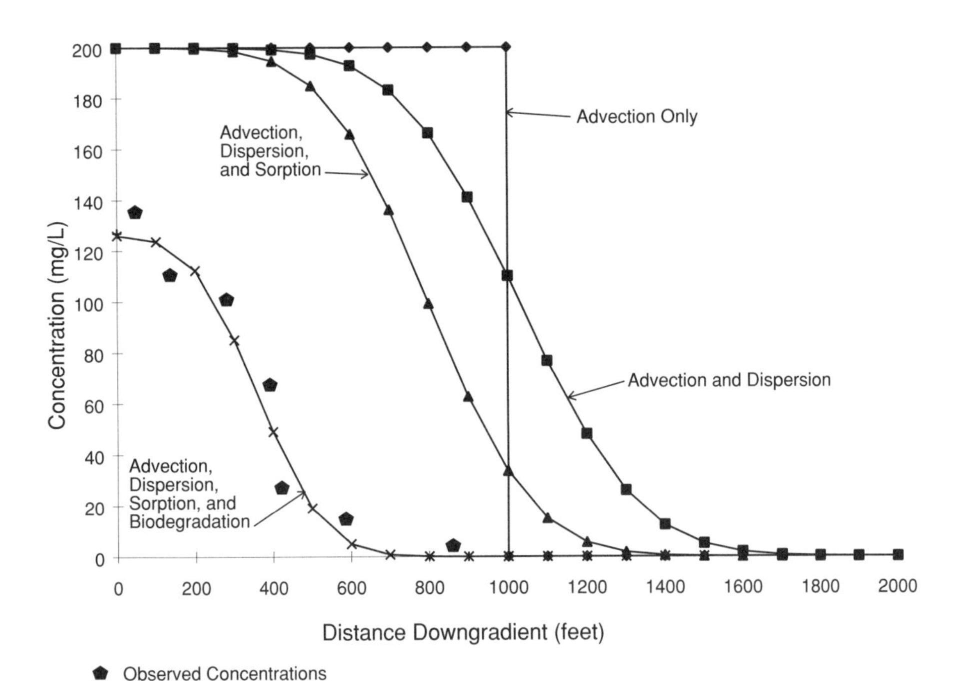

FIGURE 7.3. Use of models to evaluate natural attenuation.

7.3 DOCUMENTED LOSS OF CONTAMINANT MASS OR PLUME STABILIZATION

A statistically significant historical database showing plume stabilization and/or loss of contaminant mass over time can be used to show that natural attenuation is occurring at a site. This is perhaps the best line of evidence to have when trying to implement natural attenuation as a remediation approach. It is important to note that plume stabilization can occur with or without destructive attenuation mechanisms such as intrinsic bioremediation or hydrolysis. In rare cases, nondestructive mechanisms of natural attenuation such as dilution, dispersion, sorption, and volatilization may be sufficient to cause the dissolved contaminant plume to reach steady-state equilibrium, or even recede if the strength of the NAPL source is decreasing due to natural weathering or engineered remediation. The second line of evidence, chemical and geochemical data, can be used to help separate the component of natural attenuation attributable to destructive attenuation mechanisms, especially intrinsic bioremediation.

Both visual and statistical methods can be used to evaluate plume stability. Ultimately, it will probably be some mix of visual observations and statistical techniques that will be used to confirm plume behavior. It is important when evaluating the stability of a contaminant plume that the historical data demonstrate a clear and meaningful trend in contaminant mass and/or concentration over time at appropriate monitoring or sampling points. Methods to accomplish this are presented by the U.S. EPA (1989, 1992). In addition, two statistical methods to evaluate plume behavior are presented below.

Visual Tests for Plume Stability

There are several ways to present data showing changes in contaminant concentrations and plume configuration over time. One method consists of preparing isopleth maps of contaminant concentration over time. Figure 7.2 shows isopleth maps of total VOC concentrations in groundwater at the depth of highest contaminant concentration. Note that contaminant data were collected during the same season. This is important because seasonal variations in recharge can cause significant changes in contaminant concentrations and groundwater geochemistry and an apparent reduction in plume size and/or contaminant concentrations could simply be the result of seasonal dilution. Also plotted on Figure 7.2 is the projected extent of contamination if intrinsic bioremediation were not occurring. This projection was made using an analytical model that incorporated the effects of advection, dispersion, and sorption only; biodegradation was assumed not to occur in these simulations. Model predictions suggest that if biodegradation were not occurring at this site, the plume would grow approximately 1500 ft/yr. Chemical data show that this is not the case, so plume stabilization is probably the result of intrinsic bioremediation. This type of analysis provides good evidence for the occurrence of intrinsic bioremediation, but geochemical data can be used to provide additional confirmation of intrinsic bioremediation (see the case studies in Chapters 9 and 10).

Another method that can be used to present data showing changes in contaminant concentrations and plume configuration over time is to plot contaminant concentrations versus time for individual monitoring wells, or to plot contaminant concentrations versus distance downgradient for several wells along the groundwater flow path over several sampling events. It is important when plotting data in this matter that at least one data point be located a short distance downgradient of the contamination in the groundwater flow path. This ensures that contaminant concentrations in the aquifer as a whole are decreasing and that a pulse of contaminant is not simply migrating downgradient of the observation wells. To ensure that contaminants are not migrating downgradient, it is important that downgradient wells are located in the path of contaminated groundwater flow. Geochemical data can be used to confirm that downgradient wells are sampling groundwater that was once contaminated with organic compounds, as discussed in Chapter 11. Figure 7.4 contains a plot of contaminant concentration versus time in one well and contaminant concentrations versus distance downgradient along the flow path for several sampling events. Based on the geochemical data presented in this figure we are reasonably certain that well H is in the plume's flow path, so if the plume were migrating downgradient, it should be detected. Wells F and H are spaced 100 ft apart, and the groundwater seepage

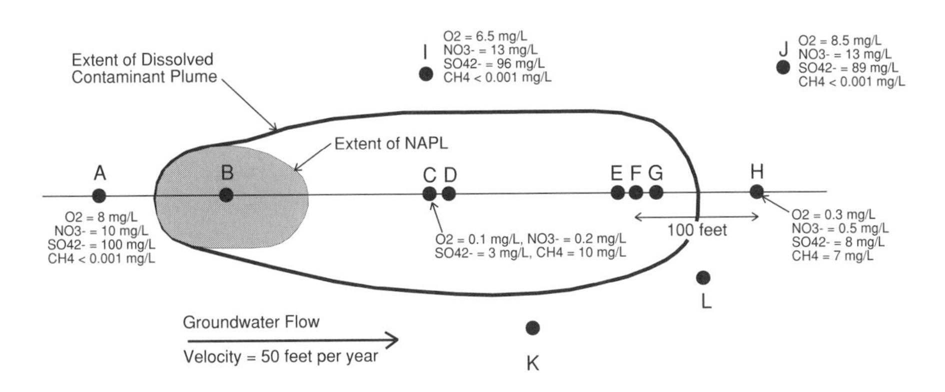

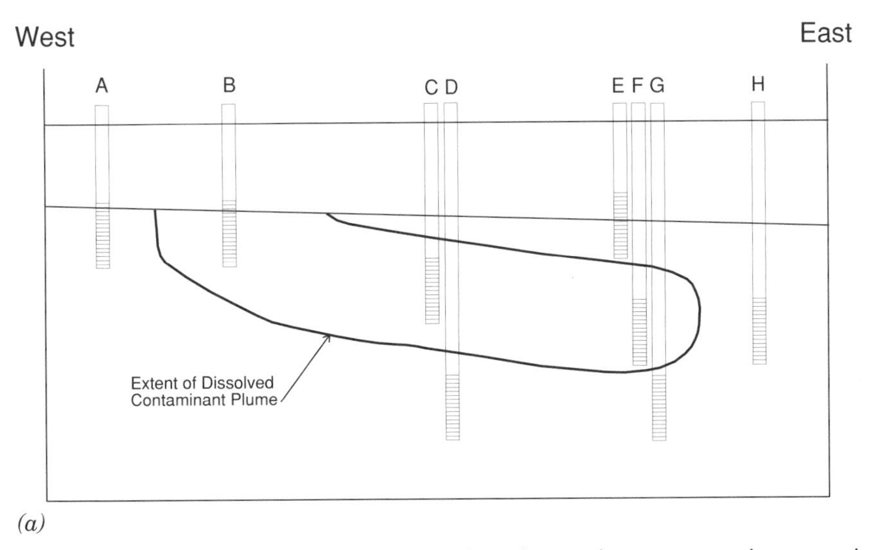

(a)

FIGURE 7.4. (*a*) Sampling locations for (*b*) plots of contaminant concentration versus time and distance downgradient.

continues

velocity is 50 ft/yr; with eight years of sampling data from the same season we can conclude with reasonable certainty that the plume is not migrating downgradient. The combination of decreasing contaminant concentrations shown by the plots in Figure 7.4 and the lack of contaminant migration provide reasonable evidence for natural attenuation and contaminant mass destruction. The chemical and geochemical data discussed in Section 7.4 can be used to show that this loss of contaminant mass is the result of intrinsic bioremediation.

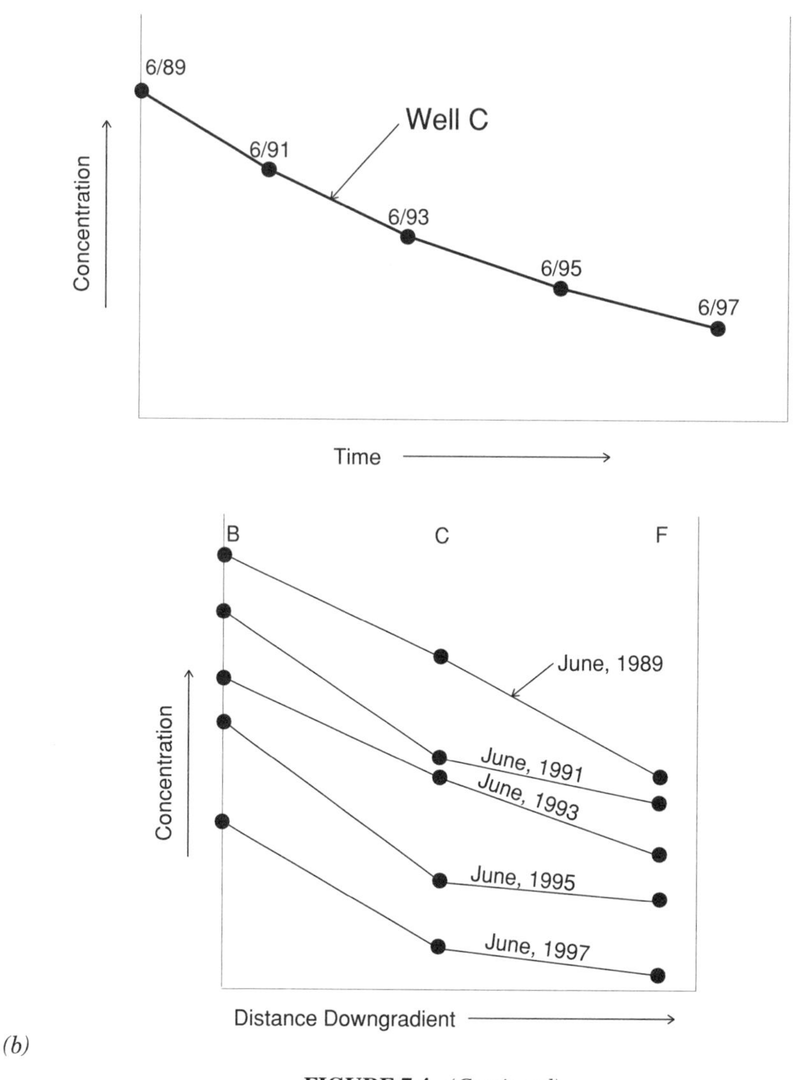

FIGURE 7.4. *(Continued)*

Statistical Tests for Plume Stability

Two different and fairly straightforward statistical approaches that can be used to evaluate plume stability are presented below. First, trends can be analyzed by plotting concentration data versus time, usually on semilog paper with log concentration being plotted against linear time. Plotting the concentration data on the log scale counters the relatively large changes in concentration data (e.g., a concentration reduction from 1 mg/L to 1 μg/L represents a 1000-fold reduction). While plotting concentration versus time data is recommended for most plume stability analyses, discerning trends in the plotted data can be a subjective process, particularly if the data do not have a uniform trend and show some variability over time (Figure 7.5). In these cases a statistical test such as the Mann–Whitney *U* test or Mann–Kendall test can be useful.

Mann–Whitney U Test. The Mann–Whitney *U* test (also called the Wilcoxon rank-sum test) is currently being used by the state of New Jersey to determine plume stability (28 N.J.R. 1143). The test is performed for every contaminant at every monitoring well at a site where this plume stability test is being applied. The test is nonparametric (Mann and Whitney, 1947), which means that the outcome of the test is not determined by the overall magnitude of the data points but depends on the ranking of individual data points.

In the New Jersey approach, eight consecutive quarters of monitoring data are divided into two groups representing the first four quarters (each point designated with an A) and the last four quarters (each point designated with an B). The Mann–Whitney *U* method tests the hypothesis that the two populations are statistically equivalent. The test is conducted by vertically ranking the eight data points from lowest to highest, with the lowest value on top and greatest value on the bottom. For each individual A concentration, the number of B concentrations that occur below the A concentration are counted. The four values (either zero or some positive number) are summed together to obtain the *U* statistic. All nondetect values are considered zero. If two or more concentrations are identical, two vertical columns are constructed. In the first column, the tying B concentration is ranked first, and in the second column the tying A concentration is ranked first. An interim *U* is calculated for each column and the average of the interim *U* values is used as the final *U* value.

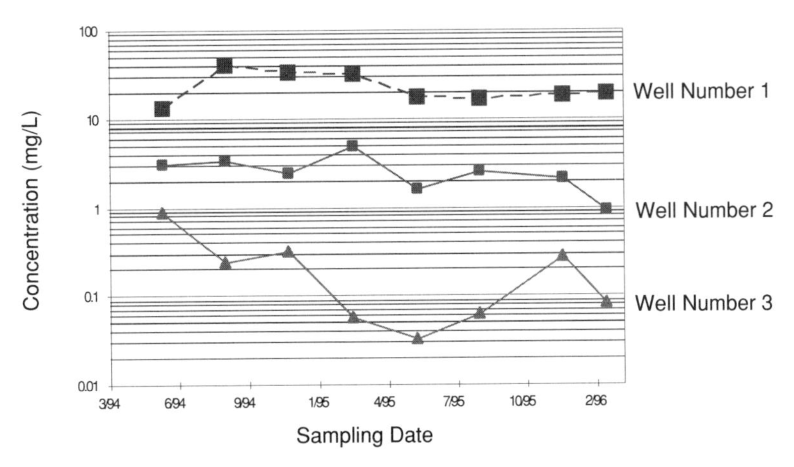

FIGURE 7.5. Example plume stability plots. Well 1 represents a stable plume, well 2 a diminishing plume, and well 3 a well with considerable scatter and no obvious trend.

If $U \leq 3$, the null hypothesis is rejected, and it is concluded with at least 90% confidence that the concentration for the individual contaminant has decreased with time at that well. If $U > 3$, the null hypothesis is accepted, and it cannot be concluded with at least 90% confidence that the concentration for the individual contaminant has decreased with time at that well. Table 7.6 presents an example of the Mann–Whitney *U* test as applied in the New Jersey methodology. Note that this is a relatively low power test, and many datasets that may appear to be declining may not yield a declining result with the *U* test.

Mann–Kendall Test. The Mann–Kendall Test is another nonparametric test (Gilbert, 1987) that can be used to define the stability of a solute plume (i.e., stable, diminishing, or expanding) based on concentration trends at individual wells. To evaluate plume stability, four or more independent sampling events are required. As with the Mann–Whitney test, the Mann–Kendall test is applied to each monitoring well located within the plume area for each contaminant of interest.

TABLE 7.6. Example of Mann–Whitney *U* Test for Plume Stability Using New Jersey Methodology[a]

Date	Concentration (mg/L)	Group
1st Q. 95	3.5	A
2nd Q. 95	2.9	A
3rd Q. 95	2.6	A
4th Q. 95	2.4	A
1st Q. 96	2.3	B
2nd Q. 96	0.9	B
3rd Q. 96	2.8	B
4th Q. 96	2.2	B

Ranked Concentration (mg/L)	Group	Is Group B Value > 2.4 mg/L?	Is Group B Value > 2.6 mg/L?	Is Group B Value > 2.9 mg/L?	Is Group B Value > 3.5 mg/L?
0.9	B	No	No	No	No
2.2	B	No	No	No	No
2.3	B	No	No	No	No
2.4	A	—	—	—	—
2.6	A	—	—	—	—
2.8	B	Yes	Yes	No	No
2.9	A	—	—	—	—
3.5	A	—	—	—	—
	Number:	1	1	0	0

[a] Data are for single contaminant at a single well. *Analysis*: number of yes values, 2; *U* statistic, 2. Since the *U* statistic is less than 3, the concentrations are stable.

However, as shown in a Mann–Kendall worksheet on Figure 7.6, the calculation approach is different.

The worksheet is used by completing the following four steps (Figure 7.6):

Step 1: Well Data. Enter contaminant concentrations for each sampling event. Include only events for which numeric or nondetect (ND) values are available.

Step 2: Data Comparisons. Complete row 1, comparing the results of events 2, 3, and so on, to event 1, as follows:

- Concentration of event x > event 1: Enter 1
- Concentration of event x = event 1: Enter 0
- Concentration of event x < event 1: Enter –1

Complete all rows in same manner until all sampling events complete.

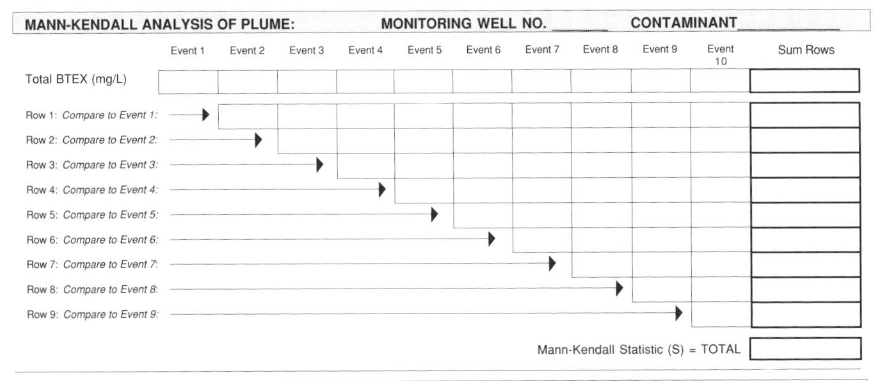

CHARACTERIZATION OF GROUNDWATER PLUME BASED ON DATA FROM THIS WELL

Use the Confidence Level Chart with the Mann-Kendall Statistic computed above (S) and the number of sampling events to estimate
confidence level in the presence of a plume trend (i.e., expanding plume or diminishing plume):

Confidence Level Chart

S Value	Total No. of Sampling Events: 4	5	6	7	8	9	10
0							
±1							
±2							
±3							
±4							
±5							
±6							
±7							
±8							
±9							
±10							
±11							
±12							
±13							
±14							
±15							
±16							
±17							
±18							
±19							
≥20							

No Trend Indicated

Trend Probably Present (≥90% Confidence)

Stability Evaluation Results

❑ **No Trend Indicated** ➊ Plume Not Dimishing or Expanding (Plume is Stable)

❑ **Trend Is Present** (≥90% Confidence)
- ❑ S < 0 ➋ Diminishing Plume
- ❑ S > 0 ➋ Expanding Plume

FIGURE 7.6. Worksheet for plume stability analysis using the Mann–Kendall test.

Step 3: Mann–Kendall Statistic. Sum across each row [e.g., 0 + 0 + (–1) + (–1) + 0 = –2] and record in far right-hand column. Sum right-hand column down to get the total sum. This total value represents Mann–Kendall statistic *S* for the data from this well.

Step 4: Results. Use the confidence level chart to determine the percent confidence in plume trend based on the *S* value and number of sampling events.

Step 5: Analysis. Compare results from all monitoring wells for all contaminants and evaluate overall plume stability.

This approach has its limitations, as a data sheet can show a tremendous amount of scatter but still return the conclusion that the plume is stable (i.e., no significant trend could be established statistically). To counter this problem, one can apply a more sophisticated analysis using Mann–Kendall by comparing the Mann–Kendall *S* statistic, a calculated confidence level, and the coefficient of variance for the sample data. Figure 7.7 is a conceptual representation of the three types of information, in which the *S* statistic shows the direction of the trend, the confidence factor shows how strong the trend is, and the coefficient of variation indicates how much scatter

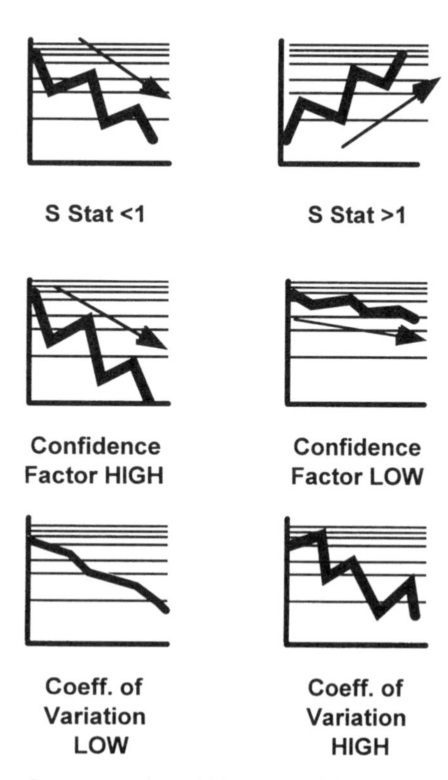

FIGURE 7.7. Conceptual representation of Mann–Kendall *S* statistic, confidence factor, and coefficient of variation.

TABLE 7.7. RNA Tool Kit Rules to Classify Plume Stability

S Statistic	Confidence Factor	Coefficient of Variation	Interpretation
< 1	> 95%	—	Declining
	90% > CF < 95%	—	Probably declining
< 1	< 90%	< 1	Stable
	< 90%	> 1	No trend
1	< 90%	—	No trend
	90% > CF < 95%	—	Probably increasing
	> 95%	—	Increasing

there is in the data. With this approach, for example, sites with confidence factors below 90% can be classified as stable if the coefficient of variation is small (e.g., < 1). One software package designed for analyzing plume stability, the RNA Tool Kit (Groundwater Services, Inc., 1998), uses these three variables together in a conservative fashion to analyze stability and will classify any data set as declining, probably declining, stable, no trend, probably increasing, or increasing. The rules used in this software package to classify plume stability are shown in Table 7.7.

7.4 ANALYTICAL DATA CONFIRMING INTRINSIC BIOREMEDIATION

The biodegradation of organic compounds such as petroleum hydrocarbons or chlorinated solvents brings about measurable changes in the chemistry of groundwater in the affected area. By measuring these changes, it is possible to document and evaluate the importance of intrinsic bioremediation. Several groundwater analytical parameters that are useful for recognizing the intrinsic bioremediation of organic contaminants are presented in this section. These analytical parameters and the associated data quality objectives are presented in Tables 7.1 through 7.4. Table 7.8 summarizes the trends in contaminant, electron acceptor, and metabolic by-product concentrations during biodegradation.

As discussed in the following sections, analytical data for evaluating intrinsic bioremediation fall into three general groups, including electron acceptors, metabolic by-products, and daughter products. It is the temporal and spatial distribution of these analytes that allows one to deduce the distribution of terminal electron-accepting processes in groundwater.

Electron Acceptors

Measurement of the electron acceptors available in an aquifer is critical in identifying the predominant microbial and geochemical processes that are occurring or have already occurred at the time of sample collection. Dissolved oxygen is the first

TABLE 7.8. Trends in Contaminant, Electron Acceptor, and Metabolic By-product Concentrations During Biodegradation

Analyte	Trend in Analyte Concentration During Biodegradation	Terminal Electron-Accepting Processes Causing Trend
Petroleum hydrocarbons	Decreases	Aerobic respiration, denitrification, manganese(IV) reduction, Fe(III) reduction, sulfate reduction, methanogenesis
Highly chlorinated solvents and daughter products	Parent compound concentration decreases, daughter products increase initially and then may decrease	Reductive dechlorination and cometabolic oxidation
Lightly chlorinated solvents	Compound concentration decreases	Aerobic respiration and Fe(III) reduction (direct oxidation) and cometabolism (indirect oxidation)
Dissolved oxygen	Decreases	Aerobic respiration
Nitrate	Decreases	Denitrification
Manganese(II)	Increases	Manganese(IV) reduction
Fe(II)	Increases	Fe(III) reduction
Sulfate	Decreases	Sulfate reduction
Methane	Increases	Methanogenesis
Chloride	Increases	Reductive dechlorination or direct oxidation of chlorinated compound
Oxidation–reduction potential	Decreases	Aerobic respiration, denitrification, reduction, Fe(III) reduction, sulfate reduction, methanogenesis, and halorespiration
Alkalinity	Increases	Aerobic respiration, denitrification, Fe(III) reduction, and sulfate reduction

electron acceptor to be utilized during the biodegradation of many organic compounds dissolved in groundwater. Nitrate, Mn(IV), Fe(III), sulfate, and carbon dioxide are found naturally in many groundwater systems and are used as electron acceptors once the system becomes anaerobic. Concentrations of these natural electron acceptors can be measured in the field using probes or colorimetric techniques. Nitrate, sulfate, and carbon dioxide concentrations can also be determined in a fixed-base, off-site analytical laboratory. Because they can change in a relatively short time after sample acquisition, dissolved oxygen concentrations should be measured while in the field.

In many cases carbon dioxide is the most abundant electron acceptor found in anaerobic environments, especially those contaminated with oxidizable organic compounds. The reasons for this are twofold: (1) carbon dioxide occurs naturally in most groundwater systems, and (2) carbon dioxide is produced during the biodegradation of organic compounds. Although carbon dioxide is used as an electron acceptor during methanogenesis, it is also produced during many biodegradation reactions. Because the concentration of this analyte typically increases in areas with elevated organic contaminant concentrations, it is discussed below under "Metabolic By-products."

Dissolved Oxygen. Dissolved oxygen is the favored electron acceptor used by microbes for the biodegradation of many forms of organic carbon. Strictly anaerobic bacteria generally cannot function at dissolved oxygen concentrations greater than about 0.5 mg/L, and hence Fe(III) reduction, sulfate reduction, methanogenesis, and reductive dechlorination cannot occur. This is why it is important to have a source of carbon in the aquifer that can be used by aerobic microorganisms as a primary substrate. During aerobic respiration, dissolved oxygen concentrations decrease. Thus dissolved oxygen concentrations below background in areas with dissolved contamination provide evidence for aerobic respiration. Changes in dissolved oxygen concentrations inside the contaminant plume versus background concentrations can be used to estimate the mass of contaminant that can be biodegraded by aerobic processes.

Dissolved oxygen measurements should be taken during well purging and immediately before sample acquisition using a direct-reading meter. Each of these measurements should be recorded. Because many well purging techniques can allow aeration of groundwater samples, it is important to minimize the potential for aeration as described in Section 7.1. Oxygen-specific probes typically work by measuring the diffusion of oxygen across a membrane. Because most dissolved oxygen probes consume oxygen, it is important that water is flowing over the membrane and that the water does not become stagnant. This is done using a flow-through cell. There are many dissolved oxygen probes on the market, including downhole (in situ) probes and probes that are designed to be used after the water has been removed from the well. The main difference between the two types of probes is the length of cord separating the dissolved oxygen probe from the meter. Care must be taken when using downhole probes because of the potential for cross-contamination of wells. Because of the extensive decontamination required between wells, the authors typically do not use down-hole probes. Regardless of the dissolved oxygen probe used, the instructions supplied with the probe should be followed. To ensure accurate measurements, the probe should be calibrated regularly and the membrane replaced regularly, or when fouling is suspected. Oxygen probes that allow a two-point calibration (oxygen-saturated water and water with no oxygen) provide more accurate readings at the low concentrations typically encountered in groundwater contaminated with organic compounds.

The aqueous solubility of dissolved oxygen is a function of temperature, salinity, and pressure. Of these variables, temperature has the greatest influence on aqueous solubility. In "fresh" (i.e., salinity = 0 parts per thousand) groundwater under water table conditions, the aqueous solubility of oxygen ranges from 12.76 mg/L at 5°C to 7.54 mg/L at 30°C. In groundwater under water table conditions with a salinity of 20 parts per thousand, the aqueous solubility of oxygen ranges from 11.18 mg/L at 5°C to 6.75 mg/L at 30°C.

The following is a troubleshooting guide for dissolved oxygen measurements.

Problem: *All dissolved oxygen readings are similar in magnitude at all locations whether upgradient of, downgradient of, or in the contaminant plume.*

Probable Cause and Solution

1. Check the membrane on the probe. Make sure that there are no bubbles, folds, tears, or films. If there are bubbles or tears, refill the probe and replace the membrane. If there are folds in the membrane, you can attempt to pull them out; however, it is best to replace it. If the membrane looks dirty, wipe it clean with a damp paper towel. Bubbles inside the membrane cause the meter to read the dissolved oxygen concentration of the bubble consistently and not the dissolved oxygen from the sample. Bubbles are not always obvious; it may be necessary to tap the probe gently (or not so gently) to have them surface.
2. Check to make sure that the meter "zeros" out in a solution of sodium sulfite and water.
   a. If the meter does "zero" (or gets really close), the measurements are correct.
   b. If the meter does not zero, return to step 1.
   c. If the meter does not zero and the probe has no bubbles, tears, folds, or films, rinse the sensor out with distilled water, refill the probe with the oxygen-free solution, and replace the membrane.
   d. If step c does not fix the problem, there is probably something mechanically wrong with the probe. Have the meter serviced by the manufacturer or a qualified technician.

Problem: *Dissolved oxygen readings are above the solubility limit for oxygen.*

Probable Cause and Solution: In general, dissolved oxygen concentrations in groundwater should not exceed about 13 mg/L. Check the probe membrane for bubbles, tears etc. and fix any problems that you observe. Check to make sure the correct conversion factor for the approximate altitude of the site has been entered. An incorrect conversion factor could lead to a consistent error (high or low) in all measurements.

Problem: *Data from a monitoring location show that there is 1 mg/L of methane and 5 mg/L of dissolved oxygen.*

Probable Cause and Solution: Methane is produced only under anaerobic (low to no oxygen) conditions. Thus groundwater data suggesting elevated methane and elevated dissolved oxygen concentrations at the same location should be evaluated critically. In the experience of the authors, accidental introduction of oxygen into a sample can occur quite readily and is a major contributor to inaccurate dissolved oxygen readings. In a simple field experiment, dissolved oxygen concentrations were observed to vary by as much as 2 mg/L, simply based on sampling technique. During the experiment a peristaltic pump was used to collect a shallow groundwater sample using the technique illustrated in Figure 7.1. Dissolved oxygen concentrations measured using this technique were less than 0.1 mg/L. Next, a disposable bailer was lowered carefully and removed from the same well and the groundwater was poured carefully from the bailer down the side of the Erlenmeyer flask. Dissolved oxygen concentrations measured using this technique were as high as 2 mg/L. The groundwater from this well contained methane concentrations on the order of 15 mg/L; clearly, this groundwater was anaerobic. This simple example illustrates that considerable care must be taken when collecting groundwater samples for dissolved oxygen measurements.

Problem: *Dissolved oxygen meter readings are erratic, jumping from high to low concentrations.*

Probable Cause and Solution: Check the batteries. However, dissolved oxygen meter erracticness is more often the result of environmental conditions such as high humidity, very low humidity, cold, and heat.

Nitrate. After dissolved oxygen has been depleted in the microbiological treatment zone, nitrate may be used as an electron acceptor for anaerobic biodegradation of organic carbon via denitrification (Chapter 5). During denitrification, nitrate concentrations measured in groundwater decrease. Thus nitrate concentrations below background in areas with dissolved contamination provide evidence for denitrification. Changes in nitrate concentrations inside the contaminant plume versus background concentrations can be used to estimate the mass of contaminant that can be biodegraded by denitrification.

Sulfate. After dissolved oxygen, nitrate, and biologically available Mn(IV) and Fe(III) have been depleted in the microbiological treatment zone, sulfate may be used as an electron acceptor for anaerobic biodegradation via sulfate reduction (Chapter 5). During sulfate reduction, sulfate concentrations measured in groundwater decrease. Thus sulfate concentrations below background in areas with dissolved contamination provide evidence for sulfate reduction. Changes in sulfate concentrations inside the contaminant plume versus background concentrations can be used to estimate the mass of contaminant that can be biodegraded by sulfate reduction. Concentrations of sulfate greater than about 20 mg/L may cause competitive exclusion of dechlorinating bacteria, so measurement of sulfate concentrations in groundwater is important when trying to determine if the prevailing groundwater geochemistry is conducive to reductive dechlorination.

Metabolic By-products

Readily measurable metabolic by-products include Mn(II), Fe(II), hydrogen sulfide, methane, ethane, ethene, chloride, hydrogen, increased alkalinity, and lowered oxidation–reduction potential. Like electron acceptors, the measurement of metabolic by-products in an aquifer is critical in identifying the predominant microbial and geochemical processes that are occurring or have already occurred at the time of sample collection. Of the readily measurable metabolic by-products, only oxidation–reduction potential must be measured in the field; the remainder of the analytes can be measured in a fixed-base laboratory.

Fe(II) and Mn(II). When Fe(III) is used as an electron acceptor during anaerobic biodegradation of organic carbon, it is reduced to Fe(II), which is soluble in water. Fe(II) concentrations can thus be used as an indicator that anaerobic degradation of organic carbon has occurred via Fe(III) reduction. Changes in Fe(II) concentrations inside the contaminant plume versus background concentrations can be used to estimate the mass of contaminant that has been biodegraded by Fe(III) reduction.

When Mn(IV) is used as an electron acceptor during anaerobic biodegradation of organic carbon, it is reduced to Mn(II). Mn(II) concentrations can thus be used as an indicator that anaerobic degradation of organic carbon has occurred via Mn(IV) reduction. Changes in Mn(II) concentrations inside the contaminant plume versus background concentrations can be used to estimate the mass of contaminant that has been biodegraded by Mn(IV) reduction.

Hydrogen Sulfide. Hydrogen sulfide (H_2S) is produced during sulfate reduction. Elevated hydrogen sulfide concentrations inside the contaminant plume versus background concentrations can be used to support the occurrence of sulfate reduction.

Methane. As implied by the name, *methanogenesis* results in the production of methane during the biodegradation of petroleum hydrocarbons. The presence of methane in groundwater is indicative of strongly reducing conditions. Changes in methane concentrations inside the contaminant plume versus background concentrations can be used to estimate the mass of contaminant that has been biodegraded by methanogenesis. In addition, the presence of methane in groundwater contaminated with chlorinated solvents suggests that the geochemistry of the water is favorable for reductive dechlorination. Analysis of methane concentrations in groundwater should be conducted by a qualified laboratory. It is important that the detection limit for methane be on the order of 0.001 mg/L, especially when evaluating the intrinsic bioremediation of chlorinated solvents.

Carbon Dioxide. Metabolic processes operating during biodegradation of petroleum hydrocarbons and other organic compounds leads to the production of carbon dioxide (CO_2). Accurate measurement of the amount of carbon dioxide produced during biodegradation is difficult because carbonate in groundwater (measured as alkalinity) serves as both a source and sink for free carbon dioxide. If the carbon dioxide produced during metabolism is not completely removed by the natural carbonate buffering system of the aquifer, carbon dioxide concentrations higher than background may be observed. Comparison of empirical data to stoichiometric calculations can provide estimates of the degree of microbiological activity and the occurrence of in situ mineralization of contaminants.

Alkalinity. Biologically active portions of a dissolved contaminant plume typically can be identified by an increase in alkalinity. This increase in alkalinity is brought about by the production of carbon dioxide during the biodegradation of organic carbon. Alkalinity results from the presence of hydroxides, carbonates, and bicarbonates of elements such as calcium, magnesium, sodium, potassium, or ammonia. These species result from the dissolution of rock (especially carbonate rocks), the transfer of carbon dioxide from the atmosphere, and respiration of microorganisms. Alkalinity is important in the maintenance of groundwater pH because it buffers the groundwater system against acids generated during both aerobic and anaerobic biodegradation. In general, areas contaminated by petroleum hydrocarbons exhibit a total alkalinity that is higher than that seen in background areas. This is expected because the microbially mediated reactions causing biodegradation of petroleum hydrocarbons cause an increase in the total alkalinity in the system. Changes in alkalinity are most pronounced during aerobic respiration, denitrification, Fe(III) reduction, and sulfate reduction, and less pronounced during methanogenesis (Morel and Hering, 1993). In addition, Willey et al. (1975) show that short-chain aliphatic acid ions, which can be produced during biodegradation of petroleum hydrocarbons as intermediates, can contribute to alkalinity in groundwater.

Carbon dioxide is produced during the biodegradation of petroleum hydrocarbons. In aquifers that have carbonate minerals as part of the matrix, the carbon dioxide forms carbonic acid, which dissolves these minerals, increasing the alkalinity of the groundwater. An increase in alkalinity (measured as $CaCO_3$) in an area with BTEX concentrations elevated over background conditions can be used to infer the amount of petroleum hydrocarbon destroyed through aerobic respiration, denitrification, Fe(III) reduction, and sulfate reduction. Assuming complete mineralization, these reactions follow the generalized stoichiometry

$$CH \rightarrow CO_2 + H_2O \rightarrow H_2CO_3 + CaCO_3 \rightarrow Ca^{2+} + 2HCO_3^- \qquad (7.1)$$

The mass ratio of alkalinity produced during oxidation of BTEX can be calculated. The molar ratio of alkalinity (as $CaCO_3$) produced during benzene oxidation via aerobic respiration, denitrification, Fe(III) reduction, and sulfate reduction is given by

$$C_6H_6 \rightarrow 6CO_2 \rightarrow 6CaCO_3 \qquad (7.2)$$

Therefore, 6 mol of $CaCO_3$ is produced during the metabolism of 1 mol of benzene. On a mass basis, the ratio of alkalinity to benzene is given by

Molecular weights: Benzene $6(12) + 6(1) = 78$ g
Alkalinity (as $CaCO_3$) $6(40) + 6(12) + 18(16) = 600$ g

$$\text{Mass ratio of alkalinity to benzene} = 600{:}78 = 7.69{:}1$$

Therefore, 7.69 mg of alkalinity (as $CaCO_3$) is produced during the metabolism of 1 mg of benzene. This means that for every 1 mg of alkalinity produced, 0.13 mg of BTEX is destroyed. Similar calculations can be made for toluene, ethylbenzene, and xylene. Table 7.9 summarizes the results of these calculations for all the BTEX compounds during aerobic respiration, denitrification, Fe(III) reduction, and sulfate reduction. Methanogenesis does not cause significant changes in alkalinity.

Oxidation–Reduction Potential. The oxidation–reduction potential of groundwater is a measure of electron activity and is an indicator of the relative tendency of a solution to accept or transfer electrons. Oxidation–reduction reactions in groundwater containing organic compounds (natural or anthropogenic) are usually biologically mediated, and therefore the oxidation–reduction potential of the groundwater system depends upon and influences biodegradation. The oxidation–reduction potential of groundwater changes with the predominant terminal electron-accepting process, with conditions becoming more reducing through the sequence oxygen, nitrate, iron, sulfate, and carbonate. Interpretation of the oxidation–reduction potential of groundwater is difficult. The potential obtained in groundwater is a mixed potential that reflects the potential of many reactions and cannot be used for quantitative interpretation (Stumm and Morgan, 1981). However, the approximate location of a fuel hydrocarbon plume can be identified in the field by measurement of the reduction (redox) potential of groundwater if background organic carbon concentrations are low.

TABLE 7.9. Mass Ratio of Alkalinity (as $CaCO_3$) Produced to BTEX Degraded During Aerobic Respiration, Denitrification, Fe(III) Reduction, and Sulfate Reduction

Alkalinity Production Reaction	Stoichiometric Mass Ratio of Alkalinity Produced to BTEX Degraded	Mass of Compound Degraded (mg) per Unit Mass of Alkalinity Produced (mg)
$C_6H_6 \rightarrow 6CO_2 \rightarrow 6CaCO_3$ Benzene oxidation	600:78	0.13
$C_7H_8 \rightarrow 7CO_2 \rightarrow 7CaCO_3$ Toluene oxidation	700:92	0.13
$C_8H_{10} \rightarrow 8CO_2 \rightarrow 8CaCO_3$ Ethylbenzene oxidation	800:104	0.13
$C_8H_{10} \rightarrow 8CO_2 \rightarrow 8CaCO_3$ Xylene oxidation	800:104	0.13

Figure 4.4 presents the redox potential range for various natural attenuation processes. The redox range observed in groundwater may be higher than the theoretical optimum redox for various electron acceptor reactions (Norris et al., 1994). This discrepancy is a common problem associated with measuring oxidizing potential using field instruments. It is likely that the platinum electrode probes are not sensitive to some of the redox couples (e.g., sulfate/sulfide, carbon dioxide/methane). Many authors have noted that field-measured redox data alone cannot be used to predict reliably the electron acceptors that may be operating at a site (Stumm and Morgan, 1981; Lovley et al., 1994). Integrating redox measurements with analytical data on reduced and oxidized chemical species allows a more thorough and reasonable interpretation of which electron acceptors are being used to biodegrade site contaminants.

To overcome the limitations imposed by traditional redox measurements, recent work has focused on the measurement of molecular hydrogen to describe accurately the predominant in situ redox reactions (Lovley and Goodwin, 1988; Lovley et al., 1994; Chapelle et al., 1995). The evidence suggests that concentrations of hydrogen in groundwater can be correlated with specific microbial processes, and these concentrations can be used to identify zones of methanogenesis, sulfate reduction, and iron reduction in the subsurface (Chapelle, 1996).

Oxidation–reduction potential measurements should be taken during well purging and immediately before sample acquisition using a direct-reading meter. Because many well-purging techniques can allow aeration of collected groundwater samples (which can affect oxidation–reduction potential measurements), it is important to minimize potential aeration by following the steps outlined in Section 7.1. The oxidation–reduction potential determined in the field using a probe is termed *Eh*. Eh can be expressed as pE, which is the hypothetical measure of the electron activity associated with a specific Eh. High pE means that the solution or redox couple has a relatively high oxidizing potential.

Dissolved Hydrogen. Concentrations of dissolved hydrogen can be used to evaluate redox processes in groundwater systems (Lovley and Goodwin, 1988; Lovley et al., 1994; Chapelle et al., 1995). Because each terminal electron-accepting process has associated with it a characteristic hydrogen concentration, hydrogen concentrations can be an indicator of predominant redox processes. These characteristic ranges are given in Table 6.5.

Oxidation–reduction potential measurements are based on the concept of thermodynamic equilibrium and, within the constraints of that assumption, can be used to evaluate redox processes in groundwater systems. The hydrogen method is based on the ecological concept of interspecies hydrogen transfer by microorganisms and, within the constraints of that assumption, can also be used to evaluate redox processes. These methods are therefore fundamentally different. A direct comparison of these methods (Chapelle, 1996) has shown that oxidation–reduction potential measurements were effective in delineating oxic from anoxic groundwater, but that oxidation–reduction potential measurements could not distinguish between nitrate-reducing, Fe(III)-reducing, sulfate-reducing, or methanogenic zones in an aquifer. In contrast, the hydrogen method could readily distinguish among anaerobic zones. For those sites at which distinguishing between different anaerobic processes is important information (such as at sites contaminated with chlorinated solvents), hydrogen measurements are an available technology for making such distinctions.

In practice, it is preferable to interpret hydrogen concentrations in the context of electron acceptor [dissolved oxygen, nitrate, Mn(IV), Fe(III), sulfate] availability and the presence of the final products [Mn(II), Fe(II), hydrogen sulfide, methane] of microbial metabolism (Chapelle et al., 1995). For example, if sulfate concentrations in groundwater are less than 0.5 mg/L, methane concentrations are greater than 0.5 mg/L, and hydrogen concentrations are in the range 5–20 nM, it can be concluded with a high degree of certainty that methanogenesis is the predominant redox process in the aquifer. Similar logic can be applied to identifying denitrification (presence of nitrate, hydrogen < 0.1 nM), Fe(III) reduction [production of Fe(II), hydrogen 0.2 to 0.8 nM]), and sulfate reduction (presence of sulfate, production of sulfide, hydrogen 1 to 4 nM).

Chapelle et al. (1997) compare three methods for measuring hydrogen concentrations in groundwater: a downhole sampler, a gas stripping method, and a diffusion sampler. The downhole sampler and gas stripping methods gave similar results. The diffusion sampler appeared to overestimate hydrogen concentrations. Of these methods, the gas stripping method is better suited to field conditions because it is faster (approximately 30 minutes for a single analysis as opposed to 2 hours for the downhole sampler and 8 hours for the diffusion sampler), the analysis is easier (less sample manipulation is required), and the data computations are more straightforward (hydrogen concentrations need not be corrected for water sample volume) (Chapelle et al., 1997).

Chloride. During biodegradation of chlorinated hydrocarbons dissolved in groundwater, chloride is released into the groundwater, resulting in the accumulation of biogenic chloride. This results in chloride concentrations in groundwater in the contaminant plume that are elevated relative to background concentrations. In aquifers with low background concentrations of inorganic constituents, the concentration of chloride in the aquifer can be seen to increase as chlorinated solvents are degraded.

Elemental chlorine is the most abundant of the halogens. Although chlorine can occur in oxidation states ranging from Cl^- to Cl^{7+}, the chloride form (Cl^-) is the only form of major significance in natural waters (Hem, 1985). Chloride forms ion pairs or complex ions with some of the cations present in natural waters, but these complexes are not strong enough to be of significance in the chemistry of fresh water (Hem, 1985). The chemical behavior of chloride is neutral. Chloride ions generally do not enter into oxidation–reduction reactions, form no important solute complexes with other ions unless the chloride concentration is extremely high, do not form salts of low solubility, are not significantly adsorbed on mineral surfaces, and play few vital biochemical roles (Hem, 1985). Thus physical processes control the migration of chloride ions in the subsurface. Kaufman and Orlob (1956) conducted tracer experiments in groundwater and found that chloride moved through most of the soils tested more conservatively (i.e., with less retardation and loss) than any of the other tracers tested. Because of the neutral chemical behavior of chloride, it can be used as a conservative tracer to estimate biodegradation rates.

Daughter Products

Concentrations of chlorinated solvents and their degradation (daughter) products give a direct indication of the presence or absence of microbial degradation (both reductive and oxidative) processes. In many cases the production of *cis*-DCE, VC, and chloride ions along aquifer flow paths is direct evidence of intrinsic bioremediation. For example, if TCE was the only contaminant disposed of at a site, any DCE or VC present at the site must have come from the degradation of the parent TCE. NAPL analyses can be helpful in establishing which compounds were disposed of at a site. At sites in which it can be demonstrated that compounds appearing in groundwater were not disposed of, these compounds are probably daughter products. It is possible that VC and some DCE isomers can be primary contaminants in some groundwater systems. However, VC is not normally present as a primary contaminant in solvent spills associated with military activities. The reasons for this are (1) VC was not used as a solvent, and (2) VC is a gas at temperatures as low as 15°C. Thus the presence of VC in groundwater associated with a chlorinated ethene spill is strong evidence of reductive dechlorination. Also, *cis*-DCE (rather then *trans*-DCE) is usually produced from the reductive dechlorination of TCE. As a rule of thumb, if the ratio of *cis*-1,2-DCE to *trans*-1,2-DCE plus 1,1-DCE is greater than about 5:1, the DCE is probably of biogenic origin. Based on these concepts, VC and *cis*-DCE are usually reliable indicators of microbial reductive dechlorination. In addition, the presence of ethene and ethane can be indicative of reductive dechlorination.

Spatial Distribution of Electron Donors, Electron Acceptors, Metabolic By-products, and Daughter Products

It has often been said that a picture is worth a thousand words. The spatial relationships between contaminants, daughter products, electron acceptors, and metabolic by-products can be used to evaluate the occurrence of a given biodegradation reaction qualitatively. Isopleth maps (isoconcentration maps) are the "pictures" that map the relationships between these parameters in both time and space and portray graphically the second line of evidence used to evaluate intrinsic bioremediation. Isopleth maps allow interpretation of data on the distribution and the relative transport and degradation rates of contaminants in the subsurface. In addition, contaminant isopleth maps allow contaminant concentrations to be gridded and used for input into a solute transport model. Isopleth maps are prepared by first plotting the concentration of a parameter on a base map prepared using surveyor's data. Lines of equal concentration (isopleths) are then drawn and labeled. It is important to ensure that each data point is honored during contouring, unless some data are suspect and there is a logical (and defendable) reason for omitting a data point.

Isopleth maps should be prepared for all contaminants of concern and their daughter products. For example, if trichloroethene and BTEX were released in the same

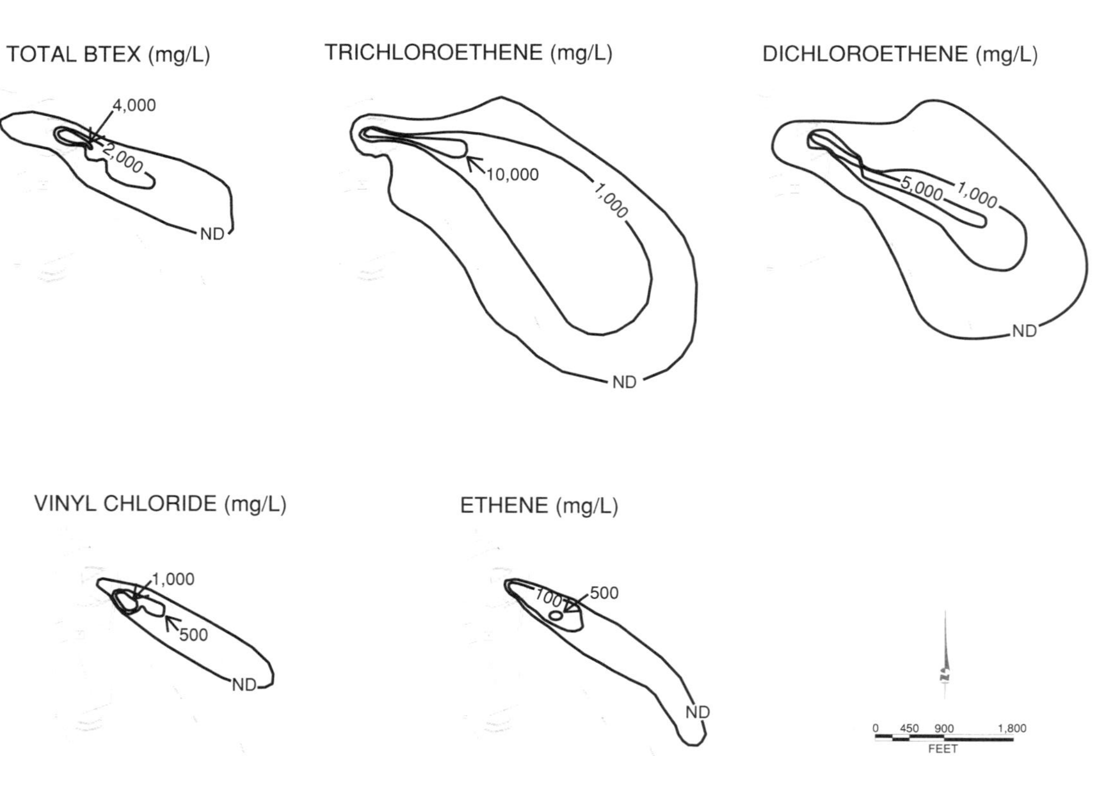

FIGURE 7.8. Total BTEX, chlorinated solvent, and daughter product isopleth maps for groundwater.

area, maps of dissolved BTEX and trichloroethene should be prepared. In addition, if the daughter products dichloroethene, vinyl chloride, and ethene are present, maps showing the distribution of these compounds should also be prepared. Figure 7.8 contains examples of contaminant and daughter product isopleth maps. Daughter product maps provide evidence for biodegradation.

Inorganic electron acceptor isopleth maps allow interpretation of data on the distribution of dissolved oxygen, nitrate, and sulfate in the subsurface. Isopleth maps for these compounds provide a visual indication of the relationship between the contaminant plume and the electron acceptors and the relative importance of each terminal electron-accepting process. Dissolved oxygen concentrations below background in areas with high organic carbon concentrations (e.g., BTEX, landfill leachaete, or naturally occurring organic carbon) are indicative of aerobic respiration. Nitrate concentrations below background in areas with high organic carbon concentrations are indicative of denitrification. Sulfate concentrations below background in areas with high organic carbon concentrations are indicative of sulfate reduction. Figure 7.9 gives examples of completed isopleth maps for dissolved oxygen, nitrate, and sulfate.

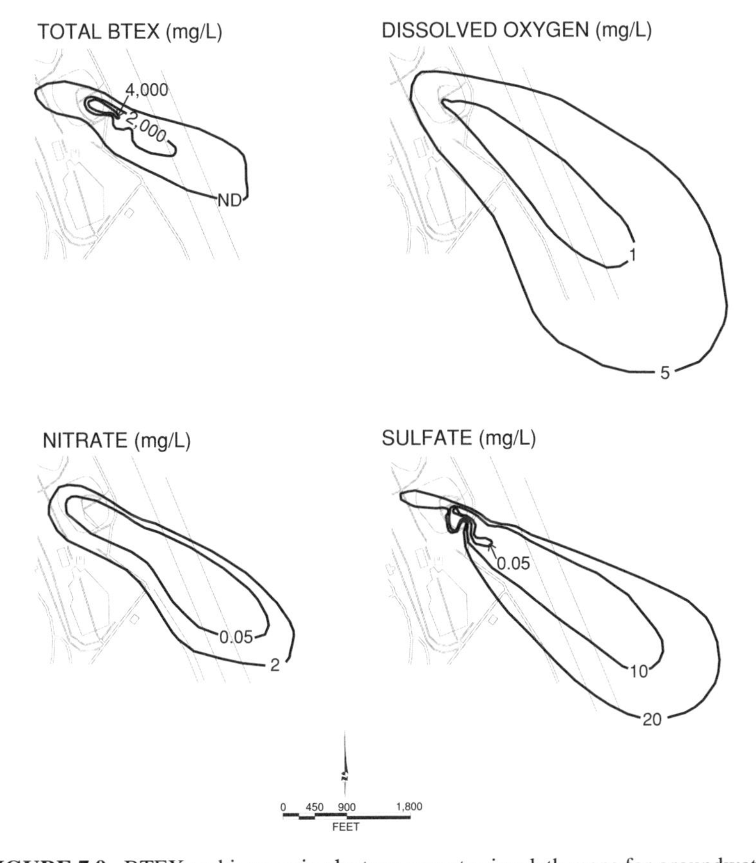

FIGURE 7.9. BTEX and inorganic electron acceptor isopleth maps for groundwater.

This figure also contains the total BTEX (electron donor) isopleth map for the site. Comparison of the total BTEX isopleth map and the electron acceptor isopleth maps shows that there is a strong correlation between areas with elevated organic carbon and depleted electron acceptor concentrations. The correlation indicates that the electron acceptor demand exerted during the metabolism of BTEX has resulted in the depletion of soluble inorganic electron acceptors. These relationships provide strong evidence that biodegradation is occurring via the processes of aerobic respiration, denitrification, and sulfate reduction.

Metabolic by-product maps should be prepared for Fe(II), methane, and if chlorinated solvents are present, chloride. In addition, an equipotential map should be prepared to show the distribution of oxidation–reduction potential in the aquifer. The Fe(II) map is prepared in lieu of an electron acceptor isopleth map for Fe(III) because the amount of bioavailable amorphous or poorly crystalline Fe(III) in an aquifer matrix is extremely hard to quantify. Fe(II) concentrations above background in areas with elevated organic carbon concentrations are indicative of anaerobic Fe(III) reduction. Methane concentrations above background in areas with

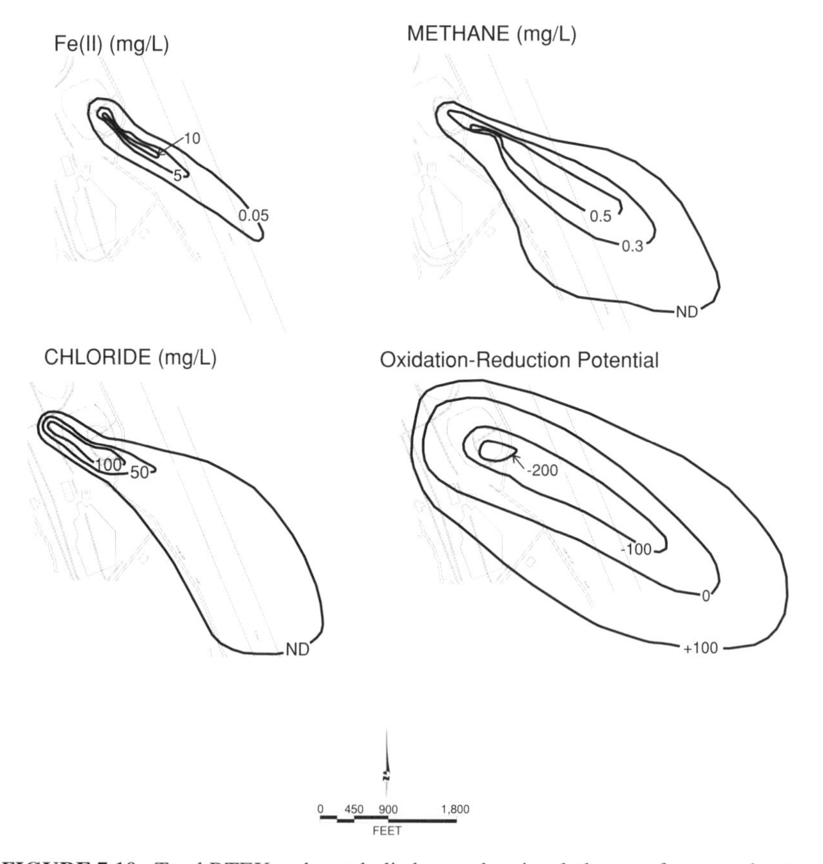

FIGURE 7.10. Total BTEX and metabolic by-product isopleth maps for groundwater.

elevated organic carbon concentrations are indicative of methanogenesis. Biodegradation of chlorinated solvents tends to increase the chloride concentration found in groundwater. Thus chloride concentrations inside the contaminant plume generally increase to concentrations above background. This map will allow visual interpretation of chloride data by showing the relationship between the contaminant plume and chloride. During biodegradation, the oxidation–reduction potential of groundwater is lowered. Thus the oxidation–reduction potential inside the contaminant plume generally becomes negative. Figure 7.10 gives examples of completed isopleth maps for Fe(II), methane, chloride, and oxidation–reduction potential. This figure shows that there is a strong correlation between areas with elevated organic carbon and elevated metabolic by-product concentrations. These relationships provide strong evidence that biodegradation is occurring via the processes of Fe(III) reduction, methanogenesis, and reductive dechlorination.

Figures 7.11, 7.12, and 7.13 show the three-dimensional relationships between total BTEX, total chlorinated solvents, Fe(II), sulfate, methane, and oxidation–reduction potential at a site where reductive dechlorination is known to be occurring.

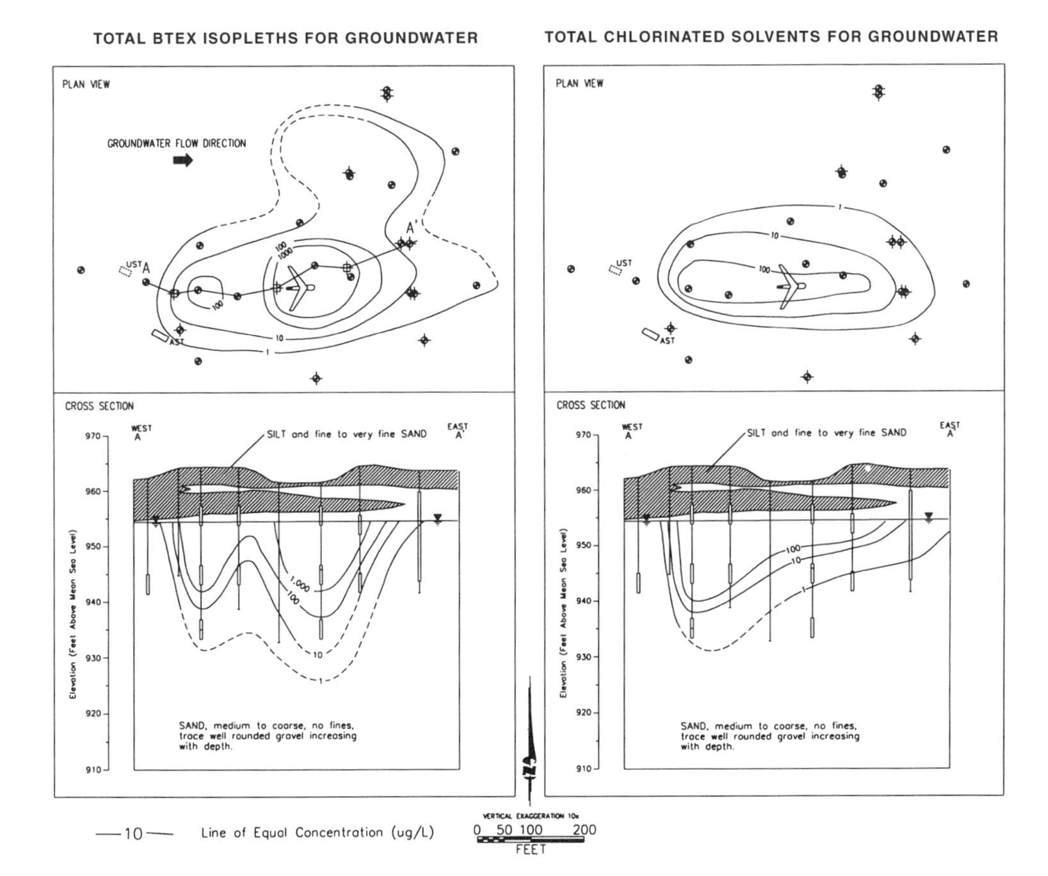

FIGURE 7.11. Three-dimensional relationships: BTEX and chlorinated solvents.

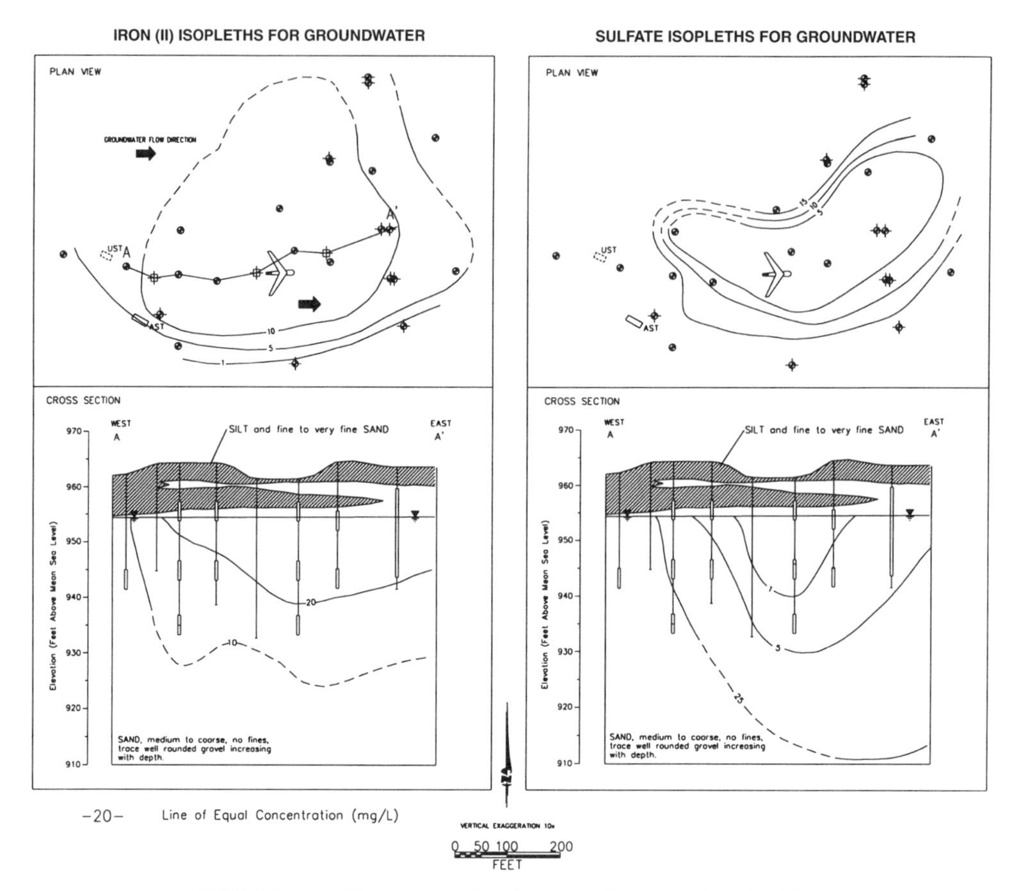

FIGURE 7.12. Three-dimensional relationships: Fe(II) and sulfate.

Such three-dimensional relationships can often be seen if sufficient data are available. Dissolved oxygen and nitrate concentrations were low both inside and outside the plume, so no trends in these analytes could be discerned.

Deducing the Distribution of Terminal Electron Accepting Processes in Groundwater

Although important for simulating the fate and transport of petroleum hydrocarbons using models such as BIOSCREEN or BIOPLUME III, accurate deduction of the distribution of terminal electron-accepting processes in groundwater is especially important for sites contaminated with chlorinated solvents. The reason for this is that the occurrence and efficiency of intrinsic bioremediation of chlorinated solvents is strongly related to the dominant terminal electron-accepting process in groundwater in the area affected. For example, the common chlorinated solvents such as TCA, PCE, TCE, and CT can be biodegraded only under the strongly reducing conditions associated with sulfate reduction or methanogenesis.

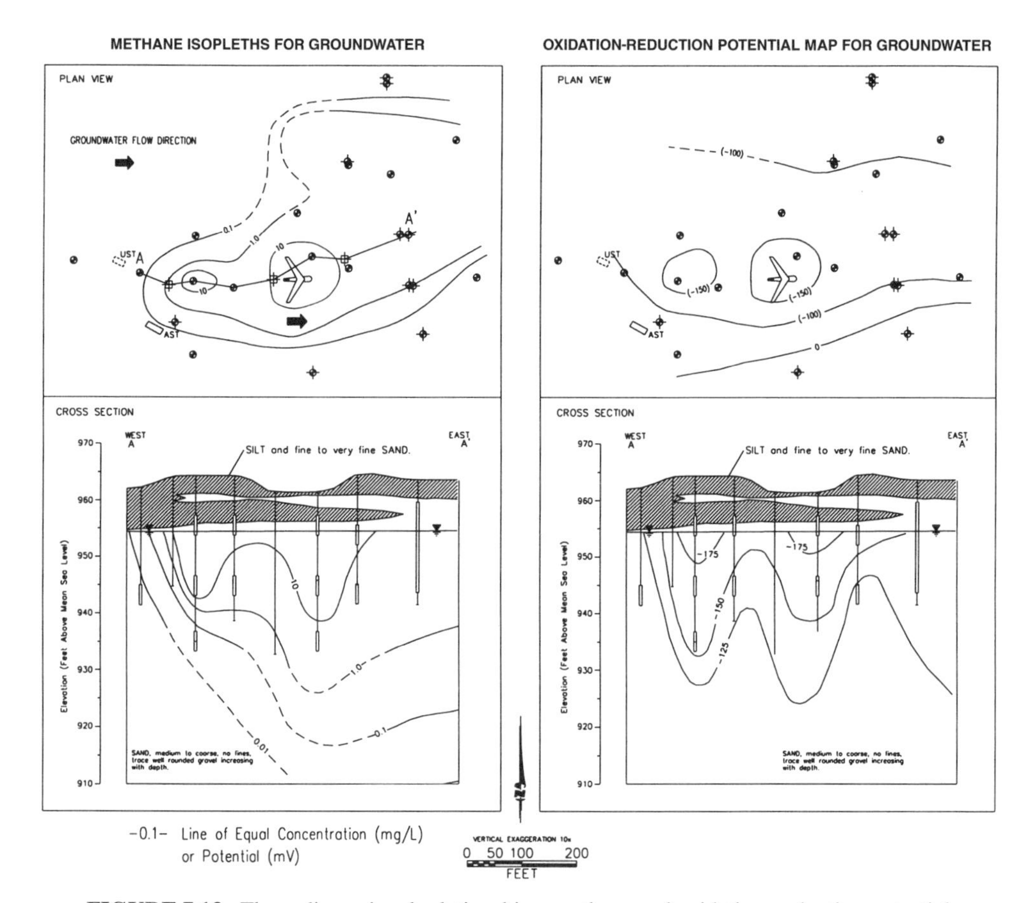

FIGURE 7.13. Three-dimensional relationships: methane and oxidation–reduction potential.

The distribution of terminal electron-accepting reactions in groundwater is caused by, and influences the intrinsic bioremediation of organic contaminants dissolved in groundwater. The most common electron-accepting processes in groundwater systems are aerobic respiration, denitrification, Fe(III) reduction, sulfate reduction, and methanogenesis. The stoichiometry of these processes can be represented by the following generalized equations (Wiedemeier and Chapelle, 1998):

$$O_2 + CH_2O \rightarrow CO_2 + H_2O \qquad (7.3)$$

$$4NO_3 + 4H^+ + 5CH_2O \rightarrow 2N_2 + 5CO_2 + 7H_2O \qquad (7.4)$$

$$4Fe(OH)_3 + CH_2O + 8H^+ \rightarrow 4Fe^{2+} + CO_2 + 11H_2O \qquad (7.5)$$

$$2CH_2O + SO_4^{2+} + H^+ \rightarrow 2CO_2 + HS^- + 2H_2O \qquad (7.6)$$

$$2CH_2O \rightarrow CH_4 + CO_2 \qquad (7.7)$$

Based on the resulting changes in groundwater geochemistry caused by these reactions, it is possible to deduce logically which microbial processes predominate in a groundwater system. Chapelle et al. (1995) present a technique for delineating the distribution of terminal electron-accepting processes in groundwater. Figure 7.14 is a modification of a flowchart for deducing the dominant terminal electron-accepting process presented by Chapelle et al. (1995). The following paragraph summarizes the logic presented by these authors.

If dissolved oxygen is present in groundwater at concentrations greater than about 0.5 mg/L, oxygen reduction will be the predominant microbial process. If dissolved oxygen concentrations are less than 0.5 mg/L but nitrate is present at concentrations

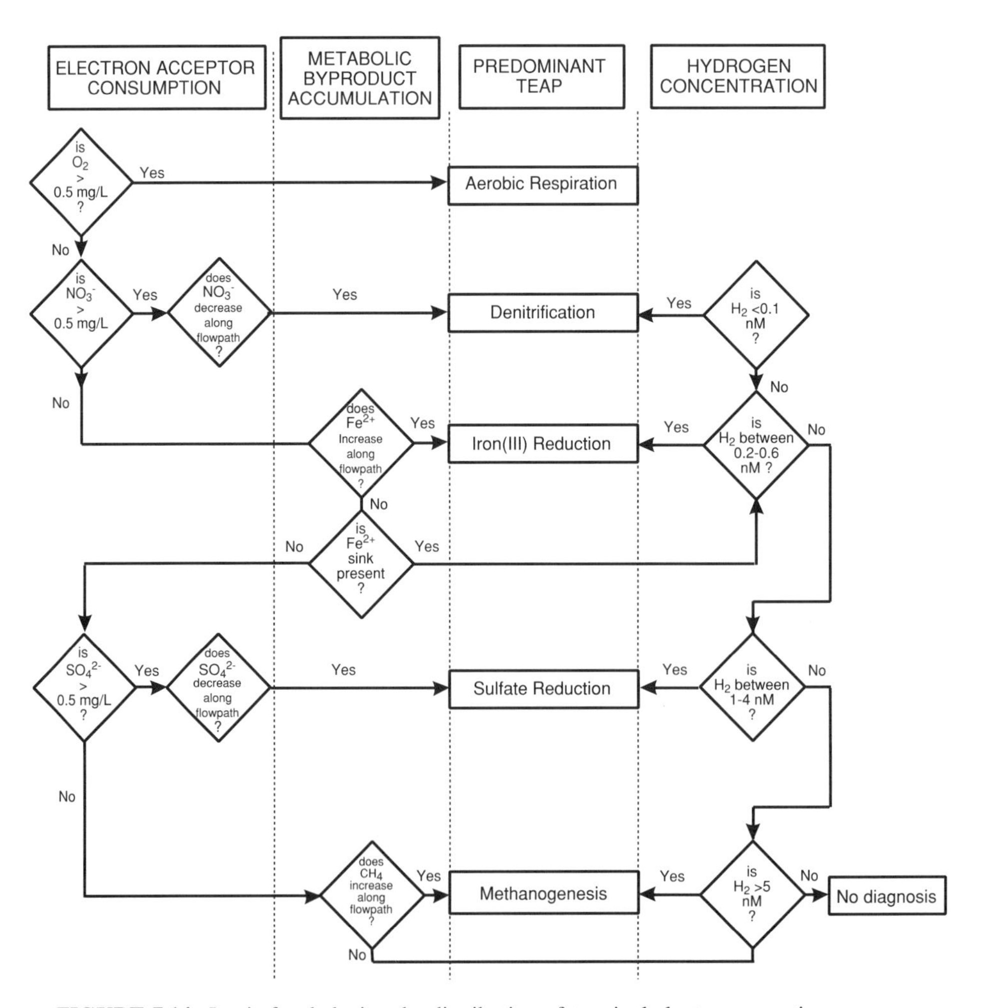

FIGURE 7.14. Logic for deducing the distribution of terminal electron-accepting processes in groundwater systems. (Modified from Chapelle et al., 1995.)

greater than 1 mg/L, nitrate reduction will be the predominant microbial process. If groundwater lacks dissolved oxygen, nitrite, or nitrate but concentrations of Fe(II) are greater than 0.5 mg/L, Fe(III) reduction will be the predominant process. If groundwater lacks Fe(II) but contains concentrations of sulfate greater than 1 mg/L and hydrogen sulfide greater than 0.05 mg/L, sulfate reduction will be the predominant process. Finally, if the water lacks dissolved oxygen, nitrate, Fe(II), sulfate, and hydrogen sulfide but contains concentrations of methane greater than 0.2 mg/L, methanogenesis will be the predominant process.

In practice, this methodology often encounters uncertainties. Many products of microbial metabolism, such as Fe(II), H_2S, and methane, can be transported by groundwater flow. In cases in which such transport is significant, it is difficult to determine the exact redox zonation with this water-chemistry information. In these cases, direct measurement of dissolved hydrogen gas (H_2) concentrations can be used to evaluate ambient redox processes (Figure 7.14). Fermentative microorganisms in groundwater systems continuously produce H_2 during anoxic decomposition of organic matter. This H_2 is then consumed by respiratory microorganisms that may use Fe(III), sulfate, or carbon dioxide as terminal electron acceptors. In microbial ecology, this process is referred to as *interspecies hydrogen transfer*. Significantly, Fe(III)-, sulfate-, and carbon dioxide–reducing (methanogenic) microorganisms exhibit different efficiencies in utilizing H_2. Fe(III) reducers are relatively efficient in utilizing H_2 and thus they maintain lower steady-state H_2 concentrations (0.2 to 0.8 nM H_2) than those of sulfate reducers (1 to 4 nM H_2) or of methanogens (5 to 15 nM H_2). Because each terminal electron-accepting process has associated with it a characteristic H_2 concentration, H_2 concentrations can be an indicator of predominant redox processes in groundwater systems.

As discussed in Chapter 5, the dominant terminal electron-accepting process can vary in both time and space. This can have a profound influence on the efficacy of intrinsic bioremediation, especially for the chlorinated solvents, which require strongly reducing conditions for degradation. Although some engineered remediation systems may be effective in removing contaminant mass from groundwater, some systems may have an adverse impact on intrinsic bioremediation. For example, the introduction of oxygen via air sparging into a system contaminated with chlorinated solvents may raise the oxidation–reduction potential and oxygen content of the groundwater to the point that reductive dechlorination can no longer occur and the natural treatment system is destroyed. A groundwater pump-and-treat system can have the same impact by flushing oxygen-rich groundwater through the contaminant plume. In light of this, it is imperative that the impacts of any proposed remediation system on naturally occurring processes be evaluated (see Chapter 11).

7.5 MICROBIOLOGICAL DATA

Although several types of microbiological data are available, the most common type of data collected for evaluating the degradation of organic contaminants in aquifer material are those from a laboratory microcosm study. If properly designed, implemented, and interpreted, microcosm studies can provide very convincing documentation of the occurrence of intrinsic bioremediation. They are the only line of evidence that allows an unequivocal mass balance on the biodegradation of environmental contaminants. If the microcosm study is properly designed, it will be easy for decision makers with nontechnical backgrounds to interpret. The results of a microcosm study are strongly influenced by the nature of the geological material submitted for study, the physical properties of the microcosm, the sampling strategy, and the duration of study.

Microcosm studies are time consuming and expensive and suffer from several limitations, as discussed below. For these reasons, microcosm studies should be used very selectively in assessing the efficiency of natural attenuation processes. Biodegradation processes are often specific to ambient redox conditions (i.e., aerobic versus anaerobic), the availability of electron donors and acceptors, and the availability of nutrients. All of these factors can be changed significantly, either intentionally or unintentionally, when sediments or groundwater from a particular site are brought to the laboratory. This is because of sediment disturbance caused by sampling and transport and because of the difficulty in reproducing in situ conditions in the laboratory. For example, if anaerobic sediments are brought back to the laboratory and incubated under aerobic conditions, the experimental results may greatly overestimate in situ rates of petroleum-hydrocarbon biodegradation, or greatly underestimate rates of in situ chlorinated solvent biodegradation. It is difficult to verify that experimental conditions in the laboratory match ambient in situ conditions, so conclusions from such studies can be difficult to interpret. To help alleviate this problem, material for a microcosm study should not be selected until the geochemical behavior of the site is well understood. Microbial degradation of organic substrates can result in the consumption of oxygen, nitrate, or sulfate, and the production of Fe(II), Mn(II), or methane. These processes usually operate concurrently in different parts of the plume. Regions in which each process prevails may be separated in directions parallel to groundwater flow by hundreds of meters, in directions perpendicular to groundwater flow by tens of meters, and vertically by only a few meters. Rate constants and constraints for petroleum-hydrocarbon biodegradation will be influenced by the prevailing geochemistry. Material from microcosms must be acquired from depth intervals and locations that have been predetermined to be representative of the prevailing geochemical milieu in the plume. Even with this information, conclusions from such studies are often difficult or impossible to interpret.

A number of existing studies pertaining to the activity and rate of anaerobic degradation processes have been conducted in a laboratory setting and/or artificial microcosms. The results of these studies must be viewed with caution because several factors can influence the results including:

- Laboratory findings cannot be translated directly to field settings.
- Anaerobic biodegradation of contaminants results from the interactions of a microbial consortia. Removing aquifer material from its original setting disrupts the balance of the consortia, which in turn inhibits biodegradation.

- Because of the difficulty in collecting, handling, transporting, and transferring aquifer samples to an anaerobic incubation chamber or microcosm, it is difficult to replicate naturally occurring, site-specific anaerobic processes in the laboratory. Any exposure to oxygen can alter the balance of a specific consortium.
- Laboratory studies may not be run for long enough after sample collection for the system to recover from the effects discussed above (e.g., sample collection, handling, etc.), resulting in a time lag before biodegradation can begin. For this reason, laboratory studies often tend to underestimate the effect of biodegradation mechanisms, especially anaerobic processes.

In some circumstances, however, laboratory studies are necessary. When specific questions are raised concerning *conditions* under which biodegradation processes occur or do not occur, controlled laboratory studies are often required. For example, if concentrations of a particular compound are observed to decrease in the field, it is often not clear whether this decrease is due to sorption, dilution, or biodegradation. Laboratory studies in which the effects of each process can be isolated and controlled (they usually cannot be controlled in the field) are the only method available for answering these questions.

7.6 ESTIMATING BIODEGRADATION RATES

Three different approaches are required to estimate rates for the three kinetic models presented in Chapter 4. For the Monod model, data are derived from carefully controlled laboratory experiments (see, e.g., Tabak et al., 1990; Desai et al., 1990) to obtain half-saturation and maximum utilization rate data. Monod parameters have been measured for only a limited data set of contaminants and reactions. For the instantaneous reaction model at petroleum-hydrocarbon sites, the reactions depend on the amount of electron acceptors that are available. This quantity, referred in this book as *expressed biodegradation capacity*, is presented in Chapter 5 along with calculation techniques and data from field sites. Expressed biodegradation capacity is also referred to as *expressed assimilative capacity and biodegradation capacity*. The BIOPLUME II and III models use the concept of expressed biodegradation capacity to simulate the fate and transport of petroleum hydrocarbons dissolved in groundwater. Because of its simple nature and utility, the first-order decay model is so widely used that to many researchers and practitioners, the term *rate* is synonymous with *first-order decay rate constants* (even though other models may provide a better fit to observed data). Considerable attention has been devoted to the calculation of these rate constants and to key calculation techniques using (1) conservative tracers, (2) methods that assume steady-state equilibrium of the plume, and (3) mass flux methods.

The two kinetic models used most often to represent petroleum-hydrocarbon biodegradation in solute fate and transport models are the instantaneous reaction and first-order decay models. The kinetic model most commonly used to simulate the fate and transport of chlorinated solvents is first-order decay. How the instantaneous reaction is used in models is described in Chapter 8. In this section we discuss methods that can be used to estimate first-order biodegradation rate constants using field data.

Calculating Biodegradation Rate Constants Using Conservative Tracers

Conservative tracers found commingled with a contaminant plume can be useful for estimating the biodegradation rates of petroleum hydrocarbons and chlorinated solvents. For example, the isomers of trimethylbenzene have been found to be biologically recalcitrant under some anaerobic conditions. Comparing the loss of trimethlybenzene to the loss of BTEX along the groundwater flow path can allow estimation of biodegradation rates. In a manner similar to petroleum hydrocarbons, the biodegradation rate for chlorinated hydrocarbons may be estimated using a conservative tracer. At sites where commingled petroleum-hydrocarbon and chlorinated solvent plumes are present, trimethylbenzene isomers can be used to estimate biodegradation rates for BTEX and chlorinated solvents.

Wiedemeier et al. (1996b) have derived a relationship to approximate biodegradation rates at sites contaminated with petroleum hydrocarbons. To determine approximate biodegradation rate constants with this method, measured concentrations of dissolved BTEX are corrected for the effects of dispersion, dilution from recharge, volatilization, and sorption using a tracer. One tracer that has proven useful in some but not all groundwater environments is trimethylbenzene (TMB). The three isomers of this compound (1,2,3-TMB, 1,2,4-TMB, and 1,3,5-TMB) have Henry's law constants and soil sorption coefficients similar to (although somewhat higher than) those of the BTEX compounds. Also, the TMB isomers are generally present in sufficient quantities in fuel mixtures to be readily detectable in groundwater in contact with a fuel spill. Finally, they often are recalcitrant to biodegradation in the anaerobic portion of a plume. Other compounds of potential use as conservative tracers are the tetramethylbenzene isomers, provided that they are detectable throughout most of the plume.

The corrected concentration of a compound is the concentration of the compound that would be expected at one point (B) located downgradient from another point (A) after correcting for the effects of dispersion, dilution from recharge, volatilization, and sorption between points A and B. One method of calculating the corrected concentration (Wiedemeier et al., 1996b) is given by

$$C_{\mathrm{B,corr}} = C_{\mathrm{B}} \frac{T_{\mathrm{A}}}{T_{\mathrm{B}}} \tag{7.8}$$

where $C_{\mathrm{B,corr}}$ = corrected concentration of compound of interest at point B
C_{B} = measured concentration of compound of interest at point B
T_{A} = measured concentration of tracer (trimethylbenzene or similar) at point A
T_{B} = concentration of tracer (trimethylbenzene or similar) at point B

Because trimethylbenzene is slightly more hydrophobic than BTEX and therefore has a higher soil sorption coefficient, there probably is preferential sorption of TMB relative to the BTEX compounds. In addition, TMB may not be entirely recalcitrant under some anaerobic conditions, and it appears to degrade rapidly under aerobic conditions. The degree of recalcitrance of TMB is site specific, and use of this compound as a tracer must be evaluated on a case-by-case basis. However, if any TMB mass is lost to the processes of biodegradation or preferential sorption, eq. (7.8) is conservative because the calculated mass losses and the attenuation rate constants calculated on the basis of those losses will be lower than the actual losses and attenuation rates.

An approximate first-order biodegradation rate can be calculated if it can be shown that the corrected contaminant distribution approximates a distribution resulting from a first-order process. To show that the process is first order, a log-linear plot of tracer-corrected dissolved BTEX concentrations versus downgradient travel time along the flow path should be made. If the points plot along a straight line, the process is first order.

Substituting the TMB-corrected concentration, $C_{\mathrm{B,corr}}$, at a downgradient point (B) for C in eq. (4.3) and the measured concentration, C_{A}, at an upgradient point (A) for C_0, this relationship becomes

$$C_{\mathrm{B,corr}} = C_{\mathrm{A}} e^{-\lambda t} \tag{7.9}$$

where $C_{\mathrm{B,corr}}$ = tracer-corrected contaminant concentration at time t at downgradient point B
C_{A} = measured contaminant concentration at upgradient point A
λ = approximate first-order biodegradation constant

The travel time between two points is given by

$$t = \frac{x}{\upsilon_c} \tag{7.10}$$

where t = travel time between two points
x = distance between two points
υ_c = retarded solute velocity (where applicable)

Use of the retarded solute velocity is appropriate in which the organic carbon content of the aquifer matrix is sufficient to retard the migration of dissolved BTEX compounds. However, in many cases, retardation of the plume will be so limited that use of the seepage velocity will approximate the retarded velocity. Background aquifer organic carbon concentrations may be used to calculate a retardation factor and a retarded solute velocity, but this may not accurately represent the solute velocity in the plume interior. However, as long as the retarded solute velocity is overestimated (i.e., retardation is underestimated), the rate constant calculated using this method will be conservatively low.

Example 7.1: Estimating First-Order Rates Using TMB as a Conservative Tracer. Given an upgradient total BTEX concentration of 1250 μg/L and trimethylbenzene concentration of 425 μg/L and a downgradient total BTEX concentration of 100 μg/L and trimethylbenzene concentration of 400 μg/L, calculate an approximate first-order biodegradation rate constant. Assume that the groundwater flow velocity is 1000 ft/yr and the two points are 1000 ft apart in the direction parallel to groundwater flow. Aquifer sediments consist of well-sorted beach sand with a total organic carbon content of 0.0005% so sorption can be neglected.

SOLUTION: Using eq. (7.8), we have

$$C_{\text{B,corr}} = 100\left(\frac{425}{400}\right) = 106 \ \mu\text{g/L}$$

Solving eq. (7.9) for λ and knowing that the travel time between points A and B is one year, the biodegradation rate is approximated by

$$\lambda = -\frac{\ln(106/1250)}{1} = 2.5 \text{ year}^{-1} = 0.007 \text{ day}^{-1}$$

Wiedemeier et al. (1996b) discuss in depth how this technique is applied and compare the results obtained using this method with results obtained using the method of Buscheck and Alcantar (1995) (see below).

Methods That Assume Steady-State Equilibrium

Buscheck and Alcantar (1995) derive a relationship that allows calculation of approximate biodegradation rate constants. An important assumption that must be made when using this method is that the contaminant plume has reached a steady-state configuration. This method involves coupling the regression of contaminant concentration (plotted on a logarithmic scale) versus distance downgradient (plotted on a linear scale) to an analytical solution for one-dimensional, steady-state, contaminant transport that includes advection, dispersion, sorption, and biodegradation. The effects of volatilization are assumed to be negligible. For a steady-state plume, the first-order decay rate is approximated by (Buscheck and Alcantar, 1995)

$$\lambda = \frac{\upsilon_c}{4\alpha_x}\left[\left(1 + 2\alpha_x\left(\frac{k}{\upsilon_x}\right)\right)^2 - 1\right] \tag{7.11}$$

where λ = first-order biological decay rate
υ_c = retarded contaminant velocity in x direction
α_x = dispersivity
υ_x = groundwater seepage velocity in x direction
k/υ_x = slope of line formed by making log-linear plot of contaminant concentration versus distance downgradient along flow path

When used with accurate estimates of dispersivity and groundwater flow and solute transport velocity, this method gives reasonable first-order biodegradation rates. Examples of how to apply this method are given in Buscheck and Alcantar (1995) and Wiedemeier et al. (1996b). This method can also be used to estimate biodegradation rates for chlorinated solvents dissolved in groundwater. Chapelle et al. (1996) present a similar method for estimating biodegradation rate constants.

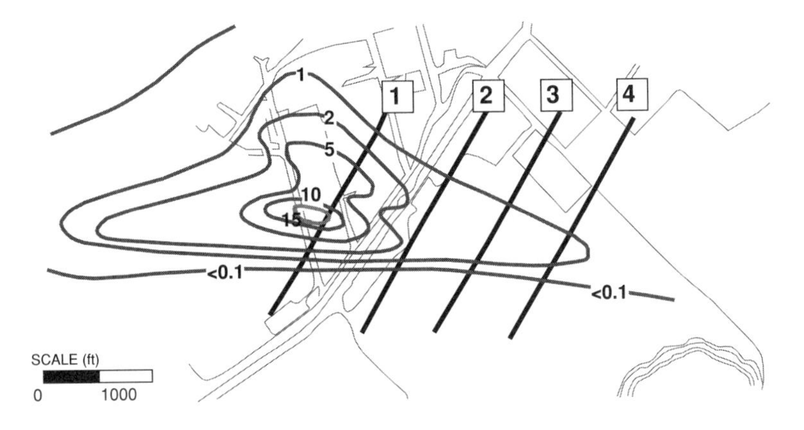

Units: mg/L Total Chlorinated Solvents

FIGURE 7.15. Total chlorinated solvents for mass balance calculation.

Mass Balance Methods

For sites in which sufficient historical data have been collected (a minimum of three sampling events), it may be possible to estimate a biodegradation rate constant by estimating the change in dissolved mass within the plume as a function of time. There are several approaches for estimating the dissolved mass DM at a given time t, including the use of an average plume concentration:

$$\mathrm{DM}_t = C_{\mathrm{avg},t}\, bnLW \tag{7.12}$$

where b = aquifer thickness
n = porosity
L = plume length
W = plume width

Rifai et al. (1988) and Chiang et al. (1989) used a variation of this method to estimate mass loss rate constants for sites in Traverse City, Michigan and at the Michigan Gas Plant Facility, respectively. Both studies estimated a rate constant of approximately 1% per day for BTEX at these sites.

The Remediation Technologies Development Forum (1997) uses a graphical method for estimating mass loss that is based on having good isoconcentration maps for the site in question. The authors draw several lines perpendicular to the flow and at various distances away from the source (see Figure 7.15) on the site isoconcentration map and using the thickness of the aquifer and the groundwater velocity, estimate the mass of groundwater per year that passes through each line. Their estimates involve using the following relationships:

$$V_w = Lbvn \tag{7.13}$$

$$M_w = Vp \tag{7.14}$$

$$M_c = M_w C_i \tag{7.15}$$

where V_w = water volume
L = length of perpendicular line contained between isoconcentration contour
b = aquifer thickness
M_w = water mass
M_c = contaminant mass
C_i = average concentration within the contour interval crossed by the perpendicular line

Table 7.10 presents an example of mass balance calculations for the site shown in Figure 7.15.

TABLE 7.10. Mass Loss Calculation Worksheet

Length of 0–1 ppm	300 ft
Length of 1–2 ppm	250 ft
Length of 2–5 ppm	330 ft
Length of 5–10 ppm	590 ft
Length of 10–15 ppm	200 ft
	1670 ft
Plume width	1670 ft

Hint: Be sure to measure both sides of the plume and then add the two concentration segment lengths.

$$M_{w(0-1)} = (300)(10)(140)(0.3)62.5 = 7{,}875{,}000 \text{ lb}$$
$$M_{w(0-2)} = (250)(10)(140)(0.3)62.5 = 6{,}562{,}500 \text{ lb}$$
$$M_{w(2-5)} = (330)(10)(140)(0.3)62.5 = 8{,}662{,}500 \text{ lb}$$
$$M_{w(5-10)} = (590)(10)(140)(0.3)62.5 = 15{,}487{,}500 \text{ lb}$$
$$M_{w(10-15)} = (200)(10)(140)(0.3)62.5 = 5{,}250{,}000 \text{ lb}$$

$$M_{c(0-1)} = \frac{M_{w(0-1)}(0.5)}{1{,}000{,}000} = 3.9 \text{ lb}$$

$$M_{c(1-2)} = \frac{M_{w(1-2)}(0.5)}{1{,}000{,}000} = 9.8 \text{ lb}$$

$$M_{c(2-5)} = \frac{M_{w(2-5)}(0.5)}{1{,}000{,}000} = 30.3 \text{ lb}$$

$$M_{c(5-10)} = \frac{M_{w(5-10)}(0.5)}{1{,}000{,}000} = 116.2 \text{ lb}$$

$$M_{c(10-15)} = \frac{M_{w(10-15)}(0.5)}{1{,}000{,}000} = 65.6 \text{ lb}$$

$$\text{Mass/year} = M_{c(0-1)} + M_{c(1-2)} + M_{c(2-5)} + M_{c(5-10)} + M_{c(10-15)} = 225.8 \text{ lb/yr}$$

Source: Remediation Technologies Development Forum (1997).

TABLE 7.11. Worksheet for Half-Life Calculation by Graphical Extrapolation

Measure distances from the most-downgradient source.

The compound whose half-life is being estimated: ________________

Starting concentration: ________________

–50% = __________ –75% = __________ –90% = __________

D_1 = distance to 50% decline = __________ ft

D_2 = distance to 75% decline = __________ ft

D_3 = distance to 90% decline = __________ ft

V_x = groundwater seepage velocity = __________ ft/yr

$$\text{Estimated half-life} = \frac{(D_1/v_x) + (D_2/v_x) + (D_3/v_x)}{6}$$

$$= \frac{(\_\_\_\_) + (\_\_\_\_) + (\_\_\_\_)}{6}$$

$$= \_\_\_\_\_\_\_\_\_\_ \text{ yr}$$

Caution: This calculation is an approximation.

Source: Remediation Technologies Development Forum (1997).

Graphical Method. The Remediation Technologies Development Forum (1997) presents a graphical method for estimating rate constants using field data. Their method estimates points along a transect within the plume where 50%, 75%, and 90% concentration reduction is observed (relative to source concentrations). These concentration reductions are associated with one, two, and three half-lives, respectively. The distances from the source area to these points are then measured and divided by the groundwater velocity. The sum of the resulting values divided by 6 will yield the half-life for the chemical at the site. These calculations are shown in Table 7.11.

7.7 SCREENING FOR NATURAL ATTENUATION OF PETROLEUM HYDROCARBONS

Experience has shown that the efficiency of natural attenuation depends on particular hydrologic and geochemical characteristics of the groundwater system into which the contaminants are introduced. This, in turn, makes it possible to assess the efficiency of natural attenuation using a few general principles. In this section we emphasize these general principles for petroleum hydrocarbons. It is not possible, nor is it desirable, to assess natural attenuation at all sites using exactly the same procedures. It is possible, however, to identify general guidelines that can generally be applied to a wide variety of hydrologic conditions. This section is intended to provide this general guidance. Ultimately, the accurate assessment of natural attenuation will rely as much on the sound professional judgment of the practitioner responsible for the assessment as on the guidelines being followed. A more detailed methodology for evaluating the natural attenuation of petroleum hydrocarbons is presented in the AFCEE protocol entitled *Technical Protocol for Implementing Intrinsic Remediation with Long-Term Monitoring for Natural Attenuation of Fuel Contamination Dissolved in Groundwater* (Wiedemeier et al., 1995).

The precise methodology for assessing redox processes in groundwater systems and evaluating the efficacy of natural attenuation of petroleum hydrocarbons will vary from site to site depending on ambient conditions. Nevertheless, the following steps can be taken in sequence to screen for natural attenuation of petroleum hydrocarbons dissolved in groundwater (after Wiedemeier and Chapelle, 1998):

Step 1. *Place a sufficient number of monitoring wells at the site to delineate the areal and vertical extent of groundwater contamination and to determine the distribution of hydrostratigraphic units.* Figure 7.16 shows extremely generalized locations for placement of monitoring wells. A considerable amount of professional judgment should be used when placing wells to characterize a site. The final number and placement of monitoring wells will depend on many factors, including but not limited to (a) the size of the spill, (b) the hydrogeologic complexity of the site, and (c) the distance and potential pathways to sensitive receptors.

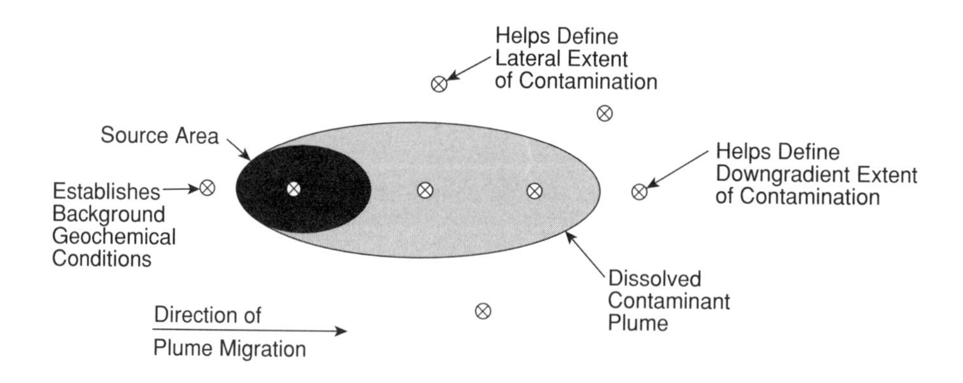

FIGURE 7.16. General areas for collection of screening data.

Step 2. *Measure water levels in monitoring wells, prepare a potentiometric surface map, determine the hydraulic gradient, and perform slug tests or pumping tests to determine the distribution of hydraulic conductivity.* Use measured hydraulic conductivity data, hydraulic gradient data, and estimates of aquifer porosity to determine the direction and velocity of groundwater flow using eq. (3.1):

$$\upsilon_x = -\frac{K}{n_c}\frac{dH}{dL}$$

where υ_x = average groundwater seepage velocity (ft/day)
K = hydraulic conductivity (ft/day)
n_e = effective porosity (dimensionless)
dH/dL = hydraulic gradient (dimensionless)

Step 3. *Measure concentrations of contaminants, inorganic electron acceptors, and metabolic by-products listed in Table 7.3.* The chemical and geochemical data presented in Table 7.3 should be obtained from four general areas, which should include (Figure 7.16): (a) the most contaminated portion of the aquifer (generally in the area in which NAPL currently is present or was present in the past), (b) downgradient from the NAPL source area but still in the dissolved contaminant plume, (c) downgradient from the dissolved contaminant plume, and (d) upgradient and lateral locations that are not affected by the plume. Keep in mind that Figure 7.16 is a simplified two-dimensional representation of a contaminant plume. Real contaminant plumes are three-dimensional objects. This must be considered during site characterization and groundwater monitoring well placement.

The sample collected in the NAPL source area provides information on the dominant terminal electron-accepting processes operating in the source area. This area is likely to be the most reducing part of the contaminant plume. In conjunction with the sample collected in the NAPL source zone, samples collected in the dissolved plume downgradient from the NAPL allow one to determine if the plume is degrading with distance along the flow path and to determine the distribution of electron acceptors and metabolic by-products along the flow path. The sample(s) collected downgradient from the dissolved plume aid in plume delineation and allow one to determine if metabolic by-products are present in an area of groundwater that has been remediated. The upgradient and lateral samples allow delineation of the plume and determination of background concentrations of electron acceptors.

The spatial relationships between contaminants, inorganic electron acceptors [dissolved oxygen, nitrate, Mn(IV), Fe(III), and sulfate], and metabolic by-products [Mn(II), Fe(II), and methane] can be used to evaluate qualitatively the occurrence of petroleum-hydrocarbon degradation. Isopleth maps (isoconcentration maps) show the relationships between these parameters in space. These maps portray graphically the first and second lines of evidence used to document natural attenuation. Figures 7.9 and 7.10 give examples of completed isopleth maps for BTEX, dissolved oxygen, nitrate, Fe(II), sulfate, and methane. The patterns illustrated by these figures show that there is a strong correlation between areas with elevated BTEX, depleted electron acceptors, and elevated metabolic by-product concentrations at the site. These relationships provide strong evidence that biodegradation is occurring via the processes of aerobic respiration, denitrification, Fe(III) reduction, sulfate reduction, and methanogenesis.

Step 4. *Estimate the biodegradation rates.* Such estimates can be made by evaluating field data (Section 7.6). Experience has shown that in many cases the biodegradation kinetics for petroleum hydrocarbons are adequately described by a first-order rate law.

Step 5. *Use the results of steps 1, 2, and 4 to compare the rate of contaminant transport to the rate of biodegradation.* Such comparisons can be made using simple analytical solutions of the advection–dispersion equation such as that presented in the BIOSCREEN model (Chapter 8). For complex problems (e.g., sites with a heterogeneous distribution of hydraulic conductivity), numerical solute transport models may be appropriate.

Step 6. *Evaluate potential receptor impacts.* If the efficiency of natural attenuation is sufficient to prevent contaminant transport to sensitive receptors, natural attenuation may be a feasible remedial option. If the efficiency of natural attenuation is insufficient to prevent contaminant transport to sensitive receptors, natural attenuation may be ruled out as a remedial option.

7.8 SCREENING FOR NATURAL ATTENUATION OF CHLORINATED SOLVENTS

Experience has shown that the efficiency of natural attenuation depends on particular hydrologic and geochemical characteristics of the groundwater system into which the contaminants are introduced. This, in turn, makes it possible to assess the efficiency of natural attenuation using a few general principles. In this section we emphasize these general principles for chlorinated solvents. It is not possible, nor is it desirable, to assess natural attenuation at all sites using exactly the same procedures. It is possible, however, to identify guidelines that can be applied generally to a wide variety of hydrologic conditions. This section is intended to provide this general guidance. Ultimately, accurate assessment of natural attenuation will rely as much on the sound professional judgment of the practitioner responsible for the assessment as on the guidelines being followed. A more detailed methodology for evaluating the natural attenuation of chlorinated solvents is presented in the AFCEE protocol entitled *Technical Protocol for Evaluating the Natural Attenuation of Chlorinated Solvents Dissolved in Groundwater* (Wiedemeier et al., 1996a, 1998).

The precise methodology for assessing redox processes in groundwater systems and evaluating the efficacy of natural attenuation of chlorinated solvents will vary from site to site depending on ambient conditions. Nevertheless, the following steps, modified from Wiedemeier et al. (1996a, 1998) and Wiedemeier and Chapelle (1998), can be taken in sequence to delineate ambient microbial redox processes in groundwater systems:

Step 1. *Place enongh PVC-cased monitoring wells at the site to delineate the areal and vertical extent of groundwater contamination and determine the distribution of hydrostratigraphic units.* Figure 7.16 shows extremely generalized locations for placement of monitoring wells. A considerable amount of professional judgement should be used when placing monitoring wells to characterize the site. The final number and placement of monitoring wells will depend on many factors, including but not limited to (a) the size of the spill, (b) the hydrogeologic complexity of the site, and (c) the distance and potential pathways to sensitive receptors.

Step 2. *Measure water levels in the wells, prepare a potentiometric surface map, determine hydraulic gradient, and perform slug tests or pumping tests to determine the distribution of hydraulic conductivity.* Use measured hydraulic conductivity data, hydraulic gradient data, and estimates of aquifer porosity to determine the direction and velocity of groundwater flow using eq. (3.1):

$$\upsilon_x = \frac{K}{n_e}\frac{dH}{dL}$$

where υ = average groundwater seepage velocity (ft/day)
K = hydraulic conductivity (ft/day)
n_e = effective porosity (dimensionless)
dH/dL = hydraulic gradient (dimensionless)

Step 3. *Measure the concentrations of indicator parameters listed in Table 7.3.* The chemical and geochemical data presented in Table 7.3 should be obtained from four general areas (Figure 7.16): (1) the most contaminated portion of the aquifer (generally in the area where NAPL currently is present or was present in the past); (2) downgradient from the NAPL source area but still in the dissolved contaminant plume; (3) downgradient from the dissolved contaminant plume; and (4) upgradient and lateral locations that are not affected by the plume. Keep in mind that Figure 7.16 is a simplified two-dimensional representation of a contaminant plume. Real contaminant plumes are three-dimensional objects. This must be considered during site characterization and groundwater monitoring well placement.

The sample collected in the NAPL source area provides information on the dominant terminal electron-accepting processes operating in the source area. This area is likely to be the most reducing part of the contaminant plume. In conjunction with the sample collected in the NAPL source zone, samples collected in the dissolved plume downgradient from the NAPL allow one to determine if the plume is degrading with distance along the flow path and to determine the distribution of electron acceptors and donors and metabolic by-products along the flow path. The sample(s) collected downgradient from the dissolved plume aid in plume delineation and will show if metabolic by-products are present in an area of groundwater that has been remediated. The upgradient and lateral samples allow delineation of the plume and determination of background concentrations of electron acceptors and donors.

Step 4. *Using the data collected in step 3, apply the procedure discussed in Section 7.4 and the logic of Figure 7.14 to deduce the distribution of ambient redox processes at the site and to screen the site to determine if biodegradation via reductive dechlorination is occurring.* In addition, use these data to document contaminant and daughter product concentration changes as groundwater flows downgradient, and concentration changes of chloride as groundwater flows downgradient. First use the logic discussed in Section 7.4 and summarized in Figure 7.14. Once the distribution of terminal electron-accepting processes is known, apply the screening process presented next to determine if biodegradation via reductive dechlorination is occurring.

The biogeochemical signature left in groundwater when organic compounds are biodegraded, in conjunction with the ambient geochemical conditions within the aquifer, can be used to assess the potential for chlorinated solvent biodegradation. This approach was first published by Wiedemeier et al. (1996a) and has been applied at many sites across the country. Because biodegradation is the most important destructive process acting to reduce contaminant mass in groundwater, an accurate estimate of the potential for natural biodegradation is important when assessing risk to human health and the environment. The screening process presented herein can be used to determine if biodegradation is probably occurring at a site contaminated with chlorinated solvents or a mixture of petroleum hydrocarbons and chlorinated solvents.

For most of the common chlorinated solvents (e.g., PCE, TCE, TCA, and CT), the initial biotransformation in the environment is reductive dechlorination. It is important to note that TCA can be chemically transformed by means of hydrolysis, as discussed in Chapter 3. The initial screening process is designed to recognize geochemical environments in which reductive dechlorination is plausible. It is recognized, however, that some of the less halogenated compounds, such as DCE, VC, the dichloroethanes, chloroethane, dichlorobenzenes, monochlorobenzene, methylene chloride, and ethylene dibromide, can be biodegraded via oxidative pathways, so this screening processes is only to be used to determine if reductive dechlorination is occurring.

The biodegradation screening process involves using the data collected in step 3 to determine if biodegradation is occurring. The right-hand column of Table 7.12 contains scoring values that can be used as a test to assess the likelihood that biodegradation via reductive dechlorination is occurring. This method relies on the fact that biodegradation will cause predictable changes in groundwater chemistry. For example, if the dissolved oxygen concentration in the area of the plume with the highest contaminant concentration is less than 0.5 mg/L, 3 points are awarded. Table 7.13 summarizes the range of possible scores and gives an interpretation for each score. If the score totals 15 or more points, it is likely that biodegradation via reductive dechlorination is occurring.

The following two examples illustrate how this simple screening process is implemented. The site used in the first example is a former fire training area contaminated with chlorinated solvents mixed with petroleum hydrocarbons. Tetrachloroethene was the only chlorinated solvent released at this site, so the

TABLE 7.12. Analytical Parameters and Weighting for Preliminary Screening for Anaerobic Biodegradation Processes

Analysis	Concentration in Most Contaminated Zone	Interpretation	Value
Oxygen	< 0.5 mg/L	Tolerated, suppresses reductive pathway at higher concentrations	3
Oxygen	> 1 mg/L	VC may be oxidized aerobically	–3
Nitrate	< 1 mg/L	May compete with reductive pathway at higher concentrations	2
Fe(II)	> 1 mg/L	Reductive pathway possible; VC may be oxidized under Fe(III)-reducing conditions	3
Sulfate	< 20 mg/L	At higher concentrations may compete with reductive pathway	2
Methane	< 0.5 mg/L	VC oxidizes	0
	> 0.5 mg/L	Ultimate reductive daughter product; VC accumulates	3
Oxidation reduction potential (ORP)	< 50 mV	Reductive pathway possible	1
	< –100 mV	Reductive pathway likely	2
PH	5 < pH < 9	Optimal range for reductive pathway	0
	5 > pH > 9	Outside optimal range for reductive pathway	–2
TOC	> 20 mg/L	Carbon and energy source; drives dechlorination; can be natural or anthropogenic	2
Temperature	> 20°C	At $T > 20$°C, biochemical process is accelerated	1
Carbon dioxide	> 2 × background	Ultimate oxidative daughter product	1
Alkalinity	> 2 × background	Results from interaction of carbon dioxide with aquifer minerals	1
Chloride	> 2 × background	Daughter product of organic chlorine	2
Hydrogen	> 1 nM[a]	Reductive pathway possible; VC may accumulate	3
Hydrogen	< 1 nM	VC oxidized	0
Volatile fatty acids	> 0.1 mg/L	Intermediates resulting from biodegradation of aromatic compounds; carbon and energy source	2
BTEX	> 0.1 mg/L	Carbon and energy source; drives dechlorination	2
Tetrachloroethene	—	Material released	0

continues

TABLE 7.12. *(Continued)*

Analysis	Concentration in Most Contaminated Zone	Interpretation	Value
Trichloroethene	—	Material released	0
		Daughter product of PCE	2[a]
DCE	—	Material released	0
		Daughter product of TCE	2[a]
		If *cis* is greater than 80% of total DCE, it is probably a daughter product of TCE	
		1,1-DCE can be chemical reaction product of TCA	
VC	—	Material released	0
		Daughter product of DCE	2[a]
1,1,1-Trichloroethane	—	Material released	0
DCA	—	Daughter product of TCA under reducing conditions	2
Carbon tetrachloride	—	Material released	0
Chloroethane	—	Daughter product of DCA or VC under reducing conditions	2
Ethene/ethane	> 0.01 mg/L	Daughter product of VC/ethene	2
	> 0.1 mg/L		3
Chloroform	—	Material released	0
		Daughter product of carbon tetrachloride	2
Dichloromethane	—	Material released	0
		Daughter product of chloroform	2

[a]Points awarded only if it can be shown that the compound is a daughter product (i.e., not a constituent of the source NAPL).

TABLE 7.13. Interpretation of Points Awarded During Screening Step 1

Score	Interpretation
0–5	Inadequate evidence for anaerobic biodegradation[a] of chlorinated organics
6–14	Limited evidence for anaerobic biodegradation[a] of chlorinated organics
15–20	Adequate evidence for anaerobic biodegradation[a] of chlorinated organics
> 20	Strong evidence for anaerobic biodegradation[a] of chlorinated organics

[a]Reductive dechlorination.

TCE, DCE, and VC could only have come from reductive dechlorination of the PCE. Groundwater within the contaminant plume has low dissolved oxygen, nitrate, and sulfate concentrations and elevated Fe(II), methane, and chloride concentrations. Table 7.14 summarizes the geochemical scoring for this site. The geochemical conditions at the site led to a score of 23 points. The presence of petroleum hydrocarbons at this site appears to have reduced the oxidation–reduction potential of the groundwater to the extent that reductive dechlorination is occurring. Although intrinsic bioremediation of contaminants is occurring at this site, a full assessment of natural attenuation should be made to ensure protection of human health and the environment. This is especially true since some of the daughter products of PCE are more toxic than the parent compound.

The second example contains data from a dry cleaning site contaminated with PCE. This site was contaminated with spent cleaning solvents that were dumped into a shallow dry well situated just above a well-oxygenated unconfined aquifer with low concentrations of organic carbon. Groundwater within the contaminant plume has high concentrations of dissolved oxygen and low concentrations of nitrate, Fe(II), sulfate, methane, and chloride. Table 7.15 summarizes the geochemical scoring for this site. The geochemical conditions at the site lead to a score of only 1 point. In this example it can be inferred that reductive dechlorination is probably not occurring or is occurring too slowly to contribute significantly to natural attenuation. In this case, other natural attenuation mechanisms, such as dispersion, dilution from recharge, sorption, and volatilization, should be critically analyzed or an engineered remediation system should be designed.

TABLE 7.14. Strong Evidence for Anaerobic Biodegradation (Reductive Dechlorination) of Chlorinated Organics

Analyte	Concentration in Most Contaminated Zone	Points Awarded
Dissolved oxygen	0.1 mg/L	3
Nitrate	0.3 mg/L	2
Fe(II)	10 mg/L	3
Sulfate	2 mg/L	2
Methane	5 mg/L	3
ORP	–190 mV	2
Chloride	3 × background	2
PCE (released)	1000 μg/L	0
TCE (none released)	1200 μg/L	2
cis-1,2-DCE (none released)	500 μg/L	2
VC (none released)	50 μg/L	2
		23

TABLE 7.15. Anaerobic Biodegradation (Reductive Dechlorination) Unlikely

Analyte	Concentration in Most Contaminated Zone[a]	Points Awarded
Dissolved oxygen	8 mg/L	–3
Nitrate	0.3 mg/L	2
Fe(II)	Not detected (ND)	0
Sulfate	10 mg/L	2
Methane	ND	0
ORP	100 mV	0
Chloride	Background	0
TCE (released)	1200 mg/L	0
cis-1,2-DCE	ND	0
VC	ND	0
		1

[a]ND, not detected.

Step 5. *Categorize the site according to the progression of redox processes and deduce the efficiency of natural attenuation according to the progression of redox conditions, mass loss of solvent concentrations, and the production and destruction of daughter products.* The most common, but certainly not the only, progressions observed in the field include:

- *Progression from methanogenic to Fe(III)-reducing to oxic conditions (Type 1 or 2 environment grading to Type 3 environment).* This scenario is characterized by methanogenic conditions in the contaminant source area, grading to Fe(III)-reducing conditions immediately downgradient of the source area, and finally grading to oxic conditions farther downgradient (Type 1 or 2 environment grading to Type 3 environment downgradient). This scenario results in rapid and efficient reductive dechlorination at the source area followed by oxidation of DCE and VC downgradient of the source area. Extensive mass loss of solvent along aquifer flow paths; production and subsequent mass loss of daughter products along aquifer flow paths. *Highly efficient natural attenuation is probable.*
- *Progression from sulfate- or Fe(III)-reducing to oxic conditions (Type 1 or 2 environment grading to Type 3 environment).* This scenario is characterized by sulfate- or Fe(III)-reducing conditions at the contaminant source area, grading to oxic conditions downgradient (Type 1 or 2 environment grading to Type 3 environment downgradient). This scenario results in reductive dechlorination at the source area followed by oxidation of DCE and VC downgradient of the source area. Partial mass loss of solvents; production and subsequent loss of daughter products along aquifer flow paths. Low concentrations of PCE and TCE transported to oxic zone, in which biodegradation processes stop but dilution continues. *Moderately efficient natural attenuation is probable.*

- *Progression from oxic to anoxic conditions (Type 3 environment grading to Type 1 or 2 environment).* Oxic conditions at contaminant source area grading to reducing [Fe(III)-reducing, sulfate-reducing, or methanogenic] conditions downgradient (Type 3 environment grading to Type 1 or 2 environment downgradient). No reductive dechlorination at the source area, some dechlorination, and mass loss of solvent downgradient. Production and degradation of daughter products. *Moderately efficient natural attenuation is probable.*
- *Uniformly oxic conditions (Type 3 environment throughout).* Uniformly oxic conditions in contaminant source area and downgradient areas (Type 3 environment throughout). No reductive dechlorination, no production of oxidizable daughter products. *Inefficient natural attenuation is probable.*

Step 6. *Estimate biodegradation rate constants.* Biodegradation is the most important process that degrades contaminants in the subsurface. Because of this, biodegradation rates are among the most important model input parameters. In many cases, the biodegradation of chlorinated solvents can commonly be approximated using a first-order decay rate. Whenever possible, use site-specific biodegradation rates estimated from field data. Calculation of site-specific biodegradation rates is discussed in Section 7.6. Biodegradation rate constants can be estimated using a conservative tracer found commingled with the contaminant plume, as described by Wiedemeier et al. (1996b) and in Section 7.6. When dealing with a plume that contains chlorinated solvents, this procedure can be modified to use chloride as a tracer. Rate constants derived from microcosm studies can also be used, but these must be used with caution, for the reasons discussed in Section 7.5.

Step 7. *Evaluate the feasibility of using monitored natural attenuation as a remedial strategy in the context of contaminant transport to receptors.* If the efficiency of natural attenuation is sufficient to prevent contaminant transport to sensitive receptors, natural attenuation may be a workable remedial option. If the efficiency of natural attenuation is insufficient to prevent contaminant transport to sensitive receptors, natural attenuation may be ruled out as a remedial option. Solute transport models that incorporate (at a minimum) advection, dispersion, sorption, and biodegradation may be useful for evaluating potential receptor impacts.

- *Efficient natural attenuation, receptors not affected.* Natural attenuation is a possible sole remedial strategy.
- *Efficient natural attenuation, receptors affected.* Natural attenuation is not a possible sole remedial strategy, but may be used in conjunction with other remedial strategies.
- *Moderate natural attenuation, receptors not affected.* Natural attenuation may be either a sole remedial strategy, or may be used in conjunction with other remedial strategies.
- *Moderate natural attenuation, receptors affected.* Natural attenuation is not a possible sole remedial strategy, but may be used in conjunction with other remedial strategies.

- *Inefficient natural attenuation.* Natural attenuation is not an appropriate remedial strategy.

The primary purpose of comparing the rate of transport to the rate of attenuation is to determine if the residence time along the flow path is adequate to be protective of human health and the environment (i.e., to estimate qualitatively if the contaminant is attenuating at a rate fast enough to allow degradation of the contaminant to acceptable concentrations before receptors are exposed). It is important to perform a sensitivity analysis to help evaluate the confidence in the preliminary screening modeling effort. If modeling shows that receptors will not be exposed to contaminants at concentrations above risk-based corrective action criteria, the screening criteria are met and a natural attenuation evaluation should be completed.

7.9 WEIGHT-OF-EVIDENCE APPROACH FOR DEALING WITH UNCERTAINTY

The terrestrial subsurface varies in complexity from relatively straightforward "sandbox" systems to extremely heterogeneous and complex systems. In addition, the biogeochemical fate of dissolved organic compounds, especially chlorinated solvents, can be quite complicated and will vary depending on site conditions. To remove some of the uncertainty associated with sampling and analyzing sediment and groundwater from complicated systems, care should be taken when collecting samples and independent and converging lines of evidence should be used to evaluate the efficiency of natural attenuation. Such a weight-of-evidence approach will greatly increase the likelihood of successful implementation of natural attenuation at sites in which natural processes are restoring the environmental quality of groundwater.

For example, methane data may suggest that the oxidation–reduction potential of groundwater is sufficiently reducing to allow reductive dechlorination, but this piece of evidence alone is insufficient to show that reductive dechlorination is occurring. Additional evidence that could be used to document reductive dechlorination includes hydrogen concentrations in the range of sulfate reduction or methanogenesis, depleted dissolved oxygen, nitrate, and sulfate concentrations, lowered oxidation–reduction potential, the presence of daughter products that were not spilled at the site, and a stable contaminant plume. In total, the sum of this evidence provides very strong weight of evidence for the occurrence of reductive dechlorination.

REFERENCES

American Petroleum Institute, 1985, Laboratory study on solubilities of petroleum hydrocarbons in groundwater, American Petroleum Institute, Publication Number 4395.

American Society for Testing and Materials (ASTM), 1998, *ASTM Guide for Remediation by Natural Attenuation at Petroleum Release Sites*, E 1943–98.

Baedecker, M. J., and Back, W., 1979, Hydrogeological processes and chemical reactions at a landfill, *Ground Water*, vol. 17, no. 5, pp. 429–437.

Buscheck, T. E., and Alcantar, C. M., 1995, Regression techniques and analytical solutions to demonstrate intrinsic bioremediation, in *Proceedings of the 1995 Battelle Inter-*

national Conference on In Situ and On Site Bioreclamation, Apr., Battelle Press, Columbus, OH.

Chapelle, F. H., 1993, *Groundwater Microbiology and Geochemistry*, Wiley, New York, 424 p.

———, 1996, Identifying redox conditions that favor the natural attenuation of chlorinated ethenes in contaminated groundwater systems, in *Proceedings of the Symposium on Natural Attenuation of Chlorinated Organics in Groundwater*, Dallas, TX, Sept. 11–13, EPA/540/R-96/509, U.S. EPA, Washington, DC.

Chapelle, F. H., McMahon, P. B., Dubrovsky, N. M., Fujii, R. F., Oaksford, E. T., and Vroblesky, D. A., 1995, Deducing the distribution of terminal electron-accepting processes in hydrologically diverse groundwater systems, *Water Resour. Res.*, vol. 31, pp. 359–371.

Chapelle, F. H., Bradley, P. M., Lovley, D. R., and Vroblesky, D. A., 1996, Measuring rates of biodegradation in a contaminated aquifer using field and laboratory methods, *Ground Water*, vol. 34, no. 4, pp. 691–698.

Chapelle, F. H., Vroblesky, D. A., Woodward, J. C., and Lovley, D. R., 1997, Practical considerations for measuring hydrogen concentrations in groundwater, *Environ. Sci. Technol.*, vol. 31, pp. 2873–2877.

Chiang, C. Y., Salanitro, J. P., Chai, E. Y., Colthart, J. D., and Klein, C. L., 1989, Aerobic biodegradation of benzene, toluene, and xylene in a sandy aquifer—data analysis and computer modeling, *Ground Water*, vol. 27, no. 6, pp. 823–834.

Cline, P. V., Delfino, J. J., and Rao, P. S. C., 1991, Partitioning of aromatic constituents into water from gasoline and other complex solvent mixtures, *Environ. Sci. Technol.*, vol. 25, pp. 914–920.

Desai, S., Govind, R., and Tabak, H., 1990, *Determination of Monod Kinetics of Toxic Compounds by Respirometry for Structure-Biodegradability Relationships*, EPA/600/D-90/134, 14 p.

Gilbert, R. O., 1987, *Statistical Methods for Environmental Pollution Monitoring*, Van Nostrand Reinhold, New York.

Groundwater Services, Inc., 1998, *Remediation by Natural Attenuation (RNA) Tool Kit Users Manual*, GSI, Houston, TX.

Hem, J. D., 1985, *Study and Interpretation of the Chemical Characteristics of Natural Water*, Water Supply Paper 2254, U.S. Geological Survey, Washington, DC, 264 pp.

Kampbell, D. H., Wilson, J. T., and Vandegrift, S. A., 1989, Dissolved oxygen and methane in water by a GC headspace equilibrium technique, *Intern. J. Environ. Ana. Chem.*, vol. 36, pp. 249–257.

Kaufman, W. J., and Orlob, G. T., 1956, Measuring groundwater movement with radioactive and chemical tracers, *Am. Water Works Assoc. J.*, vol. 48, pp. 559–572.

Lovley, D. R., and Goodwin, S., 1988, Hydrogen concentrations as an indicator of the predominant terminal electron-accepting reaction in aquatic sediments, *Geochim. Cosmochim. Acta*, vol. 52, pp. 2993–3003.

Lovley, D. R., Chapelle, F. H., and Woodward, J. C., 1994, Use of dissolved H_2 concentrations to determine distribution of microbially catalyzed redox reactions in anoxic groundwater, *Environ. Sci. Technol.*, vol. 28, no. 7., pp. 1205–1210.

Mann, H. B., and Whitney, D. R., 1947, On a test of whether one or more random variables is stochastically larger than in the other, *Ann. Math. Sci.*, vol. 18, pp. 52–54.

Morel, F. M. M., and Hering, J. G., 1993, *Principles and Applications of Aquatic Chemistry*, Wiley, New York.

National Research Council, 1993, *In Situ Bioremediation, When Does it Work?* National Academy Press, Washington, DC, 207 pp.

Norris, R. D., Hinchee, R. E., Brown, R., McCarty, P. L., Semprini, L., Wilson, J. T., Kampbell, D. H., Reinhard, M., Bouwer, E. J., Borden, R. C., Vogel, T. M., Thomas, J. M., and Ward, C. H., 1994, *Handbook of Bioremediation*, Lewis Publishers, Inc., p. 257 p.

Remediation Technologies Development Forum, 1997, Natural attenuation of chlorinated solvents in groundwater, class notes, RTDF.

Rifai, H. S., Bedient, P. B., Wilson, J. T., Miller, K. M., and Armstrong, J. M., 1988, Biodegradation modeling at aviation fuel spill site, *J. Environ. Eng.*, vol. 114, no. 5, pp. 1007–1029.

Smith, J. H., Harper, J. C., and Jaber, H., 1981, *Analysis and Environmental Fate of Air Force Distillate and High Density Fuels*, Report No. ESL-TR-81-54, Tyndall Air Force Base Engineering and Services Laboratory, FL.

Stumm, W., and Morgan, J. J., 1981, *Aquatic Chemistry*, Wiley, New York.

Tabak, H. H., Govind, R., and Desai, S., 1990, *Determination of Biodegradability Kinetics of RCRA Compounds Using Respirometry for Structure Activity Relationships*, EPA/600/D-90/136, 18 p.

U.S. Environmental Protection Agency (U.S. EPA), 1989, *Methods for Evaluating Attainment of Cleanup Standards*, Vol. 1, *Soils and Solid Media*, EPA/230/02-89-042, U.S. EPA, Washington, DC.

———, 1992, *Methods for Evaluating Attainment of Cleanup Standards*, Vol. 2, *Groundwater*, EPA/230-R-92-014, U.S. EPA, Washington, DC.

Wiedemeier, T. H., and Chapelle, F. H., 1998, *Technical Guidelines for Evaluating Monitored Natural Attenuation of Petroleum Hydrocarbons and Chlorinated Solvents in Ground Water at Navy and Marine Corps Facilities*, Naval Facilities Engineering Command, Alternative Restoration Technology Team, September, 1998.

Wiedemeier, T. H., Wilson, J. T., Kampbell, D. H., Miller, R. N., and Hansen, J. E., 1995, *Technical Protocol for Implementing Intrinsic Remediation with Long-Term Monitoring for Natural Attenuation of Fuel Contamination Dissolved in Groundwater*, U.S. Air Force Center for Environmental Excellence, San Antonio, TX.

Wiedemeier, T. H., Swanson, M. A., Moutoux, D. E., Wilson, J. T., Kampbell, D. H., Hansen, J. E., and Haas, P., 1996a, *Overview of the Technical Protocol for Natural Attenuation of Chlorinated Aliphatic Hydrocarbons under Development for the U.S. Air Force Center for Environmental Excellence*, EPA/540/R-96/509, U.S. EPA, Washington, DC, pp. 35–59.

Wiedemeier, T. H., Swanson, M. A., Wilson, J. T., Kampbell, D. H., Miller, R. N., and Hansen, J. E., 1996b, Approximation of biodegradation rate constants for monoaromatic hydrocarbons (BTEX) in groundwater, *Ground Water Monit. Remediation*, vol. 16, no. 3, Summer, pp. 186–194.

Wiedemeier, T. H., Swanson, M. A., Montoux, D. E., Gordon, E. K., Wilson, J. T., Wilson, B. H., Kampbell, D. H., Haas, P. E., Miller, R. N., Hansen, J. E., and Chapelle, F. H., 1998, *Technical Protocol for Evaluating Natural Attenuation of Chlorinated Solvents iin Ground Water*, EPA/600/R-98/128, September 1998, U.S. EPA, Washington, DC, ftp://ftp.epa.gov .pub/ada/reports/protocol.pdf.

Willey, L. M., Kharaka, Y. K., Presser, T. S., Rapp, J. B., and Barnes, Ivan, 1975, Short chain aliphatic acid anions in oil field waters and their contribution to the measured alkalinity, *Geochim. Cosmochim. Acta*, vol. 39, pp. 1707–1711.

CHAPTER 8

MODELING NATURAL ATTENUATION

A groundwater solute fate and transport model (possibly coupled with a groundwater flow model) is a useful tool for evaluating natural attenuation. The model can be used to predict the migration and degradation of the dissolved contaminant plume, to predict contaminant concentrations in a receptor well, or in some cases to provide an estimate of the time required to reach the receptor well. Although these are the most common uses of models for this type of study, models can also be used to synthesize, interpret, and present available data and in turn, to guide any additional data collecting necessitated by the need to verify remediation using natural attenuation. Some models can also be used to evaluate the effects of other remedial actions (e.g., source removal, source reduction, pump-and-treat, or hydraulic containment) on their own or in conjunction with natural attenuation.

Whether used for solute fate and transport, groundwater flow, or both, a model consists of several components. As described by Spitz and Moreno (1996), these components are:

- The natural system for which the model is designed
- A conceptual model as an idealized representation of the natural system
- A mathematical model representing controlling mechanisms in mathematical terms
- Solution of the mathematical model
- Calibration of the solution by adjusting simulated to observed responses of the natural system
- Validation of the accuracy of the model predictions
- Simulations based on the calibrated solution of the conceptual model

In general, any modeling effort will contain some or all of these components, depending on the type of study, the data available, and the type of model being used. The intent of this chapter is to provide an overall understanding of these concepts in the context of their application to natural attenuation. A *natural attenuation model* is a model that simulates solute transport affected by advection, dispersion, sorption, and biodegradation as well as the groundwater flow field in which contaminants are transported. Models used to simulate groundwater flow and solute transport can be classified according to the mathematical technique used to solve the governing partial differential equations. Analytical models provide an exact solution to the governing flow and transport equation and its associated initial and boundary conditions. Numerical models, on the other hand, compute an approximate solution to the problem. In the following sections we describe the mathematical relationships that govern one-, two-, and three-dimensional solute transport and the numerical techniques used to solve these relationships. Also included is a discussion of the merits of analytical and numerical models and some suggestions regarding model selection. Finally, consideration is given to whether or not a model is necessary to implement remediation successfully using natural attenuation at a given site.

8.1 WHY USE MODELS?

Two key questions will invariably arise during a natural attenuation assessment: (1) How far will the plume migrate? and (2) how long will the contaminant plume persist? When properly applied, most analytical and/or numerical fate and transport models along with an adequate field characterization database, can address the first question. The second question is addressed by a much more limited group of models that incorporate variables such as source strength and expected lifetime (see Chapter 2).

One of the first questions to be answered before implementing a solute transport model is whether a model is really necessary. The answer to this question will depend on several factors, including the rate of plume migration and expansion and the locations of potential receptors. For example, if abundant historical data are available for the site and the data show that the dissolved contaminant plume has reached a steady-state configuration or is receding, a solute transport model probably is not necessary to determine how long the plume may extend. However, a model of this site might allow an estimation of how long it will take for the plume to attenuate to acceptable levels. If, on the other hand, the plume is close to a potential receptor and historical data are not available, a solute transport model in conjunction with the appropriate data can be useful for predicting the impact of the plume on potential receptors.

A groundwater model can be applied in two basic ways, such as:

1. To make a priori predictions of plume length without calibration. Best estimates for site data are entered into the model, and a plume length is calculated. No plume calibration is performed because the modeling process is strictly forward looking. Because of uncertainties in various parameters

(such as seepage velocity, dispersion, biodegradation parameters), a sensitivity analysis is typically performed to indicate resolution of the prediction. There is usually considerable range in possible plume lengths using this approach.

2. To determine if a particular plume conceptual model fits the observed data. In this approach, the plume is calibrated to the observed plume conditions using one or more input variables to the model. If the model can be calibrated, the input data are evaluated to determine if they are reasonable and fit the current conceptual model of a site. For example, if the conceptual model of the site is based on no biodegradation at the site but biodegradation is needed to calibrate the model adequately, the model provides evidence that the original no-biodegradation conceptual model needs to be modified.

In the Air Force Natural Attenuation Initiative, natural attenuation models are used in the latter approach, where the modeling results are used as a line of evidence to demonstrate that biodegradation is a key process that affects site consitutents. Many *risk-based corrective action* (RBCA) programs, on the other hand, use models to make a priori plume length predictions without calibration. Because of several different types of conservative safety factors (e.g., using low-end estimates of biodegradation and assuming all receptors are on the centerline of contaminant plumes), the resulting predictions are useful indicators of the maximum risk that a site could pose to groundwater receptors. Other RBCA programs are based on calibration and use of dedicated plume calibration tools (e.g., FATE V; Nevin et al., 1997) to match the model prediction to site conditions observed.

8.2. SELECTION OF A NATURAL ATTENUATION MODEL

The partial differential equations that describe groundwater flow and/or solute transport can be solved analytically or numerically. The type of model selected (analytical or numerical) to simulate site conditions will depend on the results of data review and conceptual model development. *Analytical methods* (*models*) provide exact, closed-form solutions, whereas *numerical methods* (*models*) provide approximate solutions. Analytical models are the simplest to set up and solve, allowing the user to evaluate many scenarios in a relatively short time. Numerical methods are more efficient for complex systems. Analytical models are restricted as to the nature of the problems for which they can be used, and may be inadequate or inappropriate for some field situations. Numerical models, on the other hand, can simulate a broader range of hydrogeological systems and reactions within them.

Groundwater flow and solute transport modeling is often referred to as both an art and a science. The art involves the ability to select a reasonable set of assumptions that will yield a simple but rigorous model that can accurately represent the system being modeled. In essence, the user needs to achieve a balance between simplifying assumptions and the predominant subsurface conditions at the site. Such a balance will depend to a great extent on the intended use of the model results. As an example, a simple analytical model will, in many cases, provide the appropriate information with much less effort than would be required to produce a numerical model. The science involves model setup, parameter selection, and model calibration.

Analytical models may be useful for defining the possible magnitude of a contaminant problem. If limited data are available, or the hydrogeological conditions are simple, an analytical model can be selected to simulate contaminant fate and transport. The basic parameters typically include groundwater seepage velocity, hydraulic conductivity, saturated thickness of the aquifer, porosity, source area configuration and contaminant concentrations, leakage rates, dispersion coefficients, retardation values, and decay rates.

Analytical models provide exact, closed-form solutions to the governing advection–dispersion equation by making significant simplifying assumptions. The more closely the actual system approximates these assumptions, the more accurate the analytical model will be in predicting solute fate and transport. Analytical solutions are continuous in time and space and provide information on the temporal and spatial distribution of hydraulic head or solute concentrations for the governing initial and boundary conditions. The main advantage of analytical solutions is that they are simple to use and provide a good first approximation of solute transport in relatively simple hydrogeologic settings. Analytical solutions are generally limited to steady, uniform flow or radial flow and should not be used for groundwater flow or solute transport problems in strongly anisotropic or heterogeneous media. In some cases, such as where potential receptors are a great distance away or where the aquifer is extremely homogeneous and isotropic, an analytical solution may adequately describe contaminant fate and transport. At a minimum, analytical models are useful for conceptual model development and can aid in locating additional data collection points.

The analytical solutions of the advective–dispersive equation presented in this chapter give solute concentration as a function of time and distance from the source of contamination. Analytical solutions are sometimes used to verify the accuracy of numerical solutions. This is done by applying both the exact analytical solution and the numerical solution to the same groundwater flow system and comparing the results. Several well-documented and widely accepted analytical models are available for modeling the fate and transport of fuel hydrocarbons under the influences of advection, dispersion, sorption, and biodegradation. A number of analytical solute transport models are described in Section 8.4.

Analytical models are used to estimate the impacts of contamination on a site given the qualifying assumptions used to develop the model. Analytical models are best utilized for order-of-magnitude results because a number of potentially important processes are treated in the model in an approximate manner, or are ignored completely. For example, analytical models may include terms describing a variety of chemical and hydrological processes, but usually are not capable of incorporating subsurface heterogeneity. Because of the nature of the simplifying assumptions, analytical models may overestimate or underestimate the spread of contamination. By making assumptions which will ensure the model that will overpredict contaminant concentrations and travel distances (or at least not underpredict them), the model predictions will be conservative. The more conservative a model is, the more confidence there should be that potential receptors will not be affected by site contamination. This will aid in using natural attenuation for remediating a given site.

Numerical solutions provide approximate solutions to the advection–dispersion equation. Numerical models are less burdened by simplifying assumptions and are capable of addressing more complicated problems. Unlike analytical models, numerical models allow subsurface heterogenieties and varying aquifer parameters to be simulated as well as transient simulations (i.e., one or more properties or conditions change over time) if the requisite data are available. In numerical models, the continuous problem domain is replaced by a discretized domain. The resolution of the results provided by a model depends on the degree of discretization (in the model) of the groundwater system under investigation.

Many of the assumptions required for the analytical solutions are not necessary when numerical techniques are used to solve the governing solute transport equation. However, a greater number of site-specific data are needed to implement a numerical model, and the solutions are inexact numerical approximations. Numerical models require input parameters similar to those used for analytical models, but their spatial distribution must be known to warrant use of a numerical model. Several well-documented and widely accepted numerical models are available for modeling the fate and transport of chlorinated aliphatic hydrocarbons and fuel hydrocarbons dissolved in groundwater. Specific numerical fate and transport models are described in Section 8.5.

Numerical models require a reasonably good understanding of the three-dimensional distribution of both aquifer hydraulic properties and the contaminants. Implementation of a numerical model is much more complex than implementation of an analytical model and generally requires an experienced hydrogeologist familiar with the model code. The most commonly used numerical groundwater flow and transport models fall into one of the following four model types:

1. Finite difference
2. Finite element
3. Random walk
4. Method of characteristics (MOC)

These differing methods have been developed to address the multitude of problems presented by the number of physical and chemical processes that can affect groundwater flow and dissolved contaminant transport. Detailed descriptions of these methods may be found in the books by Anderson and Woessner (1992) and Spitz and Moreno (1996).

Figure 8.1 shows a decision process that can be used to determine if an analytical or a numerical model is most appropriate to simulate site conditions. The specific modeling objectives of the project, the available data, and the conceptual model should be the primary factors governing model selection. In addition, the user should avoid making the problem more complex than necessary. Success in solute fate and transport modeling depends on the ability to conceptualize properly the processes

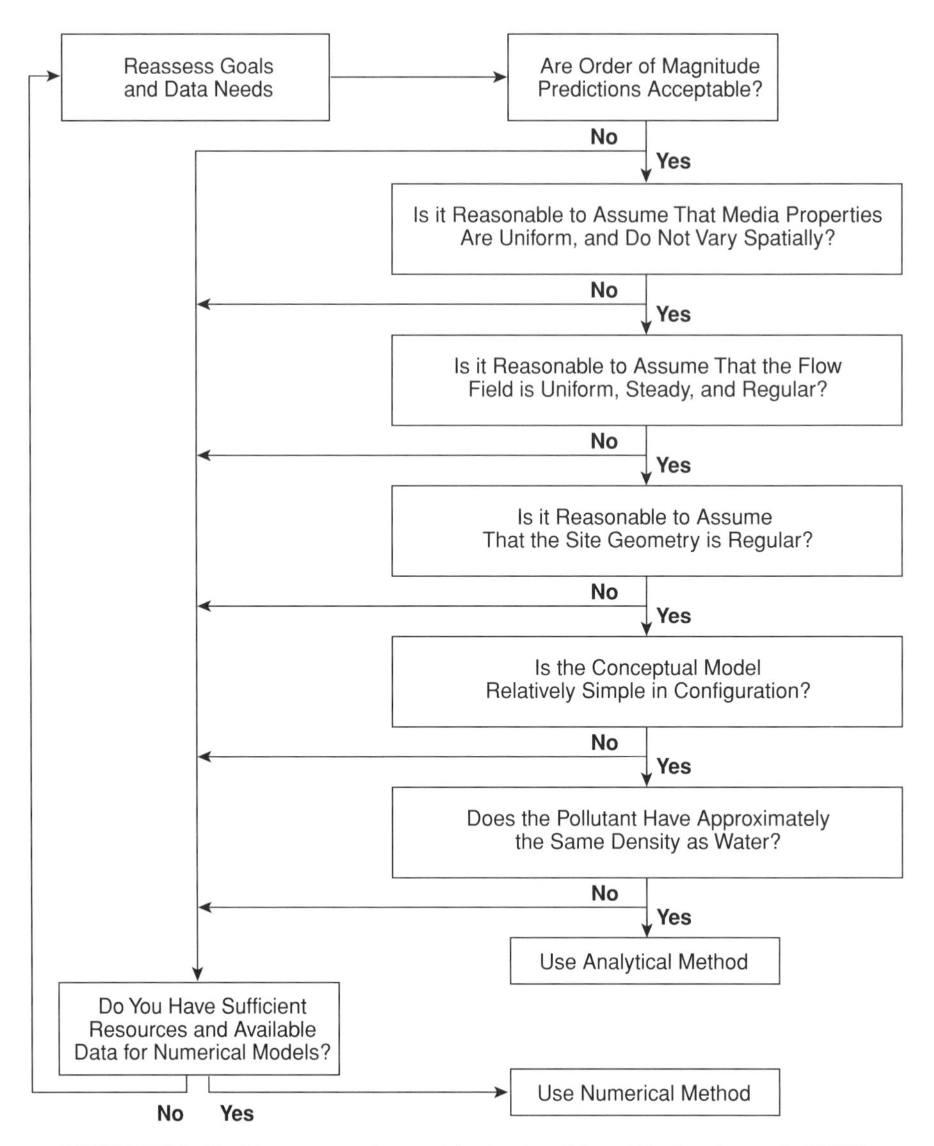

FIGURE 8.1. Decision process for model selection. (From Wiedemeier et al., 1995.)

governing contaminant transport, to select a model that simulates the most important processes at a site, and to achieve reasonable model predictions.

One final caveat should be considered before deciding what type of model to use as well as at all times during the modeling process. Models are a powerful tool but do not necessarily provide definitive answers. With a good sense of the limitations imposed by the simplifying assumptions inherent in the models and the data available, the modeler and/or model user should be able to apply the model and/or its results (as a tool) to reach reasonable conclusions and apply those conclusions appropriately. Failure to understand and work within the limitations of a particular model and data set will lead to erroneous conclusions that will hinder the application of remediation by natural attenuation. The final decision to use an analytical or numerical solute transport model should be based on the complexity of the problem, the amount of available data, and the importance of the decisions that will be based on the model.

Example 8.1 Consider a site located 5 miles from the nearest potential receptor. The database for this site consists of five sampling points, with one round of sampling data from each point. The aquifer system at the site consists of 50 ft of unconsolidated well-sorted medium-grained sand overlying a horizontal shale unit. The shallow water table is 5 ft below the surface. Such a site is an excellent candidate for an analytical model. Consider, on the other hand, a site located approximately 1000 ft from the nearest potential receptor. The database for this site consists of 40 data points for which there are five years of quarterly groundwater-quality sample analyses. The aquifer at this site consists of 10 ft of poorly sorted, silty sand, underlain by 5 ft of well-sorted medium-grained sand, underlain by 20 ft of silt. The quarterly groundwater-quality data indicate that a dissolved contaminant plume is migrating downgradient from the source area. In this situation, a numerical model may be a more appropriate tool to predict the fate and transport of the dissolved contaminant plume.

8.3 GOVERNING EQUATIONS FOR CONTAMINANT FATE AND TRANSPORT

The mathematical relationships that describe groundwater flow and solute transport are based on the equation of continuity and Darcy's law. In the following sections we present one-, two-, and three-dimensional partial differential equations that describe solute transport by the processes of advection, dispersion, sorption, and biodegradation. Several texts derive these equations (Bear, 1972, 1979; Domenico and Schwartz, 1990; Bedient et al., 1994). No discussion of groundwater flow equations is presented in this chapter. The reader is referred to the aforementioned texts, as well as those by Strack (1989) and Spitz and Moreno (1996), for a detailed discussion of the flow equation.

One-, Two-, and Three-Dimensional Reactive Solute Transport

The one-dimensional partial differential equation describing transient solute transport with first-order decay of the solute is

$$\frac{\partial C}{\partial t} = \frac{D_x}{R}\frac{\partial^2 C}{\partial x^2} - \frac{v_x}{R}\frac{\partial C}{\partial x} - \lambda C \tag{8.1}$$

where C = solute concentration
t = time
D_x = hydrodynamic dispersion along flow path
R = coefficient of retardation
x = distance along flow path
υ_x = groundwater seepage velocity in x direction
λ = first-order decay rate constant

The decay rate may be used to simulate any process that is observed to be reducing solute concentrations in a manner that approximates first-order decay, such as biodegradation, radioactive decay, or hydrolysis. Under steady-state conditions, the change in contaminant concentration with respect to time becomes zero, resulting in the following modified equation:

$$0 = \frac{D_x}{R}\frac{\partial^2 C}{\partial x^2} - \frac{\upsilon_x}{R}\frac{\partial C}{\partial x} - \lambda C \qquad (8.2)$$

The two- and three-dimensional partial differential equations describing transient solute transport with first-order biodegradation in the saturated zone are, respectively,

$$\frac{\partial C}{\partial t} = \frac{D_x}{R}\frac{\partial^2 C}{\partial x^2} + \frac{D_y}{R}\frac{\partial^2 C}{\partial y^2} - \frac{\upsilon_x}{R}\frac{\partial C}{\partial x} - \lambda C \qquad (8.3)$$

$$\frac{\partial C}{\partial t} = \frac{D_x}{R}\frac{\partial^2 C}{\partial x^2} + \frac{D_y}{R}\frac{\partial^2 C}{\partial y^2} + \frac{D_z}{R}\frac{\partial^2 C}{\partial z^2} - \frac{\upsilon_x}{R}\frac{\partial C}{\partial x} - \lambda C \qquad (8.4)$$

where C = solute concentration
t = time
D_x = hydrodynamic dispersion along flow path
x = distance along flow path
D_y = hydrodynamic dispersion transverse to flow path
y = distance transverse to flow path
υ_x = groundwater seepage velocity in x direction
R = coefficient of retardation
λ = first-order decay rate constant
D_z = vertical hydrodynamic dispersion
z = vertical distance transverse to flow path

Under steady-state conditions, the change in contaminant concentration with respect to time becomes zero, resulting in the following equations for two- and three-dimensional representations, respectively:

$$0 = \frac{D_x}{R}\frac{\partial^2 C}{\partial x^2} + \frac{D_y}{R}\frac{\partial^2 C}{\partial y^2} - \frac{\upsilon_x}{R}\frac{\partial C}{\partial x} - \lambda C \tag{8.5}$$

$$0 = \frac{D_x}{R}\frac{\partial^2 C}{\partial x^2} + \frac{D_y}{R}\frac{\partial^2 C}{\partial y^2} + \frac{D_z}{R}\frac{\partial^2 C}{\partial z^2} - \frac{\upsilon_x}{R}\frac{\partial C}{\partial x} - \lambda C \tag{8.6}$$

Initial and Boundary Conditions

An important consideration in developing solutions to the governing equations for contaminant fate and transport includes constructing a set of applicable initial and boundary conditions. Initial conditions specify the groundwater flow and contaminant concentrations at the begining of the simulation. Boundary conditions specify the value of the dependent variable (e.g., concentration), or the value of the first derivative of the dependent variable, along the boundaries of the system being modeled. Boundary conditions describe the interaction between the system being modeled and its surroundings. Three types of boundary conditions are generally utilized to describe the transport system: Dirichlet (first type), Neumann (second type), and Cauchy (third type). Table 8.1 summarizes the mathematical expressions for these different types of boundary conditions. In the remaining paragraphs in this section we discuss in more detail which initial and boundary conditions may be applicable for specific transport scenarios.

Initial Conditions. Initial conditions for solute transport models are used to specify the solute concentration, C, in the system at the beginning of the simulation (i.e., at time $t = 0$). Generally, initial conditions take the form

$$C(x,y,z,0) = 0 \tag{8.7}$$

or

$$C(x,y,z,0) = C_i \tag{8.8}$$

where x = distance downgradient of source
y = distance transverse to the source in horizontal direction
z = distance vertically down into aquifer
C_i = initial contaminant concentration

The initial condition shown in eq. (8.7) indicates that there are no contaminants in the aquifer at the beginning of the simulation, whereas eq. (8.8) implies that the aquifer contains a background concentration C_i at time = 0.

Upgradient (Inflow) Boundary Conditions. First-type (constant concentration) and third-type (constant flux) boundary conditions can be used to describe source concentrations at the upgradient (inflow) boundary of an analytical model.

TABLE 8.1. Typical Mathematical Expressions for Boundary Conditions

Boundary Condition	Boundary Type	Formal Name	Mathematical Description
Concentration specified	One	Dirichlet	$C = f(x,y,z,t)$
Flux specified	Two	Neumann	$\frac{\partial C}{\partial n} = f(x,y,z,t)$
Concentration-dependent flux	Three	Cauchy	$\frac{\partial C}{\partial n} + cC = f(x,y,z,t)$

Source: From Wiedemeier et al. (1995).

For example, if the source concentration is constant, a first type of boundary condition of the following form can be used:

$$C(0,y,z,t) = C_0 \tag{8.9}$$

where C is the contaminant concentration and C_0 is the initial contaminant concentration. Similarly, if the source concentration is declining exponentially, the following first-type boundary condition can be used:

$$C(0,y,z,t) = C_0 e^{-\alpha t} \tag{8.10}$$

where α is the source decay rate.

The third type of boundary condition is used in which the concentration gradient across the boundary is dependent on the difference between a specified concentration on one side of the boundary and the solute concentration on the opposite side of the boundary (Wexler, 1992). This boundary condition best describes solute concentrations in which a well-mixed solute enters the system by advection across the boundary. Mathematically, this type of boundary condition can be expressed as

$$v_x C - D_x \frac{\partial C}{\partial x} = v_x C_0 \tag{8.11}$$

where v_x is the velocity across the boundary, D_x is the dispersion coefficient, and x is the distance across the boundary.

Wexler (1992) indicates that the third type of boundary condition is more accurate than the first-type boundary condition. This is because the first type of boundary condition assumes that the concentration gradient across the upgradient boundary is zero the instant flow begins, which tends to overestimate the mass of solute in the system for early time (Wexler, 1992). Most analytical models, however, typically use the first-type boundary condition. Table 8.2 lists additional boundary conditions used to describe the upgradient boundary of a solute transport system for analytical models.

Downgradient (Outflow) Boundary Conditions. Solute transport systems can be simulated as systems of finite length, semi-infinite length, and infinite length. For systems in which the outflow boundary is sufficiently far from the source of

TABLE 8.2. Upgradient Boundary Conditions Used to Simulate the Addition of Contaminants to a Hydrogeologic System

Type of Source Being Simulated	Type of Boundary	One-Dimensional Form[a]
Constant concentration	1	$C(0,t) = C_0$
Pulse-type loading with constant concentration	1	$C(0,t) = C_0, 0 < t \leq t_0$ $C(0,t) = 0, t > t_0$
Decaying source, exponential decay with source concentration approaching 0	1	$C(0,t) = C_0 e^{-k_s[t - x/(V_s/R)]}$
Exponential decay with source concentration approaching C_a	1	$C(0,t) = C_a + C_b e^{-k_s[t - x/(V_s/R)]}$
Constant flux with constant input concentration	3	$\upsilon_x C - D_x \left.\frac{\partial C}{\partial x}\right\|_{x=0} = \upsilon_x C_0$
Pulse-type loading with constant input fluxes	3	$\upsilon_x C - D_x \left.\frac{\partial C}{\partial x}\right\|_{x=0} = \begin{cases} \upsilon_x C_0, & 0 < t \leq t_0 \\ 0, & t > t_0 \end{cases}$

Source: P.A. Domenico and F. W. Schwartz, *Physical and Chemical Hydrogeology*. Copyright © 1990 John Wiley & Sons, Inc. Reprinted by permission of John Wiley & Sons, Inc.

[a] t_0 = time at which concentration changes during pulse loading.

contamination that the downgradient boundary will not influence solute concentrations within the area of interest, the system can be treated as semi-infinite (Wexler, 1992). Semi-infinite systems are modeled using a first or second type of boundary condition at the downgradient boundary.

Lateral and Vertical Boundary Conditions. Lateral and vertical boundary conditions apply to two- and three-dimensional models only. One-dimensional models require only inflow and outflow boundaries. In two- and three-dimensional systems, impermeable or no-flux (no-flow) boundaries may be present at the base, top, or sides of the aquifer (Wexler, 1992). Because there is no flux across the boundary and molecular diffusion across the boundary is assumed to be negligible, the general third type of boundary condition simplifies to a second type of boundary condition, and the boundary conditions are expressed as (Wexler, 1992)

$$\frac{dC}{dy} = 0, \qquad y = 0 \quad \text{and} \quad y = W \tag{8.12}$$

and

$$\frac{dC}{dz} = 0, \qquad z = 0 \quad \text{and} \quad z = H \tag{8.13}$$

where C = contaminant concentration
y = distance in y direction
W = width of aquifer
H = height of aquifer
z = distance in z direction

In many cases, the lateral and vertical boundaries of the system may be far enough away from the area of interest that the system can be treated as being infinite along the y and z axes. If this is the case, the boundary conditions are specified as (Wexler, 1992)

$$C = 0, \quad \frac{dC}{dy} = 0, \quad y = \pm\infty \tag{8.14}$$

and

$$C = 0, \quad \frac{dC}{dz} = 0, \quad z = \pm\infty \tag{8.15}$$

8.4 ANALYTICAL NATURAL ATTENUATION MODELS

Analytical models provide exact, closed-form solutions to the governing advection–dispersion equation. The use of analytical models requires the user to make several simplifying assumptions about the solute transport system. For this reason, analytical models are most valuable for relatively simple hydrogeologic systems that are relatively homogeneous and isotropic and have uniform geometry (straight boundaries and constant thickness, width, and length). Heterogeneous and anisotropic hydrogeologic systems can be modeled using analytical models only if the system is simplified and average hydraulic characteristics are used. As an example, consider a hydrogeologic system composed of several layers with differing thicknesses and hydraulic conductivities. This system could be simulated using an analytical model by averaging the hydraulic conductivity over the entire thickness being modeled, an objective that can be accomplished by dividing the sum of the products of each layer's thickness and hydraulic conductivity by the total aquifer thickness (Walton, 1991).

Table 8.3 lists the analytical solutions presented in this section. Models based on these solutions are capable of simulating advection, dispersion, sorption, and biodegradation (or any first-order decay process). One-, two-, and three-dimensional analytical solutions that simulate a continuous source of contamination are good for determining the worst-case distribution of the dissolved contaminant plume. This is unrealistic, however, if for no other reason than source concentrations decrease over time via natural weathering processes. The models used to simulate a decaying source are especially applicable in which an engineered solution is implemented for source removal. An important model input parameter for such models is the source decay rate. This topic is addressed in detail in Chapter 2.

TABLE 8.3. Analytical Models Commonly Used to Simulate Solute Transport

Processes Simulated	Description	References
	One-Dimensional Model	
Advection, dispersion, linear sorption, and biodegradation—constant source term	Solute transport in a semi-infinite system with a continuous source of contamination. Biodegradation is simulated using a first-order decay rate constant. Solute concentration is given as a function of time and distance.	Bear (1972), van Genuchten and Alves (1982), Wexler (1992)
Advection, dispersion, linear sorption, and biodegradation—decaying source term	Solute transport in a semi-infinite system with a decaying source of contamination. Biodegradation is simulated using a first-order decay rate constant. Solute concentration is given as a function of time and distance.	van Genuchten and Alves (1982)
	Two-Dimensional Model	
Advection, dispersion, linear sorption, and biodegradation—constant source term	Solute transport in a semi-infinite system with a continuous source of contamination. Biodegradation is simulated using a first-order decay rate constant. Solute concentration is given as a function of time and distance.	Wilson and Miller (1978)
	Three-Dimensional Model	
Advection, dispersion, linear sorption, and biodegradation—constant source term	Solute transport in a semi-infinite system with a continuous source of contamination. Biodegradation is simulated using a first-order decay rate constant. Solute concentration is given as a function of distance from the source and time.	Domenico (1987)
Advection, dispersion, linear sorption, and biodegradation—decaying source term	Solute transport in a semi-infinite system with a decaying source of contamination. Biodegradation is simulated using a first-order decay rate constant. Solute concentration is given as a function of distance from the source and time.	Domenico (1987), modified for decaying source concentration

Source: From Wiedemeier et al. (1995).

One-Dimensional Analytical Models

Models presented in this section include a solution for a semi-infinite system with a constant contaminant source of constant concentration and first-order decay of the solute [modified from Bear (1972) by van Genuchten and Alves (1982) and by Wexler (1992)] and a solution for a semi-infinite system with a point source of diminishing concentration and first-order decay of solute (van Genuchten and Alves, 1982). Because they are one-dimensional, these models can represent:

- Contaminant transport down a laboratory column
- Contaminated aquifers in which a constant source extends throughout the saturated zone in both the y and z directions (such as a river channel aquifer

surrounded by impermeable material, with a source penetrating through the entire cross section of the aquifer)

- Cases in which it is appropriate to ignore dispersion in the y and z directions (such as a simple planning model that is focusing on advection or biodegradation)

Semi-infinite System with Constant Source. The one-dimensional analytical solution for eq. (8.1) under the initial and boundary conditions listed below is given by [from Wexler (1992, eq. 60, p. 18), modified from Bear (1972, p. 630), and van Genuchten and Alves (1982, p. 60)]

$$C(x,t) = \frac{C_0}{2}\left(\exp\left\{\frac{x}{2(D_x/R)}\left[\frac{\upsilon_x}{R} - \sqrt{\left(\frac{\upsilon_x}{R}\right)^2 + 4\lambda\frac{D_x}{R}}\right]\right\}\right.$$
$$\operatorname{erfc}\left[\frac{x - t\sqrt{(\upsilon_x/R)^2 + 4\lambda(D_x/R)}}{2\sqrt{D_x/R\ t}}\right]$$
$$+ \exp\left\{\frac{x}{2(D_x/R)}\left[\frac{\upsilon_x}{R} + \sqrt{\left(\frac{\upsilon_x}{R}\right)^2 + 4\lambda\frac{D_x}{R}}\right]\right\}$$
$$\left.\operatorname{erfc}\left[\frac{x + t\sqrt{(\upsilon_x/R)^2 + 4\lambda(D_x/R)}}{2\sqrt{D_x/R\ t}}\right]\right) \quad (8.16)$$

where $C(x,t)$ = contaminant concentration at a distance x downgradient from source at time t
C_0 = initial contaminant concentration at source
x = distance downgradient of upgradient boundary
t = time
D_x = longitudinal hydrodynamic dispersion coefficient
R = coefficient of retardation
υ_x = unretarded linear groundwater flow velocity
λ = first-order decay rate constant for dissolved contaminant
erfc = complementary error function (Table 3.3 and Figure 8.2)

Boundary conditions:

$$C = C_0, \qquad x = 0$$

$$C = 0, \quad \frac{\partial C}{\partial x} = 0, \qquad x = \infty$$

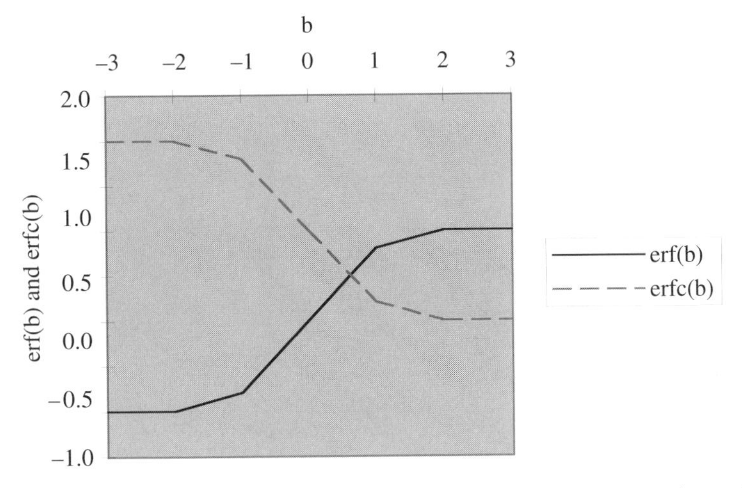

FIGURE 8.2. erf(b) and erfc(b) plotted versus b. (Adapted from P. A. Domenico and F. W. Schwartz, *Physical and Chemical Hydrogeology*. Copyright © 1990 John Wiley & Sons, Inc. Reprinted by permission of John Wiley & Sons, Inc.)

Initial condition:

$$C = 0, \quad 0 < x < \infty \qquad \text{at } t = 0$$

Assumptions:

- The fluid is of constant density and viscosity.
- Biodegradation of the solute is approximated by first-order decay.
- Flow is in the x direction only, and the velocity is constant.
- The longitudinal hydrodynamic dispersion D_x is constant.
- Sorption is approximated by the linear sorption model.

For steady-state conditions, solute transport is described by eq. (8.2) and the solution reduces to (Wexler, 1992, eq. 62, p. 20)

$$C(x) = C_0 \exp\left\{\frac{x}{2(D_x/R)}\left[\frac{v_x}{R} - \sqrt{\left(\frac{v_x}{R}\right)^2 + 4\lambda\frac{D_x}{R}}\right]\right\} \tag{8.17}$$

Example 8.2 Given the hydraulic and contaminant transport parameters below, plot the change in concentration through time at a location 30 m downgradient of the source using eq. (8.16). At what time does the concentration at this point reach steady-state equilibrium?

SOLUTION

Hydrogeologic data:

Hydraulic conductivity	κ = 3.15 m/day
Hydraulic gradient	I = 0.02 m/m
Effective porosity	n_e = 0.25
Total porosity	n = 0.35
Longitudinal dispersivity	α_x = 30 m
Initial contaminant concentration	C_0 = 12 mg/L

Retardation coefficient calculation:

Contaminant decay rate	λ = 0.01 day^{-1}
Soil sorption coefficient	κ_{oc} = 79 mL/g
Particle mass density (for quartz)	ρ_s = 2.65 g/cm^3
Bulk density	$\rho_b = \rho_s(1 - n)$ = 1.722 g/cm^3
Organic carbon content	f_{oc} = 0.8%
Retardation coefficient	$R = 1 + \rho_b \kappa_{oc} f_{oc}/n_e$ = 5.354

Groundwater hydraulics calculation:

Groundwater velocity (pore-water)	$\upsilon_x = \kappa/n_e$ = 0.252 m/day
Longitudinal dispersion coefficient	$D_x = \alpha_x \upsilon_x$ = 7.56 m^2/day
Retarded contaminant velocity	$\upsilon_c = \upsilon_x/R$ = 0.047 m/day

Change in concentration with time calculation:

$i = 1 \ldots 1000, \quad \Delta t = 1 \cdot \text{day}, \quad t_i = \Delta t \cdot_i, \quad x = 30\ \text{m}$

$$C_1 = \frac{C_0}{2}\left(\exp\left\{\frac{x}{2(D_x/R)}\left[\frac{\upsilon_x}{R} - \sqrt{\left(\frac{\upsilon_x}{R}\right)^2 + 4\lambda\frac{D_x}{R}}\right]\right\} \cdot \left\{1 - \operatorname{erf}\left[\frac{x - t_i\sqrt{\left(\frac{\upsilon_x}{R}\right)^2 + 4\lambda(D_x/R)}}{2\sqrt{\frac{D_x}{R}t_i}}\right]\right\} + \exp\left\{\frac{x}{2(D_x/R)}\left[\frac{\upsilon_x}{R} + \sqrt{\left(\frac{\upsilon_x}{R}\right)^2 + 4\lambda\frac{D_x}{R}}\right]\right\} \left\{1 - \operatorname{erf}\left[\frac{x - t_i\sqrt{\left(\frac{\upsilon_x}{R}\right)^2 + 4\lambda(D_x/R)}}{2\sqrt{\frac{D_x}{R}t_i}}\right]\right\}\right)$$

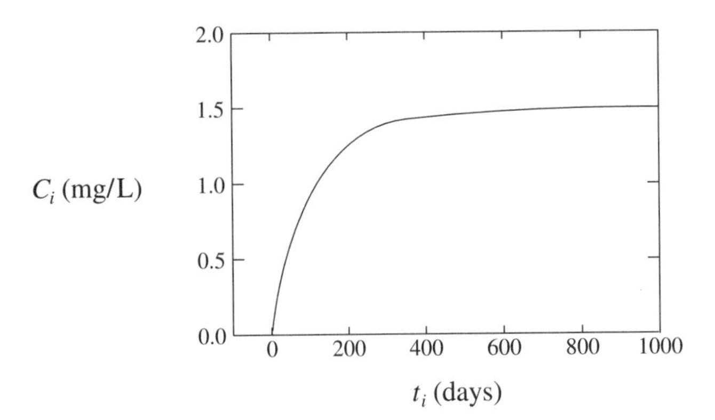

where

$$
\begin{aligned}
C_1 &= 0 \text{ mg/L} \\
C_{100} &= 0.714 \text{ mg/L} \\
C_{200} &= 1.303 \text{ mg/L} \\
C_{300} &= 1.455 \text{ mg/L} \\
C_{400} &= 1.494 \text{ mg/L} \\
C_{500} &= 1.504 \text{ mg/L} \\
C_{1000} &= 1.508 \text{ mg/L}
\end{aligned}
$$

Plume reaches steady-state equilibrium after approximately 400 days.

Semi-infinite System with Decaying Source. The analytical relationships presented in the preceding section are useful for simulating solute transport at sites with a constant source of contamination. In reality, contaminant source concentrations generally decrease over time as a result of weathering of mobile and residual LNAPL. Temporal variations in source concentrations are simulated using the third-type boundary condition. Van Genuchten and Alves (1982) give a solution to eq. (8.1) for a decaying contaminant source and a solute subject to first-order decay. For cases where the decay rate for the dissolved contaminant, λ, is not equal to the decay rate for the source, k_s:

$$C(x,t) = C_0 A(x,t) + C_s E(x,t) \tag{8.18}$$

where

$$
\begin{aligned}
A(x,t) = \exp(-\lambda t) \Bigg\{ 1 &- \frac{1}{2} \operatorname{erfc}\left(\frac{Rx - v_x t}{2\sqrt{D_x R t}} \right) \\
&- \frac{v_x^2 t}{\pi D_x R} \exp\left[- \frac{(Rx - v_x t)^2}{4 D_x R t} \right] \\
&+ \frac{1}{2} \left(1 + \frac{v_x x}{D_x} + \frac{v_x^2 t}{D_x R} \right) \exp\left(\frac{v_x x}{D_x} \right) \operatorname{erfc}\left(\frac{Rx + v_x t}{2\sqrt{D_x R t}} \right) \Bigg\}
\end{aligned} \tag{8.19}
$$

and

$$
\begin{aligned}
E(x,t) = \exp(-k_s t) & \left\{ \frac{v_x}{v_x + v_x\sqrt{1 + (4D_x R / v_x^2)(\lambda - k_s)}} \right\} \\
& \exp\left\{ \left[\frac{[v_x - v_x\sqrt{1 + (4D_x R / v_x^2)(\lambda - k_s)}]x}{2D_x} \right] \right\} \\
& \operatorname{erfc}\left[\frac{Rx - tv_x\sqrt{1 + (4D_x R / v_x^2)(\lambda - k_s)}}{2\sqrt{D_x R t}} \right] \\
+ & \frac{v_x}{v_x - v_x\sqrt{1 + (4D_x R / v_x^2)(\lambda - k_s)}} \\
& \exp\left\{ \left[\frac{[v_x + v_x\sqrt{1 + (4D_x R / v_x^2)(\lambda - k_s)}]x}{2D_x} \right] \right\} \\
& \operatorname{erfc}\left[\frac{Rx + tv_x\sqrt{1 + (4D_x R / v_x^2)\,(\lambda - k_s)}}{2\sqrt{D_x R t}} \right] \\
+ & \frac{v_x^2}{2D_x R(\lambda - k_s)} \exp\left[\frac{v_x x}{D_x} - (\lambda - k_s)t \right] \operatorname{erfc}\left(\frac{Rx + v_x t}{2\sqrt{D_x R t}} \right)
\end{aligned}
\tag{8.20}
$$

where $C(x,t)$ = contaminant concentration at a distance x downgradient from the source at time t

x = distance downgradient of upgradient boundary

t = time

C_0 = initial dissolved contaminant concentration at boundary

C_s = concentration of injected contaminant (source term)

λ = first-order decay rate constant for dissolved contaminant

R = coefficient of retardation

v_x = unretarded linear groundwater flow velocity

D_x = longitudinal hydrodynamic dispersion coefficient

k_s = first-order decay rate constant for source term

Assumptions:

- The aquifer is homogeneous and isotropic.
- The fluid is of constant density and viscosity.
- Biodegradation of dissolved constituents is approximately first order.
- Flow is in the x direction only, and the velocity is constant with a uniform flow field.
- The longitudinal hydrodynamic dispersion D_x is constant.

- There is no advection or dispersion into or out of the aquifer.
- Sorption is approximated by the linear sorption model.
- The source fully penetrates the aquifer.

Boundary conditions:

$$\left(-D_x \frac{\partial C}{\partial x} + v_x C\right)\Bigg|_{x=0} = v_x C_s \exp(-\alpha t)$$

$$\frac{\partial C}{\partial x}(\infty, t) = 0$$

where α is the solute decay coefficient across the boundary.

Initial condition:

$$C(x,0) = C_0$$

Because the source is decaying, the solute transport system will never reach truly steady-state conditions, and therefore no steady-state solution is available.

Two-Dimensional Analytical Models

The model presented in this section is for a semi-infinite system with a constant source of constant concentration and first-order decay of solute (Wilson and Miller, 1978). Equation (8.3) is the two-dimensional partial differential equation describing transient solute transport with advection, dispersion, sorption, and first-order biodegradation in the saturated zone. For large values of time when the system has reached steady-state equilibrium, solute transport with advection, dispersion, sorption, and biodegradation is described by eq. (8.4).

Wilson and Miller (1978) give the following solution for eq. (8.3) (Bedient et al., 1994, eq. 6.27, p. 136):

$$C(x,y,t) = \frac{f'_m \exp(v_x x / 2D_x)}{4\pi n_e \sqrt{D_x D_y}} W\left(u, \frac{r}{B}\right) \tag{8.21}$$

where

f'_m = continuous rate of contaminant injection per vertical unit aquifer [M/LT] (see Figure 8.3 for source configuration)

n_e = effective porosity

D_x = longitudinal dispersion coefficient (L^2/T)

D_y = transverse dispersion coefficient (L^2/T)

$W\left(u, \frac{r}{B}\right)$ = Hantush leaky well function (Hantush,1956)

u = $r^2 / 4\gamma_s D_x t$

γ_s = $1 + 2B\lambda / v_x$ $\qquad \lambda$ = solute decay coefficient (1/T)

$$r = \sqrt{\left(x^2 + \frac{D_x y^2}{D_y}\right)\gamma}$$

$$B = 2\frac{D_x}{\upsilon_x}$$

Wilson and Miller (1978) give an approximate solution to the Hantush well function. This relationship is

$$W\left(u, \frac{r}{B}\right) \approx \sqrt{\frac{\pi B}{2r}} \exp\left(-\frac{r}{B}\right) \operatorname{erfc}\left[-\frac{(r/B) - 2u}{2\sqrt{u}}\right]. \tag{8.22}$$

This approximation is reasonably accurate (within 10%) for $r/B > 1$, and more accurate (within 1%) for $r/B > 10$ (Wilson and Miller, 1978).

Three-Dimensional Analytical Models

Models presented in this section include a semi-infinite system with a constant source of constant concentration and first-order decay of solute (Domenico, 1987) and a semi-infinite system with a decaying source and first-order decay of solute. Equation (8.5) is the three-dimensional partial differential equation describing transient solute transport with advection, dispersion, sorption, and first-order biodegradation in the saturated zone. For large values of time (when the system has reached steady-state equilibrium), solute transport with advection, dispersion, sorption, and biodegradation is described by eq. (8.6).

Domenico (1987) developed an analytical solution for a finite (patch) source that incorporates one-dimensional groundwater velocity, longitudinal dispersion, transverse dispersion, vertical dispersion (in both the upward and downward directions), and first-order decay. In this model, the source is modeled as a vertical plane source, which can be visualized as a curtain of contamination that delivers dissolved-phase

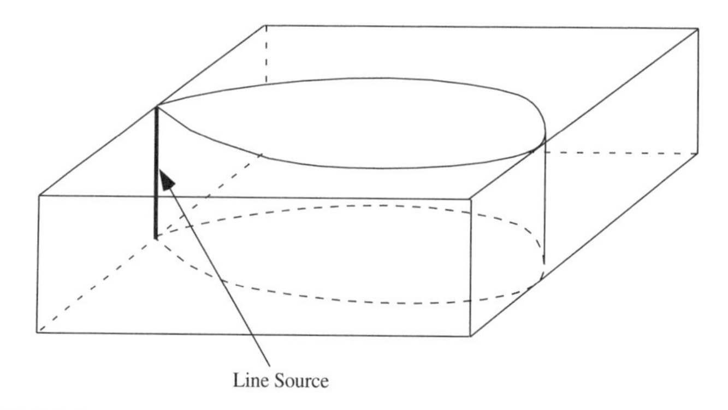

FIGURE 8.3. Line source configuration for the Wilson and Miller (1978) model.

contamination to groundwater moving through the curtain. The contaminants then spread out due to dispersion, sorb to the aquifer, and biodegrade. For transient conditions [eq. (8.5)] the Domenico (1987) solution is given as

$$\begin{aligned} C(x,y,z,t) &= \frac{C_0}{8}\exp\left[\frac{x}{2\alpha_x}\left(1-\sqrt{1+\frac{4\lambda R\alpha_x}{\upsilon_x}}\right)\right] \\ &\quad \operatorname{erfc}\left[\frac{x - t(\upsilon_x/R)\sqrt{1+4\lambda R\alpha_x/\upsilon_x}}{2\sqrt{\alpha_x(\upsilon_x/R)\,t}}\right] \\ &\quad \left[\operatorname{erf}\left(\frac{y+Y/2}{2\sqrt{\alpha_y x}}\right) - \operatorname{erf}\left(\frac{y-Y/2}{2\sqrt{\alpha_y x}}\right)\right] \\ &\quad \left[\operatorname{erf}\left(\frac{z+Z/2}{2\sqrt{\alpha_z x}}\right) - \operatorname{erf}\left(\frac{z-Z/2}{2\sqrt{\alpha_z x}}\right)\right] \end{aligned} \tag{8.23}$$

where $C(x,y,z,t)$ = contaminant concentration as a function of x, y, z, and t
C_0 = initial dissolved contaminant concentration at boundary
x = distance downgradient of upgradient boundary
α_x = longitudinal dispersivity
R = coefficient of retardation
λ = first-order decay rate constant for dissolved contaminant
υ_x = unretarded linear groundwater flow velocity
t = time
y = distance lateral to flow direction
Y = source dimension in y direction (see Figure 8.4 for source configuration)
α_y = transverse dispersivity
z = vertical distance perpendicular to flow direction
Z = source dimension in z direction
α_z = vertical dispersivity

For steady-state conditions this expression becomes (Domenico, 1987)

$$\begin{aligned} C(x,y,z,t) &= \frac{C_0}{4}\exp\left[\frac{x}{2\alpha_x}\left(1-\sqrt{1+\frac{4\lambda R\alpha_x}{\upsilon_x}}\right)\right] \\ &\quad \left[\operatorname{erf}\left(\frac{y+Y/2}{2\sqrt{\alpha_y x}}\right) - \operatorname{erf}\left(\frac{y-Y/2}{2\sqrt{\alpha_y x}}\right)\right] \\ &\quad \left[\operatorname{erf}\left(\frac{z+Z/2}{2\sqrt{\alpha_z x}}\right) - \operatorname{erf}\left(\frac{z-Z/2}{2\sqrt{\alpha_z x}}\right)\right] \end{aligned} \tag{8.24}$$

Assumptions:

- The fluid is of constant density and viscosity.
- The solute may be subject to first-order decay via biodegradation.
- Flow is in the x direction only, and the velocity is constant.
- The longitudinal dispersion D_x is constant.
- Sorption is approximated by the linear sorption model.

Dispersion in Two Directions Compared to Dispersion in One Direction. Equations (8.23) and (8.24) assume that vertical dispersion occurs in both directions as the contaminants travel away from the source zone (i.e., downward and upward). For source zones located in the middle of a thick aquifer, or in cases in which recharge produces a clean zone on top of the plume, this would be an appropriate approach. For source zones located at the top of an aquifer (as is the case at most petroleum release sites), upward vertical dispersion above the water table does not occur (unless recharge is significant), and therefore the model could overestimate the effects of dispersion. Fortunately, eqs. (8.23) and (8.24) can be easily modified to account for dispersion in a single direction by replacing $Z/2$ in the last two terms of these equations with Z (i.e, replace $z + Z/2$ with $z + Z$ and $z - Z/2$ with $z - Z$).

In summary, if vertical dispersion is used, it should be used in both directions if there is clean groundwater downgradient of the source above the top of the source, and if there is clean groundwater downgradient of the source below the bottom of the source (this may be true at some DNAPL sites but will be true at relatively few LNAPL sites). The adjustment for vertical dispersion in one direction should be used if the bulk of the source is either at the top of the water table (such as at almost all LNAPL sources) or the bottom of the source (at some DNAPL sites where most of the source is pooled at the bottom of the aquifer), and the dissolved-phase contaminants can spread in only one vertical direction. If the source extends vertically throughout the entire saturated zone, no vertical dispersion should be used, and a very low value for α_z is used in the model. Note that vertical dispersion is typically very small.

Biodegradation of the Sorbed Phase and the Dissolved Phase. Analytical models are typically run with first-order decay coefficients (λ) representing biodegradation of the dissolved phase and the sorbed phase. This can be illustrated by changing the retardation factor in a steady-state model that includes first-order decay, and observing how the steady-state plume length changes with changes in retardation. A higher λ value results in more biodegradation and a plume that stabilizes close to the source; a lower λ value will produce less biodegradation and a longer plume under steady-state conditions. In this case, the time required for the solute to travel from the source to any downgradient point has increased due to retardation, and therefore the time available for the first-order reaction has increased. This, in turn, shows that a steady-state analytical model typically considers degradation in both the sorbed and dissolved phases, as the only reason the reaction time increases is because of the increased time the solute is sorbed to the aquifer matrix. Similarly, if only the dissolved phase were considered, retardation would not affect the time that the solute is in the dissolved phase (the constituent velocity while in groundwater is not changed as the solute moves from sand grain to sand grain).

The type of biodegradation that is considered (dissolved phase only or both dissolved and sorbed phases) in an analytical model depends on the degradation constant selected. The advection–dispersion equation can be solved with different λ values for the dissolved and sorbed phases. If for the sorbed phase, λ is set to zero, retardation will have no effect on steady-state plume lengths even when first-order decay is used in the model. Note that most analytical solutions are based on combined λ values in which the first-order decay rate for the dissolved phase is equal to the first-order decay rate for the sorbed phase.

Source Term Characterization. In most analytical flow models, the contaminant source input term is characterized as a constant-flux or constant-concentration point, line, or area source. Typically, this source characterization does not match the actual variation of contaminant concentrations across a source zone. Using a simple superposition technique, however, a source zone can be characterized as a set of multiple source terms (i.e., multiple points, lines, or areas) configured to match the transverse cross-sectional concentration distribution of the actual plume.

An example of this source zone characterization for the Domenico analytical contaminant transport model is illustrated in Figures 8.4 and 8.5*A*. As shown in Figure 8.5*A*, the transverse plume cross section at the source zone location can be approximated by a stacked set of vertical plane source terms, resembling a layer cake. Steady-state plume concentrations predicted by the Domenico model for cases of single and multiple source terms are compared in Figure 8.5*B* to the output of BIOPLUME II (without biodegradation) for a hypothetical plume. As shown, the layer-cake source configuration significantly improves the agreement of the Domenico analytical model with BIOPLUME II for this steady-state no-biodegradation case.

In summary, the source term in an analytical model can best be matched to the transverse plume cross section using a multisource configuration (e.g., layer cake for the Domenico model). If NAPLs are present within the subsurface, the source term should be assumed to remain constant, with no significant attenuation over time, to obtain a conservative value for risk assessment or design.

Decaying Source. The Domenico model was modified in the BIOSCREEN model (see Newell et al., 1997) to account for a decaying source (see Section 2.8). In this approximation, the C_0 term in eq. (8.23) was replaced with the following expression:

$$C_0(x,t) = C_0 e^{-k_s[t - x/(V_s/R)]} \tag{8.25}$$

where $C_0(x,t)$ = source concentration used in eq. (8.24) for point distance x away from source at time t

C_0 = initial source concentration at $x = 0$ and $t = 0$

k_s = first-order source decay rate constant

t = time

x = distance away from source

Domenico Model with First Order Decay Algorithm

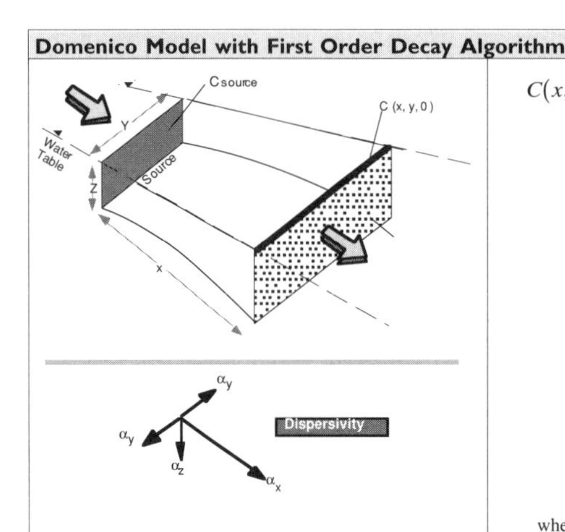

$$C(x,y,o,t) = C_o \exp[-k_s(t - x/v)]$$
$$\frac{1}{8}\exp\left[\frac{x}{\alpha_x 2}\left(1 - (1 + 4\lambda\alpha_x / v)^{1/2}\right)\right]$$
$$erfc\left[\frac{\left(x - vt(1 + 4\lambda\alpha_x / v)^{1/2}\right)}{2(\alpha_x vt)^{1/2}}\right]$$
$$\left\{erf\left[\frac{(y + Y/2)}{2(\alpha_y x)^{1/2}}\right] - erf\left[\frac{(y - Y/2)}{2(\alpha_y x)^{1/2}}\right]\right\}$$
$$\left\{erf\left[\frac{(Z)}{2(\alpha_z x)^{1/2}}\right] - erf\left[\frac{(-Z)}{2(\alpha_z x)^{1/2}}\right]\right\}$$

where: $v = \dfrac{K \cdot i}{\theta_e R}$

Domenico Model with Instantaneous Reaction Superposition Algorithm

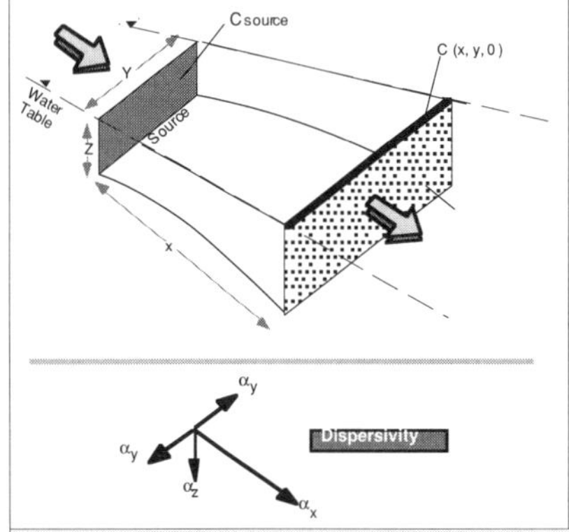

$$C(x,y,o,t) = \left(C_o \exp[-k_s(t - x/v)] + BC\right)$$
$$\frac{1}{8} erfc\left[\frac{(x - vt)}{2(\alpha_x vt)^{1/2}}\right]$$
$$\left\{erf\left[\frac{(y + Y/2)}{2(\alpha_y x)^{1/2}}\right] - erf\left[\frac{(y - Y/2)}{2(\alpha_y x)^{1/2}}\right]\right\}$$
$$\left\{erf\left[\frac{(Z)}{2(\alpha_z x)^{1/2}}\right] - erf\left[\frac{(-Z)}{2(\alpha_z x)^{1/2}}\right]\right\} - BC$$

where: $v = \dfrac{K \cdot i}{\theta_e R}$ $\quad BC = \Sigma \dfrac{C(ea)_n}{UF_n}$

Definitions

Symbol	Definition
BC	Biodegradation capacity (mg/L)
C(x,y,z,t)	Concentration at distance x downstream of source and distance y off centerline of plume at time t (mg/L)
C_s	Concentration in Source Zone (mg/L)
Co	Concentration in Source Zone at t=0 (mg/L)
x	Distance downgradient of source (ft)
y	Distance from centerline of source (ft)
z	Vertical Distance from groundwater surface to measurement point (assumed to be 0; concentration is always assumed to be at top of water table).
$C(ea)_n$	Concentration of electron acceptor (or by-product equivalent) n in groundwater (mg/L)
UF_n	Utilization factor for electron acceptor *n* (i.e., mass ratio of electron acceptor/by-product to hydrocarbon consumed in biodegradation reaction)
α_x	Longitudinal groundwater dispersivity (ft)
α_y	Transverse groundwater dispersivity (ft)
α_z	Vertical groundwater dispersivity (ft)
λ	First-order decay coefficient for dissolved contaminants (yr^{-1})
θ_e	Effective soil porosity
υ	Contaminant velocity in groundwater (ft/yr)
K	Hydraulic conductivity (ft/yr)
R	Constituent retardation factor
i	Hydraulic gradient (ft/ft)
Y	Source width (ft)
Z	Source depth (ft)
t	Time (yr)
k_s	First-order decay term for source concentration (yr^{-1})

FIGURE 8.4. Domenico analytical model used in BIOSCREEN. (From Newell et al., 1997.)

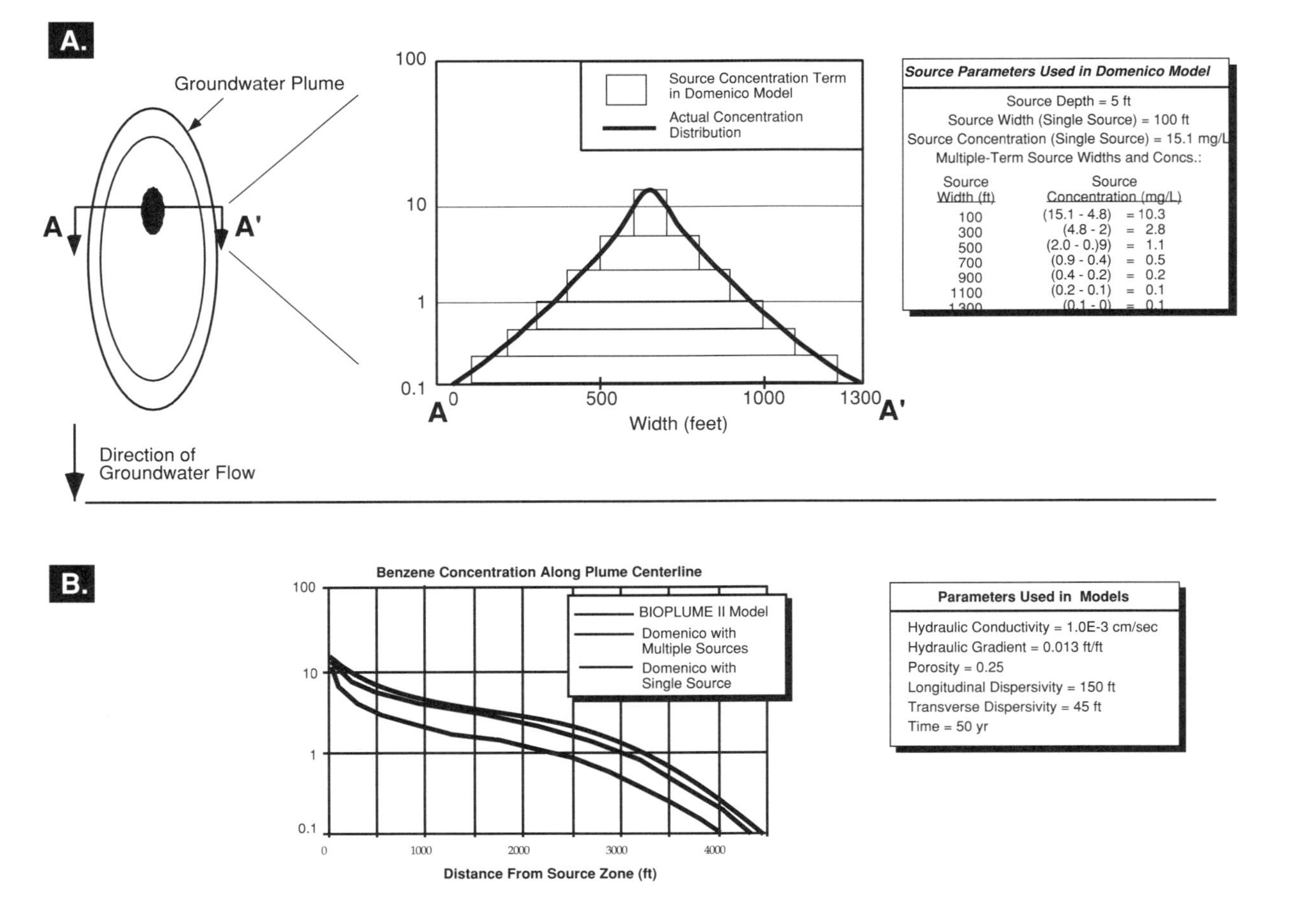

FIGURE 8.5. Schematic of "layer cake" superposition approximation for analytical models (*A*) and comparison of single-source analytical model, multiple-source analytical model, and BIOPLUME II numerical model results (*B*). (From Connor et al., 1994.)

Equation (8.25) can be used to simulate a contaminant source (typically, a zone with continuous or residual NAPL) that is slowly decaying, resulting in a reduction in source concentration over time. As discussed in Chapter 2, the most widely accepted conceptual model for source decay is due to dissolution, although biodgradation in the source zone (which is universal at fuel sites and common at chlorinated solvent sites) will accelerate the dissolution process. Therefore, a source zone with high rates of dissolution (e.g., very soluble contaminants in the NAPL that are rapidly biodegraded in the dissolved phase) will show a much higher first-order source decay constant (k_s) than that of a source zone with relatively low dissolution rates (e.g., low solubity contaminants that are not very biodegradable).

An additional feature of the decaying source term in BIOSCREEN as shown in eq. (8.25) is that the actual source concentration used in eq. (8.24) is adjusted to reflect the time that the dissolved contaminants left the source. In other words, the time used to determine the source concentration is adjusted to account for the travel time between the source and measurement point. For example, consider the case in which a declining source term is used with a source half-life of 10 years, a solute velocity of 100 ft/yr, and a retardation factor of 1. To calculate the concentration at a point 2000 ft away at a time of 30 years, BIOSCREEN follows these steps:

1. Calculates travel time from point to source: 2000 ÷ (100 ÷ 1) = 20 years.
2. Subtracts travel time from simulation time: 30 years – 20 years = 10 years (this represents that actual time that the contaminants left the source area).
3. Uses the adjusted time that the contaminants left the source area in the exponential decay relationship.

All three of these steps are incorporated into eq. (8.25). The resulting concentration, $C_0(x,t)$, is substituted for the C_0 term in the constant source equation in eq. (8.23):

$$C(x,y,z,t) = \frac{C_0 e^{-k_s[t - x/(V_s/R)]}}{8} \exp\left[\frac{x}{2\alpha_x}\left(1 - \sqrt{1 + \frac{4\lambda R\alpha_x}{\upsilon_x}}\right)\right] \operatorname{erfc}\left[\frac{x - t(\upsilon_x/R)\sqrt{1 + 4\lambda R\alpha_x/\upsilon_x}}{2\sqrt{\alpha_x(\upsilon_x/R)\,t}}\right] \times \left[\operatorname{erf}\left(\frac{y + Y/2}{2\sqrt{\alpha_y x}}\right) - \operatorname{erf}\left(\frac{y - Y/2}{2\sqrt{\alpha_y x}}\right)\right] \left[\operatorname{erf}\left(\frac{z + Z/2}{2\sqrt{\alpha_z x}}\right) - \operatorname{erf}\left(\frac{z - Z/2}{2\sqrt{\alpha_z x}}\right)\right] \tag{8.26}$$

where $C(x,y,z,t)$ = contaminant concentration as a function of x, y, z, and t
C_0 = initial dissolved contaminant concentration at boundary
k_s = first-order decay rate constant for contaminant source
t = time
x = distance downgradient of upgradient boundary
R = coefficient of retardation
α_x = longitudinal dispersivity
λ = first-order decay rate constant for dissolved contaminant
υ_x = unretarded linear groundwater flow velocity
y = distance lateral to flow direction
Y = source dimension in y direction
α_y = transverse dispersivity
z = vertical distance perpendicular to flow direction
Z = source dimension in z direction
α_z = vertical dispersivity
V_s = average groundwater seepage velocity at site

Assumptions:

- The fluid is of constant density and viscosity.
- The solute may be subject to first-order decay via biodegradation.
- The source may be subject to first-order decay via weathering or engineered remediation.
- Flow is in the x direction only, and the velocity is constant.
- The longitudinal dispersion α_x is constant.
- Sorption is approximated by the linear sorption model.

Computer Applications for Analytical Models

Depending on the needs, resources, and skills of the modeler, analytical modeling can be performed using commonly available spreadsheets, mathematical analysis applications such as MathCAD®, or codes written expressly for analytical modeling. Use of spreadsheets requires the most effort on behalf of the modeler, owing to the need to set up the sheet, enter the appropriate equations, and format the output. Mathematical analysis applications also require the user to enter the equations, but entering the equations and formatting the input and output can be much simpler and more intuitive than in a spreadsheet. However, once set up, both methods provide the modeler with a template that can be used repeatedly. For specific analytical modeling codes, the methods of input vary, as do the methods of displaying output. In general, though, these codes require the least amount of effort on behalf of the user.

A wide range of analytical solute transport modeling software is available. A partial list of codes is presented in Table 8.4. Some of the codes are proprietary and some are public domain. Proprietary codes often have graphical interfaces for processing input and output, resulting in a greater cost. Depending on the needs of the user, these extra costs may well be worth the time and labor for preparation and data entry that

TABLE 8.4. Analytical Solute Transport Models

Model Code Name	Capabilities	Distributor
AGU-10	Analytical flow, advective solute transport, advective-dispersive transport, and advective–dispersive transport with decay of source and solute. Based on American Geophysical Union's Water Resources Monograph 10 (Javandel et al., 1984).	IGWMC; documentation includes AGU Monograph 10
AT123D	Based on analytical solution for transient one-, two-, or three-dimensional transport in a homogeneous isotropic aquifer with uniform regional flow. Allows for retardation, dispersion, and first-order decay, with differing source configurations and boundary conditions. Has a shell for pre- and postprocessing. Authored by G. T. Yeh of Oak Ridge National Laboratories.	IGWMC
ONE-D	Package of five analytical solutions of the one-dimensional advective–dispersive transport equation with adsorption, dispersion, and first-order-decay options; includes a zero-order production term. Written by M. T. van Genuchten and W. J. Alves of the U.S. Department of Agriculture's Salinity Laboratory.	IGWMC
PLUME, PLUME2D	Analytical models for calculating point concentrations of solutes. Includes advection, dispersion, retardation, and first-order decay. Source terms can be varied over time (in discrete intervals). Written by P. K. M. van der Heijde of the IGWMC.	IGWMC
PRINCE	Proprietary package of 10 analytical solute transport and flow models, widely referred to as the Princeton Analytical Models. Seven solute transport models allow calculation of concentrations and breakthrough curves for one-, two-, and three-dimensional problem domains. Advection, dispersion, retardation, and first-order decay can all be simulated, along with a wide range of source terms (including multiple sources in two and three dimensions). A self-contained package with graphical user interface for pre- and postprocessing.	Waterloo Hydrogeologic Software or IGWMC
SOLUTE	Menu-driven set of five different programs that provide the user with nine different types of analytical solute transport models. The nine models include one-, two-, and three-dimensional solutions with differing boundary conditions and options for retardation and first-order decay. Displays output graphically; output can also be used with other utilities for postprocessing. Authored by M. S. Beljin and P. K. M. van der Heijde of IGWMC.	IGWMC
USGS-SOL	Analytical solutions describing advective–dispersive transport, as well as first-order decay and retardation. Includes one-, two-, and three-dimensional models with a limited number of boundary conditions. No preprocessor is provided, and users must set up their own input files. Output can be sent to a file for use in other postprocessors. Originally prepared by E. J. Wexler of the USGS.	IGWMC
WALTON35	Set of 35 analytical and numerical models for a variety of groundwater flow and solute transport problems. Input is interactive, and results can be saved to a file. Prepared by W. C. Walton in conjunction with the IGWMC.	IGWMC

Source: From Wiedemeier et al. (1996).

may be saved by using such a program. Other codes may be available; this is by no means an exhaustive list.

BIOSCREEN Model for Fuel Hydrocarbon Sites

The BIOSCREEN Natural Attenuation Decision Support System is a public-domain, spreadsheet-based screening tool for simulating the natural attenuation of dissolved hydrocarbons at petroleum fuel release sites (Newell et al., 1996, 1997). It is based on the Domenico (1987) analytical solute transport model (Figure 8.4) that simulates groundwater flow, a fully penetrating vertical plane source oriented perpendicular to groundwater flow, linear isotherm sorption, and three-dimensional dispersion. The original Domenico model was designed to incorporate first-order decay during solute transport; the BIOSCREEN model extends the Domenico solution by incorporating a decaying source concentration [eq. (8.25)] and a simple electron-acceptor-limited "instantaneous" reaction assumption. BIOSCREEN was developed by the Air Force Center for Environmental Excellence (AFCEE) and is currently being distributed by the EPA through their Web site at (http://www.epa.gov/ada/bioscreen.html).

BIOSCREEN attempts to answer both fundamental natural attenuation questions: (1) How far will the plume migrate, and (2) how long will the plume persist. The AFCEE development team and EPA reviewers intended BIOSCREEN to be used in two ways (Newell et al., 1996):

1. *As a screening model to determine if natural attenuation is a feasible remediation approach at a site.* In this case, BIOSCREEN is used early in the remedial investigation to determine if a remediation by natural attenuation (RNA) field program should be implemented to quantify the natural attenuation occurring at a site. Some data, such as electron acceptor concentrations, may not be available, so typical values are used. In addition, the model can be used to help develop long-term monitoring plans for natural attenuation remediation projects.

2. *As an additional line of evidence for demonstrating a natural attenuation groundwater model at smaller, simpler fuel hydrocarbon sites.* The Air Force intrinsic remediation protocol (Wiedemeier et al., 1995) describes how groundwater models may be used to help verify that natural attenuation is occurring and to help predict how far plumes might extend under an RNA scenario. At large, high-effort sites such as complicated Superfund and RCRA sites, a more sophisticated model such as BIOPLUME is probably more appropriate. At less complicated, lower-effort sites such as service stations, BIOSCREEN may be sufficient to serve as a line of evidence for a natural attenuation study.

The conceptual model used in BIOSCREEN is

1. Groundwater upgradient of the source contains electron acceptors.
2. As the upgradient groundwater moves through the source zone, non-aqueous-phase liquids (NAPLs) and contaminated soil release dissolvable hydrocarbons (in the case of petroleum sites, the BTEX compounds are released).
3. Biological reactions occur until the available electron acceptors in groundwater are consumed. (Two exceptions to this conceptual model are iron

reactions, in which the electron acceptor, ferric iron, dissolves from the aquifer matrix; and methane reactions, in which the electron acceptor, CO_2, is also produced as an end product of the reactions. For these two exceptions, the metabolic by-products, ferrous iron and methane, can be used as proxies for the potential amount of biodegradation that could occur from the iron-reducing and methanogenesis reactions.)

4. The total amount of electron acceptors available for biological reactions can be estimated by (a) calculating the difference between upgradient concentrations and source zone concentrations for oxygen, nitrate, and sulfate; and (b) measuring the production of metabolic by-products (ferrous iron and methane) in the source zone.
5. Using stoichiometry, a utilization factor can be developed showing the ratio of the oxygen, nitrate, and sulfate consumed to the mass of dissolved hydrocarbon degraded in the biodegradation reactions. Similarly, utilization factors can be developed to show the ratio of the mass of metabolic by-products that are generated to the mass of dissolved hydrocarbon degraded in the biodegradation reactions. Table 5.2 presented the utilization factors based on the degradation of combined BTEX constituents.
6. For a given background concentration of an individual electron acceptor, the potential contaminant mass removal or *biodegradation capacity* depends on the *utilization factor* for that electron acceptor. Dividing the background concentration of an electron acceptor by its utilization factor provides an estimate (in BTEX concentration units) of the assimilative capacity of the aquifer by that mode of biodegradation. The biodegradation capacity is then used directly in the model to simulate an instantaneous reaction. The suggested calculation approach to develop BIOSCREEN input data is

$$\text{biodegradation capacity (mg/L)}$$

$$= \frac{\text{(average upgradient oxygen conc.)} - \text{(minimum source zone oxygen conc.)}}{3.14}$$
$$+ \frac{\text{(average upgradient nitrate conc.)} - \text{(minimum source zone nitrate conc.)}}{4.9}$$
$$+ \frac{\text{(average upgradient sulfate conc.)} - \text{(minimum source zone sulfate conc.)}}{4.7}$$
$$+ \frac{\text{average observed ferrous iron conc. in source area}}{21.8}$$
$$+ \frac{\text{average observed methane conc. in source area}}{0.78}$$

where 3.14, 4.9, 4.7, 21.8, and 0.78 represent utilization factors for oxygen, nitrate, sulfate, ferrous iron, and methane, respectively (see Chapter 5).

As an analytical model, BIOSCREEN has the limitation that it assumes simple groundwater flow conditions and should not be applied in which pumping systems create a complicated flow field. In addition, the model should not be applied in which vertical flow gradients affect contaminant transport. Finally, BIOSCREEN approximates only more complicated processes that occur in the field and should not be applied in which extremely detailed, accurate results that closely match site conditions are required. More comprehensive numerical models should be applied in these cases.

BIOSCREEN output includes (1) plume centerline graphs, (2) three-dimensional color plots of plume concentrations, and (3) mass balance data showing the contaminant mass removal by each electron acceptor (instantaneous reaction option). Other features of the original BIOSCREEN model (Version 1.3) included a concentration versus time animation module and a water balance showing the volume of water in the plume and the flux of water moving through the plume. In a later release (BIOSCREEN Version 1.4), a mass flux calculator was added that shows the mass flux of contaminants at any point in the plume. With the mass flux calculator, dilution calculations can be performed for plumes that are discharging to streams.

Estimating Plume Length with First-Order Decay Kinetics. First-order decay kinetics requires a rate constant as input. In general, two different approaches can be used. First, literature values can be used directly in the model, preferably using data from other field sites rather than data from laboratory studies (see Chapters 5 and 6). This approach can provide an approximation of expected plume conditions and is useful for comparing plume lengths under no-biodegradation assumptions versus more realistic model runs that do account for biodegradation (this is particularly true for fuel hydrocarbon plumes). Tables 5.4, 5.5, 5.6, and 6.7 provide first-order decay coefficients from laboratory and field studies for fuel hydrocarbon and chlorinated solvent sites, respectively.

The second approach is to calibrate to existing plume data. If the plume is in a steady-state or diminishing condition, BIOSCREEN can be used to determine first-order decay coefficients that best match the site concentrations observed. One may adopt a trial-and-error procedure to derive a best-fit decay coefficient for each contaminant, use one of several methods provided in Chapter 7 for converting field data from sites to first-order decay coefficients, or use a plume calibration package such as FATE V (Nevin et al., 1977). For still-expanding plumes, this steady-state calibration method may overestimate actual decay-rate coefficients and contribute to an underestimation of predicted concentration levels.

Incorporating Instantaneous Reaction Kinetics into BIOSCREEN. Early biodegradation research focused on the role of dissolved oxygen in controlling the rate of biodegradation in the subsurface (Borden and Bedient, 1986; Borden et al., 1986; Lee et al., 1987). Because microbial biodegradation kinetics are relatively fast in comparison to the rate of oxygen transport in the groundwater flow system, Borden and Bedient (1986) demonstrated that the biodegradation process can be simulated as an instantaneous reaction between the organic contaminant and oxygen.

This simplifying assumption was incorporated into the BIOPLUME I numerical model, which calculated organic mass loss by superposing background oxygen concentrations onto the organic contaminant plume. In BIOPLUME II, a dual-particle mover procedure was incorporated to simulate more accurately the separate transport of oxygen and organic contaminants within the subsurface (Rifai et al., 1987, 1988).

To simulate electron-acceptor-limited biodegradation with an analytical model, Newell et al. (1995) incorporated the concept of oxygen superposition into BIOSCREEN in a manner similar to that employed in the BIOPLUME I and II models (Borden and Bedient, 1986; Rifai et al., 1988). By this method, contaminant mass concentrations at any location and time within the flow field are corrected by subtracting 1 mg/L organic mass for each 3 mg/L of background oxygen, in accordance with the instantaneous reaction assumption. Borden et al. (1986) concluded that this simple superposition technique was an exact replacement for more sophisticated oxygen-limited models, as long as the oxygen and the hydrocarbon had the same transport rates (e.g., retardation factor, $R = 1$).

In their original work, Borden et al. (1986) noted that for highly sorptive contaminants the oxygen-superposition method might erroneously characterize biodegradation due to the differing transport rates of dissolved oxygen and the organic contaminant within the aquifer matrix. However, as demonstrated by Connor et al. (1994), the oxygen superposition method and BIOPLUME II (dual-particle transport) are in reasonable agreement for contaminant retardation factors as high as 6. Therefore, the superposition method can be employed as a reasonable approximation in BIOSCREEN regardless of contaminant sorption characteristics.

BIOSCREEN employs the same superposition approach for all of the aerobic and anaerobic biodegradation reactions (based on evaluation of O_2, NO_3, SO_4, Fe^{2+}, and CH_4). Based on work reported by Newell et al. (1995), the anaerobic reactions (nitrate, ferric iron, and sulfate reduction and methanogenesis) are amenable to simulation using the instantaneous reaction assumption (see Chapter 4). Based on the expressed biodegradation capacity of electron acceptors present in the groundwater system, this algorithm will correct the nondecayed groundwater plume concentrations predicted by the Domenico model for the effects of organic constituent biodegradation.

The background research for the instantaneous reaction assumption in BIOSCREEN is summarized below.

1. The BIOPLUME I model (Borden and Bedient, 1986) used a superposition method to simulate the fast or "instantaneous" reaction of dissolved hydrocarbons with dissolved oxygen in groundwater.
2. Borden et al. (1986) reported that this version of BIOPLUME was mathematically exact for the case in which the retardation factor of the contaminant was 1.0.
3. Rifai et al. (1988) developed the BIOPLUME II model with a dual-particle tracking routine that expanded the original BIOPLUME model to handle contaminants with retardation factors other than 1.0, in addition to other improvements.

4. Connor et al. (1994) compared the superposition method with the more sophisticated BIOPLUME II model and determined that the two approaches yielded very similar results for readily biodegradable contaminants with retardation factors between 1.0 and 6.0.
5. BIOSCREEN was developed using the superposition approach to simulate the "instantaneous" reaction of aerobic and anaerobic reactions in groundwater. This mathematical approach (superposition) matches the more sophisticated BIOPLUME II model very closely for readily biodegradable contaminant retardation factors of up to 6.0. BIOSCREEN simulations using the instantaneous reaction assumption at sites with retardation factors greater than 6.0 should be performed with caution and verified using a more sophisticated model such as BIOPLUME III (Rifai et al., 1997).

Estimating Plume Length with Instantaneous Reaction Kinetics. To use the instantaneous reaction model, an estimate of the expressed biodegradation capacity of groundwater flowing through the source zone and the plume area is needed. Newell et al. (1997) reported that when the expressed biodegration capacity presented in Chapter 5 was used in BIOSCREEN, the model results tended to overpredict the amount of biodegradation occurring at the site, and therefore recommended using average observed source zone concentrations of ferrous iron and methane to calculate the expressed biodegradation capacity.

When instantaneous reaction kinetics are used, a two-step process for calibrating the BIOSCREEN model was recommended:

1. The primary calibration step (if necessary) is to manipulate the model's dispersivity. Values for dispersivity are related to aquifer scale (defined as the plume length or distance to the measurement point), and simple relationships are usually applied to estimate dispersivities. Gelhar et al. (1992) cautions that dispersivity values vary between two and three orders of magnitude for a given scale (plume length), due to natural variation in hydraulic conductivity at a particular site. Therefore, dispersivity values can be manipulated within a large range and may still be within the range of values observed at field test sites. In BIOSCREEN, adjusting the transverse dispersivity alone will usually be sufficient to calibrate the model.
2. As a secondary calibration step, the biodegradation capacity calculation may be reevaluated. There is some judgment involved in averaging the electron acceptor concentrations observed in upgradient wells; determining the minimum oxygen, nitrate, and sulfate concentrations in the source zone; and estimating the average observed ferrous iron and methane concentrations in the source zone. These values may be adjusted as a final level of calibration.

Handling Multiple Constituents Such as BTEX. BIOSCREEN requires source zone concentrations that either correspond to lumped constituents (such as the BTEX compounds together) or individual compounds (such as benzene alone).

In the BIOSCREEN manual, Wilson (1996) provided suggested rules of thumb regarding how to handle multiple constituents:

1. If the maximum plume length is desired, model lumped constituents (such as BTEX). If a risk assessment is being performed, data on individual constituents are needed.
2. If lumped constituents are being modeled (total BTEX), use either average values for the chemical-specific data (K_{oc} and λ) or the worst-case values (e.g., use the lowest of the K_{oc} and λ values from the group of constituents being modeled) to overestimate concentrations. Most modeling will be performed assuming that the ratio of BTEX at the edge of the plume is the same as that at the source. For more detailed modeling studies, Wilson (1996) has proposed the following rules of thumb to help account for different rates of reaction among the BTEX compounds:
   - If the site is dominated by aerobic degradation (most of the biodegradation capacity is from oxygen, a relatively rare occurrence), assume that the benzene will degrade first and that the dissolved material at the edge of the plume is primarily TEX.
   - If the site is dominated by nitrate utilization (most of the biodegradation capacity is from nitrate, a relatively rare occurrence), assume that benzene will degrade last and that the dissolved material at the edge of the plume is primarily benzene.
   - If the site is dominated by sulfate reduction (most of the biodegradation capacity is due to sulfate utilization, a more common occurrence), assume that the benzene will degrade at the same rate as the TEX constituents and that the dissolved material at the edge of the plume is a mixture of BTEX.
   - If the site is dominated by methane production (most of the biodegradation capacity is due to methanogenesis, a more common occurrence), assume that benzene will degrade last and that the dissolved material at the edge of the plume is primarily benzene.
3. If individual constituents are being modeled with the instantaneous reaction assumption, note that the total biodegradation capacity must be reduced to account for electron acceptor utilization by other constituents present in the plume. For example, in order to model benzene as an individual constituent using the instantaneous reaction model in a BTEX plume containing equal source concentrations of benzene, toluene, ethylbenzene, and xylene, the amount of oxygen, nitrate, sulfate, iron, and methane should be reduced by 75% to account for utilization by toluene, ethylbenzene, and xylene.

Source Term in BIOSCREEN. The original Domenico (1987) model assumes that the source is infinite (i.e., the source concentrations are constant). In BIOSCREEN, however, an approximation for a declining source concentration was added. The declining source term is based on the following assumptions:

- There is a finite mass of organics in the source zone present as a free-phase or residual NAPL. The NAPL in the source zone dissolves slowly as fresh groundwater passes through.
- The change in source zone concentration can be approximated as a first-order decay process. For example, if the source zone concentration half-life is 10 years and the initial source zone concentration is 1 mg/L, the source zone concentration will be 0.5 mg/L after 10 years and 0.25 mg/L after 20 years.
- The mass flux of contaminant leaving the source can be approximated by multiplying the source times a representative source concentration for the first-order decay model (thereby assuming no biodegradation in the source zone) or by multiplying the source times the sum of the source concentration and biodegradation capacity (thereby assuming that there is biodegradation in the source zone).

The source term in the BIOSCREEN model was described in more detail in Chapter 2.

Example 8.3: BIOSCREEN Modeling of Keesler Air Force Base In the following example, the BIOSCREEN model is used to simulate natural attenuation of the BTEX plume at the Keesler Air Force Base, SWMU 66, Mississippi (see Table 8.5 for site data and Figure 8.6 for model input). The BIOSCREEN model is used first to reproduce the movement of the plume from 1989 (the best guess for when the release occurred) to 1995. The soluble mass in soil and NAPL was estimated by integrating BTEX soil concentrations contours mapped as part of the site soil delineation program. An estimated 2000 kg of BTEX is thought to be present at the site based on GC/MS analysis of soil samples collected from both the vadose and saturated zones. This value represented a source half-life of 60 years with the instantaneous reaction, a relatively long half-life, so the 2000 kg measured in 1995 was assumed to be representative of 1989 conditions.

The instantaneous reaction model was used as the primary model to try to reproduce the plume length (about 280 ft). Because a decaying source was used, the source concentration on the input screen (representing concentrations six years ago) were adjusted so that the source concentrations on the centerline output screen (representing concentrations now) were equal to 12 mg/L. Because the source decay term is different for the first-order decay and instantaneous reaction models, this simulation focused on matching the instantaneous reaction model. The final result was a source concentration of 13.68 mg/L in the center of the source zone (note that on the centerline output the source concentration is 12.021 mg/L).

Results from the initial run of the instantaneous reaction model indicated that the plume was too long, implying that there is more mixing of hydrocarbon and electron acceptors at the site than is predicted by the model. Therefore, the longitudinal dispersivity was adjusted upward (more mixing) until the BIOSCREEN results matched the observed plume length. The final longitudinal dispersivity was 32.5 ft.

As a check, the first-order decay model was used with the BIOSCREEN default value of 2 years. This run greatly overestimated the plume length, so the amount of biodegradation was increased by decreasing the solute half-life. A good match of the

TABLE 8.5. Site Data for the Keesler Air Force Base, SWMU 66, Mississippi

Data Type	Parameter	Value				Source of Data
Hydrogeology	Hydraulic conductivity (cm/s)	1.1×10^{-2}				Slug-test results
	Hydraulic gradient (ft/ft)	0.003				Static water-level measurements
	Porosity	0.3				Estimated
Dispersion	Original					
	Longitudinal dispersivity (ft)	13.3				Based on estimated length of 280 ft and Xu–Eckstein relationship
	Transverse dispersivity (ft)	1.3				
	Vertical dispersivity (ft)	0				
	After calibration					
	Longitudinal dispersivity (ft)	32.5				Based on calibration to plume length (*Note:* this is well within the observed range for long dispersivity; remember to convert from feet to meters before using the chart)
	Transverse dispersivity (ft)	3.25				
	Vertical dispersivity (ft)	0				
Adsorption	Retardation factor	1.0				Calculated from $R = 1 + K_{oc}f_{oc}(\rho_b/n)$
	Soil bulk density, ρ_b (kg/L)	1.7				Estimated
	f_{oc} (%)	0.0057				Laboratory analysis
	K_{oc}	B: 38		T: 135		Literature—use $K_{oc} = 38$
		E: 95		X: 240		
Biodegradation	Electron acceptor	O_2	NO_3	SO_4		Based on March 1995 groundwater sampling program conducted by Groundwater Services, Inc.
	Background conc. (mg/L)	2.05	0.7	26.2		
	Minimum conc. (mg/L)	–0.4	–0	–3.8		
	Change in conc. (mg/L)	[1.65]	[0.7]	[22.4]		
	Electron acceptor	Fe	CH_4			
	Maximum conc. (mg/L)	36.1	7.4			
	Average conc. (mg/L)	[16.6]	[6.6]			
		Note: Boxed values are BIOSCREEN input values.				
General	Modeled area length (ft)	320				Based on area of affected groundwater plume
	Modeled area width (ft)	200, 50				
	Simulation time (yr)	6				Steady-state flow
Source data	Source thickness (ft)	10				Based on geologic logs and lumped BTEX monitoring data
	Source concentration	See Figure 8.9				
Actual data	Distance from source (ft)	30	60	180	280	Based on concentrations observed at site
	BTEX conc. (mg/L)	5.0	1.0	0.5	0.001	
Output	Centerline concentration	See Figure 8.11				
	Array concentration	See Figure 8.12 and 8.14				

Source: From Newell et al. (1996).

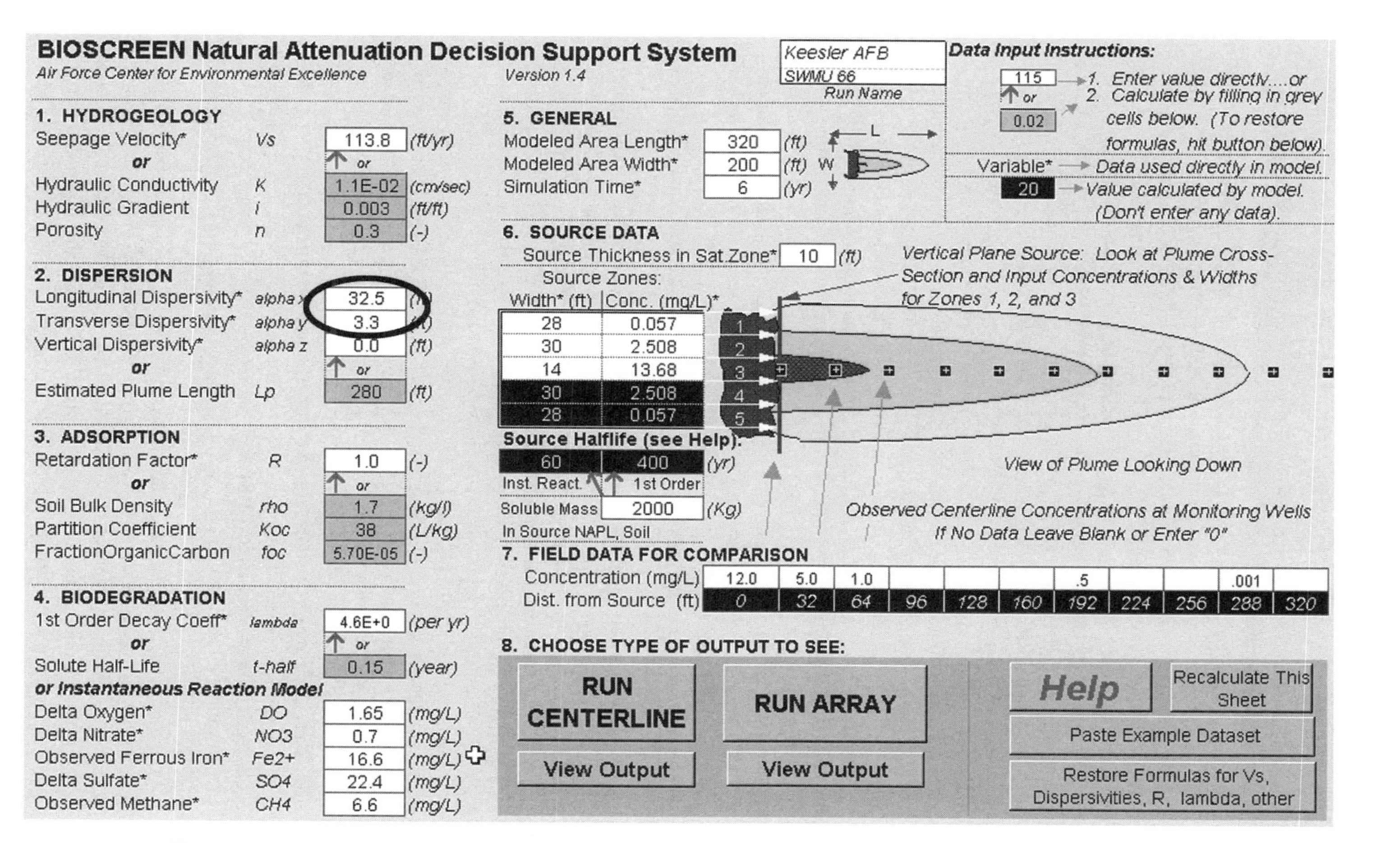

FIGURE 8.6. BIOSCREEN input screen for Keesler Air Force Base, Mississippi. (*Note:* Longitudinal dispersivity has been changed from the original computed value of 13.3 ft to 32.5 ft during calibration.) (From Newell et al., 1996.)

plume was reached with a solute half-life of 0.15 year. This is within observed ranges reported in the literature.

As shown in Figure 8.7, BIOSCREEN matches the observed plume fairly well. The instantaneous model is more accurate near the source, while the first-order decay model is more accurate near the middle of the plume. Both models reproduce the actual plume length relatively well.

As shown in Figure 8.8, the current plume is estimated to contain 7.8 kg of BTEX. BIOSCREEN indicates that the plume under a no-degradation scenario would contain 126.3 kg BTEX. In other words, BIOSCREEN indicates that 94% of the BTEX mass that has left the source since 1989 has biodegraded. Most of the source mass postulated to be in place in 1989 is still there in 1996 (2000 kg versus 1837 kg, or 92% left).

The current plume contains 1.0 acre-ft of contaminated water, with 1.019 acre-ft/yr of water being contaminated as it flows through the source. Because the plume is almost at steady state, 1.019 acre-ft of water become contaminated per year, with the same amount being remediated every year due to in situ biodegradation and other attenuation processes. This indicates that a long-term monitoring approach would probably be more appropriate for this site than active remediation, as the plume is no longer growing in size.

A hypothetical stream is assumed to be located approximately 210 ft downgradient of the source (note that no such stream exists at the actual site). Using an estimated model width of 200 ft (see Figure 8.6), a mass flux of 1500 mg/day is calculated (see Figure 8.8) at a distance of 224 ft away from the source (the closest point calculated by BIOSCREEN).

Users should be aware that the mass flux calculation is sensitive to the model width assigned in section 6 of the input screen (see Figure 8.6). A model width of 200 ft was used in the original example so that most of the no-degradation plume was in the array, allowing calculation of the plume and source. For the mass flux calculation, however, a more accurate result will be obtained by selecting a width in which most of the plume of interest (in this case the instantaneous reaction plume) appears across the array. As shown in Figures 8.9 and 8.10, a model width of 50 ft was selected so that the instantaneous reaction plume covered most of the BIOSCREEN array. With this width, a mass flux value of 860 mg/day was calculated. This is a more accurate estimate of the mass flux than the 1500 mg/day calculated above.

Comparison of First-Order versus Instantaneous Kinetics. The first-order decay and instantaneous reaction kinetics were compared by Newell et al. (1995) using a numerical version of BIOSCREEN. To compare the two models, three lines of evidence were evaluated:

1. *Available reaction time.* The maximum utilization rates observed in the laboratory were compared to the reaction time available for natural attenuation in the contaminant plumes. As discussed in Chapter 4, these microcosm data support the conclusion that microbial kinetics are very fast compared to the typical residence time of dissolved hydrocarbons in the subsurface, in which hydraulic residence times are on the order of years or tens of years, which is

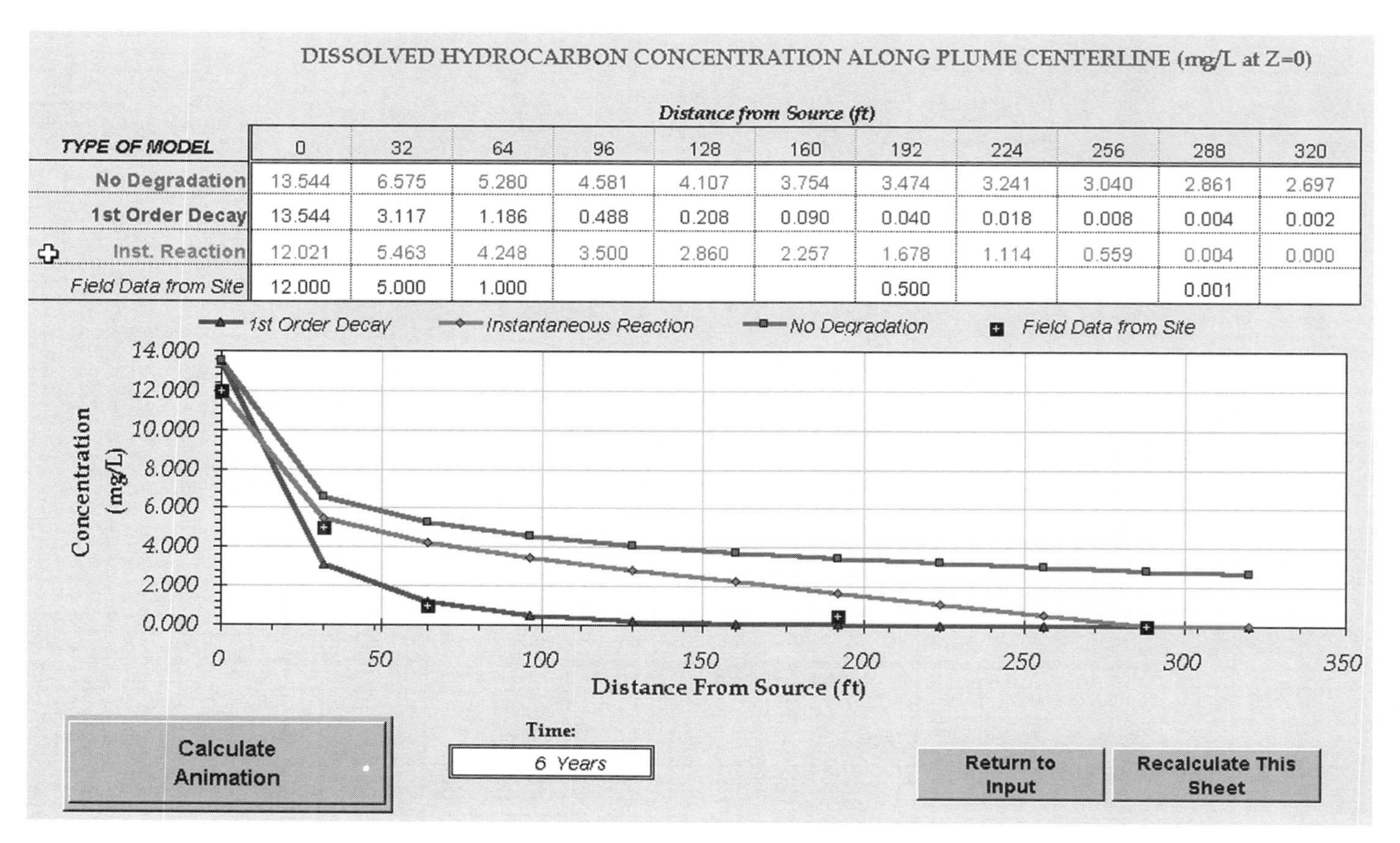

DISSOLVED HYDROCARBON CONCENTRATION ALONG PLUME CENTERLINE (mg/L at Z=0)

TYPE OF MODEL	Distance from Source (ft) 0	32	64	96	128	160	192	224	256	288	320
No Degradation	13.544	6.575	5.280	4.581	4.107	3.754	3.474	3.241	3.040	2.861	2.697
1st Order Decay	13.544	3.117	1.186	0.488	0.208	0.090	0.040	0.018	0.008	0.004	0.002
Inst. Reaction	12.021	5.463	4.248	3.500	2.860	2.257	1.678	1.114	0.559	0.004	0.000
Field Data from Site	12.000	5.000	1.000				0.500			0.001	

FIGURE 8.7. Centerline output for Keesler Air Force Base, Mississippi. (From Newell et al., 1976.)

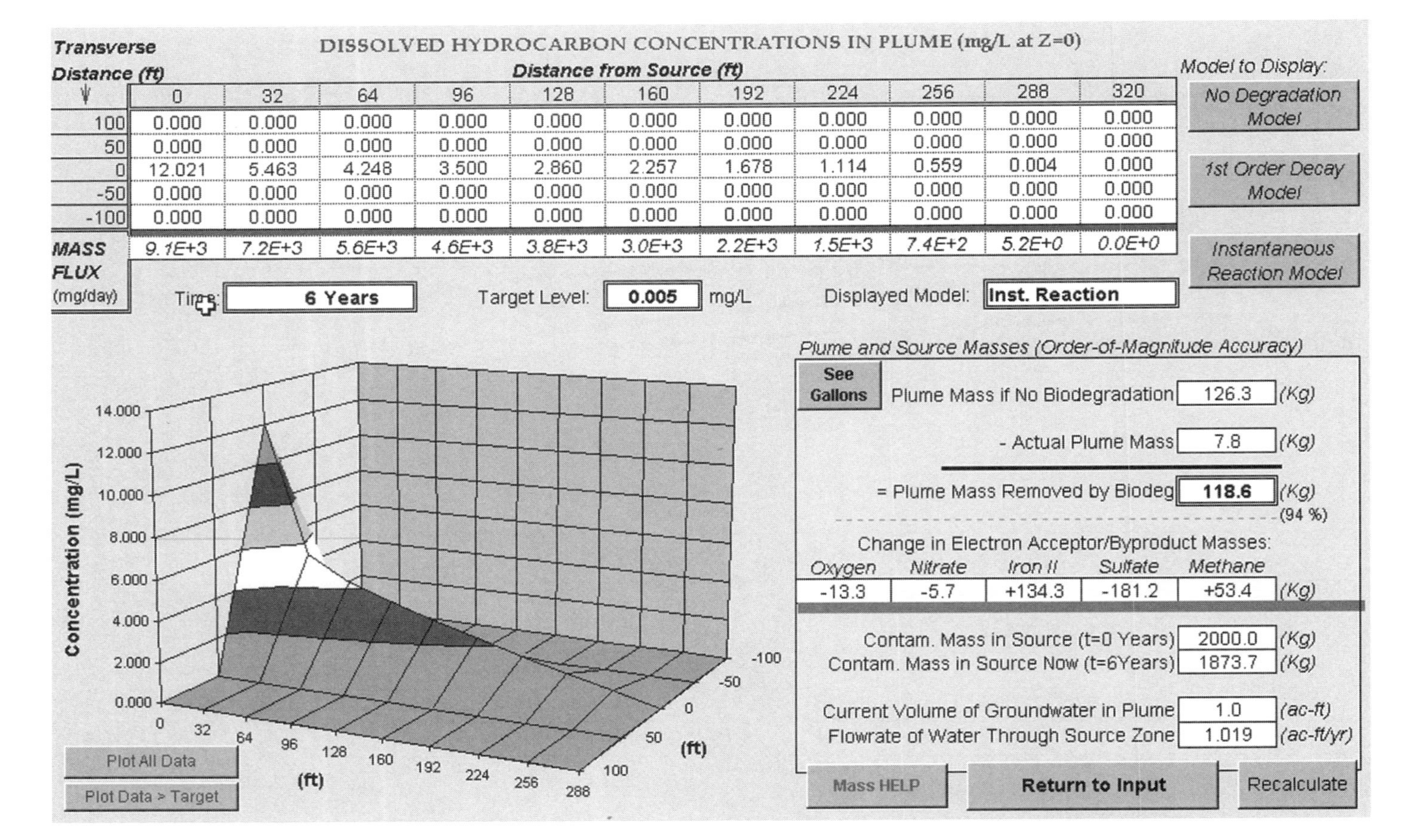

DISSOLVED HYDROCARBON CONCENTRATIONS IN PLUME (mg/L at Z=0)

Transverse Distance (ft) ↓ / Distance from Source (ft) →	0	32	64	96	128	160	192	224	256	288	320
100	0.000	0.000	0.000	0.000	0.000	0.000	0.000	0.000	0.000	0.000	0.000
50	0.000	0.000	0.000	0.000	0.000	0.000	0.000	0.000	0.000	0.000	0.000
0	12.021	5.463	4.248	3.500	2.860	2.257	1.678	1.114	0.559	0.004	0.000
-50	0.000	0.000	0.000	0.000	0.000	0.000	0.000	0.000	0.000	0.000	0.000
-100	0.000	0.000	0.000	0.000	0.000	0.000	0.000	0.000	0.000	0.000	0.000
MASS FLUX (mg/day)	9.1E+3	7.2E+3	5.6E+3	4.6E+3	3.8E+3	3.0E+3	2.2E+3	1.5E+3	7.4E+2	5.2E+0	0.0E+0

FIGURE 8.8. Array concentration output for Keesler Air Force Base, Mississippi. (From Newell et al., 1996.)

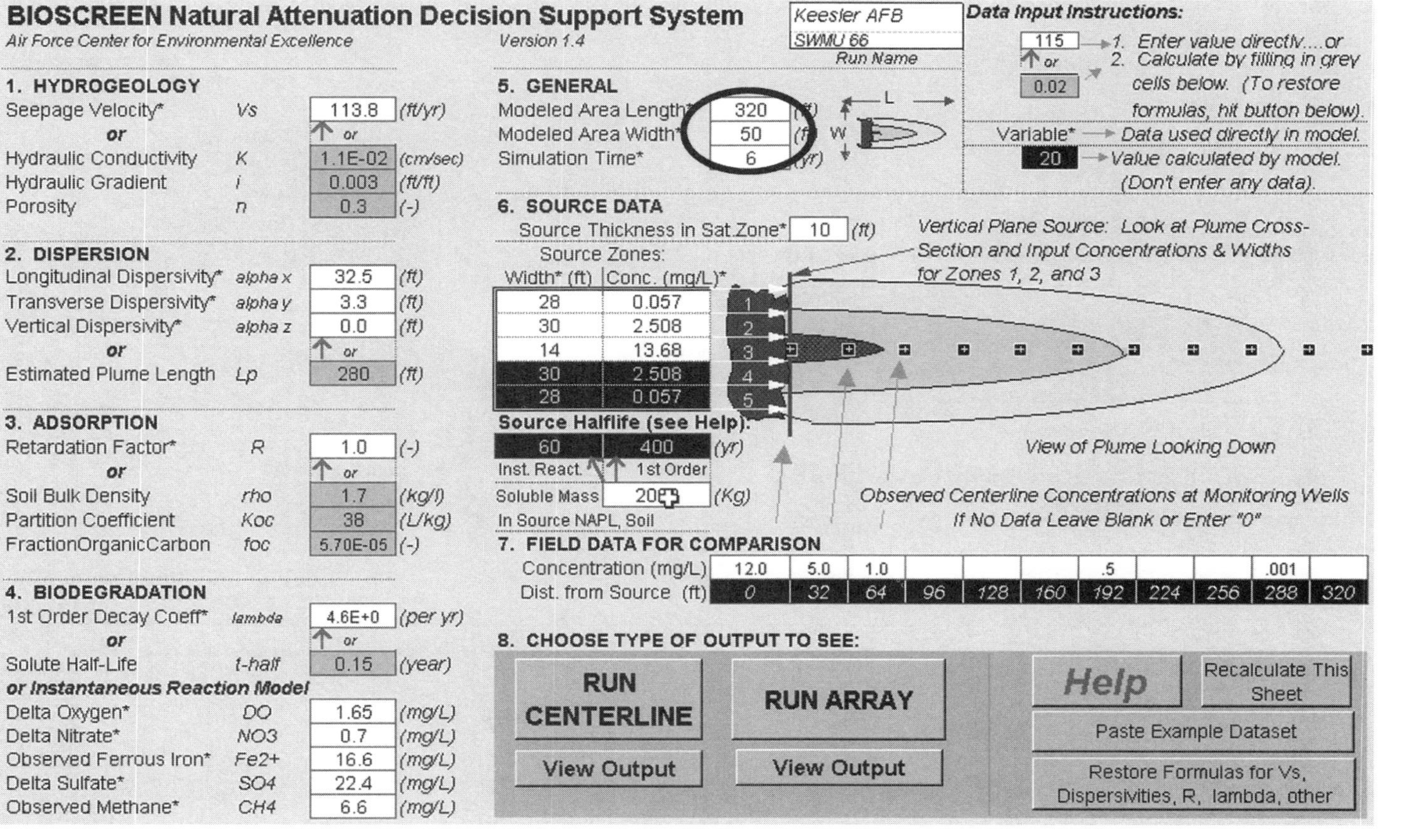

FIGURE 8.9. BIOSCREEN input screen with 50-ft modeled area width for Keesler Air Force Base, Mississippi. (From Newell et al., 1997.)

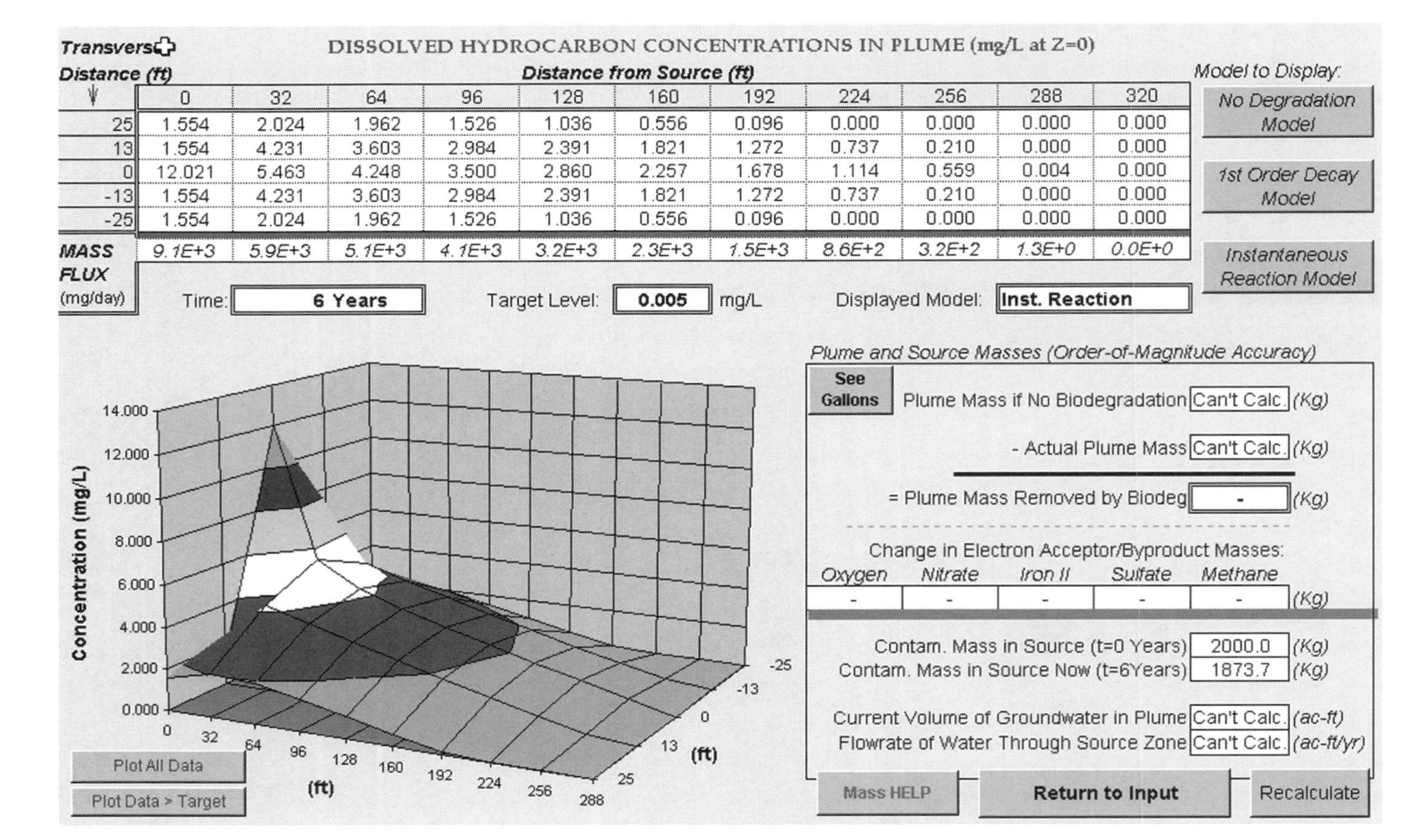

DISSOLVED HYDROCARBON CONCENTRATIONS IN PLUME (mg/L at Z=0)

Transverse Distance (ft)	0	32	64	96	128	160	192	224	256	288	320
25	1.554	2.024	1.962	1.526	1.036	0.556	0.096	0.000	0.000	0.000	0.000
13	1.554	4.231	3.603	2.984	2.391	1.821	1.272	0.737	0.210	0.000	0.000
0	12.021	5.463	4.248	3.500	2.860	2.257	1.678	1.114	0.559	0.004	0.000
-13	1.554	4.231	3.603	2.984	2.391	1.821	1.272	0.737	0.210	0.000	0.000
-25	1.554	2.024	1.962	1.526	1.036	0.556	0.096	0.000	0.000	0.000	0.000
MASS FLUX (mg/day)	9.1E+3	5.9E+3	5.1E+3	4.1E+3	3.2E+3	2.3E+3	1.5E+3	8.6E+2	3.2E+2	1.3E+0	0.0E+0

FIGURE 8.10. Array concentration output with 50-ft modeled area width for Keesler Air Force Base, Mississippi. (From Newell et al., 1997.)

much longer than the time required for the biomass to complete the biological reaction.

2. *Prediction of dissolved hydrocarbon concentrations.* A comparison of the two kinetic expressions was performed by calibrating both simulations to match the observed hydrocarbon concentrations for three sites. Both models were able to approximate the centerline plume concentrations after calibration (see Figure 8.11).
3. *Depletion of electron acceptors.* A comparison of the modeled electron acceptor consumption (and by-product generation) versus the observed electon acceptor consumption (and by-product generation) was performed for both the first-order decay kinetics and instantaneous reaction kinetics. The results indicated that the first-order decay model underpredicted the observed contaminant mass removal by a factor of 2 to 3 (see Figure 8.11). The instantaneous reaction model performs much better in predicting the mass removal by alternative electron acceptors [see Figure 8.11; note that the mass balance computed the total mass (in kilograms) of contaminant consumed per foot of saturated aquifer depth in the contaminated area]. The reason is that the instantaneous reaction model simulates vigorous aerobic and (when the oxygen is depleted) anaerobic biodegradation in the dissolved-phase groundwater in the source zones, while the first-order decay model has much lower rates of biodegradadation in source zones.

Natural Attenuation Tool Kit

Some of the analytical models described in this chapter and in other chapters have been integrated into one single natural attenuation tool kit (Groundwater Services, Inc., 1997). This package includes a simplified version of the BIOSCREEN model, the SAM model (see Section 2.5), a source mass calculator called SMaC (see Section 2.8), and Mann–Kendall plume stability analysis tools (see Chapter 7). This model-based software, originally developed for the state of Florida, allows users to develop primary and secondary lines of evidence for natural attenuation studies performed under the ASTM remediation by natural attenuation (RNA) standard (ASTM, 1997).

BIOCHLOR Natural Attenuation Model

The BIOCHLOR natural attenuation model (Aziz et al., 1999) simulates chlorinated solvent natural attenuation using an interface similar to BIOSCREEN's. However, there are significant differences in the way BIOCHLOR and BIOSCREEN handle biodegradation reactions. BIOSCREEN simulates BTEX biodegradation and natural attenuation at petroleum-hydrocarbon sites, in which biodegradation is assumed to occur as relatively fast, single-step reactions, with BTEX components biodegrading simultaneously (i.e., parallel reactions). Although this is not true in a thermodynamic sense, it is a relatively good approximation at fuel hydrocarbon sites, in which most monitoring wells show oxygen, nitrate, and sulfate depletion and ferrous iron and methane production concurrently. BIOCHLOR simulates chlorinated solvent biodegradation, in

ST-41 Site/Elmendorf AFB

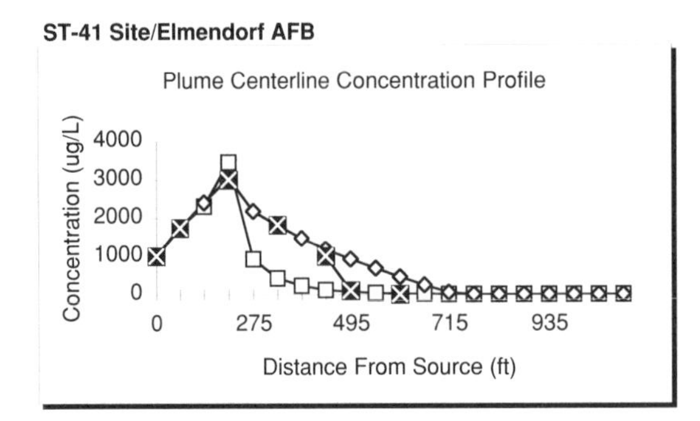

Total Mass of BTEX Degraded (kg/ft):		
Modeled		Observed
First-Order	Instantaneous	
9	260	277

ST-29 Site/Patrick AFB

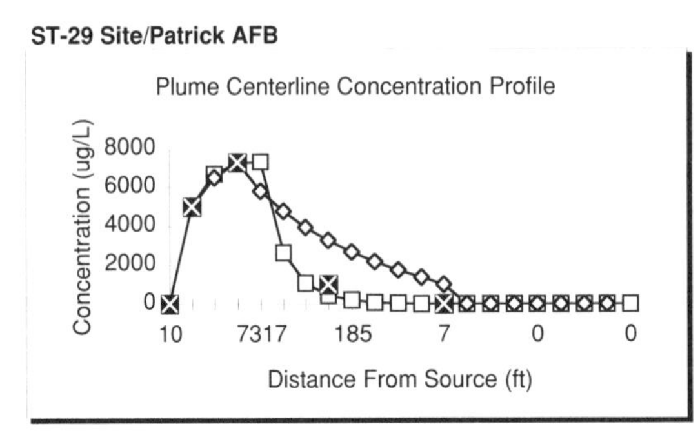

Total Mass of BTEX Degraded (kg/ft):		
Modeled		Observed
First-Order	Instantaneous	
17	27	30

POL Site/Hill AFB

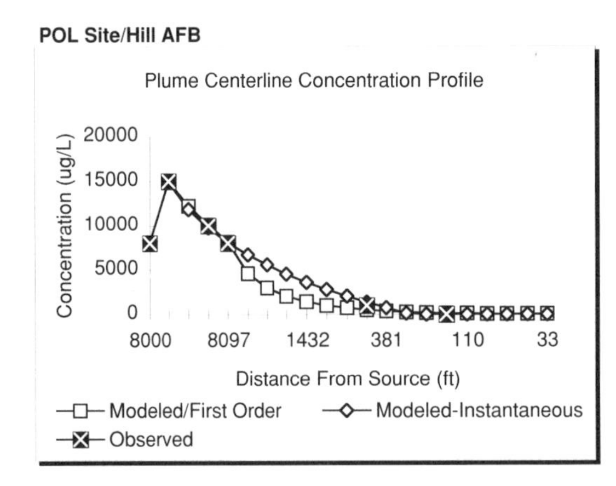

Total Mass of BTEX Degraded (kg/ft):		
Modeled		Observed
First-Order	Instantaneous	
143	780	600

FIGURE 8.11. BIOSCREEN modeling results for three Air Force sites. The total mass of BTEX degraded for the two modeled approaches was calculated by running the models with and without biodegradation, and subtracting the mass of the plume with biodegradation from the mass of the plume without biodegradation. To calculate the "observed" mass of BTEX that had been degraded, the areas of the contours associated with depleted electron acceptors (oxygen, nitrate, and sulfate) and the contours for metabolic by-products (ferrous iron and methane) and converting to mass of BTEX using utilization factors (see Chapter 5) were integrated and summed to determine masses. As can be seen, both models, after calibration, were successful at matching the observed BTEX concentrations. The instantaneous reaction model, however, provided a more accurate match to observed loss of BTEX as indicated by electron acceptor consumption and by-product production. (From Newell et al., 1995.)

which the reaction kinetics may be much slower, making the instantaneous reaction assumption less valid. In addition, chlorinated solvent biodegradation is assumed to occur through reductive dechlorination. This process involves sequential reactions, in which the parent compound biodegrades into a daughter product and that daughter product biodegrades into another daughter product, and so on. For the chlorinated ethenes, the reaction sequence is shown below for the degradation of PCE through ethene (ETH):

$$\text{PCE} \xrightarrow{k_1} \text{TCE} \xrightarrow{k_2} \text{DCE} \xrightarrow{k_3} \text{VC} \xrightarrow{k_4} \text{ETH}$$

The equations describing the sequential first-order biodegradation reaction rates are shown below for each of the components:

$$\begin{aligned}
r_{\text{PCE}} &= -k_1 C_{\text{PCE}} \\
r_{\text{TCE}} &= k_1 C_{\text{PCE}} - k_2 C_{\text{TCE}} \\
r_{\text{DCE}} &= k_2 C_{\text{TCE}} - k_3 C_{\text{DCE}} \\
r_{\text{VC}} &= k_3 C_{\text{DCE}} - k_4 C_{\text{VC}} \\
r_{\text{ETH}} &= k_4 C_{\text{VC}}
\end{aligned}$$

where k_1, k_2, k_3, and k_4 are the first-order rate constants and C_{PCE}, C_{TCE}, C_{DCE}, C_{VC}, and C_{ETH} are the aqueous concentration of PCE, TCE, DCE, vinyl chloride, and ethene, respectively. These equations assume no degradation of ethene.

To describe the transport and reaction of these compounds in the subsurface, one-dimensional advection, three-dimensional dispersion, linear adsorption, and sequential first-order biodegradation are assumed as shown in the equations below. All equations but the first are coupled to another equation through the reaction term.

$$\begin{aligned}
R_{\text{PCE}} \frac{dC_{\text{PCE}}}{dt} &= -v \frac{dC_{\text{PCE}}}{dx} + D_x \frac{d^2 C_{\text{PCE}}}{dx^2} + D_y \frac{d^2 C_{\text{PCE}}}{dy^2} + D_z \frac{d^2 C_{\text{PCE}}}{dz^2} - k_1 C_{\text{PCE}} \\
R_{\text{TCE}} \frac{dC_{\text{TCE}}}{dt} &= -v \frac{dC_{\text{TCE}}}{dx} + D_x \frac{d^2 C_{\text{TCE}}}{dx^2} + D_y \frac{d^2 C_{\text{TCE}}}{dy^2} + D_z \frac{d^2 C_{\text{TCE}}}{dz^2} \\
&\quad + k_1 C_{\text{PCE}} - k_2 C_{\text{TCE}} \\
R_{\text{DCE}} \frac{dC_{\text{DCE}}}{dt} &= -v \frac{dC_{\text{DCE}}}{dx} + D_x \frac{d^2 C_{\text{DCE}}}{dx^2} + D_y \frac{d^2 C_{\text{DCE}}}{dy^2} + D_z \frac{d^2 C_{\text{DCE}}}{dz^2} \\
&\quad + k_2 C_{\text{TCE}} - k_3 C_{\text{DCE}} \\
R_{\text{VC}} \frac{dC_{\text{VC}}}{dt} &= -v \frac{dC_{\text{VC}}}{dx} + D_x \frac{d^2 C_{\text{VC}}}{dx^2} + D_y \frac{d^2 C_{\text{VC}}}{dy^2} + D_z \frac{d^2 C_{\text{VC}}}{dz^2} \\
&\quad + k_3 C_{\text{DCE}} - k_4 C_{\text{VC}} \\
R_{\text{ETH}} \frac{dC_{\text{ETH}}}{dt} &= -v \frac{dC_{\text{ETH}}}{dx} + D_x \frac{d^2 C_{\text{ETH}}}{dx^2} + D_y \frac{d^2 C_{\text{ETH}}}{dy^2} + D_z \frac{d^2 C_{\text{ETH}}}{dz^2} + k_4 C_{\text{VC}}
\end{aligned}$$

where R_{PCE}, R_{TCE}, R_{DCE}, R_{VC}, and R_{ETH} = retardation factors
v = seepage velocity
D_x, D_y, and D_z = dispersivities in x, y, and z directions

BIOCHLOR uses a novel analytical solution to solve these coupled transport and reaction equations in an Excel spreadsheet. To uncouple these equations, BIOCHLOR employs transformation equations developed by Sun and Clement (in press). The uncoupled equations were solved using the Domenico model, and inverse transformations were used to generate concentration profiles. Details of the transformation are presented elsewhere (Sun and Clement, in press). Representative concentration profiles for biodegradation of PCE through ethene are shown in Figure 8.12. Typically, source zone concentrations of *cis*-1,2-dichloroethythene (DCE) are high because biodegradation of PCE and TCE has been occurring since the solvent release.

BIOCHLOR also simulates different first-order decay rates in two different zones at a chlorinated solvent site. For example, BIOCHLOR is able to simulate a site with high dechlorination rates in a high-carbon area near the source that becomes a zone with low dechlorination rates downgradient when fermentation substrates have been depleted.

In addition to the model, a database of chlorinated solvent sites is currently being analyzed to develop empirical rules for predicting first-order decay coefficients that can be used in BIOCHLOR. For example, at sites with evidence of considerable halorespiration, the use of higher first-order decay coefficients will be recommended. Indicators of high rates of halorespiration may include (1) high concentrations of fermentation substrates such as BTEX at the site; (2) high methane concentrations, which indicate high rates of fermentation; (3) large ratios of daughter products to parent compounds; and (4) high concentrations of source zone chloride compared to background chloride concentrations. The BIOCHLOR model and database is now available at www.gsi-net.com.

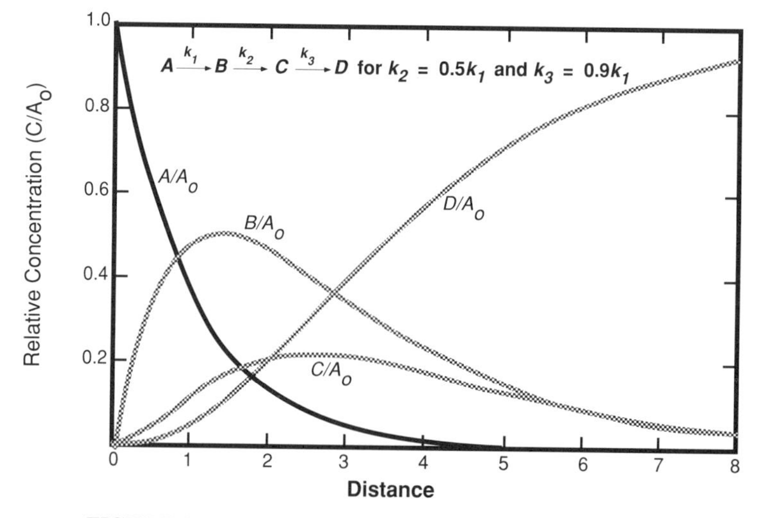

FIGURE 8.12. Concentration profiles simulated in BIOCHLOR.

8.5 NUMERICAL NATURAL ATTENUATION MODELS

Numerical models provide inexact (relative to analytical methods) and, in some cases, nonunique solutions to the governing advection–dispersion equation. As with analytical models, the use of numerical models requires the user to make some simplifying assumptions about the solute transport system. However, fewer simplifying assumptions must be made, so numerical models can simulate more complex systems. Numerical models can be used to simulate complex hydrogeologic systems or contaminant transport affected by multiple reactions for which rates or properties may vary spatially. Heterogeneous and anisotropic hydrologic systems can be modeled using numerical models, as can transient flow systems (i.e., systems in which stresses, parameters, or boundary conditions affecting or controlling groundwater flow change over time). Another advantage of numerical models is that most codes are more flexible in allowing simulation of contaminant sources that vary over time, allowing more straightforward simulation of scenarios, including source reduction through weathering or through engineered solutions.

Success in groundwater solute fate and transport modeling using numerical methods depends on the ability to conceptualize properly the processes governing contaminant transport, to select a model that simulates the most important processes at a site, to understand the limitations of the solution methods, and to present model predictions that are reasonable within those limitations. When using a numerical transport model (and an associated flow model, if applicable), remember that implementation of a numerical model is much more complex than implementation of an analytical model and generally requires at a minimum the supervision of an experienced hydrogeologist who is familiar with the model code. Keep in mind that any numerical model code or analytical solution selected for a demonstration of remediation by natural attenuation should be validated properly through sufficient previous application at a variety of field sites.

Summary of Existing Numerical Models

Numerical models used for the evaluation of natural attenuation processes in groundwater should be capable of simulating advection, dispersion, sorption, and biodegradation (or any first-order decay process). A large number of numerical codes are available for simulating groundwater flow and solute transport under the influence of these processes. Some include both flow and transport, while others are flow models alone or transport models alone (but which may be combined, i.e., flow fields from a flow model are incorporated into a transport model). The selection of a code will ultimately depend on the user's needs, the available data, the sophistication of the desired predictions, the sophistication of the user's computer hardware, and the limitations of the available model codes.

A limited number of numerical models are available for simulating the reactant-limited transport of solutes, but these codes are intended largely for simulating transport and biodegradation of petroleum compounds. Examples include BIOPLUME II (Rifai et al., 1988), BIOPLUME III (Rifai et al., 1997), and Bio1D© (Srivinsan and Mercer, 1988).

A wide range of other codes that can be used to evaluate contaminant transport with biodegradation is currently available. In these codes, biodegradation reaction rates are typically controlled by a first-order rate constant rather than by the availability of the reactants. Nonetheless, with appropriately calculated rate constants, these codes often are the most suitable means with which fate and transport of chlorinated solvents dissolved in groundwater may be simulated. Table 8.6 presents a partial list of numerical model codes that may be used for simulating the fate and transport of contaminants dissolved in groundwater. Of these codes, the combination of MODFLOW and MT3D may be one of the most commonly used, especially given the number of proprietary software packages that are designed around those programs. Table 8.6 is not a thorough review of available codes; rather, this overview is to illustrate the variety of readily available codes for personal computers that may be useful in evaluating the natural attenuation of contaminants dissolved in groundwater.

BIOPLUME II/BIOPLUME III Model

The BIOPLUME III model (Rifai et al., 1997) is a two-dimensional, finite difference model for simulating the natural attenuation of organic contaminants in groundwater due to the processes of advection, dispersion, sorption, ion exchange, and biodegradation. The model simulates the biodegradation of organic contaminants using a number of aerobic and anaerobic electron acceptors: oxygen, nitrate, iron(III), sulfate, and carbon dioxide. BIOPLUME III is based on the U.S. Geological Survey (USGS) method of characteristics model dated July 1989 (Konikow and Bredehoeft, 1989).

The BIOPLUME III code was developed primarily to model the natural attenuation of organic contaminants in groundwater. The model solves the transport equation six times to determine the fate and transport of the hydrocarbons and the electron acceptors and reaction by-products. For the case in which iron(III) is used as an electron acceptor, the model simulates the production and transport of iron(II) or ferrous iron.

Three different kinetic expressions can be used to simulate the aerobic and anaerobic biodegradation reactions. These include first-order decay, instantaneous reaction, and Monod kinetics (see Chapter 4). The principle of superposition is used to combine the hydrocarbon plume with the electron acceptor plume(s).

Model Background. Borden and Bedient (1986) developed the BIOPLUME I model based on their work at the United Creosoting Company Superfund site in Conroe, Texas. BIOPLUME I is based on the assumption that aerobic biodegradation of hydrocarbons is often limited by the availability of dissolved oxygen in groundwater aquifers. Borden and Bedient (1986) simulated the aerobic biodegradation of hydrocarbons as an instantaneous reaction between the hydrocarbon and oxygen.

Rifai et al. (1988) developed the BIOPLUME II model by incorporating the concepts developed by Borden and Bedient (1986) into the USGS two-dimensional solute transport model (Konikow and Bredehoeft, 1978). The BIOPLUME II model tracks two plumes: oxygen and hydrocarbon. The two plumes are superimposed to

TABLE 8.6. Numerical Groundwater Flow and Solute Transport Models

Model Code Name	Capabilities	Distributor
AQUA	Two-dimensional, transient groundwater flow and transport. Aquifer may be heterogeneous and anisotropic. Can simulate advection, dispersion, linear sorption, and decay. A proprietary code with initeractive/graphical interface.	Scientific Software Group
ASM	Aquifer simulation model for two-dimensional modeling of groundwater flow and solute transport. Uses random-walk method for solute transport, and can simulate advection, dispersion, linear sorption, and decay. Aquifer can be heterogeneous and anisotropic. Menu-driven, graphical interface. A proprietary program prepared by W. Kinzelbach (University of Heidelberg) and R. Rausch (University of Stuttgart).	IGWMC
BIO1D®	One-dimensional model for simulation of biodegradation and sorption of hydrocarbons. Transport of substrates and electron acceptors is considered, assuming a uniform flow field. Several reaction options are available for biodegradation and sorption. Has a preprocessor and display graphics. A proprietary code developed at GeoTrans, Inc.	GeoTrans, Inc.; IGWMC
BIOPLUME II	Two-dimensional model for simulating transport of a single dissolved hydrocarbon species under the influence of oxygen-limited biodegradation, first-order decay, linear sorption, advection, and dispersion. Aquifer may be heterogeneous and anisotropic. Based on the USGS two-dimensional MOC model (including a finite-difference flow model) by Konikow and Bredehoeft (1978). Oxygen-limited biodegradation is a reactive transport process. A public-domain code with a menu-driven preprocessor and limited postprocessing abilities. Developed by Rifai et al. (1988) at Rice University.	IGWMC
BIOPLUME III	Successor to BIOPLUME II. Two-dimensional model for reactive transport of multiple hydrocarbons under the influence of advection, dispersion, sorption, first-order decay, and reactant-limited biodegradation. Development commissioned by AFCEE. Has interactive, graphical pre- and postprocessing capabilities.	AFCEE (currently under development)
BioTrans	Proprietary two-dimensional finite-element transport code requiring flow velocity data from another code (e.g., MODFLOW). Models transport of multiple species under the influence of advection, dispersion, sorption, first-order decay, and oxygen-limited biodegradation. Allows internal computation of source terms due to dissolution of NAPL. Graphical, interactive user interface with pre- and postprocessing capabilities. Prepared by Environmental Systems and Technologies, Inc.	Environmental Systems and Technologies, Inc.

continues

TABLE 8.6. *(Continued)*

Model Code Name	Capabilities	Distributor
FEMSEEP	Set of programs for solving steady-state and transient groundwater flow and solute transport problems in simplified two- and three-dimensional systems. Transport under the influence of advection, dispersion, linear sorption, and first-order decay may be simulated using finite element methods. A proprietary program with graphical and menu-driven user interfaces and pre- and postprocessing capabilities. Prepared by D. Meiri of FEMSEEP Software, Inc.	FEMSEEP Software, Inc.; IGWMC
FEMWATER, FEMWASTE	Finite-element flow (FEMWATER) and transport (FEMWASTE) models. FEMWATER can simulate variably saturated conditions in two and three dimensions. FEMWASTE can simulate transport in one, two, and three dimensions. The system may be heterogeneous and anisotropic, and the code can account for advection, dispersion, first-order decay, and three types of sorption. Public-domain codes developed by researchers at Oak Ridge National Laboratories. Some proprietary versions of FEMWATER are available; they are based on the Department of Defense's Groundwater Modeling System (GMS) modeling and data management package.	Oak Ridge National Laboratories, NTIS, distributors of-proprietary GMS programs
FLONET®, FLOTRANS®	Two-dimensional steady-state groundwater flow (FLONET) and transient solute transport (FLOTRANS) models for cross-sectional problems. FLOTRANS is an extension of FLONET that can simulate transport under the influence of advection, dispersion, linear sorption, and first-order decay. A proprietary program with an interactive graphical user interface and extensive pre- and postprocessing capabilities. Developed by Waterloo Hydrogeologic Software, Inc.	IGWMC; Waterloo Hydrogeologic Software, Inc.
FTWORK	Block-centered finite-difference model for one-, two-, and three-dimensional flow and transport. The transport model includes advection, dispersion, first-order decay, and two types of sorption (linear and nonlinear equilibrium). A public-domain code that may be acquired with a proprietary (IGWMC) textual and menu-driven preprocessor and postprocessor. Originally developed by Faust et al. (1990) at GeoTrans, Inc.	IGWMC; GeoTrans, Inc.
HST3D	Program for simulating groundwater flow and associated heat and solute transport in three dimensions. Solute transport is for a single solute with advection, dispersion, linear sorption, and first-order decay. A public-domain code with no pre- and postprocessors. Prepared by K. L. Kipp of the USGS.	IGWMC
MOC, USGS2D-MOC	Two-dimensional model for simulation of groundwater flow and nonconservative solute transport. Derived from the original model developed by Konikow and Bredehoeft (1978). The latest version (March 1995) simulates transport under the influence of advection, dispersion, first-order decay, reversible equilibrium-controlled sorption, and reversible equilibrium-controlled ion exchange. The flow model is a finite-difference model, while transport is simulated using MOC methods. A public-domain code with an interactive preprocessor.	IGWMC

MODFLOW	Block-centered finite-difference code for steady-state and transient simulation of groundwater flow in two and three dimensions. Consists of a main program and a large number of subroutines (modules) that are used to simulate a wide variety of boundaries and stresses on the hydrogeologic sys tem. Originally coded by McDonald and Harbaugh (1988) of the USGS. Possibly the most widely used flow model in the United States and Canada, MODFLOW can be used to generate flow fields that may be coupled with a wide variety of transport models (e.g., MT3D, BioTrans®, or RAND3D). The popularity of MODFLOW is also evidenced by the great number of proprietary pre- and postprocessing programs that are available. MODFLOW is a public-domain code, although it is typically acquired in conjunction with a pre-/postprocessing package.	USGS; IGWMC; in addition, many companies have developed pre- and postprocessing programs with a wide variety of capabilities and features
MODFLOWP	Extension of MODFLOW that includes a package that uses nonlinear regression techniques to estimate model parameters under constraints given by the modeler. Model input includes statistics for analyzing the parameter estimates and the model to quantify the reliability of the resulting model, to suggest changes in model construction, and to compare results of models constructed in different ways. Prepared by M. C. Hill of the USGS. Requires a user with advanced skills.	USGS; IGWMC
MT3D	Three-dimensional transport model for simulation of advection, dispersion, linear or nonlinear sorption, and first-order decay of a single species. Uses a modular structure similar to that of MODFLOW. Intended for use with any block-centered finite-difference flow model, such as MODFLOW, on the assumption that concentration changes will not affect the flow field. MT3D uses one of three methods (all based on MOC) for solution of the transport equation. Prepared by C. Zheng (for S. S. Papadopulos & Associates, Inc.), MT3D is available in public-domain and proprietary versions. Proprietary versions are typically the most advanced in terms of pre- and postprocessing capabilities.	S. S. Papadopulos & Associates, Inc.; IGWMC; many versions available from many companies with pre- and postprocessing programs with a wide variety of capabilities and features. Often coupled with MODFLOW in such codes. Public-domain version may be acquired from USEPA.
RANDOM WALK	Code for simulation of two-dimensional groundwater flow and solute transport. Flow is simulated using either analytical solutions or a version of the PLASM finite-difference model (Prickett and Lonnquist, 1971). Transport is simulated using particle-tracking methods coupled with the random-walk technique for dispersion. The model also handles first-order decay, linear sorption, and zero-order production. A public-domain code originally produced at the Illinois State Water Survey (Prickett et al., 1981). The IGWMC version includes pre- and postprocessing utilities.	IGWMC

continues

TABLE 8.6. *(Continued)*

Model Code Name	Capabilities	Distributor
RAND3D	Three-dimensional version of the random walk algorithm developed by Prickett et al. (1981). RAND3D is designed to be coupled with MODFLOW input files for calculation of velocity vector files that are used to run the code. May be used for transient simulation of advection, dispersion, linear sorption, and zero-order, first-order, or variable-order decay. Code has some pre- and postprocessing capabilities. A proprietary code prepared by D. Koch of Engineering Technologies Associates and T. A. Prickett.	IGWMC
RT3D	Modification of MT3D, under development by researchers at Washington State University and Pacific Northwest National Laboratory. RT3D (Reactive Transport in Three Dimensions) is designed to describe multispecies transport and reactions, including attenuation of chlorinated compounds and their daughter products, and fate of solid-phase species. Also included are modules for aerobic, instantaneous BTEX reactions (similar to Bioplume II) and multiple-electron-acceptor, kinetically limited BTEX reactions (similar to Bioplume III).	Under development
SUTRA	Code for simulating two-dimensional fluid movement and transport of energy or dissolved substances. May be used for saturated systems or variably saturated systems in profile view. Can simulate advection, dispersion, sorption, and first-order decay. A public-domain code originally prepared by C. I. Voss of the USGS, IGWMC version has a graphical postprocessor.	IGWMC; USGS
SWICHA	Three-dimensional finite-element code for simulating steady-state and transient flow and transport in confined (fully saturated) aquifers. Transport includes advection, dispersion, sorption, and first-order decay. A public-domain code with no pre- or postprocessing capabilities. Authored by B. Lester of GeoTrans, Inc.	IGWMC
SWIFT, SWIFT/486	Fully three-dimensional fnite-difference model for simulating flow and transport of fluid, heat, and solutes in porous and fractured media. Includes linear and nonlinear sorption, dispersion, diffusion, and decay, as well as dissolution, leaching, and dual porosity. An advanced code developed at Sandia National Laboratories (Reeves and Cranwell, 1981); now in the custody of GeoTrans, Inc. Public-domain and proprietary versions of the code are available through IGWMC.	IGWMC
SWMS_2D	Two-dimensional model for simulating water and solute movement in variably saturated media. Includes dispersion, linear sorption, zero-order production, and first-order decay. A public-domain code prepared by researchers at the U.S. Salinity Lab. No pre- or postprocessing utilities.	IGWMC
TARGET	Code for simulating two- and three-dimensional flow and transport under a wide variety of conditions. Can simulate advection, dispersion, diffusion, sorption, and first-order decay. A proprietary code prepared by workers at Dames & Moore, Inc. Has been used for a wide variety of applications.	Dames & Moore, Inc.

Source: From Wiedemeier et al. (1996).

determine the resulting concentrations of oxygen and hydrocarbon at each time step. Anaerobic biodegradation in BIOPLUME II was simulated as a first-order decay in hydrocarbon concentrations.

Other major differences between BIOPLUME II and BIOPLUME III include:

- BIOPLUME III runs in a Windows95 environment, whereas BIOPLUME II was developed primarily in a DOS environment.
- BIOPLUME III model was integrated with a sophisticated groundwater modeling platform known as Environmental Impact System from ZEi/MicroEngineering, Inc., of Annandale, Virginia.

Conceptual Model for Biodegradation. Recent research suggests that hydrocarbons are degraded both aerobically and anaerobically in subsurface environments. The main electron acceptors include oxygen for aerobic biodegradation and nitrate, iron(III), sulfate, and carbon dioxide for anaerobic biodegradation. Manganese has also been identified as an anaerobic electron acceptor; however, manganese has not been incorporated into the current version of BIOPLUME III.

The conceptual model used in BIOPLUME III to simulate these biodegradation processes tracks six plumes simultaneously: hydrocarbon, oxygen, nitrate, iron(II), sulfate, and carbon dioxide. Iron(III) is input as a concentration matrix of ferric iron in the formation. Once ferric iron is used for biodegradation, BIOPLUME III simulates the production and transport of ferrous iron. Biodegradation occurs sequentially in the following order:

$$\text{oxygen} \rightarrow \text{nitrate} \rightarrow \text{iron(III)} \rightarrow \text{sulfate} \rightarrow \text{carbon dioxide}$$

The biodegradation of hydrocarbon in a given location using nitrate, for example, can occur only if oxygen has been depleted to its threshold concentration at that location.

Three different kinetic expressions can be utilized for the biodegradation reaction for each of the electron acceptors:

1. First-order decay
2. Instantaneous reaction
3. Monod kinetics

These kinetic expressions were presented in detail in Chapter 4. The first-order decay model implemented in BIOPLUME III for any of the electron acceptors is limited by the availability of the electron acceptor in question. In other words, the model allows the first-order reaction to take place up to the point that the electron acceptor concentration available in the aquifer has been depleted. The Monod kinetic model used in BIOPLUME III assumes a constant microbial population for each of the aerobic and anaerobic reactions and does not simulate the growth, transport, and decay of the microbial population in the subsurface.

Model Applicability and Limitations. The BIOPLUME III model has been developed primarily to simulate the natural attenuation of hydrocarbons using oxygen, nitrate, iron(III), sulfate, and carbon dioxide as electron acceptors for biodegradation. BIOPLUME III is generally used to answer a number of questions regarding natural attenuation:

1. How long will the plume extend if no engineered/source controls are implemented?
2. How long will the plume persist until natural attenuation processes dissipate the contaminants completely?
3. How long will the plume extend or persist if some engineered controls or source reduction measures are undertaken (e.g., free phase removal or residual soil contamination removal)?

The model can also be used to simulate bioremediation of hydrocarbons in groundwater by injecting electron acceptors [except for iron(III)] and can also be used to simulate air sparging for low injection airflow rates. Finally, the model can be used to simulate advection, dispersion, and sorption without including biodegradation.

As with any model, there are limitations to the use of BIOPLUME III. The assumptions used in the USGS MOC code include:

1. Darcy's law is valid, and hydraulic-head gradients are the only driving mechanism for flow.
2. The porosity and hydraulic conductivity of the aquifer are constant with time, and porosity is uniform in space.
3. Gradients of fluid density, viscosity, and temperature do not affect the velocity distribution.
4. No chemical reactions occur that affect the fluid properties or the aquifer properties.
5. Ionic and molecular diffusion are negligible contributors to the total dispersive flux.
6. Vertical variations in head and concentration are negligible.
7. The aquifer is homogeneous and isotropic with respect to the coefficients of longitudinal and transverse dispersivity.

The limitations imposed by the biodegradation expressions incorporated in BIOPLUME III include:

1. The model does not account for selective or competitive biodegradation of the hydrocarbons. This means that hydrocarbons are generally simulated as a lumped organic which represents the sum of benzene, toluene, ethylbenzene, or xylene. If a single component is to be simulated, the user would have to determine how much electron acceptor would be available for the component in question.

2. The conceptual model for biodegradation used in BIOPLUME III is a simplification of the complex biologically mediated redox reactions that occur in the subsurface.

BIOPLUME II Equation Formulation. Borden and Bedient (1986) simulated the growth of microorganisms and removal of hydrocarbon and oxygen using a modification of the Monod function where

$$\frac{dH}{dt} = -M_t k \frac{H}{K_h + H} \frac{O}{K_O + O} \tag{8.27}$$

$$\frac{dO}{dt} = -M_t kF \frac{H}{K_h + H} \frac{O}{K_O + O} \tag{8.28}$$

$$\frac{dM_t}{dt} = M_t kY \frac{H}{K_h + H} \frac{O}{K_O + O} + k_c YC - bM_t \tag{8.29}$$

where H = hydrocarbon concentration
M_t = total microbial concentration
k = maximum hydrocarbon utilization rate per unit mass microorganisms
K_h = hydrocarbon half-saturation constant
K_o = oxygen half-saturation constant
O = oxygen concentration
F = ratio of oxygen to hydrocarbon consumed
Y = microbial yield coefficient (g cells/g hydrocarbon)
k_c = first-order decay rate of natural organic carbon
C = natural organic carbon concentration
b = microbial decay rate

Equations (8.27) through (8.29) were combined with the advection–dispersion equation for a solute undergoing linear instantaneous adsorption to result in

$$\frac{\partial H}{\partial t} = \nabla \cdot \frac{D\nabla H - vH}{R_h} - M_t k \frac{H}{K_h + H} \frac{Q}{K_O + O} \tag{8.30}$$

$$\frac{\partial O}{\partial t} = \nabla \cdot (D\nabla O - vO) - M_t kF \frac{H}{K_h + H} \frac{O}{K_O + O} \tag{8.31}$$

where D = dispersion tensor
v = groundwater velocity vector
R_h = retardation factor for hydrocarbon

The movement of naturally occurring microorganisms will be limited by the tendency of the organisms to grow as microcolonies attached to the formation. Borden and Bedient (1986) assumed that the transfer of microorganisms between the solid surface and groundwater will be rapid and will follow a linear relationship with total concentration, thus allowing them to simulate the transport of microorganisms using a simple retardation factor approach:

$$\frac{\partial M_S}{\partial t} = \nabla \cdot \frac{D\nabla M_S - vM_S}{R_m} + M_S kY \frac{H}{K_h + H} \frac{O}{K_O + O} + \frac{k_c YC}{R_m} - bM_s \qquad (8.32)$$

where $M_t = M_s + M_a = (1 + K_m)M_s = R_m M_s$
M_s = concentration of microbes in solution
M_a = concentration of microbes attached to aquifer $K_m M_s$
K_m = ratio of microbes attached to microbes in solution
R_m = microbial retardation factor

Borden and Bedient (1986) conducted one-dimensional simulations with eqs. (8.30) through (8.32) and determined that there are three general regions in which different processes control the rate and extent of degradation: near the contaminant source, in the heart of the plume, and at the leading edge of the plume. Biodegradation rates will be very high near the source and will result in depleted oxygen levels. Biodegradation in the heart of the plume will be limited by the availability of oxygen. The primary mass transfer processes include horizontal mixing with oxygenated formation water, advective fluxes of oxygen, and vertical exchange with the unsaturated zone. The limited oxygen supply to the heart of the plume will result in a region of reduced oxygen and hydrocarbon concentrations. At the leading edge of the plume, oxygen is present in excess and hydrocarbons will be absent or present in trace quantities.

Sensitivity analyses with the one-dimensional model indicated that the microbial parameters had little or no effect on the hydrocarbon concentration in the body of the plume and on the time to hydrocarbon breakthrough. This led Borden and Bedient (1986) to assume that the consumption of hydrocarbon and oxygen might be approximated as an instantaneous reaction between oxygen and hydrocarbon:

$$H(t+1) = H(t) - \frac{O(t)}{F} \qquad O(t+1) = 0 \quad \text{where } H(t) > \frac{O(t)}{F} \qquad (8.33)$$

$$O(t+1) = O(t) - H(t)F \qquad H(t+1) = 0 \quad \text{where } O(t) > H(t)F \qquad (8.34)$$

where $H(t)$, $H(t + 1)$, $O(t)$, and $O(t + 1)$ are the hydrocarbon and oxygen concentrations at time t and $t + 1$.

Borden and Bedient (1986) concluded that the instantaneous reaction assumption is a close approximation to eqs. (8.30) through (8.32). Their simulations indicate that the most significant errors using this assumption occur in the region near the source area, especially when groundwater velocities are very high or for poorly degradable hydrocarbons.

Borden and Bedient (1986) used the instantaneous reaction assumption to simplify the system of eqs. (8.30) through (8.32) to

$$\frac{\partial H}{\partial t} = \frac{1}{R_h}\left(D_1\frac{\partial^2 H}{\partial x^2} + D_t\frac{\partial^2 H}{\partial y^2} - v\frac{\partial H}{\partial x}\right) - \delta \tag{8.35}$$

$$\frac{\partial O}{\partial t} = D_1\frac{\partial^2 O}{\partial x^2} + D_t\frac{\partial^2 O}{\partial y^2} - v\frac{\partial O}{\partial x} - \delta F \tag{8.36}$$

where y = coordinate orthogonal to the flow
D_t = transverse dispersion coefficient = $\alpha_t v$
α_t = transverse dispersivity
δ = $\min(H, O/F)$

Two-dimensional simulations of eqs. (8.35) and (8.36) indicate that biodegrading plumes are generally less laterally spread than their nondegrading counterparts. Simulations also indicated that transverse mixing is the major source of oxygen to the plume.

Borden (1986) examined the vertical exchange of oxygen with the unsaturated zone. His simulations indicated that the effects of gas exchange with the unsaturated zone may be approximated as a first-order decay in space and time of hydrocarbon concentrations. Borden (1986) developed a regression function at the United Creosoting Company site to determine the reaeration first-order decay coefficient:

$$K' = 2611 D_v^{0.79} \exp\left(\frac{-10.5B}{B + 1.04}\right) \tag{8.37}$$

where D_v is the vertical dispersion coefficient and B is the saturated thickness.

Development of the BIOPLUME II Model. Rifai et al. (1988) incorporated the conclusions developed by Borden and Bedient (1986) into the USGS two-dimensional solute transport model, more commonly referred as the method of characteristics (MOC) model. The MOC model was modified from a single-particle mover to a dual-particle mover model to simulate the transport of hydrocarbon and oxygen. The system of transport equations solved in BIOPLUME II is given by

$$\frac{\partial Hb}{\partial t} = \frac{1}{R_h}\left[\frac{\partial}{\partial x_i}\left(bD_{ij}\frac{\partial H}{\partial x_j}\right) - \frac{\partial}{\partial x_i}(bHV_i)\right] - \frac{H'W}{n} \tag{8.38}$$

$$\frac{\partial Ob}{\partial t} = \left[\frac{\partial}{\partial x_i}\left(bD_{ij}\frac{\partial O}{\partial x_j}\right) - \frac{\partial}{\partial x_i}(bOV_i)\right] - \frac{O'W}{n} \qquad (8.39)$$

where
- H = concentration of hydrocarbon
- b = saturated thickness
- t = time
- R_h = retardation factor for hydrocarbon
- x_i/x_j = Cartesian coordinates
- D_{ij} = coefficient of hydrodynamic dispersion
- V_i = seepage velocity in the direction of x_i
- H' = concentration of hydrocarbon in source or sink fluid
- W = volume flux per unit area
- n = effective porosity
- O = concentration of oxygen
- O' = concentration of oxygen in source or sink fluid

The hydrocarbon and oxygen plumes are combined using the principle of superposition (Rifai et al., 1988) and eqs. (8.38) and (8.39).

BIOPLUME III Equation Formulation. Much like the approach used in developing BIOPLUME II, the 1989 version of the MOC model was modified to become a six-component particle mover model to simulate the transport of hydrocarbon, oxygen, nitrate, iron(II), sulfate, and carbon dioxide. Since the biodegradation of hydrocarbon uses iron(III) as an electron acceptor, iron(III) concentrations are simulated as an initial concentration of ferric iron that is available in each cell. Once the iron(III) is consumed, hydrocarbon concentrations are reduced and ferrous iron is produced. The resulting ferrous iron is then transported in the aquifer. The BIOPLUME III equations include

$$\frac{\partial Hb}{\partial t} = \frac{1}{R_h}\left[\frac{\partial}{\partial x_i}\left(bD_{ij}\frac{\partial H}{\partial x_j}\right) - \frac{\partial}{\partial x_i}(bHV_i)\right] - \frac{H'W}{n} \qquad (8.40)$$

$$\frac{\partial Ob}{\partial t} = \left[\frac{\partial}{\partial x_i}\left(bD_{ij}\frac{\partial O}{\partial x_j}\right) - \frac{\partial}{\partial x_i}(bOV_i)\right] - \frac{O'W}{n} \qquad (8.41)$$

$$\frac{\partial Nb}{\partial t} = \left[\frac{\partial}{\partial x_i}\left(bD_{ij}\frac{\partial N}{\partial x_j}\right) - \frac{\partial}{\partial x_i}(bNV_i)\right] - \frac{N'W}{n} \qquad (8.42)$$

$$\frac{\partial Fb}{\partial t} = \left[\frac{\partial}{\partial x_i}\left(bD_{ij}\frac{\partial F}{\partial x_j}\right) - \frac{\partial}{\partial x_i}(bFV_i)\right] - \frac{F'W}{n} \qquad (8.43)$$

$$\frac{\partial Sb}{\partial t} = \left[\frac{\partial}{\partial x_i}\left(bD_{ij}\frac{\partial S}{\partial x_j}\right) - \frac{\partial}{\partial x_i}(bSV_i)\right] - \frac{S'W}{n} \tag{8.44}$$

$$\frac{\partial Cb}{\partial t} = \left[\frac{\partial}{\partial x_i}\left(bD_{ij}\frac{\partial C}{\partial x_j}\right) - \frac{\partial}{\partial x_i}(bCV_i)\right] - \frac{C'W}{n} \tag{8.45}$$

where N = concentration of nitrate
N' = concentration of nitrate in source or sink fluid
F = concentration of iron(II)
F' = concentration of iron(II) in source or sink fluid
S = concentration of sulfate
S' = concentration of sulfate in source or sink fluid
C = concentration of carbon dioxide
C' = concentration of carbon dioxide in source or sink fluid

All other parameters are as defined previously.

The biodegradation of hydrocarbon using the aerobic and anaerobic electron acceptors is simulated using the principle of superposition and the following equations:

$$H(t+1) = H(t) - R_{HO} - R_{HN} - R_{HFe} - R_{HS} - R_{HC} \tag{8.46}$$

$$O(t+1) = O(t) - R_{OH} \tag{8.47}$$

$$N(t+1) = N(t) - R_{NH} \tag{8.48}$$

$$\mathrm{Fe}(t+1) = \mathrm{Fe}(t) - R_{FeH} \tag{8.49}$$

$$F(t+1) = R_{FeH} \tag{8.50}$$

$$S(t+1) = S(t) - R_{SH} \tag{8.51}$$

$$C(t+1) = C(t) - R_{CH} \tag{8.52}$$

where R_{HO}, R_{HN}, R_{HFe}, R_{HS}, and R_{HC} are the hydrocarbon concentration losses due to biodegradation using oxygen, nitrate, ferric iron, sulfate, and carbon dioxide as electron acceptors, respectively. The terms R_{OH}, R_{NH}, R_{FeH}, R_{SH}, and R_{CH} are the corresponding concentration losses in the electron acceptors. These reaction terms are computed using one of the three biodegradation expressions: first-order, instantaneous, or Monod. For example, and for the instantaneous model, the reaction terms are computed as follows:

$$
\begin{aligned}
R_{\text{HO}} &= \frac{O(t)}{F_{\text{O}}} \\
R_{\text{HN}} &= \frac{N(t)}{F_{\text{N}}} \\
R_{\text{HFe}} &= \frac{\text{Fe}(t)}{F_{\text{Fe}}} \\
R_{\text{HS}} &= \frac{S(t)}{F_{\text{S}}} \\
R_{\text{HC}} &= \frac{C(t)}{F_{\text{C}}}
\end{aligned} \tag{8.53}
$$

and

$$
\begin{aligned}
R_{\text{OH}} &= H(t)F_{\text{O}} \\
R_{\text{NH}} &= H(t+1)^1 F_{\text{N}} \\
R_{\text{FeH}} &= H(t+1)^2 F_{\text{Fe}} \\
R_{\text{SH}} &= H(t+1)^3 F_{\text{S}} \\
R_{\text{CH}} &= H(t+1)^4 F_{\text{C}}
\end{aligned} \tag{8.54}
$$

where F_{O}, F_{N}, F_{Fe}, F_{S}, and F_{C} are the stoichiometric ratios for each of the electron acceptors, respectively, and $H(t + 1)$, $H(t + 1)^2$, $H(t + 1)^3$, and $H(t + 1)^4$ are the hydrocarbon concentrations modified by loss due to the reaction with oxygen; oxygen and nitrate; oxygen, nitrate, and iron; and oxygen, nitrate, iron, and sulfate, respectively, in the given time step.

For each of the electron acceptors, the following constraints are applied:

$$
\begin{aligned}
H(t+1) &= 0 \quad \text{where } O(t) > H(t)F_{\text{O}} \\
O(t+1) &= 0 \quad \text{where } H(t) > \frac{O(t)}{F_{\text{O}}}
\end{aligned} \tag{8.55}
$$

$$
\begin{aligned}
H(t+1)^2 &= 0 \quad \text{where } N(t) > H(t+1)^1 F_{\text{N}} \\
N(t+1) &= 0 \quad \text{where } H(t+1)^1 > \frac{N(t)}{F_{\text{N}}}
\end{aligned} \tag{8.56}
$$

$$
\begin{aligned}
H(t+1)^3 &= 0 \quad \text{where Fe}(t) > H(t+1)^2 F_{\text{Fe}} \\
\text{Fe}(t+1) &= 0 \quad \text{where } H(t+1)^2 > \frac{\text{Fe}(t)}{F_{\text{Fe}}}
\end{aligned} \tag{8.57}
$$

$$
\begin{aligned}
H(t+1)^4 &= 0 \qquad \text{where } S(t) > H(t+1)^3 F_S \\
S(t+1) &= 0 \qquad \text{where } H(t+1)^3 > \frac{S(t)}{F_S}
\end{aligned} \tag{8.58}
$$

$$
\begin{aligned}
H(t+1) &= 0 \qquad \text{where } C(t) > H(t+1)^4 F_C \\
C(t+1) &= 0 \qquad \text{where } H(t+1)^4 > \frac{C(t)}{F_C}
\end{aligned} \tag{8.59}
$$

Furthermore, these reaction terms are subject to additional constraints. For first-order decay, instantaneous, and Monod kinetic models:

$$R_{HN} = 0 \qquad \text{if} \quad O(t+1) > \mathrm{O} \tag{8.60}$$

$$
\begin{aligned}
R_{HFe} = 0 \qquad &\text{if} \quad O(t+1) > \mathrm{O}_{min} \\
&\text{or} \quad N(t+1) > \mathrm{N}_{min}
\end{aligned} \tag{8.61}
$$

$$
\begin{aligned}
R_{HS} = 0 \qquad &\text{if} \quad O(t+1) > \mathrm{O}_{min} \\
&\text{or} \quad N(t+1) > \mathrm{N}_{min} \\
&\text{or} \quad \mathrm{Fe}(t+1) > \mathrm{Fe}_{min}
\end{aligned} \tag{8.62}
$$

$$
\begin{aligned}
R_{HC} = 0 \qquad &\text{if} \quad O(t+1) > \mathrm{O}_{min} \\
&\text{or} \quad N(t+1) > \mathrm{N}_{min} \\
&\text{or} \quad \mathrm{Fe}(t+1) > \mathrm{Fe}_{min} \\
&\text{or} \quad S(t+1) > \mathrm{S}_{min}
\end{aligned} \tag{8.63}
$$

where O_{min}, N_{min}, Fe_{min}, S_{min}, and C_{min} are the threshold concentrations for the corresponding electron acceptor below which no biodegradation will take place.

For the first-order decay and Monod kinetic models the reaction terms are compared to the concentration of the electron acceptor:

$$R_{HO} \le \frac{O(t)}{F_O} \tag{8.64}$$

$$R_{HN} \le \frac{N(t)}{F_N} \tag{8.65}$$

$$R_{HFe} \le \frac{\mathrm{Fe}(t)}{F_{Fe}} \tag{8.66}$$

$$R_{HS} \leq \frac{S(t)}{F_S} \tag{8.67}$$

$$R_{HC} \leq \frac{C(t)}{F_C} \tag{8.68}$$

Example 8.4: Sensitivity Analysis with BIOPLUME III The purpose of a sensitivity analysis is to quantify the effects of uncertainty in the estimates of model parameters on model results. During a sensitivity analysis, calibrated values for transmissivity, thickness, recharge, dispersivity, and so on, are systematically changed within a prescribed range of applicable values. The magnitude of change in heads and concentrations from the calibrated model is a measure of the sensitivity of the model results to the particular parameter. The results of this analysis are expressed as the effects of the parameter change on the spatial distribution of heads and concentrations.

The sensitivity of BIOPLUME III model results to the input parameters is a key analysis that the user should perform for each site application. In this section we present in general the relative sensitivity of the model to various input parameters using a hypothetical case study scenario. The hypothetical site base-case scenario used in the sensitivity simulations was set up as follows:

Grid size	9×10
Cell size	900 ft $\times$ 900 ft
Aquifer thickness	20 ft
Transmissivity	0.1 ft^2/sec
Porosity	0.3
CELDIS	0.5
Longitudinal dispersivity	100 ft
Transverse dispersivity	30 ft
Simulation time	2.5 years
Source of contamination	1 injection well at 0.1 ft^3/sec and 100 mg/L source concentration
Recharge	0 ft^3/sec
Boundary conditions	Constant head, upgradient and downgradient
Chemical reactions	None
Biodegradation reactions	None

Three categories of parameters were analyzed: hydrogeologic, chemical, and biodegradation model parameters. In each category and for each parameter analyzed, the value of the parameter was changed by a factor of up to one order of magnitude from the base-case scenario. The associated model results were then analyzed to determine the impact of the changed parameter values on the contaminant plume shape, size, and concentrations.

Hydrogeologic parameters. Five hydrogeologic parameters were evaluated: porosity, aquifer thickness, transmissivity, and longitudinal and transverse dispersivity.

Overall, model results were most sensitive to changes in porosity, thickness, and transmissivity. This is to be expected since the three parameters affect the seepage velocity for the aquifer. The data in Table 8.7 indicate that model results are most sensitive to changes in the transmissivity and aquifer thickness.

Chemical parameters. Two variables, linear sorption and radioactive decay, were used in this analysis to illustrate the sensitivity of the model to selected chemical parameters. The user is encouraged to determine the sensitivity of model results to the remaining chemical parameters (Langmuir and Freundlich sorption parameters and ion exchange) if they apply to their site.

Both linear sorption and radioactive decay have a substantial impact on the model results, as can be seen in Table 8.8. A retardation factor of 2 caused plume concentrations to decline by 27% from the base-case scenario and a half-life of 2×10^7 s or 231 days caused plume concentrations to decline by over 50%.

Biodegradation parameters. The BIOPLUME III model simulates biodegradation using two basic methods. The first method involves specifying an overall first-order decay rate to simulate both aerobic and anaerobic processes. The second method involves specifying the background electron acceptor concentrations in the aquifer and selecting an associated kinetic model for the analysis.

The sensitivity analyses conducted for the biodegradation parameters involved simulating the impact of using an overall first-order decay parameter as well as the impact of specifying electron acceptor concentrations with instantaneous kinetics.

TABLE 8.7. Sensitivity of BIOPLUME III Model Results to Changes in Hydrogeologic Parameters

Variable		Maximum Plume Concentration (mg/L)	Plume Length (ft)	Plume Width (ft)
Porosity	0.15	75	5400	4500
	0.3[a]	67	3600	2700
	0.45	80	3600	2700
Thickness (ft)	10	75	5400	4500
	20[a]	67	3600	2700
	40	47	1800	1800
Transmissivity (ft^2/sec)	0.01	90	2700	2700
	0.1[a]	67	3600	2700
	0.2	57	4500	2700
Longitudinal dispersivity (ft)	10	70	2700	2700
	50	69	3600	2700
	100[a]	67	3600	2700
Transverse dispersivity (ft)	10	68	3600	2700
	30[a]	67	3600	2700
	60	66	3600	2700

Source: From Rifai et al. (1997).

[a] Base case.

TABLE 8.8. Sensitivity of BIOPLUME III Model Results to Linear Sorption and Radioactive Decay

Variable		Maximum Plume Concentration (mg/L)	Plume Length (ft)	Plume Width (ft)
R	1[a]	67	3600	2700
	2	49	2700	1800
	5	28	1800	900
THALF (sec)	0[a]	67	3600	2700
	1.00E + 07	20	1800	1800
	2.00E + 07	33	1800	2700

Source: From Rifai et al. (1997).

[a]Base case.

The results from the analyses are shown in Table 8.9. The data in the table illustrates that model results are very sensitive to biodegradation parameters. Regardless of the modeling methodology and biodegradation kinetics, the simulated concentrations using biodegradation are likely to differ substantially from their counterparts without biodegradation.

Example 8.5: BIOPLUME III Modeling of Patrick Air Force Base Patrick AFB is located on a barrier island off the east coast of Florida near Cocoa Beach. Site ST-29, at the northeastern end of the base's northeast–southwest runway, consists of a gasoline dispensing area, a car wash facility, and a plume of groundwater contamination extending to the west. The land surface is composed of beach deposits and is approximately 7 ft above mean sea level with a slight slope westward. The site is shown in Figure 8.13.

TABLE 8.9. Sensitivity of BIOPLUME III Model Results to First-Order Decay and Instantaneous Reaction Kinetics

Variable		Maximum Plume Concentration (mg/L)	Plume Length (ft)	Plume Width (ft)
First-order decay(1^{-1})	0[a]	67	3600	2700
	1.16E – 08	58	3600	2700
	1.16E – 07	43	1800	900
Oxygen (mg/L)	10	67	3600	2700
	20[a]	67	3600	2700
	40	66	1800	2700
Oxygen, nitrate, iron, sulfate, carbon dioxide	0 for all[a]	67	3600	2700
	3 for all[b]	62	1800	1800

Source: From Rifai et al. (1997).

[a]Base case.

[b]Threshold concentration = 0.5 mg/L for all. F = 3, 5, 22, 5, and 2, respectively.

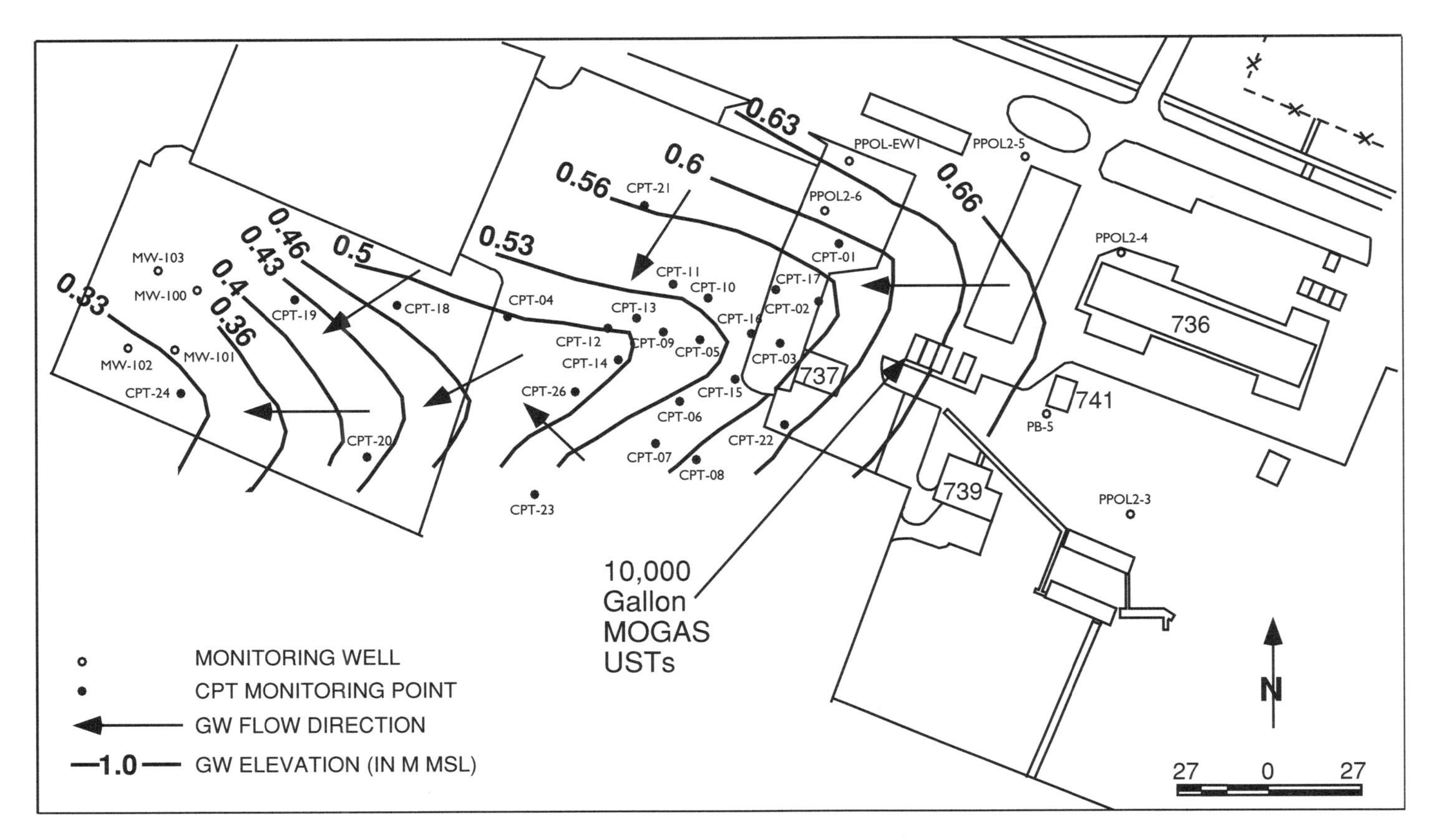

FIGURE 8.13. ST-29 site map and location of sampling points.

The closest bodies of surface water are the Atlantic Ocean, lying 750 ft to the east, and the Banana River, which runs roughly north–south 2400 ft west of the site. Limestone layers form several localized aquifers. The shallowest aquifer is composed of undifferentiated marine sands mixed with shell fragments that reach approximately 25 ft below the ground surface. Rising-head slug tests conducted in March 1994 indicate a hydraulic conductivity for the shallow aquifer ranging from 0.023 to 0.089 ft/min (Parsons Engineering Science, 1994). A groundwater divide is present in the vicinity of the site. Groundwater lies at depths 4 to 5 ft below ground surface, and flow is to the west with a hydraulic gradient ranging from 0.001 to 0.003 ft/ft (Environmental Science and Engineering, 1991; O'Brien and Gere Engineers, 1992). Table 8.10 gives water table elevation data, and the corresponding well locations are shown on Figure 8.13. A porosity of 0.35 was assumed, yielding a seepage velocity estimate of 0.43 ft/day.

The Caloosahatchee Marl Formation underlying the shallow aquifer is composed of sandy clay deposits. It separates the shallow aquifer from a bedrock aquifer containing limestones, silty sands, and clay. Migration of groundwater from the shallow aquifer deeper into the subsurface is unlikely due to the higher pressure head in the deep aquifers. The probable source of contamination is a 10,000-gallon UST and its product lines, which were discovered leaking in 1986. An estimated 700 gallons of product were released. Soil contamination is concentrated in areas near and downgradient from the site USTs, located north of the car wash. The highest observed BTEX concentration was 1236 mg/kg at a soil sample taken 120 ft west–northwest of the suspected source. Table 8.11 shows all the soil sample analyses for fuel hydrocarbons; the majority are below 0.1 mg/kg. The soil BTEX contamination appears to extend 220 ft downgradient of the source and is 90 ft wide at its greatest width.

Groundwater in the site vicinity has displayed unusual odors and characteristics. Laboratory analyses indicate that the groundwater is contaminated by fuel hydrocarbons; analysis results are displayed in Table 8.12. BTEX compounds were present in 38 of 48 groundwater samples collected in March 1994. Trimethylbenzene was found in 11 of 48 samples; 2 of these samples had concentrations greater than 25 mg/L. The BTEX plume, shown in Figure 8.14, is approximately 560 ft in length and 200 ft wide at its widest part, as defined by the 5-mg/L isopleth, with a maximum total BTEX concentration of 7304 μg/L.

Data from groundwater samples taken in March 1994 are shown in Table 8.13. Background dissolved oxygen concentrations are 3.7 mg/L, with the areas of elevated BTEX concentrations showing decreased oxygen levels. Nitrate concentrations are low at this site; virtually all observed concentrations are below 0.29 mg/L. Ferrous iron and methane levels increase in the source BTEX area with maxima of 1.9 and 14.95 mg/L, respectively. No clear trends are apparent for sulfate. The data indicate that biodegradation mechanisms involving oxygen, iron, and methane are occurring at ST-29.

The BIOPLUME III model was used to predict the fate and transport of the BTEX contamination. The shallow contaminated zone was modeled as an unconfined

TABLE 8.10. Water-Level Data for Patrick AFB Site ST-29[a]

Sample Location	Sample Date	Datum Elevation (ft MSL)	Ground Elevation (ft MSL)	Total Depth (ft BTOC)	Depth to Water (ft BTOC)	Water Elevation (ft MSL)
CPT-02S	3/27/94	6.61	7.07	7.58	4.93	1.68
CPT-03D	3/27/94	6.23	6.55	19.13	4.59	1.64
CPT-03S	3/27/94	6.26	6.55	7.9	4.57	1.69
CPT-04D	3/27/94	6.01	6.57	11.36	4.46	1.55
CPT-04S	3/27/94	5.99	6.57	NR	4.46	1.53
CPT-09S	3/27/94	6.13	6.35	7.76	4.35	1.78
CPT-12S	3/27/94	5.9	6.43	7.56	4.4	1.5
CPT-18S	3/27/94	6.11	6.58	7.4	4.69	1.42
CPT-22D	3/27/94	6.32	6.75	9.77	4.49	1.83
CPT-22S	3/27/94	6.35	6.75	7.5	4.5	1.85
CPT-23D	3/27/94	5.84	6.57	12.7	4.2	1.64
CPT-23S	3/27/94	5.94	6.57	6.31	4.3	1.64
CPT-24D	3/27/94	5.44	5.89	12.49	4.46	0.98
CPT-24S	3/27/94	5.49	5.89	6.06	4.52	0.97
CPT-25D	3/27/94	6.49	6.6	12.72	5.72	0.77
CPT-25S	3/27/94	6.43	6.6	6.31	5.66	0.77
CPT-26D	3/27/94	5.79	6.47	12.73	4.13	1.66
CPT-26S	3/27/94	5.89	6.47	6.29	4.25	1.64
PB5	3/23/94	10.86	7.39	15.45	8.75	2.11
PPOL2-1	3/25/94	7.47	6.92	NA	5.33	2.14
PPOL2-5	3/25/94	7.48	NA	12.51	5.36	2.12
PPOL2-6	3/25/94	6.64	6.39	NA	4.93	1.71

Source: Parsons Engineering Science (1994).

[a]NA, not analyzed; NP, not reported.

TABLE 8.11. Fuel-Hydrocarbon Compounds Detected in Soil at Patrick AFB[a]

Sample Location	Sample Depth (ft)	Compound (mg/kg)											
		Benzene	Toluene	Ethylbenzene	*p*-Xylene	*m*-Xylene	*o*-Xylene	Total Xylenes	Total BTEX	1,3,5-TMB	1,2,4-TMB	1,2,3-TMB	TPH
CPT-02-A17	4	BLQ	BLQ	0.0345	0.0249	0.054	0.0296	0.1085	0.143	0.018	0.514	0.017	ND
CPT-02-A18	5	0.197	0.0605	0.514	1.44	0.976	0.703	3.119	3.8905	3.1	3.28	2.65	140
CPT-02-A19	6	6.99	8.08	191	257	542	231	1,030	1,236	274	786	180	17,100
CPT-02-A20	6.45	0.68	0.345	16.1	20.3	45.9	17.9	84.1	101.2	26	77.4	17.5	1,660
CPT-02-A21	7	0.339	0.106	2.26	2.91	6.77	2.74	12.42	15.13	3.75	11.7	2.59	289
CPT-03-A22	3	0.164	0.0142	0.21	0.253	0.523	0.243	1.019	1.407	0.319	0.844	0.227	20.9
CPT-03-A23	4.5	ND	BLQ	0.00838	0.0102	0.0209	0.0107	0.0418	0.0502	0.0155	0.0461	0.013	0.2
CPT-03-A24	5.5	BLQ	0.462	BLQ	39.8	45.7	85.8	171.3	171.8	253	332	123	11,700
CPT-03-A25	6.5	2.33	29.5	120	186	428	20	82.3	974.8	181	675	124	9,300
CPT-03-A26	7	0.366	0.532	2.02	3.48	7.58	4.4	15.46	18.38	3.97	12.7	2.87	287
CPT-5-A1	3.5	BLQ	BLQ	BLQ	BLQ	BLQ	BLQ	BLQ	BLQ	BLQ	0.0075	BLQ	ND
CPT-5-A2	4.5	BLQ	BLQ	BLQ	BLQ	BLQ	BLQ	BLQ	BLQ	BLQ	0.0117	BLQ	ND
CPT-5-A3	5	BLQ	BLQ	BLQ	0.00725	BLQ	BLQ	0.00725	0.00725	BLQ	0.0088	BLQ	ND
CPT-5-A4	5.5	BLQ	BLQ	BLQ	BLQ	BLQ	BLQ	BLQ	BLQ	BLQ	0.0149	BLQ	ND
CPT-9-A4	2.5	BLQ	BLQ	ND	BLQ	BLQ	BLQ	BLQ	BLQ	BLQ	ND	ND	2
CPT-9-A5	4.5	BLQ	BLQ	BLQ	BLQ	BLQ	N	BLQ	BLQ	BLQ	BLQ	BLQ	47.1
CPT-9-A6	5	BLQ	0.034	BLQ	0.0311	0.0229	0.0109	0.0649	0.0989	0.0121	BLQ	0.00976	2,740
CPT-9-A7	6	BLQ	BLQ	BLQ	BLQ	BLQ	BLQ	BLQ	BLQ	BLQ	0.00818	BLQ	15.2
CPT-13	5.5	BLQ	BLQ	BLQ	BLQ	BLQ	BLQ	BLQ	BLQ	BLQ	BLQ	BLQ	ND
CPT-15-A8	3	BLQ	BLQ	ND	BLQ	BLQ	BLQ	BLQ	BLQ	BLQ	0.016	BLQ	ND
CPT-15-A9	4	BLQ	BLQ	ND	BLQ	BLQ	BLQ	BLQ	BLQ	BLQ	0.0128	BLQ	0.01
CPT-15-A10	5	BLQ	0.012	0.00978	BLQ	BLQ	0.0226	0.0226	0.0444	BLQ	0.0114	BLQ	526
CPT-15-A11	5.5	ND	ND	BLQ	BLQ	BLQ	BLQ	BLQ	BLQ	BLQ	0.0179	BLQ	ND
CPT-16-A1	4.5	0.00752	0.00761	BLQ	BLQ	BLQ	BLQ	BLQ	0.0151	BLQ	0.0066	BLQ	0.04
CPT-16-A2	6	ND	BLQ	ND	BLQ	BLQ	BLQ	BLQ	BLQ	BLQ	0.0104	BLQ	ND
CPT-16-A12	2	ND	BLQ	BLQ	BLQ	BLQ	BLQ	BLQ	BLQ	BLQ	0.0127	BLQ	ND
CPT-16-A13	3	ND	BLQ	ND	BLQ	BLQ	BLQ	BLQ	BLQ	BLQ	0.0248	BLQ	ND
CPT-16-A14	4	ND	BLQ	BLQ	BLQ	0.00779	BLQ	0.00779	0.00779	BLQ	0.0176	BLQ	ND
CPT-16-A15	5	ND	BLQ	ND	BLQ	BLQ	BLQ	BLŒ	BLQ	BLQ	0.0172	BLQ	ND

Source: Parsons Engineering Science (1994).

[a]BLQ = below limit of quantification; NA, not analyzed; ND, not detected.

TABLE 8.12. Fuel-Hydrocarbon Compounds Detected in Groundwater at Patrick AFB[a]

Sample Location	Sample Date	Compound (μg/L)										
		Benzene	Toluene	Ethylbenzene	*p*-Xylene	*m*-Xylene	*o*-Xylene	Total Xylenes	Total BTEX	1,3,5-TMB	1,2,4-TMB	1,2,3-TMB
CPT-01	3/24/98	BLQ	2.42	BLQ	BLQ	BLQ	BLQ	BLQ	2.42	BLQ	BLQ	BLQ
CPT-02S	3/24/98	375	18.9	165	166	353	119	638	1197	71.2	NA	86.3
CPT-02D	3/24/98	1.5	1.7	1.1	1.4	2.6	0.4	5.4	9.7	BLQ	NA	1.3
CPT-03S	3/24/98	724	737	82.3	1220	2410	1390	5020	7304	347	NA	403
CPT-03M	3/24/98	207	15.6	40.5	42.2	24	7.5	73.7	336.8	2.8	NA	16
CPT-03D	3/24/98	1.8	1.1	BLQ	BLQ	1.4	BLQ	1.4	4.3	BLQ	NA	BLQ
CPT-04S	3/24/98	BLQ	6	BLQ	BLQ	BLQ	BLQ	BLQ	6	BLQ	BLQ	BLQ
CPT-04D	3/26/98	BLQ	3.7	BLQ	BLQ	BLQ	BLQ	BLQ	3.7	BLQ	BLQ	BLQ
CPT-05S	3/26/98	BLQ	1.2	BLQ	BLQ	BLQ	BLQ	BLQ	1.2	BLQ	NA	BLQ
CPT-06S	3/26/98	BLQ	11.1	BLQ	1.3	2.3	BLQ	3.6	14.7	BLQ	NA	BLQ
CPT-07S	3/26/98	BLQ	3.9	BLQ	BLQ	1	BLQ	1	4.9	BLQ	NA	BLQ
CPT-08S	3/26/98	BLQ	2.8	BLQ	BLQ	BLQ	BLQ	BLQ	2.8	BLQ	BLQ	BLQ
CPT-09S	3/26/98	2	4	BLQ	BLQ	BLQ	BLQ	BLQ	6	BLQ	BLQ	BLQ
CPT-09D	3/26/98	427	14.1	2.9	11.7	12.1	9.7	33.5	477.5	BLQ	NA	9.4
CPT-10S	3/26/98	BLQ	3.1	BLQ	BLQ	BLQ	BLQ	BLQ	3.1	BLQ	BLQ	BLQ
CPT-11S	3/26/98	BLQ	1	BLQ	BLQ	BLQ	BLQ	BLQ	1	BLQ	BLQ	BLQ
CPT-12S	3/26/98	BLQ	1.1	BLQ	BLQ	1.2	BLQ	1.2	2.3	BLQ	NA	BLQ
CPT-12D	3/26/98	93.5	5.9	BLQ	8.4	7	3.7	19.1	119	BLQ	NA	4.1
CPT-13S	3/26/98	BLQ	8.4	BLQ	10.4	5.4	3.1	18.9	56.4	BLQ	NA	6.7
CPT-16S	3/27/98	1	1.9	BLQ	BLQ	BLQ	BLQ	BLQ	2.9	BLQ	BLQ	BLQ
CPT-16DD	4/1/98	BLQ	1.9	BLQ	BLQ	BLQ	BLQ	BLQ	1.9	BLQ	BLQ	BLQ
CPT-18S	3/25/98	BLQ	1.7	2.3	3.8	6.6	4	14.4	18.4	1.4	NA	1.6
CPT-18DD	3/25/98	8.3	2.1	BLQ	BLQ	BLQ	BLQ	BLQ	10.4	BLQ	BLQ	BLQ
CPT-18DD	4/1/98	BLQ	3.5	BLQ	BLQ	BLQ	BLQ	BLQ	3.5	BLQ	BLQ	BLQ
CPT-19S	3/25/98	BLQ	BLQ	1.1	1.5	2.7	BLQ	4.2	5.3	BLQ	NA	BLQ
CPT-19D	3/25/98	1.5	BLQ	BLQ	BLQ	BLQ	BLQ	BLQ	1.5	BLQ	BLQ	BLQ

continues

TABLE 8.12. *(Continued)*

Sample Location	Sample Date	Compound (μg/L)										
		Benzene	Toluene	Ethylbenzene	*p*-Xylene	*m*-Xylene	*o*-Xylene	Total Xylenes	Total BTEX	1,3,5-TMB	1,2,4-TMB	1,2,3-TMB
CPT-20S	3/25/98	BLQ	2.1	BLQ	1.5	2.7	1.8	6	8.1	BLQ	NA	1
CPT-20D	3/25/98	BLQ	BLQ	BLQ	BLQ	BLQ	BLQ	BLQ	BLQ	BLQ	BLQ	BLQ
CPT-21S	3/26/98	NA	5.1	BLQ	BLQ	BLQ	BLQ	BLQ	5.1	BLQ	NA	BLQ
CPT-21D	3/26/98	NA	BLQ	BLQ	BLQ	BLQ	BLQ	BLQ	BLQ	BLQ	BLQ	BLQ
CPT-22S	3/27/98	NA	1.9	BLQ	BLQ	BLQ	BLQ	BLQ	1.9	BLQ	BLQ	BLQ
CPT-22D	3/27/98	1.5	BLQ	BLQ	BLQ	BLQ	BLQ	BLQ	1.5	BLQ	BLQ	BLQ
CPT-23S	3/26/98	BLQ	BLQ	BLQ	BLQ	1.7	BLQ	1.7	1.7	BLQ	NA	BLQ
CPT-23D	3/26/98	BLQ	BLQ	BLQ	BLQ	BLQ	BLQ	BLQ	BLQ	BLQ	BLQ	BLQ
CPT-24S	3/27/98	BLQ	BLQ	BLQ	BLQ	BLQ	BLQ	BLQ	BLQ	BLQ	BLQ	BLQ
CPT-24D	3/27/98	BLQ	BLQ	BLQ	BLQ	BLQ	BLQ	BLQ	BLQ	BLQ	BLQ	BLQ
CPT-25S	3/27/98	BLQ	2.1	2.3	4.1	6.9	3.5	14.5	18.9	1.7	NA	1.5
CPT-25D	3/27/98	BLQ	1	BLQ	BLQ	1.4	BLQ	1.4	2.4	BLQ	NA	BLQ
CPT-26S	3/27/98	BLQ	1.1	BLQ	BLQ	BLQ	BLQ	BLQ	1.1	BLQ	BLQ	BLQ
CPT-26D	3/27/98	BLQ	1	BLQ	BLQ	BLQ	BLQ	BLQ	1	BLQ	BLQ	BLQ
CPT-14D	3/26/98	960	16.6	11.5	39.2	36.8	44.2	120.2	1108.3	15.3	NA	23
MW-100	3/24/98	4.1	BLQ	BLQ	1.1	1	BLQ	2.1	6.2	BLQ	BLQ	BLQ
PB5	3/27/98	BLQ	BLQ	BLQ	BLQ	BLQ	BLQ	BLQ	BLQ	BLQ	BLQ	BLQ
MW-101	3/24/98	BLQ	BLQ	BLQ	BLQ	BLQ	BLQ	BLQ	BLQ	BLQ	BLQ	BLQ
MW-102	3/24/98	BLQ	BLQ	BLQ	BLQ	BLQ	BLQ	BLQ	BLQ	BLQ	BLQ	BLQ
MW-103	3/24/98	BLQ	BLQ	BLQ	BLQ	BLQ	BLQ	BLQ	BLQ	BLQ	BLQ	BLQ
PPOL2-6	3/27/98	BLQ	BLQ	BLQ	BLQ	BLQ	BLQ	BLQ	BLQ	BLQ	BLQ	BLQ
PPOL2-1	3/27/98	BLQ	BLQ	BLQ	BLQ	1.4	BLQ	1.4	BLQ	BLQ	BLQ	BLQ

Source: Adapted from Parsons Engineering Science (1994).

[a] BLQ = below limit of quantification; NA, not analyzed.

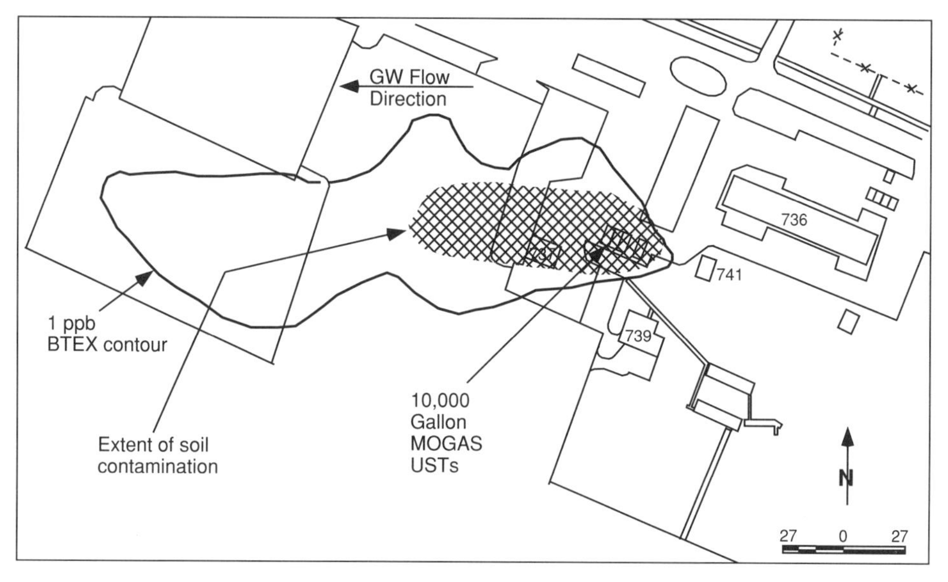

FIGURE 8.14. Extent of soil contamination and BTEX plume at ST-29 site.

aquifer composed of fine to coarse sand with shell fragments. Continuing contamination sources were included due to the presence of residual phase contamination. Figure 8.15 shows the 20 × 30 grid designed for the Patrick site. Each grid cell is 50 ft wide by 100 ft long. Constant head boundaries were set on the top and bottom rows of the grid, the top corresponding to the groundwater divide and the bottom corresponding to the Banana River. Groundwater divide levels were estimated at 2.3 to 2.4 ft above mean sea level, and the river head level was assumed to be 0.1 ft above mean sea level.

Known fuel releases occurred in 1986; thus the plume calibrations were compared with observed 1994 data after eight years of simulation time incorporating contaminant injection into the subsurface. Using the calibrated model as a base, steady-state predictive simulations were made. Initial concentrations and stoichiometric ratios are required for iron and carbon dioxide (the oxidized carbon compound generating methane). The highest concentration of ferrous iron measured was 1.9 mg/L, signifying that at least that much ferric iron is available for use as an electron acceptor. A background concentration of 2 mg/L was selected. For methane, readings as high as 14.9 mg/L were recorded; later modeling results required that a more conservative level of 2.7 mg/L be used, which equates to 7.5 mg/L of CO_2. The stoichiometric ratio for the biodegradation of xylene is 21.8 g iron/g xylene for iron and 2.14 g carbon dioxide/g xylene. These parameters and other data are shown in Tables 8.14 and 8.15.

TABLE 8.13. Geochemical Data Collected at Patrick AFB[a]

Sample Location	Temperature (°C)	Dissolved Oxygen (mg/L)	Redox Potential (mV)	Total Alkalinity (mg/L)	Conductivity (μS)	pH	Hydrogen Sulfide (mg/L)	Chloride (mg/L)	Sulfate (mg/L)	Ferrous Iron (mg/L)	$NO_2 + NO_3$ Nitrogen (mg/L)	Methane (mg/L)	TOC[b] (mg/L)
CPT-01	24.7	0.4	NA	NA	NA	NQ	NA	44.4	4.37	NA	0.13	4.99	14
CPT-02D	24.7	0.6	–190	330	771	7.1	0.4	45.7	<0.5	0.4	0.13	5.953	6.8
CPT-02S	24.7	0.2	–156	498	1,061	6.7	0.2	42.6	<0.5	1.6	0.12	14.95	16.9
CPT-03D	NA	NA	–255	315	721	7.3	1	41.6	<0.5	0.4	0.12	1.63	5.4
CPT-03M	26.4	0.2	–50	398	898	7.1	0	40.7	2.52	0.3	0.11	3.164	10.9
CPT-03S	26.4	0.1	–208	520	1,733	6.7	0.1	132	118	1.2	14.8	14.02	63.1
CPT-04D	26.1	0.2	–266	212	457	7.2	0.2	12.4	1.47	0.6	0.09	3.76	5.6
CPT-04S	26.9	0.3	–286	215	469	6.9	0.5	12.5	<0.5	0.6	0.19	7.66	6.6
CPT-05S	26.4	1.1	–160	215	488	7.3	0.1	23.6	6.86	0.1	0.17	4.86	12
CPT-06S	25.1	0.2	–278	148	437	7.6	1.5	47.8	7.03	0.3	0.13	6.60	3.8
CPT-07S	25.3	0.2	–250	254	577	7.2	1.2	30.2	2.52	1	0.12	6.34	3.4
CPT-08S	25	0.2	–60	420	974	7.1	0	44.7	8.51	1.9	0.1	1.74	10.1
CPT-09D	27.8	0.3	–200	422	938	7.1	0.6	34.7	15.3	0.2	0.11	4.24	12
CPT-09S	27.3	0.2	–24	340	530	7.3	0	14.3	6.64	0.2	0.1	3.80	10.2
CPT-10S	26	0.1	–60	192	460	7.3	0	26.6	9.5	0.2	0.13	3.49	21.3
CPT-11S	25.9	0.1	–35	210	508	7.2	0	12.7	15.9	0.4	0.15	4.24	NA
CPT-12D	27.1	0.4	10	329	715	7.2	0.1	28.1	3.86	0.1	0.12	0.98	8.1
CPT-12S	27.3	0.9	30	266	564	7	0	15.2	8.38	0.1	0.1	5.37	10.5
CPT-13S	25.7	0.1	–230	362	801	7.3	0.6	35.5	6.94	0.3	0.12	2.04	7.2
CPT-14D	25.5	0.3	–240	460	906	7	0.6	34.6	3.68	0.3	0.11	8.79	12.8
CPT-16DD	26.7	2.7	NA	NA	NA	NA	0.3	NA	NA	NA	NA	NA	NA
CPT-16S	25.5	0.1	–190	231	776	7	0.1	37.9	8.23	0.4	0.13	0.781	9.4
CPT-18D	26.1	0.3	–60	294	620	7.1	0.2	15	1.85	0.3	0.11	4.56	5.6

CPT-18DD	27.3	2.3	NA	NA	NA	NA	0	NA	NA	NA	NA	NA	NA
CPT-18S	26.6	2	25	286	834	6.9	0	36.6	86	0.5	0.12	NA	7.8
CPT-19D	26.6	0.2	–50	328	744	7.1	0.2	33.7	1.51	0.2	0.1	2.14	7.3
CPT-19S	NA	NA	41	335	800	7	0	37.4	8.85	0.1	0.11	0.92	10.3
CPT-20D	25.7	0.3	–44	380	842	7.1	0.1	52	<0.5	0.2	0.07	1.11	8.6
CPT-20S	25	1.5	23	148	368	7	0	9.83	25.5	0.3	0.1	1.28	3.6
CPT-21D	26.4	0.2	–20	304	716	7.2	0.1	29.8	13.3	0.2	0.14	0.46	6.1
CPT-21S	26	3.2	20	245	610	7.1	0	26.6	25.5	0.2	0.29	2.41	7.6
CPT-22D	25.4	0.1	–287	415	936	6.9	5	NA	NA	0.2	0.12	0.87	11.4
CPT-22S	25.5	0.3	–153	450	1,271	6.8	0	66.6	128	1.2	0.07	3.22	10
CPT-23D	26.7	0.4	–167	332	779	7.1	0.2	36.1	1.49	0.2	0.1	2.28	8.2
CPT-23S	26.5	3.5	54	346	757	6.9	0	23.4	<0.5	0.2	0.12	1.99	2.8
CPT-24D	26	0.3	–60	192	376	7.5	0.1	5.46	3.61	0.1	0.1	0.69	2.8
CPT-24S	25.7	1.7	30	190	358	7	0	6.63	<0.5	0.1	0.12	2.2	5.6
CPT-25D	NA	NA	62	371	892	7.1	0	54.7	6.16	<0.05	0.12	1.56	15.7
CPT-25S	25	3.7	53	157	664	7.3	0	28	51.9	<0.05	0.12	0.15	15.7
CPT-26D	26.2	0.2	–293	311	751	7	3	44.9	19.8	0.4	0.11	2.96	8.3
CPT-26S	26	2.2	–20	264	558	7.6	0	15.1	1.22	0.3	0.12	3.57	5
MW-100	25.9	0.5	–241	331	607	7.2	0.2	24.9	16.3	0.1	0.12	2.82	18.6
MW-101	26.5	0.3	–247	287	533	7.2	0.8	21.6	5.75	0.1	0.13	2.31	9
MW-102	25.8	0.2	–281	250	523	7.2	0.5	17.9	3.51	0.1	0.12	3.26	7.6
MW-103	25.4	0.1	–271	209	445	7.4	1	12.5	4.69	0.1	0.11	5.29	2.6
PB5	24.9	0.2	–140	291	743	7.1	NA	51.4	4.45	<0.05	0.1	4.41	NA
PPOL2-6	26.3	0.1	–220	305	747	7.1	0.7	44	3.2	0.6	<0.05	5.33	NA
PPOL2-1	27.2	0.2	–230	334	30,100	7	0.8	10,200	1,150	<0.05	<0.05	0.034	NA

Source: Adapted from Parsons Engineering Science (1994).

[a]NA, not analyzed.

[b]TOC, total organic carbon.

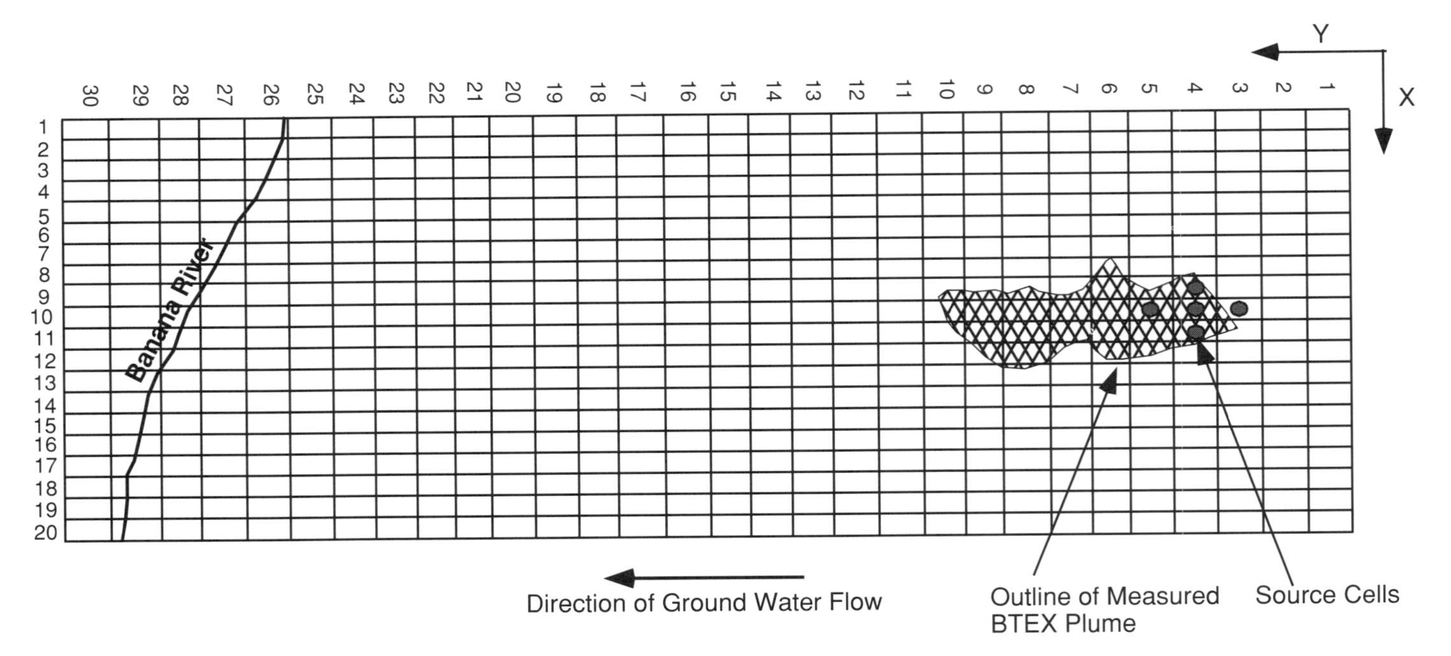

FIGURE 8.15. Model grid.

TABLE 8.14. BIOPLUME II Model Input Parameters for Patrick AFB

Parameter	Description	Calibrated Model	No Source Removal
NTIM	Maximum number of times steps in a pumping period	8	58
NPMP	Number of pumping periods	1	1
NX	Number of nodes in the x direction	20	20
NY	Number of nodes in the y direction	30	30
NPMAX	Maximum number of particles	5250	5250
NPNT	Time-step interval for printing data	1	1
NITP	Number of iteration parameters	7	7
NUMOBS	Number of observation points	0	0
ITMAX	Maximum number of iterations	200	200
NREC	Number of pumping or injection wells	5	5
NPTPND	Initial number of particles per node	9	9
NCODES	Number of node identification codes	3	3
NPNTMV	Particle movement interval	0	0
NREACT	Option for biodegradation, retardation, and decay	0	0
PINT	Pumping period (yrs)	8	58
TOL	Convergence criteria	0.001	0.001
POROS	Effective porosity	0.35	0.35
BETA	Beta (ft)	20	20
S	Storage coefficient	0	0
TIMX	Time increment multiplier for transient flow	—	—
TINIT	Size of initial time step (secs)	0	0
XDEL	Width of finite-difference cell (ft)	50	50
YDEL	Length of finite-difference cell (ft)	100	100
DLTRAT	Ratio of transverse to longitudinal dispersivity	0.3	0.3
CELDIS	Maximum cell distance per particle move	0.5	0.5
ANFCTR	Ratio of T_{yy} to T_{xx} (1 = isotropic)	1	1
DK	Distribution coefficient	0.35	0.35
RHOB	Solid bulk density (glc.c.)	1.6	1.6
THALF	Solute half-life	—	—
DEC1	Anaerobic decay coefficient (day^{-1})	0	0
DEC2	Reaeration coefficient (day^{-1})	0.0009	0.0009
F	Stoichiometric ratio of hydrocarbon to oxygen	—	—

Source: Adapted from Parsons Engineering Science (1994).

TABLE 8.15. Electron Acceptor Input Data for Patrick AFB

Parameter	Description	Value
CONC1	Background concentration of oxygen (μg/L)	3500
IRECO	Biodegradation kinetics specifier for oxygen	2
DCO	First-order decay rate for aerobic biodegradation (day^{-1})	—
FO	Stoichiometric ratio of oxygen to contaminant	3.14
DOMIN	Threshold concentration of oxygen (μg/L)	0
CONC2	Background concentration of nitrate (μg/L)	—
IRECN	Biodegradation kinetics specifier for nitrate	0
DCN	First-order decay rate for denitrification (day^{-1})	—
FN	Stoichiometric ratio of nitrate to contaminant	—
NMIN	Threshold concentration of nitrate (μg/L)	0
CONC3	Background concentration of iron (μg/L)	2000
IRECF	Biodegradation specifier for iron	2
DCF	First-order decay rate for iron reduction (day^{-1})	—
FF	Stoichiometric ratio of iron to contaminant	21.8
FMIN	Threshold concentration of iron (μg/L)	0
CONC4	Background concentration of sulfate (μg/L)	—
IRECS	Biodegradation specifier for sulfate	0
DCS	First-order decay rate for sulfate reduction (day^{-1})	—
FS	Stoichiometric ratio of sulfate to contaminant	—
SMIN	Threshold concentration of sulfate (μg/L)	0
CONC5	Background concentration of carbon dioxide (μg/L)	7500
IRECC	Biodegradation specifier for carbon dioxide	2
DCC	First-order decay rate for methanogenesis (day^{-1})	—
FC	Stoichiometric ratio of oxygen to contaminant	2.14
CMIN	Threshold concentration of carbon dioxide (mg/L)	0

Figure 8.16 shows the BTEX calibration plume from the model with contour level values in μg/L. This calibration plume was generated by running the simulation for eight years. The extent and concentrations of the calibration plume are in excellent agreement with the data of Figure 8.14. The maximum measured concentration was 7300 μg/L, with the next-highest point at 1200 mg/L; the highest simulated concentration was 3600 μg/L over an entire grid cell. A 1-μg/L contour of the simulated plume is very comparable to the 560- × 220-ft extent observed for the plume. The 50-year prediction resulting from the model shows that the plume has spread an additional 200 ft downgradient. Concentration levels have increased slightly, with a maximum of 5200 μg/L predicted at the plume source. Based on these simulation results, it appears that a steady-state equilibrium will be reached, resulting in a plume approximately 750 ft long and maximum concentration of approximately 5000 μg/L.

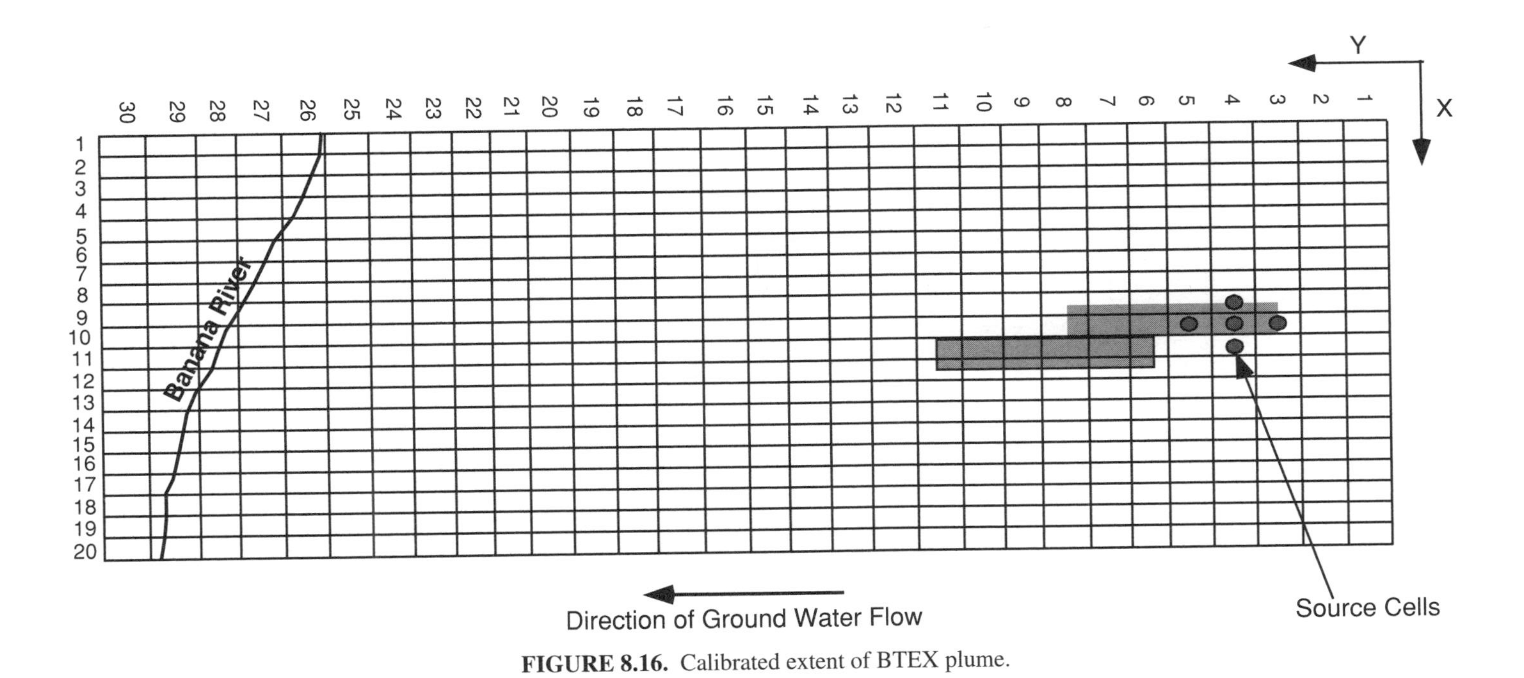

FIGURE 8.16. Calibrated extent of BTEX plume.

8.6 MODELING NATURAL ATTENUATION OF CHLORINATED SOLVENTS

Very few models exist (analytical or numerical) that are designed specifically for simulating the natural attenuation of chlorinated solvents in groundwater. Ideally, a model for simulating natural attenuation of chlorinated solvents would be able to track the degradation of a parent compound through its daughter products and allow the user to specify differing decay rates for each step of the process. This may be referred to as a *reactive transport model*, in which transport of a solute may be tracked while it reacts, its properties change due to those reactions, and the rates of the reactions change as the solute properties change. Moreover, the model would also be able to track the reaction of those other compounds that react with or are consumed by the processes affecting the solute of interest (e.g., electron donors and acceptors).

At this time, at least two natural attenuation models for chlorinated solvents are under development. The BIOCHLOR model, under development by the authors of the BIOSCREEN model was described earlier.

Researchers at Battelle Pacific Northwest Laboratories are developing a numerical model referred to as RT3D. RT3D (Reactive Transport in Three Dimensions, Sun et al., 1996) is a FORTRAN 90–based model for simulating three-dimensional multispecies, reactive transport in groundwater. This model is based on the 1997 version of MT3D (DOD Version 1.5), but has several extended reaction capabilities. RT3D can accommodate multiple sorbed and aqueous-phase species with any reaction framework that the user needs to define. RT3D can simulate different scenarios, since a variety of preprogrammed reaction packages are already provided and users have the ability to specify their own reaction kinetic expressions. This allows, for example, natural attenuation processes or an active remediation to be evaluated. Simulations can be applied to modeling contaminants such as heavy metals, explosives, petroleum hydrocarbons, and/or chlorinated solvents.

RT3D's preprogrammed reaction packages include

1. Two-species instantaneous reaction (hydrocarbon and oxygen)
2. Instantaneous hydrocarbon biodegradation using multiple electron acceptors (O_2, NO_3^-, Fe_2^+, SO_4^-, CH_4)
3. Kinetically limited hydrocarbon biodegradation using multiple electron acceptors (O_2, NO_3^-, Fe_2^+, SO_4^-, CH_4)
4. Kinetically limited reaction with bacterial transport (hydrocarbon, oxygen, and bacteria)
5. Nonequilibrium sorption/desorption (can also be used for non-aqueous-phase liquid dissolution)
6. Reductive anaerobic biodegradation of PCE, TCE, DCE, and VC
7. Reductive anaerobic biodegradation of PCE, TCE, DCE, and VC combined with aerobic biodegradation of DCE and VC
8. Combination of reactions 3 and 7

RT3D represents a remarkable breakthrough in the development and solving of optimization models of bioremediation design. It is a modular three-dimensional simulator capable of predicting multispecies (solutes and microbes) bioreactive transport while helping us to understand natural attenuation and active bioremediation processes.

8.7 APPLYING NATURAL ATTENUATION MODELS TO SITES

Sufficient field data are essential when using models for simulating existing flow and/or contaminant conditions at a site or when using the model for predictive purposes. However, it may be desirable to model a site even when few data exist. The modeling in this case may serve as a method for identifying those areas in which detailed field information needs to be collected.

Applying an appropriate modeling methodology will increase confidence in the modeling results. Anderson and Woessner (1992) propose the following general modeling protocol that can be applied to any site:

1. Establish the purpose of the model.
2. Develop a conceptual model of the system.
3. Calibrate the site model.
4. Determine the effects of uncertainty on model results.
5. Verify the calibrated model.
6. Predict results based on the calibrated model.
7. Determine the effects of uncertainty on model predictions.
8. Present modeling results.
9. Postaudit and update the model as necessary.

Stating the purpose of the modeling effort with natural attenuation models helps focus the study and determine the expectations from the analysis. Typical objectives include

1. Determining the effectiveness of natural attenuation for remediating a given site
2. Designing a groundwater pump-and-treat and/or bioremediation system

Formulating a conceptual model of the site is essential to the success of a natural attenuation modeling effort. A conceptual model is a pictorial representation of the groundwater flow and transport system, frequently in the form of a block diagram or a cross section. The nature of the conceptual model will determine the dimensions of the site model and the design of the grid.

Formulating a conceptual model includes (1) defining the hydrogeologic features of interest (i.e., the aquifers that will be modeled); (2) defining the flow system (including boundary and initial conditions) and sources and sinks of water in the system, such as recharge from infiltration, and pumping; and (3) defining the transport system (velocity, dispersion, sorption, and biodegradation) and sources and sinks of chemicals in the system (including boundary and initial conditions).

Calibration, Verification, and Prediction

Calibrating a site model is the process of demonstrating that the model is capable of producing field-measured values of the head and concentrations at the site. For the case of groundwater flow, for example, calibration is accomplished by finding a set of parameters, boundary and initial conditions, and stresses that produce simulated values of heads that match measured values within a specified range of error.

The procedure for calibrating the BIOPLUME III model, for example, is by manual trial-and-error selection of parameters. The main parameters that are used for calibrating the flow at a site include transmissivity, thickness, recharge, and boundary conditions. The main parameters that are used to calibrate the transport and fate of chemicals at a site include source definition, dispersion, sorption, and biodegradation parameters. In addition, the transmissivity, thickness, and recharge data used in calibrating the flow solution determine the transport velocity and should be checked for accuracy against observed field velocities.

The FATE V groundwater model calibration tool (Nevin et al., 1997) has been developed to assist users with determining site-specific first-order biodegradation rates for organic constituents dissolved in groundwater. FATE V represents an enhancement to the Domenico analytical groundwater transport model (Domenico, 1987). Using the Solver routine built into Excel, FATE V is able to calibrate the attenuation rate used by the Domenico model to match site-specific data. By calibrating the decay rate to site-specific measurements, FATE V can yield accurate predictions of long-term natural attenuation processes within a groundwater plume. In addition, FATE V includes a formulation of the transient Domenico solution used to help the user determine if the steady-state assumptions used by the model are appropriate.

Obtaining the information necessary for the natural attenuation model is a process that involves interpreting field data to estimate the values for the model parameters. This process, although not straightforward in some cases, is crucial to the modeling effort. In general, the site hydrogeologic and water quality data are analyzed with the objective of predicting trends and estimating the parameter values for the site model. The subsurface geologic data are usually interpreted to yield values for transmissivity, thickness, and porosity. The water level or potentiometric surface data are analyzed to determine the direction of groundwater flow and the water level contours. Water-quality data are analyzed to determine the spatial and temporal trends in contaminant distributions at the site.

An emerging tool in spatial data analysis that should be mentioned here is *geostatistics*. Geostatistics can be viewed as a set of statistical procedures for describing the correlation of spatially distributed random variables and for performing interpolation and aerial estimation of these variables (Cooper and Istok, 1988). *Kriging*, for example, is one of the most widely used geostatistical methods to determine spatial distributions of the hydraulic conductivity (or transmissivity and thickness) at a site. Contouring data using other statistical methods can also be used as an alternative to kriging.

A quantitative evaluation of the calibration process involves an assessment of the *calibration error*. The calibration error is determined by comparing model predicted values to observed values of the heads and concentrations. Two equations are commonly used for this purpose:

$$\text{mean error} = \frac{1}{n}\sum_{i=1}^{n}(x_m - x_s)i \tag{8.69}$$

$$\text{root mean squared (RMS) error} = \left[\frac{1}{n}\sum_{i=1}^{n}(x_m - x_s)_i^2\right]^{0.5} \tag{8.70}$$

where x_m and x_s are the measured and simulated values, respectively. It should be noted that the calibration error is very different and distinct from the *computational error*, which is a result of the numerical approximation procedures used in numerical models.

Verifying the calibrated site model is the process of using the calibrated model to predict a second set of measured data from the site. The purpose of this step is to ensure that the calibrated model is indeed capable of simulating observed site conditions. If the modeling results for the verification step do not match within reasonable error the observed field data, the model might require fine tuning and recalibration.

Prediction is the process of using the calibrated and verified model to determine site conditions in response to an anticipated set of future events. The prediction process is often associated within a sensitivity analysis similar to that completed with the model after calibration. This is necessary to determine which parameters specifically affect the results predicted.

8.8 SOURCE TERM IN NATURAL ATTENUATION MODELING

If one desires to estimate a source term for a contaminant fate and transport model, one can attempt to estimate the mass loading rate and use that estimate as an input parameter. However, this often does not yield model concentrations (dissolved) that are similar to observed concentrations. As a result, the source in the model often becomes a calibration parameter (Mercer and Cohen, 1990; Spitz and Moreno, 1996). This is because the effects of the source (i.e., the dissolved contaminant plume) are easier to quantify than the actual flux from the source. The frequent need for such a blackbox source term has been borne out during modeling associated with evaluations of natural attenuation of fuel hydrocarbons [following the AFCEE technical protocol (Wiedemeier et al., 1995)] at over 30 U.S. Air Force sites. Use of other methods to calculate source loading for those models often produced model concentrations that differed from observed concentrations by as much as an order of magnitude. From the model, the flux estimate can then be used for estimating source lifetimes or other such calculations.

One can estimate flux using several methods, as summarized in Chapter 2. For bodies of mobile LNAPL, this may be more practical, because the area of NAPL in contact with groundwater can be estimated from pool dimensions. In contrast, where most NAPL is residual, the surface area can be highly variable and cannot be measured in the field. Laboratory studies to understand and quantify mass transfer from residual NAPL in porous media are in the early stages, and when such mass transfer is modeled, surface area is a calibration parameter with great uncertainty (Abriola, 1996). Some methods of estimating NAPL dissolution rates require an estimate of the contact area and therefore will contain a great deal of uncertainty. This is one of the main reasons why for purposes of modeling, the blackbox source term is more commonly used.

One reason that practitioners want to estimate mass transfer rates is to provide a basis for estimating contaminant source lifetimes, which can affect regulatory decisions and remedial designs. To determine how long it will take for a dissolved contaminant plume to attenuate fully, it is necessary to estimate how fast the contaminants are being removed from the NAPL. In general, it is difficult to estimate cleanup times, so conservative estimates should be made based on NAPL dissolution rates. Predicting the cleanup time for sites with mobile NAPL is especially difficult because residual NAPL will remain after the recoverable mobile NAPL has been removed. Of course, this is all complicated by the many factors that affect dissolution rates as discussed above. Moreover, most methods do not account for changing dissolution rates as a result of NAPL volume loss (and subsequent surface area decrease), preferential partitioning from mixed NAPLs, and the change in porosity (and therefore groundwater velocity) resulting from NAPL dissolution. Finally, the mass of the NAPL present in the subsurface must also be estimated, lending further uncertainty to any calculation of source lifetime.

If estimating mass flux rates is less important, one can use direct measurement or equilibrium concentration calculations to estimate contaminant source zone concentrations. The first method involves directly measuring the concentration of dissolved contaminants in groundwater near the NAPL-containing zone. The second method involves the use of partitioning calculations. These approaches are described in the following sections. This type of data can be useful if it can be demonstrated that the source is not capable of introducing concentrations of compounds of concern that exceed regulatory limits, or that with slight weathering the same results can be expected. Source zone concentrations, whether measured or calculated, may also be used to provide calibration targets for transport models in which a blackbox source term is used.

A discussion of estimating source terms for sites contaminated solely with fuel hydrocarbons is presented by Wiedemeier et al. (1995). In general, estimating dissolution rates of individual compounds from fuels is simpler than estimating rates of dissolution from other NAPL mixtures because there are a great many experimental data regarding partitioning and equilibrium solubilities of individual compounds from common fuel mixtures.

Typical uses of chlorinated solvents (e.g., degreasing or parts cleaning) and past disposal practices that generally mixed different waste solvents or placed many types of waste solvents in close proximity have resulted in complex and greatly varying NAPL mixtures being released at sites. For mixtures containing other compounds (e.g., either DNAPLs containing multiple chlorinated compounds, or fuel LNAPLs containing commingled chlorinated compounds), the equilibrium solubility of the individual compounds of interest must first be calculated; then that information can be used in the common mass transfer rate calculations. Except in the case of pure solvent spills, therefore, the estimation of dissolution rates is then further complicated by this need to estimate equilibrium solubilities from the mixture.

REFERENCES

Abriola, L. M., 1996, Organic liquid contaminant entrapment and persistence in the subsurface: interphase mass transfer limitation and implications for remediation, 1996 Darcy Lecture, National Ground Water Association, presented at Colorado School of Mines, Golden, CO, Oct. 25.

Anderson, M. P., and Woessner, W. W., 1992, *Applied Groundwater Modeling*, Academic Press, San Diego, CA.

American Society for Testing and Materials (ASTM), 1997, *Standard Guide for Risk-Based Corrective Action Applied at Petroleum Release Sites*, ASTM E-1739-95, ASTM, Philadelphia.

Aziz, C. E., Newell, C. J., Gonzales, A. R., Haas, P., Clement, T. P., and Sun, Y., 1999, BIOCHLOR Natural Attenuation Decision Support System User's Manual. Prepared for the Air Force Center for Environmental Excellence, Brooks AFB, San Antonio, TX. (see also www.gsi-net.com)

Bear, J., 1972, *Dynamics of Fluids in Porous Media*, Dover Publications, New York, 764 pp.

Bear, J., 1979, *Hydraulics of Groundwater*, McGraw-Hill, New York, 569 pp.

Bedient, P. B., Rifai, H. S., and Newell, C. J., 1994, *Ground Water Contamination, Transport and Remediation*, PTR Prentice Hall, Upper Saddle River, NJ.

Borden, R. C., 1986, *Influence of Adsorption and Oxygen Limited Biodegradation on the Transport and Fate of a Creosote Plume: Field Methods and Simulation Techniques*, Houston, TX, Ph.D. thesis, Rice University, 193 p.

Borden, R. C., and Bedient, P. B., 1986, Transport of dissolved hydrocarbons influenced by oxygen-limited biodegradation: 1. Theoretical development, *Water Resour. Res.*, vol. 13, pp. 1973–1982.

Borden, R. C., Bedient, P. B., Lee, M. D., Ward, C. H., and Wilson, J. T., 1986, Transport of dissolved hydrocarbons influenced by oxygen-limited biodegradation: 2. Field application, *Water Resour. Res.*, vol. 13, pp. 1983–1990.

Connor, J. A., Newell, C. J., Nevin, J. P., and Rifai, H. S., 1994, Guidelines for use of groundwater spreadsheet models in risk-based corrective action design," *Proceedings of the National Ground Water Association Petroleum Hydrocarbons and Organic Chemicals in Ground Water Conference*, Houston, TX, Nov., pp. 43–55.

Cooper, R. M., and Istok, J. D., 1988, Geostatistics applied to groundwater contamination: I, Methodology, *J. Environ. Eng.*, vol. 114, no. 2, pp. 270–286.

Domenico, P. A., 1987, An analytical model for multidimensional transport of a decaying contaminant species. *J. Hydrol.*, vol. 91, pp. 49–58.

Domenico, P. A., and Schwartz, F. W., 1990, *Physical and Chemical Hydrogeology*, Wiley, New York, 824 pp.

Environmental Science and Engineering, Inc., 1991, *Installation Restoration Program, Phase II, Stage 2: Remedial Investigation/Feasibility Study*, Vol. I–X, Patrick Air Force Base, FL.

Faust, C. R., Sims, P. N., Spalding, C. P., Anderson, P. F., and Stephenson, D. E., 1990, FTWORK, A Three-Dimensional Groundwater Flow and Transport Code, Westinghouse Savannah River Company Report WSRC-RP-89-1085, Aiken, South Carolina.

Gelhar, L. W., Welty, C., and Rehfeldt, K. R., 1992, A critical review of data on field-scale dispersion in aquifers. *Water Resour. Res.*, vol. 28, no. 7, pp. 1955–1974.

Groundwater Services, Inc., 1998, Remediation by natural attenuation (RNA) tool kit, *User's Manual*, Houston, TX.

Hantush, M. S., 1956, Analysis of data from pumping tests in leaky aquifers. *J Geophys. Res.*, vol. 69, pp. 4221–4235.

Javandel, I., Doughty, C., and Tsang, C., 1984, Groundwater Transport Handbook of Mathematical Models: American Geophysical Union Water Resources Monograph Series 10, Washington, DC, 288 p.

Konikow, L. F., and Bredehoeft, J. D., 1978, Computer model of two-dimensional solute transport and dispersion in ground water, in *Techniques of Water Resources Investigation of the United States Geological Survey*, Book 7, U.S. Geological Survey, Reston, VA.

———, 1989, Computer model of two-dimensional solute transport and dispersion in ground water, in *Techniques of Water Resources Investigation of the United States Geological Survey*, Book 7, U.S. Geological Survey, Reston, VA.

Lee, M. D., Jamison, V. W., and Raymond, R. L., 1987, Applicability of in-situ bioreclamation as a remedial action alternative, in *Proceedings of the National Ground Water Association Petroleum Hydrocarbons and Organic Chemicals in Ground Water Conference*, Houston, TX, Nov., pp. 167–185.

McDonald, G. and Harbaugh, A. W., 1988, A modular three-dimensional finite-difference ground water flow model, U.S. Geological Survey Techniques of Water Resources, Investigations, Book 6, Chapter A1.

Mercer, J. W., and Cohen, R. M., 1990, A review of immiscible fluids in the subsurface: properties, models, characterization and remediation, *J. Contam. Hydrol.*, vol. 6, pp. 107–163.

Nevin, J. P., Connor, J. A., Newell, C. J., Gustafson, J., and Lyons, K., 1997, FATE V: a natural attenuation calibration tool for groundwater fate and transport modeling, *NGWA Petroleum Hydrocarbons Conference*, Houston, TX, Nov. (see also www.gsi-net.com)

Newell, C. J., Winters, J. W., Rifai, H. S., Miller, R. N., Gonzales, J., and Wiedemeier, T. H., 1995, Modeling intrinsic remediation with multiple electron acceptors: results from seven sites, *Proceedings of the National Ground Water Association Petroleum Hydrocarbons and Organic Chemicals in Ground Water Conference*, Houston, TX, Nov., pp. 33–48.

Newell, C. J., McLeod, R. K., and Gonzales, J. R., 1996, *BIOSCREEN Natural Attenuation Decision Support System User's Manual*, Version 1.3, EPA/600/R-96/087, Robert S. Kerr Environmental Research Center, Ada, OK, Aug.

———, 1997, *BIOSCREEN Natural Attenuation Decision Support System*, Version 1.4 revisions. *Groundwater Services, Inc.*, Houston, TX.

O'Brien and Gere Engineers, Inc., 1992, *ST-29 (PPOL-2) Work Plan Draft Final Report*, Patrick Air Force Base, FL.

Parsons Engineering Science, Inc., 1994, *Intrinsic Remediation Engineering Evaluation/Cost Analysis for Site ST-29*, Patrick Air Force Base, FL.

Prickett, T. A., and Lonnquist, G., 1971, Selected Digital Computer Techniques for Groundwater Resource Evaluations, Illinois State Water Survey Bulletin 65, 103 pp.

Prickett, T. A., Naymik, T. G., and Lonnquist, C. G., 1981, A "Random Walk" Solute Transport Model for Selected Groundwater Quality Evaluation, Illinois State Water Survey Bulletin 65, 103 pp.

Rifai, H. S., Bedient, P. B., Borden, R. C., and Haasbeek, J. F., 1987, *BIOPLUME II: Computer Model of Two-Dimensional Transport Under the Influence of Oxygen Limited Biodegradation in Ground Water, User's Manual*, Version 1.0, Rice University, Houston, TX.

Rifai, H. S., Bedient, P. B., Wilson, J. T., Miller, K. M., and Armstrong, J. M., 1988, Biodegradation modeling at aviation fuel spill site, *J. Environ. Eng.*, vol. 114, no. 5, pp. 1007–1029.

Rifai, H. S., Newell, C. J., Gonzales, J. R., Dendrou, S., Kennedy, L., and Wilson, J., 1997, *BIOPLUME III Natural Attenuation Decision Support System Version 1.0 User's Manual*, prepared for the U.S. Air Force Center for Environmental Excellence, Brooks Air Force Base, San Antonio, TX.

Spitz, K., and Moreno, J., 1996, *A Practical Guide to Groundwater and Solute Transport Modeling*, Wiley, New York, 461 pp.

Srivinsan, P. and Mercer, J. W., 1988, Simulation of biodegradation and sorption processes in groundwater, *Ground Water*, vol. 26, no. 4, pp. 475–487.

Strack, O. D. L., 1989, *Groundwater Mechanics*, Prentice Hall, Upper Saddle River, NJ, 732 pp.

Sun, Y., and Clement, T.P., A decomposition method for solving coupled multi-species reactive transport problems, *Transport in Porous Media Journal* (in press).

Sun, Y., Petersen, J. N., Clement, T. P., and Hooker, B. S., 1996, A modular computer model for simulating natural attenuation of chlorinated organics in saturated ground-water aquifers, in *Proceedings of the Symposium on Natural Attenuation of Chlorinated Organics in Ground Water*, Dallas, TX, Sept. 11–13, EPA/540/R-96/509, U.S. EPA, Washington, DC.

van Genuchten, M. Th., and Alves, W. J., 1982, *Analytical Solutions of the One-Dimensional Convective–Dispersive Solute Transport Equation*, Technical Bulletin 1661, U.S. Department of Agriculture, Washington, DC, 151 pp.

Walton, W. C., 1991, *Principles of Groundwater Engineering*, Lewis Publishers, Boca Raton, FL, 546 pp.

Wexler, E. J., 1992, Analytical solutions for one-, two-, and three-dimensional solute transport in ground-water systems with uniform flow, *Techniques of Water-Resources Investigations of the United States Geological Survey*, Book 3, Chapter B7, U.S. Geological Survey, Washington, DC, 190 pp.

Wiedemeier, T. H., Wilson, J. T., Kampbell, D. H., Miller, R. N., and Hansen, J. E., 1995, *Technical Protocol for Implementing Intrinsic Remediation with Long-Term Monitoring for Natural Attenuation of Fuel Contamination Dissolved in Groundwater*, Vol. 1, U.S. Air Force Center for Environmental Excellence, Technology Transfer Division, Brooks Air Force Base, San Antonio, TX.

Wiedemeier, T. H. et al., 1996, *Technical Protocol for Evaluating Natural Attenuation of Chlorinated Solvents in Groundwater*, U.S. Air Force Center for Environmental Excellence, Technology Transfer Division, Brooks Air Force Base, San Antonio, TX.

Wilson, J. L., and Miller, P. J., 1978, Two-dimensional plume in uniform ground-water flow, ASCE, *J. Hydraul. Div.*, vol. 104, no. HY4, pp. 503–514.

Wilson, J. T., 1996, personal communication. He may be reached at the Subsurface Protection and Remediation Division of the National Risk Management Laboratory, Ada, OK.

CHAPTER 9

CASE STUDIES: FUEL HYDROCARBONS

Because of the diversity inherent in the terrestrial subsurface, we have asked additional authors to contribute sections to Chapters 9 and 10. The four case studies presented in this chapter pertain to the natural attenuation of fuel hydrocarbons dissolved in groundwater. Case studies pertaining to the natural attenuation of chlorinated solvents dissolved in groundwater, are presented in Chapter 10.

The first case study presented in this chapter illustrates the effect of a rising groundwater table on the natural attenuation of a fuel-hydrocarbon plume. In this case, the rising groundwater table has caused increased dissolution of BTEX from mobile LNAPL as it becomes trapped below the water table. Although dissolved BTEX concentrations have been increasing steadily at this site, the dissolved BTEX plume is still at steady-state equilibrium and has in fact receded somewhat since the early 1990s. The second case study, of a former gasoline station, illustrates a natural attenuation evaluation at a relatively small site. The next case study is for a site at Elmendorf AFB in Alaska. This case study illustrates the efficacy of natural attenuation in cold groundwater environments. The final case study presents the findings of a natural attenuation case study conducted at Hill Air Force Base, Utah. From this case study it is apparent that even in areas with fast-flowing groundwater, dissolved fuel-hydrocarbon plumes reach steady-state equilibrium only a short distance downgradient from the NAPL source.

9.1 NATURAL ATTENUATION OF FUEL HYDROCARBONS: FORMER LIQUID FUELS STORAGE AREA, SITE ST-12, WILLIAMS AIR FORCE BASE, ARIZONA

Activities at the former liquid fuels storage area (site ST-12) at Williams Air Force Base (AFB), Arizona resulted in contamination of subsurface sediments and groundwater.

A soil vapor extraction (SVE) system currently is in operation at the site for the remediation of vadose zone sediment contamination. Mobile LNAPL, probably a weathered mixture of JP-4 jet fuel and AVGAS (aviation gasoline), has been observed at the site. This LNAPL is being removed as the result of volatilization caused by the SVE system. A pilot-scale groundwater extraction and treatment system installed as part of a treatibility study achieved poor results. Because of the poor results from the extraction test, an evaluation of natural attenuation was undertaken. The results of the natural attenuation evaluation indicate that natural attenuation of fuel hydrocarbons is occurring at ST-12. Historical data show that the dissolved BTEX plume is not migrating downgradient, despite a rising water table that is causing an increase in dissolved BTEX concentrations. In fact, the dissolved BTEX plume appears to have receded between 1992 and 1995. Geochemical data strongly suggest that intrinsic bioremediation of fuel hydrocarbons is occurring via aerobic respiration and the anaerobic processes of denitrification, iron(III) reduction, sulfate reduction, and methanogenesis. When coupled with sorption, dispersion, and dilution, the estimated rates of biodegradation were found to be more than sufficient to prevent downgradient migration of the dissolved BTEX contamination. With source remediation in place, intrinsic bioremediation should begin to decrease the amount of dissolved contaminant mass in the system.

Facility Background

Military operations at Williams AFB began in 1942. Throughout its history, the primary mission of Williams AFB was pilot training. As a result of Department of Defense downsizing, Williams AFB was closed on September 30, 1993. Surrounding land uses include the General Motors Desert Proving Ground, irrigated agricultural land, and increasingly, residential development.

Site ST-12 is located in the former liquid fuels storage area, which consists of several buildings, aboveground storage tanks (ASTs), underground storage tanks (USTs), fuel distribution lines, fences, and paved and unpaved areas. Liquid fuels have been stored at a variety of facilities in the immediate area around ST-12 since 1942.

Site ST-12 (Figure 9.1) includes areas of contaminated sediment and groundwater affected primarily by releases of JP-4 and AVGAS. These releases included multiple documented fuel spills and leaks between 1977 and 1989, and probably other undocumented fuel spills and leaks that occurred since Williams AFB initiated operations. Unsaturated sediment from a depth of about 25 ft BGS to the water table, which is approximately 210 ft BGS, is contaminated with weathered JP-4 and AVGAS.

In August 1990, pumping was initiated to remove free-phase floating JP-4/AVGAS from the water table beneath the site. During this period, JP-4/AVGAS was skimmed from eight wells located in the vicinity of reported spills. As of December 1994, approximately 24,000 gallons of mobile LNAPL and grossly contaminated groundwater had been extracted.

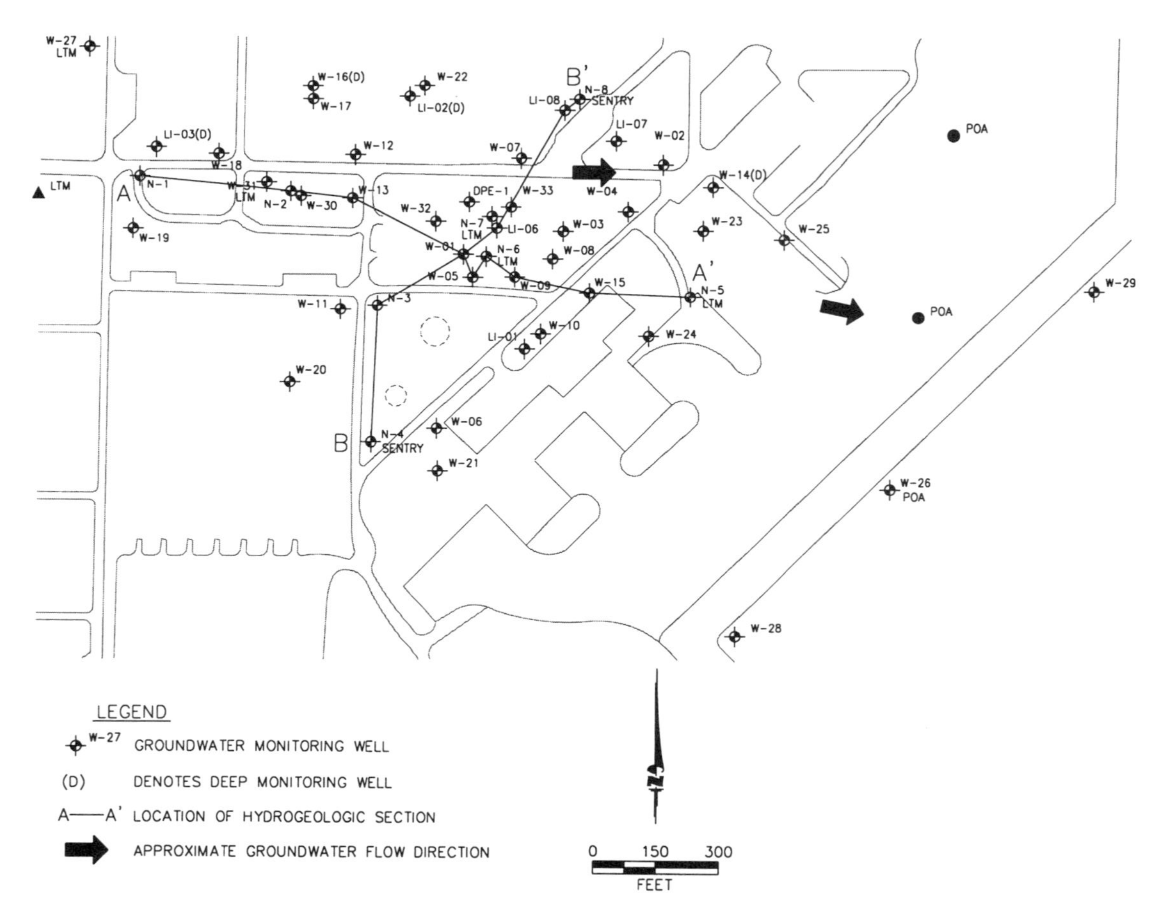

FIGURE 9.1. Site map and location of hydrogeologic cross sections.

Physical Characteristics of the Study Area

Williams AFB is located 30 miles southeast of Phoenix, in the south-central Arizona lowlands within the eastern portion of the Basin and Range Physiographic Province. The local topography is the result of large-scale normal faulting that has created broad, flat, alluvium-filled valleys bounded by steep mountain ranges. The eastern Salt River Valley, where the base is located, is bounded on the north by the Usery Mountains and on the east by the Superstition Mountains. The topography at Williams AFB slopes gently to the west with an approximate surface grade of 0.4%. Ground surface elevations at the Base range from 1390 to 1326 ft above mean sea level (MSL). The area around Williams AFB historically was agricultural but now is becoming urbanized, especially north and northwest of the base.

Regional Geology and Hydrogeology

Six geologic units underlie Williams AFB (Laney and Hahn, 1986). They are, from oldest to youngest, crystalline rocks of Precambrian to Mesozoic age, extrusive rhyolitic and basaltic rocks of Middle to Late Tertiary age, and consolidated and unconsolidated sedimentary rocks of Tertiary through Cenozoic age. The sediments are divided into the red unit, the lower unit, the middle unit, and the upper unit. The youngest sediments are fluvial and lacustrine deposits of Cenozoic age (Eberly and Stanley, 1978).

Because of the pumping of groundwater for agricultural purposes, an extensive vadose zone has been produced in the vicinity of the base. Groundwater beneath the base is encountered at depths of between 180 to 250 ft and flows to the north and east on a base-wide scale. The hydrogeology beneath the base is presumed to be a complex, stratified aquifer system, which is interconnected both vertically and horizontally to varying degrees across the base.

A general rise in groundwater elevations has been observed since December 1989. The rises in groundwater levels may be attributed to decreased local pumping, increased recharge from additional agricultural irrigation, and increased recharge from unusually rainy periods over the past 10 to 15 years. General seasonal fluctuations in groundwater elevation over the base indicate highest groundwater levels in late winter and early spring and lowest groundwater levels in summer and fall. This fluctuation coincides with the pumping schedules of agricultural wells and groundwater recharge. Pumping is at a minimum during the winter and spring when recharge is at a maximum. Conversely, pumping is at a maximum in the summer and fall when recharge is at a minimum.

Site Geology and Hydrogeology

As discussed above, the six geologic units identified in the area include crystalline Precambrian to Mesozoic rocks, extrusive rhyolitic and basaltic rocks of Middle to Late Tertiary age, and consolidated and unconsolidated sedimentary rocks. Two regional geologic strata were observed during past field sampling efforts at site ST-12 to a depth of approximately 350 ft. These strata correlate to the upper and middle units.

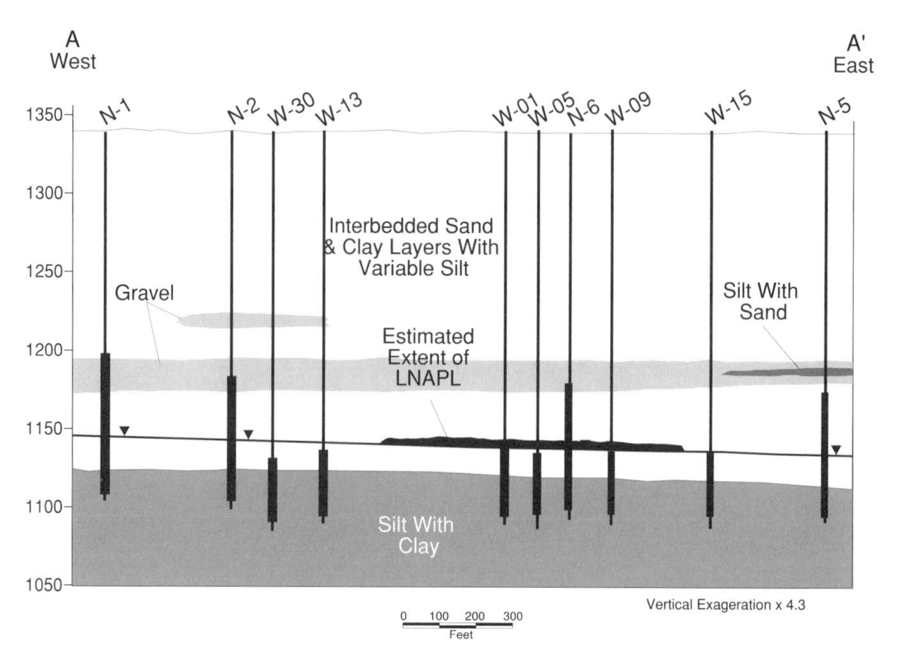

FIGURE 9.2. Hydrogeologic cross section A–A′.

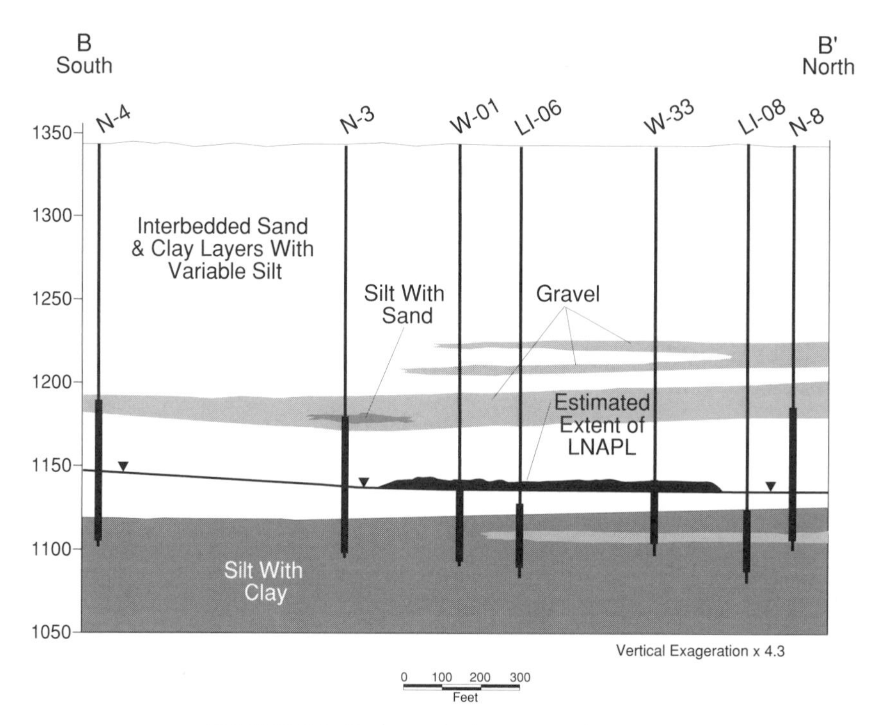

FIGURE 9.3. Hydrogeologic cross section B–B′.

The youngest unit in the stratigraphic sequence, referred to as the upper unit, consists of channel, floodplain, terrace, and alluvial fan deposits of mostly unconsolidated gravel, sand, silt, and clay. This unit is about 150 ft thick. The upper unit received its sediment from the Salt River and Queen Creek drainages. The middle unit is comprised of playa, alluvial fan, and fluvial deposits with no associated evaporites. The middle unit received its sediments primarily from the Salt River. The middle unit is about 150 ft thick at the base.

To illustrate these stratigraphic relationships at site ST-12, hydrogeologic cross sections have been developed using logs of the newly installed and previously installed monitoring wells. The locations of these cross sections are shown on Figure 9.1. Figure 9.2 presents hydrogeologic section A–A′, which is approximately parallel to the direction of groundwater flow. Figure 9.3 presents hydrogeologic section B–B′, which is approximately perpendicular to the direction of groundwater flow. As shown by these figures, the sediments within the vadose zone at ST-12 (to about 215 ft BGS) consist primarily of fine-grained subunits that typically consist of varying percentages of silt and clay with minor amounts of very fine sand. The observed thickness of these subunits ranges from 2.5 to 20 ft. These subunits exhibit lateral continuity across the site. Well-sorted, clean sand units were not observed, although some beds of moderately sorted, medium-grained sand with minor silt and clay were observed. A cobble zone, containing sand and gravel, is present between 135 and 175 ft BGS. Below 215 ft BGS, a more homogeneous layer of silt with clay was observed.

There are two aquifers at site ST-12 that are separated by a competent aquitard. These are referred to as the shallow (unconfined) and deep (confined) aquifers. The shallow aquifer extends from the water table (currently between 200 and 210 ft BGS) to approximately 245 ft BGS. The shallow aquifer is separated from the deep aquifer by a fine-grained aquitard approximately 20 ft thick.

Shallow groundwater elevations have steadily increased since 1989, due to the decrease in agricultural use of the surrounding area. Figure 9.4 presents a hydrograph for three monitoring wells screened in the shallow aquifer at site ST-12. Data

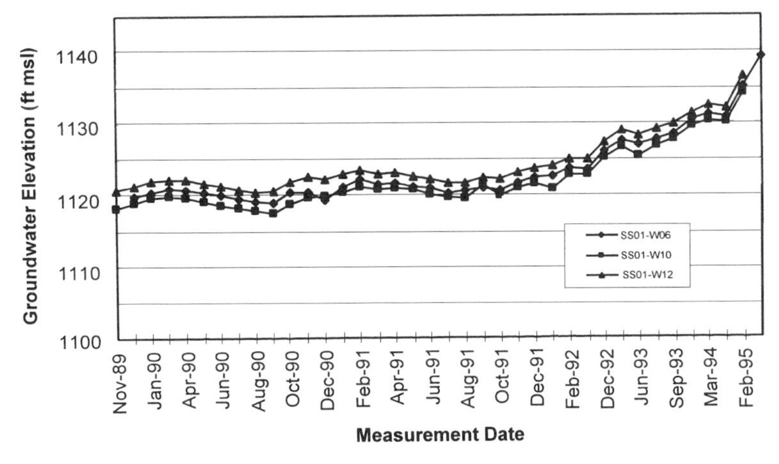

FIGURE 9.4. Shallow well hydrograph.

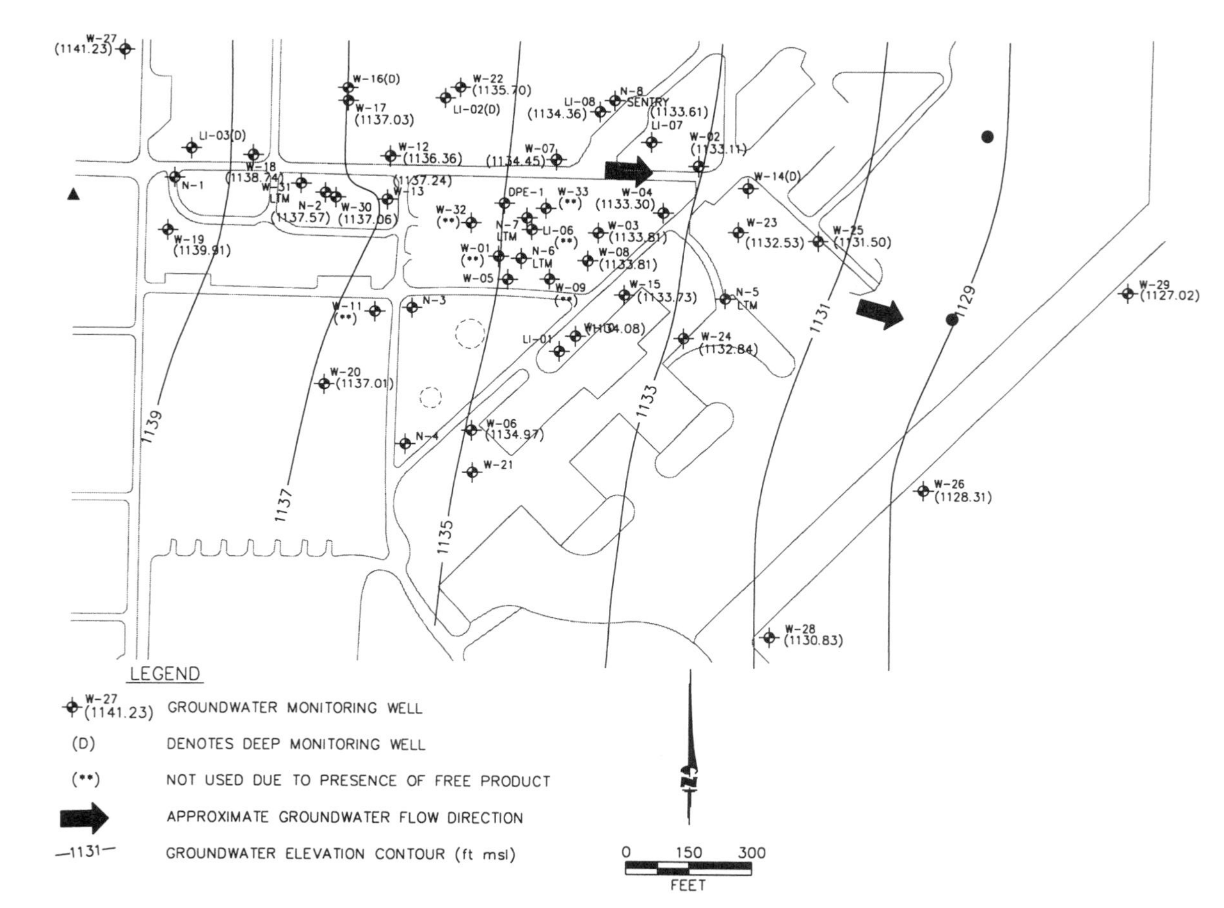

FIGURE 9.5. Groundwater elevation contours.

presented in this figure suggest that groundwater elevations will continue to increase by about 3 ft/yr during the near future. Figure 9.5 is a water table elevation map for the site. This figure shows that groundwater flow is to the east with local variations to the southeast. This is consistent with the shape of the BTEX plume and the distribution of groundwater geochemical parameters. The horizontal groundwater gradient is approximately 0.0055 ft/ft.

Based on several individual well pump tests across ST-12, a mean hydraulic conductivity value of 15 ft/day was estimated. Because of the difficulty involved in accurately determining effective porosity, accepted literature values for the type of sediment making up the aquifer were used. Walton (1988) gives a range of effective porosity for silt of 0.01 to 0.30. An average effective porosity of 0.25 was assumed for this project.

The seepage velocity of groundwater in the direction parallel to groundwater flow is given by eq. (3.1):

$$\bar{v} = \frac{K}{n_e}\frac{dH}{dL}$$

where $\bar{v}$ = average seepage velocity of groundwater [L/T]
K = hydraulic conductivity [L/T] (15 ft/day)
n_e = effective porosity (0.25)
dH/dL = gradient [L/L] (0.0055 ft/ft)

Using this relationship in conjunction with site-specific data, the average seepage velocity of groundwater at ST-12 is 0.33 ft/day, or approximately 120 ft/yr.

Contaminant Sources and Sediment and Groundwater Chemistry

Free Product. Figure 9.6 shows the estimated extent of mobile LNAPL as observed in October 1996. Figure 9.7 presents plots of apparent mobile LNAPL thickness versus water table elevation. From these plots it appears that as the water table rises, mobile LNAPL is trapped below the water table. This increases the surface area of NAPL in contact with groundwater and has resulted in increasing dissolved BTEX concentrations at the site. Figure 9.8 presents a conceptual model of the site. This figure shows conceptually how product has been trapped below the water table, forming a smear zone that exposes a large cross section of the aquifer to NAPL contamination.

The maximum observed mobile LNAPL thickness measured at site ST-12 was 14.70 ft in well W-01, observed in September 1989, and LNAPL thicknesses greater than 9 ft were observed until August 1990; however, due to the rising water table, LNAPL thicknesses measured in W-01 have been less than 1 ft since that time. In 1996, the maximum LNAPL thickness observed at the site was 11.39 ft, in monitoring well W-05. According to Mercer and Cohen (1990), measured mobile LNAPL thickness in wells is typically 2 to 10 times greater than the actual mobile LNAPL thickness in the formation.

Samples of LNAPL were collected from wells W-05, W-33, and N-7 in December 1996 for analysis of BTEX content, as well as for TMB isomers and 1,2,3,4-

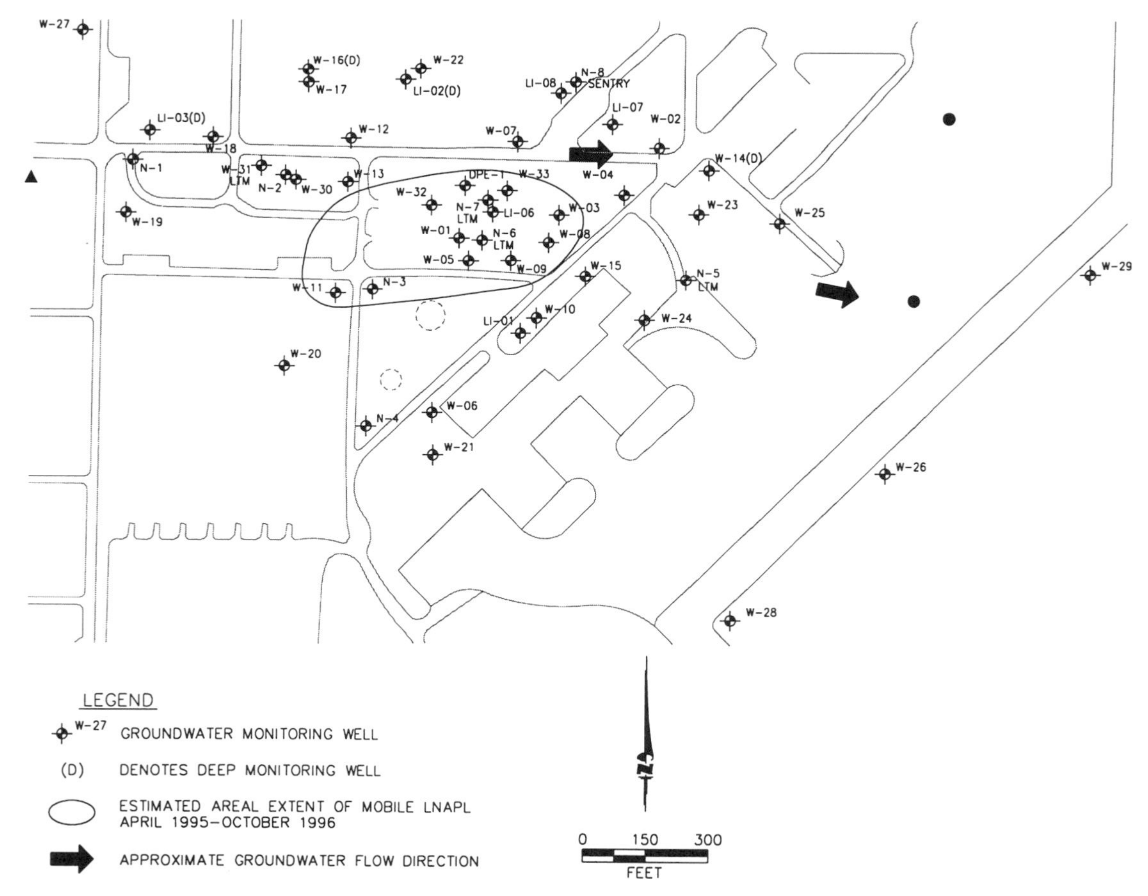

FIGURE 9.6. Estimated extent of mobile LNAPL, October 1996.

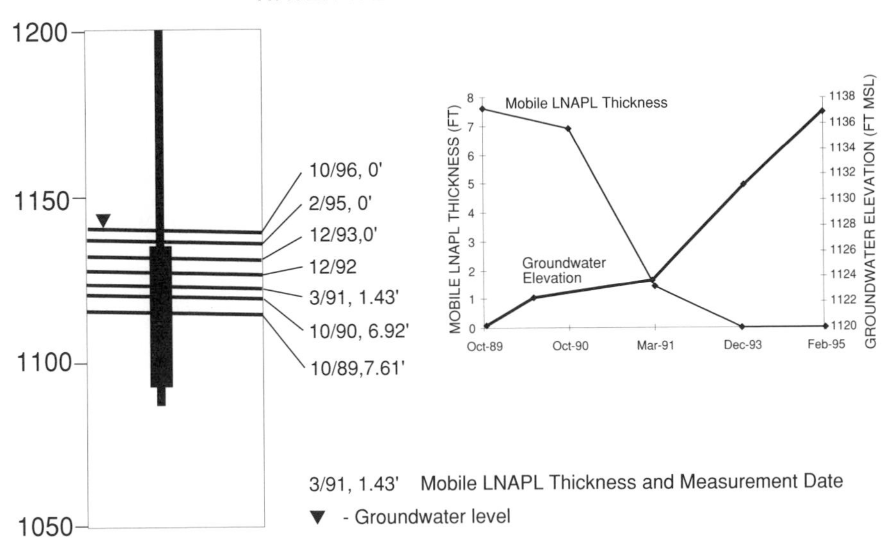

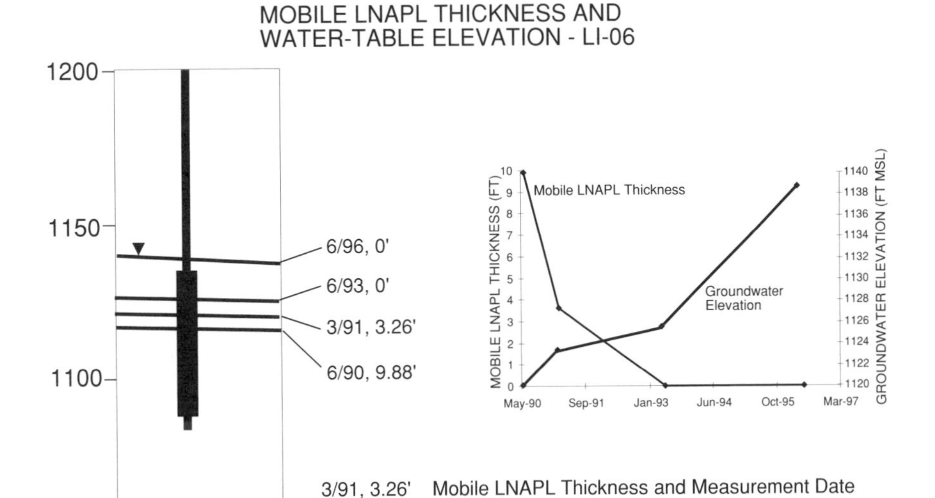

FIGURE 9.7. Plots of mobile LNAPL thickness and water table elevation.

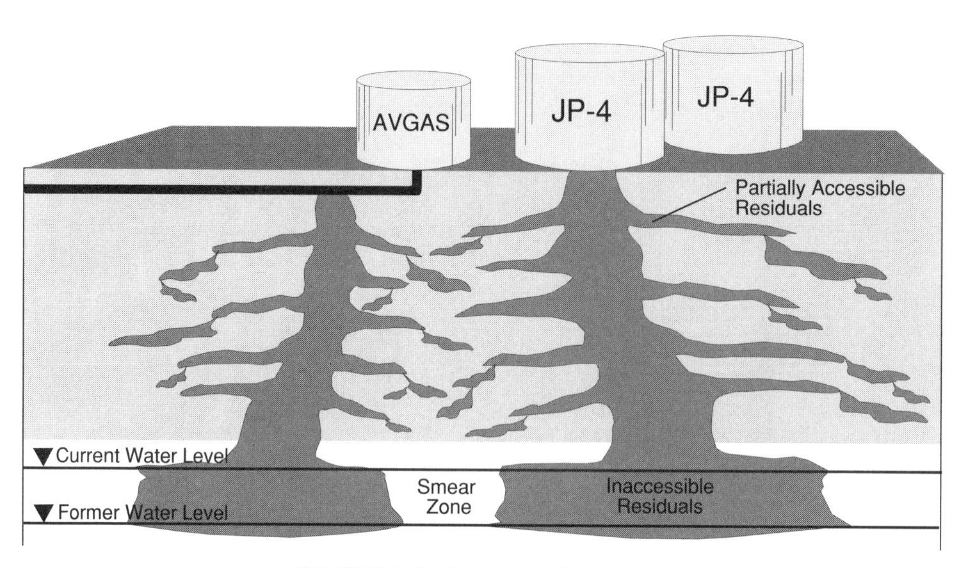

FIGURE 9.8. Conceptual site model.

tetramethylbenzene. Results of these analyses are presented in Table 9.1. Data suggest that LNAPL present at site ST-12 is most likely slightly weathered AVGAS, with some JP-4. Partitioning calculations performed using data collected from LNAPL samples suggest that groundwater in contact with the LNAPL at site ST-12 could contain up to 64.0 mg/L (64,000 μg/L) of BTEX.

Residual Contamination. *Residual NAPL* is defined as the NAPL that is trapped in the aquifer by the processes of cohesion and capillarity, and therefore will not flow within the aquifer and will not flow from the aquifer matrix into a well under the influence of gravity (see Chapter 2). At this site, the residual LNAPL consists primarily of fuel hydrocarbons. In general, sediment contamination exists from the ground surface to the water table at the site. However, discrete areas with much greater contamination are observed with consistency, corresponding with more clayey subsurface units. During installation of monitoring wells N-1 through N-8,

TABLE 9.1. Calculated versus Observed BTEX Concentrations

Compound	Average Concentration in Site LNAPL (mg/L)	Calculated Concentration in Groundwater (mg/L)	Measured Concentration in Groundwater (mg/L)
Benzene	6,525	28.2	30
Toluene	24,000	26.8	24
Ethylbenzene	8,250	2.4	2.9
Xylene	21,225	6.6	6.2
		64	63.1

field screening techniques [e.g., photoionization detector (PID) readings, appearance, and smell] performed on sediment cores confirmed these areas with high residual contamination.

Total Organic Carbon in Sediment. Sediment TOC concentrations were measured in 24 samples from eight monitoring well boreholes installed in October 1996. Sediment samples were collected from below the water table (>240 ft BGS) up to 35 ft BGS. The average TOC from sediment samples taken at or below 200 ft BGS was 0.027%. In general, the sediment TOC data for site ST-12 suggest that there is minimal organic carbon in the saturated zone and that contaminant retardation will be limited.

BTEX in Groundwater. The distributions of BTEX measured in groundwater at site ST-12 in 1990, 1991, 1992, 1995, March 1996, and October 1996 are shown on Figure 9.9. Where detected, BTEX concentrations ranged from 1.3 to 26,000 μg/L in

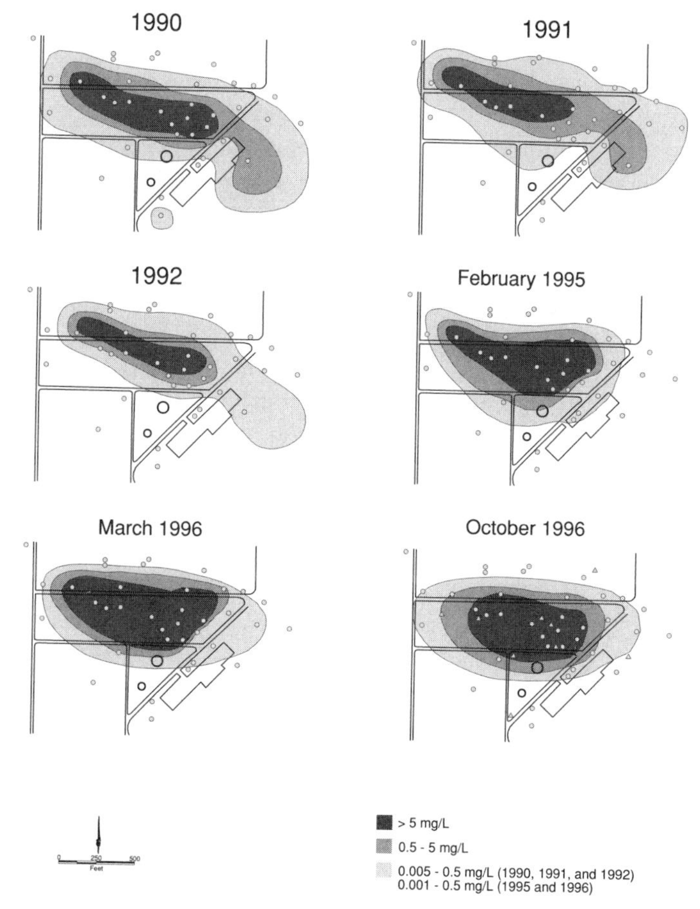

FIGURE 9.9. Benzene concentrations in groundwater.

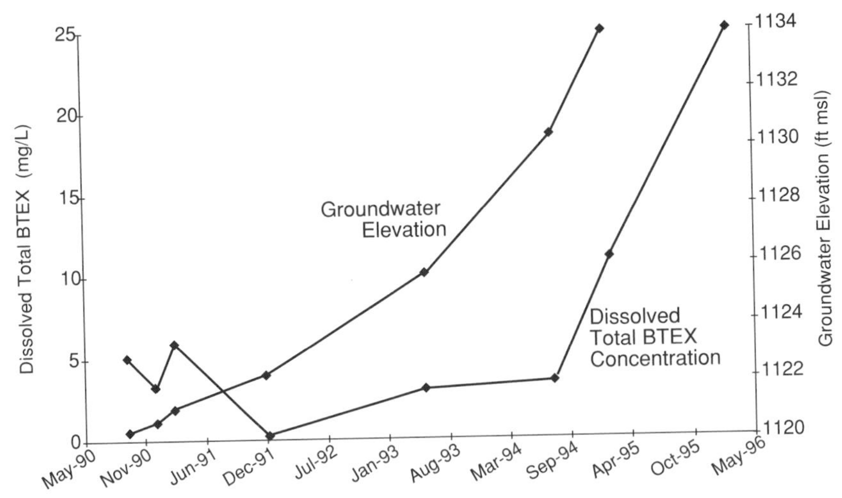

FIGURE 9.10. Dissolved BTEX concentration versus groundwater elevation: W-03.

February 1995; from 20 to 63,010 μg/L in March 1996, and from 16 to 42,080 μg/L in October 1996. Partitioning calculations using data collected from LNAPL samples suggest that groundwater in contact with the LNAPL at ST-12 could contain up to 64,000 μg/L of BTEX (Table 9.1). This is consistent with dissolved total BTEX concentrations observed in March 1996. Figure 9.10 is a plot of dissolved BTEX concentration versus groundwater elevation for the period May 1990 through May 1996. As can be seen from this figure, as the water table rises, dissolved total BTEX concentrations have risen. This is to be expected, as an increasing amount of NAPL becomes exposed to groundwater. Figure 9.11 shows conceptually how NAPL trapped below the groundwater table will increase mass

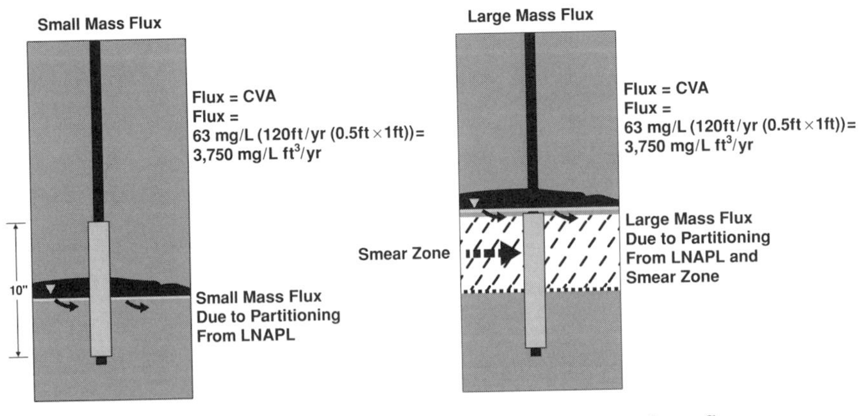

FIGURE 9.11. Conceptualization of increasing contaminant flux.

flux of BTEX into groundwater as the cross-sectional area of NAPL exposed to groundwater increases. Although dissolved total BTEX concentrations have increased at this site due to the rising water table, the downgradient extent of the dissolved BTEX plume has not increased; in fact, available data suggest that the plume receded from 1992 and has achieved steady-state equilibrium.

Biodegradation of Fuel Hydrocarbons

Site-specific data for electron acceptors in groundwater at site ST-12 indicate that intrinsic bioremediation of fuel hydrocarbons in the water table aquifer is occurring by aerobic oxidation, denitrification, iron(III) reduction, sulfate reduction, and methanogenesis. This is evidenced by detectable changes in groundwater geochemistry in areas with BTEX contamination relative to background conditions. Areas of the site that show the greatest changes in concentrations of geochemical indicators (relative to background) generally correspond well with areas of low ORP and high BTEX concentrations.

Dissolved Oxygen. Dissolved oxygen concentrations were measured at monitoring wells at the time of groundwater sampling in February 1995, March 1996, and October 1996. Figure 9.12 presents isopleth maps showing the distribution of dissolved oxygen in groundwater during these sampling events. Relatively high background dissolved oxygen concentrations have been measured in the groundwater near site ST-12. For example, the dissolved oxygen concentrations at monitoring well W-27 in February 1995 and March 1996 were 8.04 mg/L and 5.69 mg/L, respectively. In October 1996, the highest dissolved oxygen concentrations outside the plume were 6.86 mg/L at N-4 and 8.04 mg/L at well LI-01. In wells

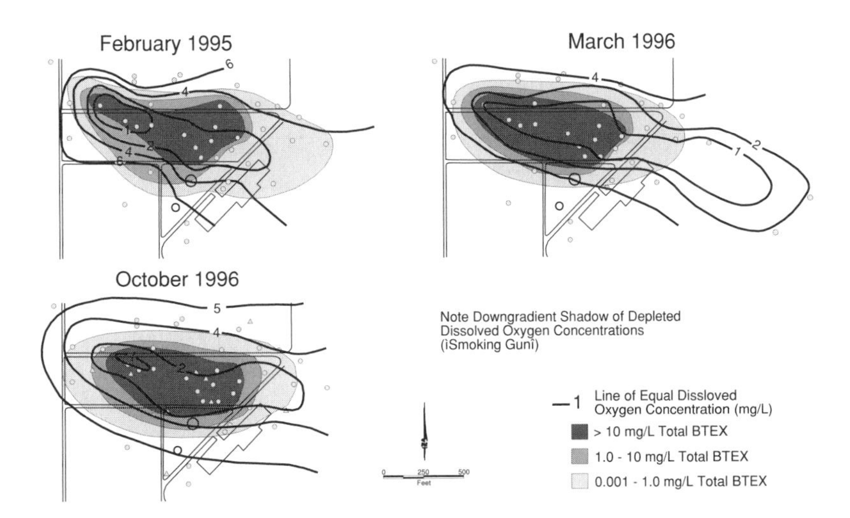

FIGURE 9.12. Total BTEX and dissolved oxygen.

containing dissolved BTEX, dissolved oxygen concentrations are lower than background levels, and typically are less than 1 mg/L. These data provide strong evidence that aerobic biodegradation of fuel hydrocarbons is occurring at site ST-12.

Nitrate/Nitrite. Concentrations of nitrate and nitrite [as nitrogen (N)] were measured in groundwater samples collected in February 1995, March 1996, and October 1996. Figure 9.13 presents isopleth maps that show the distribution of nitrate in groundwater. The data from all three sampling events indicate reduced nitrate concentrations (relative to upgradient, background data) within the dissolved BTEX plume, which suggests that nitrate is an important electron acceptor at this site. Nitrate was detected in site groundwater at concentrations ranging from 1.2 to 19.6 mg/L, <0.1 to 22.0 mg/L, and <0.1 to 26.7 mg/L in February 1995, March 1996, and October 1996, respectively.

Iron(II). Iron(II) (Fe^{2+}) concentrations were measured in groundwater samples collected in February 1995, March 1996, and October 1996. Figure 9.14 presents isopleth maps showing the distribution of iron(II) in groundwater at the site. This figure indicates that iron(II) is being produced in the anaerobic portions of the BTEX plume due to the reduction of iron(III) during anaerobic biodegradation of BTEX compounds. In March 1996, the highest concentration of iron(II), 1.95 mg/L, was detected at W-18, which is located in the western part of the observed BTEX plume. In October 1996, the highest iron(II) concentrations were observed at the center and western side of the plume at monitoring wells LI-06 and W-31 (3.65 and 3.06 mg/L, respectively).

Sulfate. Sulfate concentrations were measured in groundwater samples collected in February 1995, March 1996, and October 1996. Sulfate concentrations at the site

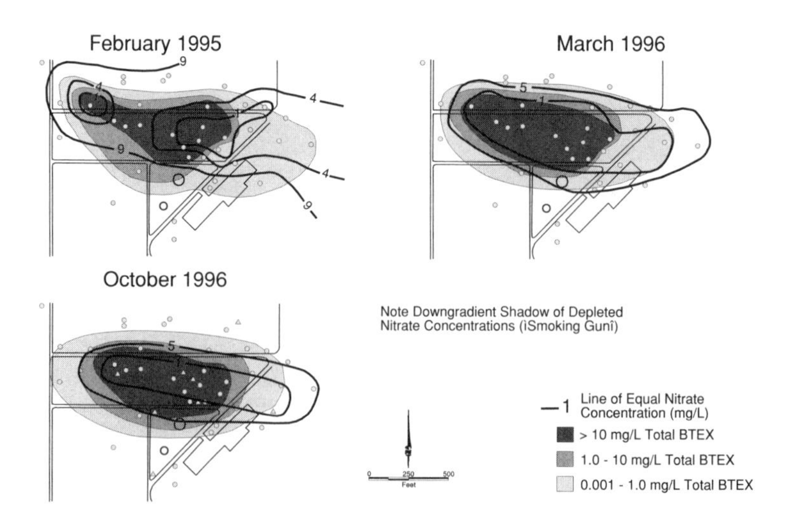

FIGURE 9.13. Total BTEX and nitrate.

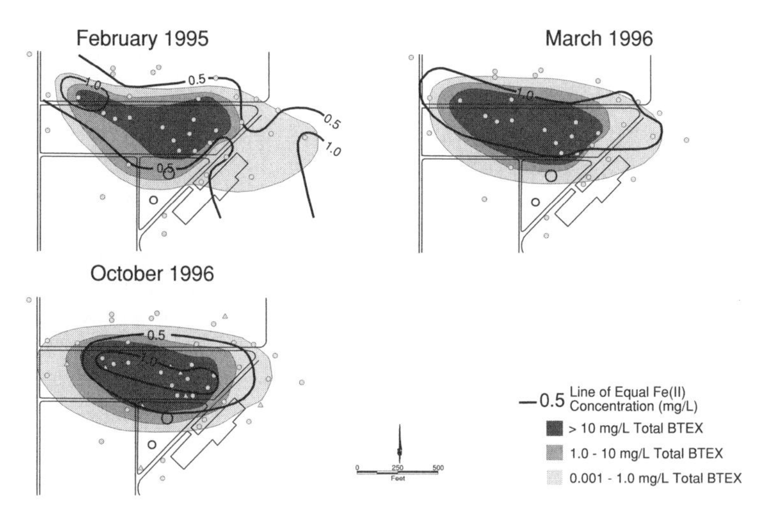

FIGURE 9.14. Total BTEX and Fe(II).

ranged from 0.4 to 250 mg/L, 2.3 to 350 mg/L, and 5.8 to 388 mg/L in February 1995, March 1996, and October 1996, respectively. Figure 9.15 presents isopleth maps that show the distribution of sulfate in groundwater for these three sampling events. This figure shows graphically that areas of depleted sulfate concentrations correspond to areas of elevated BTEX concentrations. This is a strong indication that anaerobic biodegradation of the BTEX compounds is occurring at the site via sulfate reduction.

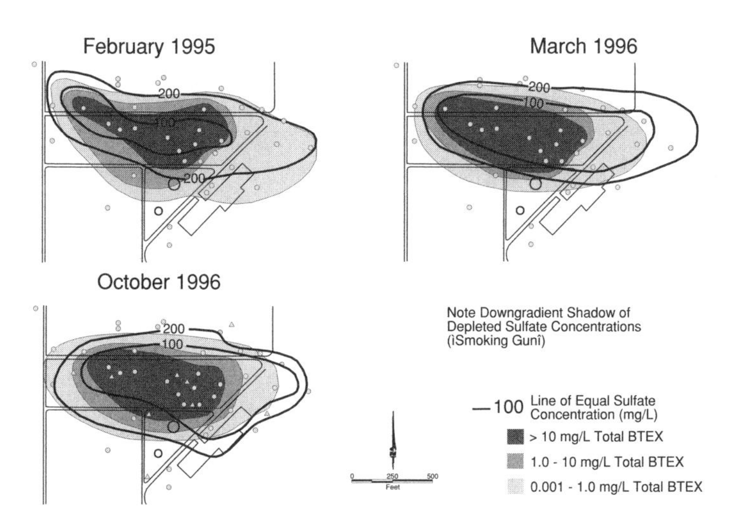

FIGURE 9.15. Total BTEX and sulfate.

Methane. Methane concentrations were measured in groundwater samples collected in February 1995, March 1996, and October 1996. Methane concentrations ranged from not detected to 0.4 mg/L, not detected to 2.42 mg/L, and not detected to 0.511 mg/L in February 1995, March 1996, and October 1996, respectively. Figure 9.16 presents isopleth maps showing the distribution of methane in groundwater for all three sampling events. This figure illustrates that the areas with elevated total BTEX concentrations correlate with elevated methane concentrations. The lack of consistently elevated methane concentrations above 1.0 mg/L at ST-12 suggests that methanogenesis is not a significant BTEX degradation pathway at this site.

Oxidation–Reduction Potential. ORP was measured at groundwater monitoring wells in February 1995, March 1996, and October 1996. The ORP of site groundwater ranged from –305 to 276 mV, –328 to 174 mV, and –462 to 181 mV in February 1995, March 1996, and October 1996, respectively. As expected, areas at the site with low ORP coincide with areas of high BTEX contamination; low dissolved oxygen, nitrate, and sulfate concentrations; and elevated iron(II) and methane concentrations. In general, the ORP of site groundwater decreased from February 1995 to October 1996. This general decrease in ORP may correspond to a shift toward a more reducing environment. However, different sampling techniques and ORP probes may also explain the temporal and spatial difference in observed ORP.

Alkalinity. In February 1995, March 1996, and October 1996, total alkalinity (as calcium carbonate) was measured in groundwater samples. Total alkalinity at the site ranges from 80 to 280 mg/L in February 1995, 40.5 to 177.8 mg/L in March 1996, and 54 to 180 mg/L. Elevated alkalinity correlates with the dissolved BTEX

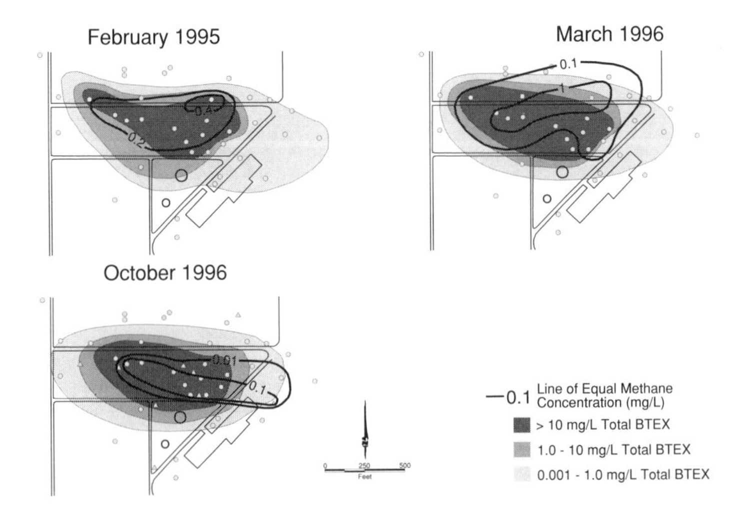

FIGURE 9.16. Total BTEX and methane.

plume. The increase in alkalinity, over background, in areas with dissolved BTEX contamination is in response to increased carbon dioxide levels that result from BTEX biodegradation. Increasing alkalinity acts to buffer the pH of the groundwater from potential changes brought about by an increase in carbon dioxide.

pH. pH was measured for groundwater samples collected from groundwater monitoring locations in February 1995, March 1996, and October 1996. Groundwater pH measured at the site for all three sampling events fell within the range 6.0 to 7.5. This range of pH is within the optimal range for BTEX-degrading microbes.

Temperature. Temperature affects the types and growth rates of bacteria that can be supported in the groundwater environment, with high temperatures generally resulting in higher growth rates. Groundwater temperature measurements made in February 1995, March 1996, and October 1996. Observed temperatures in the aquifer were all within 3° of 25°C. These are moderate temperatures for groundwater, suggesting that bacterial growth rates should not be inhibited.

Shadowing of Electron Acceptor Concentrations

At site ST-12, zones of groundwater with low electron acceptor (i.e., dissolved oxygen, nitrate, and sulfate) concentrations were observed downgradient from the observed dissolved BTEX plume in all three sampling events. These downgradient zones of depleted electron acceptors are referred to as *shadows*, as they appear to foreshadow the dissolved fuel hydrocarbon plume. For electron acceptor shadows to develop, two groundwater criteria must be met. First, dissolved contamination must either be degraded rapidly, or contaminant velocity must be slower than the seepage velocity of groundwater (i.e., high contaminant retardation). Second, electron acceptors must be depleted in the zone of dissolved contamination (biodegradation is occurring) and must be transported at nearly the same velocity as groundwater (i.e., no retardation). Although contaminant retardation at site ST-12 is probably minimal, degradation processes destroy the dissolved BTEX mass quickly enough so that the dissolved BTEX plume does not migrate downgradient. However, groundwater is moving through the contaminated area, and electron acceptors (e.g., dissolved oxygen, nitrate, and sulfate) are being consumed. The groundwater moves downgradient from the dissolved hydrocarbon plume and produces the observed shadow of depleted electron acceptors.

This shadowing effect provides further evidence of the effectiveness of biodegradation at site ST-12. Groundwater with no contamination and low depleted electron acceptors, which once contained both fuel hydrocarbons and high dissolved oxygen, nitrate, and sulfate concentrations, has moved downgradient from the source and dissolved hydrocarbon plumes (Figures 9.12, 9.13, and 9.15). While such a shadowing effect could be partially attributed to retardation of fuel hydrocarbons, the apparent lack of sorption (i.e., low TOC) at site ST-12 suggests that contaminants are being rapidly biodegraded. If biodegradation was negligible, fuel-hydrocarbon compounds would be detected in those downgradient wells within the electron-acceptor shadows.

Approximation of Biodegradation Rates

The method of Buscheck and Alcantar (1995) was used to estimate first-order biodegradation rate constants for BTEX compounds at site ST-12. TMB could not be used as a conservative tracer at this site because it appears to degrade as rapidly as BTEX. Decay rates computed using the method of Buscheck and Alcantar (1995) using data from March 1996 and October 1996 are 0.010 and 0.009 day^{-1}, respectively. The equivalent BTEX half-lives are 69 and 77 days, respectively. Historical site contaminant data support the relatively high calculated biodegradation rates for site ST-12; dissolved BTEX data from as early as 1989 show that the current plume has not migrated downgradient during the past seven years. High dissolved contaminant concentrations in the source area, limited plume migration, and the presence of electron-acceptor shadows all suggest that natural attenuation processes are capable of degrading large amounts of dissolved hydrocarbons at site ST-12 in relatively short periods of time.

Long-Term Groundwater Monitoring

To implement monitored natural attenuation for remediation of groundwater contaminated with fuel hydrocarbons at site ST-12, a long-term groundwater monitoring plan was developed. This long-term monitoring plan identifies the location of two separate groundwater monitoring networks and outlines a groundwater sampling and analysis strategy to demonstrate the continuing effectiveness of natural attenuation. The strategy described in this section is designed to monitor plume migration (or lack thereof) over time and to verify that natural attenuation continues to occur at rates sufficient to protect potential receptors. In the event that data collected under the

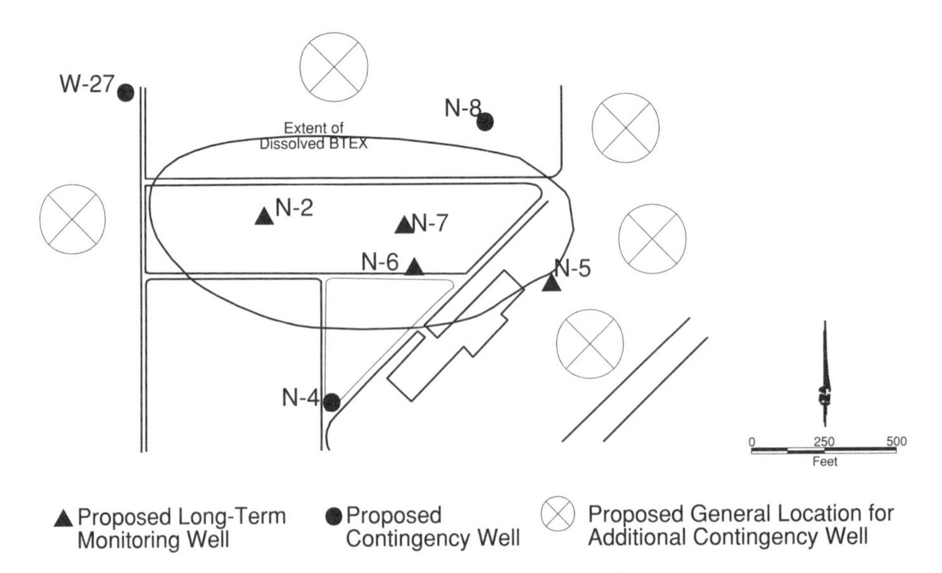

FIGURE 9.17. Proposed long-term monitoring locations.

long-term monitoring program indicate that natural processes are insufficient to protect human health and the environment, contingency controls to augment the beneficial effects of natural attenuation would be necessary.

Two separate sets of wells were recommended at the site as part of the monitored natural attenuation alternative. These wells are shown on Figure 9.17. The first set will consist of four long-term monitoring wells located upgradient from, within, and downgradient from the observed contaminant plume. The second set of wells will consist of three existing contingency wells located cross gradient from the source area, and five proposed downgradient contingency wells. The purpose of the contingency wells is to verify that benzene concentrations exceeding the federal MCL of 5 mg/L do not migrate beyond the area under institutional control. Because of the shallow groundwater gradient at this site, contingency wells were placed to surround the dissolved contaminant plume. Historical data analysis and calculated site-specific biodegradation rates suggest that in the future benzene concentrations are not expected to exceed federal standards at any of the contingency wells. Furthermore, source reduction using SVE technology should help reduce future contaminant concentrations in groundwater. The long-term monitoring and contingency wells will be sampled and analyzed for the parameters listed in Tables 9.2 and 9.3.

In addition to groundwater sampling, two mobile LNAPL samples will be collected annually for five years to monitor the fraction of BTEX (using USEPA Method SW8020) remaining in the LNAPL and to help determine the remediation time required for natural attenuation to achieve groundwater cleanup criteria. The fraction of BTEX remaining in the mobile LNAPL should be reduced by natural attenuation and weathering processes, and the operation of the SVE system, which will enhance volatilization of VOCs from the LNAPL. In addition to annual monitoring of the percent BTEX in the mobile LNAPL, mobile LNAPL thickness measurements will be conducted quarterly to assess the effectiveness of SVE at reducing the volume of mobile LNAPL.

Conclusions

The following conclusions can be inferred from the analysis of data at site ST12:

- Natural biodegradation of dissolved hydrocarbons is occurring and preventing the migration of the dissolved BTEX plume.
- Sulfate reduction is the predominant natural biodegradation pathway currently occurring in site groundwater. Sulfate concentrations from background and cross-gradient locations exceeded 200 mg/L, whereas sulfate concentrations in highly contaminated areas were less than 100 mg/L and often less than 50 mg/L. Sulfate reduction has been observed to dominate biodegradation processes at other sites in similar dry climates, suggesting that groundwater throughout the region favors sulfate reduction over other natural attenuation mechanisms.
- Analysis of field data suggests that degradation of BTEX through aerobic respiration, nitrate reduction, iron reduction, manganese reduction, and methanogenesis also is occurring at ST-12.

TABLE 9.2. Long-Term Groundwater Monitoring Analytical Protocol

Analyte	Method/Reference	Comments	Data Use	Recommended Frequency of Analysis	Sample Volume, Sample Container, Sample Preservation	Field or Fixed-Base Laboratory
Temperature	E170.1	Field only	Metabolism rates for microorganisms depend on temperature	Every year for 10 years	Not applicable	Field
Dissolved oxygen	Dissolved oxygen meter	Refer to method A4500 for a comparable laboratory procedure	The oxygen concentration is an indicator of biodegradation conditions; concentrations less than 1 mg/L generally indicate an anaerobic pathway	Every year for 10 years	Collect 300 mL of water in biochemical oxygen demand bottles; analyze immediately; alternatively, measure dissolved oxygen in situ	Field
pH	E150.1/SW9040, direct-reading meter	Protocols/handbook methods	Aerobic and anaerobic processes are pH-sensitive	Every year for 10 years	Collect 100–250 mL of water in a glass or plastic container; analyze immediately	Field
Conductivity	E120.1/SW9050, direct-reading meter	Protocols/handbook methods	General water quality parameter used as a marker to verify that site samples are obtained from the same groundwater system	Every year for 10 years	Collect 100–250 mL of water in a glass or plastic container	Field

Redox potential	A2580 B	Measurements are made with electrodes; results are displayed on a meter; samples should be protected from exposure to atmospheric oxygen	The ORP of groundwater influences and is influenced by biologically mediated reactions; the redox potential of groundwater may range from more than 200 mV to less than −400 mV	Every year for 10 years	Collect 100–250 mL of water in a glass container, filling container from bottom; analyze immediately	Field
Sulfate (SO_4^{2-})	IC method E300 or method SW9056 or Hach SulfaVer 4 method	Method E300 is a handbook method; method SW9056 is an equivalent procedure; the Hach method is photometric	Substrate for anaerobic microbial respiration	Every year for 10 years	Collect up to 40 mL of water in a glass or plastic container; cool to 4°C	Fixed-base or field (for Hach method)
Methane	RSKSOP-114 modified to analyze water samples for methane by headspace sampling with dual thermal conductivity and flame ionization detection	Method published and used by the U.S. EPA National Risk Management Research Laboratory	The presence of methane suggests BTEX degradation via an anaerobic pathway utilizing carbon dioxide (carbonate) as the electron acceptor (methanogenesis)	Every year for 10 years	Collect water samples in 40-mL volatile organic analysis (VOA) vials with butyl gray/Teflon-lined caps (zero headspace); cool to 4°C	Fixed-base
Aromatic hydrocarbons (BTEX)	Purge and trap GC method SW8020	Handbook method; analysis may be extended to higher-molecular-weight alkylbenzenes	BTEX is the primary target analyte for monitoring natural attentuation; BTEX concentrations must also be measured for regulatory compliance	Every year for 10 years	Collect water samples in a 40-mL VOA vial with zero headspace; cool to 4°C; add hydrochloric acid to pH ≤2	Fixed-base

TABLE 9.3. Contingency Well Groundwater Monitoring Analytical Protocol

Analyte	Method/Reference	Comments	Data Use	Recommended Frequency of Analysis	Sample Volume, Sample Container, Sample Preservation	Field or Fixed-Base Laboratory
Temperature	E170.1	Field only	Well development	Every year for 10 years	Not applicable	Field
Dissolved oxygen	Dissolved oxygen meter	Refer to method A4500 for a comparable laboratory procedure	The oxygen concentration is an indicator of biodegradation conditions; concentrations less than 1 mg/L generally indicate an anaerobic pathway	Every year for 10 years	Collect 300 mL of water in biochemical oxygen demand bottles; analyze immediately; alternatively, measure dissolved oxygen in situ	Field
pH	E150.1/SW9040, direct-reading meter	Protocols/handbook methods	Aerobic and anaerobic processes are pH sensitive	Every year for 10 years	Collect 100–250 mL of water in a glass or plastic container; analyze immediately	Field
Conductivity	E120.1/SW9050, direct-reading meter	Protocols/handbook methods	General water quality parameter used as a marker to verify that site samples are obtained from the same groundwater system	Every year for 10 years	Collect 100–250 mL of water in a glass or plastic container	Field

Redox potential	A2580 B	Measurements are made with electrodes; results are displayed on a meter; samples should be protected from exposure to atmospheric oxygen	The redox potential of groundwater influences and is influenced by biologically mediated reactions; the redox potential of groundwater may range from more than 200 mV to less than −400 mV	Every year for 10 years	Collect 100–250 mL of water in a glass container, filling container from bottom; analyze immediately	Field
Sulfate (SO_4^{2-})	IC method E300 or method SW9056 or Hach SulfaVer 4 method	Method E300 is a handbook method; method SW9056 is an equivalent procedure; the Hach method is photometric	Substrate for anaerobic microbial respiration	Every year for 10 years	Collect up to 40 mL of water in a glass or plastic container; cool to 4°C	Fixed-base or field (for Hach method)
Methane	RSKSOP-114 modified to analyze water samples for methane by headspace sampling with dual thermal conductivity and flame ionization detection	Method published and used by the U.S. EPA, Robert S. Kerr Laboratory	The presence of methane suggests BTEX degradation via an anaerobic pathway utilizing carbon dioxide (carbonate) as the electron acceptor (methanogenesis)	Every year for 10 years	Collect water samples in 40-mL volatile organic analysis (VOA) vials with butyl gray/Teflon-lined caps (zero headspace); cool to 4°C	Fixed-base
Aromatic hydrocarbons (BTEX)	Purge and trap GC method SW8020	Handbook method; analysis may be extended to higher-molecular-weight alkylbenzenes	BTEX are the primary target analytes for monitoring natural attentuation; BTEX concentrations must also be measured for regulatory compliance	Every year for 10 years	Collect water samples in a 40-mL VOA vial with zero headspace; cool to 4°C; add hydrochloric acid to pH ≤2	Fixed-base

- Calculated BTEX biodegradation rates for March 1996 and October 1996 were 0.009 and 0.010 day^{-1}, respectively. BTEX half-lives of 60 and 70 days, respectively, were estimated using data from these two sampling events.
- Historical dissolved BTEX data from February 1995 to October 1996 show that the extent of dissolved BTEX contamination has remained steady and that significant downgradient migration of dissolved contamination has not occurred. Based on partitioning calculations using mobile LNAPL sample data, the dissolved BTEX concentration at W-09 of 63.01 mg/L in March 1996 represents the maximum dissolved BTEX concentrations possible at site ST-12. Concentrations as high as 54.1 mg/L total dissolved BTEX have been exhibited at site ST-12 since 1989. Even though dissolved contaminant concentrations have not decreased in recent years, the lack of plume migration suggests that natural attenuation, and perhaps most important, intrinsic bioremediation at site ST-12 contain the dissolved contaminant plume.
- As site mobile LNAPL becomes more weathered and as source removal through SVE reduces the BTEX available for dissolution, reductions in dissolved BTEX through natural attenuation should be enhanced.
- Long-term groundwater monitoring will continue to verify the effects of natural attenuation and help predict the effects of LNAPL weathering and source reduction on dissolved contamination.

9.2 NATURAL ATTENUATION OF FUEL HYDROCARBONS: FORMER GASOLINE STATION, SWMU 66 SITE, KEESLER AIR FORCE BASE, BILOXI, MISSISSIPPI

Robert Balcells
Groundwater Services, Inc.
Houston, Texas

Facility Background

Keesler AFB is located in Biloxi, Mississippi, in Harrison County, on a peninsula that stretches east along the Mississippi Gulf coast. The base, including housing, runways, and aircraft maintenance facilities, covers 1610 acres. The base was activated in 1941 as a training center for aircraft mechanics. Keesler AFB currently provides technical training, medical care, and flying operations and support. Technical training operations include electronics communications and personnel training schools. Flying support operations are performed at industrial shops that repair and fabricate parts for aircraft and ground equipment.

The SWMU 66 site is located 60 ft northeast of Building 4038, a military vehicle service station, and approximately 1800 ft north of the shoreline of the Gulf of Mexico. The SWMU 66 site is the location of a former 8000-gallon gasoline UST that was taken out of service and removed in December 1987. Three additional USTs, a 5000-gallon gasoline tank, an 8000-gallon gasoline tank, and an 8000-gallon diesel fuel tank, were removed from service and excavated during November 1996 (see Figure 9.18).

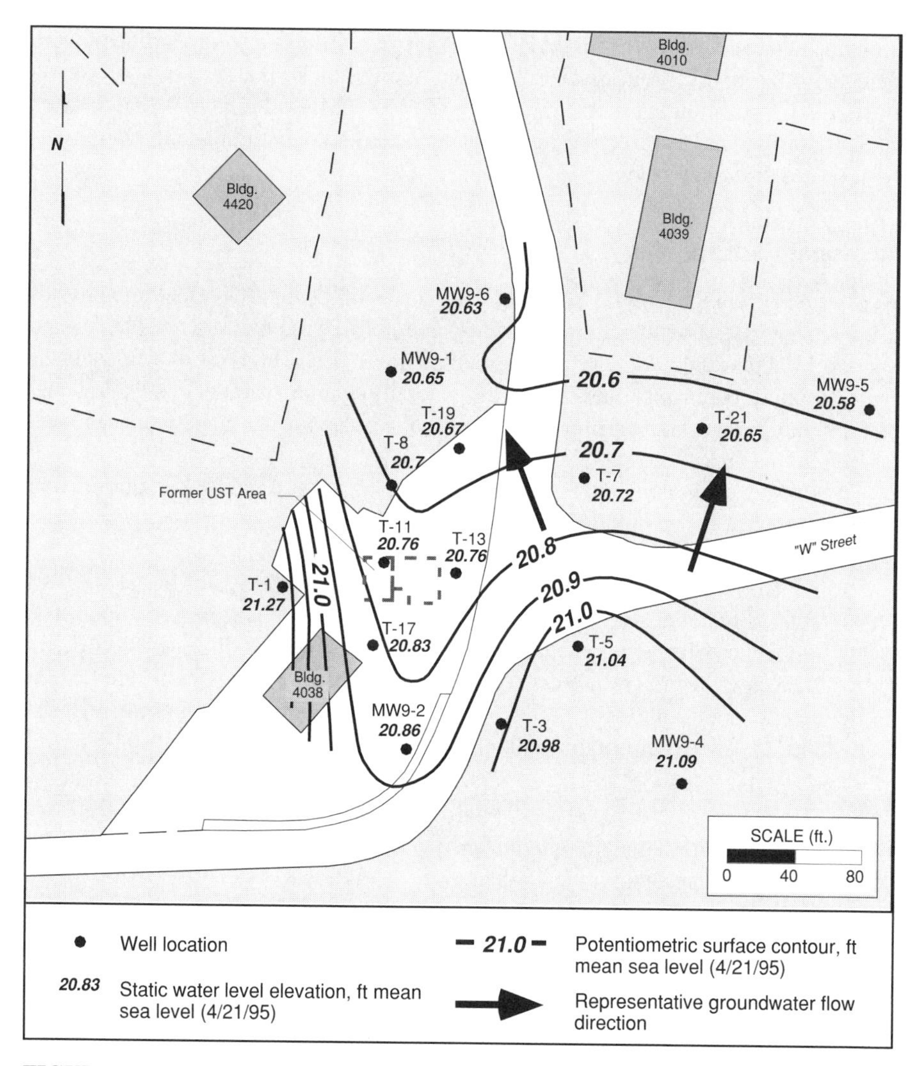

FIGURE 9.18. Potentiometric surface contour map for Shallow Coastal Sand interval, April 21, 1995, SWMU 66 site, Keesler Air Force Base, Mississippi.

Three groundwater monitoring wells (MW9-1, MW9-2, and MW9-3, the latter subsequently destroyed) were installed and sampled in October 1989. Water samples were tested for total petroleum hydrocarbons (TPH), lead, and purgeable aromatics. Elevated TPH concentrations were detected at MW9-3, and lead was detected in all three wells. Purgeable aromatics were not detected in any of the groundwater samples.

Four soil borings were drilled, and one temporary monitoring well was installed in the vicinity of the former USTs in October 1992. Groundwater in the temporary well and existing wells was sampled and tested during this program. Elevated levels of BTEX (benzene, toluene, ethylbenzene, and xylenes) and TPH were detected in all soil borings; and BTEX and PAHs (polynuclear aromatic hydrocarbons) were detected in groundwater immediately adjacent to the tank area (i.e., in well T66-1) but not in the surrounding monitoring wells (MW9-1, MW9-2, and MW9-3). Details of the sampling and testing program conducted during the natural attenuation investigation are discussed below.

Regional Hydrogeology

Keesler AFB is in the Coastal Lowlands subsection of the Gulf Coastal Plain physiographic province, a region characterized by geomorphic remnants of the Pleistocene sea level fluctuations. These remnants include linear accumulations of sand, such as barrier islands, peninsulas, and sand spits, generally oriented parallel to the coast, and shallow bays and estuaries. Elevations on the Biloxi peninsula range from sea level to 40 ft above mean sea level (MSL).

The subsurface in the vicinity of Keesler AFB consists of the following principal units, in order of increasing depth:

1. *Coastal Deposits and Citronelle Formation.* An average of about 75 ft of relatively permeable sand and clayey sand overlie the deeper formations. The Pliocene Citronelle Formation and the overlying Holocene–Pleistocene coastal deposits are not differentiated in the Biloxi area, due to the similar characteristics of the two units. The Surficial Aquifer, subject of this investigation, occurs within this unit.
2. *Graham Ferry and Pascagoula Formations.* The interval from approximately 75 ft to more than 2000 ft below ground surface (BGS) consists of Pliocene and Miocene clay with sand layers. The Miocene Aquifer includes the water-yielding zones within the Graham Ferry and Pascagoula Formations.

Groundwater in the Surficial Aquifer is generally brackish or saline in the coastal areas of Mississippi. Fresh water is obtained from the upper part of the Miocene Aquifer between 600 and 2000 ft BGS. Water in the lower part of the Miocene Aquifer below a depth of about 2000 ft, which includes the Hattiesburg Formation and Catahoula Sandstone, is saline in the Biloxi area.

The Surficial Aquifer is usually under water table conditions and is recharged by infiltration from the ground surface and by artesian flow from the underlying Miocene Aquifer, indicating a general upward gradient from the deeper aquifer to the shallow Surficial Aquifer. The Miocene Aquifer is recharged by infiltration in the outcrop zones of the member formations more than 25 miles north of the site.

Site Stratigraphy of Shallow Soils

The subsurface investigations indicate that the shallow soils underlying the SWMU 66 site at Keesler AFB generally consist of fine to coarse loose quartzose sand with minor amounts of clay and silt. The shallow soils from the surface to a depth of 26 ft BGS (exclusive of a surface fill layer 1 to 2 ft thick) can be divided into two units based on color, composition, and texture. The distribution of shallow soils is shown on the cross sections, Figures 9.19 and 9.20, and is described below in order of increasing depth.

1. *Olive gray and black clayey sand:* a dark clayey or silty fine-grained sand with abundant organic material (e.g., limbs and roots) and a strong organic odor, encountered in boring locations in the vicinity of the USTs. The thickness of this unit ranges from 2.5 ft to a maximum of 9 ft.
2. *Brown beach and dune sand:* a brown medium-grained beach and dune sand encountered in all borings. This unit extends to the maximum depth penetrated during this investigation (i.e., 26 ft BGS).

The Coastal Deposit sands underlying the SWMU 66 site from the ground surface to a depth of more than 26 ft BGS represent the uppermost water-bearing unit beneath

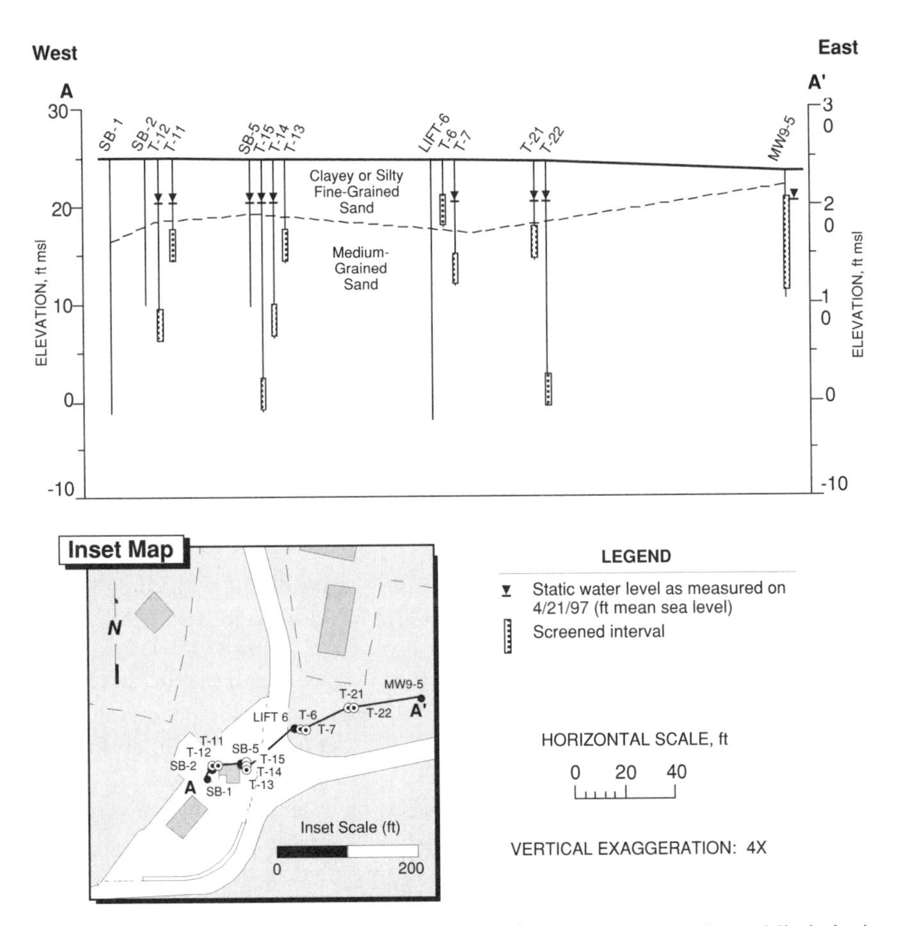

FIGURE 9.19. Hydrogeologic cross section A–A′, Keesler Air Force Base, Mississippi.

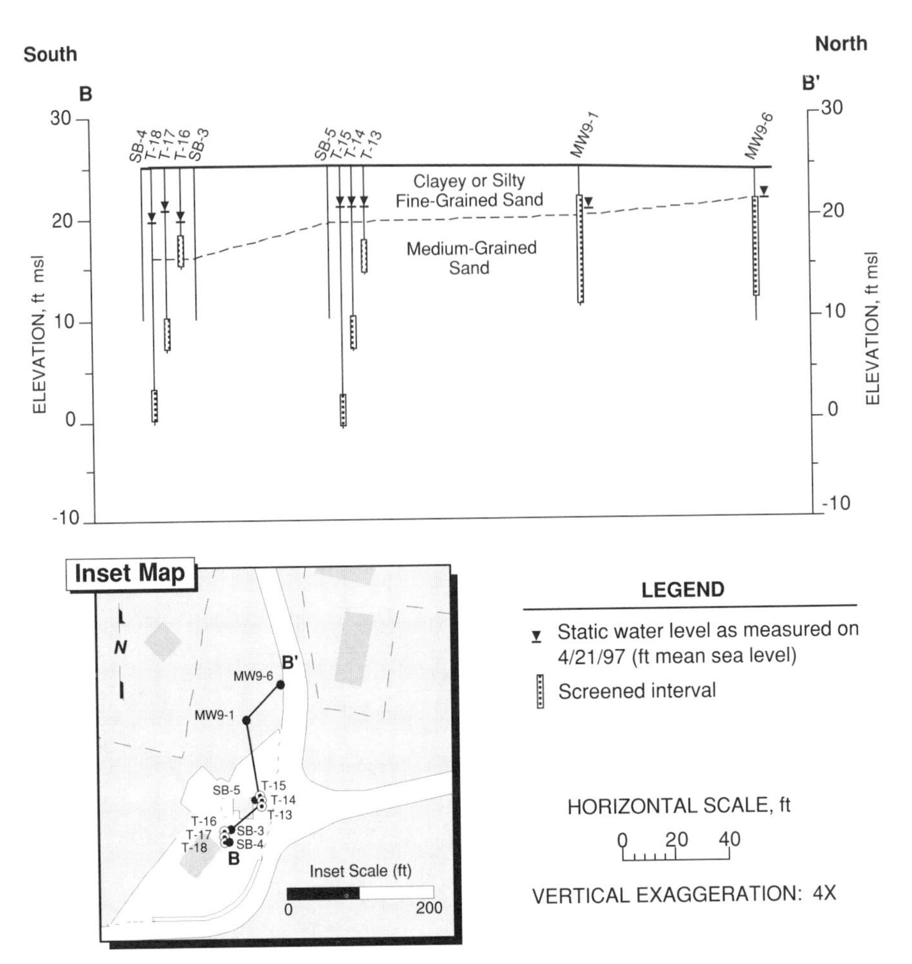

FIGURE 9.20. Hydrogeologic cross section B–B′, Keesler Air Force Base, Mississippi.

this site. The depth to the water table, which varies seasonally and with rainfall events at this site, is relatively shallow. Static water levels measured in April 1995 ranged from 2.63 to 4.36 ft BGS. The groundwater level in the immediate vicinity of the SWMU 66 site has been measured as low as 6.44 ft BGS at well MW9-2 in November 1989, compared to 3.79 ft BGS in April 1995. Fluctuation in groundwater levels on the order of ±2 ft can result from infiltration of water from the ground surface following rainfall.

The potentiometric surface contour map, based on water levels measured in April 1995, indicate that groundwater within the upper water-bearing unit is generally moving in a northern direction at an average lateral flow gradient of 0.003 ft/ft (see Figure 9.18). However, water levels measured in December 1992 indicate that groundwater within the sand unit was generally moving in a east–southeast direction. The variation of the groundwater flow direction is reflected in the dissolved contaminant plume outline, as shown on Figure 9.21. The groundwater appears to be recharged from the surface west of the SWMU 66 site. During the April 1995 field investigation a large pool of standing water was observed in the grassy area immediately west of the site, near piezometer location T-1, which exhibited the highest water level elevation.

To investigate the hydraulic characteristics of the upper water-bearing unit at the site, falling-head and rising-head slug tests were performed at three groundwater monitoring wells screened in the interval 3.0 to 13.0 ft BGS. The results of rising- and falling-head slug test calculations indicate an average hydraulic conductivity of 1.1×10^{-2} cm/s for the water-bearing unit at the site. The hydraulic conductivity at

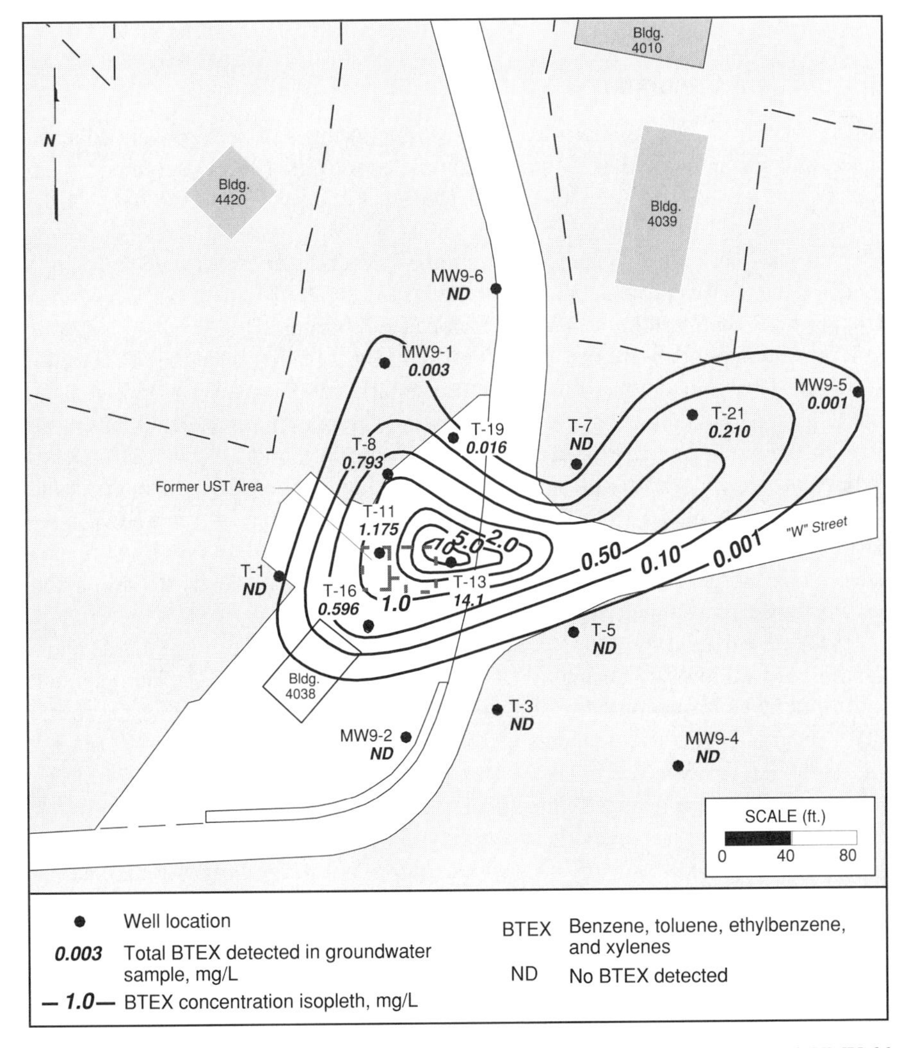

FIGURE 9.21. Concentration isopleth map: BTEX in Coastal Sand groundwater, SWMU 66 site, Keesler Air Force Base, Mississippi.

monitoring wells MW9-1 and MW9-6, which were screened in the beach and dune sand, was calculated to be 1.28×10^{-2} cm/s. Monitoring well MW9-3, which was partially screened in the clayey sand, exhibited a lower hydraulic conductivity of 7.25×10^{-3} cm/s. Based on an average hydraulic conductivity of 1.1×10^{-2} cm/s, a hydraulic gradient of 0.003 ft/ft, and an assumed porosity of 0.30, the calculated seepage velocity within the upper water-bearing zone is 114 ft/yr.

No consistent upward or downward gradients have been observed at the SWMU 66 site. For example, on March 24, 1995, an upward gradient of 0.3 ft was observed at the T-8/T-9/T-10 well cluster, which changed to a 0.09-ft downward gradient by April 21, 1995. Similarly, the T-13/T-14/T-15 well cluster showed a 0.65-ft downward gradient on March 24, 1995, changing to a lower gradient (0.01 ft) on April 21, 1995.

Affected Soil Conditions

Soil samples from five soil borings and three monitoring well borings were collected and tested during the field investigation completed in April 1995. A summary of laboratory analyses of soil samples is presented on Figure 9.22. On the basis of this investigation and previous work conducted at SWMU 66, a 70- by 70-ft area of affected soils has been delineated (see Figure 9.22). The deepest affected soil sample collected during this program (0.339 mg/kg BTEX and 0.03 mg/kg PAHs) is from a depth of 14 ft BGS within the saturated zone at soil boring SB-2.

BTEX compounds were detected in soil samples collected from the fill placed in the former UST excavation, from the northeastern corner of the former UST area, and along the piping leading to the fuel dispensers. The maximum detected concentrations of benzene (24 mg/kg), ethylbenzene (50 mg/kg), and xylenes (240 mg/kg) were measured at a depth of 4 ft BGS in soil boring location SB-3. The maximum concentration of toluene (92 mg/kg) was measured at a depth of 5.5 ft BGS in soil boring location SB-2. In general, the highest hydrocarbon concentrations were measured at the top of the saturated zone (i.e., 4 to 5.5 ft BGS). Soil hydrocarbon concentrations decreased with depth within the saturated zone.

PAHs were detected at four soil boring locations (i.e., SB-2 through SB-5) and at monitoring well MW9-4 during this investigation (see Figure 9.22). The maximum total PAH level was recorded in soil from boring SB-3, located south of the former UST. The principal constituent detected at this location was naphthalene (17 mg/kg). Phenanthrene was detected in one boring, SB-3, at a concentration of 0.074 mg/kg. No other PAHs were detected in soil from borings in the UST area. Concentrations of fluoranthene, chrysene, and benzo[*b*]fluoranthene (total PAHs of 0.167 mg/kg), which were detected in soils from well location MW9-4, are probably unrelated to contamination from the UST site and may be related to construction activities near the well area. Soil testing indicated that PAH concentrations decreased with depth below the top of the saturated zone (i.e., 4.0 to 5.5 ft).

The maximum concentration of petroleum extractables detected in the soil, 1900 mg/kg, was detected at a depth of 5.5 ft BGS at soil boring SB-2 and at a depth of 4.0 ft BGS at well MW9-4. Lead levels measured in the soil samples tested ranged from 11 mg/kg at soil boring SB-1 (5.5 ft BGS) to 26 mg/kg at soil boring SB-3 (4 ft BGS).

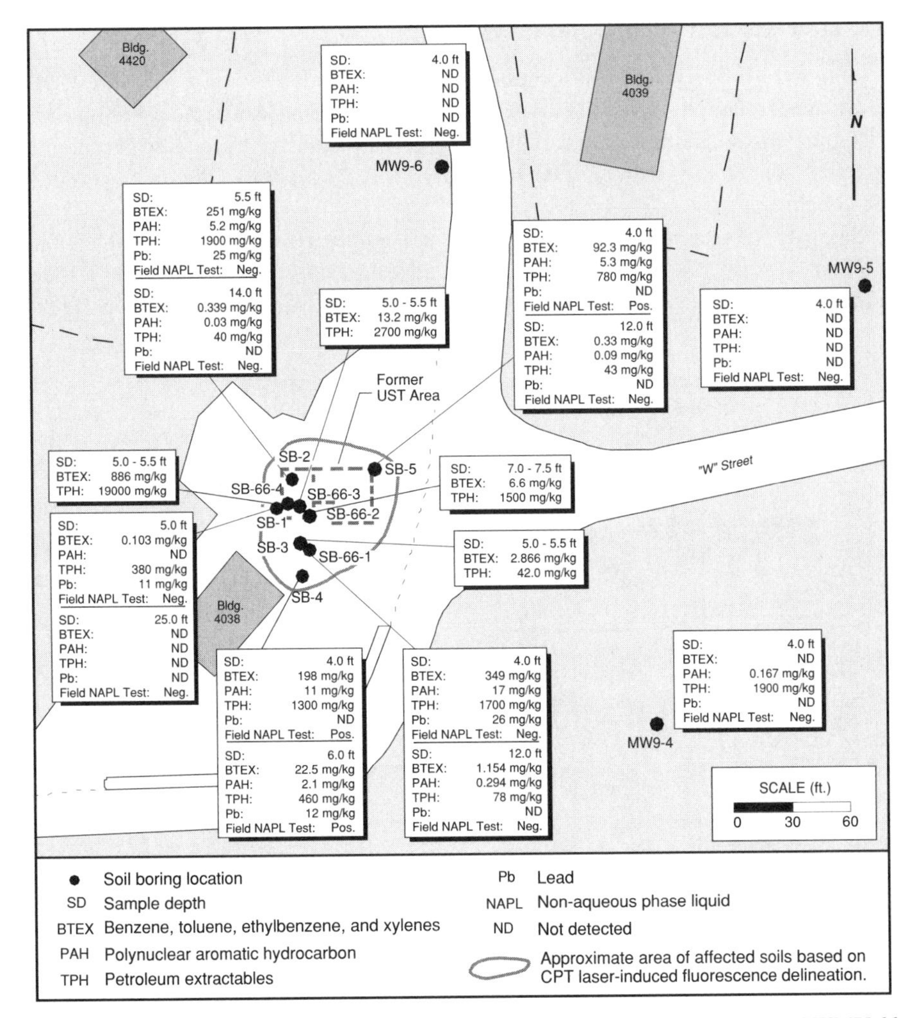

FIGURE 9.22. Results of soil testing: BTEX, PAHs, TPH, lead, and field NAPL; SWMU 66 site, Keesler Air Force Base, Mississippi.

Field testing using shortwave ultraviolet light and Sudan IV dye (a hydrophobic dye) detected the presence of residual (i.e., nonmobile) NAPL at a depth of 4.5 ft BGS at soil boring SB-5 and at a depth of 4.0 to 6.0 ft at soil boring SB-4.

Sixteen cone penetrometer test/laser-induced fluorescence (CPT/LIF) locations were used to supplement the stratigraphic information at the SWMU 66 site as well as to delineate qualitatively areas of apparently affected soils. This delineation was accomplished by identifying stratigraphic intervals having strong overall LIF response (generally, >500 counts). The affected area identified corresponds to the locations of the former USTs and the soils adjacent to the piping leading to dispensing units. The LIF-delineated area generally corresponds to the analytically defined BTEX-affected soil area.

Affected Groundwater Conditions

A total of 20 groundwater test locations were sampled during April and May 1995. Groundwater samples were obtained from five monitoring wells and 15 temporary CPT piezometers. Sampling locations and summary results of groundwater laboratory analyses are shown on Figures 9.21 and 9.23. A discussion of the findings is provided below.

Samples collected from monitoring wells and temporary cone piezometer locations show an area of affected groundwater that measures approximately 41,000 ft^2 and extends approximately 320 ft in length by 140 ft in width (see Figure 9.21). During the

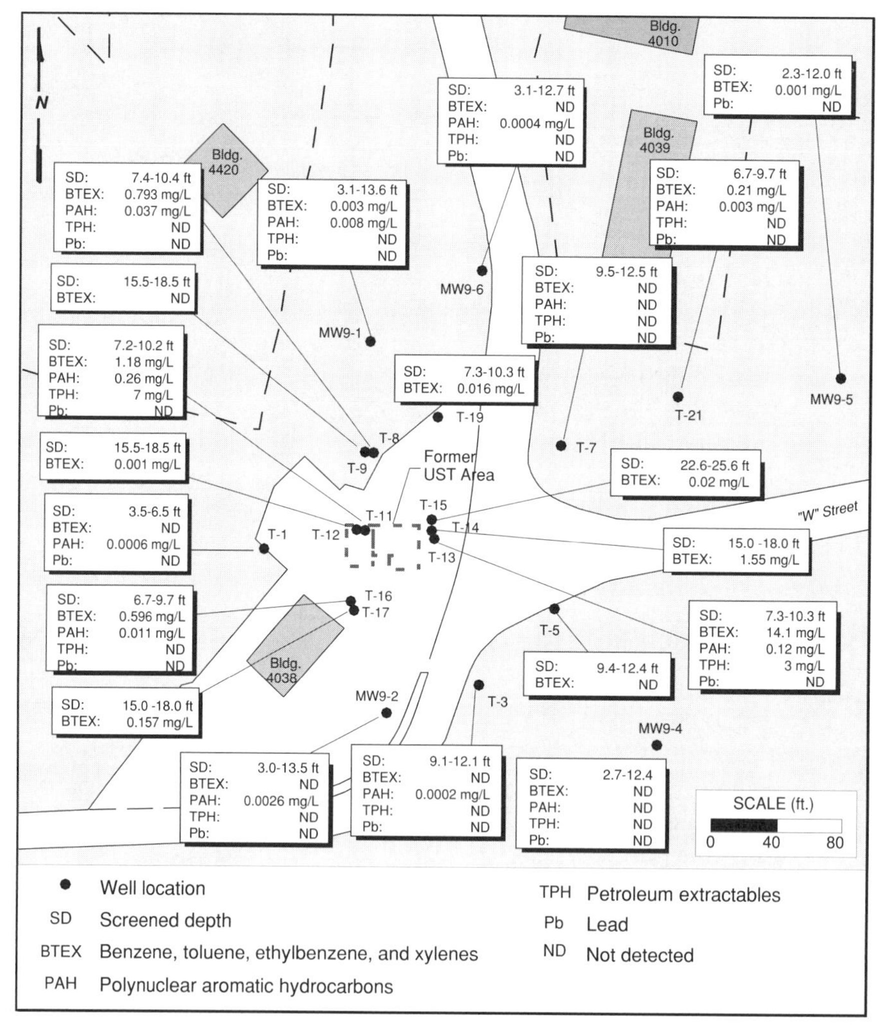

FIGURE 9.23. Results of groundwater testing: BTEX, PAHs, and TPH; SWMU 66 site, Keesler Air Force Base, Mississippi.

investigation of the shallow water-bearing sand zone, no mobile NAPL was detected in any sampling locations. The highest concentration of BTEX compounds (benzene, 3.55 mg/L; ethylbenzene, 0.75 mg/L; toluene, 3.45 mg/L; and xylene, 6.35 mg/L) was reported in groundwater from piezometer T-13 (see Figure 9.23). BTEX concentrations decreased significantly with depth. For example, at the three-well cluster location of T-13/T-14/T-15, BTEX concentration decreased from 14.1 mg/L at the 7.3- to 10.0-ft depth, to 1.55 mg/L at the 15- to 18-ft depth, to 0.02 mg/L at the 22.6- to 25.6-ft depth. Trimethylbenzenes (TMBs) were detected in groundwater from eight monitoring wells, with the maximum concentration (1.7 mg/L of 1,2,3-TMB) recorded at piezometer T-13.

The maximum level of petroleum extractables measured in the groundwater samples was 7 mg/L at piezometer T-11. Total organic carbon levels measured in these wells ranged from 13 to 292 mg/L. PAHs were detected in 10 of 12 groundwater samples analyzed. The only PAH detected, with one exception, was naphthalene, which was measured at a maximum concentration of 0.28 mg/L in piezometer T-13. Trace levels of acenaphthene (i.e., 0.0002 mg/L), the only other PAH constituent detected, were measured at piezometer location T-3. Low levels of naphthalene (ranging from 0.0004 to 0.0026 mg/L) were detected in some outlying locations (i.e., wells MW9-2, MW9-6, and piezometer T-1) where no BTEX was measured in groundwater.

Laboratory results of groundwater testing for lead did not report concentrations greater than the analytical detection limit for this metal at any location. Significant depletion of electron acceptors (oxygen and sulfate) and generation of metabolic byproducts [most notably Fe(II) and methane] indicate active bioremediation is occurring at this site. These data are discussed below.

Estimated Mass of Contaminants in Groundwater and Soil

Groundwater and soil sampling results were used to estimate the distribution of BTEX compounds in the aqueous phase and in the soil at the site. To estimate the total BTEX mass of all the unsaturated and saturated soils, the sampling data were divided into depth intervals, and concentration contours were drawn across the affected soil area. This approach yielded a total mass estimate of 610 kg of BTEX material in the affected soils, most of which (608 kg) was in the saturated soils at the SWMU 66 site.

A separate analysis was performed for the area of affected groundwater. Estimated groundwater concentration contours were used to calculate the total mass of dissolved BTEX at the site. Using this approach, the total dissolved BTEX mass was estimated to be 3.2 kg.

Evidence of Natural Attenuation: Analysis of Monitoring Data

Plume Characteristics. Two methods were used to determine if the groundwater monitoring data indicated active natural attenuation:

1. *Use of organic tracers.* At fuel spill sites, organic tracers, consisting of three TMB compounds that are typically present in petroleum fuels, have the same mobility

characteristics as benzene but do not biodegrade as readily. The extent of migration of these tracers is compared to the extent of benzene migration. If the travel distance and plume width for the TMB constituents is greater than for the BTEX compounds, natural attenuation is indicated.

At the SWMU 66 site, BTEX and TMBs are found in the source area at concentrations in the range 1 to 10 mg/L. At the farthest extent of the BTEX plume (well MW9-5 located 240 ft from the source) only xylene and 1,2,3-TMB are found. These observations support an active ongoing natural attenuation scenario, as both xylene and the TMBs are more recalcitrant to biodegradation than benzene or toluene.

A more detailed evaluation of the MW9-5 data provides further evidence of active in situ biodegradation. The concentration ratios between the source zone well (T-13) and the edge of the plume (MW9-5) indicate that xylene exhibited a larger relative drop in concentration than the less biodegradable 1,2,3-TMB compound:

- Xylene concentrations decreased from 6.8 mg/L to 0.001 mg/L (6800-fold reduction).
- 1,2,3-TMB concentrations decreased from 1.5 mg/L to 0.0014 mg/L (1286-fold reduction).

Therefore, the TMB tracer indicates the presence of ongoing intrinsic remediation of more biodegradable BTEX compounds at the SWMU 66 site.

2. *SWMU 66 site plume over time.* Historical information was available for only four wells. Data collected in 1991 from a temporary well (T66-1) were compared with data from a nearby well (T-11) tested in 1995. These nearly adjacent wells are both located in the northeastern corner of the former UST area. Total BTEX concentrations in this area were 1.505 mg/L in 1991 and 1.175 mg/L in 1995, indicating a slowly shrinking plume.

Available hydrogeologic data from the SWMU 66 site also suggest that groundwater is migrating at a rate of more than 100 ft/yr. Based on this seepage velocity, the dissolved BTEX plume should have migrated at least 600 ft between 1989 (when gasoline product leakage is assumed to have occurred) and 1995. (This calculation ignores longitudinal dispersion, which would increase the length of the plume, and retardation that could decrease the plume length.) However, contaminant data collected in 1995 indicate that the plume extends only 260 ft downgradient of the source area. Because sorption alone cannot account for the discrepancy between the predicted and actual distance of contaminant plume migration, these data indicate that natural biodegradation has removed dissolved BTEX from the site.

Electron Acceptor Consumption / By-product Generation

Geochemical data indicate that intense aerobic and anaerobic biodegradation is occurring in the dissolved contaminant plume at the SWMU 66 site. As with most natural attenuation sites, areas with high total BTEX concentrations show depleted dissolved oxygen concentrations. Dissolved oxygen concentrations fall from background levels of over 2.0 mg/L outside the plume area to approximately 0.4 mg/L inside the plume (see Figure 9.24). Aerobic degradation plays an important role in subsurface attenuation at the SWMU 66 site.

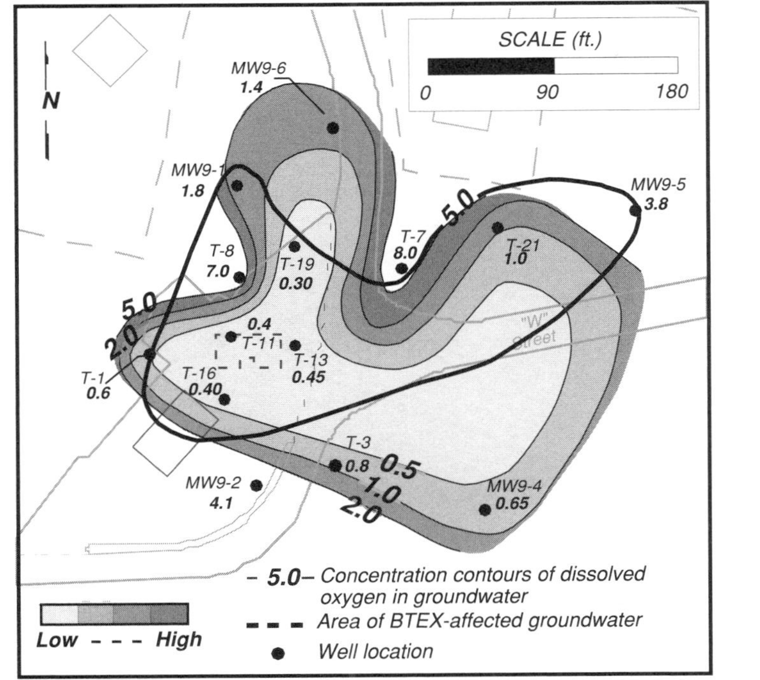

Dissolved oxygen in groundwater (mg/L)

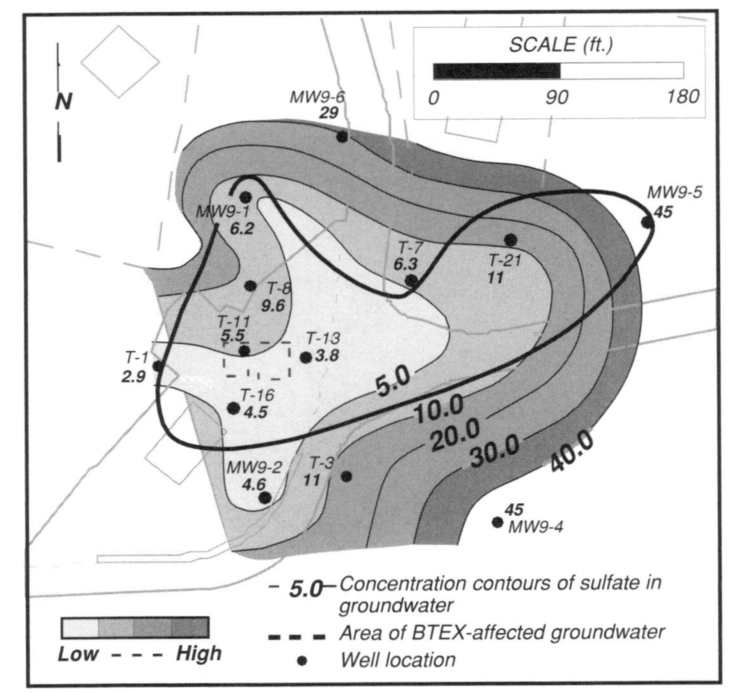

Sulfate in groundwater (mg/L)

FIGURE 9.24. Distribution of electron acceptors in groundwater, April 1995.

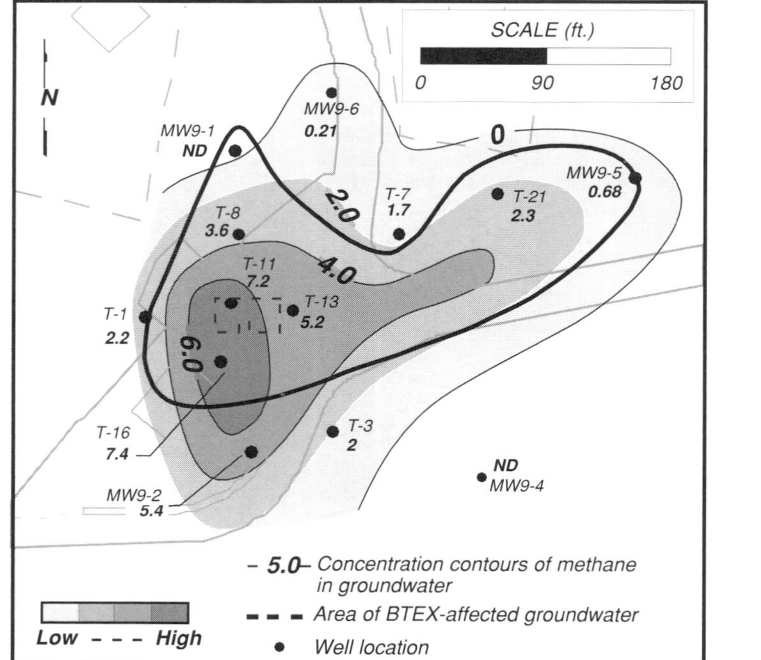

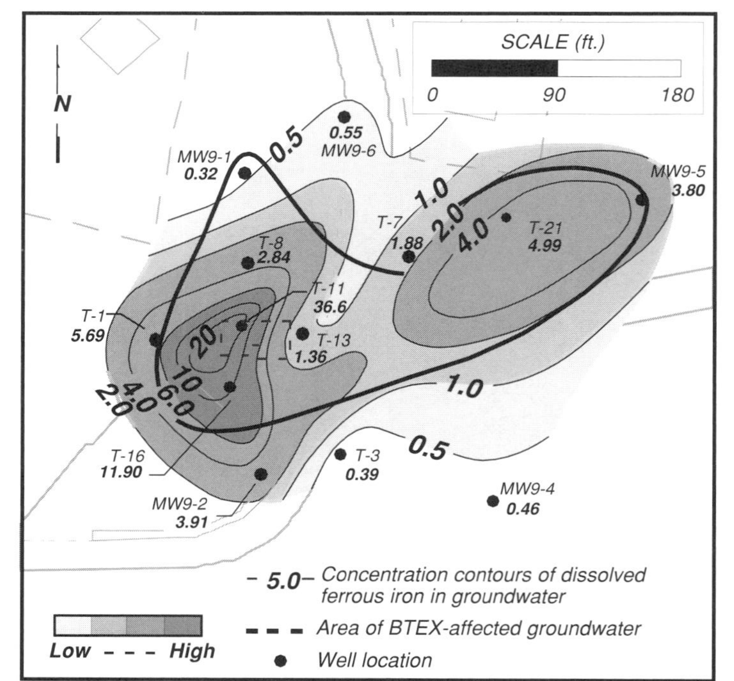

FIGURE 9.25. Metabolic by-products in groundwater, April 1995.

Sulfate data show considerable depletion of this electron acceptor in the center of the plume, with background levels of 26.2 mg/L falling to 3.8 mg/L inside the plume (see Figure 9.24). This decrease in sulfate concentration is a strong indication that anaerobic biodegradation of the BTEX compounds is occurring throughout the site. The by-product data for Fe(II) and methane (shown on Figure 9.25) show strong correlations with the dissolved plume. The methane reading of 7.4 mg/L at well T-16 (compared to a background level of <0.12 mg/L) indicates that methanogens are active at this site. Measured Fe(II) concentrations of 36.6 mg/L at T-11 indicate that Fe(III) is being reduced to Fe(II) during biodegradation of BTEX compounds.

Based on the monitoring data, the groundwater contains significant biodegradation capacity, defined as the mass of contaminant that can be removed by 1 L of groundwater moving from a clean background zone through the plume area. The calculated biodegradation capacity at the SWMU 66 site is about 16.7 mg/L, with most of this capacity associated with sulfate reduction and methanogenesis.

9.3 NATURAL ATTENUATION OF FUEL HYDROCARBONS: PETROLEUM, OIL, AND LUBRICANT FACILITY, SITE ST41, ELMENDORF AIR FORCE BASE, ALASKA

MATT SWANSON
PARSONS ENGINEERING SCIENCE, INC.
DENVER, COLORADO

Facility Background

Elmendorf AFB is located in south-central Alaska, immediately north of the city of Anchorage. Site ST41 is located at the crest of the Elmendorf Moraine, a low east-to-west-trending ridge in the southwestern portion of the base. Figure 9.26 shows the site and its features. In the early 1940s, four 1,000,000-gallon AVGAS USTs were installed at this site. The USTs are known to have leaked, and numerous aboveground spills also have occurred at this site. Spills suspected or known to have occurred include a 60,000-gallon AVGAS spill in the 1960s and a 33,000-gallon spill in 1964. Several hundred thousand gallons of JP-4 jet fuel were reportedly spilled in this area between 1975 and 1984 (Engineering-Science, Inc., 1983; Black and Veatch, 1990).

In the 1970s and early 1980s, a concrete dam and an oil–water separator were installed in an effort to intercept fuel discharging from seeps on the hillside immediately south of the USTs. In late 1990, testing indicated that all four USTs were leaking, and all the tanks and piping were emptied and taken out of service in early 1991. During site investigation and remediation work, two wood stave condensate pipes were uncovered. These pipes allowed condensed fuel and water to drain away from tanks 601 and 604 and discharge at the ground surface. These pipes may have been connected to sump pumps that were used to remove water from the bottoms of the tanks (Jacobs Engineering Group, 1994).

A body of mobile LNAPL consisting of petroleum hydrocarbons is present in the vicinity of the former USTs. The full extent of suspected mobile LNAPL contamination

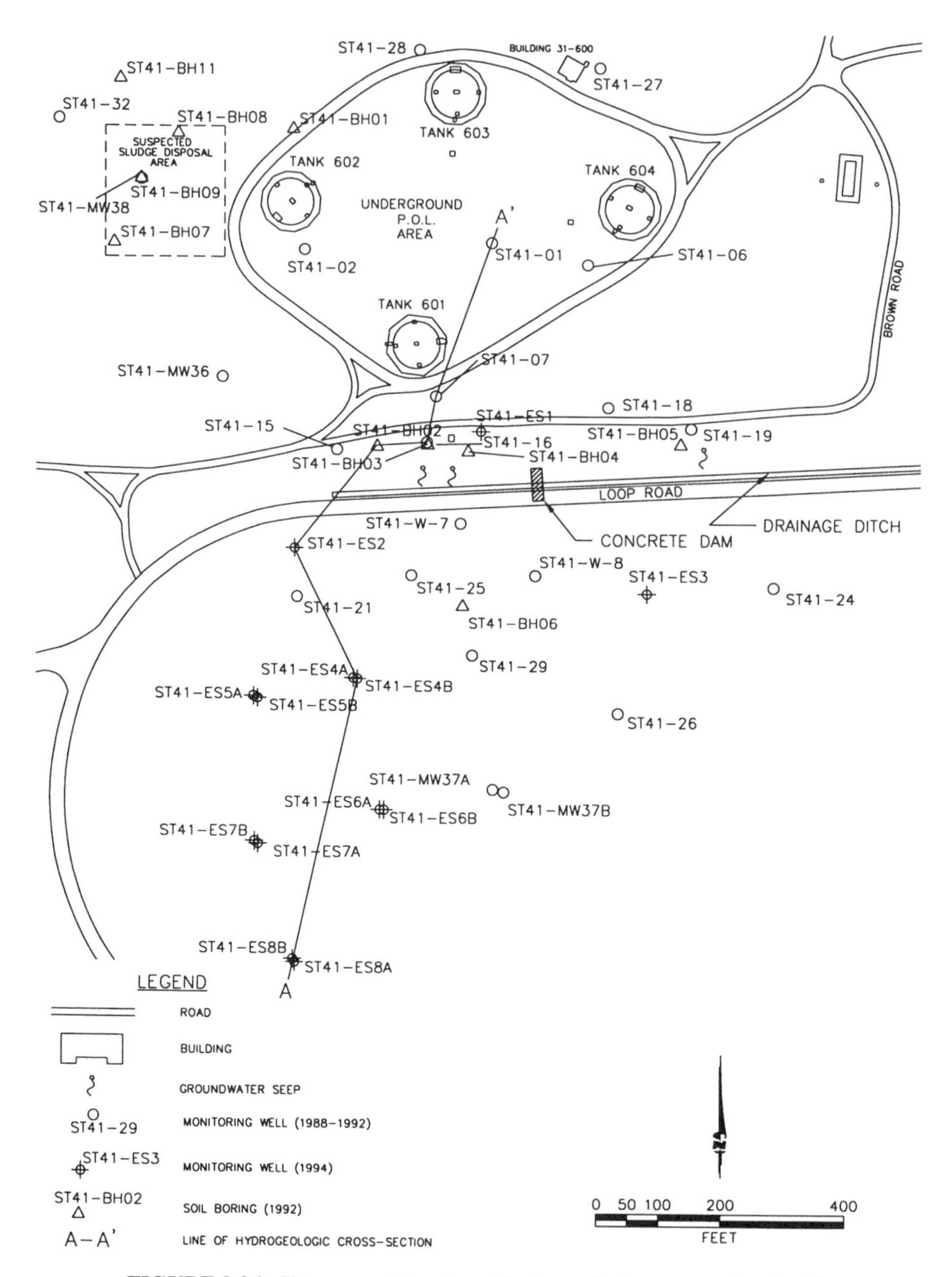

FIGURE 9.26. Site map and location of hydrogeologic cross section A–A′.

cannot be determined accurately from the available measurements. However, because LNAPL is present in only one well (immediately south of tank 601), the extent of the LNAPL body appears limited. A plume of dissolved fuel hydrocarbons extends south from the vicinity of tank 601. This plume consists largely of BTEX.

In October 1993, two groundwater extraction wells and three groundwater extraction trenches were installed in the area on either side of the groundwater divide as an interim remedial action (IRA). The IRA system was constructed to intercept mobile LNAPL and contaminated water. A limited amount of product and contaminated groundwater was extracted, and at the time of sampling for this work (June 1994), the system was not in operation.

As part of the Air Force Center for Environmental Excellence's ongoing natural attenuation initiative, this work was undertaken to collect data on natural attenuation of fuel hydrocarbons and to evaluate the potential use of remediation by natural attenuation (RNA) as a component of overall site remediation plans for site ST41. The results of this work were also used in support of a record of decision (ROD) for the site. This ROD included natural attenuation as part of the recommended alternative for the operable unit containing ST41.

Site Geology and Hydrogeology

Site ST41 is underlain by glacial till of the Elmendorf Moraine. The moraine consists of unconsolidated deposits of silt, clay, and sand, with occasional intervals containing up to 15% coarse sand and gravel. As is typical of glacial till, these deposits are massive (nonstratified), poorly sorted, and heterogeneous. Near the center of the hill, the till is directly underlain by a clayey unit known as the Bootlegger Cove Formation. At the flanks of the hill, the till is not present, and the Bootlegger Cove Formation is overlain by a silty sand to sandy silt unit, known as *cover sand*. This unit is described as well sorted, possibly of alluvial origin, and older than the till but younger than the Bootlegger Cove Formation (Jacobs Engineering Group, 1994). Soils south of the moraine flank largely consist of sand and silty sand, covered by a thin layer of topsoil. Boreholes south of the moraine reach the top of the Bootlegger Cove Formation, which is a dense blue-gray clay, silty clay, or clayey silt unit encountered at depths ranging from 8 to 22.5 ft BGS. The upper surface of the unit generally slopes to the south, away from the moraine. These stratigraphic relationships are illustrated on cross section A–A′ in Figure 9.27.

Depth to groundwater in the vicinity or site ST41 ranges from 1 to 35 ft BGS. In general, depths are greatest along the moraine crest and shallowest along the limbs. A groundwater divide is present in the vicinity of site ST41, running northeast to southwest along the crest of the moraine and between the USTs. Groundwater flow, generally following the local topography, is to the northwest on the northern limb of the moraine and to the south or southeast on the southern limb of the moraine. Basewide groundwater elevation data indicate that away from the moraine, the flow trends to the south and eventually southwest. Groundwater flow patterns in the area of tank 601 are shown on Figure 9.28. Flow in this area is generally to the south and

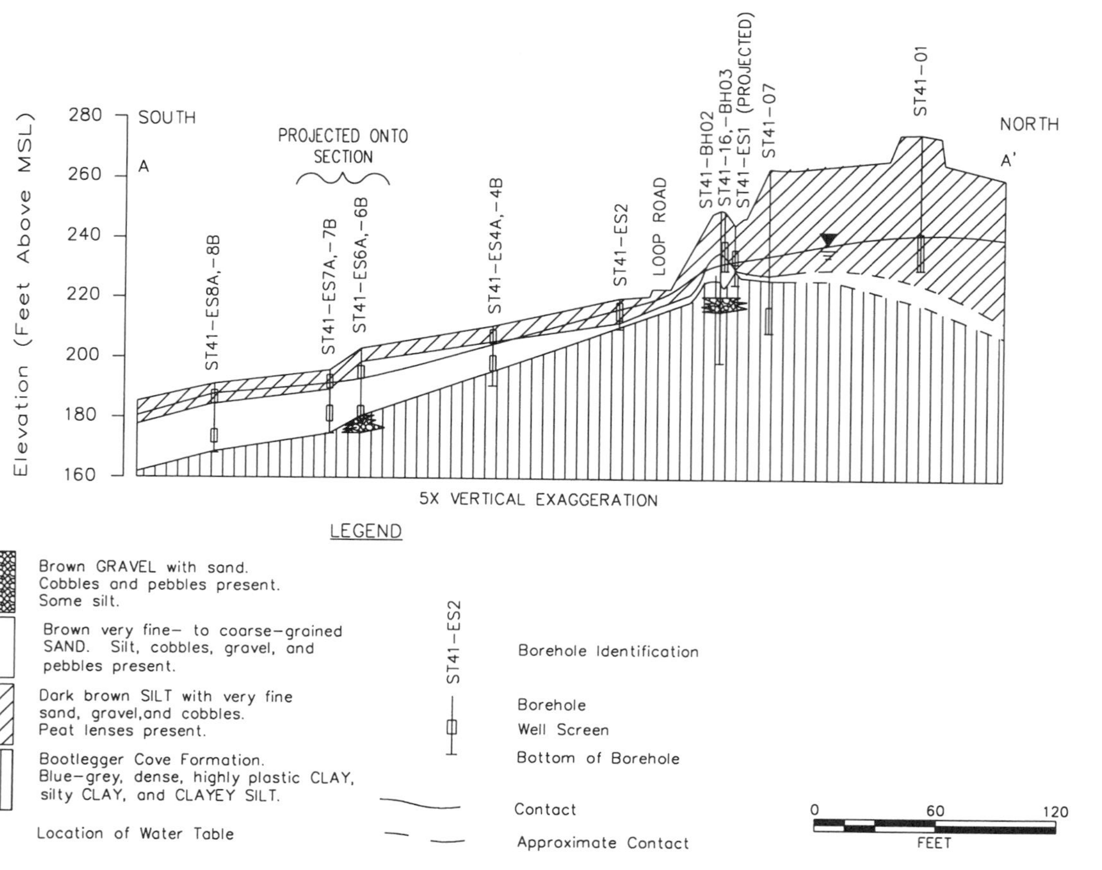

FIGURE 9.27. Hydrogeologic cross section A–A′.

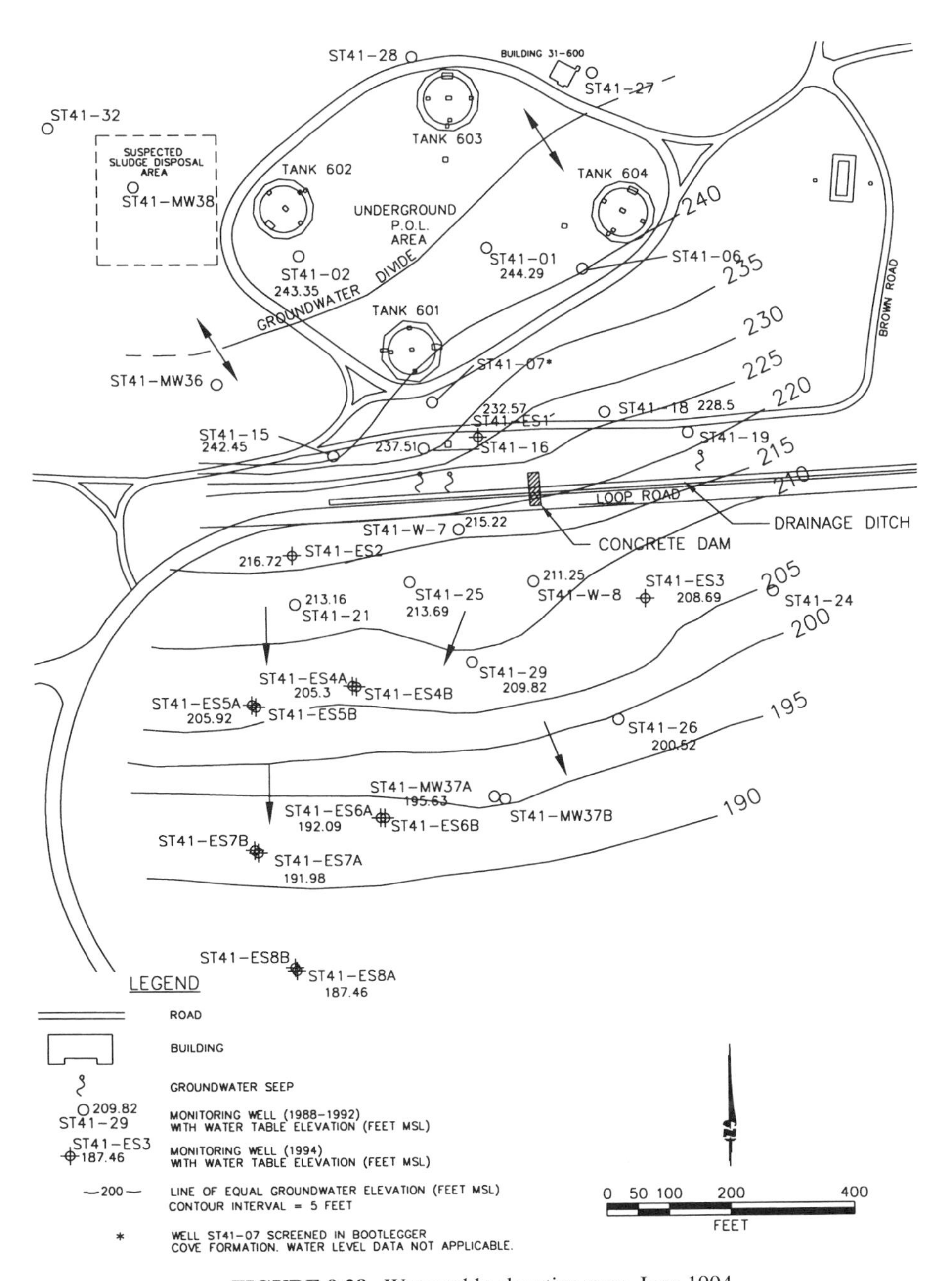

FIGURE 9.28. Water table elevation map, June 1994.

TABLE 9.4. Retardation Factors and Contaminant Velocities

Compound	Average Site-Specific Retardation Factor[a]	Transport Velocity (ft/day)
Benzene	1.9	0.38
Toluene	3.1	0.23
Ethylbenzene	6.2	0.12
Xylenes	5.5	0.13

[a] Calculated using an average f_{oc} value of 0.0024.

south–southeast. In the flatter area south of the moraine, the average horizontal flow gradient is about 0.06 ft/ft. On the moraine, the gradient ranges from 0.05 to 0.16 ft/ft. Flow patterns are generally consistent through the year, with slight seasonal variations in water levels.

Hydraulic conductivity of the saturated units at the site ranges from about 1×10^{-5} to 6×10^{-3} ft/min. The average hydraulic conductivity for the saturated zone at ST41 is about 2.9×10^{-3} ft/min. On the basis of site data and literature values, effective porosity was assumed to be 0.35. Using these data yields an average groundwater seepage velocity for the unconfined sand aquifer of 0.72 ft/day, or about 260 ft/yr.

Background TOC concentrations in soil at the site are high enough that the transport velocities of organic contaminants will be retarded relative to the seepage velocity of groundwater. The TOC content of the aquifer material at this site ranges from 0.071 to 25.9%. Site-specific retardation factors for BTEX dissolved in site groundwater were estimated using the TOC data and used to calculate contaminant transport velocities (Table 9.4).

Groundwater and LNAPL Chemistry

Contaminants. LNAPL at the site appears to be a weathered mixture of AVGAS and JP-4 jet fuel. Mobile LNAPL was detected in only one well (ST41-16, with 0.67 ft of LNAPL), so the extent of mobile LNAPL appears to be limited. Residual LNAPL is also present in site soils, particularly those near tank 601 and on the hillside immediately south of the tank. Groundwater beneath and downgradient from the LNAPL is contaminated with dissolved BTEX and other fuel-related compounds. The dissolved contaminant plume is about 700 ft long and 500 ft wide (Figure 9.29) and extends south from tank 601, consistent with the groundwater flow pattern indicated in Figure 9.28. Where detected, total BTEX concentrations range from 1.1 to 43,280 μg/L. The maximum observed concentration of 43,280 μg/L was detected at monitoring well ST41-16. This concentration may be unrealistically high. Using mass-fraction data from analysis of an LNAPL sample from ST41-16 together with the fuel–water partitioning model of Bruce et al. (1991), the

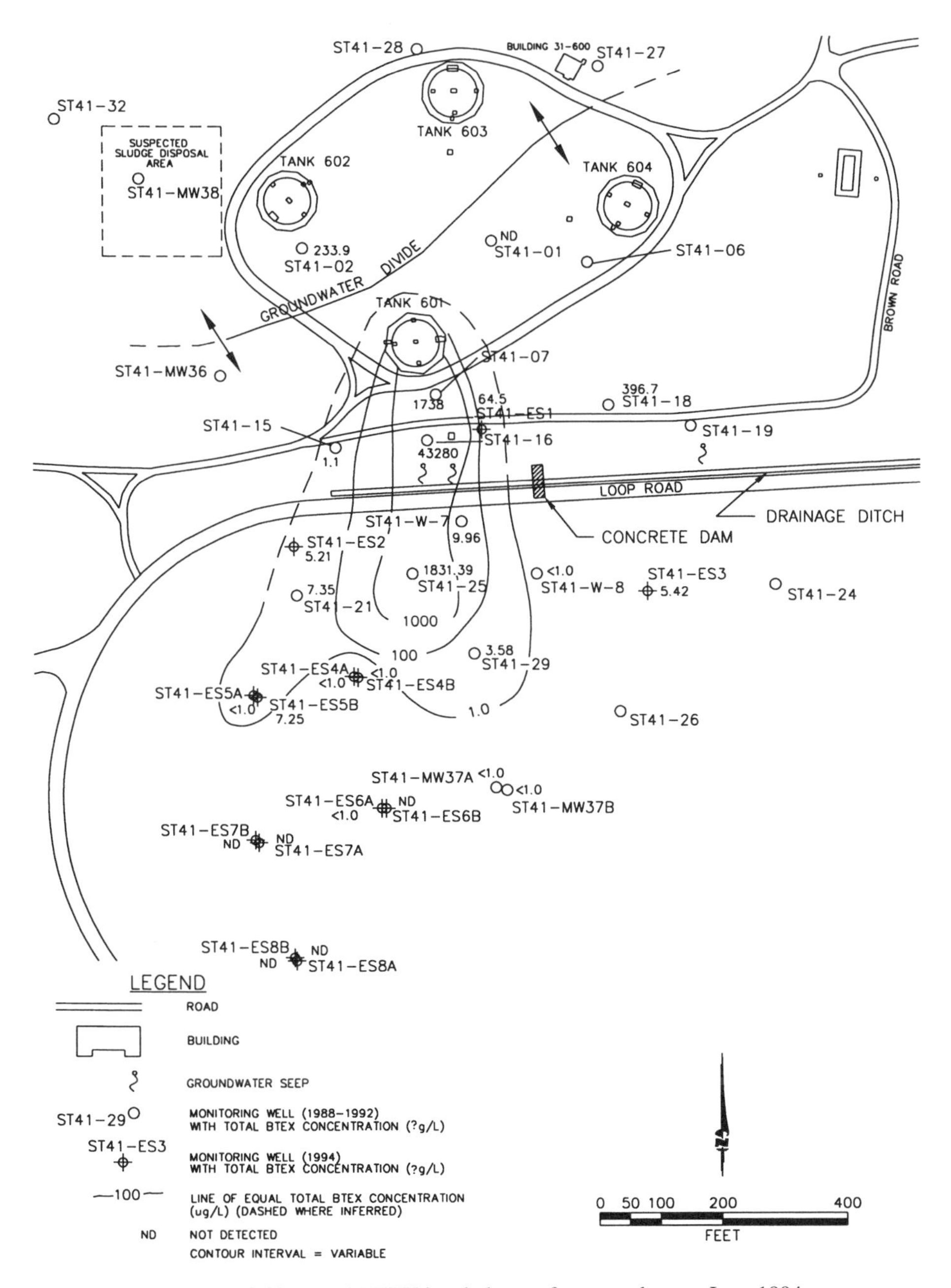

FIGURE 9.29. Total BTEX isopleth map for groundwater, June 1994.

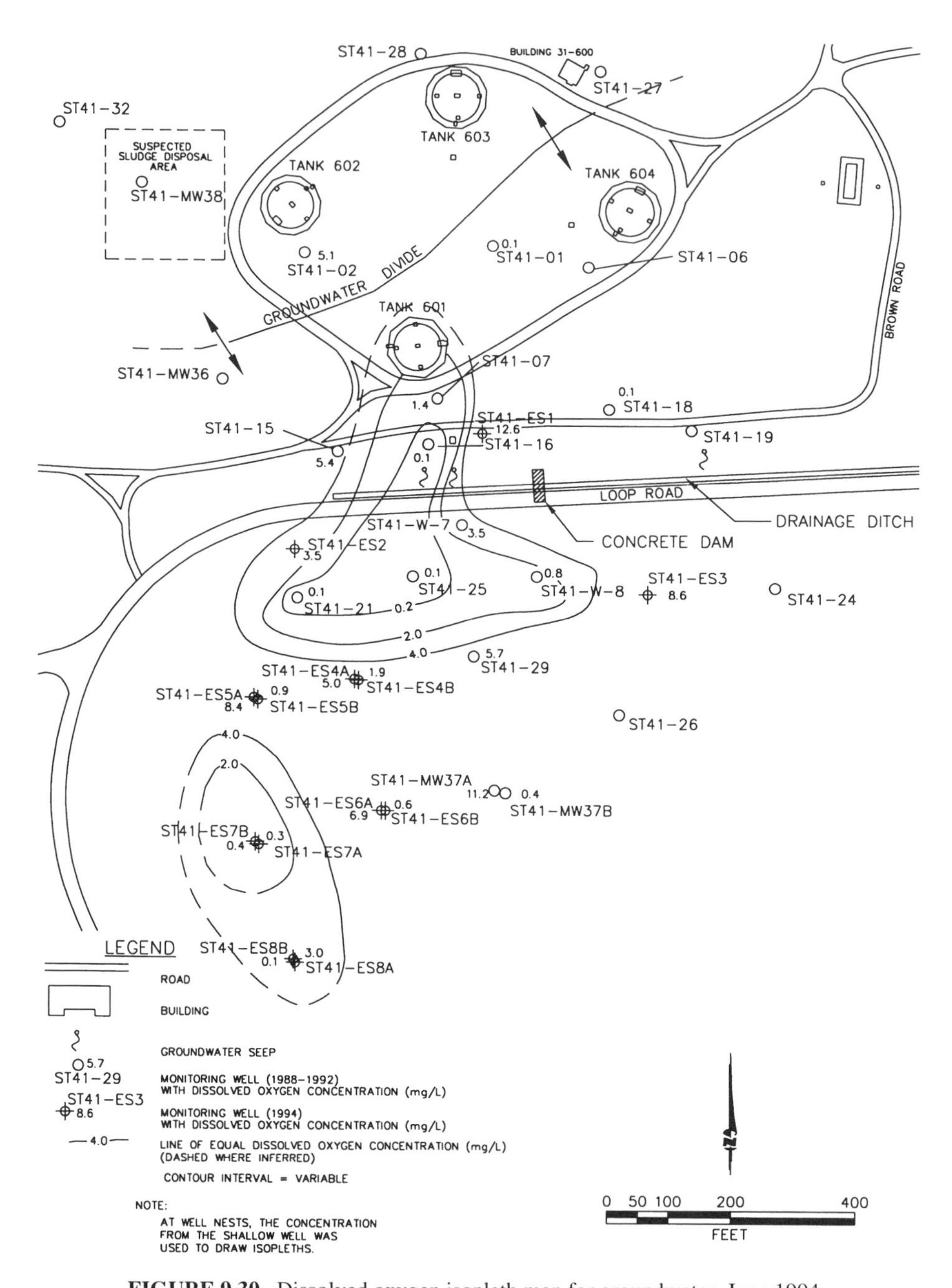

FIGURE 9.30. Dissolved oxygen isopleth map for groundwater, June 1994.

maximum dissolved benzene, toluene, ethylbenzene, and total xylenes concentrations expected in groundwater would be approximately 30,600 μg/L.

At the time of sampling in 1994, historical data on the BTEX plume at ST41 were not available. However, the observed extent of BTEX contamination suggests that natural attenuation mechanisms are limiting plume migration. Given that releases of petroleum hydrocarbons have occurred in the site vicinity since the early 1960s (and probably as early as the mid-1940s), and given the average seepage velocity of groundwater (261 ft/yr), it is reasonable to expect that the plume would be much longer than was observed in 1994. To limit plume movement to the extent observed, biodegradation would be necessary to remove BTEX mass from the groundwater.

Geochemical Parameters. Data for electron acceptors and biodegradation by-products further confirm that biodegradation is ongoing. At ST41, there is a correlation between elevated BTEX concentrations and depleted dissolved oxygen concentrations (Figure 9.30). Background dissolved oxygen concentrations are as high as 12.6 mg/L, with the solubility of dissolved oxygen enhanced by low groundwater temperatures, which range from 4.1 to 11.7°C. Within the plume, dissolved oxygen concentrations are typically less than 1.0 mg/L. Aerobic biodegradation of fuel hydrocarbons is thus likely to consume a significant mass of hydrocarbons, particularly at the margins of the plume.

Geochemical evidence of anaerobic biodegradation of fuel hydrocarbons is also available. Nitrate concentrations are also below 1 mg/L in the vicinity of the BTEX plume, while background concentrations are greater than 10 mg/L (Figure 9.31). Iron(II) was detected in groundwater contaminated with BTEX, with concentrations within the plume ranging from 0.7 to 40.5 mg/L (Figure 9.32). Outside the plume, iron(II) was not detected at a detection limit of 0.05 mg/L. Sulfate concentrations suggest that sulfate is also used as an alternative electron acceptor for fuel biodegradation (Figure 9.33). Outside the area contaminated by BTEX, sulfate concentrations in site groundwater range from 15 to 59 mg/L, while within the plume, sulfate concentrations range from less than 0.05 mg/L to about 10 mg/L. Finally, methane is present within the boundaries of the BTEX plume (Figure 9.34), indicating that methanogenesis is occurring. Methane concentrations within the plume range from 0.044 to 1.48 mg/L, and outside the plume methane concentrations range from less than 0.001 to 0.006 mg/L.

The evidence for use of alternative electron acceptors during BTEX biodegradation at Site ST41 is supported by the distribution of ORPs in groundwater (Figure 9.35). Background ORPs are over 100 mV, while in the center of the plume the ORPs are in the range –53 to about 15 mV. Further evidence of BTEX biodegradation is provided by volatile fatty acids analyses performed by EPA researchers on samples from ST41-16 and ST41-25, which are in the center of the BTEX plume. The test is a GC/MS method wherein the samples are compared to a standard mixture containing 13 phenols, 25 aliphatic acids, and 19 aromatic acids. Compounds in

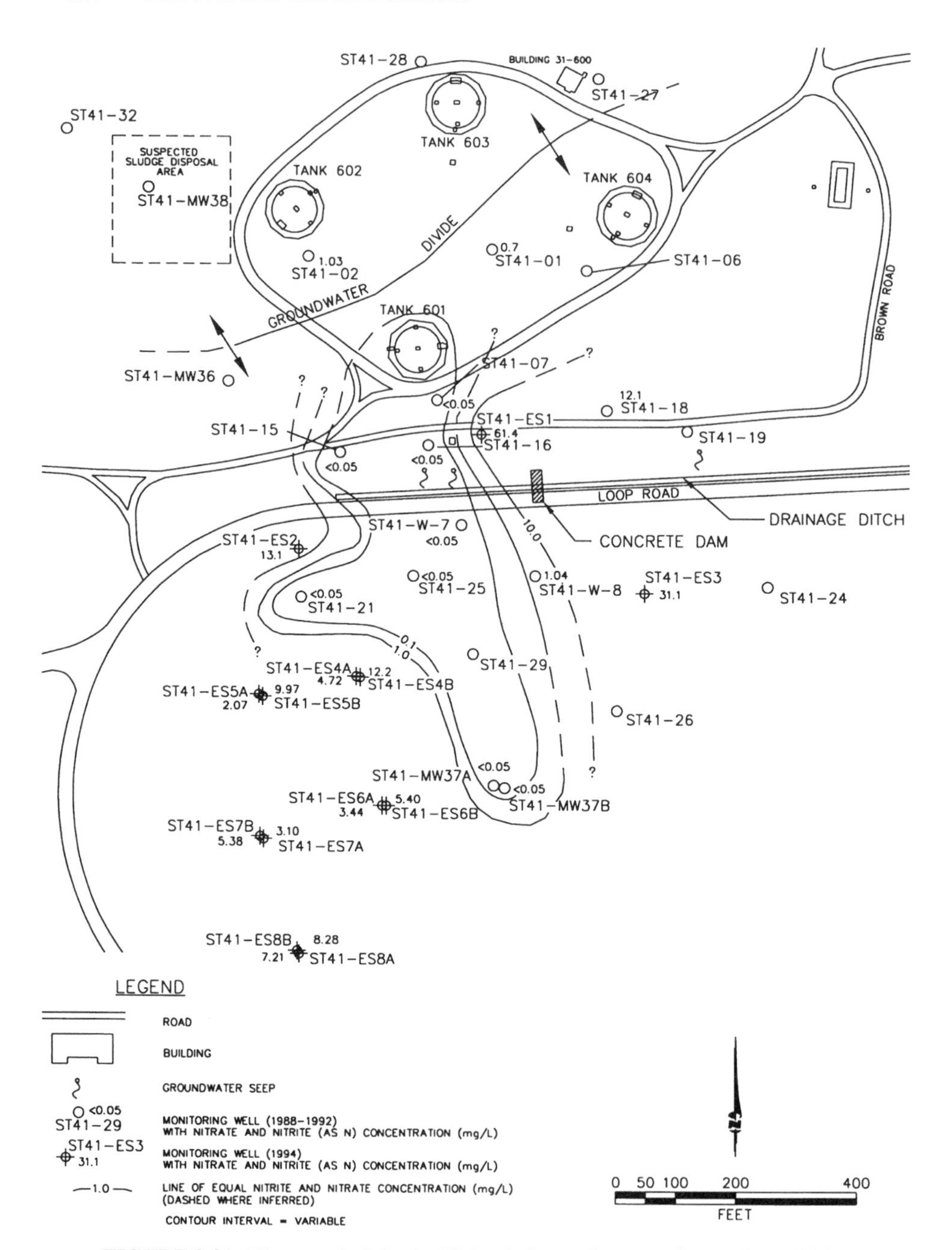

FIGURE 9.31. Nitrate and nitrite (as N) isopleth map for groundwater, June 1994.

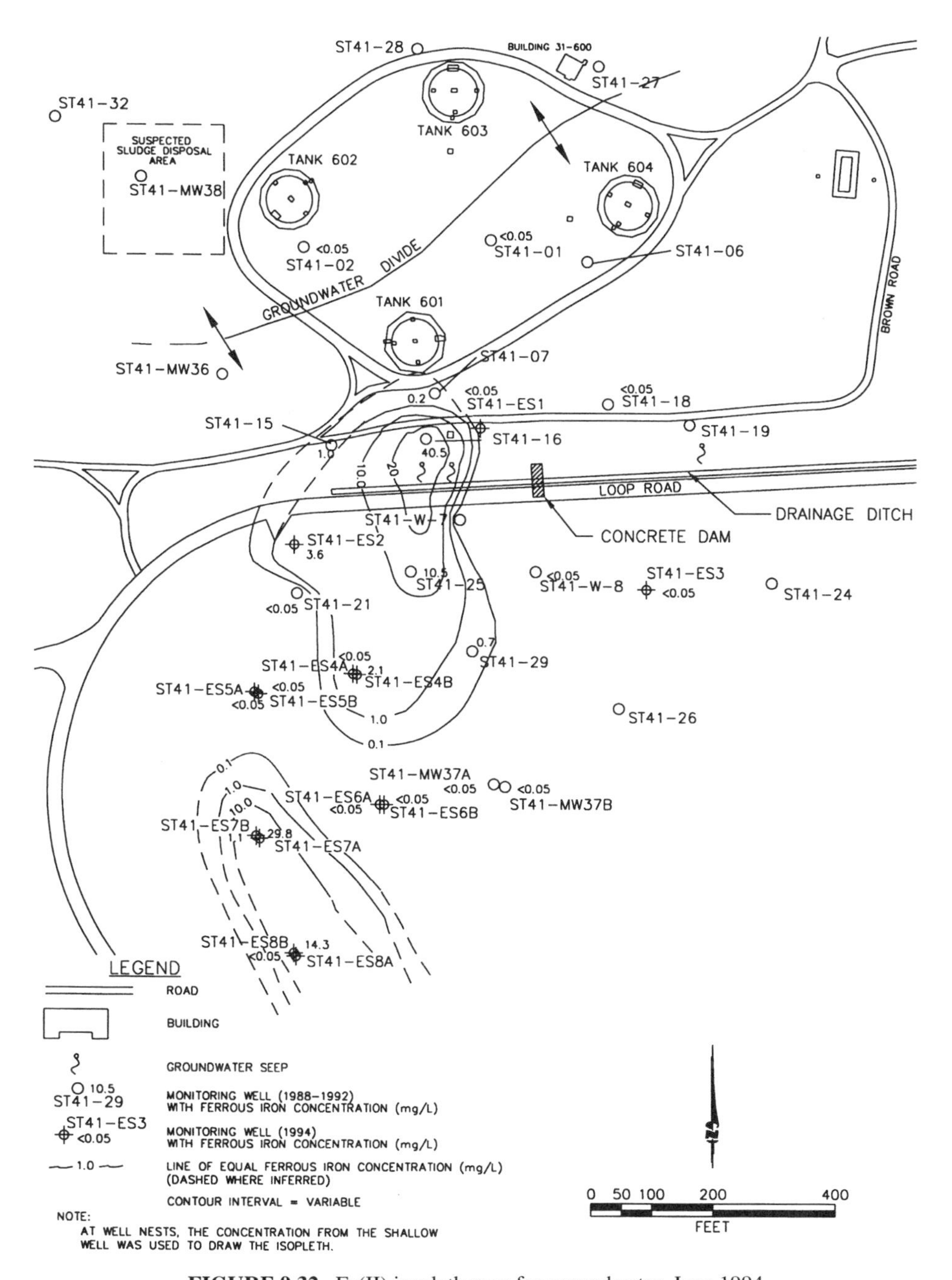

FIGURE 9.32. Fe(II) isopleth map for groundwater, June 1994.

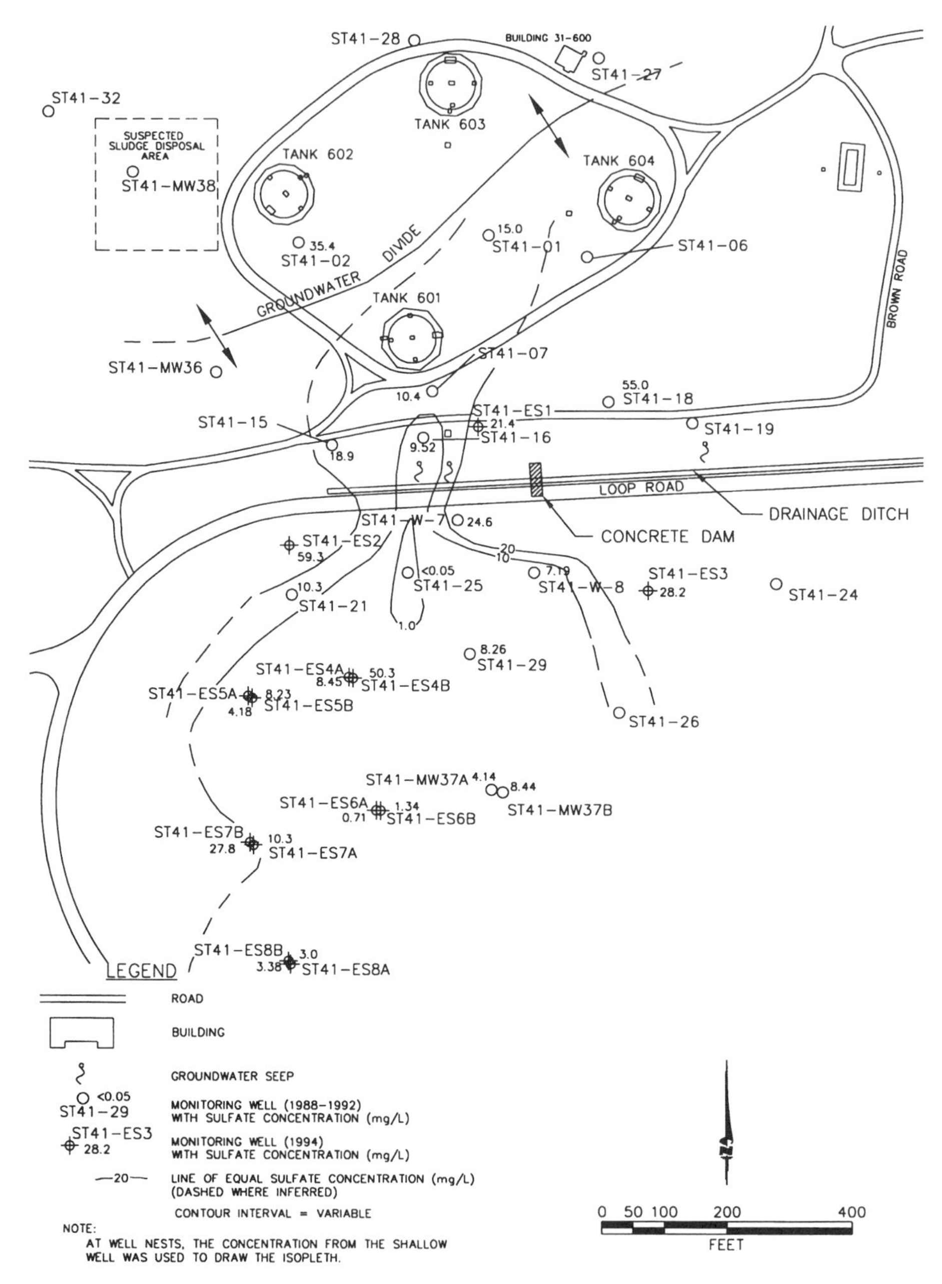

FIGURE 9.33. Sulfate isopleth map for groundwater, June 1994.

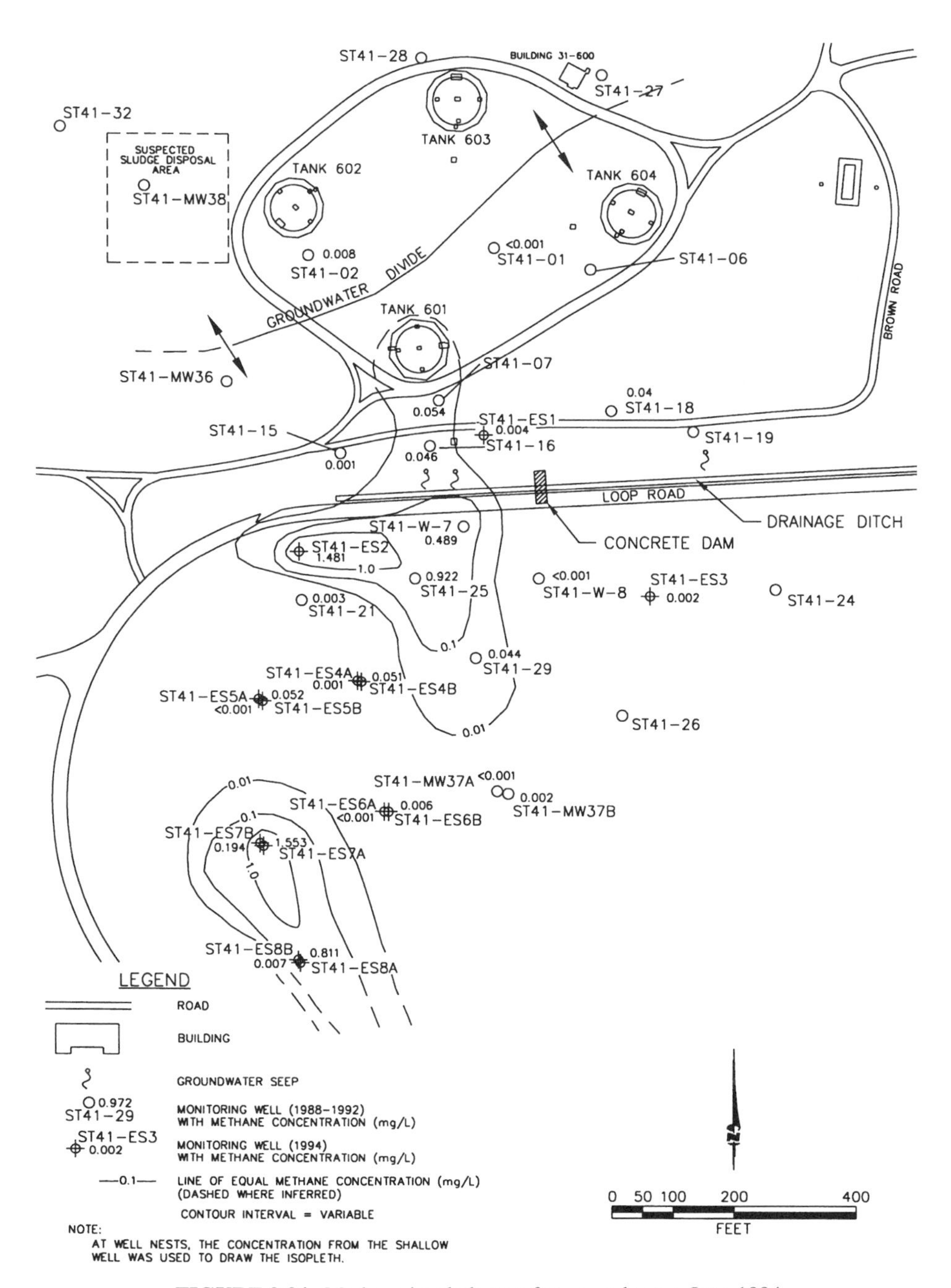

FIGURE 9.34. Methane isopleth map for groundwater, June 1994.

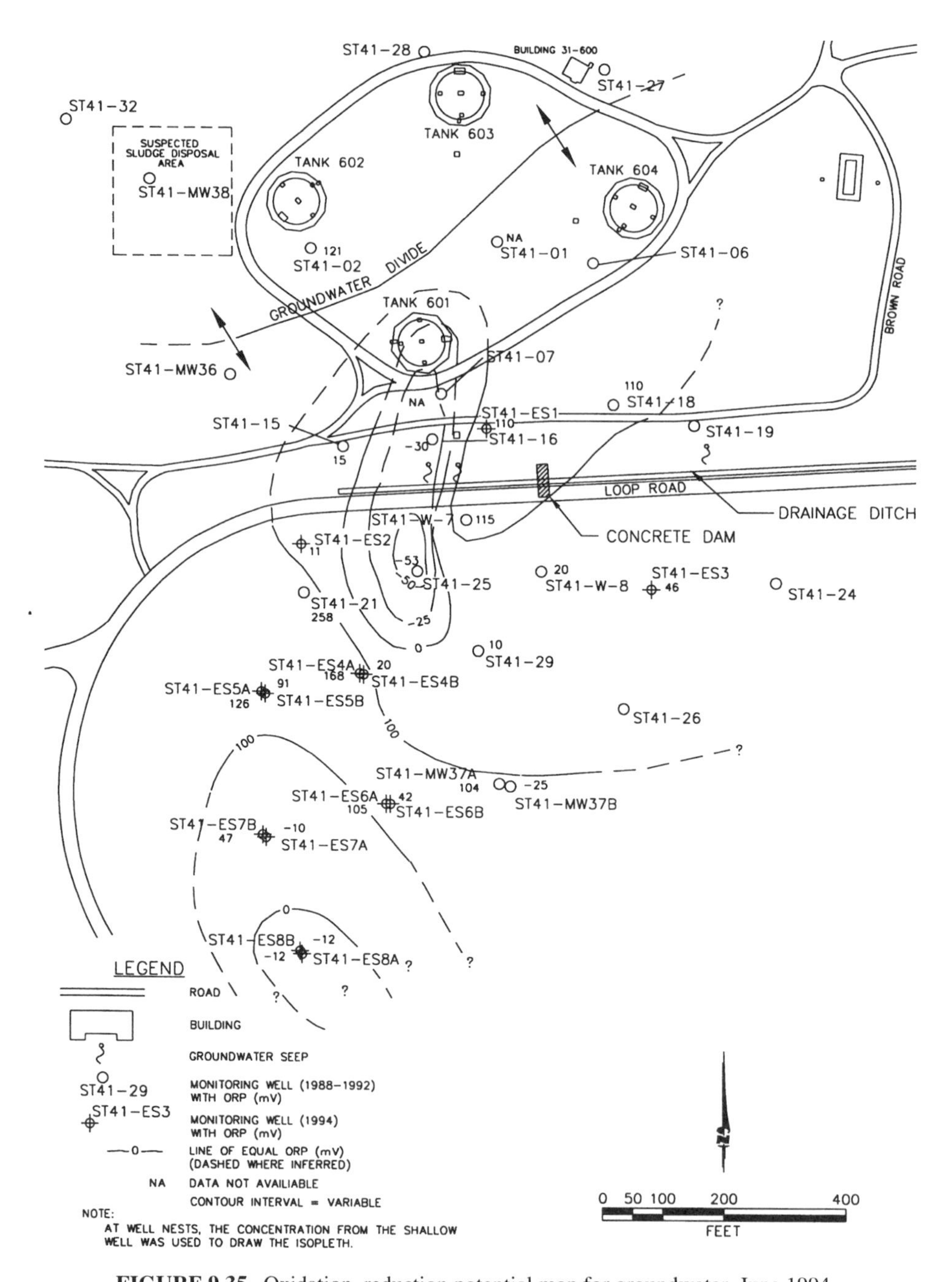

FIGURE 9.35. Oxidation–reduction potential map for groundwater, June 1994.

the standard mixture are generally associated with microbial processes that break down petroleum hydrocarbons. The EPA researchers reported that the sample from ST41-16 contained 43 of the 57 compounds in the standard mixture, and that the chromatogram for the sample from ST41-25 was similar. In both samples, benzoic acid was the compound detected in the greatest concentrations detected.

Biodegradation Rate Constant Calculations

A field-scale first-order biodegradation rate constant for the anaerobic portion of the BTEX plume was estimated using the method presented by Wiedemeier et al. (1996). This method uses TMB isomer as a tracer to calculate an anaerobic biodegradation rate along a flow path. Calculated rate constants for site ST41 range from 0.003 to 0.03 day^{-1}. One reason that rate estimates were needed was that a groundwater flow and contaminant transport model (using Bioplume II) was to be used to evaluate the plume fate. During model calibration, the rate was varied within the range calculated, with the model ultimately incorporating a value of 0.004 day^{-1}.

Groundwater Flow and Contaminant Transport Model

To evaluate the fate of the BTEX plume and to assess the effects of differing remedial strategies, a numerical groundwater flow and solute transport model was developed for the site using BIOPLUME II. BIOPLUME II was used to simulate the transport of dissolved BTEX under the influence of advection, dispersion, sorption, and biodegradation. Model input parameters were obtained from field data where possible; parameters not measured at the site were estimated using reasonable literature values. A 20- by 20-cell grid was used for the BIOPLUME II model; Figure 9.36 shows the grid and the calibrated water table map. To calibrate the transport model, BTEX was introduced to the model in four source-area cells (Figure 9.37), and the model was run for 30 years to reproduce the 1994 BTEX concentrations. The 30-year time frame was selected because the earliest recorded spills occurred in the mid-1960s. Figure 9.37 shows the calibrated BTEX plume and the location of the source cells.

After calibration and sensitivity analyses, the model was used to evaluate plume fate for three scenarios. The first scenario was a worst-case scenario which assumed that the mobile and residual LNAPL would not weather and would provide a continuous and undiminishing source of BTEX. The second scenario assumed that the IRA system would operate for five years and reduce the BTEX mass entering groundwater by 70%, and the remaining mass would weather at a rate of about 8% per year (0.0002 day^{-1}). Finally, the third scenario assumed that the source LNAPL would only be affected by the weathering rate of 8% per year. The weathering rate selected was chosen to be about one-half the source weathering rate observed

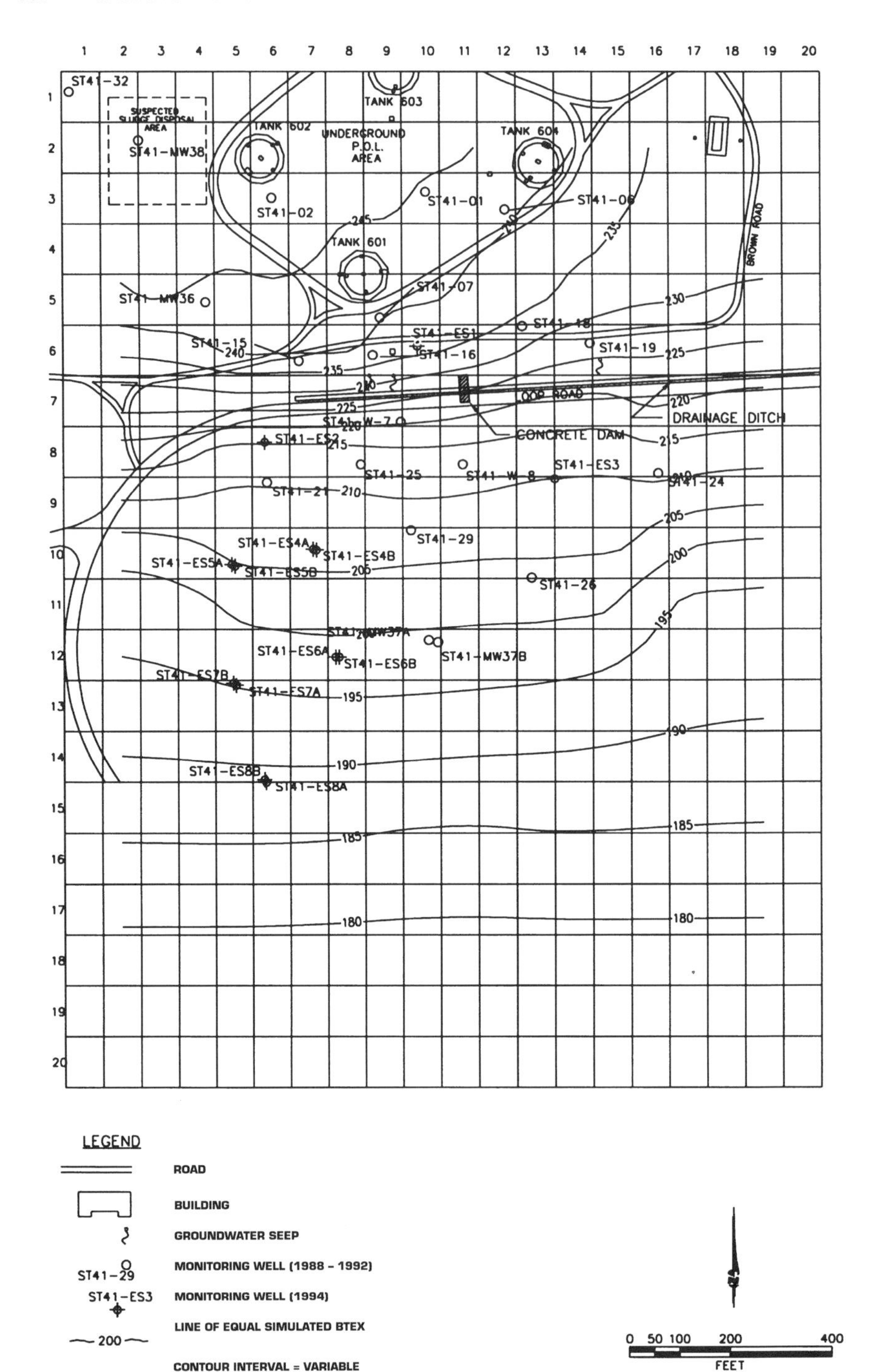

FIGURE 9.36. Calibrated water table map.

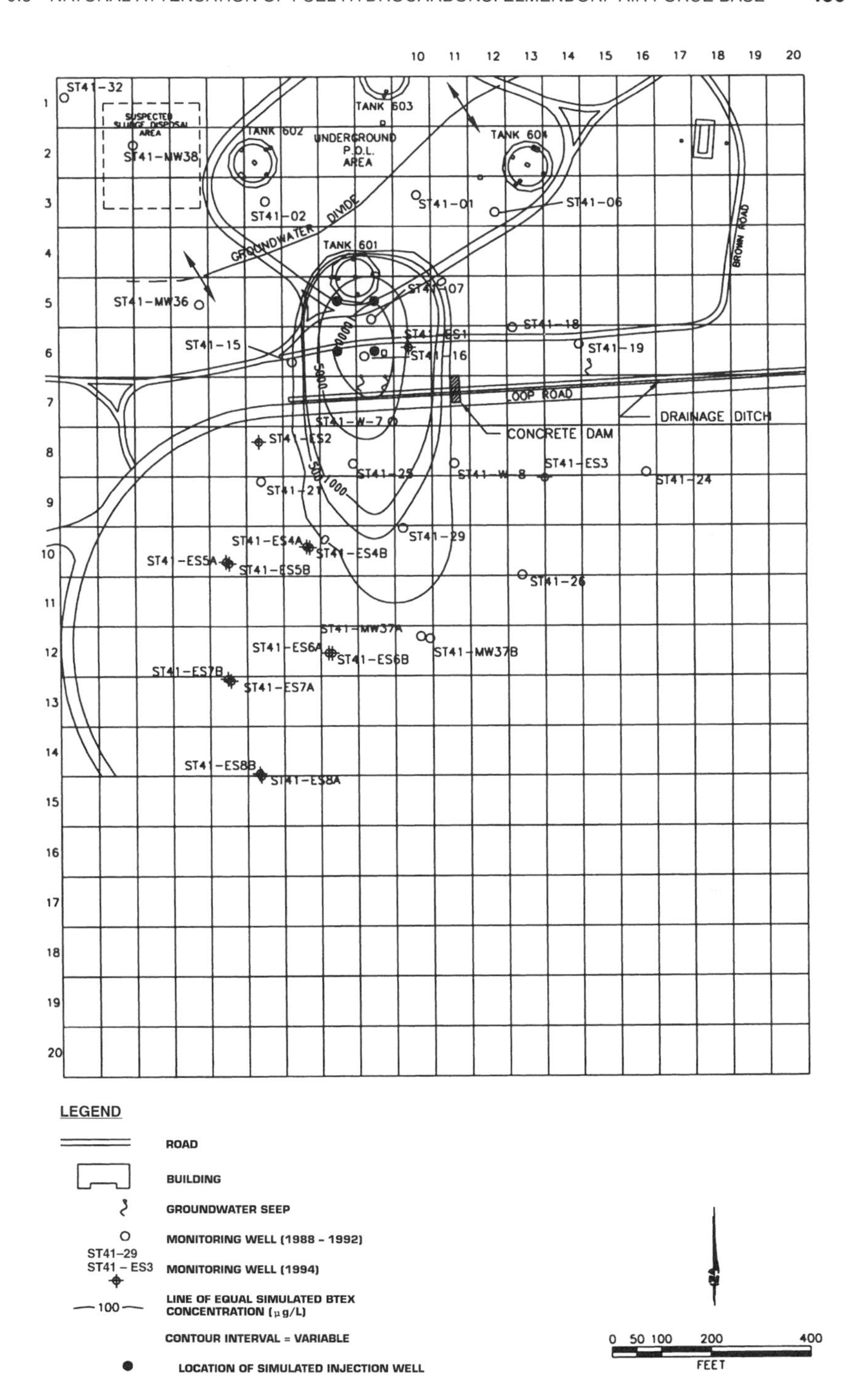

FIGURE 9.37. Calibrated BTEX plume.

Eglin AFB, where the LNAPL history was better known (Parsons Engineering Science, 1995).

Figure 9.38 shows the predicted plume configurations for the first scenario (the worst-case simulation). This model suggests that the plume will lengthen for five years, then stabilize in a configuration varying between those shown for after five years and after 20 years (Figure 9.38). The maximum BTEX concentration for the stable plume is about 20,000 μg/L. For the second scenario, the plume shrinks considerably within the first 10 years, then diminishes more slowly (Figure 9.39). After a total of 22 years of model prediction time, the predicted plume concentrations drop below zero. Results of the third scenario suggest that the plume will gradually shrink from the extents observed in 1994 (Figure 9.40), with model concentrations dropping to zero after 30 years of prediction time. With no likely receptor exposure points present within 2500 ft downgradient from the site, it is very unlikely that receptor exposure to harmful concentrations of BTEX will occur under any scenario.

Discussion

Contaminant and geochemical data collected from site ST41 indicate that microbial degradation of fuel hydrocarbons is limiting plume migration. The limited plume extent relative to the age of the site suggested that mass was being removed from the system, and the geochemical data provided additional evidence of biodegradation. Further supporting this evidence was the volatile fatty acid data provided by the EPA, which confirmed that biodegradation processes were ongoing within the BTEX plume.

Site data were also used to construct a BIOPLUME II model, the results of which indicate that RNA with long-term monitoring is an appropriate remedial strategy for site ST41. While operation of the IRA could shorten the time frame, the added cost and limited initial effectiveness of the IRA system made monitored RNA a more attractive option. Weathering of the source area will ultimately reduce BTEX loading rates enough that concentrations will drop below levels that pose a risk to potential receptors, and the time frame for this is not much longer than if the IRA system were used in an attempt to remove more mass from the source area.

A long-term monitoring plan was recommended to confirm the effectiveness of RNA and to provide historical data supporting the other evidence of biodegradation. Sampling locations for this plan are shown on Figure 9.41. The monitoring plan recommends that in addition to sampling wells within the plume and upgradient from the plume, two contingency wells downgradient from the plume front should be sampled. Should BTEX concentrations be detected and persist in these wells, site remedial options may need to be reconsidered.

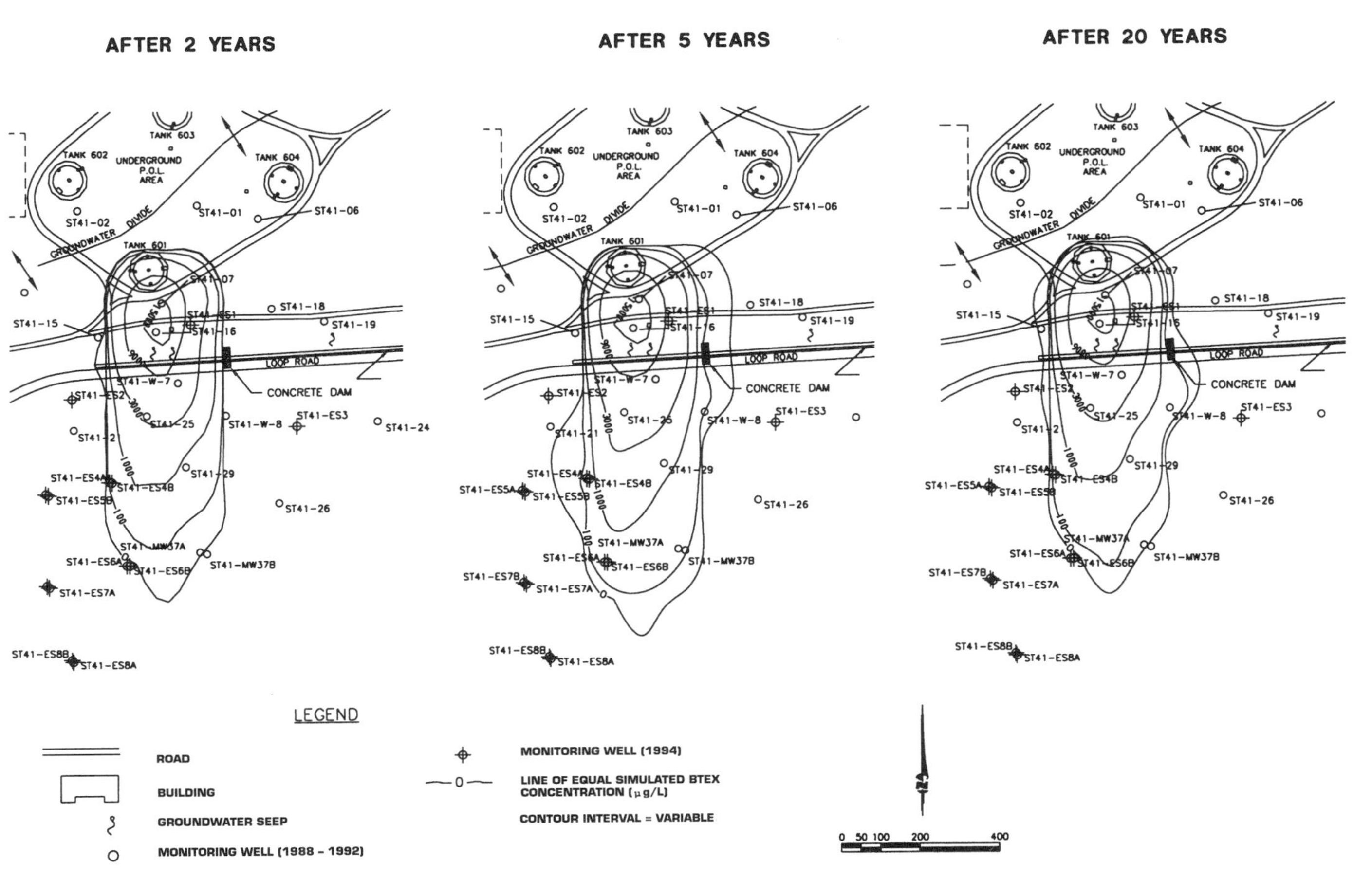

FIGURE 9.38. Predicted BTEX plume (no source reduction).

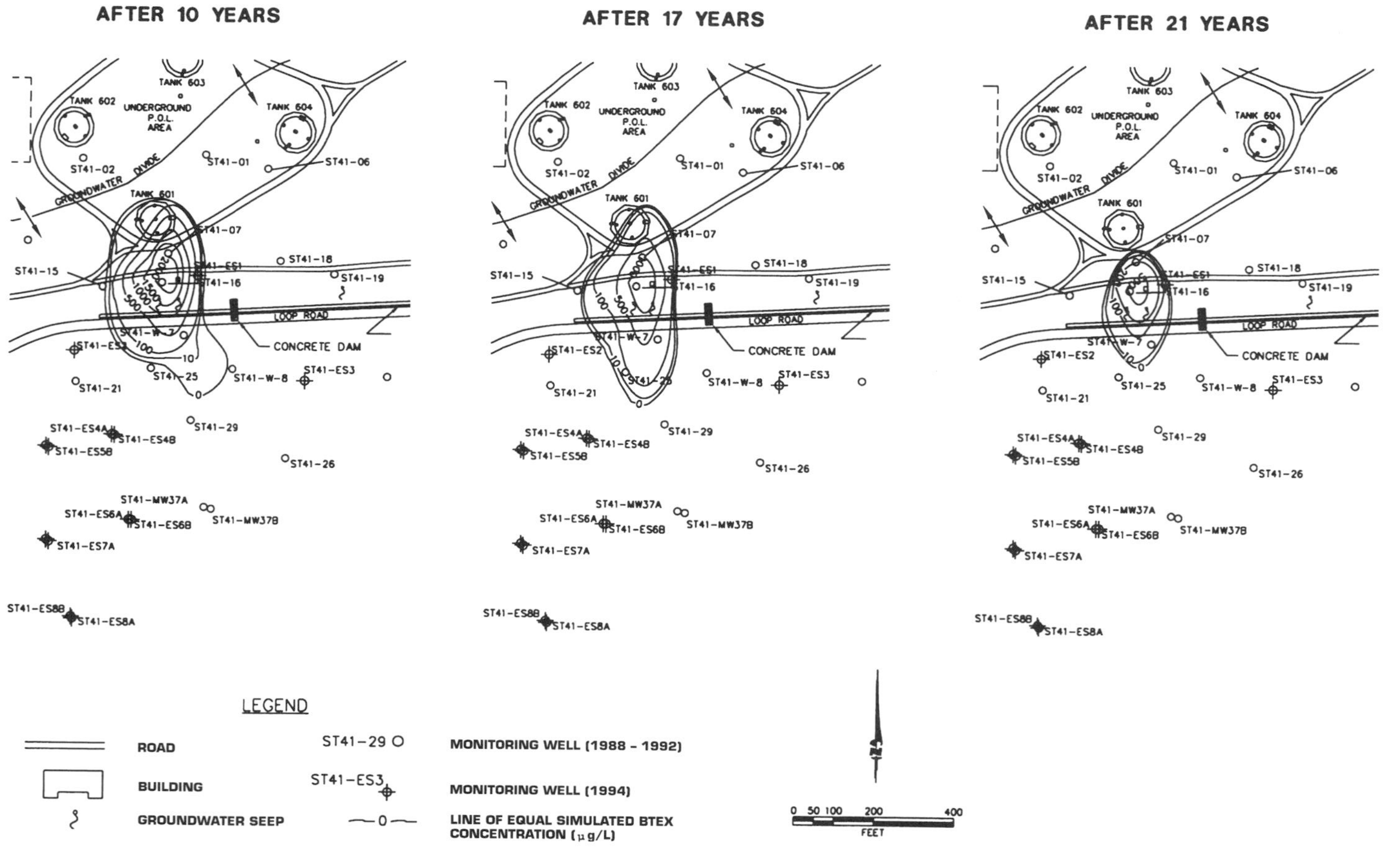

FIGURE 9.39. Predicted BTEX plume (with source reduction).

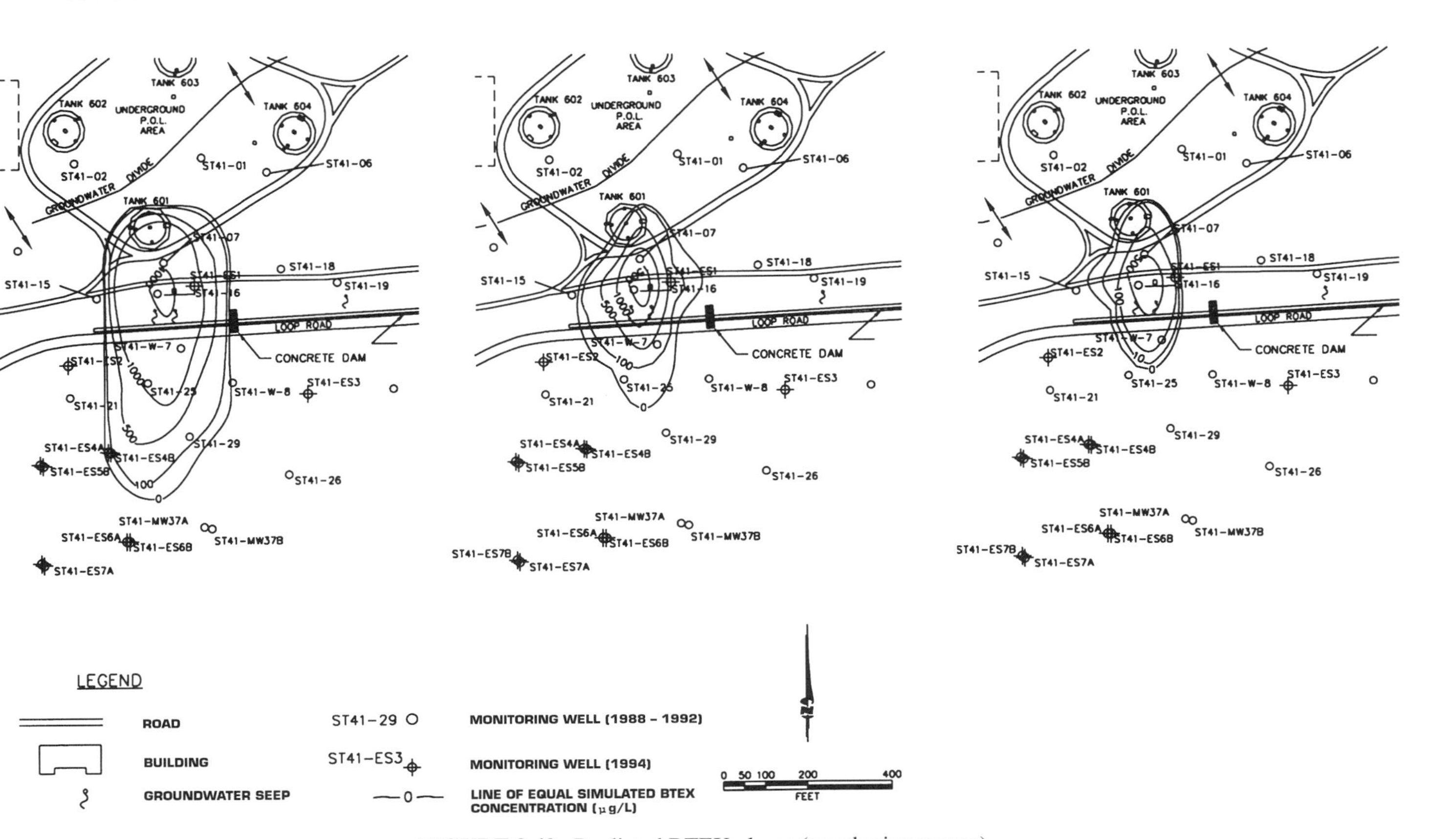

FIGURE 9.40. Predicted BTEX plume (weathering source).

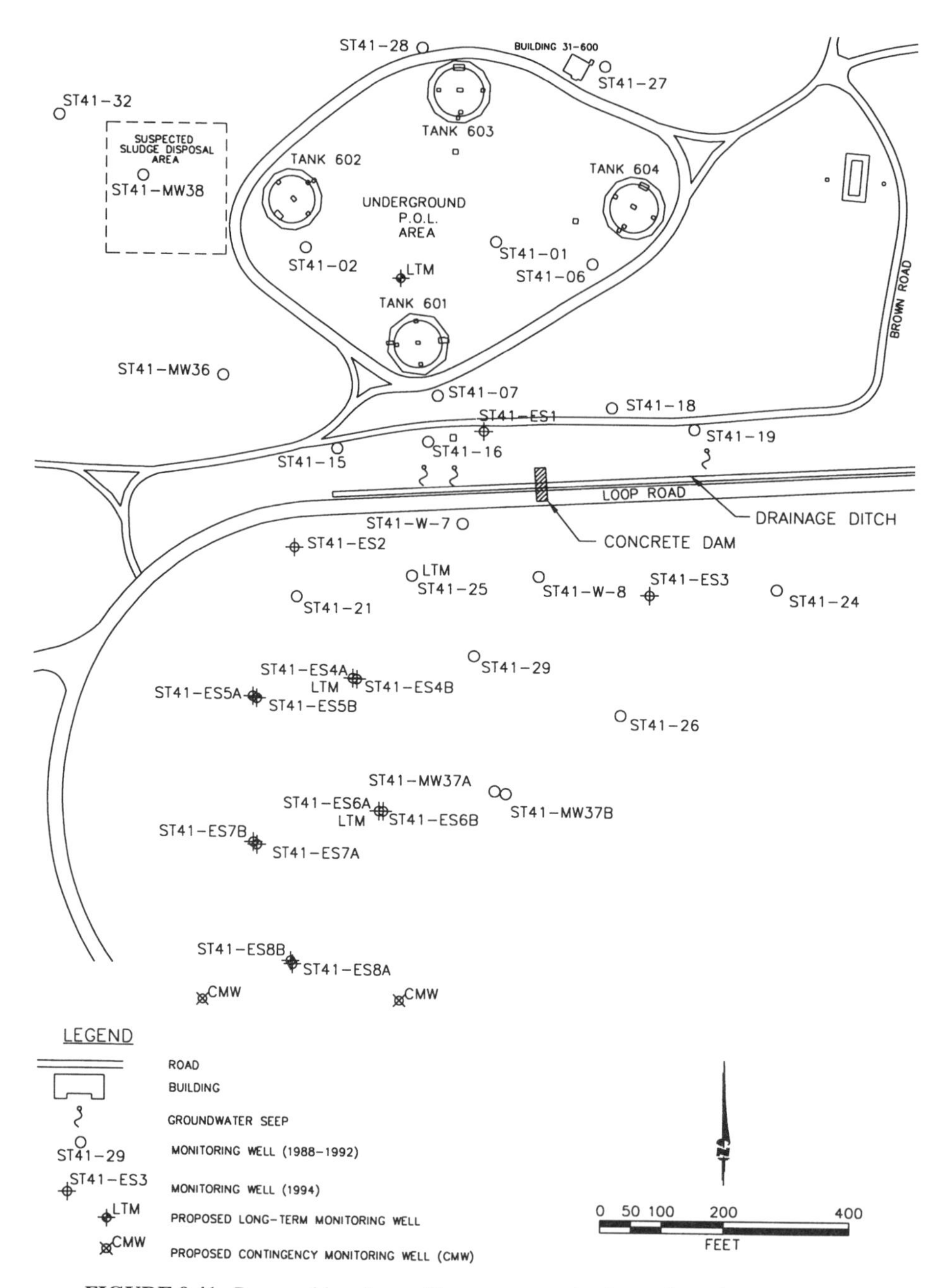

FIGURE 9.41. Proposed locations of long-term monitoring and contingency wells.

9.4 NATURAL ATTENUATION OF FUEL HYDROCARBONS: PETROLEUM, OIL, AND LUBRICANTS FACILITY, HILL AIR FORCE BASE, UTAH

Facility Background

Investigations at the petroleum, oils, and lubricants (POL) facility at Hill AFB, Utah, determined that JP-4 jet fuel had been released into the soil and shallow groundwater. The chronology of the JP-4 spill or spills is not known, but the facility began operating in the early 1950s. Site characterization methods used to evaluate natural attenuation included Geoprobe® sampling of groundwater, cone penetrometer testing (CPT), soil borehole drilling, soil sample collection and analysis, monitoring well installation, sampling and analysis of groundwater from monitoring wells, and aquifer testing. Although the velocity of groundwater at this site is on the order of 1600 ft/yr, the dissolved plume appears to be close to steady-state equilibrium and extends only about 400 ft downgradient from the leading edge of the mobile LNAPL body. Geochemical evidence suggests that aerobic respiration, denitrification, iron(III) reduction, sulfate reduction, and methanogenesis all are contributing to intrinsic bioremediation of dissolved BTEX. Sulfate reduction appears to be the dominant biodegradation mechanism at this site.

Site Geology and Hydrogeology

Hill AFB is located on a bench of the Wasatch Mountains on the edge of the Great Salt Lake Basin. Surface topography at the site slopes to the southwest. Shallow sediments consist of light reddish-brown to dark gray cohesive clayey silts to silty clays. This unit is 4 to 15 ft (1.2 to 4.6 m) thick and is underlain by poorly to moderately sorted, yellowish-brown to reddish-brown silty fine-grained sands that coarsen downward into a 3- to 22-ft (0.9- to 6.7 m)-thick sequence of moderately sorted medium- to coarse-grained sands. Underlying the sands is a sequence of competent thinly interbedded clay to silty clay and fine- to very fine-grained clayey sand and silt of unknown thickness. This sequence of interbedded clay and fine-grained sand and silt acts as an effective barrier to the downward migration of water and contaminants, as indicated by geochemical data. Upward hydraulic gradients in the area also prevent downward migration of the contaminant plume. Figure 9.42 shows the location of hydrogeologic cross section A–A′. Figure 9.43 is hydrogeologic cross section A–A′, which is oriented approximately parallel to the direction of groundwater flow.

The water table aquifer is present in the medium- to coarse-grained sands described above. The water table is present between 5 and 20 ft (1.5 and 6.1 m) BGS, and groundwater flow is to the southwest with an average horizontal gradient of 0.046 ft/ft (0.046 m/m). Available data suggest that there is almost no seasonal variation in groundwater flow direction or gradient at the site. Based on slug tests and pumping tests, the average hydraulic conductivity for the shallow medium- to coarse-grained sands of the shallow saturated zone is 0.0084 cm/s. Effective porosity for the types of sediments observed in the saturated zone ranges from 0.15 to 0.35

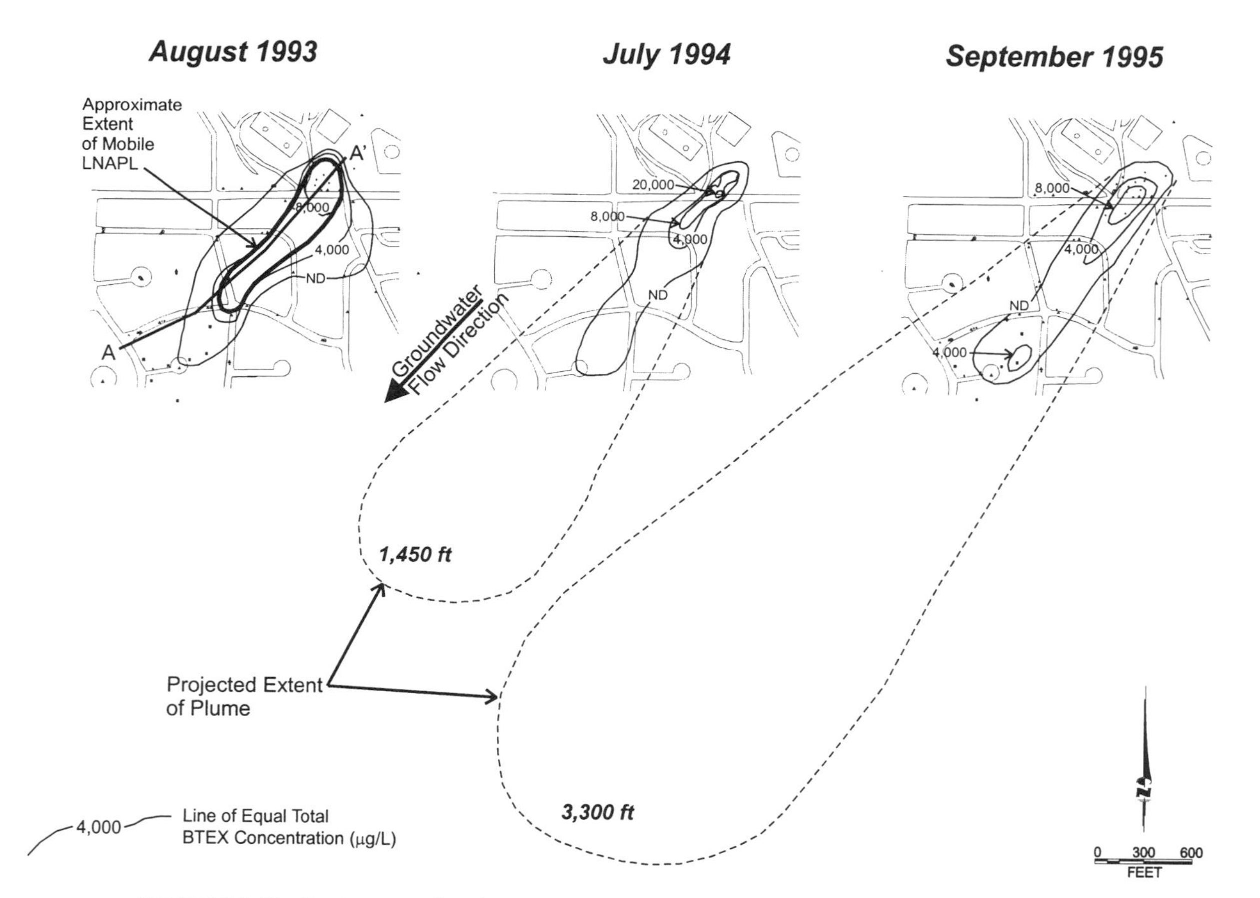

FIGURE 9.42. Site layout and projected extent of dissolved BTEX plume with advection, dispersion, and sorption only (biodegradation omitted).

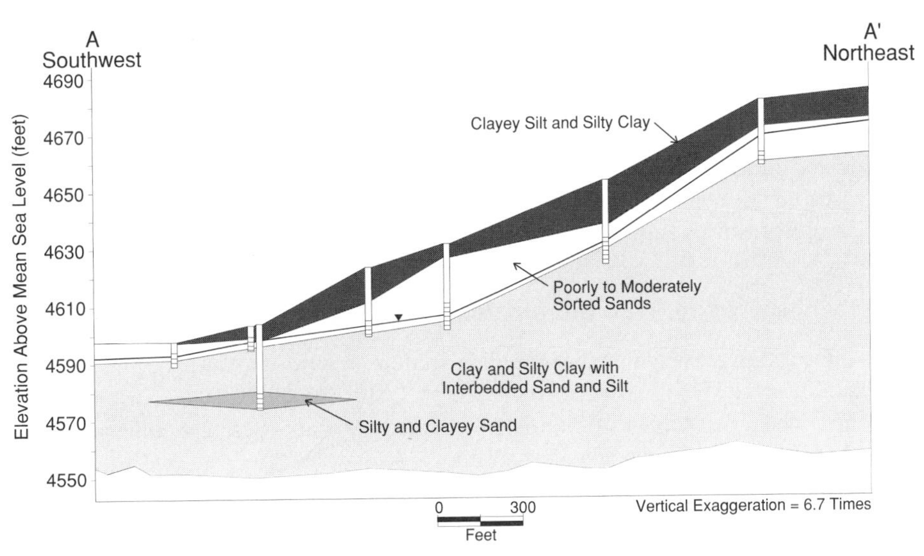

FIGURE 9.43. Hydrogeologic section A–A′.

(Domenico and Schwartz, 1990). Assuming an effective porosity of 0.25, the average groundwater seepage velocity is 4.4 ft/day (1.34 m/day) or approximately 1600 ft/yr (488 m/yr). The concentration of TOC measured in four soil samples ranged from 0.069 to 0.094%. Because of the low TOC content and clay mineral content observed in the shallow saturated zone, retardation of the BTEX compounds is not considered to be an important process affecting solute transport at this site.

Nature and Extent of Contamination

Figure 9.42 shows the approximate extent of mobile LNAPL at the site, which is the source of BTEX dissolved in groundwater. The LNAPL plume is composed of weathered JP-4 released from the POL facility. Measured residual total BTEX concentrations in soil decrease rapidly outside the area of mobile LNAPL contamination. The highest dissolved BTEX concentration observed in groundwater during this study was 21,475 μg/L. Figure 9.42 also contains isopleth maps that show the distribution of total BTEX dissolved in groundwater in August 1993, July 1994, and September 1995. Dissolved BTEX contamination is migrating to the southwest, in the direction of groundwater flow.

Evidence of Plume Stabilization

Dissolved BTEX data collected in 1993, 1994, and 1995 indicate that the dissolved BTEX plume has reached approximately steady-state conditions, and although fluctuations in contaminant concentrations were observed during this period, the plume is not significantly expanding or migrating further downgradient. Based on a two-dimensional analytical model incorporating advection and dispersion, the contaminant plume should have migrated approximately 3300 ft between August 1993 and September 1995 (Figure 9.42). The relative stability of the dissolved BTEX plume represents the first line of evidence used to document and support natural attenuation, as discussed in Chapter 7. Available geochemical data suggest that stabilization of the BTEX plume is primarily the result of intrinsic bioremediation.

Mechanisms of Intrinsic Bioremediation

The information presented in the preceding section suggests that intrinsic bioremediation of dissolved BTEX is occurring at the POL site. Geochemical data are used to estimate which mechanisms of intrinsic bioremediation are operating at the site and the relative importance of each mechanism.

General Groundwater Geochemistry. General ambient groundwater geochemistry at the site is conducive to intrinsic bioremediation. Total alkalinity (as calcium carbonate) at the site is fairly high and ranges from 349 to 959 mg/L. This amount of alkalinity is sufficient to buffer potential changes in pH caused by biologically mediated BTEX oxidation reactions. The pH of groundwater at the POL facility ranges from 6.3 to 8.3 standard units. This range of pH is optimal for BTEX-degrading microbes. The average temperature of groundwater is 18°C. The ORP of groundwater ranges from 274 to –190 mV. Areas at the site with low ORPs coincide with areas of BTEX contamination, low dissolved oxygen, nitrate, and sulfate concentrations, and elevated Fe(II) and methane concentrations. This suggests that dissolved BTEX at the site is subject to a variety of biodegradation processes, including aerobic respiration, denitrification, iron reduction, sulfate reduction, and methanogenesis.

Aerobic Respiration. Dissolved oxygen concentrations in unimpacted groundwater at the site are on the order of 6 mg/L. In the portions of the plume with the highest BTEX concentrations, dissolved oxygen concentrations are less than 0.1 mg/L. Figure 9.44 shows the distribution of dissolved oxygen in groundwater in July 1994. This figure shows that areas with elevated total BTEX concentrations have depleted dissolved oxygen concentrations. This is an indication that aerobic biodegradation of BTEX is occurring at the site.

Denitrification. Nitrate concentrations in unimpacted groundwater at the site are on the order of 17 mg/L. In the portions of the plume with the highest BTEX concentrations, nitrate concentrations are less than 0.1 mg/L. Figure 9.44 shows the distribution of nitrate in groundwater in July 1994. This figure shows that areas with elevated total BTEX concentrations have depleted the nitrate present in groundwater. These relationships provide strong evidence that anaerobic biodegradation of the BTEX compounds is occurring at the site through the microbially mediated process of denitrification.

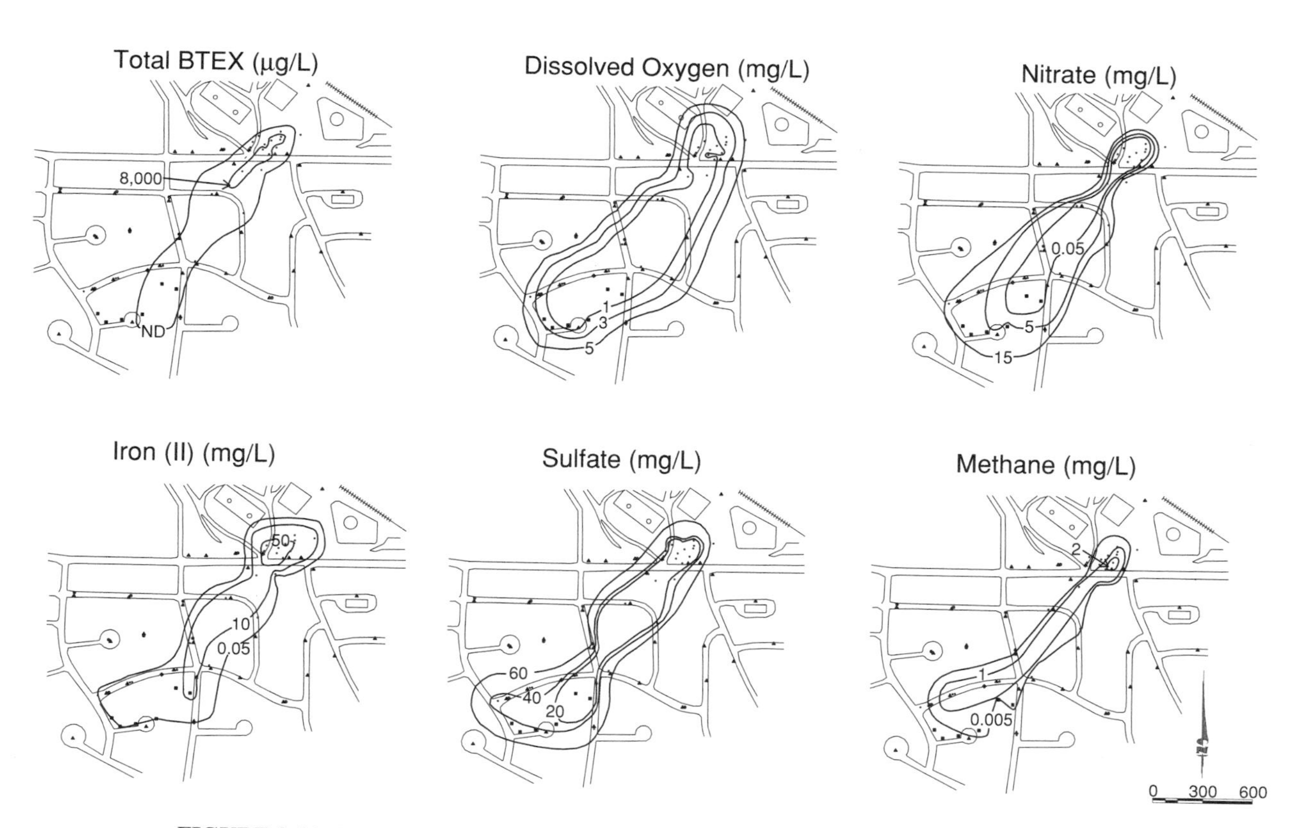

FIGURE 9.44. Isopleth maps showing electron acceptor and metabolic by-product concentrations.

Iron(III) Reduction. Iron(II) concentrations in unimpacted groundwater at the site are less than 0.05 mg/L. In the portions of the plume with the highest BTEX concentrations, iron(II) concentrations as high as 51 mg/L were observed. Figure 9.44 shows the distribution of iron(II) in groundwater in July 1994. This figure shows that areas with elevated total BTEX concentrations have elevated iron(II) concentrations. This is an indication that iron(III) is being reduced to iron(II) during biodegradation of BTEX compounds.

Sulfate Reduction. Sulfate concentrations in unimpacted groundwater at the site are on the order of 100 mg/L. In the portions of the plume with the highest BTEX concentrations, sulfate concentrations are less than 0.05 mg/L. Figure 9.44 shows the distribution of sulfate in groundwater in July 1994. This figure shows that areas with elevated total BTEX concentrations have depleted sulfate concentrations. This is an indication that anaerobic biodegradation of the BTEX compounds is occurring at the site through the microbially mediated process of sulfate reduction.

Methanogenesis. Methane concentrations in unimpacted groundwater at the site are less than 0.001 mg/L. In the portions of the plume with the highest BTEX concentrations, methane concentrations are on the order of 2 mg/L. Figure 9.44 shows the distribution of methane in groundwater in July 1994. This figure shows that areas with elevated total BTEX concentrations have elevated methane concentrations. This is an indication that anaerobic biodegradation of the BTEX compounds is occurring at the site through the microbially mediated process of methanogenesis, with carbon dioxide serving as the electron acceptor. This is consistent with other electron acceptor data collected at the site, with the area having elevated methane concentrations being confined to areas with depleted dissolved oxygen, nitrate, and sulfate concentrations and elevated iron(II) concentrations.

Biodegradation Rate Constants

Wiedemeier et al. (1996) calculate approximate first-order rate constants for each of the BTEX compounds and for total BTEX at the Hill AFB site using two independent methods. The first method utilizes a biologically recalcitrant tracer with physical and chemical properties similar to the BTEX compounds (TMB in this case) to correct for the effects of dispersion, dilution, sorption, and volatilization. The second method utilizes a modification of the advection-dispersion equation to solve for biodegradation rates and was presented by Buscheck and Alcantar (1995). Table 9.5 presents the results of these calculations.

The approximate site-specific biodegradation rates presented in Table 9.5 are within the ranges published in recent literature. Calculated degradation rates along the centerline of the Hill AFB plume are generally slower than those calculated along a flow path near the periphery of the plume. This is probably due to the continuing introduction of dissolved BTEX in the vicinity of the source and the greater availability of electron acceptors near the margins of the BTEX plume.

TABLE 9.5. Biodegradation Rates for Hill Air Force Base

Compound	Approximate Biodegradation Rate Constant Calculated Using Trimethylbenzene as a Conservative Tracer (day^{-1})	Approximate Biodegradation Rate Constant Calculated Using the Method of Buscheck and Alcantar (1995) (day^{-1})
	Flow Path Down Plume Centerline	
Benzene	0.028	0.038
Toluene	0.023	0.031
Ethylbenzene	0.009	0.010
Xylene	0.006	0.008
Total BTEX	0.010	0.012
	Flow Path Along Plume Periphery	
Benzene	0.025	0.027
Toluene	0.026	0.029
Ethylbenzene	0.020	0.024
Xylene	0.027	0.024
Total BTEX	0.021	0.025

Benzene degradation rates at Hill AFB generally were similar to the rates calculated for the other BTEX compounds. This is not consistent with experience at other sites where intrinsic bioremediation of BTEX has been documented. Barker et al. (1987), Cozzarelli et al. (1990), and Wilson et al. (1993) found benzene to be more recalcitrant to biodegradation than the other BTEX compounds at the sites they studied. This suggests that pathways and patterns of BTEX biodegradation are site specific, as are the apparent rates of biodegradation. The fate of benzene may be controlled by the dominant terminal electron-accepting process. The dominant electron-accepting process at Hill AFB was sulfate reduction, accounting for 59% of the total expressed electron-accepting activity (Wiedemeier et al., 1995b). Methanogenesis accounted for 7%, iron reduction for 7%, denitrification for 21%, and aerobic respiration for 6% of the total expressed electron-accepting activity.

Long-Term Monitoring Plan

The recommended remedial alternative for the fuel-hydrocarbon contamination present in groundwater at the POL facility was mobile LNAPL recovery and bioventing for mobile- and residual-phase LNAPL contamination, and monitored natural attenuation for contaminated groundwater. The purpose of long-term monitoring is to assess site conditions over time, to confirm the effectiveness of naturally occurring processes at reducing contaminant mass and minimizing contaminant migration, to validate a solute transport model, and to evaluate the need for additional remediation. The long-term monitoring plan identifies the locations of two separate groundwater monitoring networks and develops a groundwater and stormwater discharge point sampling and analysis strategy.

The first groundwater monitoring network consists of nine long-term monitoring wells located upgradient, within, and downgradient from the observed total BTEX plume (Figure 9.45). The purpose of the long-term monitoring well network is to provide confirmation and verification of natural attenuation. Long-term monitoring wells are sampled and analyzed to monitor trends in groundwater chemistry and to

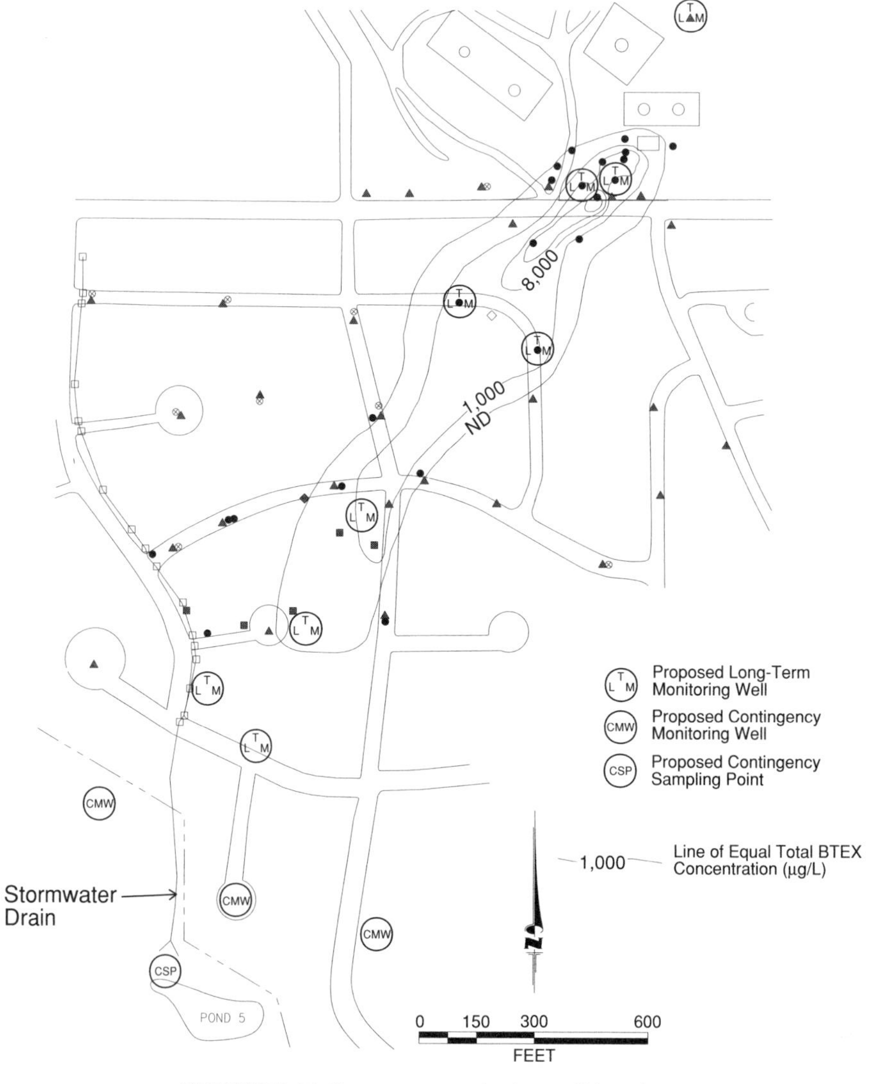

FIGURE 9.45. Long-term monitoring well locations.

verify the effectiveness of intrinsic remediation at the site. In this case, it was recommended that all groundwater samples from long-term monitoring wells be analyzed for dissolved oxygen, nitrate, iron(II), sulfate, methane, temperature, pH, conductivity, ORP, and aromatic hydrocarbons. Based on the results of natural attenuation demonstrations conducted at numerous sites around the country, such an extensive list of analytical parameters probably is not necessary. Instead, it may be necessary to analyze only for aromatic hydrocarbons, dissolved oxygen, pH, temperature, conductivity, and oxidation–reduction potential. If the results of these analyses suggest that a significant change in groundwater geochemistry has occurred, it may be beneficial to analyze the groundwater for the entire list of parameters.

The second network of groundwater monitoring points consists of three contingency wells and a contingency sampling point at the pond 5 outfall of a stormwater sewer located downgradient from the dissolved BTEX plume (Figure 9.45). The purpose of this monitoring network is to verify that no BTEX compounds in concentrations exceeding risk-based corrective action goals migrate beyond the area under institutional control. Should BTEX compounds be detected through contingency sampling in the stormwater sewer discharge in excess of risk-based corrective action goals, a stormwater treatment system will be installed. Should BTEX compounds be detected in contingency wells at concentrations above risk-based corrective action goals, other remedial options will be evaluated. Contingency sampling points will be sampled and analyzed to monitor trends in groundwater chemistry, to verify the effectiveness of intrinsic remediation at the site, and to demonstrate protection of human health and the environment and compliance with site-specific numerical remediation goals. All groundwater samples from contingency wells should be analyzed for aromatic hydrocarbons, dissolved oxygen, pH, temperature, conductivity, and ORP.

For this site it was recommended that each of the long-term monitoring and contingency sampling points be sampled twice annually for 13 years. If the data collected during this time period support the anticipated effectiveness of the monitored natural attenuation alternative at this site, the sampling frequency can be reduced to once every year for all wells in the LTM program, or eliminated.

Conclusions

Historical data and geochemical evidence show that intrinsic bioremediation of dissolved BTEX is occurring at the POL facility at Hill AFB at rates sufficient to be protective of human health and the environment. Geochemical evidence suggests that aerobic respiration, denitrification, iron(III) reduction, sulfate reduction, and methanogenesis are the primary processes contributing to microbial degradation of BTEX at Hill AFB. The approximate site-specific biodegradation rates calculated along the centerline of the plume are generally slower than those calculated along a flow path near the periphery of the plume. This is probably due to the continuing introduction of dissolved BTEX in the vicinity of the source and the greater availability of electron acceptors near the margins of the BTEX plume. Because of the extensive mobile LNAPL contamination at the site, recommendations were made to remove as much of this source as practicable. Long-term monitoring was recommended to ensure that natural attenuation continues to be protective of human health and the environment.

REFERENCES

Black and Veatch, 1990, *Installation Restoration Program Stage 3 Remedial Investigation/ Feasibility Study*, Elmendorf Air Force Base, AK, May.

Bruce, L., Miller, T., and Hockman, B., 1991, Solubility versus equilibrium saturation of gasoline compounds: a method to estimate fuel/water partition coefficient using solubility or K_{oc}, in A. Stanley, ed., *Proceedings of the NWWA/API Conference on Petroleum Hydrocarbons in Groundwater*, NWWA/API, pp. 571–582.

Buscheck, T. E., and Alcantar, C. M., 1995, Regression techniques and analytical solutions to demonstrate intrinsic bioremediation, in *Proceedings of the 1995 Battelle International Symposium on In Situ and On Site Bioreclamation*, Apr., Battelle Press, Columbus, OH.

Cozzarelli, I. M., Eganhouse, R. P., and Baedecker, M. J., 1990, Transformation of monoaromatic hydrocarbons to organic acids in anoxic groundwater environment, *Environ. Geol. Water Sci.*, vol. 16.

Cozzarelli, I. M., Baedecker, M. J., Eganhouse, R. P., and Goerlitz, D. F., 1994, The geochemical evolution of low-molecular-weight organic acids derived from the degradation of petroleum contaminants in groundwater, *Geochim. Cosmochim. Acta*, vol. 58, no. 2, pp. 863–877.

Domenico, P. A., and Schwartz, F. W., 1990, *Physical and Chemical Hydrogeology*, Wiley, New York, 824 pp.

Eberly, L. D., and Stanley, T. B., 1978, Cenozoic stratigraphy and geologic history of southwestern Arizona, *Geol. Soc. Am. Bull.*, vol. 89, pp. 921–940.

Engineering-Science, Inc., 1983, *Installation Restoration Program Phase I: Record Search*, Sept.

Jacobs Engineering Group, 1994, *Environmental Restoration Program: Operable Unit 2 Remedial Investigation/Feasibility Study Report*, Elmendorf Air Force Base, AK, Mar.

Laney, R. L., and Hahn, M. E., 1986, *Hydrogeology of the Eastern Part of the Salt River Valley Area, Maricopa and Pinard Counties, Arizona*, Water Resources Investigations Report 86-4147, U.S. Geological Survey, Washington, DC.

Mercer, J. W., and Cohen, R. M., 1990, A review of immiscible fluids in the subsurface: properties, models, characterization and remediation, *J. Contam. Hydrol.*, vol. 6, pp. 107–163.

Parsons Engineering Science, Inc., 1995, *Draft Treatability in Support of Intrinsic Remediation for POL Site SS-36*, Eglin Air Force Base, FL, July.

Wiedemeier, T. H., Wilson, J. T., Kampbell, D. H., Miller, R. N., and Hansen, J. E., 1995a, *Technical Protocol for Implementing Intrinsic Remediation with Long-Term Monitoring for Natural Attenuation of Fuel Contamination Dissolved in Groundwater*, U.S. Air Force Center for Environmental Excellence, San Antonio, TX.

Wiedemeier, T. H., Wilson, J. T., and Miller, R. N., 1995b, Significance of anaerobic processes for the intrinsic bioremediation of fuel hydrocarbons, in *Proceedings of the Petroleum Hydrocarbons and Organic Chemicals in Groundwater: Prevention, Detection, and Restoration Conference*, Houston, TX, Nov. 29–Dec. 1, NWWA/API.

Wiedemeier, T. H., Swanson, M. A., Wilson, J. T., Kampbell, D., Miller, R. N., and Hansen, J. E., 1996, Comparison of two methods to estimate biodegradation rates of monoaromatic hydrocarbons (BTEX) in groundwater; *Ground Water Monit. Remed.*, vol. 16, no. 3, pp. 186–195.

Wilson, J. T., Kampbell, D. H., and Armstrong, J., 1993, Natural bioreclamation of alkylbenzenes (BTEX) from a gasoline spill in methanogenic groundwater, *Proceedings of the Environmental Restoration Technology Transfer Symposium*, San Antonio, TX.

CHAPTER 10

CASE STUDIES: CHLORINATED SOLVENTS

As with Chapter 9, additional authors have contributed to the case studies presented in this chapter to demonstrate the variable effects of subsurface conditions on the natural attenuation of chlorinated aliphatic hydrocarbons (CAHs) dissolved in groundwater. In particular, the case studies reflect the potential for different governing biochemical conditions in different portions of the same contaminant plume.

10.1 NATURAL ATTENUATION OF CHLORINATED SOLVENTS: FT-17, CAPE CANAVERAL, FLORIDA

MATT SWANSON
PARSONS ENGINEERING SCIENCE, INC.
DENVER, COLORADO

Facility Background

Cape Canaveral Air Station is located in east-central Florida, on the Atlantic coast. Site CCFTA-2 (FT-17) is a former fire training area on the western edge of the station, about 1100 ft east of the Banana River and about 1000 ft northeast of a canal that empties into the river. Figure 10.1 is a map of the site. The site was used to train Air Force firefighting personnel from about 1965 until 1985. These activities resulted in contamination of shallow soils and groundwater with a mixture of chlorinated solvents and fuel hydrocarbons and the formation of a body of mobile LNAPL containing petroleum hydrocarbons with cosolvenated CAHs. This LNAPL body (Figure 10.1) is the main source of dissolved contamination.

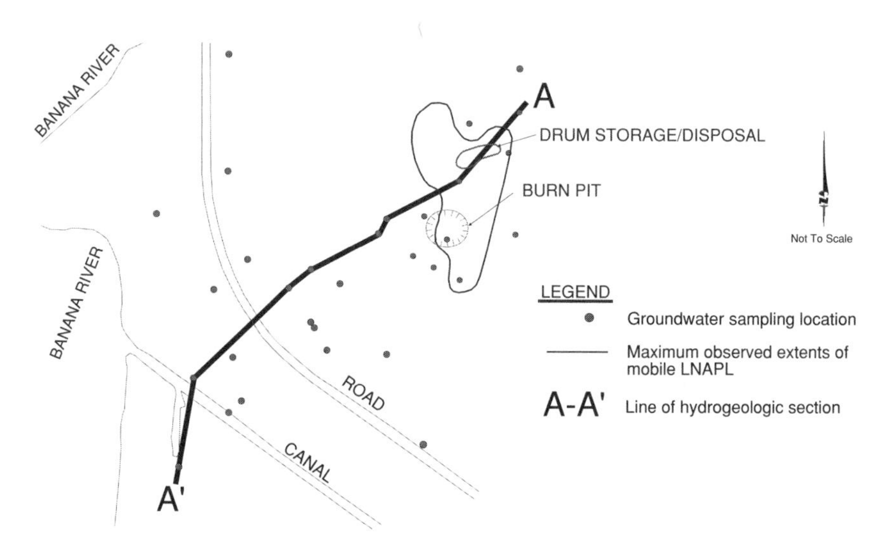

FIGURE 10.1. Site layout, FT-17, Cape Canaveral Air Station, Florida.

A plume of dissolved CAHs and petroleum hydrocarbons is present beneath and downgradient from the LNAPL. Groundwater contaminants include TCE, *cis*-1,2-DCE, *trans*-1,2-DCE, vinyl chloride (VC), and BTEX. Present in lower concentrations are PCE, 1,1,1-trichloroethane (1,1,1-TCA), and 1,1-DCA. VC is the compound of greatest concern at the site because it is discharging to surface water in the drainage canal.

Engineered remedial actions for addressing some aspects of the dissolved contaminant plume have been implemented to minimize environmental impacts resulting from the presence of LNAPL and the discharge of VC to the canal. A horizontal air sparging system (HASS) has been installed near the canal to reduce, through volatilization, concentrations of VC entering the canal. In addition, soil containing LNAPL was excavated, processed through an onsite soil-washing system, and returned to the excavation. This action has helped eliminate the primary source of dissolved contaminants and ultimately will reduce the time needed for site remediation. As part of the Air Force Center for Environmental Excellence's ongoing natural attenuation initiative, data on natural attenuation of chlorinated solvents and fuel hydrocarbons were collected and the potential use of monitored natural attenuation as a component of overall site remediation plans for CCFTA-2 (FT-17) was evaluated.

Site Geology and Hydrogeology

Shallow sediment at the site consists of an undifferentiated marine sand unit about 10 to 15 ft thick containing fine to coarse sand and shell fragments, with occasional clay lenses and peat stringers. The sand unit is underlain by the Caloosahatchee Marl; in this area, the upper part of this formation is a unit of generally fine-grained calcareous sand, unconsolidated shells, and shell fragments, with some interbedded clay units and

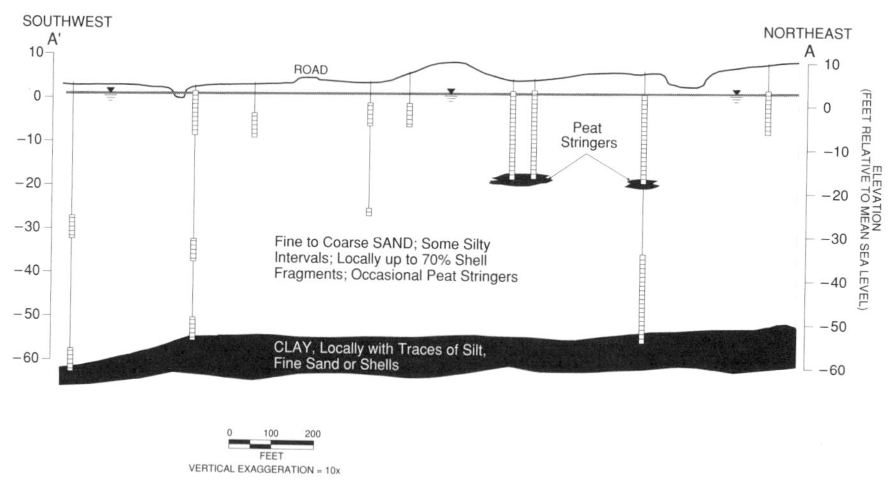

FIGURE 10.2. Hydrogeologic cross section, FT-17, Cape Canaveral Air Station, Florida.

peat stringers. About 60 BGS, a clay layer is present within the marl. This unit appears to be present throughout the site area and probably is a lower confining unit for the shallow groundwater flow system. Figure 10.2 presents a hydrogeologic cross section of the site.

Depth to groundwater ranges from 2 ft BGS in wells near surface water bodies to about 11 ft BGS in wells northeast of the fire training pit. Groundwater flow is to the south–southwest (Figure 10.3). The average horizontal gradient is approxi-

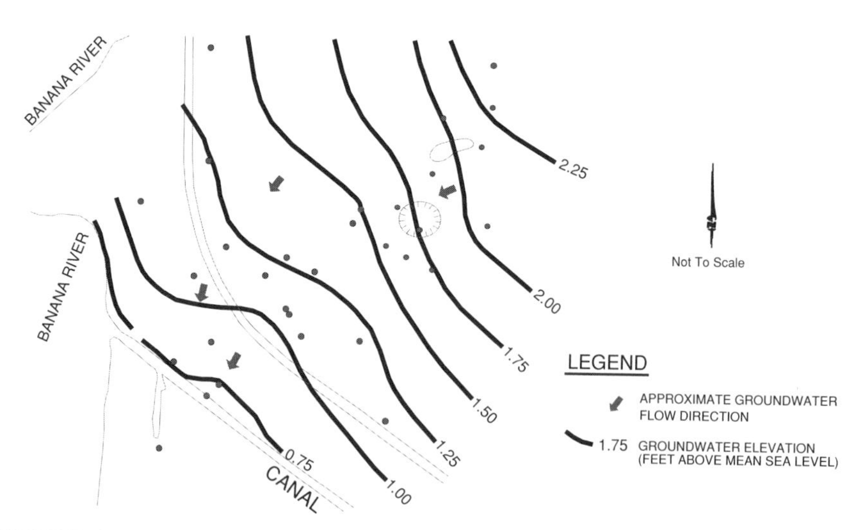

FIGURE 10.3. Water table elevations, January 1996, FT-17, Cape Canaveral Air Station, Florida.

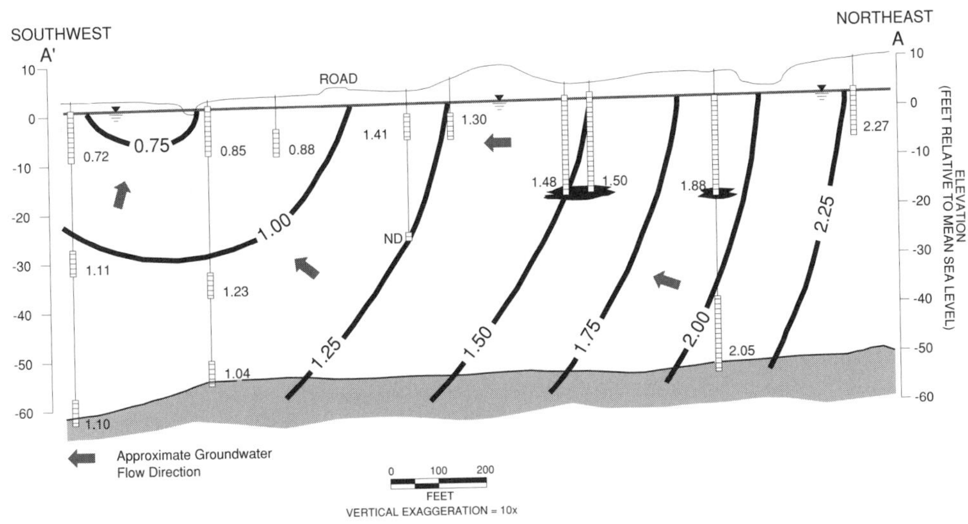

FIGURE 10.4. Vertical groundwater flow profile, FT-17, Cape Canaveral Air Station, Florida.

mately 0.0012 ft/ft, and vertical gradients are upward, with groundwater discharging to the canal and the Banana River (Figure 10.4). Hydraulic conductivity values for the unconfined sand aquifer underlying the site range from about 25 ft/day to about 115 ft/day, with an average of about 45 ft/day. Effective porosity for this site was assumed to be 0.20. Using these data, an average advective groundwater velocity for the unconfined sand aquifer of 0.27 ft/day, or about 99 ft/year, was estimated.

Background TOC concentrations in soil at the site are high enough that the transport velocities of organic contaminants will be retarded relative to the advective groundwater velocity. In addition, the peat stringers at the site probably provide localized areas where sorption of organic contaminants will be greatly enhanced. Site-specific retardation factors for chlorinated ethenes dissolved in site groundwater were estimated using the TOC data and used to calculate contaminant transport velocities (Table 10.1).

TABLE 10.1. Retardation Factors and Contaminant Velocities

Compound	Average Site-Specific Retardation Factor[a]	Transport Velocity (ft/day)
TCE	2.58	0.10
1,2-DCE	2.18	0.12
VC	1.04	0.27

[a]Calculated using an average f_{oc} value of 0.0018.

Groundwater and LNAPL Chemistry

Contaminants and Daughter Products. LNAPL at the site is a mixture of fuels and waste solvents from which petroleum hydrocarbons (including BTEX), TCE, and lesser amounts of PCE and DCE partition into groundwater. However, the chlorinated compounds are found entirely in the portion of the LNAPL body north of the burn pit, in the vicinity of the drum storage and disposal area (Figure 10.5). Analysis of the LNAPL from north of the pit shows that TCE is the CAH present in the greatest concentrations. DCE and VC were not present in measurable concentrations in one sample, while one sample contained concentrations of *cis*-1,2-DCE and PCE at about 1/10 to 1/100 the TCE concentration, respectively.

Groundwater beneath and downgradient from the LNAPL is contaminated with dissolved chlorinated solvents and fuel-related compounds consistent with those

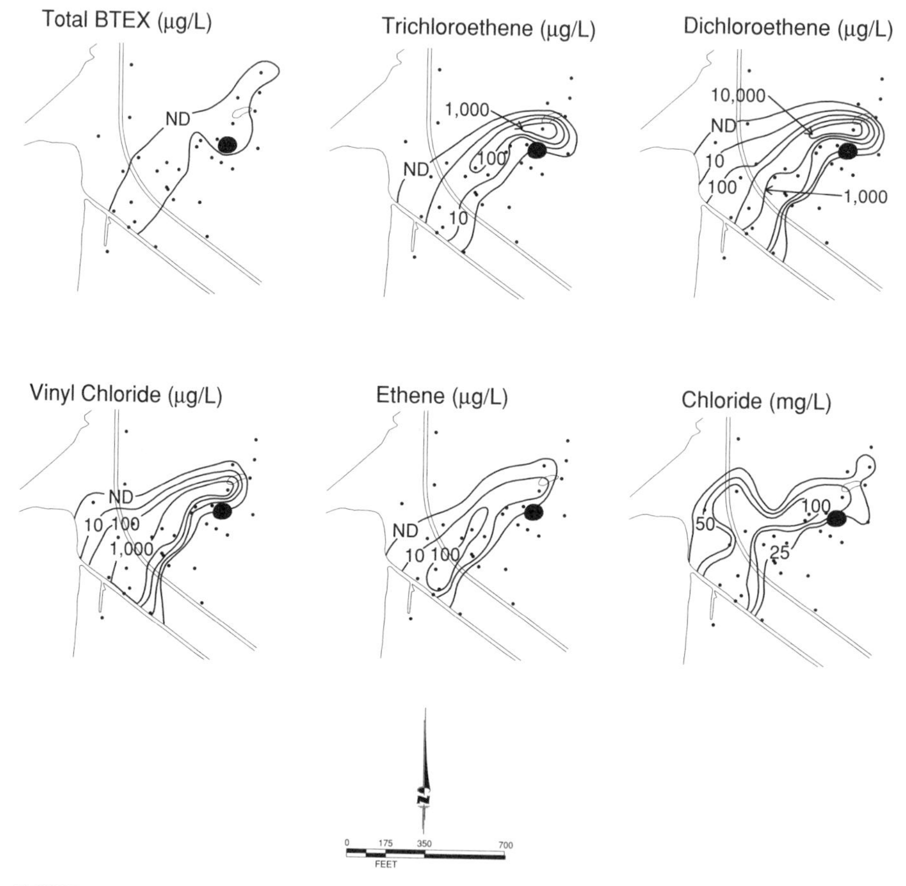

FIGURE 10.5. Chlorinated solvents and by-products in groundwater, January 1996, FT-17, Cape Canaveral Air Station, Florida.

identified in the LNAPL. The dissolved contaminant plume extends along a curved path from the source area to the canal, a distance of about 1300 ft (Figures 10.5 and 10.6). The highest total dissolved BTEX concentration (measured in a sample collected beneath the LNAPL body) was about 330 μg/L. Downgradient from the source area, total BTEX concentrations ranged from <1 to 46 μg/L. The low dissolved BTEX concentrations suggest that the LNAPL has weathered significantly since fire training activities ceased, and that natural attenuation has further reduced dissolved petroleum hydrocarbon concentrations.

Concentrations of TCE, DCE, and VC as high as 15,800, 98,500, and 6520 μg/L, respectively, have been observed at the site. Figure 10.5 shows the distribution of chlorinated ethenes at CCFTA-2 (FT-17). Beneath the LNAPL, dissolved DCE concentrations are higher than TCE concentrations, and VC is present. In addition, Figure 10.7 shows that the concentrations of *cis*-1,2-DCE detected in site groundwater are about 5 to 400 times the *trans*-1,2-DCE and 1,1-DCE concentrations, and at only three locations are these ratios less than 20:1. The elevated ratio of DCE to TCE, the predominance of the *cis*-1,2-DCE isomer, and the presence of VC beneath the LNAPL strongly suggest that TCE and DCE are being reductively dechlorinated. VC concentrations are greatest in the vicinity of the canal, as would be expected due to the production of VC from the biodegradation of DCE, coupled with the faster transport velocity for VC.

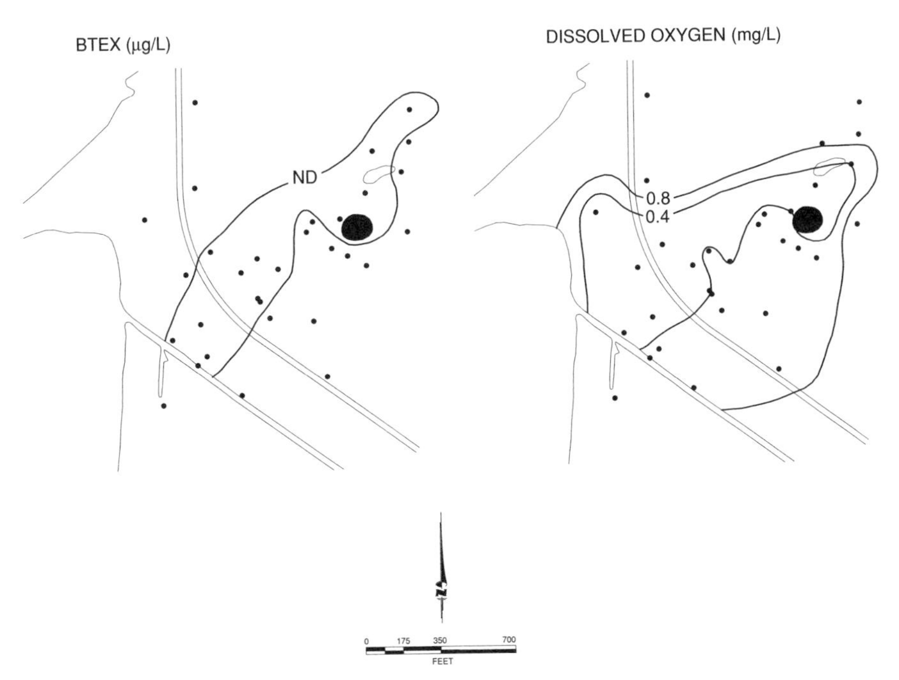

FIGURE 10.6. BTEX and dissolved oxygen, FT-17, Cape Canaveral Air Station, Florida.

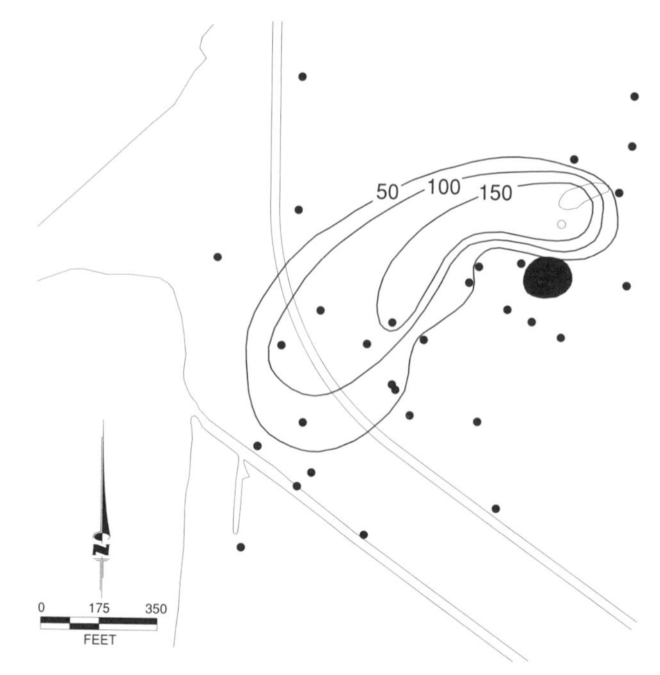

FIGURE 10.7. Ratio of *cis*-1,2-DCE to *trans*-1,2-DCE, FT-17, Cape Canaveral Air Station, Florida.

Ethene and chloride (shown on Figure 10.5) are also produced as a result of reductive dechlorination. The highest ethene concentrations generally coincide with the highest VC concentrations. This indicates that a portion of the VC is being reductively dechlorinated to ethene. However, because this process is slow relative to the dechlorination of TCE, VC is reaching the canal and discharging to surface water. Chloride is a by-product of the dechlorination process, and chloride concentrations within the CAH plume are elevated well above background concentrations.

Geochemical Parameters. At CCFTA-2 (FT-17), there is a correlation between elevated BTEX concentrations and depleted DO concentrations (Figure 10.6). However, low upgradient and cross-gradient DO concentrations observed at the site suggest that DO is being consumed before it enters the plume area, most likely as microbes use native organic carbon (e.g., peat) as a substrate. Nitrate concentrations in groundwater are generally below 0.1 mg/L; thus nitrate is not likely to be used as an electron acceptor at this site. Sulfate concentrations in shallow groundwater at the site ranged from 7.85 to 842 mg/L. Given the amount of sulfate dissolved in site groundwater, it is likely that sulfate reduction is an ongoing anaerobic biodegradation process at CCFTA-2 (FT-17), but the data indicate no clear pattern. Dissolved hydrogen data (discussed later) indicate that sulfate reduction was occurring in the vicinity of the dissolved CAH plume source area at the time of sampling.

By-products of microbial degradation of organic matter detected in site groundwater include Fe(II), methane, and ammonia. There is a strong correlation between areas with elevated BTEX concentrations and areas with elevated Fe(II), methane, and ammonia (Figure 10.8), suggesting that Fe(III) reduction and methanogenesis are working to biodegrade fuel hydrocarbons at the site. Background Fe(II) and methane concentrations are <0.1 and <0.005 mg/L, respectively. The presence of ammonia in the absence of nitrate suggests that conditions are sufficiently reducing to allow microbial fixation of atmospheric nitrogen.

The pE of groundwater was also measured to give an indication of redox conditions. Areas of lowest pE correspond to areas with contamination (Figure 10.9). This is an indication that biologically mediated oxidation–reduction reactions are occurring

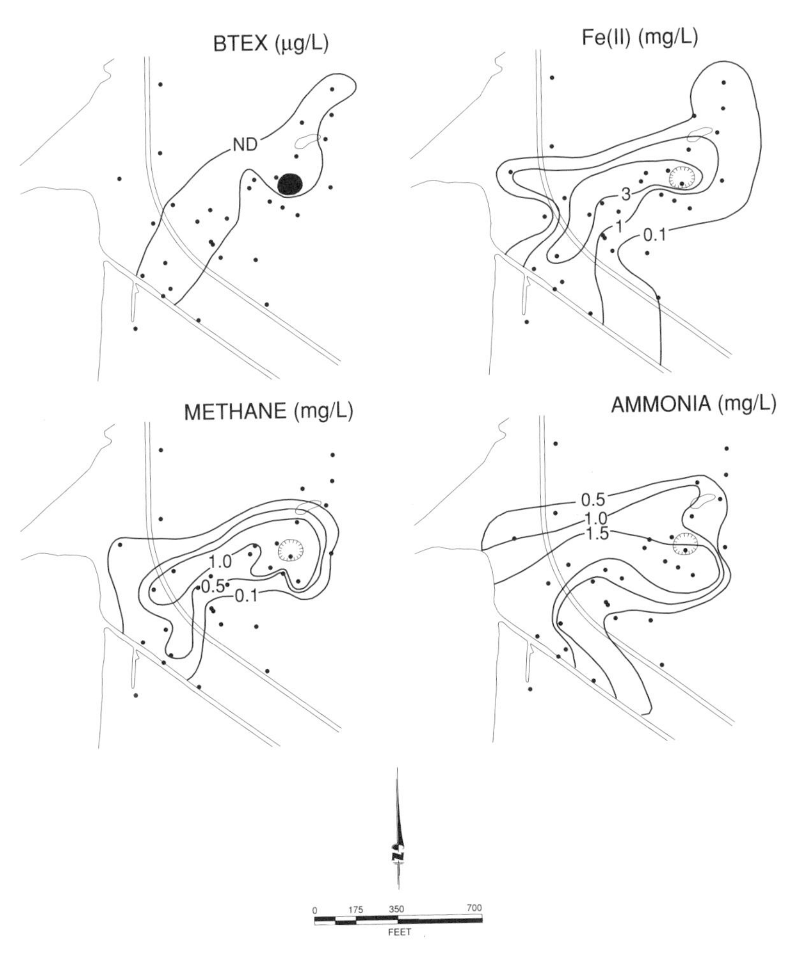

FIGURE 10.8. BTEX and metabolic by-products, FT-17, Cape Canaveral Air Station, Florida.

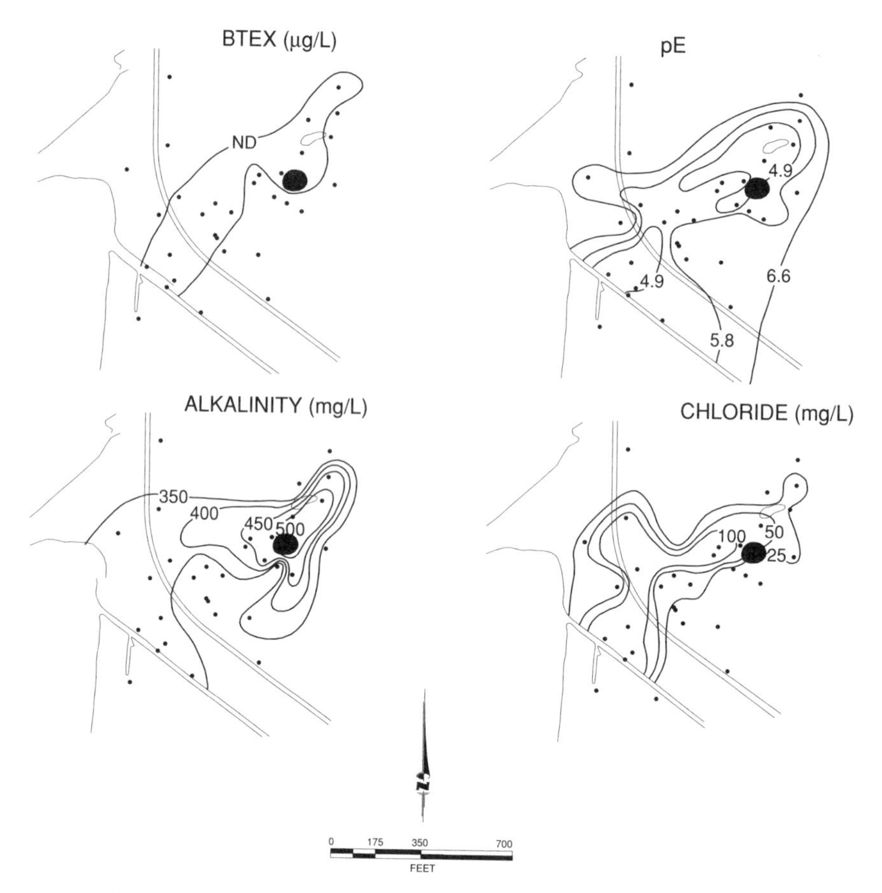

FIGURE 10.9. BTEX and other geochemical parameters, FT-17, Cape Canaveral Air Station, Florida.

in the area with groundwater contamination. Background pE values are relatively low, a further indication that microbial consumption of native organic matter is taking place outside the contaminant plume.

Alkalinity (as $CaCO_3$) also was measured at CCFTA-2 (FT-17). An increase in alkalinity in an area with BTEX concentrations elevated above background conditions can be used to infer that petroleum hydrocarbons (or native organic carbon) has been destroyed through aerobic and anaerobic microbial respiration. The areas with increased alkalinity generally coincide with the BTEX and CAH plumes (Figure 10.9). This is further evidence that biodegradation utilizing BTEX and native organic carbon compounds as substrates is ongoing at the site.

As noted previously, there is a strong correlation between areas with contamination and areas with elevated chloride concentrations relative to measured background concentrations (Figures 10.5 and 10.9). Background chloride concentrations at the site are in the range 5 to 20 mg/L, while chloride concentrations in the CAH plume area are as high as 373 mg/L. The presence of elevated concentrations of chloride in contaminated groundwater provides further evidence that TCE, DCE, and VC are being reductively dechlorinated.

Dissolved native organic carbon also can act as a source of electron donors during the reductive dechlorination of CAHs. Background dissolved TOC concentrations at upgradient and cross-gradient wells ranged from 1.1 to 6.3 mg/L. The background concentrations probably represent compounds dissolved from peat units and other organic matter dispersed throughout the aquifer. In addition to the soil TOC, this native carbon source should provide a continuing source of electron donors to be used in microbial redox reactions. Consumption of this matter and the TOC in soil probably has resulted in the depletion of DO and the relatively low pE values observed outside the contaminant plume.

Dissolved hydrogen (H_2) concentrations can be used to determine the dominant terminal electron-accepting process (TEAP) in an aquifer. Much research has been done on the topic of using hydrogen measurements to delineate TEAPs (Lovley and Goodwin, 1988; Lovley et al., 1994; Chapelle et al., 1995). Groundwater samples for analysis of dissolved H_2 were collected from 17 wells at CCFTA-2 (FT-17). Measured H_2 concentrations range from <0.01 to 1.80 nmol/L, suggesting that Fe(III) reduction and sulfate reduction are the dominant TEAPs at the site. The highest hydrogen concentrations were measured in the vicinity of the northern LNAPL body.

Plume Behavior

As discussed in Chapter 6, contaminant plumes formed by CAHs dissolved in groundwater can exhibit three types of behavior based on the amount and type of primary substrate present in the aquifer. The dissolved CAH plume at CCFTA-2 (FT-17) exhibits characteristics of Type 1 and Type 2 behaviors. Dissolved petroleum hydrocarbons are acting as a carbon source, and the geochemical data indicate that some microbial consumption of native organic matter is taking place outside the plume and probably within the plume. The introduction of petroleum hydrocarbons due to fire training activities probably stimulated additional microbial activity and made the groundwater system reducing enough to foster reductive dechlorination of CAHs, including VC.

Biodegradation Rate Constant Calculations

Field-scale first-order biodegradation rate constants were estimated using methods presented by Buscheck and Alcantar (1995) and Moutoux et al. (1996). Data from a flow path roughly parallel to cross section A–A′ (Figure 10.2) were used in the calculations, the results of which are summarized in Table 10.2. The method of Moutoux et al. (1996) yields a dechlorination rate for all CAHs (averaged across the entire flow path and for all dechlorination reactions) of about 0.00007 day^{-1}, or a half-life of about 27 years. This includes the slowest transformations, such as DCE to VC, and VC to ethene, and probably represents a minimum decay rate.

TABLE 10.2. Estimated First-Order Reductive Dechlorination Rates

Compound(s)	Method	Estimated Dechlorination Rate (day^{-1})	Approximate Half-Life (yr)
Total CAHs	Moutoux et al. (1996)	0.00007	27
Total CAHs	Buscheck and Alcantar (1995)	0.0003	6.3
TCE	Buscheck and Alcantar (1995)	0.0005	3.8
cis-1,2-DCE	Buscheck and Alcantar (1995)	0.0005	3.8
VC	—	VC is produced.	

The method of Buscheck and Alcantar (1995) can be applied to individual compounds as well as to total CAHs. For both TCE and DCE, this method yielded first-order rates of about 0.0005 day^{-1}, or half-lives of about 3.8 years. This probably underestimates the DCE degradation rate, because DCE is being produced (by the dechlorination of TCE) in addition to being degraded. A total CAH degradation rate of about 0.0003 day^{-1} (a half-life of about 6.3 years) was also estimated. Given that VC is accumulating, this probably represents a rate of dechlorination from TCE to VC.

Groundwater Flow and Contaminant Transport Model

To assess the time required to attain surface water and groundwater standards, a numerical groundwater flow and solute transport model was developed for the site using MODFLOW and MT3D. After calibration and sensitivity analyses, the combined model was used to evaluate the fate and transport of dissolved CAHs in site groundwater under the influence of advection, dispersion, sorption, and biodegradation. Model input parameters were obtained from field data where possible; parameters not measured at the site were estimated using reasonable literature values.

Model predictions were made for two scenarios. The first scenario was a baseline model constructed with the assumption that the contaminant source (i.e., mobile and residual LNAPL) would not be removed but that CAH dissolution from the source area would naturally decrease (due to weathering, volatilization, and dissolution) by a rate of 10% per year (i.e., each year's source loading rate was 90% of the previous year's). The second scenario assumed that due to excavation and soil washing, 90% of the source would be removed instantly and the remaining source would weather at the same rate of 10% per year.

Because the HASS has been installed, model output was evaluated by tracking the maximum total CAH concentrations predicted to pass through the HASS. The object was to determine the time until CAH concentrations were low enough (less than about 50 μg/L) so that surface water standards would not be exceeded due to discharge of contaminated groundwater. Results for the first scenario suggest that CAH concentrations migrating beyond the HASS would drop below the target concentration in about 46 to 53 years (from 1997) and that source area concentrations would not drop below the target concentration for at least 70 years. Results for the second scenario suggest that source removal will reduce the time needed to reach the target concentrations by about 20 years. In addition, source concentrations would drop below the target levels in about 60 years under this scenario.

Discussion

Several lines of contaminant and geochemical data indicate that microbial degradation of fuel hydrocarbons and native organic matter is fostering reductive dechlorination of CAHs dissolved in groundwater at CCFTA-2 (FT-17). DCE concentrations in groundwater beneath the source area are six times greater than TCE concentrations, despite the fact that DCE concentrations in the LNAPL are 10 to 100 times lower than the associated TCE concentrations. In addition, nearly all the DCE detected in groundwater is *cis*-1,2-DCE, which is preferentially produced during biological dechlorination of TCE. VC also is present in groundwater but was not detected in the mobile LNAPL. Geochemical data confirm that conditions are sufficiently reducing for biological reductive dechlorination and that chloride is being produced as a by-product of the process.

Both Type 1 and Type 2 behavior are exhibited by the CAH plume. Some of the TCE entering groundwater from the source area LNAPL is being fully dechlorinated to ethene. Most of the TCE is converted rapidly to *cis*-1,2-DCE in the vicinity of the source area. The DCE is dechlorinated relatively rapidly to VC; however, because of the slow rate of transformation of VC to ethene, VC is accumulating and is entering surface water at the canal. Biodegradation (reductive dechlorination) rates calculated using site data suggest that the half-life for the transformation of TCE to ethene is about 6 to 27 years. These calculations also suggest that the rates of TCE and DCE dechlorination (to DCE and VC, respectively) are more rapid, which is consistent with the observed contaminant data.

Results of groundwater flow and solute transport modeling suggest that natural attenuation is best used to reduced dissolved CAH concentrations to desired levels in conjunction with source removal and operation of the HASS. The HASS will reduce, through volatilization, concentrations of VC entering the canal, and the addition of oxygen may also foster aerobic biodegradation of VC and DCE. Source area excavation will reduce greatly the mass of TCE entering site groundwater, although some of the electron donors present in the LNAPL will also be removed. Together, all of these components should bring about site restoration in a cost-effective and timely manner.

While reductive dechlorination is unlikely to cease due to removal of the source, the rates may change as the nature of the substrates and electron donors changes. In the future, as dissolved BTEX concentrations continue to decrease, it is likely that microbes will rely more on heavier petroleum hydrocarbons and native organic matter as sources of energy and carbon. If and when the system might run out of substrates is difficult to predict, especially because the suitability and availability of the native organic carbon for a microbial population is not readily determined with current practices. However, if the system eventually becomes aerobic, it is possible that VC and DCE will be consumed as substrates (e.g., Davis and Carpenter, 1990; Bradley and Chapelle, 1998). This is important to note because most of the dissolved contaminant mass consists of DCE and VC, and once the mobile and residual LNAPL are removed, the relative fractions of DCE and VC in site groundwater will increase relative to TCE.

10.2 INTRINSIC ANAEROBIC BIOTRANSFORMATION OF CHLORINATED SOLVENTS AT A MANUFACTURING PLANT

TIM BUSCHECK AND KIRK O'REILLY
CHEVRON RESEARCH AND DEVELOPMENT COMPANY

AND

GARY HICKMAN
CH2M HILL

Facility Background

Groundwater monitoring data collected since 1986 provide evidence that intrinsic biotransformation of PCE and TCE is occurring in groundwater within the two shallowmost water-bearing zones (A- and C-zones) beneath a manufacturing plant in the San Francisco Bay Area (Buscheck et al., 1997). Low DO measurements, low (negative) redox potentials, and the presence of other indicator parameters suggest that the anaerobic conditions required for reductive dechlorination of chlorinated solvents exist in groundwater at the site. The presence of intermediate and final breakdown products, such as 1,2-DCE, VC, and ethene, demonstrates that reductive dechlorination of PCE and TCE is occurring.

Natural attenuation relies on processes that reduce the concentration of subsurface contaminants without artificial enhancement. Specific attenuation processes include dispersion, sorption, volatilization, and biotic and abiotic transformation processes. Biotransformation offers the potential for mass removal and contaminant destruction and is therefore of particular significance with respect to natural attenuation.

Manufacturing Plant History. The subject of this case study is a manufacturing plant in the San Francisco Bay area. This plant has operated for more than 60 years, producing a variety of agricultural chemicals and fuel additives. Agricultural chemical manufacturing ceased at the end of 1996, and additives manufacturing ceased in early 1997. The plant continues to conduct additives blending/terminaling, warehousing, and laboratory activities. Most of the areas not covered with buildings are either paved or covered with landscape materials.

Site Hydrogeology

The site hydrogeology is characterized by alluvial fan and estuary deposits within 130 ft of ground surface. The alluvial fans consist of clays with layers of silts and sands. The estuary deposits are low-permeability clays. There are three water-bearing zones within 130 ft of ground surface, named in the order in which they were identified. The A-zone is a shallow layer consisting mostly of placed fill over bay mud. Groundwater in the A-zone occurs under water table conditions within 10 ft of the ground surface. Groundwater velocities in the A-zone are approximately 10 to 100 ft/yr.

The C-zone consists of alluvial fan deposits, approximately 20 to 75 ft BGS. Groundwater velocities in the C-zone are 10 to 100 ft/yr. The B-zone is a deeper, coarser-grained layer, more than 85 ft BGS, 10 to 15 ft thick. Groundwater within the C- and B-zones occurs under confined conditions. In general, groundwater gradients in the three water-bearing zones tend to slope to the west. Figure 10.10 illustrates the monitoring well locations.

Contaminant Delineation

Subsurface investigations have been conducted at the plant since the early 1980s. From 1986 to the present, a groundwater monitoring program has been implemented at the plant. Contaminants detected in the A-zone include metals, pesticides, BTEX, and chlorinated solvents (PCE and TCE). The transformation products, 1,2-DCE and VC, were never used at the facility. The contaminants in the C-zone are limited primarily to chlorinated solvents. No contamination has been detected in the B-zone.

Evidence of Natural Attenuation

In 1996 the existing groundwater monitoring program was augmented to include geochemical parameters to evaluate the potential for anaerobic biotransformation of PCE/TCE. This monitoring program, similar to that described by Wiedemeier et al. (1996a), will be conducted for a minimum of two years. The second quarter 1996 analytical results for selected monitoring wells in the A- and C-zone are summarized in Tables 10.3 and 10.4, respectively.

In this section we describe several lines of evidence that demonstrate the occurrence of reductive dechlorination. These lines of evidence are

- Loss of parent contaminants and appearance of breakdown products
- Redox geochemistry to support anaerobic conditions
- Availability of electron donor

Loss of Parent Contaminants and Appearance of Breakdown Products. Monitoring well GW-12A, completed in the A-zone, is located in the northwestern corner of the plant (Figure 10.10). Evidence of reductive dechlorination is illustrated

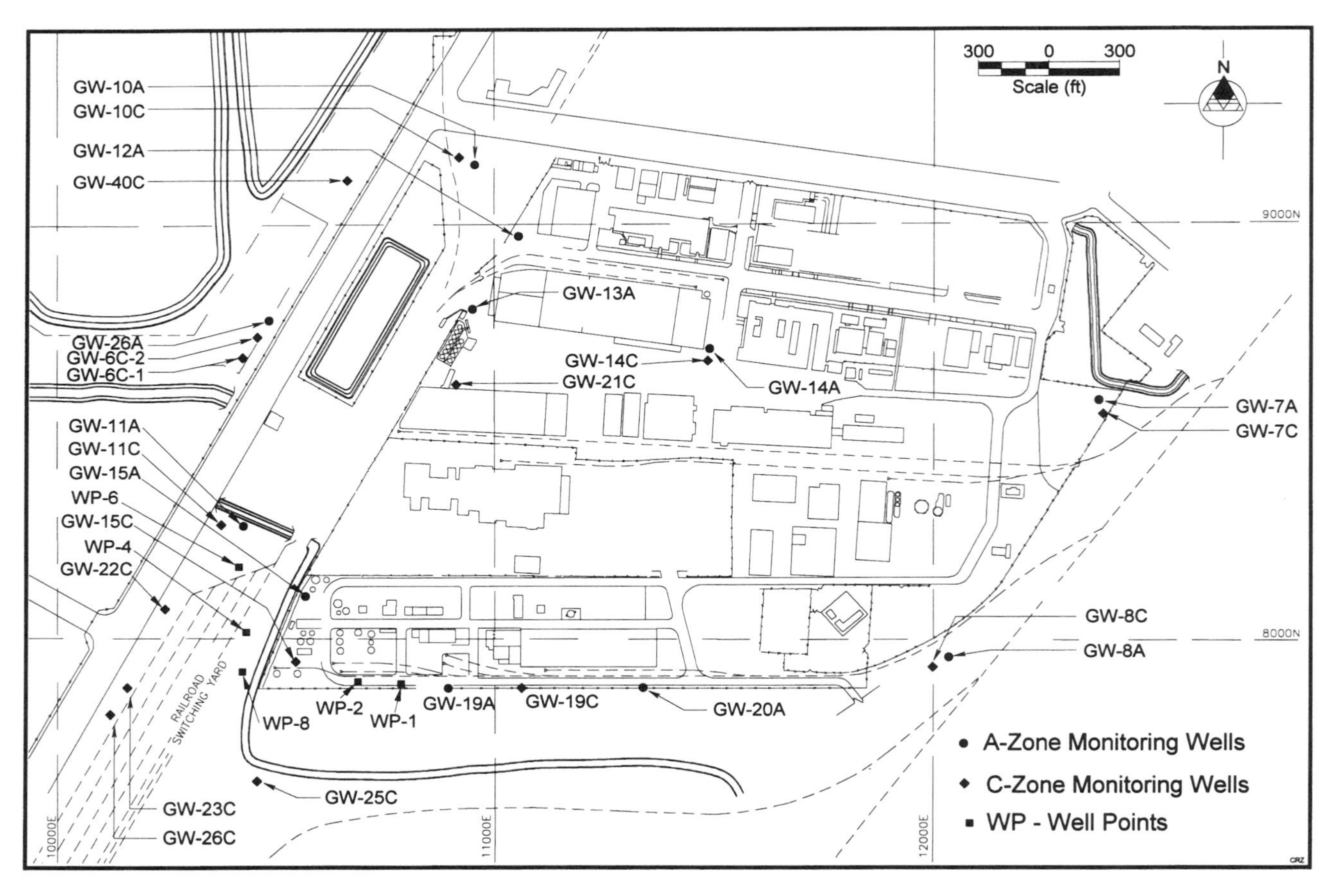

FIGURE 10.10. Site map with monitoring well and well-point locations, manufacturing plant site, California.

TABLE 10.3. Indicator Parameters for Natural Attenuation in A-Zone Monitoring Wells at the California Manufacturing Plant Site[a]

	Monitoring Location					
Analyte	GW-7A	GW-8A	GW-12A	GW-13A	GW-15A	GW-19A
TCE (μg/L)	ND	ND	ND	ND	ND	21
1,2-DCE (μg/L)	ND	ND	79	3.1	4.8	485
VC (μg/L)	ND	ND	367	ND	3.1	43
Ethene (μg/L)	ND	ND	210	ND	ND	ND
Ethane (μg/L)	ND	ND	ND	240	ND	ND
Methane (μg/L)	9.6	11	3000	1900	2000	52
DO (mg/L)	0.21	0.03	1.6	0.65	0.71	0.71
ORP (mV)	98	82	−65	−47	−84	15
TOC (mg/L)	5.47	1.8	24.7	29.2	17	1.65
BTEX (μg/L)	ND	ND	43	ND	315	109

[a] Samples obtained second quarter of 1996. ND, not detected.

TABLE 10.4. Indicator Parameters for Natural Attenuation in C-Zone Monitoring Wells at the California Manufacturing Plant Site[a]

	Monitoring Location						
Analyte	GW-23C	GW-25C	GW-26C	GW-10C	GW-11C	GW-15C	GW-40C
PCE (μg/L)	ND	ND	ND	3446	336	27	4060
TCE (μg/L)	ND	ND	ND	879	5476	34,961	943
1,2-DCE (μg/L)	ND	ND	ND	364	672	102	ND
VC (μg/L)	ND	ND	ND	9.2	30	ND	ND
DO (mg/L)	0.09	0.18	0	0.09	0.04	0.03	0.05
ORP (mV)	−17	−27	−117	−49	−113	36	−80
COD (mg/L)	472	400	449	ND	67.4	157	58.4
Sulfate (mg/L)	1170	683	1330	199	151	48.6	237

[a] Samples obtained second quarter of 1996. ND, not detected.

by concentration versus time data for well GW-12A in Figure 10.11, which includes TCE, 1,2-DCE (*cis* and *trans* isomers), VC, and ethene. The plots in Figure 10.11 illustrate the sequential appearance, accumulation, and depletion of the breakdown products along the reductive dechlorination pathway. Note, that ethene was first included in the sampling program, during the third quarter of 1995. The concentrations of TCE breakdown products at GW-12A exceed TCE concentrations for the period of record. However, historical TCE concentrations (prior to 1986) probably were higher than those shown on Figure 10.11.

Table 10.3 data suggest reductive dechlorination of parent contaminants to VCVC in monitoring wells GW-15A and GW-19A. Table 10.4 provides data for seven C-zone wells (three uncontaminated and four contaminated wells). Although VC was detected in two of the four contaminated wells during the second quarter of 1996, it has historically been measured in each of these wells, as evidence of reductive dechlorination.

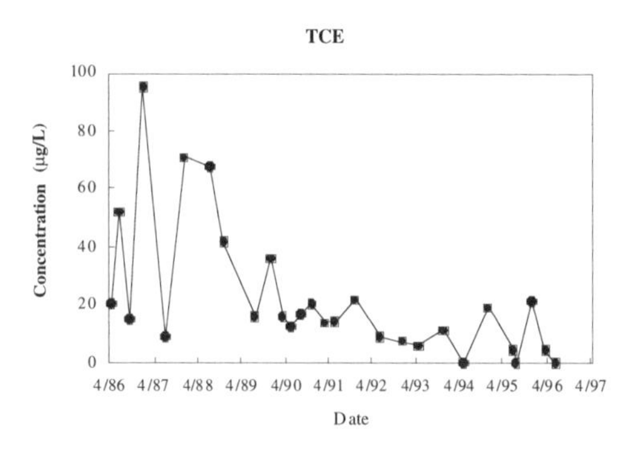

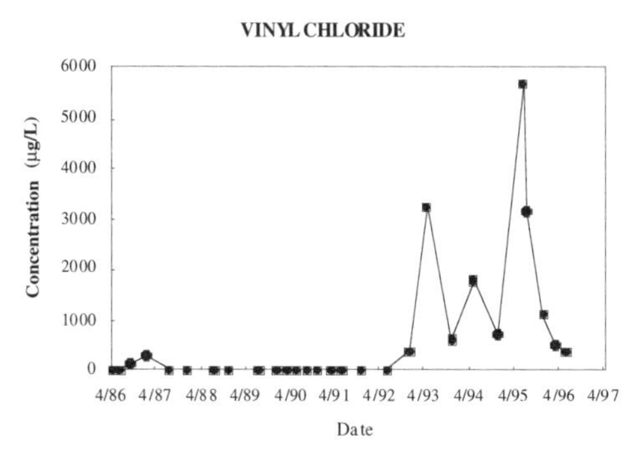

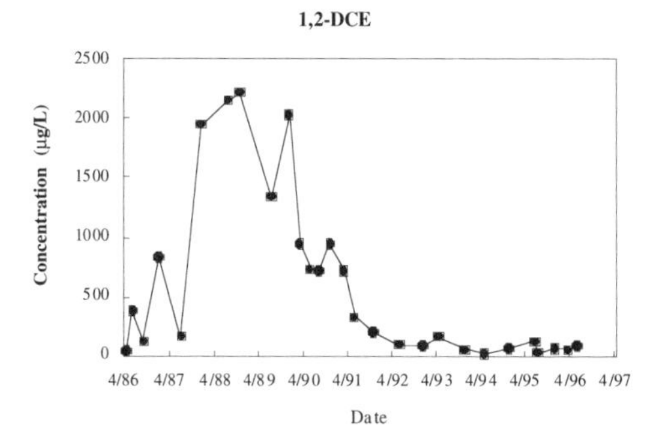

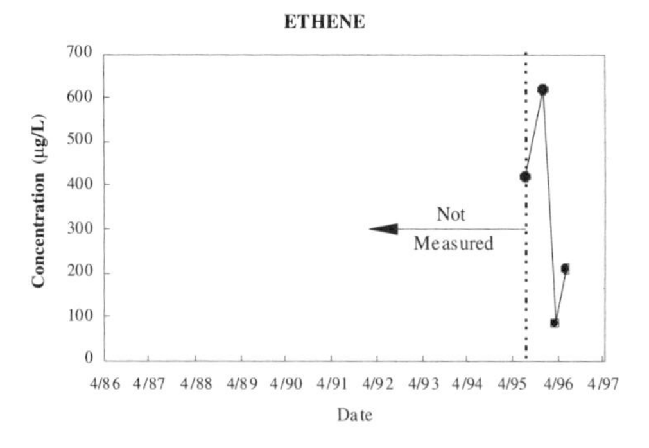

FIGURE 10.11. Concentration versus time for well GW-12A, manufacturing plant site, California.

Redox Geochemistry. Reductive dechlorination will occur under the appropriate geochemical conditions. Redox geochemistry can be used to infer aerobic versus anaerobic conditions. Some of the indicator parameters, such as methane and sulfate, suggest the redox environment. DO concentrations below 1 to 2 mg/L suggest the anaerobic conditions necessary for reductive dechlorination. In Table 10.3, DO concentrations in upgradient wells GW-7A and GW-8A were 0.21 and 0.03 mg/L, respectively. In the second quarter, 22 wells were sampled, and only one well had a DO concentration greater than 2 mg/L (2.1 mg/L). Groundwater in both the A- and C-zones is predominantly anaerobic.

Oxidation–reduction potential (ORP) measurements below +100 mV suggest reducing conditions, expected for anaerobic groundwater. During the second sampling quarter, only two ORP measurements exceeded +100 mV. ORP measurements appear to validate DO measurements.

In three of four monitoring wells with historical measurements of TCE, methane concentrations range from 1900 to 3000 μg/L (Table 10.3). These concentrations suggest methanogenic conditions, generally required for complete reductive dechlorination of TCE to ethene. While ethene has been detected in only one of these monitoring wells, further investigation described below demonstrates that the occurrence of ethene is more widespread than monitoring well data would suggest.

Table 10.4 sulfate concentrations in three uncontaminated wells (GW-23C, 25C, and 26C) ranged from 683 to 1330 mg/L, while concentrations in four contaminated wells were 237 mg/L or less, suggesting sulfate reduction. Reductive dechlorination also occurs under sulfate-reducing conditions.

Electron Donors. Organic compounds other than the chlorinated solvents are required to serve as the electron donor in reductive dechlorination. Several indicators for the occurrence of electron donor were measured, including TOC and chemical oxygen demand (COD). Historically, TOC concentrations in the A-zone have been as high as 30 to 40 mg/L in certain wells. BTEX in the A-zone also serves as an electron donor. TOC concentrations in the C-zone are not as high as A-zone concentrations.

COD levels measured in the C-zone may serve as a better indicator of electron donors. COD in three uncontaminated wells (GW-23C, 25C, and 26C) varied between 400 and 472 mg/L during the second quarter monitoring. COD concentrations in four contaminated wells (GW-10C, 11C, 15C, and 40C) varied from not detected to 157 mg/L during the second quarter. The reduction in COD levels in contaminated wells suggests that this may be an indicator of organics that are serving as the electron donor in reductive dechlorination.

Well-Point Investigation

In late 1996, 0.75-in.-diameter monitoring wells (well points) were installed in the southwestern portion of the plant using direct-push techniques. A total of 11 multilevel well-point "clusters" were installed at discrete depths. The locations of six of these well-point clusters are shown in Figure 10.10. Data from this investigation provide additional evidence for reductive dechlorination of PCE and TCE in the A- and C-zones. Table 10.5 includes five A-zone and two C-zone well points where ethene was detected. These results demonstrate the occurrence of ethene and ethane in the A- and C-zones in this portion of the site, where there were previously no monitoring wells. Well points WP-1C and WP-4C, completed in the C-zone, have low but detectable ethene concentrations. Ethene has not been detected in any C-zone monitoring well. Prior to this well-point investigation, ethene and ethane had been measured in only one A-zone monitoring well (GW-12A).

The Table 10.5 results also provide evidence for redox geochemistry to support anaerobic conditions. Although it was not possible to take DO measurements, well-point monitoring did include ORP measurements, which were negative with the exception of WP-1C, where ORP was 10 mV. Most of the A-zone well-point methane concentrations demonstrate methanogenic conditions, which are generally required for complete reductive dechlorination of CAHs to ethene. TOC concentrations obtained from A-zone well points varied from 11 to 72 mg/L, suggesting the presence of adequate electron donor for reductive dechlorination.

Summary

The groundwater monitoring program at the California manufacturing plant provides good evidence for reductive dechlorination of the PCE and TCE. The breakdown products of PCE and TCE are present in the A- and C-zones. The appearance of ethene and ethane in well GW-12A and in recently installed well points in the southwestern portion of the site indicates that complete dechlorination of chlorinated solvents to innocuous products is occurring. Field measurements for DO and ORP indicate anaerobic conditions in groundwater, as is required for reductive dechlorination. Methane concentrations in A-zone monitoring wells and well points

TABLE 10.5. Indicator Parameters for Natural Attenuation in Wells Points at the California Manufacturing Plant Site[a]

	Monitoring Location						
Analyte	WP-1A	WP-1C	WP-2A	WP-4A	WP-4C	WP-6A	WP-8A
PCE (μg/L)	ND	1,040	ND	ND	ND	ND	18.5
TCE (μg/L)	0.8	225,286	ND	31,900	297	237	23
1,2-DCE (μg/L)	26	ND	131,385	10,025	36	1,687	117
VC (μg/L)	255	ND	16,487	1,050	3.5	128	570
Ethene (μg/L)	2,100	16	220	220	5.2	4.1	840
Ethane (μg/L)	ND	8.5	ND	12	ND	4.5	ND
Methane (μg/L)	1,500	11	200	31	9.8	86	2,400
ORP (mV)	−176	10	−59	NM	−119	−46	−296
TOC (mg/L)	71.7	11	47.8	11.4	2.2	19.4	21.8
COD (mg/L)	331	1,220	568	97.4	315	109	255
Sulfate (mg/L)	40.3	622	2,830	222	170	1,270	31.3

[a] Samples obtained fourth quarter of 1996. ND, not detected; NM, not measured.

demonstrate methanogenic conditions, which generally are required for complete reductive dechlorination of TCE to ethene. Sulfate-reducing conditions appear to exist in the C-zone. Other indicator parameters, such as TOC (A-zone) and COD (C-zone), suggest sufficient concentrations of organic compounds to serve as electron donors.

10.3 NATURAL ATTENUATION OF THE BUILDING 301 TRICHLOROETHENE PLUME AT OFFUTT AIR FORCE BASE, NEBRASKA

David Moutoux and John Hicks
Parsons Engineering Science, Inc.
Denver, Colorado

Facility Background

Offutt AFB is located in eastern Nebraska adjacent to the town of Bellevue, several miles south of Omaha. Building 301 (B301) is situated in the northwestern corner of the base, along the rim of a dissected Pleistocene alluvial terrace remnant of the Missouri River. To the west of B301, the ground slopes steeply downward into the Papillion Creek alluvial valley. A small seepage area and drainage ditch lie at the base of the slope near the base boundary. Papillion Creek lies an additional 3000 ft to the west across the alluvial valley.

The B301 complex was originally constructed in 1941 for the manufacture of bombers during World War II. Manufacturing operations ceased after the war but were resumed briefly between 1959 and 1965 for the assembly of guided missiles. Currently, B301 is the home of the 55th Strategic Reconnaissance Wing as well as several squadrons, assorted base services, and recreational facilities. Solvents used during the varied manufacturing operations are believed to be the source of groundwater contamination identified at the site. Figure 10.12 is a map of B301 showing potential historical source areas.

A plume of dissolved CAHs extends westward approximately 3000 ft from the northwestern corner of B301. Groundwater contaminants include (but are not limited to) PCE, TCE, *cis*-1,2-DCE, *trans*-1,2-DCE, 1,1-DCE, VC, and 1,1-dichloroethane (DCA). TCE, *cis*-1,2-DCE, and 1,1-DCE are the only contaminants detected at concentrations exceeding the federal safe drinking water standards. TCE is the compound of most concern at the site because concentrations exceeding regulatory standards have repeatedly been detected at off-base groundwater monitoring locations.

A low-vacuum dual-phase extraction (LVDPE) system pilot test was conducted near the northwestern corner of B301 in order to evaluate the efficacy of this technology at eliminating subsurface contaminant mass through removal of both soil gas and groundwater. On the basis of the results of this treatibility study, it was suggested that a high-vacuum dual-phase extraction (HVDPE) system could be a more effective remedy; however, a decision to install a full-scale HVDPE system has not yet been

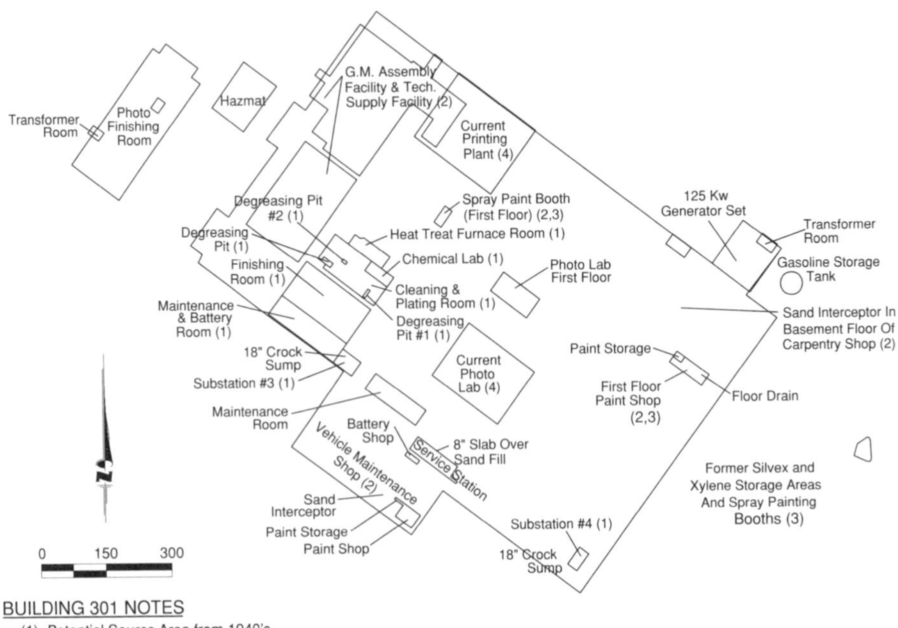

FIGURE 10.12. Building 301 layout, Offutt Air Force Base, Nebraska.

made. As part of the Air Force Center for Environmental Excellence's ongoing natural attenuation initiative, the work summarized in the following few pages was undertaken to collect data on natural attenuation of chlorinated solvents and to evaluate the potential use of natural attenuation as a component of site remediation plans for the B301 TCE plume.

Site Geology and Hydrogeology

The leveled area upon which B301 was constructed consists of up to approximately 13 ft of clayey fill soil overlying a clayey loess. The combined thickness of these units is approximately 60 ft beneath B301, pinching out to the west as the land surface slopes down. The loess overlies a sequence of glaciofluvial sediments and glacial outwash that is up to at least 50 ft thick and consists of well-sorted sand and silty sand. Relatively thin layers of clay-rich glacial till are present within the coarser-grained outwash deposits. The glaciofluvial sediments overlie clay-rich glacial till, which is in turn underlain by limestone.

As described above, the loess and glacial outwash deposits pinch out near the Offutt AFB boundary and the base of the upland terrace. The subsurface west of B301 in the Papillion Creek valley consists primarily of clays and silts of alluvial origin, with intermittent sand lenses. Available data indicate that the alluvial section west of the base boundary is approximately 50 to 70 ft thick. The alluvium is underlain by glacial till. Figure 10.13 identifies the location of a simplified hydrogeologic cross section provided on Figure 10.14. The section runs the length of the TCE plume parallel to the direction of groundwater flow.

The depth to groundwater in the immediate vicinity of B301 is approximately 45 to 60 ft BGS. The depth to groundwater decreases toward the west as a result of the steep decline in ground surface elevation between the upland terrace and the alluvial valley. Near and west of the base boundary, depth to groundwater is 3 to 10 ft BGS. The groundwater flow direction is westerly toward Papillion Creek (Figure 10.15). The lateral hydraulic gradient beneath the upland terrace is approximately 0.002 ft/ft, increasing to 0.03 ft/ft beneath the terrace slope to the west of B301. Beneath the alluvial valley west of Fort Crook Road, the gradient decreases to approximately 0.01 ft/ft. Vertical hydraulic gradients appear to be upward in the top 25 ft of the Surficial Aquifer and slightly downward deeper in the aquifer. Hydraulic conductivity and effective porosity values vary considerably between hydrogeologic units. Estimated values for these parameters and the resultant groundwater velocities are provided in Table 10.6. Groundwater flow in the alluvial silt and clay unit in the valley may be enhanced by the presence of desiccation fractures and/or sand seams.

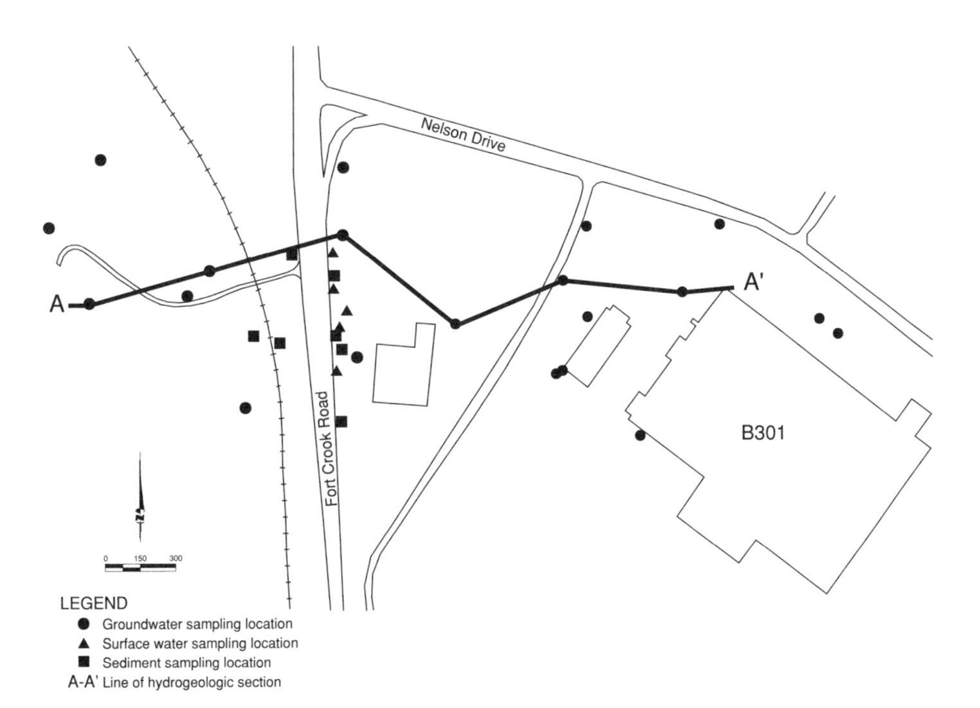

FIGURE 10.13. Site location map, Offutt Air Force Base, Nebraska.

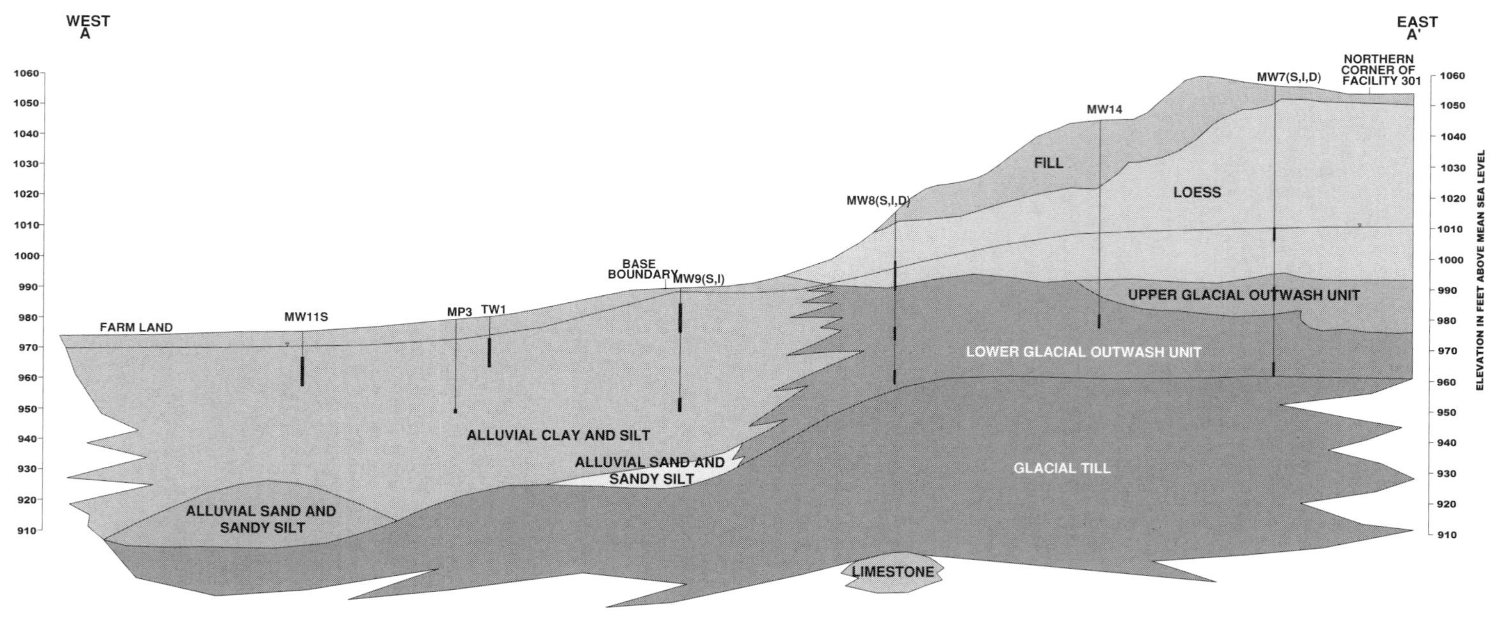

FIGURE 10.14. Hydrogeologic cross section, Offutt Air Force Base, Nebraska.

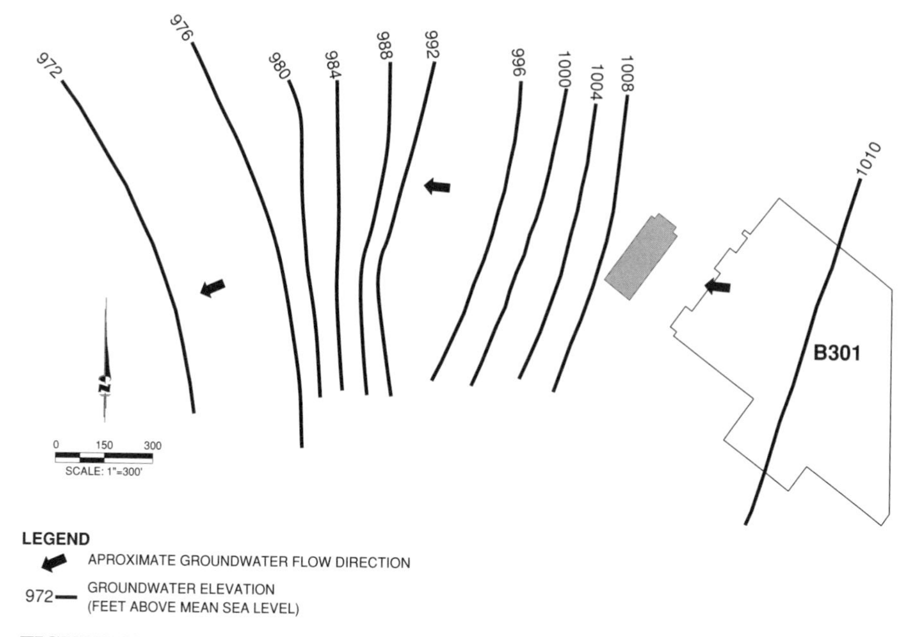

FIGURE 10.15. Groundwater table elevations, July 1996, Offutt Air Force Base, Nebraska.

Background TOC concentrations in soil also vary between hydrogeologic units. Specifically, the TOC concentrations are higher within the alluvial valley than within the upland terrace deposits. Some of the measured TOC concentrations in the alluvial valley deposits are sufficient to retard the transport of organic contaminants significantly relative to the advective groundwater velocity. Average retardation factors for each hydrogeologic unit were estimated using the TOC data. These factors, and the estimated retarded velocity for TCE, also are provided in Table 10.6.

TABLE 10.6. Hydrogeologic Parameters and Resultant Velocities by Lithologic Unit at Building 301

Lithologic Unit	Mean Hydraulic Conductivity (ft/day)	Estimated Effective Porosity	Estimated Groundwater Velocity Range (ft/yr)	Average TOC-based Retardation Factor	Retarded Velocity for TCE (ft/yr)
Loess	0.3	0.2	0.20–58	1.18	0.17–49
Upper glacial outwash	1.1	0.2	1.1–7.3	1.14	0.96–6.4
Lower glacial outwash	16.9	0.25	7.3–1971	1.14	6.4–1729
Alluvial silt and clay	3.5	0.15	55–124	3.49	16–36

Groundwater Chemistry

Distribution of CAHs and Metabolites. Groundwater beneath and downgradient from the B301 site is contaminated with dissolved chlorinated solvents, principally TCE, that were introduced into the groundwater as a result of past activities at B301. The source of TCE contamination appears to be located beneath the northwestern corner of B301, as evidenced by an elevated TCE concentration (25,000 μg/L) in a groundwater sample collected in this area. The source has not been located or characterized. The TCE plume extends westward approximately 2800 ft from the suspected source area, terminating beneath a soybean field (Figure 10.16). Vertical contaminant profiling indicates that the TCE migrates through the glacial outwash beneath the upland terrace and through the alluvial clay and silt within the adjacent valley (Figure 10.17).

In addition to TCE, lesser concentrations of PCE, DCE, and VC also were detected at the site at maximum concentrations of 1.9, 1237, and 1.3 μg/L, respectively. Figure 10.16 shows the distribution of chlorinated ethenes at the B301 site. PCE was detected at three locations in and immediately downgradient from the suspected source area; no PCE has been detected beyond or even close to the base boundary. All three isomers of DCE were detected at the site. The *cis*-1,2-DCE isomer was detected over the largest area and at the highest concentrations (ranging up to 1230 μg/L). Relatively

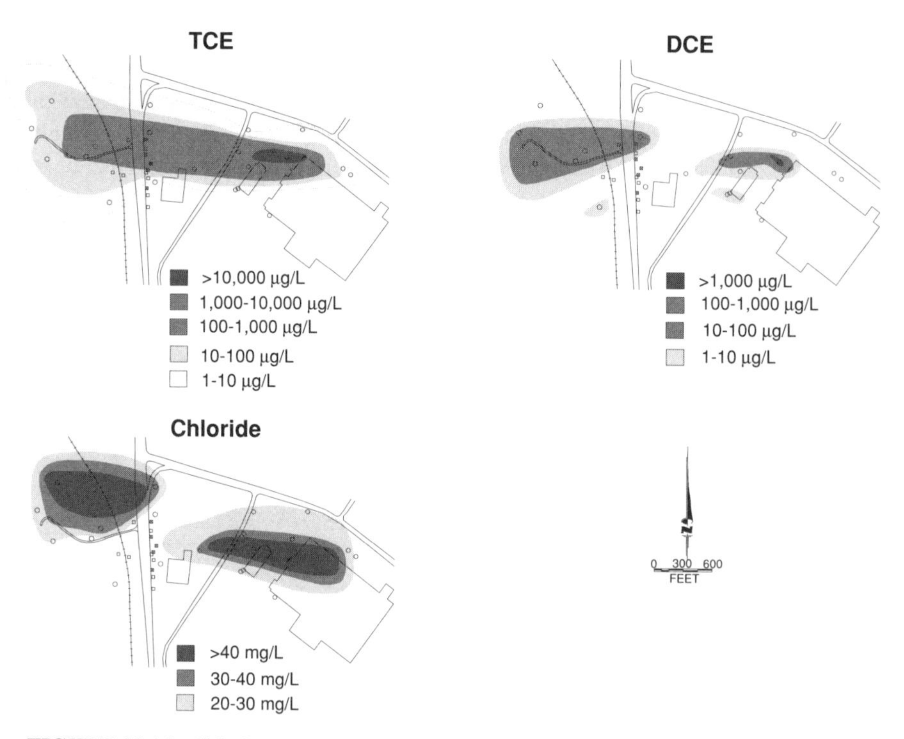

FIGURE 10.16. Chlorinated solvents and by-products, Offutt Air Force Base, Nebraska.

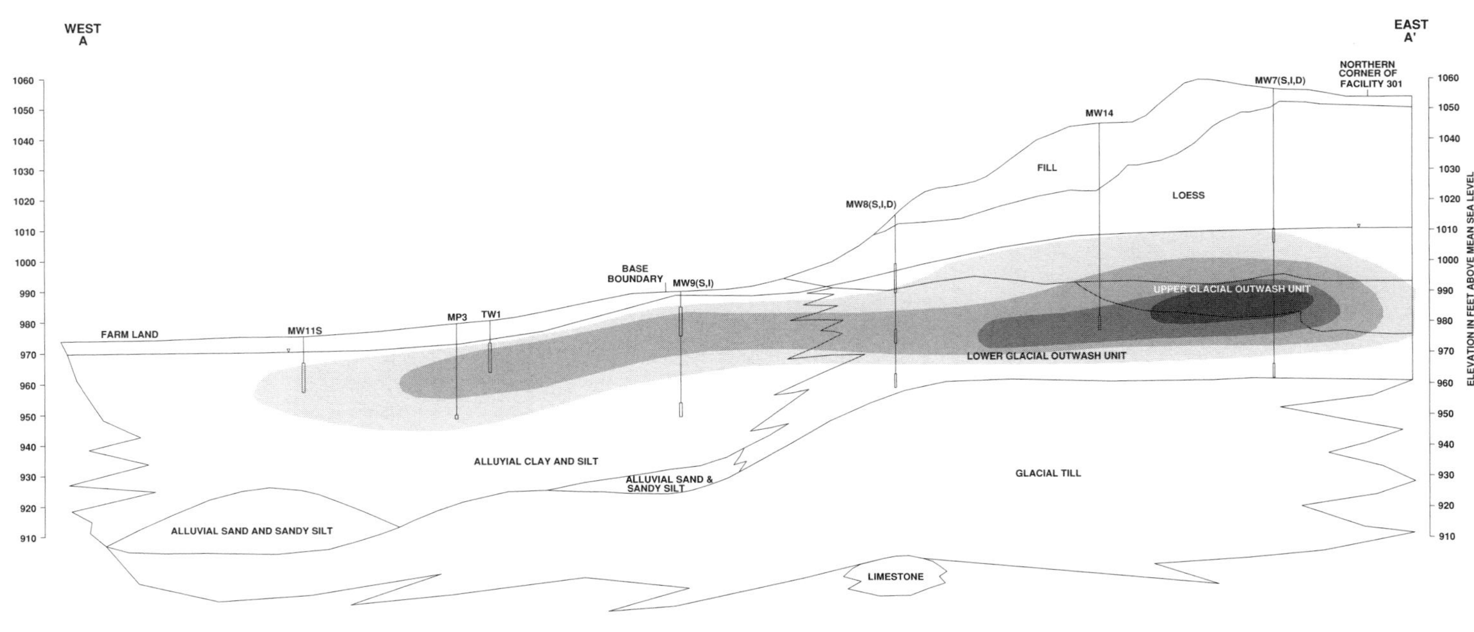

FIGURE 10.17. Vertical TCE distribution, Offutt Air Force Base, Nebraska.

low concentrations of *trans*-1,2-DCE and 1,1-DCE (ranging up to 9.4 and 28.6 μg/L, respectively) also were detected during the sampling event. The low magnitude of detected *trans*-1,2-DCE and 1,1-DCE concentrations relative to those of *cis*-1,2-DCE is common in environments where reductive dechlorination (halorespiration) of TCE occurs. VC was detected at two locations in the downgradient portion of the plume in the Papillion Creek alluvial valley.

Seven surface water samples were collected at the site, TCE was detected in the two samples collected from a westerly-to-southerly flowing drainage ditch west of the base boundary; however, TCE was not detected in surface water samples collected from other drainage ditches and areas of standing water in the vicinity of the plume. The surface water sample with the higher observed TCE concentration (75 μg/L) also contained the reductive dechlorination daughter product *cis*-1,2-DCE at 3.7 μg/L.

Natural Attenuation Analysis. In the B301 plume, limited biodegradation of the TCE appears to be occurring, primarily by reductive dechlorination of TCE to DCE. The areas of most active reductive dechlorination appear to be characterized by Type II conditions and coincide with the Papillion Creek alluvial valley and possibly also the source area. Between the source area and the Papillion Creek alluvial valley Type III conditions prevail and may be accompanied by DCE oxidation. These observations are supported by the distribution of TCE and metabolites, the oxidation–reduction potential (ORP) of the groundwater, electron donors, and alternative electron acceptors and their metabolic by-products.

Results from three groundwater monitoring events performed since 1991 do not indicate overall decreases in dissolved TCE concentrations. Rather, concentrations have remained relatively uniform or have varied erratically. These observations suggest that the rate of natural attenuation is not sufficient to overcome the rate at which TCE partitions from the source into groundwater. The inability to locate a definite source area, and the supposition that the source is at least 30 years old, imply (but do not confirm) that the rate of natural attenuation is not rapid.

As already described, the presence of TCE together with its daughter products *cis*-1,2-DCE and VC is strong evidence that reductive dechlorination is occurring at the site; however, the low concentrations of these metabolites in relation to the TCE concentrations at the same locations is a strong indication that the process is limited. Also, the presence of daughter products only in the source area and the Papillion Creek alluvial valley suggests that reductive dechlorination is localized. Chloride, another metabolite of reductive dechlorination, occurs at elevated concentrations within and upgradient from the source area and in the Papillion Creek alluvial valley (Figure 10.16).

Additional evidence regarding biodegradation processes at the site can be inferred by evaluating the TCE/DCE ratio at several points along a flow path. At the B301 site, TCE/DCE ratios were computed for eight points along a flow path (Figure 10.18). Within the source area, the ratio is approximately 10:1; however, downgradient from the source area, the ratio steadily increases to more than 1000:1. This trend implies that DCE concentrations are attenuating faster than TCE concentrations. Because VC is not present within this portion of the plume, it can

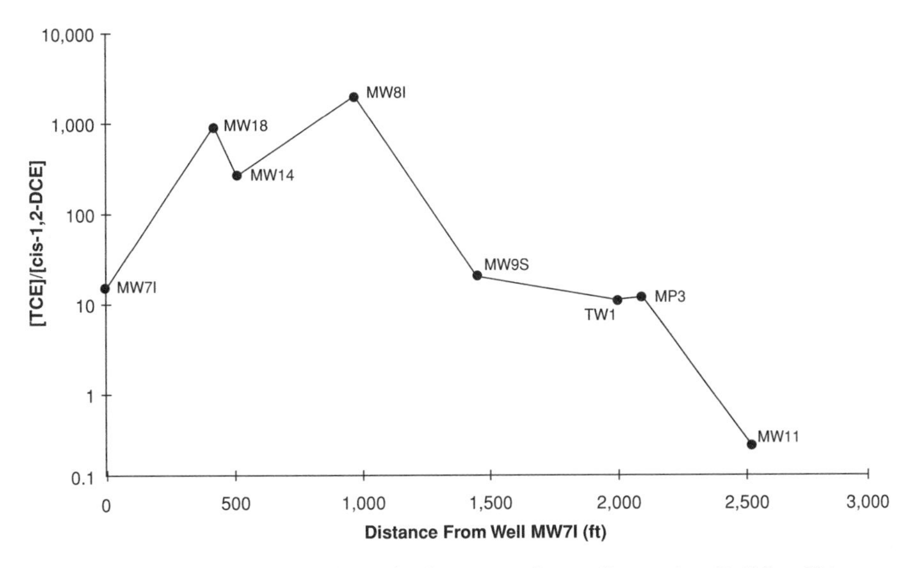

FIGURE 10.18. TCE to DCE ratio along groundwater flow path at Building 301.

be inferred that the DCE is degraded through biological oxidation. Current research also suggests that conditions favorable for DCE oxidation are unfavorable for reductive dechlorination, implying that reductive dechlorination does not occur where oxidation is inferred. As the plume enters the Papillion Creek alluvial valley, the TCE/DCE ratio begins to decrease. In fact, by the end of the flow path, the ratio falls below 1:1. This implies that reductive dechlorination resumes within the alluvial valley. In fact, the presence of VC along with the lowest TCE/DCE ratios suggest that the most aggressive reductive dechlorination at the site occurs within the alluvial valley.

The ORP of groundwater, as well as dissolved hydrogen concentrations, can provide insight into preferred microbiological reactions within a given area. In particular, reducing conditions are required for reductive dechlorination to occur. Conversely, relatively oxidizing conditions are required for DCE oxidation. At the B301 site, the relatively reducing conditions required for reductive dechlorination appeared to be limited to the downgradient portion of the plume within the Papillion Creek alluvial aquifer; more oxidizing conditions were present in the glaciofluvial deposits nearer the source area. Concentrations of dissolved hydrogen (H_2) also can be used to evaluate the groundwater ORP. H_2 is produced continuously in anaerobic groundwater systems by fermenting microorganisms that decompose natural and anthropogenic organic matter. H_2 is then consumed by respiratory microorganisms, with the least efficient H_2 users (including halorespirators) operating under the most reducing conditions. Although exact ranges for specific halorespirators are still being determined, H_2 concentrations higher than 1 nmol/L would probably be required to support substantial reductive dechlorination. Such conditions were observed at over half of the measurement locations within the downgradient, anaerobic portion of the contaminant plume, suggesting that conditions in that area are favorable for reductive dechlorination.

When investigating the biodegradation of CAHs, it is necessary to examine the distribution of other compounds that are used in the microbially mediated reactions that facilitate CAH degradation. The presence of an anthropogenic carbon source such as BTEX could create favorable conditions for reductive dechlorination, because BTEX compounds can function as electron donors and facilitate microbial reactions that drive down the local groundwater ORP. BTEX compounds, however, were detected in only two B301 groundwater samples at low levels during the sampling event and do not appear to be a factor in the biodegradation of CAHs within the sampled area at B301. Dissolved native organic carbon (measured as TOC) also can act as a source of electron donors during the reductive dechlorination of CAHs. At the B301 site, background dissolved TOC concentrations averaged approximately 2.5 mg/L. Although this native carbon source could provide a continuing source of electron donors, Wiedemeier et al. (1996) report that dissolved TOC concentrations less than 20 mg/L may not be sufficient to drive dechlorination reactions significantly and sustainably.

Biodegradation of organic compounds, whether natural or anthropogenic, brings about measurable changes in the chemistry of groundwater in the area affected. Concentrations of compounds used as electron acceptors (e.g., DO, nitrate, and sulfate) are depleted, and by-products of electron-acceptor reduction [e.g., Fe(II), methane, and

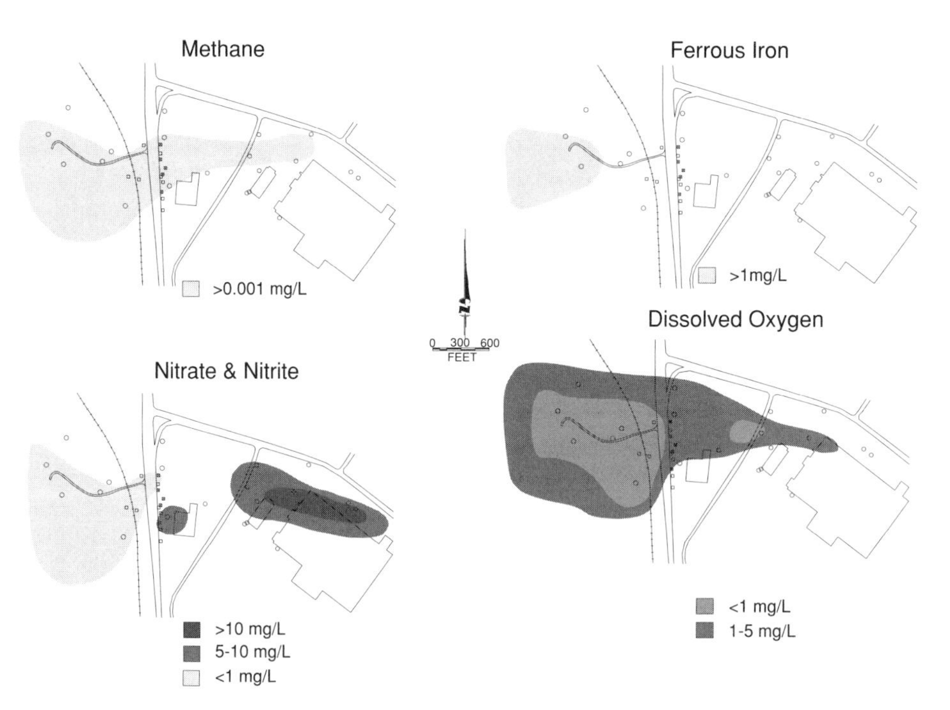

FIGURE 10.19. Electron acceptors and metabolic by-products, Offutt Air Force Base, Nebraska.

sulfide] are produced. As explained in Chapter 5, microbial reactions typically proceed from most energetically favorable to least energetically favorable as follows: aerobic respiration, denitrification, Fe(III) reduction, sulfate reduction, and methanogenesis. When used as electron acceptors, chlorinated solvents are in competition with these native electron acceptors. Typically, reductive dechlorination becomes favorable under sulfate-reducing or methanogenic conditions. At B301, DO and nitrate concentrations are depleted west of the base boundary, Fe(II) concentrations are elevated downgradient beneath the soybean field, sulfate concentrations are inconclusive and may be sufficiently high to inhibit reductive dechlorination, and low methane concentrations (0.001 to 0.184 mg/L) are present along the entire length of the TCE plume with an isolated high methane concentration (12.3 mg/L) detected in the downgradient portion of the plume (Figure 10.19). Of these acceptors and by-products, the methane distribution clearly supports the inference of limited and localized occurrence of reductive dechlorination. The low methane concentrations along the plume length may represent a methane "shadow," assuming that at some time in the past (or perhaps currently beneath B301) conditions were (are) methanogenic and therefore conducive to reductive dechlorination. The 12.3-mg/L methane concentration in the downgradient portion of the plume suggests that conditions are stongly methanogenic at that location and therefore very conducive to reductive dechlorination.

Numerical Natural Attenuation Screening. Wiedemeier et al. (1996b) present a worksheet to allow an initial assessment of the prominence of reductive dechlorination at a site. The worksheet, including the point values determined for B301 near the source area and within the western portion of the contaminant plume within the alluvial valley, are included in Table 10.7. The score for the western portion of the contaminant plume in the Papillion Creek alluvial valley is 16, indicating that adequate evidence for reductive dechlorination of chlorinated organics is present. The score for the upgradient portion of the plume nearest the source area is –1, indicating inadequate evidence for reductive dechlorination of chlorinated organics is present, although "limited" may be a better description for reductive dechlorination in the source area based on the presence of a relatively high concentration of *cis*-1,2-DCE and the uncertainty regarding the precise location and physical–chemical characteristics of the continuing source, as well as the absence of data on the possible past presence of anthropogenic electron donors.

Biodegradation Rate Constant Calculations

Field-scale first-order biodegradation rate constants were estimated in order to simulate the fate and transport of contaminants dissolved in groundwater using segments of the flow path presented in cross section A–A′ (Figure 10.13). A method proposed by Buscheck and Alcantar (1995) was used to estimate a TCE biodegradation rate of 4×10^{-4} to 7×10^{-4} day^{-1}, or a half-life of 2.7 to 4.7 years. This method involves coupling the regression of contaminant concentration versus distance downgradient to an analytical solution for one-dimensional steady-state contaminant transport that includes advection, dispersion, sorption, and biodegradation. For the B301 TCE

TABLE 10.7. Analytical Parameters and Weighting for RNA Screening at Building 301[a]

Analysis	Concentration in Most Contaminated Zone	Interpretation	Value	Western Portion Score	Source Area Score
Oxygen	<0.5 mg/L	Tolerated, suppresses the reductive pathway at higher concentrations	3	2	0
	>1 mg/L	VC may be oxidized aerobically	–3	0	–3
Nitrate	<1 mg/L	At higher concentrations may compete with reductive pathway	2	2	0
Fe(II)	>1 mg/L	Reductive pathway possible	3	3	0
Sulfate	<20 mg/L	At higher concentrations may compete with reductive pathway	2	0	0
Sulfide	>1 mg/L	Reductive pathway possible	3	—	—
Methane	<0.5 mg/L	VC oxidizes	0	0	0
	>0.5 mg/L	Ultimate reductive daughter product; VC accumulates	3	1[b]	0
ORP	<50 mV	Reductive pathway possible	1	10	
	<–100 mV	Reductive pathway likely	2	0	0
pH	5 < pH < 9	Optimal range for reductive pathway	0	0	0
	5 > pH > 9	Outside optimal range for reductive pathway	–2	0	0
TOC	>20 mg/L	Carbon and energy source; drives dechlorination; can be natural or anthropogenic	2	0	0
Temperature	>20°C	At T > 20°C, biochemical processes are accelerated	1	0	0
Carbon dioxide	>2 × background	Ultimate oxidative daughter product	1	—	—
Alkalinity	>2 × background	Results from interaction of carbon dioxide with aquifer minerals	1	0	0
Chloride	>2 × background	Daughter product of organic chlorine	2	1[c]	0
Hydrogen	>1 nM/L	Reductive pathway possible; VC may accumulate	3	3	0
Hydrogen	<1 nM/L	VC oxidized	0	0	0
Volatile fatty acids	>0.1 mg/L	Intermediates resulting from biodegradation of aromatic compounds; carbon and energy source	2	—	0
BTEX	>0.1 mg/L	Carbon and energy source; drives dechlorination	2	0	0
PCE		Material released	0	0	0
TCE		Material released	0	0	0
		Daughter product of PCE	2[d]	0	0

TABLE 10.7. *(Continued)*

Analysis	Concentration in Most Contaminated Zone	Interpretation	Value	Western Portion Score	Source Area Score
DCE		Material released	0	0	0
		Daughter product of TCE	2[d]	2	2
		If *cis* is greater than 80% of total DCE, it is probably a daughter product of TCE			
VC		Material released	0	0	0
		Daughter product of DCE	2[d]	1	0
Ethene/ethane	>0.01 mg/L	Daughter product of VC/ethene	2	0	0
	>0.1 mg/L		3	0	0
Chloroethane		Daughter product of VC under reducing conditions	2	—	—
1,1,1-Trichloroethane		Material released	0	0	0
1,2-Dichlorobenzene		Material released	0	—	—
1,3-Dichlorobenzene		Material released	0	—	—
1,4-Dichlorobenzene		Material released	0	—	—
Chlorobenzene		Material released or daughter product of dichlorobenzene	2[d]	—	—
1,1-DCE		Daughter product of TCE or chemical reaction of 1,1,1-TCA	2[d]	0	0[e]
				16	−1

[a]— indicates that the analysis was not completed.

[b]Point awarded because of the methane detection in monitoring point MP3.

[c]Partial points awarded because the only well in the flow path with a chloride concentration higher than 2 × the average background chloride concentration is TW1.

[d]Points awarded only if it can be shown that the compound is a daughter product (i.e., not a constituent of the source NAPL).

[e]The low ratio of *cis*-1,2-DCE to 1,1-DCE in groundwater from MW18 indicates that 1,1-DCE is not a daughter product.

plume, approximation using this method is viewed as an upper bound on the biodegradation rate because the method assumes a steady-state plume, while the B301 plume may in fact be expanding. A typical expanding plume exhibits decreasing source area concentrations, increasing downgradient concentrations, or both. Over time, these changes result in a decreasing slope on a log-linear plot, and consequently a decreasing biodegradation rate.

Another method for estimating dechlorination rates of CAHs is described by Moutoux et al. (1996). This method can be used to estimate the theoretical contaminant concentration resulting from reductive dechlorination alone for every point along a flow path on the basis of the measured contaminant concentration at the point of plume origin and the ratio of molar mass decrease to the molar chlorination decrease between consecutive points along the flow path. Using the method of Moutoux et al. (1996) and data from the downgradient portion of the plume where significant reductive dechlorination is believed to occur produces an estimated dechlorination rate of 3×10^{-5} to 4×10^{-5} day^{-1} or a half-life of 47 to 63 years. These rates are considered to be lower bounds for the TCE degradation rate because they provide a total dechlorination rate for all dechlorination steps. In other words, all rates, including the rapid TCE-to-DCE rate and the slow VC-to-ethene rate, are averaged into a single first-order rate. The method also assumes that each of the CAH compounds respond identically to all physical, chemical, and biologic processes. At the B301 site it is possible that oxidative reactions, abiotic degradation, or volatilization preferentially reduce DCE mass with respect to TCE. Such a loss would be reflected as a decrease in the calculated total reductive dechlorination rate.

Groundwater Flow and Contaminant Transport Model

To help predict the future migration of CAHs dissolved in groundwater at the B301 site, a numerical groundwater flow and solute transport model was developed for the site using MODFLOW (McDonald and Harbaugh, 1988) and MT3D$^{96®}$ (S.S. Papadopulos and Associates, Inc., 1996). After calibration and sensitivity analyses, the model was used to evaluate the fate and transport of dissolved CAHs in site groundwater under the influence of advection, dispersion, sorption, and biodegradation. Model input parameters were obtained from field data where possible; parameters not measured at the site were estimated using reasonable literature values.

Model predictions were made for three scenarios. The first scenario was a baseline model which assumed that the contaminant source would not be removed. Because the source has not been identified, the continuing source used to calibrate the model was left unchanged for the first 100 years of the model simulation to be conservative. Thereafter, TCE dissolution from the source was assumed to decrease naturally at a rate of 1% per year (i.e., each year's source loading rate was 99% of the previous year's). The second scenario assumed that due to dual-phase extraction (DPE), approximately 80% of the source would be removed after 10 years of system operation, and the remaining source would weather at the same rate of 1% per year. The third scenario mimicked the second in the source area; however, the effect of a hypothetical groundwater extraction (GWE) trench that would intercept most of the TCE plume at the base boundary was also simulated. Because of the large range of possible TCE biodegradation rates in the alluvial silts and clays, two simulations were run for each of the three remedial scenarios: the first with a 24-year TCE half-life (intermediate between the upper- and lower-bound half-lives computed for the alluvium) and the second with a 4.5-year TCE half-life (best-case scenario). The TCE half-life assigned to the upgradient portion of the plume area for each simulation was 316 years (decay rate of 6×10^{-6} day^{-1}). This value was obtained during calibration of the fate and transport model and is consistent with the evidence that significant biodegradation of TCE is not occurring between B301 and the base boundary. Results of the six simulations are summarized in Table 10.8.

Discussion

Several lines of chemical and geochemical evidence indicate that although dissolved CAHs at B301 are undergoing biologically facilitated reductive dechlorination, the occurrence of this process is limited and localized. As a result, the parent CAH (TCE) still comprises the majority of the contamination present in groundwater throughout most of the plume. The evidence supporting the localized occurrence of TCE biodegradation is summarized below.

- The presence of *cis*-1,2-DCE in groundwater in the Papillion Creek alluvial valley and in the source area is a direct indication that TCE is being dechlorinated reductively in these areas, but the limited extent of elevated *cis*-1,2-DCE concentrations and the changing TCE/DCE ratio indicates that this is a localized process.
- The presence of elevated chloride concentrations is very localized (source area and western portion of plume), indicating that reductive dechlorination reactions are not prevalent enough in many portions of the plume to increase chloride concentrations significantly.
- DO, ORP, and dissolved H_2 data indicate that the groundwater is relatively anaerobic and reducing west of the base boundary and therefore more conducive to the occurrence of reductive dechlorination. East of the base boundary, conditions are generally less reducing and more aerobic and are therefore not within the optimal range for reductive dechlorination.

TABLE 10.8. Summary Results from Model Simulations for the Building 301 Plume

Simulation	TCE Half-Life in Alluvium (years)	Reduction in Source Strength	Maximum Downgradient Extent of TCE Plume
Alternative 1: RNA	24	0% after 100 years/ 63% after 200 years	Papillion Creek (8 μg/L at 200 years)
	4.5	0% after 100 years/ 63% after 200 years	1450 ft updgradient from Papillion Creek
Alternative 2: DPE	24	80% after 10 years/ 97% after 200 years	Papillion Creek (4 μg/L at 200 years)
	4.5	80% after 10 years/ 97% after 200 years	1500 ft updgradient from Papillion Creek
Alternative 3: DPE + GWE	24	80% after 10 years/ 97% after 200 years	Papillion Creek (2 μg/L at 200 years)
	4.5	80% after 10 years/ 97% after 200 years	1550 ft updgradient from Papillion Creek

- Soil TOC concentrations are elevated in the alluvial silts and clays of the Papillion Creek alluvial valley, providing an ample electron donor supply to fuel reductive dechlorination in this portion of the plume.
- Reduced nitrate/nitrite, elevated Fe(II), and elevated methane concentrations in groundwater within the western portion of the contaminant plume (alluvial valley) indicate that conditions are reducing enough to support localized reductive dechlorination.

An important component of this study was an assessment of the potential for contamination in groundwater to migrate from the source area to potential receptor exposure points at concentrations above regulatory levels intended to be protective of human health and the environment. To accomplish this objective, three remedial alternatives were simulated using the calibrated numerical model and a range of potential decay rates for the alluvial deposits. The lower decay rate (TCE half-life of 24 years) is a reasonable value for the alluvial deposits on the basis of site-specific data collected to date. The higher decay rate (TCE half-life of 4.5 years) is optimistic, although still potentially realistic.

Model simulations performed using the lower decay rate all indicate that substantial plume migration could occur within the Papillion Creek valley and that the plume could potentially migrate to Papillion Creek within the next 200 years even if the evaluated remedial actions are implemented. However, model results suggest that Nebraska drinking water criteria and aquatic life standards are unlikely to be exceeded in the creek. Model simulations performed using the higher decay rate indicate that the plume may only migrate several hundred feet farther west of its current downgradient extent. All simulations suggest that dissolved TCE concentrations substantially greater than 5 μg/L will persist in groundwater for more than 200 years.

Implementation of natural attenuation, institutional controls, and long-term monitoring (LTM) of groundwater and surface water is recommended, at least in the short term. Available data indicate that the CAH plume is slowing and biodegrading west of the base boundary and is mostly contained within an easement area where land use is controlled by the Air Force. The degree to which RNA will prevent further downgradient migration of the plume is not known with certainty but will become more apparent as successive LTM events are performed and contaminant decay rates and fate and transport predictions are refined. In addition, the shallow groundwater in the Papillion Creek valley near the plume is not being used as a drinking water or irrigation source and may not represent a reliable groundwater source due to the typically low permeability of the alluvial deposits. To assess the effectiveness of naturally occurring processes at reducing contaminant mass and minimizing contaminant migration, groundwater and surface water sampling should continue on an annual basis for approximately five years, followed by less frequent (e.g., once every five years) sampling. The LTM plan should be reevaluated periodically and modified as necessary on the basis of newly obtained data. If data collected at any time during the monitoring period indicate the need for additional remedial activities at the site, or if shortening of the LTM period is desired, sampling frequency should be adjusted accordingly and the appropriate remedial actions (e.g., DPE and/or GWE) should be implemented.

Source remediation is not recommended at this time because the contaminant source has not been located and characterized; however, implementation of a source reduction technology could provide significant benefits in reducing overall remediation time. Therefore, if performance of source location activities beneath B301 is administratively feasible, a minimally intrusive soil gas survey beneath and adjacent to the building should be considered to pinpoint subsurface TCE "hotspots" and facilitate source reduction system design.

10.4 NATURAL ATTENUATION OF CHLORINATED SOLVENTS: PLATTSBURGH AIR FORCE BASE, NEW YORK

Facility Background

Activities at a former fire training area (site FT-002) at Plattsburgh AFB, New York resulted in contamination of shallow soils and groundwater with a mixture of chlorinated solvents and fuel hydrocarbons. Groundwater contaminants include TCE, *cis*-1,2- DCE, VC, and BTEX. Table 10.9 contains contaminant data for selected wells at the site.

Plattsburgh AFB is located in northeastern New York State, approximately 26 miles south of the Canadian border and 167 miles north of Albany, New York. Site FT-002 is located in the northwestern corner of the base and is approximately 700 ft wide and 800 ft long. The site is located on a land surface that slopes gently eastward toward the confluence of the Saranac and Salmon rivers, which is located approximately 2 miles east of the site. Site FT-002 was used to train base and municipal firefighting personnel from the mid- to late-1950s until the site was permanently closed to fire training activities in May 1989. Figure 10.20 is a map of the site.

Four distinct stratigraphic units underlie the site: sand, clay, till, and carbonate bedrock. Figure 10.21 shows three of the four stratigraphic units at the site. The sand unit consists of well-sorted fine- to medium-grained sand with a trace of silt, and generally extends from ground surface to as much as 90 ft BGS in the vicinity of the site. A 7-ft-thick clay unit has been identified on the eastern side of the site. The thickness of the clay on the western side of the site has not been determined. A 30- to 40-ft-thick clay till unit is also present from 80 to 105 ft BGS in the vicinity of the site. Bedrock is present approximately 105 ft BGS.

Groundwater Hydraulics

The depth to groundwater in the sand aquifer ranges from 45 ft BGS on the west side of the site to zero on the east side of the runway, where groundwater discharges to a swamp (Figure 10.21). Groundwater flow at the site is to the southeast. The average gradient is approximately 0.010 ft/ft. Hydraulic conductivity of the upper sand aquifer was measured using constant drawdown tests and rising-head tests. Hydraulic conductivity values for the unconfined sand aquifer underlying the site range from 0.059 to 90.7 ft/day. The average hydraulic conductivity for the site is 11.6 ft/day.

TABLE 10.9. Groundwater Analytical Data, Plattsburgh Air Force Base[a]

Point[b]	Date	Distance from Source (ft)	TMB (μg/L)	BTEX (μg/L)	TCE (μg/L)	Total DCE[c] (μg/L)	Vinyl Chloride (μg/L)	Methane (μg/L)	Ethene (μg/L)	Chloride (mg/L)	Dissolved Oxygen (mg/L)	Nitrate (mg/L)	Fe(II) (mg/L)	Sulfate (mg/L)	Hydrogen (nM)	TOC (mg/L)
A	8/95	0	1,757	16,790	25,280	51,412	ND	1,420	<0.001	63	0.1	0.2	4.0	5.5	6.70	80
	5/96		828	6,598	580	12,626	ND	1,600	<0.001	82	0.5	<0.05	45.6	1.0	2.00	94
B	8/95	970	491	3,060	2	14,968	897	305	35.00	48	0.5	0.2	15.3	<0.1	1.66	30
	5/96		463	4,198	1	9,376	1,520	339	13.00	43	0.1	<0.05	16.0	<0.1	1.40	31
C	8/95	1,240	488	3,543	3	10,035	1,430	1,010	182.00	46	0.4	0.2	13.8	<0.1	NA	21
	5/96		509	3,898	1	10,326	1,050	714	170.00	57	0.2	<0.05	19.3	<0.1	11.13	24
D	8/95	2,050	NA	NA	NA	NA	NA	NA	NA	NA	NA	NA	NA	NA	NA	NA
	5/96		9	89	0	1,423	524	617	4.00	14	0.2	0.1	2.5	1.5	NA	14
E	8/95	2,560	ND	40	24	2,218	8	3,530	<0.001	20	0.9	0.3	0.7	0.5	NA	8
	5/96		ND	40	17	1,051	12	1,800	<0.001	18	0.1	<0.05	<0.05	1.0	0.81	8
F	8/95	3,103	ND	2	1	226	5	115	<0.001	3	0.4	10.4	<0.05	14.7	0.22	NA
	5/96		ND	2	ND	177	4	44	<0.001	3	0.2	9.5	0.1	14.4	0.25	NA

[a] NA, not analyzed; ND, not detected.

[b] Point A, MW-02-108; B, MW-02-310; C, 84DD; D, 84DF; E, 34PLTW12; F, 35PLTW13.

[c] Greater than 99% of DCE is *cis*-1,2-DCE.

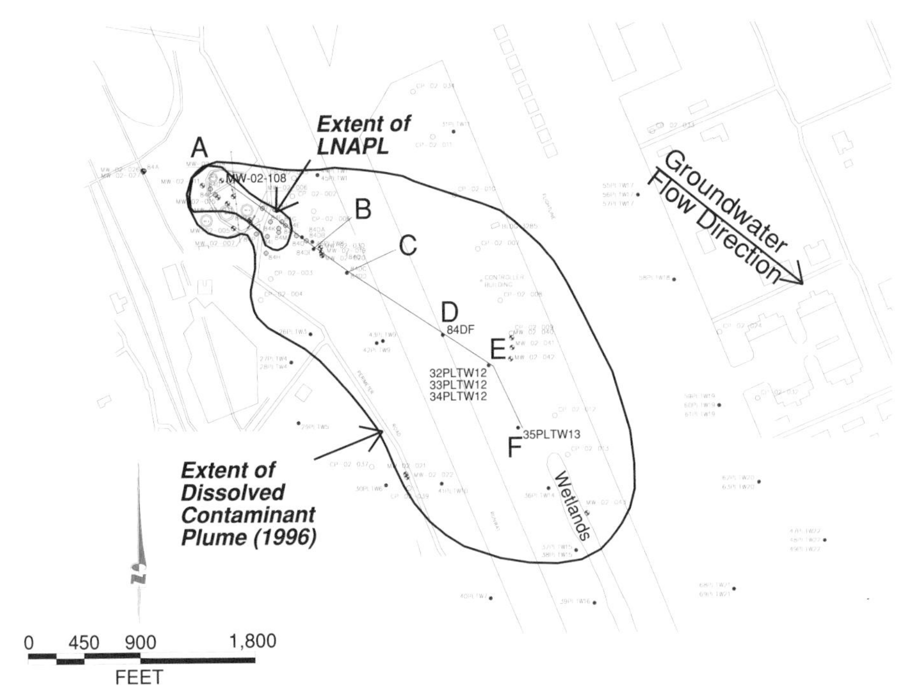

FIGURE 10.20. Site layout, Plattsburgh Air Force Base, New York.

Freeze and Cherry (1979) give a range of effective porosity for sand of 0.25 to 0.50. Effective porosity was assumed to be 0.30. Using the horizontal gradient of 0.010 ft/ft, the average hydraulic conductivity value of 11.6 ft/day, and an effective porosity of 0.30, an average advective groundwater velocity for the unconfined sand aquifer of 0.39 ft/day, or approximately 142 ft/yr, was calculated. Because of low background TOC concentrations at the site, retardation is not considered to be an important transport parameter.

Groundwater and LNAPL Chemistry

Contaminants. Figure 10.20 shows the approximate distribution of LNAPL at the site. This LNAPL is a mixture of jet fuel and waste solvents from which BTEX and TCE partition into groundwater. Analysis of the LNAPL shows that the predominant chlorinated solvents are PCE and TCE; DCE and VC are not present in measurable concentrations. For the most part, groundwater beneath and downgradient from the LNAPL is contaminated with dissolved fuel-related compounds and solvents consistent with those identified in the LNAPL. The most notable exceptions are the presence of *cis*-1,2-DCE and VC, which because of their absence in the LNAPL, probably have been formed through reductive dechlorination of PCE/TCE.

The dissolved BTEX plume currently extends approximately 2000 ft downgradient from the site and has a maximum width of about 500 ft. Total dissolved BTEX

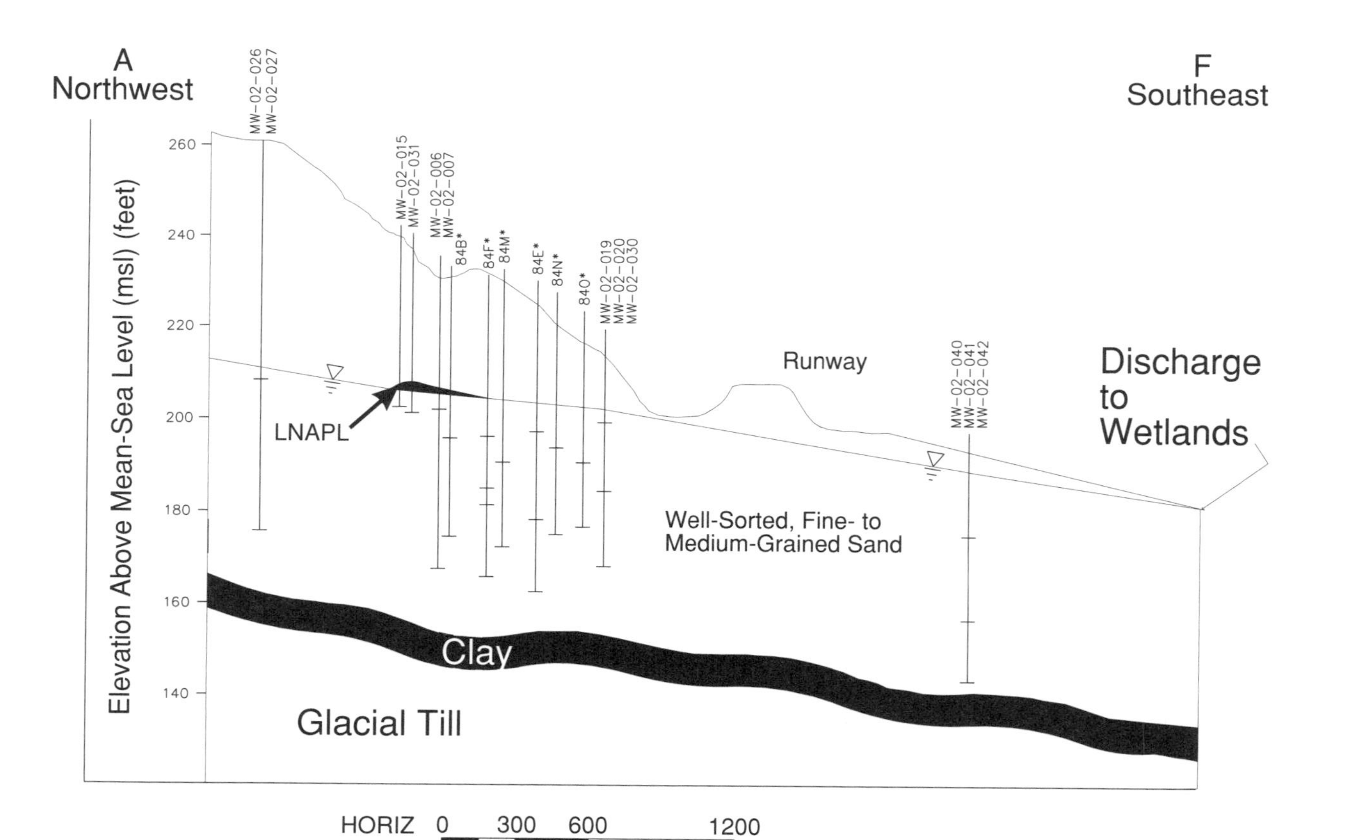

FIGURE 10.21. Hydrogeologic section, Plattsburgh Air Force Base, New York.

concentrations as high as 17 mg/L have been observed in the source area. Figure 10.22 shows the extent of BTEX dissolved in groundwater. As indicated on this map, dissolved BTEX contamination is migrating to the southeast in the direction of groundwater flow. Five years of historical data for the site show that the dissolved BTEX plume is at steady-state equilibrium and is no longer expanding.

Detectable concentrations of dissolved TCE, DCE, and VC currently extend approximately 4000 ft downgradient from FT-002 (Figure 10.22). Concentrations of TCE, DCE, and VC as high as 25, 51, and 1.5 mg/L, respectively, have been observed at the site. As stated previously, no DCE was detected in the LNAPL plume at the site, and more than 99% of the DCE found in groundwater is the *cis*-1,2-DCE isomer. Figure 10.22 shows the extent of CAH compounds dissolved in groundwater at the site. Five years of historical data for the site show that the dissolved CAH plume also is at steady-state equilibrium and is no longer expanding.

Indicators of Biodegradation. Figure 10.23 shows the distributions of electron acceptors used in microbially mediated redox reactions. Electron acceptors displayed in this figure include DO, nitrate, and sulfate. There is a strong correlation between areas with elevated BTEX concentrations and areas with depleted DO,

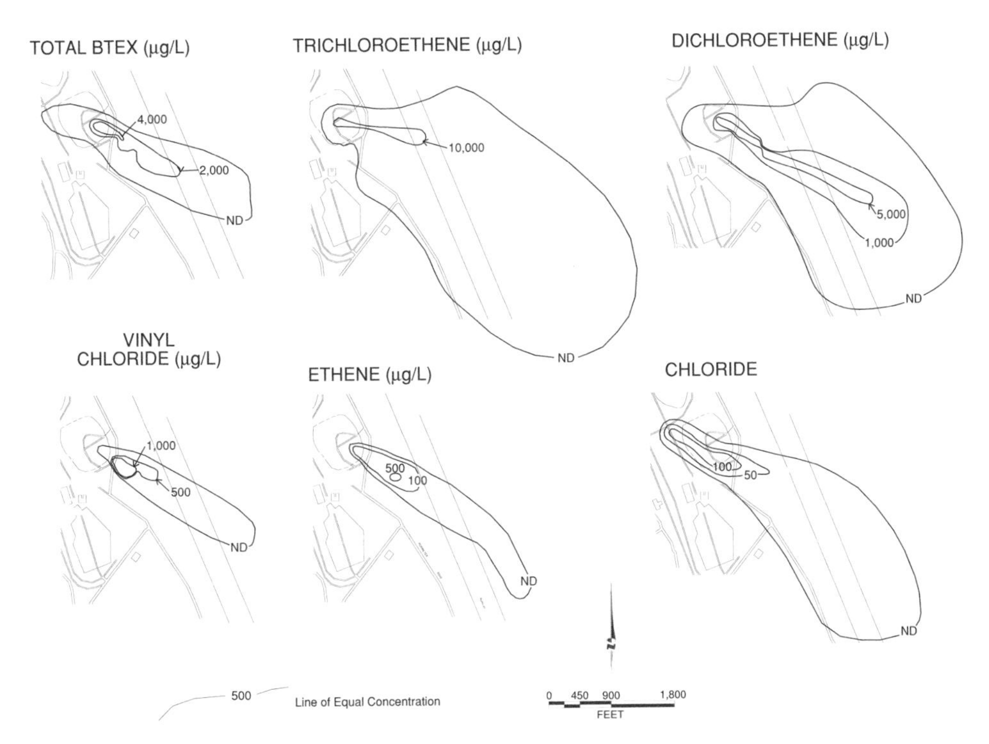

FIGURE 10.22. Chlorinated solvents and by-products, 1995, Plattsburgh Air Force Base, New York.

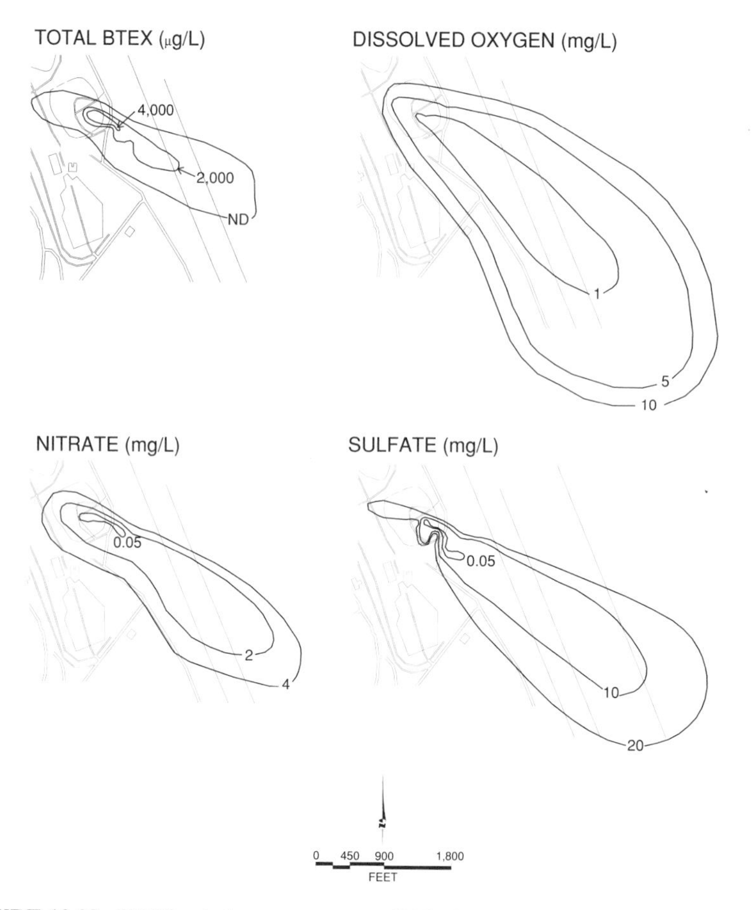

FIGURE 10.23. BTEX and electron acceptors, 1995, Plattsburgh Air Force Base, New York.

nitrate, and sulfate. The absence of these compounds in contaminated groundwater suggests that aerobic respiration, nitrate reduction, and sulfate reduction are working to biodegrade fuel hydrocarbons at the site. Background dissolved oxygen, nitrate, and sulfate concentrations are on the order of 10, 10, and 25 mg/L, respectively.

Figure 10.24 shows the distribution of metabolic by-products produced by microbially mediated redox reactions that biodegrade fuel hydrocarbons. Metabolic by-products displayed in this figure include Fe(II) and methane. There is a strong correlation between areas with elevated BTEX concentrations and areas with elevated Fe(II) and methane. The presence of these compounds in concentrations above background in contaminated groundwater suggests that Fe(III) reduction and methanogenesis are working to biodegrade fuel hydrocarbons at the site. Background Fe(II) and methane concentrations are <0.05 and <0.001 mg/L, respectively. The pE of groundwater also is shown on Figure 10.24. Areas of low pE correspond to areas with contamination. This is a further indication that biologically mediated reactions are occurring in the area with groundwater contamination.

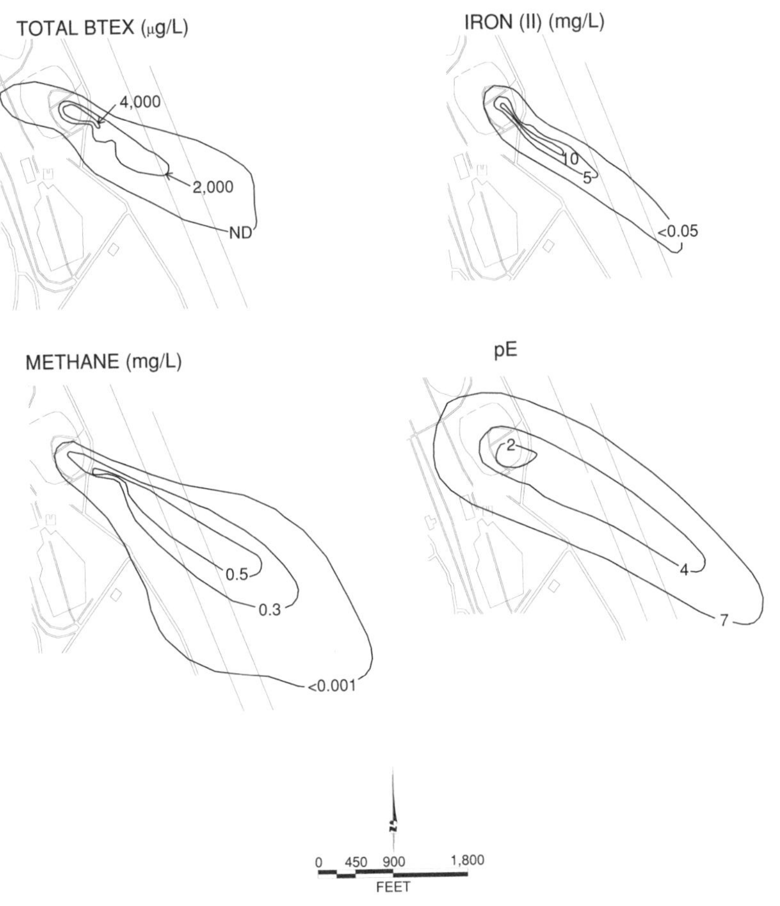

FIGURE 10.24. BTEX and metabolic by-products, 1995, Plattsburgh Air Force Base, New York.

The distribution of chloride in groundwater is shown in Figure 10.22. This figure also compares measured concentrations of total BTEX and CAHs in the groundwater with chloride. There is a strong correlation between areas with contamination and areas with elevated chloride concentrations relative to measured background concentrations. The presence of elevated concentrations of chloride in contaminated groundwater suggests that TCE, DCE, and VC are being biodegraded. Background chloride concentrations at the site are approximately 2 mg/L. Figure 10.22 also compares measured concentrations of total BTEX and CAHs in the groundwater with the distribution of dissolved ethene. There is a strong correlation between areas with contamination and areas with elevated ethene concentrations relative to measured background concentrations. The presence of elevated concentrations of ethene in contaminated groundwater suggests that TCE, DCE, and VC are being biodegraded. Background ethene concentrations at the site are <0.001 mg/L. As discussed in Chapter 7, dissolved H_2 concentrations can be used to determine the dominant terminal electron-accepting process in an aquifer. Table 10.9 presents H_2 data for the site.

Biodegradation Rate Constant Calculations

Apparent biodegradation rate constants were calculated using trimethylbenzene as a conservative tracer (Chapter 7). A modified version of this method that takes into account the production of chloride during biodegradation also was used to calculate approximate biodegradation rates. Table 10.10 presents the results of these rate constant calculations.

Primary Substrate Demand for Reductive Dechlorination

For reductive dechlorination to occur, a carbon source that can be used as a primary substrate must be present in the aquifer. This carbon substrate can be in the form of anthropogenic carbon (e.g., fuel hydrocarbons) or native organic material.

Reductive Dechlorination Supported by Fuel Hydrocarbons (Type 1 Behavior). Fuel hydrocarbons are known to support reductive dechlorination in aquifer material. The following equation describes the oxidation of BTEX compounds (approximated as CH) to carbon dioxide during reduction of carbon-to-chlorine bonds (represented as C–Cl) to carbon-to-hydrogen bonds (represented as C–H):

$$CH + 2H_2O + 2.5C{-}Cl \rightarrow CO_2 + 2.5H^+ + 2.5Cl^- + 2.5C{-}H \qquad (10.1)$$

Based on this equation, each 1.0 mg of BTEX that is oxidized via reductive dechlorination requires the consumption of 6.8 mg of organic chloride and the liberation of 6.8 mg of biogenic chloride. TCE loses two C–Cl bonds while being reduced to VC.

TABLE 10.10. Approximate First-Order Biodegradation Rate Constants[a]

Compound	Correction Method	A–B 0–970 ft (year^{-1})	B–C 970–1240 ft (year^{-1})	C–E 1240–2560 ft (year^{-1})
TCE	Chloride	1.27	0.23	–0.30
	TMB	1.20	0.52	NA
	Average	1.24	0.38	–0.30
DCE	Chloride	0.06	0.60	0.07
	TMB	0.00	0.90	NA
	Average	0.03	0.75	0.07
VC	Chloride	0.00	0.14	0.47
	TMB	0.00	0.43	NA
	Average	0.00	0.29	0.47
BTEX	Chloride	0.13	0.30	0.39
	TMB	0.06	0.60	NA
	Average	0.10	0.45	0.39

[a] NA, not analyzed.

Based on eq. (10.1), 1.25 mol of TCE would have to be reduced to VC to oxidize 1.0 mol of BTEX to carbon dioxide. Therefore, 1.0 mg of BTEX oxidized would consume 12.6 mg of TCE. If DCE were reduced to VC, each 1.0 mg of BTEX oxidized would consume 18.6 mg of DCE. To be more conservative, these calculations should be completed assuming that TCE and DCE are reduced to ethene. However, because the amount of ethene produced is trivial compared to the amount of TCE and DCE destroyed, this step has been omitted here.

Reductive Dechlorination Supported by Natural Organic Carbon (Type 2 Behavior). Wershaw et al. (1994) analyzed dissolved organic material in groundwater underneath a dry well that had received TCE discharged from the overflow pipe of a degreasing unit. The dissolved organic material in groundwater exposed to the TCE was 50.57% carbon, 4.43% hydrogen, and 41.73% oxygen. The elemental composition of this material was used to calculate an empirical formula for the dissolved organic matter and to estimate the number of moles of C−Cl bonds required to reduce 1 mol of dissolved organic carbon in this material:

$$C_{1.0}H_{1.051}O_{0.619} + 1.38H_2O + 1.91C{-}Cl \rightarrow CO_2 + 1.91Cl^- + 1.91C{-}H + 1.91H^+ \quad (10.2)$$

Based on this equation, each 1.0 mg of dissolved organic carbon that is oxidized via reductive dechlorination requires the consumption of 5.65 mg of organic chloride and the liberation of 5.65 mg of biogenic chloride. Using eq. (10.2), 0.955 mol of TCE would have to be reduced to VC to oxidize 1.0 mol of organic carbon to carbon dioxide. Therefore, 1.0 mg of organic carbon oxidized would consume 10.5 mg of TCE. If DCE were reduced to VC, each 1.0 mg of organic carbon oxidized would consume 15.4 mg of DCE.

Table 10.11 compares the electron-donor demand required to dechlorinate the alkenes remaining in the plume with the supply of potential electron donors. Table 10.10 reveals that removal of TCE and *cis*-1,2-DCE slows or ceases between points C and E. This correlates with the exhaustion of BTEX in the plume. Over this interval, the supply of BTEX is a small fraction of the theoretical demand required for dechlorination. There are adequate supplies of native organic matter, suggesting that

TABLE 10.11. Comparison of the Estimated Electron Donor Demand to Support Reductive Dechlorination to the Supply of BTEX and Native Organic Carbon

Point	Chloride (mg/L)	Organic Chloride (mg/L)	BTEX Available (mg/L)	BTEX Demand (mg/L)	Total Organic Carbon Supply (mg/L)	Organic Carbon Demand (mg/L)
A	63	58.1	16.8	8.5	80.4	10.3
B	43	7.72	4.2	1.13	31.1	1.37
C	57	8.26	3.9	1.21	24.3	1.46
D	13.6	1.34	0.09	0.20	13.8	0.24
E	18.4	0.78	0.04	0.114	8.2	0.14

native organic matter may not be of sufficient nutritional quality to support reductive dechlorination in this aquifer.

Discussion and Conclusions

Available geochemical data indicate that the geochemistry of groundwater in the FT-002 source area and about 1500 ft downgradient is significantly different from the groundwater found between 1500 and 4000 ft downgradient from the source. Near the source the plume exhibits Type 1 behavior. At about 1500 ft downgradient from the source, the plume reverts to Type 3 behavior. Figure 10.25 shows the zones of differing plume behavior at the site.

Type 1 Behavior. In the area extending to approximately 1500 ft downgradient from the former fire training pit (source area), the dissolved contaminant plume consists of commingled BTEX and TCE and is characterized by anaerobic conditions that are strongly reducing (i.e., Type 1 behavior). Dissolved oxygen concentrations are on the order of 0.1 mg/L (background = 10 mg/L), nitrate concentrations are on the order of 0.1 mg/L (background = 10 mg/L), Fe(II) concentrations are

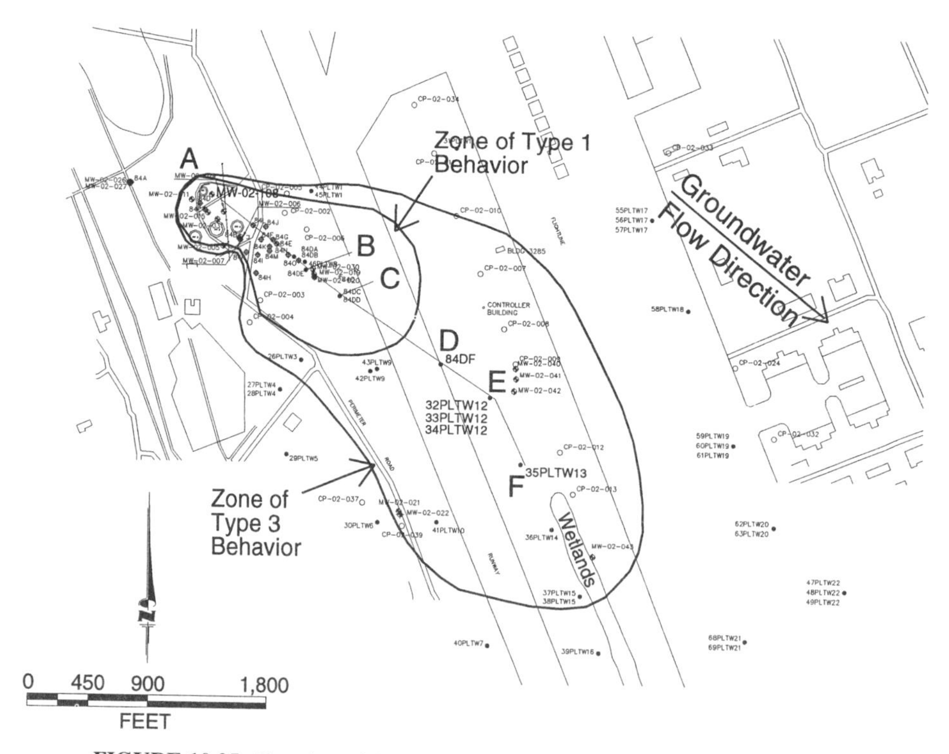

FIGURE 10.25. Zonation of CAH plume, Plattsburgh Air Force Base, New York.

on the order of 15 mg/L (background = <0.05 mg/L), sulfate concentrations are <0.05 mg/L (background = 25 mg/L), and methane concentrations are on the order of 3.5 mg/L (background = <0.001 mg/L). Hydrogen concentrations in the source area range from 1.4 to 11 nM. These H_2 concentrations are indicative of sulfate reduction and methanogenesis, even though there is no sulfate available and relatively little methane is produced. Thus reductive dechlorination may be competitively excluding these processes. In this area BTEX are being used as a primary substrate, and TCE is being dechlorinated reductively to *cis*-1,2-DCE and VC. This is supported by the fact that no detectable DCE or VC was found in the LNAPL present at the site and is strong evidence that the dissolved DCE and VC at the site are produced by the biogenic reductive dechlorination of TCE. Furthermore, the dominant isomer of DCE found at the site is *cis*-1,2-DCE, the isomer preferentially produced during reductive dechlorination. Average calculated first-order biodegradation rate constants in this zone are as high as 1.24, 0.75, and 0.29 $year^{-1}$ for TCE, *cis*-1,2-DCE, and VC, respectively. Figure 10.25 shows the approximate extent of this type of behavior. Because reductive dechlorination of VC is slower than direct oxidation, VC and ethene are accumulating in this area (Figure 10.26).

Type 3 Behavior. Between 1500 and 2000 ft downgradient from the source area, the majority of the BTEX has been biodegraded and the system begins to exhibit Type 3 behavior. DO concentrations are on the order of 0.5 mg/L

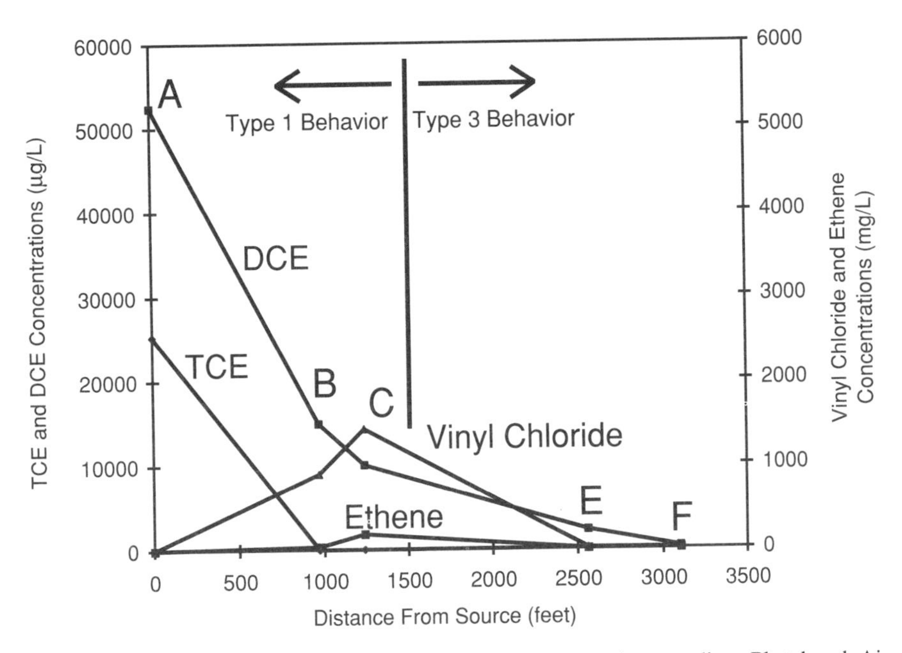

FIGURE 10.26. Plot of TCE, DCE, and ethene versus distance downgradient, Plattsburgh Air Force Base, New York.

(background = 10 mg/L). Nitrate concentrations start increasing downgradient from where Type 3 behavior begins and are near background levels of 10 mg/L at the downgradient extent of the CAH plume. Fe(II) concentrations have significantly decreased and are on the order of 1 mg/L (background = <0.05 mg/L). Sulfate concentrations start increasing to 15 mg/L at the downgradient extent of the CAH plume. Methane concentrations are the highest in this area, but could have migrated from upgradient locations. The H_2 concentrations at points E and F are 0.8 and 0.25 nM, respectively, suggesting that the dominant terminal electron-accepting process in this area is Fe(III) reduction. These conditions are not optimal for reductive dechlorination, and it is likely that VC is being oxidized via Fe(III) reduction or aerobic respiration. Average calculated rate constants in this zone are 0.3, 0.07, and 0.47 year^{-1} for TCE, *cis*-1,2-DCE, and VC, respectively. The biodegradation rates of TCE and DCE slow because reductive dechlorination stops when the plume runs out of primary substrate (i.e., BTEX). The rate of VC biodegradation in this area increases, probably because VC is being oxidized. Because biodegradation of VC is faster under Type 3 geochemical conditions than the biodegradation of other CAH compounds, the accumulation of VC ceases and the accumulated VC degrades rapidly. Ethene concentrations also begin to decrease because ethene is no longer being produced from the reductive dechlorination of VC (Figure 10.26).

REFERENCES

Buscheck, T. E., and Alcantar, C. M., 1995, Regression techniques and analytical solutions to demonstrate intrinsic bioremediation, in *Proceedings of the 1995 Battelle International Conference on In Situ and On Site Bioreclamation*, Apr., Battelle Press, Columbus, OH.

Buscheck, T. E., O'Reilly, K. T., and Hickman, G., 1997, Intrinsic anaerobic biodegradation of chlorinated solvents at a manufacturing plant, in *In Situ and On-Site Bioremediation*, Vol. 3, Battelle Press, Columbus, OH, pp. 149–154.

Chapelle, F. H., McMahon, P. B., Dubrovsky, N. M., Fujii, R. F., Oaksford, E. T., and Vroblesky, D. A., 1995, Deducing the distribution of terminal electron-accepting processes in hydrologically diverse groundwater systems, *Water Resour. Res.*, vol. 31, pp. 359–371.

Davis, J. W., and Carpenter, C. L., 1990, Aerobic biodegradation of vinyl chloride in groundwater samples, *Appl. Environ. Microbiol.*, vol. 56, pp. 3878.

Lovley, D. R., and Goodwin, S., 1988, Hydrogen concentrations as an indicator of the predominant terminal electron-accepting reaction in aquatic sediments, *Geochim. Cosmochim. Acta*, vol. 52, pp. 2993–3003.

Lovley, D. R., Chapelle, F. H., and Woodward, J. C., 1994, Use of dissolved H_2 concentrations to determine distribution of microbially catalyzed reactions in anoxic groundwater, *Environ. Sci. Technol.*, vol. 28, pp. 1255–1210.

McCarty, P. L., 1996, Biotic and abiotic transformations of chlorinated solvents in ground water, in *Proceedings of the Symposium on Natural Attenuation of Chlorinated Organics in Ground Water*, Dallas, TX, Sept. 11–13, EPA/540/R-96/509, U.S. EPA, Washington, DC, pp. 35–59.

McDonald, G., and Harbaugh, A. W., 1988, A modular three-dimensional finite-difference groundwater flow model, *U.S. Geological Survey Techniques of Water Resources Investigations*, Book 6, Chapter A1, U.S. Geological Survey, Washington, DC.

Moutoux, D. E., Benson, L. A., Lenhart, J., Wiedemeier, T. H., Wilson, J. T., and Hansen, J. E., 1996, Estimating the changing rate of anaerobic reductive dehalogenation of chlorinated aliphatic hydrocarbons in the presence of petroleum hydrocarbons, in *Proceedings of the Conference on Petroleum Hydrocarbons and Organic Chemicals in Ground Water: Prevention, Detection, and Remediation*, Houston, TX, Nov. 13–15.

S. S. Papadopulos and Associates, Inc., 1996, *MT3D*[96]*: A Modular Three-Dimensional Transport Model for Simulation of Advection, Dispersion, and Chemical Reactions of Contaminants in Ground-Water Systems*, Papadopulos and Assoc., Bethesda, MD.

Wiedemeier, T. H., Swanson, M. A., Moutoux, D. E., Wilson, J. T., Kampbell, D. H., Hansen, J. E., and Haas, P., 1996a, Overview of technical protocol for natural attenuation of chlorinated aliphatic hydrocarbons in ground water under development for the U.S.

Wiedemeier, T. H., Swanson, M. A., Moutoux, D. E., Gordon, E. K., Wilson, J. T., Wilson, B. H., Kampbell, D. H., Hansen, J. E., Haas, P., and Chapelle, F. H., 1996c, *Technical protocol for evaluating natural attenuation of chlorinated solvents in groundwater*, (Draft, Revision 1, U.S. Air Force Center for Environmental Excellence, San Antonio, TX, Nov.

Wu, W.-M., Nye, J., Hickey, R. F., Jain, M. K., and Zeikus, J. G., 1995, Dechlorination of PCE and TCE to ethene using an anaerobic microbial consortium, in *Bioremediation of Chlorinated Solvents*, R. E. Hinchee, A. Leeson, and L. Semprini, eds., Battelle Press, Columbus, OH, pp. 45–52.

CHAPTER 11

DESIGN OF LONG-TERM MONITORING PROGRAMS

As discussed in Chapter 1, many recent studies, including those conducted by the Lawrence Livermore National Laboratory (Rice et al., 1995) and the Texas Bureau of Economic Geology (Mace et al., 1997), suggest that most fuel-hydrocarbon plumes are stable or receding. In all of these studies, only a small fraction of the fuel-hydrocarbon plumes evaluated were observed to be expanding, and in the majority of these cases, the solutes were moving at rates slower than the seepage velocity of groundwater. These examples demonstrate the efficacy of natural attenuation for fuel hydrocarbons at many sites. Even with the vast amounts of data supporting the occurrence and protectiveness of natural attenuation for these types of contaminants, it is common practice to develop a long-term monitoring plan to monitor the behavior of the contaminant plume over time to ensure that potential receptors are not affected.

In contrast to the large amount of data showing that most fuel-hydrocarbon plumes are not migrating an appreciable distance downgradient, as of early 1998 there is not a large enough database to assess the stability of chlorinated solvent plumes. Based on the size of many dissolved chlorinated solvent plumes, it appears that these plumes are likely to migrate further downgradient than fuel-hydrocarbon plumes before reaching steady-state equilibrium or before receding. Indeed, many chlorinated solvent plumes discharge to surface water bodies before reaching steady-state equilibrium. For this reason a rigorous long-term monitoring plan should be developed for sites contaminated with chlorinated solvents.

In this chapter we discuss the design of long-term monitoring plans. Long-term monitoring of a contaminant plume can provide an empirical demonstration of the effectiveness of natural attenuation as a remedy. A properly designed long-term monitoring plan will help ensure protection of human health and the environment by allowing detection of plume movement before potential receptors are affected. Properly designed long-term monitoring plans will also provide a historical database of the trends in contaminant concentrations in the plume over time and distance, will allow detection of changes in background water quality, and will detect changes in the groundwater flow pattern.

Designing an effective groundwater monitoring plan involves locating groundwater monitoring wells and developing a groundwater sampling and analysis strategy. Long-term monitoring and contingency plans are site specific. Because of this, all available site-specific data and information developed during site characterization, conceptual model development, premodeling calculations, biodegradation rate constant calculations, groundwater modeling, and a receptor exposure pathways analysis should be used when preparing a long-term monitoring plan. The results of a solute fate and transport model can aid in the placement of wells and in determining a sampling frequency.

Over the past several years many regulatory agencies have come to recognize the importance of natural processes of contaminant attenuation. In November 1997 the Office of Solid Waste and Emergency Response (OSWER) of the U.S. Environmental Protection Agency (U.S. EPA) came out with a draft interim final policy on the use of natural attenuation entitled *Use of Monitored Natural Attenuation at Superfund, RCRA Corrective Action, and Underground Storage Tank Sites*. As implied by the title of this policy document, performance monitoring will be required to evaluate the effectiveness of natural attenuation and to ensure protection of human health and the environment. For natural attenuation to remain as the selected remedial approach it will be necessary to (1) demonstrate that natural attenuation is occurring according to expectations, (2) identify any potentially toxic transformation products resulting from biodegradation, (3) determine if the plume is expanding, (4) ensure no impact to downgradient receptors, (5) detect new releases of contaminants or changes in environmental conditions that could affect the efficacy of natural attenuation, and (6) verify attainment of cleanup objectives. In addition, a contingency remedy must be specified as a backup remedy in the event that natural attenuation fails to perform as anticipated.

The protocol *Designing Monitoring Programs to Effectively Evaluate the Performance of Natural Attenuation* being developed by the U.S. Air Force Center for Environmental Excellence (AFCEE), Technology Transfer Division, describes how to specify effectively and efficiently the location, frequency, and types of samples and analyses required to meet these objectives (Wiedemeier and Haas, 1999). Chemical and hydrogeological data should be used to site monitoring wells. For example, geochemical data such as dissolved oxygen, nitrate, Fe(II), sulfate, and methane can be used in conjunction with contaminant data to site downgradient contingency wells in locations with treated groundwater. This approach ensures that the downgradient monitoring network is in the flow path of the contaminant plume. The frequency of sampling will depend on the location of potential receptors and the seepage velocity of groundwater. To evaluate the behavior of the dissolved contaminant plume over time and to estimate cleanup time frames, statistical methods can be employed. Overall, the

AFCEE protocol will lower performance monitoring costs by reducing the number of monitoring wells and the frequency of sampling and number of analytes required to demonstrate the continuing efficacy of natural attenuation.

In the following sections we provide potential approaches for designing and implementing long-term monitoring programs.

11.1 PLACEMENT OF LONG-TERM MONITORING AND CONTINGENCY WELLS

A typical long-term monitoring plan consists of locating monitoring wells and developing a sampling and analysis strategy. It is important that the placement of monitoring wells and the frequency of sampling is such that they will yield useful data and be appropriate to detect migration of the plume and define trends in contaminant concentrations over time. In many cases it may be possible to utilize some of the existing monitoring wells at a site. This will reduce the cost of implementing the long-term monitoring plan. It is important, however, that these wells are located in appropriate locations. Keep in mind that all wells installed during site characterization may not be appropriate or necessary for long-term monitoring. Also remember that a monitoring well installed for site characterization purposes will not necessarily give meaningful long-term monitoring data, so it is important to be selective in determining which of the existing wells to sample.

For plumes that do not discharge to a surface water body, the long-term monitoring plan includes two types of monitoring wells. Long-term monitoring wells are intended to monitor trends in contaminant concentrations over time and space and to determine if the behavior of the plume is changing through time. Contingency wells are intended to detect movements of the plume outside the negotiated perimeter of containment and to trigger an action to manage potential plume expansion. These wells should be located and designed to protect potential receptors. Figure 11.1 illustrates a hypothetical long-term monitoring network for a site contaminated with dissolved organic compounds emanating from a light non-aqueous-phase liquid. Figure 11.2 illustrates a hypothetical long-term monitoring network for a site contaminated with dissolved organic compounds emanating from a dense non-aqueous-phase liquid. These figures depict (1) upgradient and side-gradient wells in unaffected groundwater, (2) a well in the NAPL source area, (3) wells downgradient of the NAPL source area, (4) a well located downgradient from the plume where contaminants are not detectable and soluble electron acceptors are depleted with respect to unaffected groundwater, and (5) contingency wells. Table 11.1 summarizes sampling locations. The upgradient and side-gradient wells (if needed) are intended to monitor for changes in background water quality that can provide an indication of changing conditions that could affect natural attenuation. The well(s) in the NAPL source area is (are) intended to monitor changing NAPL composition over time and to give an indication of the changing strength of the NAPL. Monitoring wells downgradient from the NAPL source area located along the centerline of

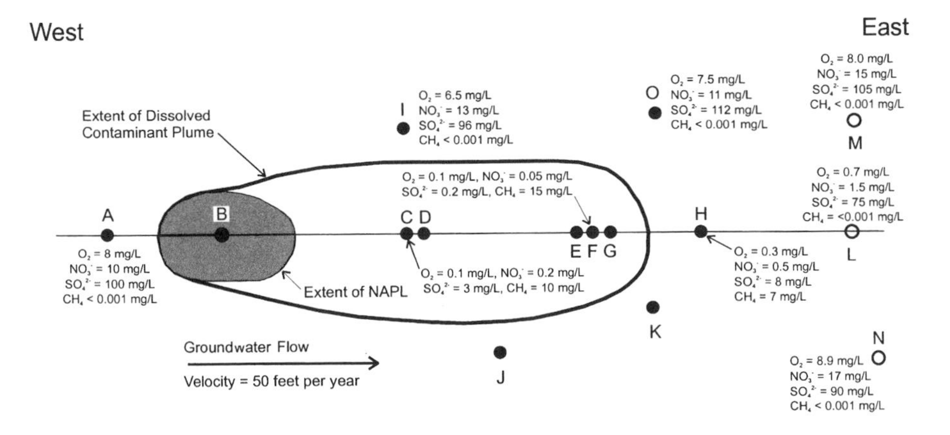

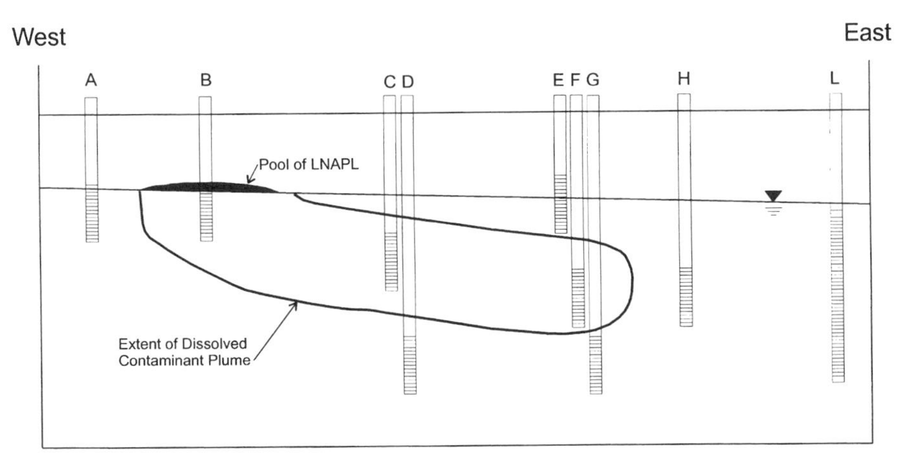

FIGURE 11.1. Locating long-term and point-of-action monitoring wells using contaminant and geochemical data (LNAPL).

the plume are intended to monitor plume behavior and changing contaminant concentrations over time.

Ideally, these wells will be aligned parallel to the direction of groundwater flow along the centerline of the plume. It should be kept in mind that this requires good definition of the plume and fairly uniform (unchanging) hydraulic gradients. The monitoring well located downgradient of the dissolved contaminant plume is intended to provide early detection of contaminant plume migration toward a contingency monitoring well. This well should be located in the flow path of the contaminant plume in treated groundwater. This is the "smoking gun" well discussed

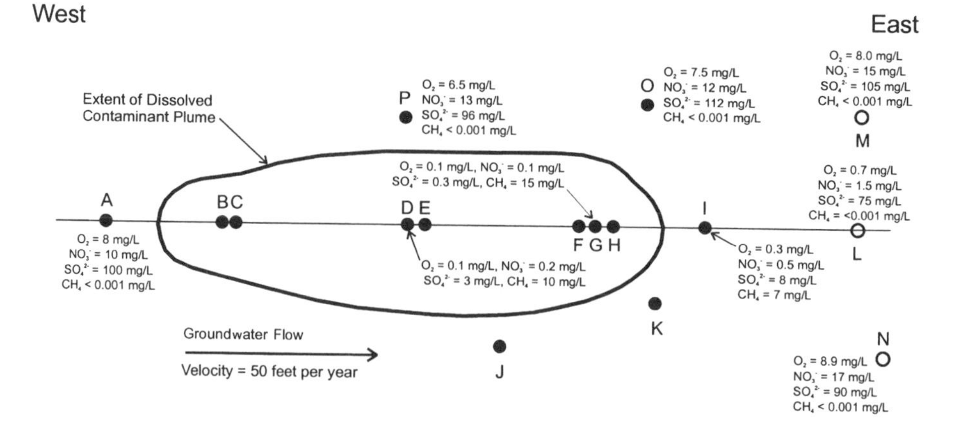

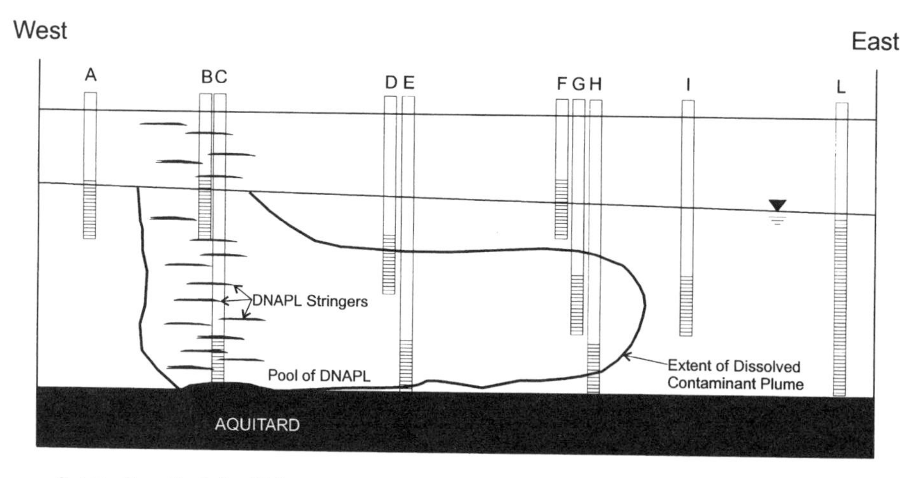

FIGURE 11.2. Locating long-term and point-of-action monitoring wells using contaminant and geochemical data (DNAPL).

below. The contingency monitoring wells are intended to monitor for plume migration and to trigger implementation of the contingency plan. If a site exhibits a significant change in the direction of groundwater flow over time (due to seasonal or other effects), one or more side-gradient wells would increase the reliability of the monitoring system.

The locations and screened intervals of long-term monitoring and contingency wells should be based on stratigraphy and the behavior of the plume as revealed during site characterization. This requires a detailed understanding of the codistribution between contaminants and hydraulic conductivity. It is important that the three-dimensional nature of hydrogeologic systems be taken into account when placing monitoring wells

TABLE 11.1. Potential Well Locations and Purpose and Analytical Parameters for Long-Term Monitoring of Groundwater

		Analytical Parameters	
Location	Purpose	Initial Sampling Rounds	Subsequent Sampling Rounds
Upgradient/side gradient	Monitor background water quality	Contaminants of concern and full suite of geochemical parameters[a]	No upgradient or side-gradient sampling or limited suite of geochemical parameters[b,c] and/or indicator contaminants[d]
NAPL source area	Monitor changing NAPL composition/ source strength	Contaminants of concern and full suite of geochemical parameters[a]	Indicator contaminants[d] in NAPL and groundwater beneath NAPL and limited suite of geochemical parameters[b,c]
Downgradient of NAPL source area along plume centerline	Monitor plume behavior over time	Contaminants of concern and full suite of geochemical parameters[a]	Indicator contaminants[c,d]
Downgradient of plume	Early detection of migration	Contaminants of concern and full suite of geochemical parameters[a]	Indicator contaminants[c,d]
Contingency wells	Monitor for plume migration toward a potential receptor and trigger contingency plan	Contaminants of concern and full suite of geochemical parameters[a]	Indicator contaminants[c,d]
Surface water	Determine surface water impacts	Contaminants of concern	Indicator contaminants[d]

[a]For fuel hydrocarbon plumes, the full suite of geochemical parameters could include dissolved oxygen, nitrate, iron(II), sulfate, methane, temperature, pH, conductivity, alkalinity, and oxidation–reduction potential. For chlorinated solvent plumes chloride, total organic carbon, and hydrogen could be added to the full suite of geochemical parameters recommended for fuel-hydrocarbon plumes. These analyses are described in Table 7.3.

[b]The limited suite of geochemical parameters could include dissolved oxygen, oxidation–reduction potential, temperature, and pH.

[c]If plume behavior changes or is suspected of changing, analyze for contaminants and the full suite of geochemical parameters.

[d]Indicator contaminants are a subset of the total list of contaminants of concern that represent contaminants with high concentrations, large contributors to risk, and high mobility.

for long-term monitoring. Detailed stratigraphic analysis is required to ensure that long-term monitoring and contingency wells are screened in the same hydrogeologic unit as the contaminant plume and that they are in the path of contaminated groundwater flow. The geologic complexity of the site will ultimately dictate the density of the sampling network. In some cases the results of a solute fate and transport model can be used to help locate long-term monitoring and contingency wells.

Geochemical data can be used to confirm that downgradient wells are sampling groundwater that was once contaminated with organic compounds. Wells downgradient of a contaminant plume completed in the same stratigraphic horizon that do not contain organic compounds but have depleted electron acceptor (e.g., dissolved oxygen, nitrate, sulfate) and/or elevated metabolic by-product concentrations [e.g., iron(II), methane, chloride, alkalinity] relative to background provide good evidence that the groundwater being sampled flowed through the contaminant plume and has been treated. Such wells have been termed "smoking guns" because they provide fairly conclusive evidence that the groundwater was contaminated at one time and has since been treated (Wiedemeier et al., 1996, 1998). Because concentrations of electron acceptors and metabolic by-products typically will return to background at some distance downgradient from the contaminant plume, it is important to locate at least one well close to the downgradient edge of the contaminant plume. This will also allow better resolution of the behavior of the leading edge of the plume to determine if the plume is at steady-state equilibrium or is expanding. For example, if wells F and H in Figure 11.1 are 500 ft apart and groundwater is flowing at 50 ft/yr, it will take at least 10 years of monitoring data to show that the contaminant plume is not migrating at the *seepage* velocity of the groundwater; it will take even longer to show that the contaminant plume is not migrating downgradient at some *retarded* solute transport velocity. If, on the other hand, wells F and H in Figure 11.1 are only 100 ft apart, it will take only about two years of monitoring data to show that the contaminant plume is not migrating at the *seepage* velocity of the groundwater and is thus being controlled by natural attenuation.

Figures 11.1 and 11.2 illustrate how geochemical data can be used to place long-term monitoring wells and contingency monitoring wells. The case study for the fuel hydrocarbon plume at Williams Air Force Base presented in Chapter 8 provides a good real-world example of wells located downgradient of a dissolved contaminant plume that have depleted electron acceptors and elevated metabolic byproducts.

The distance between downgradient long-term monitoring and contingency monitoring wells and the density of the monitoring network should be based on the groundwater seepage velocity, solute transport velocity, and the distance to potential receptors. Contingency wells should be placed a sufficient distance upgradient from potential receptors in the flow path of the affected groundwater to ensure that a contingency plan can be implemented before potential receptors are affected. To be conservative, these distance calculations should be made based on the seepage velocity of the groundwater, not the solute transport velocity.

For sites where contaminated groundwater discharges to surface water, the philosophy of monitoring is not well developed. Figure 11.3 is a hypothetical long-term monitoring strategy for a contaminant plume discharging to a body of surface water. This figure depicts (1) upgradient and side-gradient wells in unaffected groundwater, (2) a well in the NAPL source area, (3) a well downgradient of the NAPL source area in the zone of anaerobic treatment, and (4) surface water collection points. The purpose of the first three sampling locations is discussed above in relation to contaminant plumes that do not discharge to a surface water body.

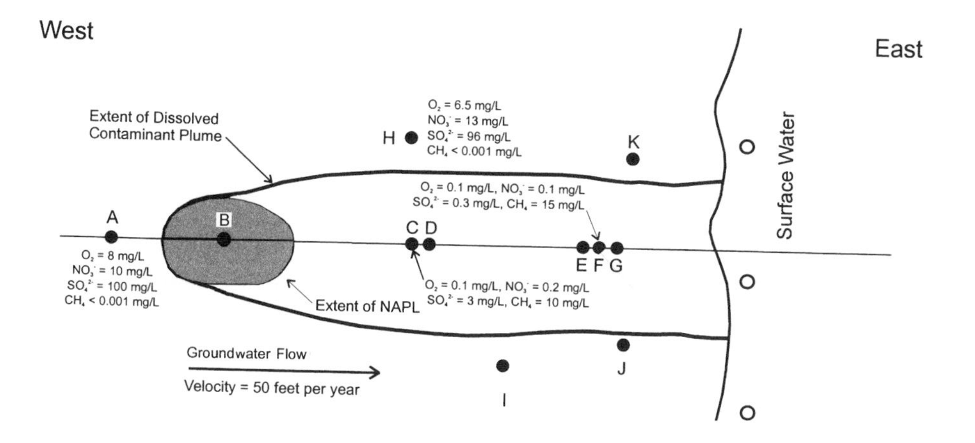

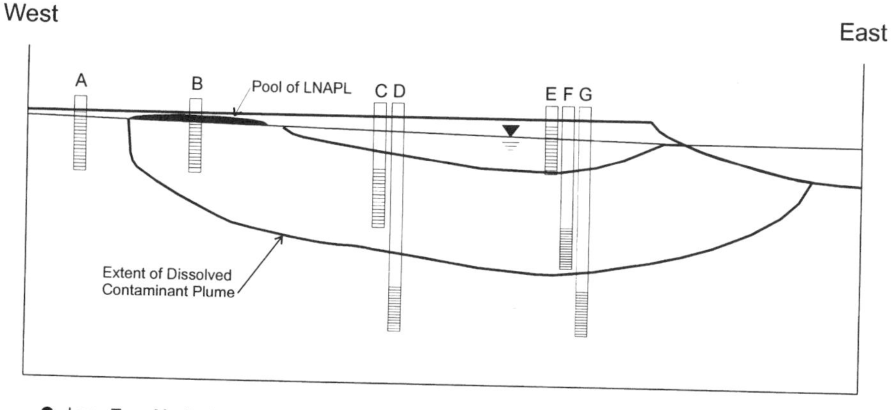

● Long-Term Monitoring Well

O Surface-Water Sampling Location

FIGURE 11.3. Locating long-term monitoring wells and surface water sampling locations for a plume discharging to a surface water body.

The fourth type of sampling location is intended to provide information on the impact of the contaminant plume on the surface water body. Mass flux calculations can be completed to estimate the amount of contamination entering the surface water body and the resultant contaminant concentrations in the surface water (see, e.g., Newell et al., 1997). Surface water samples should also be collected to help quantify the impact to surface water. In many cases the relationship between mass flux into the surface water and dilution will be such that the contamination is not detectable in the surface water body or is quickly diluted to nondetectable concentrations a short distance from the point of discharge.

11.2 GROUNDWATER SAMPLING AND ANALYSIS FOR LONG-TERM MONITORING

A groundwater sampling and analysis plan that specifies a sampling frequency and a list of analytes should be prepared in conjunction with contingency and long-term monitoring well placement. The sampling frequency should be appropriate to detect migration of the plume over time to protect potential receptors and to define trends in analyte concentrations and should account for groundwater flow and solute transport rates and monitoring well spacing. Groundwater analytical parameters for the various types of sampling locations described above are summarized in Table 11.1. The final sampling frequency and analytical parameters should be determined in conjunction with the appropriate regulatory agencies, and any analytes required for regulatory compliance that are not listed in Table 11.1 should be collected. One potential list of analytes presented in Table 11.1 include contaminants and two lists of geochemical parameters. Note that the initial round of sampling includes all the contaminants of concern (COCs) originally observed at the site, but subsequent sampling rounds could truncate the list to include a subset of the COCs that have the highest concentration, represent the highest risk, and are the most mobile. A second key point is that the geochemical analyses suggested for plumes of chlorinated solvents are particularly important because changes in groundwater geochemistry such as depletion of organic carbon or increasing dissolved oxygen concentrations at these sites can cause reductive dechlorination of chlorinated solvents to cease. Any federal or state-specific analytical requirements not listed in Table 11.1 should also be addressed in the sampling and analysis plan to ensure that all data required for regulatory decision making are collected. Water level measurements should be made during each sampling event to ensure that the groundwater flow direction has not changed.

In the past, the monitoring of dissolved contaminant plumes typically was needlessly time and location intensive and in many cases involved the quarterly sampling of every monitoring well at a site. Based on our current understanding of the behavior of dissolved contaminant plumes, this may not be necessary in many cases. However, quarterly sampling of long-term monitoring wells during the first year or sampling may be useful to help confirm the direction of plume migration and to establish baseline conditions. If significant variability is encountered during the first year, additional quarterly sampling may be required. Based on the results of the first year's sampling, the sampling frequency may be reduced to annual sampling or less in the quarter showing the highest contaminant concentrations or the greatest extent of the plume. Ideally, the number of wells to be sampled and the frequency of sampling will be based on plume behavior and the natural variability in contaminant concentrations, the distance and time of contaminant travel between long-term monitoring wells, and the distance and time of contaminant travel between long-term monitoring wells and contingency wells. Sampling frequency should be determined by the final placement of the long-term monitoring and contingency monitoring wells and the groundwater seepage/contaminant transport velocity.

One method that can be used to estimate sampling frequency is to divide the distance between the leading edge of the contaminant plume and a downgradient monitoring well located in the plume's flow path by the seepage velocity of groundwater, giving the hydraulic residence time of groundwater in the plume. For example, consider the contaminant plume depicted by Figure 11.1. If the distance between the leading edge of the plume and well H is 100 ft and the seepage velocity of groundwater is 50 ft/yr, a sampling frequency of 100 ft/50 ft per year = 2 years may be appropriate for this site. Because the exact location of the leading edge of a dissolved contaminant plume generally is not known, some professional judgment may be required when making these calculations. If this type of calculation is used to suggest relatively long intervals between sampling, a staggered sampling approach can be employed to increase the reliability of the monitoring system (i.e., for very slow sites where a five-year sampling interval may be appropriate, sample at years 1, 2, 5, 10, 15, etc.).

11.3 COST OF LONG-TERM MONITORING PLANS

The cost of long-term monitoring programs can vary significantly depending on the number of wells to be monitored and the amount of data required. A general rule of thumb that has been used for planning purposes is that sampling a well once, conducting the lab analysis, and then reporting the results costs approximately $1000 per well. When long-term monitoring plans are data intensive, the costs can be significant.

Note that the concept of net present value (NPV) is very important when evaluating the costs of long-term monitoring plans, as the present value of future expenditures is smaller than the total dollars expended through the lifetime of the project. NPV is a method to compare different payment or income streams over long time periods and is a common analysis technique in both the engineering and finance fields. The NPV can be assumed to be the amount of money one would have to borrow to make a constant set of payments over a certain time period by using both the principal and the interest (and accounting for inflation).

For example, if 10 wells are being sampled at a total cost of $1000/well (this includes sampling, analysis, and data reduction), the total annual cost is $10,000. If this sampling program is extended over a 30-year period, the total cost is $300,000. However, if a NPV approach is used, the NPV of this payment stream will be less than $300,000. The NPV calculation requires that a discount rate be used to determine the "cost" of money. Typically, the discount rate might vary between 2 and 3% (the historical real interest rate on long-term bonds) up to 10 to 15%, reflecting the required rate of return of capital for a particular company. The NPV of the hypothetical $300,000 payment stream over 30 years for four different discount rates is shown below.

Discount Rate (%)	Total Expenditure (thousands)	NPV (thousands)	NPV/Total (%)
2	$300	$224	75
4	$300	$173	58
6	$300	$137	46
8	$300	$113	38
10	$300	$94	31

Note that the NPV values shown above are only for a 30-year payment stream, and that savings from a NPV analysis will be less for shorter time periods and more for longer time periods. NPV calculations can be performed using financial tables, financial calculators, or with the functions included in most computer spreadsheets.

11.4 CONTINGENCY PLANS

Changing site conditions can result in variable plume behavior over time. To circumvent potential problems caused by such plume behavior, contingency plans should be an integral part of the long-term monitoring plan (U.S. EPA, 1997). These plans are used to help ensure protection of human health and the environment should a contaminant plume begin migrating further or faster than predicted based on the natural attenuation evaluation, and typically involve some kind of engineered remediation. It may be prudent to update the contingency plan on a periodic basis as the plume attenuates or as new remediation technologies are developed. Although some engineered remediation systems may be effective in removing contaminant mass from groundwater, it should be kept in mind when developing the contingency plan that some remediation systems may have an adverse impact on intrinsic bioremediation. Table 11.2 presents potential interactions between active remediation technologies and natural attenuation. For example, the introduction of oxygen via air sparging into a system contaminated with chlorinated solvents may raise the oxidation–reduction potential of the groundwater to the point that reductive dechlorination can no longer occur and the natural treatment system is destroyed. A groundwater pump and treat system can have the same impact by flushing unimpacted, oxygen-rich groundwater through the contaminant plume. Because of these potential adverse affects, the impacts of any proposed remediation system on naturally occurring processes should be evaluated when developing a contingency plan.

U.S. EPA (1997) provides the following discussion concerning contingency remedies:

> A contingency remedy is a cleanup technology or approach specified in the site remedy decision document that functions as a "backup" remedy in the event that the "selected" remedy fails to perform as anticipated. A contingency remedy may specify a technology (or technologies) that is (are) different from the selected remedy, or it may simply call for modification and enhancement of the selected technology, if needed. Contingency remedies should generally be flexible—allowing for the incorporation of new information about site risks and technologies.
>
> Contingency remedies are not new to OSWER programs. Contingency remedies should be employed where the selected technology is not proven for the specific site application, where there is significant uncertainty regarding the nature and extent of contamination at the time the remedy is selected, or where there is uncertainty regarding whether a proven technology will perform as anticipated under the particular circumstances of the site.
>
> It is also recommended that one or more criteria ("triggers") be established, as appropriate, in the remedy decision document that will signal unacceptable performance of

TABLE 11.2. Interactions Between Active Remediation Technologies and Natural Attenuation

Technology	Possible Benefits		Possible Detriments	
	Petroleum Hydrocarbons	Chlorinated Solvents	Petroleum Hydrocarbons	Chlorinated Solvents
Bioslurping	Volatilization, enhanced oxygen delivery/aerobic biodegradation	Volatilization, enhanced oxidation of DCE and VC, possible enhanced aerobic co-metabolism	None	Enhanced oxygen delivery/ decreased reductive dechlorination
Pump-and-treat	Plume containment, enhanced oxygen delivery/aerobic biodegradation	Plume containment, enhanced oxidation of DCE and VC, possible enhanced aerobic co-metabolism	None	Enhanced oxygen delivery/ decreased reductive dechlorination
Air sparging	Volatilization, enhanced oxygen delivery/aerobic biodegradation	Volatilization, enhanced oxidation of DCE and VC, possible enhanced aerobic co-metabolism	None	Enhanced oxygen delivery/ decreased reductive dechlorination
In-well circulation/stripping	Volatilization, enhanced oxygen delivery/aerobic biodegradation	Volatilization, enhanced oxidation of DCE and VC, possible enhanced aerobic co-metabolism	None	Enhanced oxygen delivery/ decreased reductive dechlorination
Landfill caps	Source containment/ isolation	Source containment/isolation, reduced oxygen delivery through elimination of recharge/stimulation of reductive dechlorination	Reduced oxygen delivery/ aerobic biodegradation	Decreased oxidation of DCE and VC, decreased aerobic co-metabolism

continues

TABLE 11.2. *(Continued)*

Technology	Possible Benefits		Possible Detriments	
	Petroleum Hydrocarbons	Chlorinated Solvents	Petroleum Hydrocarbons	Chlorinated Solvents
Phytoremediation	Plant-specific transpiration/enzymatically mediated degradation, enhanced biodegradation in the rhizosphere, and plume containment	Plant-specific transpiration/enzymatically mediated degradation, enhanced biodegradation in the rhizosphere, and plume containment	None	Unknown
Excavation/backfilling	Source removal, enhanced oxygen delivery/aerobic biodegradation	Source removal, enhanced oxidation of DCE and VC, possible enhanced aerobic cometabolism	None	Enhanced oxygen delivery/decreased reductive dechlorination
Chemical oxidation (e.g., Fenton's reagent, potassium permanganate, etc.)	Enhanced oxidation	Enhanced oxidation	None	Enhanced oxygen delivery/decreased reductive dechlorination through oxidation and removal of fermentable carbon substrates; lowered pH possibly inhibits microbial activity

Chemical reduction (e.g., sodium dithionate)	Unknown	Scavenges inorganic electron acceptors/enhanced reductive dechlorination	Scavenges inorganic electron acceptors/ decreased oxidation	Decreased oxidation of DCE and VC, decreased aerobic co-metabolism
Oxygen-releasing materials	Enhanced oxygen delivery/ aerobic biodegradation	Enhanced oxidation of DCE and VC	None	Enhanced oxygen delivery/ decreased reductive dechlorination through oxidation and removal of fermentable carbon substrates
Carbon substrate addition	None	Stimulation of reductive dechlorination	Competing carbon source	Decreased oxidation of DCE and VC, decreased aerobic co-metabolism at injection point
Zero-valent iron barrier walls	Unknown	Enhanced reductive dechlorination	Unknown	None
Biological barrier walls	Unknown	Enhanced reductive dechlorination	Unknown	None

Source: From Wiedemeier and Chapelle (1998).

the selected remedy and indicate when to implement contingency measures. Such criteria might include the following:

- Contaminant concentrations in soil or groundwater at specified locations exhibit an increasing trend.
- Near-source wells exhibit large concentration increases indicative of a new or renewed release.
- Contaminants are identified in sentry/sentinel wells located outside of the original plume boundary, indicating renewed contaminant migration.
- Contaminant concentrations are not decreasing at a sufficiently rapid rate to meet the remediation objectives.
- Changes in land and/or groundwater use will adversely affect the protectiveness of the monitored natural attenuation remedy.

In establishing triggers or contingency remedies, however, care is needed to ensure that sampling variability or seasonal fluctuations do not set off a trigger inappropriately. For example, an anomalous spike in dissolved concentration(s) at a well(s), which may set off a trigger, might not be a true indication of a change in trend.

EPA recommends that remedies employing monitored natural attenuation be evaluated to determine the need for including one or more contingency measures that would be capable of achieving remediation objectives. EPA believes that a contingency measure may be particularly appropriate for a monitored natural attenuation remedy which has been selected based primarily on predictive analysis (second and third lines of evidence discussed previously) as compared to natural attenuation remedies based on historical trends of actual monitoring data (first line of evidence).

11.5 WHEN TO STOP MONITORING

There are two general approaches for determining the length of time required to monitor: (1) monitor until the entire site is remediated, a process that might take decades at many sites; or (2) monitor until risk-based goals are met, where a "safe" level of contaminants are left in the subsurface for natural attenuation processes to manage. There is considerable debate about which approach is appropriate for natural attenuation sites. The EPA's draft interim final policy states that

> performance monitoring is required as long as contamination levels remain above required cleanup levels on any portion of the site. Typically, monitoring is continued for a specified period (e.g., one to three years) after cleanup levels have been achieved to ensure that concentration levels are stable and remain below target levels. The institutional and financial mechanisms for maintaining the monitoring program should be clearly established in the remedy decision or other site documents, as appropriate.

It is unclear how this policy will ultimately be implemented because the time required for natural attenuation to clean up to background levels, MCLs, and so on, are very different from the times required to reach a risk-based levels or alternative concentration levels (ACLs). It is the authors' opinion that long-term monitoring until all the contamination reaches background levels or MCLs may not be necessary at many sites. For example, 30 years of additional long-term monitoring would not appear to provide any beneficial information at a fuel-hydrocarbon site with (1) monitoring data that show shrinking plume for a time period equal to two or three times the hydraulic residence time of groundwater in the plume, and (2) a risk assessment that indicates no risk to human health or the environment from the current extent of plume migration. In this case, an appropriate course of action might be one year of confirmation monitoring, and then no additional monitoring would be required.

As with any remedial option for sites contaminated with organic compounds, remediation goals should be established early in the negotiation process. This will help establish a purpose for long-term monitoring and should help define the length of time that monitoring will be required. As indicated by the EPA guidance, confirmation monitoring is typically required to confirm that the conclusion that the site requires no further action. For example, many states implementing RBCA programs are considering one- to five-year *confirmation monitoring* periods at sites when remediation goals are first achieved. Unlike some long-term monitoring plans, the confirmation monitoring would probably include analysis of all of the original contaminants of concern observed at a site.

Continuing a monitoring program beyond the point where the data being gathered are useful is counterproductive and should be avoided. In many cases removal of NAPL will speed up the remediation process and can drastically reduce the amount of time required to meet remediation goals, and thus the length of time required for long-term monitoring.

REFERENCES

Mace, R. E., Fisher, R. S., Welch, D. M., and Parra, S. P., 1997, *Extent, Mass, and Duration of Hydrocarbon Plumes from Leaking Petroleum Storage Tank Sites in Texas*, Geological Circular 97-1, Bureau of Economic Geology, University of Texas at Austin, Austin, TX.

Newell, C. J., McCleod, R. K., and Gonzales, J. R., 1997, *BIOSCREEN Natural Attenuation Decision Support System, Version 1.4*, U.S. Air Force Center for Environmental Excellence, Brooks AFB, TX, and U.S. EPA R.S. Kerr Laboratories, Ada, OK (see www.epa.gov/ada/kerrlab.html).

Rice, D. W., Grose, R. D., Michaelsen, J. C., Dooher, B. P., MacQueen, D. H., Cullen, S. J., Kastenberg, W. E., Everett, L. G., and Marino, M. A., 1995, *California Leaking Underground Fuel Tank (LUFT) Historical Case Analyses*, California State Water Resources Control Board, Sacramento, CA.

Wiedemeier, T. H., and Chapelle, F. H., 1998, *Technical Guidelines for Evaluating Monitored Natural Attenuation of Petroleum Hydrocarbons and Chlorinated Solvents in Ground Water at Naval and Marine Corps Facilities*, Naval Facilities Engineering Command, Alternative Restoration Technology Team, September, 1998.

Wiedemeier, T. H., and Haas, P. E., 1999, *Designing Monitoring Programs to Effectively Evaluate the Performance of Natural Attenuation*, U.S. Air Force Center for Environmental Excellence, San Antonio, TX.

Wiedemeier, T. H., Swanson, M. A., Moutoux, D. E., Gordon, E. K., Wilson, J. T., Wilson, B. H., Kampbell, D. H., Hansen, J. E., Haas, P., and Chapelle, F. H., 1996, *Technical Protocol for Evaluating Natural Attenuation of Chlorinated Solvents in Groundwater*, Draft, Revision 1, U.S. Air Force Center for Environmental Excellence, San Antonio, TX.

Wiedemeier, T. H., Swanson, M. A., Montoux, D. E., Gordon, E. K., Wilson, J. T., Wilson, B. H., Kampbell, D. H., Haas, P. E., Miller, R. N., Hansen, J. E., and Chapelle, F. H., 1998, *Technical Protocol for Evaluating Natural Attenuation of Chlorinated Solvents in Ground Water*, EPA/600/R-98/128, September 1998, U.S. EPA, Washington, DC, ftp://ftp.epa.gov.pub/ada/reports/protocol.pdf.

U.S. Environmental Protection Agency (U.S. EPA), 1997, *Monitored Natural Attenuation at Superfund, RCRA Corrective Action, and Underground Storage Tank Site, Draft Interim Final Policy*, Office of Solid Waste and Emergency Response (OSWER), Washington, DC.

APPENDIX A

GEOCHEMICAL AND PHYSICAL DATA

TABLE A.1. Geochemistry of Groundwater at Fuel-Hydrocarbon-Contaminated Sites[a]

Site	Maximum Observed Total BTEX (μg/L)	Dissolved Oxygen Plume (mg/L)	Dissolved Oxygen Background (mg/L)	Nitrate (as N) Plume (mg/L)	Nitrate (as N) Background (mg/L)	Iron (II) Plume (mg/L)	Iron (II) Background (mg/L)	Sulfate Plume (mg/L)	Sulfate Background (mg/L)	Methane Plume (mg/L)	Methane Background (mg/L)	ORP Plume (mV)	ORP Background (mV)	Total Alkalinity Range (mg/L)	Average pH	Average Temperature (°C)
Hill AFB, UT, site 870	21,475	<0.5	5.9	<0.05	17	51	<0.05	<0.5	99	2.1	<0.001	−190	272	349–959	7.3	18.7
Battle Creek ANGB, MI, site 3	3,552	0.09	6.9	<0.05	3.42	12	<0.05	2.7	27.3	8.4	<0.001	−335	255	132–670	7.5	14.9
Madison ANGB, WI	28,000	0.8	7.1	<0.05	10	15.6	0.01	0.51	89.2	9.9	<0.005	−87	180	200–540	6	11.1
Elmendorf AFB, AK, hangar 10	447	<0.1	0.8	<0.5	15	9.0	<0.05	2.12	26.7	8.5	<0.001	−120	257	94–467	6.9	6.9
Elmendorf AFB, AK, ST-41	43,280	0.1	12.6	<0.5	31	40.5	<0.05	<0.05	59.3	1.48	<0.001	−53	258	47–1210	7	7.1
King Salmon AFB, AK, SS-12	5,261	0.1	10.9	<0.05	0.3	44	<0.05	<0.5	8.54	5.61	<0.001	−50	255	12–256	6.6	7.5
Eglin AFB, FL, police facility	4,030	<0.1	3.8	<0.05	0.2	10.5	<0.05	<0.05	31	16.82	0.6	−253	−30	4–94	6.1	26
Patrick AFB, FL, gas station	14,096	<0.1	3.7	<0.05	0.29	1.9	<0.05	0.52	1,000	15.53	0.034	−293	54	148–520	7.1	26.1
MacDill AFB, FL, site 56	29,636	0.02	2.4	<0.05	1.32	5.08	0.02	0.4	110	13.57	0.03	−246	−4.9	120–520	6.8	25.2
MacDill AFB, FL, pumphouse 75	676	0.08	2.1	<0.05	0.99	20.9	0.02	1.22	65.8	14.47	0.035	−169	35	10–380	6.1	26.3
Myrtle Beach, SC, police facility	18,270	0.04	1.9	<0.05	1.36	37.5	<0.05	<0.5	84.1	17.13	0.001	−255	260	3–570	6	17.2
Langley AFB, VA, site SS04	1,806	0.09	5	<0.05	1.4	44.9	<0.007	<0.25	865	7.8	<0.002	−332	185	125–648	7.1	19.5
Langley AFB, VA, site SS16	123	0.22	6.5	<0.06	5.2	9.3	0.01	<0.25	237	8	<0.004	−158	204	120–480	6.6	26.7
Griffis AFB, NY, pumphouse 5	12,840	0.1	6.5	0.06	2.7	24.5	0.1	<0.37	20	7	0.02	−158	206	120–480	6.4	14.9
Pope AFB, NC, FPTA no. 4	8,180	0	8.6	<0.05	1.95	56.3	<0.05	<0.5	13.3	48.4	<0.002	−160	393	5–251	5.8	16.8
Seymour Johnson AFB, NC, bldg. 470	13,800	0.18	9	<0.06	1.7	31.63	0.04	0.7	47	2.7	<0.004	−93	312	10–130	5.9	17.4
Fairchild AFB, WA, bldg. 1212	13,118	0.7	9.3	0.08	8.42	3.2	<0.1	<0.37	60	1.87	<0.001	−121	149	340–1160	7.1	12.1
Eaker AFB, AR, gas station	84,900	0.4	8.2	0.003	0.46	33.8	<0.01	0.32	89	3.8	<0.001	−235	222	60–780	5.5	14.5
Dover AFB, DL, site SS27/XYZ	22,900	0.1	8.3	<0.05	18.4	2.05	0.09	4	64.1	NA	NA	−431	380	5–198	6	14
Bolling AFB, DC, car care center	49,300	0.05	7.5	0.06	17.2	20.3	<0.1	50	83	9.1	0.003	−154	176	20–1100	6	20.3
Offutt AFB, NE, tank 349	90,210	0.8	6.8	<0.22	7.67	<0.05	<0.05	4.13	75.5	0.008	<0.001	153	283	80–536	7.5	22.2
Westover AFB, MA, Christmas tree fire training area	469	0.18	11.2	<0.05	7.31	23.5	<0.05	4.15	436	0.006	<0.001	−159	300	20–134	7	13.4
Columbus AFB, MS, ST-24	20,950	0.19	8.52	<0.1	2.5	63.75	<0.01	<0.1	14.6	2.06	<0.01	−234	338	1–168	5.5	20.6
Shaw AFB, SC, bldg. 1613	4,246	0	7.9	<0.05	3.45	45	<0.01	<0.1	51.6	0.99	<0.001	−149	278	10–110	5.5	22.6
Travis AFB, CA, gas station	67,000	0.1	3.7	<0.06	9.5	14.85	<0.01	4.2	2,000	5.43	<0.009	−32	325	180–1140	6.6	21.7
Beale AFB, CA, UST site	71.9	0.15	8.4	0.083	4.3	6.84	<0.01	1.1	27	3.5	<0.002	56	426	53–1220	6.8	17.6
Chanute AFB, IL, 952 site	1,540	2.2	5.5	<0.1	0.63	9.18	<0.008	0.5	146	6.60	<0.0025	−7.1	228.7	330–650	7.51	13.6
Grissom AFB, IN, bldg. 735	268	3.4	11.6	<0.6	15.4	2.8	<0.03	7	115	1.80	<0.12	180.7	315.4	71–510	6.89	13.4
Keesler AFB, MS, SWMU 66 site	14,100	0.4	2.1	<0.5	0.7	36.6	0.46	26.2	3.8	7.40	<0.12	−261.4	175.8	30–648	6.00	26.6
Tyndall AFB, FL, police B site	1,037	0.02	1.9	<0.3	0.9	1.57	0.16	7.5	13.4	4.60	<0.12	−283.5	252.5	2–152	6.18	20.6
Gulf Coast site A	405,622	1.8	3.4	<5	<5	4.09	<0.02	<5	150	7.00	<0.0012	NM	NM	212–675	6.77	24.5
Gulf Coast site B	26	0.7	4.8	<5	<5	<0.02	<0.02	21	114	0.091	<0.0012	NM	NM	353–590	7.18	25.0
Gulf Coast site C	6,267	1.4	5.6	<5	<5	0.845	<0.02	1,280	114	9.30	<0.0012	NM	NM	338–4150	7.55	25.2
Gulf Coast site D	19,400	4.2	4.2	<0.1	3.0	2.91	<0.03	445	111	9.40	<0.0012	NM	NM	282–800	6.98	22.6
Gulf Coast site E	110	2.2	3.8	<0.1	12.3	56.2	<0.03	2.65	170	4.92	<0.0012	NM	NM	380–1650	7.02	21.4
Gulf Coast site F	3,860	1.6	6.4	<5	<5	<0.02	<0.02	4	380	1.45	<0.0012	NM	NM	70–932	6.97	24.1
Gulf Coast site G	337,100	0.9	7.2	0.14	0.47	299	<0.03	<10	742	0.097	<0.0012	NM	NM	<10–996	7.15	24.0
Houston site A	83,900	2.5	3.9	<0.1	5.5	0.3	<0.02	5.2	66	1.3	<0.0012	NM	NM	NM	7.44	27.6

[a]NA, not analyzed; NM, not measured.

TABLE A.2. Size of Fuel-Hydrocarbon Plumes[a]

Site	Maximum Mobile LNAPL Length (ft)	Maximum Mobile LNAPL Width (ft)	Maximum Mobile LNAPL Thickness (ft)	Maximum Observed Total BTEX (μg/L)	Dissolved Plume Length (ft)	Dissolved Plume Width (ft)	Dissolved Plume Thickness (ft)	Hydrogeologic Setting	Groundwater Velocity (ft/yr)		
									High	Low	Average
Hill AFB, UT, site 870	1,200	350	4	21,475	1,650	750	25	Poorly to moderately sorted fine- to medium-grained sand (fluvial-deltaic)	2,113	945	1,600
Battle Creek, ANGB, MI, site 3	ND	ND	ND	3,552	900	375	30	Fine- to coarse-grained sand and silty sand with interbedded gravel and cobbles	126	38	72
Madison ANGB, WI	ND	ND	ND	28,000	750	400	10	Fine- to medium-grained sand with occasional silty or gravelly intervals	90	45	54
Elmendorf AFB, AK, hangar 10	ND	ND	ND	447	3,000[b]	1,500[b]	20	Glaciofluvial sand and gravel	2,400	800	1,600
Elmendorf AFB, AK, ST-41	100	100	0.7	43,280	700	500	20	Glacial silt sand and gravel	538	4.5	260
King Salmon AFB, AK SS-12	Trace	Trace	Sheen	5,261	850	900	10	Fine- to medium-grained sand	1,051	315	683
Eglin AFB, FL, police facility	ND	ND	ND	4,030	300[c]	150	25	Well-sorted medium-grained sand with Interbedded peat	4,380	263	657
Patrick AFB, FL, gas station	ND	ND	ND	14,096	480	120	15	Fine- to medium-grained moderately well-sorted sand	220	110	156
MacDill AFB, FL, site 56	ND	ND	ND	29,636	250	120	10	Fine- to medium-grained sand	183	33	51
MacDill AFB, FL, pumphouse 75	ND	ND	ND	676	350	150	25	Well-sorted fine-grained sand	128	7	22
Myrtle Beach, SC, police facility	500	500	5	18,270	1,150[c]	750	35	Fine- to medium-grained sand with thin clay stringers	145	5	51
Langley AFB, VA, site SS04	ND	ND	ND	1,806	NA	NA	NA	—	NA	NA	NA
Langley AFB, VA site SS16	ND	ND	ND	123	140	120	10	Fine- to very fine-grained sand	275	92	183
Griffis AFB, NY, pumphouse 5	220	70	7	12,840	360	260	10	Fine- to medium-grained sand with gravel and rock fragments	63	1.6	14.6
Pope AFB, NC, FPTA no. 4	350	140	0.6	8,180	720	380	25	Fine to coarse sand with minor silt and clay seams	840	105	473

continues

TABLE A.2. *(Continued)*

Site	Maximum Mobile LNAPL Length (ft)	Maximum Mobile LNAPL Width (ft)	Maximum Mobile LNAPL Thickness (ft)	Maximum Observed Total BTEX (μg/L)	Dissolved Plume Length (ft)	Dissolved Plume Width (ft)	Dissolved Plume Thickness (ft)	Hydrogeologic Setting	Groundwater Velocity (ft/yr)		
									High	Low	Average
Seymour Johnson AFB, NC, bldg. 470	90	30	1.95	13,800	315	250	10	Medium- to coarse-grained sand	258	15	91
Fairchild AFB, WA, bldg. 1212	ND	ND	ND	13,118	175	125	20	Gravelly sand	94	17	40
Eaker AFB, AR, gas station	80	30	4.74	84,900	420	330	15	Very hetrogeneous; discontinuous sand lenses in clay and silty clay	285	18	77.4
Dover AFB, DL, site SS27/XYZ	1,300	300	7	22,900	3,000[d]	750	25	Sand with gravel	314	131	161
Bolling AFB, DC, car care center	140	60	0.4	49,300	630	300	30	Silty clay and silty sand	14.6	1.5	7.3
Offutt AFB, NE, tank 349	ND	ND	ND	90,210	200	175	10	—	11	0.4	5.7
Westover AFB, MA, Christmas tree fire training area	ND	ND	ND	469	200	150	15	Glaciofluvial sand and gravel	219	62	63
Columbus AFB, MS, ST-24	ND	ND	ND	20,950	350	200	15	Silty to sandy gravel	978	422	700
Shaw AFB, SC, bldg. 1613	ND	ND	ND	4,246	NA	NA	NA	—	NA	NA	402
Travis AFB, CA, gas station	ND	ND	ND	67,000	680	320	20	Fine- to coarse-grained sand	60	20	40
Beale AFB, CA, UST site	ND	ND	ND	72	220	180	10	Silt and clay with interbedded sand	120	80	100

[a] NA, not analyzed; ND, not detected.

[b] Large dilute plume probably resulting from two or more commingled plumes.

[c] Plume discharges to surface water body.

[d] Groundwater flow in multiple directions due to groundwater divide.

TABLE A.3. Contaminants and Daughter Products at Sites Exhibiting Type I Behavior[a]

	Maximum Observed Concentration															
Site	Total BTEX (μg/L)	PCE (μg/L)	TCE (μg/L)	1,1-DCE (μg/L)	*trans*-1,2-DCE (μg/L)	*cis*-1,2-DCE (μg/L)	VC (μg/L)	TCA (μg/L)	1,1-DCA (μg/L)	1,2-DCA (μg/L)	CA (μg/L)	Ethene Plume (mg/L)	Ethene Back-ground (mg/L)	Ethane Plume (mg/L)	Ethane Back-ground (mg/L)	Plume Behavior (I/II/III)
Shaw AFB, SC, OU-4	3,295	90.7	718	748	2.1	4,590	416	23,000	8,700	2.1	NA	0.005	<0.003	0.01	<0.002	NA
Wurtsmith AFB, MI, OT-41/SS-42	2,421	6.1	8.1	ND	ND	29	ND	NA	NA	NA	NA	ND	ND	NA	NA	I
MacDill AFB, FL, site OT-24	2,840	1.1	2.6	ND	ND	6.8	20.8	NA	NA	NA	NA	NA	NA	NA	NA	I
Offutt AFB, NE, fire training area	3,230	ND	8.6	ND	2.4	273	817	3.3	NA	ND	NA	0.895	<0.003	NA	NA	I
Fairchild AFB, WA, FT-1	5,221	ND	63.4	<1.0	3.2	124	37.9	NA	NA	NA	NA	0.017	<0.003	<0.002	<0.002	I
Rickenbacker ANGB, OH, HWSA	440	ND	9	ND	ND	1,500	ND	NA	NA	ND	NA	0.0057	ND	NA	NA	I
Westover ARB, MA, current fire training area	26,125	ND	1600	3.7	3.2	434	2,098	NA	NA	NA	NA	0.008	ND	NA	NA	I
Westover ARB, MA, zone 1	15,601	13	28	0.51	ND	390	1.4	ND	ND	NA	NA	NA	NA	NA	NA	I

[a]NA, not analyzed; ND, not detected.

TABLE A.4. Contaminants and Daughter Products at Sites Exhibiting Mixed Behavior[a]

	Maximum Observed Concentration														
Site	Total BTEX (μg/L)	PCE (μg/L)	TCE (μg/L)	1,1-DCE (μg/L)	*trans*-1,2-DCE (μg/L)	*cis*-1,2-DCE (μg/L)	VC (μg/L)	TCA (μg/L)	1,1-DCA (μg/L)	1,2-DCA (μg/L)	CA (μg/L)	Ethene Plume (mg/L)	Ethene Background (mg/L)	Ethane Plume (mg/L)	Ethane Background (mg/L)
Plattsburgh AFB, NY, FT-002	60,100	ND	25,280	ND	233	51,400	2,080	NA	NA	NA	NA	0.474	<0.001	NA	NA
	2	ND	1	ND	ND	226	5	—	—	—	—	<0.001	—	—	—
Offutt AFB, NE, bldg. 301	1	1.9	17,500	28.6	4.8	1,230	ND	ND	1	ND	NA	<1	ND	NA	NA
III Upgradient/II downgradient	ND	ND	438	1.4	9.4	69.4	1.3	ND	ND	ND	NA	ND	ND	NA	NA
Cape Canaveral AS, FL, FT-17	331	56	15,800	200	389	98,500	3,080	ND	443	ND	NA	0.188	<0.003	55	<2
I Upgradient/II downgradient	12.5	ND	42	21	25	1,200	6,520	ND	69.6	ND	NA	0.225	<0.003	13	<2
Cape Canaveral AS, FL, bldg. 1381	6	3.3	39,400	238	145	4,120	1,350	130	26.4	ND	NA	0.114	<0.003	NA	NA
I Upgradient/II downgradient	ND	ND	ND	ND	ND	763	510	ND	ND	ND	NA	<0.003	<0.003	NA	NA
Altus AFB, OK, landfill 04	43	ND	12,700	4.3	33.3	1,340	ND	NA	NA	NA	NA	ND	ND	ND	ND
I Upgradient/III downgradient	ND	ND	13	ND	ND	ND	ND	ND	ND	ND	ND	ND	ND	ND	ND
Columbus AFB, MS, LF-06	48	ND	7.4	ND	ND	15	10	ND	1.2	ND	2.1	NA	NA	NA	NA
I Upgradient/III downgradient	ND	ND	ND	2.7	ND	ND	2.4	3.8	ND	ND	ND	NA	NA	NA	NA
Hill AFB, UT, OU-1	1,557	ND	446	12.7	8.3	7,083	469	257	187	166	NA	0.274	<0.003	NA	NA
I Upgradient/III downgradient	ND	ND	29.8	1.5	ND	648	1.4	1.6	4.4	1.4	NA	<0.005	<0.003	NA	NA

[a]NA, not analyzed; ND, not detected; —, [info TK].

TABLE A.5. Contaminants and Daughter Products at Sites Exhibiting Type III Behavior[a]

	Maximum Observed Concentration														
Site	Total BTEX (μg/L)	PCE (μg/L)	TCE (μg/L)	1,1-DCE (μg/L)	*trans*-1,2-DCE (μg/L)	*cis*-1,2-DCE (μg/L)	VC (μg/L)	TCA (μg/L)	1,1-DCA (μg/L)	1,2-DCA (μg/L)	CA (μg/L)	Ethene Plume (mg/L)	Ethene Background (mg/L)	Ethane Plume (mg/L)	Ethane Background (mg/L)	Plume Behavior (I/II/III)
Hill AFB, UT, OU-5	ND	253	355	2.3	<1.0	15.6	ND	64.2	5.6	ND	NA	<0.003	<0.003	NA	NA	III
King Salmon AFB, AK, FT01	8,620	ND	636	ND	ND	ND	ND	8.3	ND	ND	NA	NA	NA	NA	NA	III

[a]NA, not analyzed; ND, not detected.

TABLE A.6. Chemical Characteristics of Groundwater at Sites Exhibiting Type I Behavior[a]

Site	Dissolved Oxygen		Nitrate (as N)		Iron (II)		Sulfate		Methane		ORP		Hydrogen		Chloride		Alkalinity Range (mg/L)	Average pH	Average Temperature (°C)
	Plume (mg/L)	Background (mg/L)	Plume (mg/L)	Background (mg/L)	Plume (mg/L)	Background (mg/L)	Plume (mg/L)	Background (mg/L)	Plume (mg/L)	Background (mg/L)	Plume (mV)	Background (mV)	Plume (nM)	Background (nM)	Plume (mg/L)	Background (mg/L)			
Shaw AFB, SC, OU-4	0	9	<0.05	1.2	8.2	<0.05	<0.1	20.4	6.952	<0.001	−201	319	19.02	0.1	177	2.3	35–425	5.4	18.8
Wurtsmith AFB, MI, OT-41/SS-42	0.1	9.9	<0.05	6.02	15	<0.1	3.02	13.6	3.55	<0.001	−73	146	<0.1	<0.1	25.3	1.05	459–598	7.1	14.5
MacDill AFB, FL, site OT-24	0.02	1.54	1.09(D)	2.46(D)	3.48	0.2	1.04	77	9.89	0.008	−238.3	−18.7	NA	NA	115	6.51	200–260	6.6	25.1
Offutt AFB, NE, fire training area	0.1	3.3	<0.05	1.52	26.3	<0.05	<0.5	391	22.45	<0.01	−170	90	NA	NA	213	3.6	412–784	7.1	12.9
Fairchild AFB, WA, FT-1	0.46	7.95	<0.05	2.4	22.3	<0.1	<0.05	13.1	19.06	<0.001	−127	200	NA	NA	94	1.4	143–1260	7.3	10.1
Rickenbacker ANGB, OH, HWSA	0	13.4	<0.05	9.1	16.5	<0.05	6.57	938	19.2	<0.01	−136	212	0.648	0.001	53.3	1.5	212–426	7.4	12.1
Westover ARB, MA, current fire training area	0.15	8.93	<0.05	5.57	45.3	<0.05	<0.5	76.7	14.63	<0.001	−125	280	NA	NA	131	<0.5	4–238	6.9	13
Westover ARB, MA, zone 1	0.21	10.75	<0.05	9.5	288	<0.01	1.3	36.8	2.2	<0.001	−207	313	NA	NA	57.6	0.9	10–145	5.9	14.9

[a] NA, not analyzed.

TABLE A.7. Chemical Characteristics of Groundwater at Sites Exhibiting Mixed Behavior[a]

Site	Type of Behavior	Dissolved Oxygen		Nitrate		Iron (II)		Sulfate		Methane		ORP		Hydrogen		Chloride		Alkalinity Range (mg/L)	Average pH	Average Temperature (°C)
		Plume (mg/L)	Back-ground (mg/L)	Plume (mg/L)	Back-ground (mg/L)	Plume (mg/L)	Back-ground (mg/L)	Plume (mg/L)	Back-ground (mg/L)	Plume (mg/L)	Back-ground (mg/L)	Plume (mV)	Back-ground (mV)	Plume (nM)	Back-ground (nM)	Plume (mg/L)	Back-ground (mg/L)			
Plattsburgh AFB, NY, FT-002	I	0.1	10	<0.05	30	10.7	<0.5	0.08	20.3	1.5	<0.001	pE 1.08	pE 6.66	11	<0.1	222	1	96–1590	7.8	9
I Upgradient/III downgradient	III	0.4	—	10	—	<0.05	2	14.7	—	0.12	—	pE 4	—	0.22	—	3	—	NA	—	—
Offutt AFB, NE, bldg. 301	III	3.8	7	6.91	23.2	<0.05	<0.05	40[a]	25[b]	0.18	<0.001	193	312	0.26	0.26	65.7	1.68	129–488	7.3	16
III Upgradient/ II downgradient	II	0.2	—	<0.05	—	5.3	—	40[a]	25[b]	12.29	—	–98	—	9.55	—	76.2	—	287–487	—	—
Cape Canaveral AS, FL, FT-17	I	0.2	1.2	<0.05	0.72	39	<0.05	15–452[b]	7.9–441[b]	6.97	0.002–0.04	–250	100	10	<0.01	373	49	387–527	7.2	23
I Upgradient/ II downgradient	II	0.2	—	<0.05	—	3.6	—	16–3500[b]	—	0.4	—	93	—	0.29	—	169	—	188–427	—	—
Cape Canaveral AS, FL, bldg. 1381	I	0.1	0.3	<0.05	0.2	1.6[b]	4	45–71	3–25	4	0.07–3.4[c]	–136	–172 to 277	6	1.4–5.1[c]	207	13	239–354	7.6	26
I Upgradient/ II downgradient	II	0.1	—	<0.05	—	8.5	—	<1–68	—	0.08–1.23	—	–340 to 172	—	1.2–4.1	—	242	—	—	—	—
Altus AFB, MS, landfill 04	I	<0.1	4.3	<0.05	7.8	<0.05	<0.05	1440–3620[a]	680–2010	0.08	<0.001	–200	165	2.24	NA	1,500	384	205–446	6.9	14
I Upgradient/ III downgradient	III	0.3	—	1	—	<0.05	—	1500–2480[a]	—	<0.001	<0.001	70	—	0.19	—	811	—	157–504	—	—
Columbus AFB, MS, LF-06	I	0.15	5.1	0.001	6.7	95.5	0.01	ND	50	6.5	<0.005	–300	284	NA	NA	183	3	5–425	5.4	19
I Upgradient/ III downgradient	III	2	—	1.5	—	2	—	2	—	ND	—	100	—	—	—	10	—	3–68	—	—
Hill AFB, UT, OU-1	I	<0.1	8	<0.05	21	35	<0.05	<0.1–38[a]	43–366[b]	10.4	<0.001	–250	262	1.4	NA	NA	NA	190–825	7.3	11
I Upgradient/ III downgradient	III	1	—	7	—	<0.05	—	36–83[a]	—	0.17	—	90	—	NA	—	—	—	150–450	—	—

[a]Upgradient concentrations listed first followed by downgradient concentrations; NA, not analyzed; —, [info TK].

[b]No discernible trend in concentrations outside the plume versus inside the plume.

[c]Intermediate and deep wells (below the dissolved contaminant plume, groundwater gradients are upward) have methane concentrations ranging from 20 to 30 mg/L and hydrogen concentrations ranging from 5.3 to 7.1. These trends are probably caused by biodegradation of native organic matter.

TABLE A.8. Chemical Characteristics of Groundwater at Sites Exhibiting Type III Behavior

Site	Dissolved Oxygen		Nitrate (as N)		Iron (II)		Sulfate		Methane		ORP		Hydrogen		Chloride		Alkalinity Range (mg/L)	Average pH	Average Temperature (°C)
	Plume (mg/L)	Background (mg/L)	Plume (mg/L)	Background (mg/L)	Plume (mg/L)	Background (mg/L)	Plume (mg/L)	Background (mg/L)	Plume (mg/L)	Background (mg/L)	Plume (mV)	Background (mV)	Plume (nM)	Background (nM)	Plume (mg/L)	Background (mg/L)			
Hill AFB, UT, OU-5	0.3	5.39	0.53	0.96	1	<0.1	14.6	95.7	0.428	<0.001	–170	216	5	<0.1	196	16.1	70–>500	7.1	17.5
King Salmon AFB, AK, FT01	0.3	10.4	<0.05	2.52	15	<0.1	<0.5	5.9	0.39	<0.001	–65	260	NA	NA	6.02	2.07	23–177	6.8	6.3

TABLE A.9. Plume Size

Site	Dissolved Plume Length (ft)	Dissolved Plume Width (ft)	Plume Thickness (ft)	Hydrogeologic Setting	Groundwater Velocity[a] (ft/yr)		
					High	Low	Average
Plattsburgh, AFB, NY, FT-002	4,200[b]	2,050	60	Well-sorted fine- to medium-grained sand	1,102	1	139
Wurtsmith AFB, MI, OT-41/SS-42	375	100	10		1,825	110	292
MacDill AFB, FL, site OT-24	600	350	20	Poorly to well-sorted fine- to medium-grained sand	253	3	36
Offutt AFB, NE, fire training area	650	450	30	Very fine- to fine-grained sand	10.4	3.8	6.7
Offutt AFB, NE, bldg. 301	3,500	1,400	50	Loess	1,971	0.2	152
Rickenbacker ANGB, OH, HWSA	100	60	10		3.2	68	25
Westover ARB, MA, current fire training area	750	250	50	Fine-grained silty sand	6.2	38	20.8
Westover ARB, MA, zone 1	1,800 750 1,650	1,200 200 450	~50	Fine- to medium-grained sand with a trace of silt	370	39	106
Cape Canaveral AS, FL, FT-17	1,575[b]	400	15	Fine- to coarse-grained sand with occasional peat stringers	254	57	113
Shaw AFB, SC, OU-4	750	550	5		NA	NA	1,600
Hill AFB, UT, OU-5	5,000	1,400	40	Silty sand to sandy silt	110	11	60
Columbus AFB, MS, LF-06	750	550	5		12,000	750	1,500
Fairchild AFB, WA, FT-1	550	300	10	Sand and gravel	42.4	31.1	32.9
King Salmon AFB, AK, FT01	1,100	250	25	Fine- to medium-grained sand	447	69	260

[a]NA, not analyzed.

[b]Discharges to surface water body.

APPENDIX B

RCBA CHEMICAL DATABASE*

*Appendix B is from Groundwater Services, Inc. (GSI), 1995–1997. All rights reserved.
Reprinted with permission.

Physical Properties

Constituent	Abbreviation or Synonym	CAS Number	Molecular Weight (g/mole) MW	ref	Diffusion Coefficients in Air (cm^2/s) Dair	ref	Diffusion Coefficients in Water (cm^2/s) Dwat	ref	log (koc) (@20–25 C) log(l/kg) partition	ref
BTEX AND MTBE										
Benzene	B	71-43-2	78.1	5	9.30E-02	A	1.10E-05	A	1.58	A
Ethylbenzene	E	100-41-4	106.2	5	7.60E-02	A	8.50E-06	A	1.98	A
Toluene	T	108-88-3	92.4	5	8.50E-02	A	9.40E-06	A	2.13	A
Xylene (mixed isomers)	X	1330-20-7	106.2	5	7.20E-02	A	8.50E-06	A	2.38	A
Xylene, m-	X	108-30-3	106.16	5	7.00E-02	4	7.80E-06	4	3.20	29
Xylene, o-	X	95-47-6	106.2	5	8.70E-02	4	1.00E-05	4	2.11	29
Methyl t-Butyl Ether	MTBE	1634-04-4	88.146	5	7.92E-02	6	9.41E-05	7	1.08	A
CHLORINATED COMPOUNDS										
Bromodichloromethane		75-27-4	163.8	4	2.98E-02	4	1.06E-05	4	1.85	4
Carbon tetrachloride	CT	56-23-5	153.8	4	7.80E-02	4	8.80E-06	4	2.67	4
Chlorobenzene		108-90-7	112.6	4	7.30E-02	4	8.70E-06	4	2.46	4
Chloroethane		75-00-3	64.52	4	1.50E-01	4	1.18E-05	4	1.25	4
Chloroform	Trichloromethane	67-66-3	119.4	4	1.04E-01	4	1.00E-05	4	1.93	4
Chloromethane	Methyl Chloride	74-87-3	51	5	1.28E-01	4	1.68E-04	7	1.40	29
Chlorophenol, 2-		95-57-8	128.6	4	5.01E-02	4	9.46E-06	4	2.11	4
Dibromochloromethane		124-48-1	208.29	4	1.99E-02	4	1.03E-05	4	2.05	4
Dichlorobenzene, (1,2) (-o)		95-50-1	147	4	6.90E-02	4	7.90E-06	4	3.32	4
Dichlorobenzene, (1,4) (-p)		106-46-7	147	4	6.90E-02	4	7.90E-06	4	3.33	4
Dichlorodifluoromethane		75-71-8	120.92	4	5.20E-02	4	1.05E-05	4	2.12	4
Dichloroethane, 1,1-	1,1 DCA	75-34-3	98.96	4	7.42E-02	4	1.05E-05	4	1.76	4
Dichloroethane, 1,2-	1,2 DCA, EDC	107-06-2	99	4	1.04E-01	4	9.90E-06	4	1.76	4
Dichloroethene, cis-1,2-	1,2 cis DCE	156-59-2	96.936	4	7.36E-02	4	1.13E-05	4	1.38	8
Dichloroethene, 1,2,-trans-	1,2 trans DCE	156-60-5	96.936	4	7.07E-02	4	1.19E-05	4	1.46	4
Methylene chloride	Dichloromethane	75-09-2	85	4	1.01E-01	4	1.17E-05	4	1.23	4
Tetrachloroethane, 1,1,2,2-		79-34-5	168	4	7.10E-02	4	7.90E-06	4	0.00	4
Tetrachloroethene	PCE, Perc	127-18-4	165.83	4	7.20E-02	4	8.20E-06	4	2.43	29

Trichlorobenzene, 1,2,4-		120-82-1	181.5	4	3.00E-02	4	8.23E-06	4	3.91	4
Trichloroethane, 1,1,1-		71-55-6	133.4	4	7.80E-02	4	8.80E-06	4	2.45	4
Trichloroethane, 1,1,2-	TCA	79-00-5	133.4	4	7.80E-02	4	8.80E-06	4	1.75	29
Trichloroethene	TCE	79-01-6	131.4	23	8.18E-02	6	1.05E-04	7	1.26	11
Trichlorofluoromethane		75-69-4	137.4	4	8.70E-02	4	9.70E-06	4	2.49	4
Vinyl chloride	VC	75-01-4	62.5	4	1.06E-01	4	1.23E-05	4	0.39	29
PAH COMPOUNDS										
Acenaphthene		83-32-9	154.21	4	4.21E-02	4	7.69E-06	4	3.85	4
Acenaphthylene		208-96-8	152.21	4	4.39E-02	4	7.53E-06	4	4.00	4
Anthracene		120-12-7	178.23	4	3.24E-02	4	7.74E-06	4	4.15	4
Benzo (b)Fluoranthene		205-99-2	252	5	2.26E-02	6	5.56E-06	7	5.74	25
Benzo (g,h,i)Perylene		191-24-2	276	5	4.90E-02	6	5.65E-05	7	6.20	11
Benzo (k) Fluoranthene		207-08-9	252.32	4	2.26E-02	4	5.56E-06	4	5.74	4
Benzo(a)Anthracene		56-55-3	228.3	4	5.10E-02	4	9.00E-06	4	6.14	4
Benzo(a)Pyrene		50-32-8	252.3	5	5.00E-02	A	5.80E-06	A	5.59	A
Chrysene		218-01-9	228.2	4	2.48E-02	4	6.21E-06	4	5.30	4
Dibenzo(a,h) Anthracene		53-70-3	278.35	4	2.00E-02	4	5.24E-06	4	5.87	4
Fluoranthene		206-44-0	202	4	3.02E-02	4	6.35E-06	4	4.58	4
Fluorene		86-73-7	166	4	3.63E-02	4	7.88E-06	4	3.86	4
Indeno(1,2,3,c,d)Pyrene		193-39-5	276.34	4	2.33E-02	4	4.41E-06	4	7.53	4
Naphthalene		91-20-3	128.2	4	7.20E-02	A	9.40E-06	A	3.11	A
Phenanthrene		85-01-8	178.22	4	3.33E-02	4	7.47E-06	4	4.15	4
Pyrene		129-00-0	202.3	4	2.72E-02	4	7.24E-06	4	4.58	4
OTHER COMPOUNDS										
Acetone		67-64-1	58.08	4	1.24E-01	4	1.14E-05	4	−0.24	4
Benzoic acid		65-85-0	122.13	4	5.36E-02	4	7.97E-06	4	1.83	4
Butanol, n-		71-36-3	74.12	4	8.00E-02	4	9.30E-06	4	0.74	4
Carbon disulfide		75-15-0	76.1	4	1.04E-01	4	1.00E-05	4	2.47	29
Ethylene glycol		107-21-1	62.07	4	1.08E-01	4	1.22E-05	4	−0.90	4
Hexane, n-		110-54-3	86.2	5	2.00E-01	4	7.77E-06	4	2.68	4
Methanol		67-56-1	32	4	1.50E-01	4	1.64E-05	4	−0.69	4
Methyl ethyl ketone	MEK	78-93-3	72.1	4	8.08E-02	4	9.80E-06	4	0.28	4
Phenol		108-95-2	94.1	4	8.20E-02	4	9.10E-08	4	1.44	4

continues

Physical Properties *(Continued)*

Constituent	Henry's Law Constant (@20–25 C)			Vapor Pressure (@20–25 C)		Solubility (@20–25 C)	
	$\frac{(atm\text{-}m^3)}{mol}$	(unitless)	ref	(mm Hg)	ref	(mg/L)	ref
BTEX AND MTBE							
Benzene	5.29E-03	2.20E-01	A	9.52E + 01	4	1.75E + 03	A
Ethylbenzene	7.69E-03	3.20E-01	A	1.00E + 01	4	1.52E + 02	5
Toluene	6.25E-03	2.60E-01	A	3.00E + 01	4	5.15E + 02	29
Xylene (mixed isomers)	6.97E-03	2.90E-01	A	7.00E + 00	4	1.98E + 02	5
Xylene, m-	5.20E-03	2.16E-01	4	8.00E + 00	4	1.58E + 02	29
Xylene, o-	5.27E-03	2.19E-01	4	7.00E + 00	4	1.75E + 02	29
Methyl t -Butyl Ether	5.77E-04	2.40E-02		2.49E + 02		4.80E + 04	A
CHLORINATED COMPOUNDS							
Bromodichloromethane	2.05E-01	8.53E + 00	4	5.92E + 01	4	6.22E + 01	4
Carbon tetrachloride	3.00E-02	1.25E + 00	4	1.13E + 02	4	7.62E + 02	4
Chlorobenzene	3.93E-03	1.63E-01	4	1.18E + 01	4	4.45E + 02	4
Chloroethane	5.10E-03	2.12E-01	4	1.20E + 03	4	2.00E + 04	4
Chloroform	3.39E-03	1.41E-01	4	2.08E + 02	4	9.64E + 03	4
Chloromethane	8.82E-03	3.67E-01	29	3.80E + 03	5	4.00E-03	5
Chlorophenol, 2-	1.78E-05	7.40E-04	4	3.00E + 00	4	2.85E + 04	4
Dibromochloromethane	7.83E-04	3.26E-02	4	1.50E + 01	4	5.25E + 03	4
Dichlorobenzene (1,2) (-o)	1.94E-03	8.07E-02	4	1.50E + 00	4	1.50E + 02	4
Dichlorobenzene (1,4) (-p)	1.60E-03	6.65E-02	4	1.20E + 00	4	1.45E + 02	4
Dichlorodifluoromethane	4.01E-01	1.67E + 01	4	5.00E + 03	4	1.98E + 03	4
Dichloroethane, 1,1-	1.54E-02	6.41E-01	4	5.91E + 02	4	5.50E + 03	5
Dichloroethane, 1,2-	1.20E-03	4.99E-02	4	8.00E + 01	4	8.69E + 03	5
Dichloroethene, cis-1,2-	3.19E-02	1.33E + 00	4	2.00E + 02	5	8.00E + 02	5
Dichloroethene, 1,2-trans-	5.32E-03	2.21E-01	4	3.31E + 02	4	6.00E + 02	5
Methylene chloride	3.19E-03	1.33E-01	4	4.38E + 02	4	1.54E + 04	4
Tetrachloroethane, 1,1,2,2-	2.00E-03	8.32E-02	4	6.50E + 00	4	7.18E + 02	4
Tetrachloroethene	2.90E-02	1.21E + 00	4	1.90E + 01	4	1.43E + 02	4

Trichlorobenzene, 1,2,4-	1.42E-03	5.91E-02	4	1.80E-01	4	3.03E + 01	4
Trichloroethane, 1,1,1-	1.72E-02	7.15E-01	4	1.23E + 02	4	1.26E + 03	4
Trichloroethane, 1,1,2-	7.40E-04	3.08E-02	4	2.50E + 01	4	5.93E + 03	4
Trichloroethene	1.00E-02	4.17E-01	10	5.80E + 01	23	1.00E + 03	23
Trichlorofluoromethane	5.83E-02	2.42E + 00	4	7.96E + 02	4	2.47E + 03	4
Vinyl chloride	8.60E-02	3.58E + 00	4	2.66E + 03	4	2.54E + 03	4
PAH COMPOUNDS							
Acenaphthene	7.71E-03	3.21E-01	4	5.00E-03	4	3.93E + 00	29
Acenaphthylene	1.14E-04	4.74E-03	4	8.51E-10	4	3.93E + 00	29
Anthracene	6.75E-02	2.81E + 00	4	1.30E-06	4	4.50E-02	5
Benzo (b)Fluoranthene	2.01E-05	8.36E-04	25	6.67E-07	25	1.47E-02	25
Benzo (g,h,i)Perylene	1.40E-07	5.82E-06	30	1.00E-09	10	7.00E-04	5
Benzo (k) Fluoranthene	1.07E-08	4.45E-07	4	9.59E-10	4	4.30E-03	4
Benzo(a)Anthracene	1.38E-08	5.74E-07	4	1.50E-07	4	5.70E-03	5
Benzo(a)Pyrene	1.39E-09	5.80E-08	A	5.68E-04	4	1.20E-03	5
Chrysene	1.18E-08	4.91E-07	4	5.76E-09	4	1.80E-03	5
Dibenzo(a,h) Anthracene	3.81E-07	1.58E-05	4	5.20E-10	4	5.00E-04	4
Fluoranthene	6.70E-02	2.79E + 00	4	1.77E-02	4	2.06E-01	5
Fluorene	1.17E-04	4.87E-03	4	1.70E-02	4	1.69E + 00	5
Indeno(1,2,3,c,d)Pyrene	5.07E-12	2.11E-10	4	1.00E-09	4	7.17E + 02	4
Naphthalene	1.18E-03	4.90E-02	A	2.30E-01	4	3.29E + 01	4
Phenanthrene	6.05E-03	2.52E-01	4	2.10E-04	4	1.60E + 00	5
Pyrene	7.00E-09	2.91E-07	4	4.20E-08	4	1.60E-01	5
OTHER COMPOUNDS							
Acetone	2.50E-05	1.04E-03	4	2.66E + 02	4	1.00E + 06	10
Benzoic acid	1.82E-07	7.57E-06	4	7.04E-03	4	6.22E + 04	4
Butanol, n-	8.90E-06	3.70E-04	4	7.02E + 00	4	7.70E + 04	31
Carbon disulfide	1.68E-02	6.99E-01	4	3.66E + 02	4	2.30E + 03	5
Ethylene glycol	1.03E-07	4.28E-08	4	1.26E-01	4	1.00E + 06	31
Hexane, n-	1.22E-01	5.07E + 00	4	1.50E + 02	4	1.30E + 01	5
Methanol	2.70E-06	1.12E-04	4	1.14E + 02	4	1.00E + 06	
Methyl ethyl ketone	4.35E-05	1.81E-03	4	1.00E + 02	4	2.18E + 05	4
Phenol	4.54E-07	1.89E-05	4	3.41E-01	4	9.30E + 04	4

Constituent	Reference Dose (mg/kg/day)				Slope Factors 1/(mg/kg/day)				EPA Weight of Evidence	Is Constituent Carcinogen ?
	Oral RfD_oral	ref	Inhalation RfD_inhal	ref	oral SF_oral	ref	Inhalation SF_inhal	ref		
BTEX AND MTBE										
Benzene	—		1.70E-03	R	2.90E-02	A	2.90E-02	A	A	TRUE
Ethylbenzene	1.00E-01	A	2.86E-01	A	—		—		D	FALSE
Toluene	2.00E-01	A,R	1.14E-01	A,R	—		—		D	FALSE
Xylene (mixed isomers)	2.00E + 00	A,R	2.00E + 00	A	—		—		D	FALSE
Xylene, m-	2.00E + 00	R	2.00E-01	R	—		—			FALSE
Xylene, o-	2.00E + 00	R	2.00E-01	R	—		—			FALSE
Methyl t-Butyl Ether	5.00E-03	R	8.57E-01	R	—		—			FALSE
CHLORINATED COMPOUNDS										
Bromodichloromethane	2.00E-02	R	—		6.00E-02	R	—		B2	TRUE
Carbon tetrachloride	7.00E-04	R	5.71E-04	R	1.30E-01	R	5.25E-02	R	B2	TRUE
Chlorobenzene	2.00E-02	R	5.71E-03	R	—		—		D	FALSE
Chloroethane	4.00E-01	R	2.86E + 00	R	—		—			FALSE
Chloroform	1.00E-02	R	—		6.10E-03	R	8.05E-02	R	B2	TRUE
Chloromethane	—		—		1.30E-02	R	6.30E-03	R	C	TRUE
Chlorophenol, 2-	5.00E-03	R	—		—		—			FALSE
Dibromochloromethane	2.00E-02	R	—		8.40E-02	R	—		C	TRUE
Dichlorobenzene, (1,2) (-o)	9.00E-02	R	4.00E-02	R	—		—		D	FALSE
Dichlorobenzene, (1,4) (-p)	—		2.29E-01	R	2.40E-02	R	—		C	TRUE
Dichlorodifluoromethane	2.00E-01	R	5.71E-02	R	—		—			FALSE
Dichloroethane, 1,1-	1.00E-01	R	1.43E-01	R	—		—		C	FALSE
Dichloroethane, 1,2-	—		2.86E-03	R	9.10E-02	R	9.10E-02	R	B2	TRUE
Dichloroethane, cis-1,2-	1.00E-02	R	—		—		—		D	FALSE
Dichloroethene, 1,2-trans-	2.00E-02	R	—		—		—			FALSE
Methylene chloride	6.00E-02	R	8.57E-01	R	7.50E-03	R	1.64E-03	R	B2	TRUE
Tetrachloroethane, 1,1,2,2-	—		—		2.00E-01	R	2.03E-01	R	C	TRUE
Tetrachloroethene	1.00E-02	R	—		5.20E-02	R	2.03E-03	R	C-B2	TRUE

Trichlorobenzene, 1,2,4-	1.00E-02	R	5.71E-02	R	—		—		D	FALSE
Trichloroethane, 1,1,1-	9.00E-02	R	2.86E-01	R	—		—		D	FALSE
Trichloroethane, 1,1,2-	4.00E-03	R	—		5.70E-02	R	5.60E-02	R	C	TRUE
Trichloroethene	6.00E-03	R	—		1.10E-02	R	6.00E-03	R		TRUE
Trichlorofluoromethane	3.00E-01	R	2.00E-01	R	—		—			FALSE
Vinyl chloride	—		—		1.90E + 00	R	3.00E-01	R	A	TRUE
PAH COMPOUNDS										
Acenaphthene	6.00E-02	R	—		—		—			—
Acenaphthylene	4.00E-03		4.00E-03		—		—		D	FALSE
Anthracene	3.00E-01	A	—		—		—		D	FALSE
Benzo (b)Fluoranthane	—		—		7.30E-01	R	6.10E-01	R	B2	TRUE
Benzo (g,h,i)Perylene	4.00E-03		4.00E-03						D	FALSE
Benzo (k) Fluoranthene	—		—		7.30E-02	R	6.10E-02	R	B2	TRUE
Benzo(a) Anthracene	—		—		7.30E-01	R	6.10E-01	R	B2	TRUE
Benzo(a)Pyrene	—		—		7.30E + 00	R	6.10E + 00	R	B2	TRUE
Chrysene	—		—		1.15E + 00	A	1.15E + 00	A	B2	TRUE
Dibenzo(a,h) Anthracene	—		—		7.30E + 00	R	6.10E + 00	R	B2	TRUE
Fluoranthene	4.00E-02	A	—		—		—		D	FALSE
Fluorene	4.00E-02	A	—		—		—		D	FALSE
Indeno(1,2,3,c,d)Pyrene	—		—		7.30E-01	R	6.10E-01	R	B2	TRUE
Naphthalene	4.00E-03	A	—		—		—		D	FALSE
Phenanthrene	4.00E-03		4.00E-03	R					D	FALSE
Pyrene	3.00E-02	R	—		—		—		D	FALSE
OTHER COMPOUNDS										
Acetone	1.00E-01	R	—		—		—		D	FALSE
Benzoic acid	4.00E + 00	R	—		—		—		D	FALSE
Butanol, n-	1.00E-01	R	—		—		—		D	FALSE
Carbon disulfide	1.00E-01	R	2.86E-03	R	—		—			FALSE
Ethylene glycol	2.00E + 00	R	2.00E + 00	R	—		—			FALSE
Hexane, n-	6.00E-02	A	5.71E-02	R	—		—			FALSE
Methanol	5.00E-01	R	—		—		—			FALSE
Methyl ethyl ketone	6.00E-01	R	2.86E-01	R	—		—		D	FALSE
Phenol	6.00E-01	R	—		—		—		D	FALSE

Regulatory and Miscellaneous Data

	Maximum Contaminant Level	Permissible Exposure Limit PEL/TLV		Detection Limits				Half Life (First-Order Decay) (days)			
				Groundwater		Soil					
Constituent	MCL (mg/L)	(mg/m^3)	ref	(mg/L)	ref	(mg/kg)	ref	Saturated	Unsat.	ref	Ref
BTEX AND MTBE											
Benzene	5.00E-03	3.20E + 00	OSHA	0.002	C	0.005	S	720	720	H	R
Ethylbenzene	7.00E-01	4.34E + 02	ACGIH	0.002	C	0.005	S	228	228	H	S
Toluene	1.00E + 00	1.47E + 02	ACGIH	0.002	C	0.006	S	28	28	H	H
Xylene (mixed isomers)	1.00E + 01	4.34E + 02	ACGIH	0.005	C	0.006	S	360	360	H	A
Xylene, m-	1.00E + 01			0.005	C	0.005		360	360	H	3
Xylene, o-	1.00E + 01			0.005	C	0.005		360	360	H	4
Methyl t-Butyl Ether		1.44E + 02	ACGIH					360	180	H	5
CHLORINATED COMPOUNDS											
Bromodichloromethane	1.00E-01			0.001	C	0.005	S				8
Carbon tetrachloride	5.00E-03	6.20E + 01	OSHA	0.001	C	0.005	S	360	360	H	9
Chlorobenzene	1.00E-01	4.60E + 01	ACGIH	0.002	C	0.005	S	300	300	H	10
Chloroethane		2.60E + 03	OSHA	0.005	C	0.01	S	56	56	H	11
Chloroform	1.00E-01	4.90E + 01	ACGIH	0.0005	C	0.005	S	1800	1800	H	12
Chloromethane		1.03E + 02	ACGIH	0.001	C	0.01	S				13
Chlorophenol, 2-				0.005	C	0.66	S				14
Dibromochloromethane	1.00E-01			0.001	C			180	180	H	15
Dichlorobenzene, (1,2) (-o)	6.00E-01	1.50E + 02	ACGIH	0.002	C	0.66	S	360	360	H	16
Dichlorobenzene, (1,4) (-p)	7.50E-02	4.50E + 02	OSHA	0.002	C	0.66	S	360	360	H	17
Dichlorodifluoromethane		4.95E + 03	OSHA	0.005	C			360	360	H	18
Dichloroethane, 1,1-		4.00E + 0.2	OSHA	0.001	C	0.005	S	360	360	H	19
Dichloroethane, 1,2-	5.00E-03	4.00E + 00	NIOSH	0.0005	C	0.005	S	360	360	H	20
Dichloroethene, cis-1,2-	7.00E-02			0.001	C	0.005	S				23
Dichloroethene, 1,2-trans-	1.00E-01			0.001	C	0.005	S				24
Methylene chloride	5.00E-03	1.74E + 02	ACGIH	0.005	C	0.005	S	56	56	H	26
Tetrachloroethane, 1,1,2,2-		3.50E + 01	OSHA	0.0005	C	0.005	S	45	45	H	
Tetrachloroethene	5.00E-03	1.70E + 02	ACGIH	0.0005	C			720	720	H	27

Trichlorobenzene, 1,2,4-	7.00E-02		ACGIH	0.01	C	0.66	S	360	360	H	28
Trichloroethane, 1,1,1-	2.00E-01	1.90E + 03	OSHA	0.005	C	0.005	S	546	546	H	29
Trichloroethane, 1,1,2-	5.00E-03	4.50E + 01	OSHA	0.0002	C	0.005	S	730	730	H	
Trichloroethene	5.00E-03	2.69E + 02	ACGIH	0.001	C	0.005	S	1653	1653	H	
Trichlorofluoromethane		5.60E + 03	OSHA	0.005	C			720	720	H	
Vinyl chloride	2.00E-03	1.30E + 01	ACGIH	0.002	C	0.01	S	2875	2875	H	
PAH COMPOUNDS											
Acenaphthene				0.01	C	0.66	S	204	204	H	
Acenaphthylene				0.01	C	0.66	S	120	120	H	
Anthracene				0.01	C	0.66	S	920	920	H	
Benzo (b)Fluoranthene			ACGIH	0.01	C	0.66	S	1220	1220	H	
Benzo (g,h,i)Perylene				0.01	C	0.66	S	1300	1300	H	
Benzo (k) Fluoranthene				0.01	C	0.66	S	4280	4280	H	
Benzo(a)Anthracene			ACGIH	0.01	C	0.66	S	1360	1360	H	
Benzo(a)Pyrene	2.00E-04		ACGIH	0.01	C	0.66	S	1060	1060	H	
Chrysene	2.00E-04			0.01	C	0.66	S				
Dibenzo(a,h) Anthracene				0.01	C	0.66	S	1880	1880	H	
Fluoranthene				0.01	C	0.66	S	880	880	H	
Fluorene				0.01	C	0.66	S	120	120	H	
Indeno(1,2,3,c,d)Pyrene				0.01	C	0.66	S	1460	1460	H	
Naphthalene		5.00E + 01	OSHA	0.01	C	0.01	S	258	258	H	
Phenanthrene				0.01	C	0.66	S	400	400	H	
Pyrene				0.01	C	0.66	S	3800	3800	H	
OTHER COMPOUNDS											
Acetone		1.78E + 03	ACGIH	0.1	C	0.1	S	14	14	H	
Benzoic acid						3.3	S				
Butanol, n-								54	54	H	
Carbon disulfide				0.005	C	0.1	S				
Ethylene glycol								24	24	H	
Hexane, n-											
Methanol		2.60E + 02	OSHA					7	7	H	
Methyl ethyl ketone		5.90E + 02	OSHA	0.01	C			14	14	H	
Phenol		1.90E + 01	OSHA	0.001	C	0.66	S	7	10	H	

References

R	EPA Region III Risk Based Concentration Table, EPA Region 3, March 7, 1995.
S	USEPA, Test Methods for Evaluating Solid Waste, SW-846, Third Edition, OSWER, November 1986.
H	Howard, Handbook of Environmental Degradation Rates, Lewis Publishers, Chelsea, MI, 1989.
A	Emergency Standard Guide for Risk-Based Corrective Action Applied at Petroleum Release Sites, ASTM, ES 38-94.
3	Based on Kow from (2) and DiToro, D. M., 1985: "A Particle Interaction Model of Reversible Organic Chemical Sorption," Chemosphere, 14(10), 1505–1538. $\log(Koc) = 0.00028 + 0.983 \log(Kow)$
4	USEPA, 1989: Hazardous Waste Treatment, Storage, and Disposal Facilities (TSDF)-USEPA, OAQPS, Air Emission Models, (EPA-450/3-87-026).
5	Verschueren, Karel, 1983: Handbook of Environmental Data on Organic Chemicals, Second Ed., (Van Nostrand Reinhold Company Inc., New York), ISBN 0-442-28802-6.
6	Calculated diffusivity using the method of Fuller, Schettler, and Giddings from (9).
7	Calculated diffusivity using the method of Hayduk and Laudie and the reference from (9).
8	Calculated using Kenaga ang Goring Kow/solubility regression equation reference (9) and Kow data from (2), $\log(S, mg/l) = 0.922 \log(Kow) + 4.184$.
9	Handbook of Chemical Property Estimation Methods, 1982, W. J. Lyman, (McGraw-Hill, New York), ISBN 0-07-039175-0.
10	Calculated from (Pv/Patm)/(solubility/mol wt).
11	Back calculated from solubility, Note (8) and (3).
12	Aldrich Chemical Catalog, 1991.
13	Calculated using Modified Watson Correlation from (9) and normal boiling point.
14	USEPA, 1979: Water Related Environmental Fate of 129 Priority Pollutants, Vol. 1, USEPA, OWOPS, (EPA-4404-79-029a).
15	The Agrochemicals Handbook, (The Royal Society of Chemistry, The University, Nottingham, England), ISBN 0-85186-406-6.
16	Vapor pressure specified at elevated temperature, adjustments to 25C using methods presented by (9), Wauchope, R. D., T. M. Butler, A. G. Hornsby, P. W. M. Augustijn-Beckers, and J. P. Burt, 1992: "The SCS/ARS/CES Pesticide Properties Database for Environmental Decision Making," Reviews of Environmental Contamination and Toxicology, vol. 123, 1-155.
18	Farm Chemicals Handbook 91, C. Sine, ed. (Meister Publishing Company, Willoughby, Ohio).
19	Structure and Nomenclature Search System, (Version 7.00/7.03) December, 1992.
20	From Syracuse Research Corporation Calculated Value from pcchem-pcgems, 1988, ref no. 255435 in Enirofate database, Accession no. 105543.

23 NIOSH, 1990: Pocket Guide to Chemical Hazards, (U.S. Dept. of Health & Human Services, Public Health Service. Centers for Disease Control, National Institute for Occupational Safety and Health).

24 Buchter, B. et al., 1989: Correlation of Greundlich Kd and N retention Parameters with Soils and Elements, Soil Science, 148, 370–379.

25 USEPA, 1993: Air/Superfund National Technical Guidance Study series: Estimation of Air Impacts for Thermal Desorption Units Used at Superfund Sites, US Environmental Protection Agency, Office of Air Quality Planning and Standards, EPA-451/R-93-005, NTIS Accession No. PB93-215630, April 1993.

27 Based on salt solubilities in Table 3-120, R. H. Perry and D. W. Green, "Perry's Chemical Engineering Handbook" Sixth Edition, McGraw-Hill, New York, 1973.

28 Based on salt solubilities in Table of Physical Constants for Inorganic Compounds, Weast, R. C., CRC Handbook of Chemistry and Physics, 67th edition, (CRC Press, Inc., Boca Raton), 1987.

29 Montgomery and Welkom, "Groundwater Chemicals Desk Reference," Lewis Publishers, Chelsea, MI, 1990.

Note: This database was originally compiled by Shell Development Company and later modified for use in the GSI RBCA Tool Kit software system (Groundwater Services, Inc., Houston, Texas).

INDEX